Climate Change Law:
Emissions Trading in the EU and the UK

Climate Change Law:
Emissions Trading in the EU and the UK

by
J. Robinson, J. Barton, C. Dodwell,
M. Heydon and L. Milton

CAMERON MAY
INTERNATIONAL LAW & POLICY

TABLE OF CONTENTS

List of Contributors

Jonathan Robinson
Chief Legal Advisor
Ministry of Social Development
New Zealand

Jane Barton
Clean Air and Transport Unit, DG Environment
European Commission

Christopher Dodwell
Head of International Climate Change and Ozone
Department for Environment, Food and Rural Affairs, UK

Matthew Heydon
Solicitor
Legal Group
Department for Environment, Food and Rural Affairs, UK

Laura Milton
Solicitor
Treasury Solicitor's Department, UK

AUTHORS' PREFACE

The idea for this book came when we worked together in the Legal Services team of the UK Department for Environment, Food and Rural Affairs advising on various legal issues relating to climate change. Between 2001 and 2007 we advised on the development of market based measures to tackle climate change, drafting and refining the rules of the UK ETS, the world's first economy wide greenhouse gas emissions trading scheme, advising the UK delegation in the negotiations for the EU Emissions Trading Directive, Linking Directive and Registries Regulation, drafting legislation to transpose the requirements of the Directives into UK law, advising on the broader implementation of the Directive in the UK including the development of the UK's national allocation plans, and refining and interpreting the UK regime governing Climate Change Agreements.

In the process we acquired a body of knowledge and experience which could be of benefit to others. This book is our attempt to distil it into a form which is useful for those advising on these measures, as well as for those involved in the continuing development of policy on climate change.

We would like to thank our various employers for allowing us to publish. Needless to say, the views expressed are our own, and not those of the organisations we work for. A particular thank you goes to Nick May, our patient publisher, and his colleagues at Cameron May, who maintained throughout the perfect balance between supporting and tactfully cajoling. Thanks also to Damien Meadows, Martin Nesbit, Marie Pender and Ambrose Franceschini who reviewed chapters in draft and offered helpful comments and to Susan Readhead who patiently typed parts of the early drafts.

The book is very much a joint effort. Primary responsibility for chapters was as follows: Introduction, Jane Barton and Chris Dodwell; chapter 1, Chris Dodwell; chapters 2 to 9, Jane Barton and Jonathan Robinson; chapter 10, Jane Barton and Chris Dodwell; chapters 11 to 18, Jane Barton; chapters 19, Chris Dodwell; chapter 20 to 22, Matthew Heydon (who also made contributions to chapters 2 to 18); and chapter 23, Laura Milton.

The text was edited jointly by Jonathan Robinson and Jane Barton.

FOREWORD

The scientific evidence that climate change is a serious global threat is overwhelming. Climate change will affect the basic elements of life for people around the globe and hundreds of millions of people could be affected by food and water shortages, flooding and extreme weather events.

Whilst some climate change is inevitable, there is still time to avoid the worst impacts of climate change, if we take strong action now. A key conclusion of the review of the economics of climate change which my team and I completed for the UK Government in 2006 was that the benefits of taking action heavily outweigh the economic costs of not acting. If no action is taken, the overall costs and risks of climate change will be equivalent to losing at least 5 per cent of global GDP each year, possibly much more. Conversely the costs of taking action that will avoid most of these damages can be limited to around 1 per cent of global GDP each year, provided strong and deliberate policy choices are taken. Moreover, the transition to a low-carbon energy economy will create new markets and opportunities worth at least $500bn per year by 2050. Indeed it is possible that the new technologies likely to be developed and deployed could provide a boost to world growth.

A global problem such as climate change demands a global response in which all countries play their part building on mutually reinforcing approaches at national, regional and international levels. The Stern Review concluded that the first element of that policy response is the pricing of carbon. This can be implemented through taxation, regulation and emissions trading. Economic efficiency points to the advantages of a common global carbon price whereby emissions reductions will then take place wherever they are cheapest.Emissions trading provides a particularly effective way to equalise carbon prices across countries and sectors. It can also generate private sector flow of carbon finance that can help countries participate in emissions reductions.

Following in the wake of earlier experience in the UK, the EU Emissions Trading Scheme is now the centrepiece of European efforts to cut emissions. The scheme is the world's largest multi-sector, company level emissions trading scheme and its evolution highlights a number of vital lessons for the future of the carbon market.

This book provides a unique insight into the development and legal framework of the EU ETS and related UK measures. It is essential reading for anyone needing to understand how these schemes currently operate - in particular regulated industries and their advisers - and for a growing audience of policy makers and businesses who wish to understand the lessons to be learnt from the EU scheme. These lessons will not only inform decisions about the future shape of the EU scheme but more significantly will also influence the next stage in the evolution of the global carbon market, including the establishment of domestic trading schemes in other countries and the expansion of the Kyoto Protocol flexible mechanisms.

If we are to stabilise stocks of greenhouse gases at an acceptable level, the next 10 to 20 years will be a critical period of transition, from a world where carbon-pricing schemes are in their infancy, to one where carbon pricing is universal and is automatically factored into investment decisions. By shedding light on the details of how emissions trading has been implemented in the UK and EU, this book will play a valuable part in that transition.

INTRODUCTION

I.   Purpose and Structure of this Book

Climate change is one of the greatest threats facing the world today. The average surface temperature of the earth increased by around 0.6°C in the last half century and is predicted to continue rising. This increase in temperature will have significant impacts not only on the world's environment but also on future prospects for economic growth. There is increasing evidence that the change in temperature is attributable to emissions of greenhouse gases, in particular from the burning of fossil fuels and the destruction of forests.

As awareness of the need to act to reduce greenhouse gas emissions increases, new measures are being introduced. The UK and European Union (EU) have been at the forefront of efforts to address climate change and in recent years have experimented with a number of innovative market mechanisms to reduce greenhouse gas emissions. This book explains the most significant of these measures. It is hoped that it will be of value to anyone interested in the regulation of greenhouse gas emissions, from policy makers within and outside the EU to regulated industry and their legal advisers. Its analysis of existing measures also provides background information for anyone interested in the future development of the global carbon market and in particular the EU's innovative emissions trading scheme.

This introduction provides a summary of the challenge of climate change (section II) and the international, EU and UK responses (section III). It then provides an introduction to the economic theory of emissions trading and its development as a regulatory model (section IV).

The remainder of this book provides:

- a full account of the EU Emissions Trading Scheme (EU ETS) (chapters 1 to 10);
- a full account of the implementation of the EU ETS into UK law (chapters 11 to 21);
- a full account of the related UK policy instruments: the UK Emissions Trading Scheme (UK ETS) and Climate Change Agreements (CCAs)(chapters 22 and 23);

- a comprehensive glossary of terms and abbreviations (a necessary companion for newcomers to this area of law and policy); and
- the key materials necessary to understand the EU ETS and its implementation in the UK, the CCA regime, and the UK ETS (appendices).

The law is as stated as at 28 February 2007.

II. The Challenge of Climate Change[1]

The accumulation of greenhouse gases such as carbon dioxide, methane, nitrous oxide, ozone and perfluorocarbons in the atmosphere prevent radiation from the sun being reflected back into space. This effect, known as the greenhouse effect, is a natural and vital part of the earth's climate system. However, during the 1980s evidence began to emerge that increases in the levels of greenhouse gas emissions resulting from human activities were increasing this effect and resulting in changes to the earth's climate.

In 1988, the Intergovernmental Panel on Climate Change (IPCC) was established by the United Nations Environment Program and the World Meteorological Organization to collect and assess scientific, technical and other information relevant to the understanding of climate change, its potential impacts and options for adaptation and mitigation. Since 1990, the IPCC has published three assessment reportsand the fourth report is due in 2007.[2] In its Third Assessment Report, which was published in 2001, the IPCC concluded that there was compelling evidence that most of the global warming observed over the past 50 years was attributable to human-induced increases in atmospheric concentrations of greenhouse gases. More recent studies support those conclusions and acknowledge that the sensitivity of the climate system to emissions of greenhouse gases may be greater than previously thought.[3] This was confirmed in preparatory reports for the IPCC's Fourth Assessment Report due to be published in 2007 which noted that recent observations from around the world demonstrated that the pace of global warming and sea level rise had further increased and

---

[1] A detailed explanation of the science of climate change is beyond the scope of this book. Those interested in further general reading on this topic will find the following recent publications of interest: Maslin, M, (2004) *Global Warming: a Very Short Introduction*, New York, Oxford University Press; King, D, 'Science Informing Policy on Climate Change', in Helm, D (ed.), (2005) *Climate Change Policy*, Oxford, Oxford University Press.

[2] IPCC Third Assessment Report: Climate Change 2001 and IPCC Fourth Assessment Report: Climate Change 2007 (due 2007, in the meantime see summary for policy makers of the Contribution of Working Group I to the 4th Assessment Report.) Available at http://www.ipcc.ch/.

[3] Schellnhuber, H.J, Cramer, W. Nakicenovic, N et al. (eds.), (2006), *Avoiding Dangerous Climate Change*, Cambridge, Cambridge University Press.

further analysis had increased confidence that man-made emissions are the primary cause of 20th century climate change.

The earth has warmed by around 0.76°C over the last 100 years. Globally, the fifteen warmest years on record have occurred during the last 20 years and further scientific analysis has confirmed that the second half of the 20th century was the warmest phase during at least the last 1300 years in the Northern hemisphere. Moreover, it is now accepted that further temperature rises are inevitable. The IPCC concluded that global average temperatures are projected to rise between 1.1 and 6.4°C by the end of the 21st century depending on the scenario, with a best estimate range from 1.8°C to 4.0°C. The IPCC considered that the current atmospheric concentrations of carbon dioxide and methane are higher than at any other time in the last 650,000 years at least. Increases in temperature could themselves trigger powerful feedbacks (such as release of methane from wetlands and thawing permafrost and the weakening of the ability of oceans and forests to absorb carbon dioxide) which could in turn lead to further increases in temperature.

The impacts of these changes in climate will be severe and widespread: increasing risks of flooding during wet seasons and water shortages in dry seasons, reducing crop yields especially in Africa and rising sea levels. As well as impacting on people and the environment by displacing millions of people worldwide and devastating vulnerable ecosystems, the impacts of climate change will also have a huge financial cost. In October 2006, Sir Nicholas Stern, former chief economist at the World Bank and head of the UK Government's economic service, published the most comprehensive review to date of the economics of climate change.[4] The report described climate change as the greatest and widest-ranging market failure ever seen, concluding that the total cost of unabated 'business as usual' climate change to 2200 would be at least 5 per cent and up to 20 per cent of global GDP.

Although it is now inevitable that we will experience further rises in atmospheric concentrations of greenhouse gases and that this will lead to further increases in temperature and sea levels, the worst of the predicted impacts can still be avoided if urgent action is taken to reduce greenhouse gas emissions. The Stern Review concluded that the costs of action to reduce greenhouse gas emissions to avoid the worst of climate change can be limited to manageable levels if action is taken now, but that delay would result in much higher costs.

There is still no consensus as to the level of atmospheric concentration of CO_2 which will result in dangerous climate change. To date the IPCC has

---

[4] Stern, N et al, (2007) *Stern Review: The Economics of Climate Change*, Cambridge, Cambridge University Press.

declined to define what a safe level would be. The present atmospheric concentration of greenhouse gases is 430 parts per million (ppm), compared to about 280 ppm CO_2 in the pre-industrial era. In 2000, the UK Royal Commission on Environmental Pollution defined 550 ppm CO_2 as an upper limit which should not be exceeded.[5] At the current rate of increase of 2 ppm per year, we would reach this limit by 2050. More recently, the Stern Review concluded that the risks of the worst impacts of climate change can be substantially reduced if greenhouse gas levels in the atmosphere can be stabilised between 450 and 550ppm CO_2. Central estimates of the annual costs of achieving stabilisation between 500 and 550ppm CO_2 are around 1 per cent of global GDP, if we start to take strong action now, but delay in taking action would mean that the opportunity to stabilise at this level would slip away.

III.   The Response to the Challenge of Global Climate Change

A.   The International Response

In response to the growing body of evidence on global warming the UN General Assembly resolved in 1988 that climate change was a common concern of mankind. In 1990 negotiations began on an international treaty to tackle the issue and culminated in the agreement of the UN Framework Convention on Climate Change (UNFCCC) which was signed by 154 states at the UN Conference on Environment and Development in Rio de Janeiro in 1992.[6] The ultimate objective of the Convention is to achieve the 'stabilization of greenhouse gas concentrations in the atmosphere at a level that would prevent dangerous anthropogenic interference with the climate system.'[7] The Convention provides that such a level 'should be achieved within a time frame sufficient to allow ecosystems to adapt naturally to climate change, to ensure that food production is not threatened and to enable economic development to proceed in a sustainable manner.'

At the heart of the UNFCCC lies the principle of 'common but differentiated responsibilities'. This principle recognises the historical contributions of states to the problem of climate change and the differing abilities of states to respond to it. Whilst all states contribute to climate

---

[5] Royal Commission on Environmental Pollution, *Energy – The Changing Climate* (2000). Available at http://www.rcep.org.uk/newenergy.htm This upper limit was also endorsed by the UK in its 2003 Energy White Paper and by the Council of Environment Ministers of the European Union in 2005.

[6] United Nations Framework Convention on Climate Change adopted in New York on 9 May 1992. The Convention entered into force on 1 March 1994, 3 months after the fiftieth ratification document had been submitted (article 23 UNFCCC).

[7] Article 2 UNFCCC.

change and may suffer from the impacts of climate change, developed countries developed their economies at a time when there were no restrictions on emissions and they continue to emit a much greater quantity of greenhouse gases than developing countries. It is therefore argued that whilst all states should take some action to prevent climate change, developed countries should take the lead. In accordance with this principle, the UNFCCC places greater obligations on the 41 developed countries listed in Annex I to the Convention than on developing countries. In concrete terms this means that whilst all signatories are required to develop inventories of greenhouse gas emissions and national programmes to mitigate the causes and effects of climate change,[8] Annex I countries had a further obligation to take the lead in addressing climate change by developing policies and measures aimed at stabilising greenhouse gas emissions at 1990 levels by 2000.[9]

The commitments adopted by Annex I countries under the UNFCCC are strengthened by the Kyoto Protocol which was adopted in 1997.[10] Under the Protocol developed countries commit to reducing their collective emissions of six key greenhouse gases[11] by at least 5 percent below 1990 levels in the period 2008 to 2012.[12] Within this overall targets, the target for individual Annex I countries are differentiated to reflect their particular circumstances, including their state of economic development and access to low-carbon technology.[13] The individual targets range from a reduction of 8 per cent for the European Union and a number of other parties to an increase of 10 per cent for Iceland. In terms of absolute emissions, the most significant are a 7 per cent reduction target for the United States and a 0 per cent target (i.e. stabilisation at 1990 levels) for the Russian Federation.

---

[8] Article 4(1) UNFCCC.

[9] See Article 4(2) UNFCCC.

[10] The Kyoto Protocol to the United Nations Framework Convention on Climate Change adopted in Kyoto on 11 December 1997. For a more detailed account of the provisions of the Protocol relevant to the EU ETS, see section I of chapter 8. A full analysis of the Kyoto Protocol, however, is beyond the scope of this book. For fuller analysis, see Grubb, M., Vrolijk, C and Brack, D, (1999) *The Kyoto Protocol: a Guide and Assessment*, London, Royal Institute of International Affairs; Oberthür, S. and Ott, H.E, (1999), *The Kyoto Protocol: International Climate Policy for the 21st Century*, Berlin, Springer-Verlag; Yamin, F (ed), (2005) *Climate Change and Carbon Markets: a Handbook of Emission Reduction Mechanisms*, London, Earthscan.

[11] Annex A to the Kyoto Protocol lists the following greenhouse gases: carbon dioxide, methane, nitrous oxide, hydrofluorocarbons, perfluorocarbons and sulphur hexafluoride.

[12] Article 3(1) to the Kyoto Protocol.

[13] Article 3(7) of and Annex B to the Kyoto Protocol. The story of how these targets were negotiated is well worth reading, particularly for those with an interest in how future targets will be set (see sources cited in note 10).

By only setting legally binding emissions targets for Annex I countries, the Kyoto Protocol maintains the concept of differentiated responsibilities established by the UNFCCC.[14]

One of the most significant aspects of the Protocol is the inclusion of a ground-breaking set of 'flexible market-based mechanisms' designed to assist Annex I countries in complying with their targets. The development of these mechanisms can be traced back to the provisions of the UNFCCC which envisaged that parties could achieve reductions by acting jointly.[15] These flexible mechanisms allow Annex I countries to use reductions made by other parties to offset their own emissions in order to comply with their targets. There are three flexible mechanisms set out in the Protocol:

- International emissions trading, which allows Annex I countries which have overachieved their targets to sell surplus national emissions quotas ('assigned amount units' or 'AAUs') to Annex I countries with difficulties in meeting their targets through domestic action.[16]

- Joint implementation ('JI') whereby an Annex I country can receive credits ('emission reduction units' or 'ERUs') in return for providing support for emission reduction projects in another Annex I country.[17]

- The Clean Development Mechanism ('CDM'), the only mechanism open to countries without targets, which allows Annex I countries to receive credits ('certified emission reductions' or 'CERs') for financing projects in non-Annex I countries.[18]

---

[14] The Kyoto Protocol does, however, reaffirm the commitments of non-Annex I parties under the UNFCCC by requiring all parties to take measures to limit emissions and promote adaptation to future climate change impacts; submit information on their national climate change programmes and inventories; promote technology transfer; cooperate on scientific and technical research; and promote public awareness, education, and training (articles 10 and 11 of the Kyoto Protocol).

[15] Article 4(2) UNFCCC. During the discussions leading up to the agreement of the Kyoto Protocol, the US was active in seeking to convert other parties to the potential merits of using market-based mechanisms as part of the international climate change regime. Initially, the EU and developing countries were sceptical that such mechanisms would simply allow rich countries to buy their way out of their targets. However, over time the developing countries, led by Brazil, realised that the CDM could facilitate the flow of much needed foreign capital into infrastructure projects. Ironically given its subsequent enthusiasm for market-based instruments, the EU initially insisted on the paramount importance of domestic action, but eventually gave ground in order to ensure that final agreement was reached at Kyoto. For a detailed analysis of the evolution of the Kyoto mechanisms see Yamin (supra note 10), pages 4-10.

[16] Article 17 of the Kyoto Protocol.

[17] Article 6 of the Kyoto Protocol.

[18] Article 12 of the Kyoto Protocol.

In order to enter into force the Kyoto Protocol had to be ratified by at least 55 Parties to the Convention, including developed countries representing at least 55 per cent of the total 1990 carbon dioxide emissions from this group. Following the decisions by the US and Australia not to ratify, the Protocol finally entered into force on 16 February 2005 following ratification by the Russian Federation. In November 2005, at the eleventh meeting of the conference of the parties to the UNFCCC (which also served as the first meeting of the parties to the Kyoto Protocol), the detailed rules for the implementation of the Kyoto Protocol including the flexible mechanisms and the regime for monitoring compliance with the Protocol were formally adopted.[19]

At the same time, negotiations began on international commitments in relation to climate change after 2012.[20] These negotiations followed a twin track: negotiations under the Kyoto Protocol for establishing future commitments after 2012 and a dialogue under the UNFCCC itself, which will include the US and Australia, on strategic approaches for long term global cooperation on the future shape of commitments. These negotiations were furthered at the twelfth conference of the parties (also serving as the second meeting of the parties to the Kyoto Protocol) in Nairobi in November 2006 with the agreement of a comprehensive work programme on future Kyoto commitments and consideration of how to broaden efforts to address climate change under the so-called Convention Dialogue. The parties also agreed to carry out a comprehensive review of the effectiveness of the Kyoto Protocol in 2008.

B.   European Climate Change Policy

The European Union has taken a leading role in international and European efforts to combat climate change. The European Community has extensive competence to take action in the field of the environment including to address climate change. Since the entry into force of the Treaty of Amsterdam in 1999, one of the tasks of the Community is to promote 'a high level of protection and improvement of the quality of the environment' and to integrate environmental protection requirements into the definition and implementation of other Community policies and activities.[21] The environment chapter of the

---

[19] For further information on the flexible mechanisms, see chapter 8.

[20] Article 3(9) of the Kyoto Protocol required parties to start considering commitments for a new commitment period at least seven years before the end of the first commitment period (i.e. 2005).

[21] Article 3(l) EC includes 'policy in the sphere of the environment' in the list of activities of the Community and article 6 EC requires environmental protection requirements to be integrated into the definition and implementation of the Community policies and activities referred to in article 3 EC. Measures to address climate change may also be taken under the following articles of the Treaty: 37 (agriculture); 70 or 80 (transport); 93 (taxation); 95 (internal market) and article 133 (trade).

EC Treaty is set out in articles 174 to 176.[22] This includes competence to cooperate with third countries and competent international organisations and to negotiate and conclude agreements.[23] In recent years the Community has increasingly exercised its competence to adopt measures to address climate change. In doing so it has moved away from traditional command and control measures involving the prohibition or constraint of particular activities[24] and experimented with market-based measures designed to achieve environmental objectives in the most cost effective and efficient way.

Climate change was first recognised as an issue to be addressed by EC environmental policy in the EC's Fourth Environmental Action Programme which covered the period 1987-1992.[25] Acknowledging that the science was not sufficiently advanced at that point to confirm the impact of anthropogenic emissions of greenhouse gases on the climate, the Programme recognised that the Community should be thinking about possible responses and alternative energy strategies 'if the build-up of atmospheric carbon dioxide and the 'greenhouse effect' are shown (as scientists fear) to have serious impacts on climate and agricultural productivity worldwide'. In 1989, the Commission adopted its first Communication on the greenhouse effect[26] and in 1992 the Commission adopted its first Community Strategy to limit carbon dioxide emissions.[27]

By the time the Community adopted its Fifth Environmental Action Programme, covering the period 1993-2000, action on climate change was identified as one of seven priority themes for Community environmental policy and one which would require a new approach in

---

[22] For a detailed analysis of the environmental provisions of the EC Treaty see Kramer, L., (2006), *EC Environmental Law,* 6th rev ed., London, Sweet and Maxwell and chapter 1 of Jans, J, (2000), *European Environmental Law,* Groningen, Europa Law Publishing.

[23] Article 174(4) EC. Such agreements must be negotiated and concluded in accordance with article 300 EC. Under this procedure, it is the European Commission which conducts the negotiations on behalf of the European Community, having first received a negotiating mandate from the Council. It is also the Council which takes the ultimate decision as to whether to conclude the agreement which the Commission has negotiated. The Council, both in granting the Commission a negotiating mandate and in deciding whether to conclude the agreement, acts by qualified majority.

[24] Command and control regimes have been sub-divided into three categories: target standards (i.e. ambient quality targets for concentrations of particular pollutants); performance standards (i.e. limits or prohibitions on emissions operating at a plant or national level); and specification standards (i.e. restrictions on inputs or processes used to produce products). For a detailed appraisal of command and control regulation see chapter 2 of Scott, J, (1998) *EC Environmental Law,* New York, Longman.

[25] OJ C 328, 7.12.1987, p.5.

[26] Communication to the Council, The Greenhouse Effect and the Community, COM (88) 656 final.

[27] Communication, A Community Strategy to limit carbon dioxide emissions and improve energy efficiency, COM(1992) 246 final.

environmental policy making.[28] The programme acknowledged that its objectives would only be achievable by moving away from the command and control approach to regulation to the application of a wider range of instruments including market-based instruments aimed at the 'internalising of external environmental costs'.

In December 1993, the Community adopted a Decision ratifying the UNFCCC[29] and the Community and its Member States played an active role in the negotiations leading to the agreement of the Kyoto Protocol in 1997.

During the negotiations of the Kyoto Protocol, the EC pushed hard for the Protocol to include a mechanism which would enable its Member States to meet their obligations under the Protocol jointly. Article 4 of the Protocol therefore provides that any parties which have reached an agreement to meet their commitments jointly will be deemed to have met their commitments if their combined emissions do not exceed their combined required emissions reductions.

On 25 April 2002 the Council adopted a decision approving the Kyoto Protocol and providing for the Community to fulfil its commitments jointly in accordance with article 4 of the Protocol.[30] Annex II of the Decision set out the quantified emission limitation and reduction commitments agreed by Member States for the purpose of determining the respective emission levels allocated to each of them for the 2008–12 period. The contributions to the Community's target required from each Member State are differentiated to take account of *inter alia* expectations for economic growth, the energy mix and the industrial structure of the respective Member State and reflect the agreement on 16 June 1998 regarding the contribution of each Member State to the overall Community reduction commitment.[31] The UK's target under the agreement is to reduce emissions by 12.5 per cent compared to 1990 levels.[32] The Community and the Member States are therefore jointly

[28] The Fifth Environment Action Programme 'Towards Sustainability: A European Community programme of policy and action in relation to the environment and sustainable development' OJ C 138, 17.5.1993, p.5.
[29] Council Decision 94/69/EC of 15 December 1993 concerning the conclusion of the United Nations Framework Convention on Climate Change, OJ L 33, 7.2.1994, p.11.
[30] Council Decision 2002/358/EC concerning the approval, on behalf of the European Community, of the Kyoto Protocol to the United Nations Framework Convention on Climate Change and the joint fulfilment of commitments thereunder, OJ L 130, 15.05.2002, p1.
[31] Doc. 9702/98 of 19 June 1998 of the Council of the European Union reflecting the outcome of proceedings of the Environment Council of 16-17 June 1998, Annex I.
[32] The targets for other Member States are: Austria –13%, Belgium -7.5%, Denmark –21%, Finland 0%, France 0%, Germany –21%, Greece 25%, Ireland 13%, Italy -6.5%, Luxembourg –28%, Netherlands –6%, Portugal 27%, Spain 15%, Sweden 4% (the percentages are from 1990 levels).

responsible, under article 4(6) and in accordance with article 24(2) of the Protocol, for the fulfilment by the Community of its quantified emission reduction commitment under article 3(1) of the Protocol.[33] Consequently, and in accordance with article 10 of the EC Treaty, Member States individually and collectively have the obligation to take all appropriate measures, whether general or particular, to facilitate compliance with the Community's quantified emission reduction commitment under the Protocol.

The Community's target under the Kyoto Protocol is shared among the 15 Member States of the EU at the time of the adoption of the Kyoto Protocol.[34] Of the ten Member States which joined the EU on 1 March 2004, eight have individual targets under the Protocol.[35] The remaining two, Malta and Cyprus do not have targets under the Protocol as they are not included in Annex I to the UNFCCC. Romania and Bulgaria, which joined the EU on 1 January 2007, both have targets of minus 8 per cent from 1990 levels.

Although by 2000 the EU had taken some significant steps to reduce greenhouse gas emissions,[36] it recognised the need to take further measures in order to meet its obligations under the Kyoto Protocol. In order to identify the most environmentally effective and cost-effective policies and measures to reduce greenhouse gas emissions, the Commission launched, in March 2000, the European Climate Change Programme,[37] a consultative process bringing together national experts,

---

[33] Article 24(2) of the Kyoto Protocol provides that in the case of a regional economic integration organisation which becomes a Party to this Protocol, one or more of whose member States is party to the Protocol, 'the organisation and its member States shall decide on their respective responsibilities for the performance of their obligations under this Protocol'.

[34] Article 4(4) of the Protocol provides that 'if parties acting jointly do so in the framework of, and together with a regional economic integration organization, any alteration in the composition of the organization after adoption of [the] Protocol shall not affect existing commitments under the Protocol'.

[35] Czech Republic, Estonia, Latvia, Lithuania, Slovakia and Slovenia have a target of –8% from 1990 levels and Hungary and Poland have a target of –6% from 1990 levels.

[36] For example in the field of renewable energy (adoption of a white paper Energy for the future: renewable energy sources, COM(97)599 final , 26.11.1997); voluntary agreements with car manufacturers to reduce the average CO_2 emissionsof new passenger cars sold in the EU, agreed in 1999 and 2000 (Commission Recommendation 1999/125/EC on the reduction of CO_2 emissions from passenger cars, OJ L 40, 13.2.1999, p49; Commission Recommendation 2000/303/EC on the reduction of CO_2 emissions from passenger cars (KAMA), OJ L 100, 20.4.2000, p55 and Commission Recommendation 2000/304/EC on the reduction of CO_2 emissions from passenger cars (JAMA), OJ L 100, 20.4.2000, p57) and the proposal in 1997 on the taxation of energy products later to be adopted as Council Directive 2003/96/EC restructuring the Community Framework for the taxation of energy products and electricity, L 283, 31.10.2003, p.51.

[37] Commission Communication on EU policies and measures to reduce greenhouse gas emissions: Towards a European Climate Change Programme (ECCP), COM (2000) 88.

industry and environmental groups. The goal of the ECCP is to identify and develop all the necessary elements of an EU strategy to implement the Kyoto Protocol. Eleven working groups were established under the ECCP to consider different policy areas and instruments.[38]

The Commission's drive to take action to combat climate change was further underlined by the Sixth Environment Action Programme 'Environment 2010: Our Future, Our Choice'[39] which identified four priority areas for action including climate change and identified a long term objective of limiting global increases in temperature to a maximum of 2 degrees celsius over pre-industrial levels. Article 5 of the Programme set out the details of the objectives and priority areas for action on climate change, including the entry into force and implementation of the Kyoto Protocol. As well as strengthening existing policies and measures to address climate change, the Programme identified the establishment of an emissions trading system in the EU as one of the main means of achieving the long-term objective.

The EU ETS, which began operation on 1 January 2005, is the flag ship of EU action to address climate change and the largest multi-country, multi-sector trading scheme in the world. It covers about 11,500 installations which account for about 45 per cent of the EU's carbon dioxide emissions. The scheme has got off to a strong start and has also had an impact beyond Europe's borders. Trading under the EU ETS is a driver of the international carbon market, representing over sixty per cent of the total volume of global trades[40] and through the recognition of credits generated from the CDM the scheme also encourages developing countries to take action to reduce emissions. The scheme is also an important working example for other states of how carbon trading can operate and a number of regions, including in the US and Australia, are considering introducing schemes.[41]

---

[38] The following working groups were set up under the first ECCP: emissions trading, Joint Implementation and the Clean Development Mechanism; energy supply; energy demand; energy efficiency in end-use equipment and industrial processes; transport; industry (including sub-groups on fluorinated gases, renewable raw materials and voluntary agreements); research; agriculture; sinks in agricultural soils and forest-related sinks.

[39] Decision 1600/2002/EC of the European Parliament and of the Council, OJ L 242, 10.9.2002, p.1. The Sixth Action Programme, the first since the Treaty of Amsterdam came into force, was adopted as a formal Commission Decision under article 175(3) of the Treaty. Earlier Action Programmes had no legal status, being adopted by non-legally binding resolutions of the EU's Council of Ministers and of the Member States meeting in the Council. It is arguable whether the adoption of the Sixth Action Programme under article 175(3) gives any of its provisions any legal effect: but in the present context this remains academic, as none of its provisions relevant to the subject matter of this book appear capable in themselves of having any legal effect.

[40] Point Carbon Outlook for 2006: Mid-year Update, Carbon Market Analyst, 3 August 2006.

[41] See chapter 8.

Having established its leadership role in relation to climate change action, the EU is now focusing on action after the first Kyoto commitment period ends in 2012. This includes both consideration of how domestic action can be improved including through the review of the EU ETS[42] and the second phase of the ECCP,[43] and active participation in negotiations on international action post-2012 under the UNFCCC and Kyoto Protocol. A Commission Communication adopted in 2005[44] outlined key elements for the EU's strategy for a post-2012 international agreement. These include: participation by all major emitters and economic sectors; greater innovation in low-carbon technologies; continued and expanded use of cost-effective market-based instruments and the implementation of strategies for adapting to the level of climate change that is already unavoidable. The Spring 2005 European Council reaffirmed the objective of ensuring that global surface temperatures do not rise by more than 2°C compared with pre-industrial levels and provided that reduction pathways for the group of developed countries in the order of 15-30 per cent compared to the baseline envisaged by the Kyoto Protocol should be considered.[45]

On 10 January 2007, the Commission adopted a Communication entitled 'Limiting Global Climate Change to 2°C: The way ahead for 2020 and beyond'.[46] The Communication is, part of a comprehensive package of measures to establish a new Energy Policy for Europe[47] and is a major contribution to the ongoing discussions at international level on a future global agreement to combat climate change after 2012.

The Communication proposes a set of actions by developed and developing countries that would enable the world to limit global warming to meet the 2°C objective. It concludes that limiting global warming to 2°C is both technically feasible and economically affordable if the international community acts swiftly. To do this a 30 per cent cut in developed country emissions by 2020 is required as an essential step towards the longer term objective of reducing global emissions by as much as 50 per cent below 1990 levels by 2050. It proposes that the EU

---

[42] See chapter 10.

[43] A second phase of the European Climate Change Programme (ECCP II) was launched in October 2005. New working groups have been set up to consider the implementation of ECCP I policies, carbon capture and geological storage, CO_2 emissions from light duty vehicles, emissions from aviation and adaptation to climate change.

[44] Commission Communication on Winning the Battle Against Global Climate Change, COM (2005) 35 final.

[45] Presidency Conclusions of the Brussels European Council 22-23 March 2005. See also European Parliament resolution on 'Winning the Battle Against Global Climate Change' (2005/2049(INI)).

[46] COM(2007) 2 final.

[47] COM(2007) 1 final.

should continue to take the lead by committing autonomously to reduce its own emissions by at least 20 per cent below 1990 levels by 2020. This figure should be increased to 30 per cent as part of a satisfactory global agreement on reducing worldwide emissions after 2012. Developing countries should also start to slow the rate of growth in their emissions as soon as possible and then reduce their emissions in absolute terms from 2020–2025 onwards Deforestation should be halted within the next two decades and then be reversed through reforestation and afforestation.

The EU ETS will deliver an important part of the EU's reduction and has the potential to form the basis for a global carbon trading network.

C.   National Measures in the UK

The UK Government has been a strong supporter of action to address climate change and put in place a number of innovative domestic measures at an early stage. This commitment and the experience gained from the early implementation of domestic measures has enabled the UK to take a leadership role on climate change issues both internationally and within the EU. As in Europe, the development of specific policies aimed at the reduction of greenhouse gas emissions in the UK took off following the negotiation and agreement of the Kyoto Protocol.

In 1997, the Labour Government was elected with a manifesto aim of reducing CO_2 emissions by 20 per cent from 1990 levels by 2010. Although this target referred to carbon dioxide only rather than the full 'basket' of six greenhouse gases covered by the Kyoto Protocol, it went well beyond the target of a 12.5 per cent reduction below 1990 levels by 2008-12 subsequently agreed as the UK share of the EC's target under the Kyoto Protocol in the Burden Sharing Agreement.[48]

In March 1998, the Chancellor of the Exchequer, Gordon Brown, asked Lord Marshall, Chairman of British Airways, to consider whether, and if so how best, new economic instruments could play a role in the UK in improving the business use of energy, and reducing greenhouse gas emissions. Lord Marshall, assisted by a task force of officials from the Department of Trade and Industry (DTI), the Department of the Environment, Transport and the Regions (DETR) and Her Majesty's Treasury, submitted his report in November 1998.[49]

---

[48] Supra note 30.
[49] *Economic Instruments and the Business Use of Energy.* Government Task Force on the Industrial Use of Energy, November 1998. Available at: http://archive.treasury.gov.uk/pub/html/prebudgetNov98/marshall.pdf.

Lord Marshall concluded that:

–   all sectors of the economy need to contribute to the achievement of the UK's Kyoto target, which was likely to be only the first of the commitments required to combat climate change;

–   a range of economic instruments would be necessary to ensure that the Kyoto target was met;

–   emissions trading would have a role in helping the UK to meet its targets but, given that it would not be practical to introduce a fully-fledged mandatory UK scheme at that time, a pilot scheme should be introduced as soon as possible; and

–   there was a role for a tax if businesses of all sizes and sectors were to contribute to improved energy efficiency and to help meet the UK's emissions targets.

The Government took forward the recommendations of the Marshall Report in its Climate Change Programme[50] published in November 2000 setting out policies and measures to meet the UK's target under the Kyoto Protocol and move towards its domestic goal. In order to meet these targets the UK Government put in place innovative market-based mechanisms to tackle climate change, including the Climate Change Levy and associated agreements and the UK domestic emissions trading scheme.[51] These are discussed in turn below.

In 2000, to encourage energy efficiency and thereby reduce greenhouse gas emissions, the Finance Act 2000 introduced the Climate Change Levy: a tax on non-domestic use of certain types of energy.[52] In order to protect the most energy-intensive sectors of UK industry from the full consequences of the introduction of the Climate Change Levy, the Act made provision for Climate Change Agreements.[53] These agreements are made between energy intensive sectors of industry and the Secretary of State for Environment, Food and Rural Affairs and contain negotiated energy efficiency targets in return for an 80 per cent discount on the Climate Change Levy.

---

[50] Climate Change: The UK Programme, November 2000, Cm 4913. Available at: http://www.defra.gov.uk/environment/climatechange/uk/ukccp/2000/index.htm.
[51] Other policies include the Renewables Obligation and the Energy Efficiency Commitment.
[52] Section 30 and schedules 6 and 7 of the Finance Act 2000.
[53] Schedule 6.

In 2002 the UK introduced its own Emissions Trading Scheme (UK ETS). This was the first ever economy wide emissions trading scheme to tackle greenhouse gas emissions and cover all six greenhouse gases identified in the Kyoto Protocol.[54] As well as allowing direct participants in the scheme to trade emission allowances, the scheme is also linked to the Climate Change Agreements and allows parties to such agreements to trade any over-achievement or under-achievement of their energy efficiency targets. Operators of facilities that have not met their targets for the preceding target period may buy and retire UK ETS allowances to make up any shortfall in their target to remain eligible for the Climate Change Levy discount. Operators of facilities that overachieve their target (i.e. they use less energy or emit less carbon during a target period than permitted by their CCA target) are able to convert their surplus into UK ETS allowances and either bank their allowances or sell them on the UK ETS market.

The UK's experience of these policies helped it to foster greater action at the international level and in particular informed the development of the EU ETS. The pre-existing regimes of CCAs and the UK ETS were adapted and continue to operate alongside the EU ETS.

Following an extensive review launched in September 2004 of the 2000 Climate Change Programme, a new UK Climate Change Programme was published on 28 March 2006.[55] The new programme sets out the UK's policies and priorities for action both in the UK and internationally. A key aspect of the policy is extending and strengthening the EU ETS. These policies are expected to reduce the UK's emissions of the basket of greenhouse gases to 23-25 per cent below 1990 levels and is also expected to reduce the UK's carbon dioxide emissions to 15-18 per cent below 1990 levels by 2010. The review, and policies introduced already, could reduce carbon emissions by 7-12MtC by 2010. The programme confirms that the EU ETS will continue to be a central element of the business sector's contribution to this national goal and commits to working towards extending and strengthening the EU ETS to make it a key tool for emissions reductions beyond 2012, including by pushing for the inclusion of aviation in the scheme as soon as possible.

The 2006 Climate Change Programme also stated that the UK Government would consider the potential for a new mandatory, auction-based UK emissions trading scheme that would target energy use by large non-energy intensive businesses and public sector organisations not covered by the EU ETS or the Climate Change Agreements. In

---

[54] Supra note 11.
[55] Climate Change: The UK Programme 2006, March 2006, Cm 6764. Available at http://www.defra.gov.uk/environment/climatechange/uk/ukccp/index.htm.

November 2006, the Government launched a formal consultation on detailed proposals for this new emissions trading scheme (known as the Energy Performance Commitment) as well as the principal alternative option of voluntary benchmarks and reporting.[56]

IV.   Emissions Trading: a Win-Win Solution to the Challenge of Climate Change?

This section considers the economic theory behind emissions trading (or, put more accurately, the use of marketable allowances or permits), briefly reviews its application to date and considers its suitability for the regulation of greenhouse gases.

A.   Economic Theory behind Emissions Trading[57]

The use of marketable permits as a tool for achieving cost-effective environmental regulation was first proposed by Dales in Canada in the late 1960s as an alternative to the use of environmental fees or taxes.[58] According to economic theory, the regulation of a particular pollutant is cost-effective where the marginal cost of control for a polluter is equal to the marginal damage caused by the pollution which he emits.[59] Under the correct conditions, a system of marketable permits can result in this optimal situation whereby the cost of reaching a particular environmental outcome is minimised.

There are two basic approaches to the use of marketable permits: 'cap and trade' schemes and 'baseline and credit' schemes. The EU ETS is a cap and trade scheme. The UK ETS, as applied to direct participants, is also a cap and trade scheme but the CCA regime with which it is linked is a baseline and credit scheme.

1.   Cap and Trade Schemes

A cap and trade scheme (also referred to as an allowance system) is closest to the theoretical ideal and consists of the following steps:

---

[56] Details of these proposals are available at: http://www.defra.gov.uk/environment/climatechange/trading/epc/index.htm.
[57] There are a number of sources which explain the economic theory behind emissions trading in more detail. One of the most accessible is Haites, E, (2002), *An Emerging Market for the Environment: an Introduction to Emissions Trading*, Denmark, UNEP, UCCEE and UNCTAD. Available at www.unctad.org.
[58] Dales, J.H,(1968) *Pollution, Property and Prices*, Toronto, University of Toronto Press.
[59] Tietenberg, T, (1996) *Environmental and Natural Resource Economics*, New York, Harper Collins (at p. 326-332).

- An environmental authority issues a fixed number of allowances which are denominated in a quantity of pollutant to be emitted over a specified time period. The result of this process is to establish a 'cap' on emissions which is exactly equal to the desired emissions level.

- The allowances are then distributed to individual sources. The basic options for allocation are free allocation (either by 'grandfathering' on the basis of historic emissions, or by benchmarks based on products or technologies), sale (ideally via auction), or a combination of the two.

- Obligations are then imposed on affected sources (i) to monitor and report their emissions and (ii) to comply with an emission limitation obligation whereby at the end of the period each source must surrender one allowance for each unit of pollution emitted.

- A scheme providing for the unrestricted transfer of allowances is established through which allowances may be bought and sold by both affected sources and third parties.

According to economic theory, the natural operation of the emissions allowance market will ensure that the cap is achieved at minimum cost. Sources included in the scheme may comply with their legal obligations by limiting their emissions to the level covered by the allowances allocated to them; exceeding that level and buying allowances from other sources; or reducing emissions more than required in order to sell allowances (or bank them for later use if that is permitted under the scheme in question). Provided that the cap is tight enough to impose a binding constraint on emissions (typically because the total quantity of allowances is smaller than the aggregate level of emissions from all sources in the pre-control years), there will be scarcity rents attached to the possession of an allowance, at least until technological progress catches up with the cap so that industry can work within it. In other words, because demand for allowances exceeds supply, allowances carry a positive value.

As the market for allowances develops, the market price indicates the costs of each unit of pollution. In deciding whether on not to emit additional pollution, each source must consider the cost of either purchasing an additional allowance or incurring the 'opportunity cost' of not selling an allowance it already holds which would otherwise be surplus. Through trading, a firm with high abatement costs can reduce costs by purchasing allowances from a firm with lower abatement costs. Conversely a firm with low abatement costs can take advantage of that fact by reducing more than required and selling allowances for more than the abatement measures cost. As such cost-minimising behaviour is encouraged by the market, marginal abatement cost is equalised among the sources covered by the scheme.

A remarkable aspect of this apparently straightforward approach is that overall costs of compliance will be minimised regardless of the method of allocation (e.g. by auction or free of charge) because their subsequent transferability means that allowances will ultimately be used where their value is highest. In theory at least, it is not necessary for the control authority to have detailed knowledge about control costs prior to making the initial allocation decisions.[60] Perhaps more significantly, this means that the initial distribution of allowances may be designed in order to address other objectives, (such as gaining industry support for a trading scheme or broader equity issues), without undermining the cost-effectiveness of the scheme.[61]

2.   Baseline and Credit Schemes

The basic alternative to a cap and trade scheme is a 'baseline and credit' scheme, whereby credits are earned if reductions exceed a specified baseline. The baseline or 'target' can be either relative or absolute. The key distinction between this approach and cap and trade is that allowances are not distributed *ex ante* but only when an installation has over-achieved its target. Similarly, an installation which has under-achieved its target can then purchase allowances to make up any shortfall.

Unlike cap and trade schemes, baseline and credit schemes can be based on either absolute or relative targets and occasionally both.[62] Under a relative target regime, there is no absolute cap on the emissions from the sectors covered by the scheme. Instead, the target is expressed in terms of emissions intensity per unit of production, input or activity. Companies in sectors which are experiencing growth in output and therefore emissions are highly likely to favour a relative target regime, not least because economies of scale mean that they may well be able to improve their environmental efficiency as production levels increase. However, as a regime based on relative targets does not set an absolute

---

[60] Tietenberg T, supra note 59, at p.338.
[61] Tietenberg, T, *The Tradable Permits Approach to Protecting the Commons* (in Helm (2005) (supra note 1)).
[62] Uniquely, emissions trading in the UK incorporates in the UK ETS both a cap and trade system (with absolute targets) and through CCAs a baseline and credit system (with both absolute and relative targets). See chapters 22 and 23 for more details.

limit on the level of emissions, policy makers and environmental NGOs favour regimes based on absolute targets.

Baseline and credit schemes have generally not been as successful in achieving environmental benefits as cap and trade schemes when used to date.[63] This appears to be largely the result of uncertainties involved in setting baselines, in particular when reductions are to be measured against business as usual emissions[64] and the high transaction costs in particular those involved in obtaining certification of emissions reductions. Baseline and credit schemes are also not compatible with the distribution of 100 per cent of allowances by auction. However, it does not follow that such schemes cannot be effective tools for reducing emissions: the UK system of Climate Change Agreements is an example of a baseline and credit system which not only incorporates both absolute and relative targets but has also achieved substantial emissions reductions.[65]

> 3.   The Advantages and Disadvantages of Emissions Trading as a Policy Tool

Given the characteristics discussed above, an appropriately designed emissions trading system can provide advantages over traditional command and control regulation and other economic instruments such as taxation[66] for policy makers, regulators, the regulated industry and the environment.

For policy makers, emissions trading enables the total level of emissions to be set globally in advance and reduces the need to regulate individual plants. Providing the scheme is properly enforced, emissions trading can ensure that a specific environmental outcome (defined in advance) is met in the most economically efficient way. Once the regulatory framework for the scheme is established and systems for tracking and surrendering allowances are in place, policy makers can concentrate resources on setting the emissions cap at an appropriate level and

---

[63] See Tietenberg, T, Grubb, M et al. (1999). *International Rules for Greenhouse Gas Emissions Trading*; Geneva, United Nations. (Executive Summary available at http://r0.unctad.org/ghg/download/other/intl_rules_execsum.pdf.) and Tietenberg, T, (supra note 59).

[64] For example, under the project mechanisms under the Kyoto Protocol credits are issued for reductions which are additional to the level of emissions which would have otherwise occurred. The difficulties of demonstrating this counter-factual concept of 'additionality' has been one of the major challenges faced by those working on the development of the Kyoto mechanisms and there is a real risk that credits may be issued which are not fully justified.

[65] See chapter 23.

[66] Baumol and Oates, (1995) *The Theory of Environmental Policy*, Cambridge, Cambridge University Press. For a discussion of the advantages of tradable permits over taxes, see Stavins, R.N 'What Can we Learn from the Grand Policy Experiment? Lessons from SO$_2$ Allowance trading', *Journal of Economic Perspectives*, (1998) 12(3), 69-88.

allocating allowances to affected sources in a suitable way.[67] Trading is particularly well-suited where the environmental objective is to reduce emissions to a specific level or achieve a specific level of ambient environmental quality.[68] Whilst taxation can deliver the same end result, determining the appropriate level to set the tax in order to achieve this outcome requires a time consuming process of trial and error.[69]

Further, emissions trading can help deliver policy objectives by placing an economic value on emissions reductions. Industry is encouraged to invest in new plant and technology as it can profit from the resulting reduction in emissions by selling surplus allowances. Emissions trading can also raise the profile of environmental issues in companies and change the way that they address environmental risk. The need to account for allowances as assets and emissions as potential liabilities in annual reports means that environmental pollution and investment in its abatement is more likely to become an issue of interest to those at the heart of a business and its strategic direction rather than simply the responsibility of an environmental officer.

For the regulator, emissions trading reduces the need for detailed evaluation to set and enforce emission limit values and other controls on each individual source. Enforcement is also simplified as the key task of checking that allowances are surrendered to cover the reported emissions from each source can be carried out automatically. The task of ensuring that reported emissions are accurate can also to a large extent be outsourced to third party verifiers.[70]

The regulated industry, on the other hand, is provided with much greater flexibility in how to meet its legal obligations. The costs of complying with an emissions trading scheme as a whole are typically less than under traditional command and control regimes, as the trading aspect ensures that reductions take place where they can be achieved most cost-effectively.[71] Temporal flexibility can also be included if the scheme allows surplus allowances to be banked for use in future years or borrowed from future allocations.

[67] It must be borne in mind that experience with the EU ETS to date suggests that these tasks are far from straightforward or uncontroversial.
[68] Stewart, R.A. 'Economic Incentives for Environmental Protection: Opportunities and Obstacles', in Revesz, R.L., Sands, P. and Stewart, R.A (ed.), (2000) *Environmental Law, the Economy and Sustainable Development*, Cambridge, Cambridge University Press.
[69] Tietenberg, T, supra note 59, p.336 and Baumol and Oates, supra note 66.
[70] Indeed recent systems do provide for the partial outsourcing of these activities to third party verifiers (see chapter 3, section III.B).
[71] Of course, the costs for any particular source (rather than the scheme as a whole) will be driven by the amount of allowances allocated to that source, which will determine whether it is a buyer or a seller in the market. This is the reason why allocation has proved to be the most controversial aspect of emissions trading in practice.

Although some NGOs remain opposed to what they see as the commoditisation of the environment, emissions trading can even offer advantages to environmental NGOs. While the existence of sellers in a trading regime was initially seen as evidence that tougher targets should have been set, caps imposed under trading schemes can be lower than the aggregate of command and control limits. The lower compliance costs for achieving a specific environmental objective may enable regulators to impose more stringent targets or earlier deadlines for reductions. In addition there is no longer any need to build headroom for emergencies and breakdowns into individual site-specific limits as operators are able to insure against the risk of having to purchase additional allowances in the event of an emergency. Emissions trading schemes may also be more transparent and accessible than traditional command and control schemes: anyone wishing to challenge the environmental effectiveness of the trading regime can question directly the level of the overall cap rather than having to unravel the, often complex, relationship between the specific controls applied to an individual plant and an ambient environmental quality standard.[72] These advantages, together with the increased likelihood of achieving the environmental objective embodied in the cap, mean that many NGOs have now largely endorsed emissions trading as an appropriate regulatory mechanism to require operators to internalise environmental costs and thereby move consumer demand towards less polluting companies.

A possible drawback of emissions trading as compared to taxation is that it does not set a limit on compliance costs. If the cap set is too stringent, the marginal abatement costs and therefore the price of allowances will be high and the market may break down. By contrast, under a taxation scheme, the costs of abatement are naturally capped at the level of the tax. One way of ensuring that allowance prices do not rise too high is to include a mechanism in the trading scheme to prevent allowance prices from rising above a certain level. To date, there are no examples of how a 'price cap' of this type might operate in practice.[73] However, in theory, a trading scheme with a price cap suffers from the same problems as a pure taxation regime: there is no absolute cap on emissions and the same difficulties arise in setting the level of the maximum allowance price. Alternative ways of dealing with the risk of high allowance prices may achieve the same results as a 'price cap'

---

[72] Stewart, supra note 68.

[73] To date, Canada is the only jurisdiction to have committed to the use of a price cap (although the change of government in 2005 has now thrown their plans to legislate in this area into doubt). The Government guaranteed that the cost of compliance to industry regulated under its proposed Large Final Emitters system would not be more than $15 per tonne of carbon dioxide equivalent for the period 2008-12. The options under consideration include: the issuance of special credits, payment into a special fund and a rebate on verified costs which exceed $15 (see Notice of Intent to Regulate at: http://www.ec.gc.ca/CEPARegistry/notices/NoticeText.cfm?intNotice=318&intDocument=2156).

without undermining the other advantages of trading. These include the use of less stringent caps in the early years of a trading scheme coupled with the freedom to bank surplus allowances for use in later years with more stringent caps[74] and allowing regulated sources to offset their emissions against reductions which have taken place outside the scheme.[75]

One major theoretical disadvantage of emissions trading is that it does not regulate the geographical location of emissions since the only cap on emissions from an individual sources is the number of allowances it can acquire from the market. This makes emissions trading most suited to regulate problems such as climate change where the source of the emission does not effect the atmospheric concentration of a given greenhouse gas.[76] Considerable efforts have been expended on designing methods of accommodating the geographical dimensions of pollution problems into the design of trading schemes.[77] However, these solutions tend to result in additional regulatory restrictions which adversely impact on the liquidity of the market and thus undermine the economic efficiency of the scheme.[78]

B.   Early Experience of Emissions Trading

The popularity of trading schemes as a tool for environmental regulation has increased dramatically over the last two decades. Although the theory was originally promoted in the 1960s, it is only during recent years that the concept of emissions trading has evolved from being a theoretical model for minimising the costs of regulation favoured by economists to becoming the policy instrument of choice for both regulators and the regulated. Trading schemes have been advanced as a solution for a wide variety of environmental problems including regulating air pollutants from sources ranging from power plants to

[74] Jacoby, H.A. and Ellerman, A.D, (2002), *The 'Safety Valve' and Climate Policy*, MIT Joint Program on the Science and Policy of Global Climate Change, Report No.83.
[75] This was the approach taken under the EUETS which allows operators to use credits from the Kyoto Protocol project mechanisms for compliance. See further chapter 8 below.
[76] However, greenhouse gases other than carbon dioxide may also have local impacts and therefore the potential impact on local pollution cannot be ignored.
[77] For a theoretical illustration, see Baumol and Oates, supra note 66. For practical examples see: the RECLAIM scheme regulating air pollution in Los Angeles where the number of allowances which a buyer must purchase in order to offset a unit of pollution varies according to the location of the seller (Stewart, supra note 68); the proposed Clear Skies Nitrogen Oxides Program in the United States, where there are two trading zones, East and West, and no trading between the two zones (see www.epa.gov./clearskies/clearskiessummary04-11.pdf).
[78] In practice, studies suggest that emissions trading does not lead to increased incidence of pollution 'hot spots' – see Swift, B., (2000), *Allowance Trading and Potential Hot Spots – Good News from the Acid Rain Program*. Environment Reporter, Vol. 31, No. 19, p.954-959; Farrell, A., Carter, R. and Raufer, (1999), R. *The NO_x Budget: market-based control of tropospheric ozone in the northeastern United States*. Resource and Energy Economics, Vol. 21(2), p.103-124.

domestic wood-burning stoves, controlling the discharge of effluent into rivers or waste to landfill and regulating exploitation of common resources such as fish stocks.[79]

The current popularity of trading schemes can in part be traced back to the successful use of the mechanism in the United States Acid Rain Program (ARP) to control emissions of sulphur dioxide from power plants. Initial experiences with emissions trading in the United States in the 1970s were important as first steps but were modest and largely unsatisfactory, not least because they tended to be credit systems employed as an afterthought on top of traditional command and control regimes. However, US regulators built upon this early experience and in 1990 amendments were made to the Clean Air Act to establish the US Acid Rain Program which uses trading as a mechanism to address sulphur dioxide emissions from electricity utilities.[80]

The ARP, which is generally acknowledged to be the most successful emissions trading scheme introduced to date, was designed to cut emissions of sulphur dioxide by approximately 50 per cent below 1980 levels by the year 2000. The first phase, which ran from 1995 to 1999, regulated emissions from the 263 highest emitting units. Phase II, which commenced in the year 2000 and set a permanent cap of 8.95 million allowances, extended restrictions to smaller units and brought the total number of units regulated to over 2000.

Observers have concluded that the ARP regime clearly demonstrated that the theory of large scale tradeable permit regimes can work in practice.[81] The ARP more than achieved its environmental objective: 1996 emissions were at 5.43m tonnes, well below the cap of 8.12m, and it did so on time, without excessive litigation, at lower costs than expected, and with 100 per cent compliance.

---

[79] A 1999 OECD survey noted that marketable permits systems were operational in the following areas of regulation: air pollution (9); fisheries (75); water resources (3); water pollution (5); and land-use control (5). Since the date of that survey, many more schemes relating to climate change have been established.

[80] The amendments introduced a new Title IV into the Clean Air Act. Under the scheme, each allowance permits a plant to emit one tonne of sulphur dioxide in or after a specified year. There is no restriction on the banking of allowances but allowances could not be borrowed from future years. Plants are required to install continuous emission monitoring equipment to record their emissions. The penalty for excess emissions is currently US$2,000 per tonne (adjusted for inflation). Full details of the ARP are available on the US Environmental Protection Agency's website http://www.epa.gov/airmarkets/index.html.

[81] Ellerman conducted a comprehensive analysis of the ARP. See Ellerman, A.D., Joskow, P.L., Schmanlensee, R., Montero J-P. and Bailey, E. (2000) *Markets for Clean Air: The US Acid Rain Program*, Cambridge, Cambridge University Press.

The following high-level lessons can be drawn from experience with the ARP and other trading schemes:[82]

- The initial distribution of allowances is the most controversial aspect of the design of a trading scheme. The price for acceptance of a trading scheme by established industry will usually be distribution of allowances for free to existing installations. There will be winners and losers whatever the method of free allocation of allowances and regulators should expect lobbying to be heavy. Even in trading schemes where allocation appears to be transparent, detailed analysis shows that allocations have been the result of hard-nosed negotiations.[83] The most successful schemes have managed to separate completely the negotiations over the distribution of allowances from the setting of the total cap, so that the distribution of allowances becomes a zero-sum game and does not affect the environmental outcome.[84]

- Effective enforcement is fundamental to the success of a scheme: there will be no motivation to buy allowances if operators do not believe that they will actually be required to surrender an allowance for each unit of pollution. The most successful schemes benefit from high, automatic and well-enforced penalties for failure to hold allowances equal to emissions.[85]

- The emissions controlled by the scheme must be capable of measurement with sufficient accuracy and at a cost which is not excessive. Moreover, trading regimes are particularly effective at improving records of emissions and bringing errors in pre-existing data to light.

---

[82] It should not, however, be assumed that the ARP provides a simple template for the establishment of an environmental trading scheme - the complicated rules of the ARP were essential to its success and regulators should extrapolate with care.

[83] For a detailed account of allocation under the ARP, see chapter 3 of Ellerman (supra note 81).

[84] E.g. the ARP (Ellerman, supra note 81), the New Zealand ITQ system (OECD Annex I Expert Group, *Lessons from Existing Trading Systems for International Greenhouse Gas Emission Trading*, (1998), available at: http://www.oecd.org/env/cc/mechanisms.htm) and the NOx OTC (Farrell, A, (2001), 'Multilateral emissions trading: lessons from inter-State NOx control in the United States', *Energy Policy* Vol.29, p1061-76).

[85] Cap and trade schemes to date have generally operated on the basis of seller, as opposed to buyer, liability. Under seller liability, a seller is responsible for holding allowances equal to his emissions and any non-compliance on his part does not affect the validity of allowances which he has already sold. Buyer liability is more complex, as buyers are accountable for the non-compliance of the person who has sold them the allowance.

–   There is a need to strike a careful balance between flexibility and stability. US experience of air quality trading schemes demonstrates that new trading regimes rarely function perfectly from the outset but instead need to be developed and improved over time as problems are identified. It is therefore important to build in review mechanisms and to retain flexibility to change the overall cap in order to respond to improvements in the understanding of the environmental impacts of the regulated activity. However, this flexibility needs to be carefully balanced against the need for regulatory certainty to enable participants to make informed investment decisions. The introduction of a scheme is likely to be easier where the policy makers and regulators have a good understanding of the industry to be regulated.

C.   Application of the Theory to the Problem of Greenhouse Gas Emissions

Cap and trade systems are exceptionally well suited to dealing with the challenge of limiting emissions of greenhouse gases:[86]

–   Climate change is the result of a uniform mix of gases in the atmosphere as a whole rather than in any one location. Therefore, the location of an emissions source is not significant to its climate impact;

–   Greenhouse gases are emitted from a wide variety of sources in many countries, with large differences in marginal abatement costs both between sectors and countries. Trading should mean that abatement takes place in the sectors and countries where the most cost-effective options are possible;

–   The costs of abating global greenhouse gas emissions are potentially very high, so every effort should be made to limit these costs so far as possible. Moreover, it is likely that the limitation of greenhouse gases to levels necessary to avoid damaging climate change will not be possible without achieving dramatic changes in both consumption attitudes and technology. A system of marketable permits is much more likely to stimulate such activity than more traditional forms of environmental regulation.

---

[86] See in particular Tietenberg, supra note 59 and Stewart, supra note 68.

However, the application of emissions trading to greenhouse gases also raises particular challenges:

–   Emissions trading schemes have in the past been employed only where traditional forms of regulation (such as command and control quantitative limits) have been tried and failed. Generally speaking, emissions of greenhouse gases have not been regulated to date. This places policy makers at a double disadvantage. Firstly, there is limited practical experience among regulators and industry of methods to control greenhouse gas emissions. Secondly, and possibly more significantly, there is no baseline experience against which the costs of complying with the trading scheme can be compared. While compliance with a trading scheme is very likely to cost less than other forms of regulation, it will cost more (in the short term) than no regulation, leaving policy makers with a difficult job when it comes to winning the hearts and minds of the regulated industry.

–   Further, the effect of emissions of greenhouse gases on climate change can only be fully addressed if all countries take action to reduce emissions. The consequences of damage caused by emissions of greenhouse gases are neither immediate nor linked to a particular territory but will be born by future generations and other countries. Concerns about free-riders act as a deterrent to unilateral national action. There is also an issue of 'leakage' where the geographical coverage of a pollutant is initially only partial. Unless and until there is global regulation of greenhouse gas emissions, there is a theoretical risk that production facilities will move outside the regulated zone, potentially into areas where technology is more carbon intensive.

Although evidence suggests that this is unlikely to happen in practice because of other more significant factors detemining the location of production[87] the apparent force of this argument means that it cannot be ignored by regulaions.

–   Finally, long-term certainty is likely to be of particular importance in relation to greenhouse gas emission trading schemes. A large percentage of greenhouse gas emissions

---

[87] See Stern, supra note 4.

result from the burning of fossil fuels for electricity generation. The length of the investment cycle involved in constructing power stations means that it will be important to ensure that incentives to make reductions and investments are engaged at the earliest possible stage and to reinforce such incentives by allowing early reductions to be banked for use in later years and guaranteeing that the allocation methodology for later years does not disadvantage those who take early action.

Practical experience in the EU and UK of the use of emissions trading to regulate emissions of greenhouse gases will undoubtedly provide many lessons for the development of trading schemes. These lessons are already being applied in Europe, both in the future development of the EU ETS[88] and in the proposals for a new domestic trading scheme in the UK[89] and will be of interest to other countries developing similar schemes.

The remainder of this book sets out how the challenge of employing emissions trading as a tool for addressing climate change has been tackled in practice in the EU and in the UK. Chapter 1 summarises the key stages in the evolution of the EU ETS. Chapters 2 to 10 provide a detailed analysis of the Emissions Trading Directive and chapters 11 to 21 the legal and policy instruments which implemented the Emissions Trading Directive in the UK. Chapters 22 and 23 provide a detailed explanation of the domestic emissions trading instruments established in the UK before the emergence of the EU ETS, namely the UK ETS and CCA regimes.

---

[88] See further chapter 10.
[89] See section III.C above and chapter 22.

Part I

The EU Emissions Trading Scheme

CHAPTER 1

KEY ELEMENTS AND THE EVOLUTION OF THE EU EMISSIONS TRADING SCHEME

This chapter identifies the key EU legislation governing the EU ETS (section I). It then provides an account of the early policy discussions which paved the way for the adoption of the Emissions Trading Directive (section II) and discusses the formal negotiation of that Directive (section III) and of the so-called Linking Directive (section IV).

I.   Legislative Basis and Key Elements of the EU ETS

The basic instrument governing the EU ETS is the Emissions Trading Directive (the Directive) which was adopted in 2003.[1] The Directive was amended in 2004 by the so-called 'Linking Directive' which provided for a link between the EU ETS and the flexible mechanisms under the Kyoto Protocol.[2] The Directive is further supplemented by delegated legislation adopted by the Commission under the Directive in particular the Registries Regulation (adopted in 2004) which governs the establishment and operation of electronic registries to track emission allowances.[3]

The key elements of the scheme can be summarised as follows:

–   The scheme began operation on 1 January 2005.
–   The first phase covers a three-year period from 2005-2007. The second and subsequent phases cover five year periods (2008-2012, 2013-2017...).
–   It initially covers carbon dioxide emissions from the big industrial emitters but it is envisaged that other activities and gases will be included at a later stage.

---

[1] Directive 2003/87/EC of the European Parliament and of the Council of 13 October 2003 establishing a scheme for greenhouse gas emission allowance trading within the Community and amending Council Directive 96/61/EC, OJ L275, 25.10.2003, p.32.
[2] Directive 2004/101/EC of the European Parliament and of the Council of 27 October 2004 amending Directive 2003/87/EC establishing a scheme for greenhouse gas emission allowance trading within the Community, in respect of the Kyoto Protocol's project mechanisms, OJ L338, 13.11.2004, p.18.
[3] Commission Regulation No. 2216/2004 for a standardised and secured system of registries pursuant to Directive 2003/87/EC of the European Parliament and of the Council and Decision 280/2004/EC of the European Parliament and of the Council, OJ L386, 29.12.2004, p.1.

–   Operators of installations covered by the scheme are required to monitor their emissions each year and report them to the competent authority by 31 March the following year and to surrender allowances to cover those emissions by 30 April.

–   A financial penalty is applied to operators which fail to surrender sufficient allowances.

–   Allowances are allocated for each phase of the scheme on the basis of national allocation plans prepared by each Member State.

–   Operators may also use credits from the Kyoto flexible mechanisms although from 2008 a limit will apply to such use.

–   Allowances can be traded freely within the EU and can be used by operators in any Member State.

–   Each Member State operates an electronic registry to record allowance transactions.

II.   Background to the Emissions Trading Directive

A.   Commission Communications on the Implementation of the Kyoto Protocol (1998 and 1999)

The origins of the EU's adoption of market-based instruments for environmental regulation can be traced back to the European Community's Fifth Environmental Action Programme.[4] However the concept of a European emissions trading scheme for greenhouse gases only really took off following the agreement on flexible mechanisms as part of the Kyoto Protocol. Once the Protocol was adopted in December 1997 progress in the EU was swift and political momentum for setting up a pilot emissions trading scheme in advance of the first Kyoto commitment period soon gathered.

In June 1998, the Commission published a Communication entitled 'Climate Change - Towards an EU Post-Kyoto Strategy'[5] in which it sought to set out its first analysis of how to shape the European Community's strategy for meeting its commitments under the Protocol. The Communication highlighted the Community's role in 'complementing, reinforcing and supporting Member States' actions with common and co-ordinated policies and measures'. It also recognised the important role which the flexible mechanisms under the Kyoto Protocol could play in allowing Member States to reduce the costs of

[4] Supra introduction, note 28.
[5] Communication from the Commission to the Council and the European Parliament, Climate change – Towards an EU post-Kyoto strategy, COM(1998) 353 final.

meeting their commitments, thereby safeguarding the competitiveness of EU industry. The Communication proposed that the Community could set up its own internal trading regime by 2005 in order to provide early experience of trading and the necessary accompanying monitoring. It justified Community action in this field on the grounds of avoiding distortion of competition and discrimination between the Member States. Recognising that practical experience of emissions trading was limited, it proposed a step-by-step approach, rather than immediately opting for 'a comprehensive internal emissions trading scheme covering all gases and all economic sectors'.

In May 1999, the Commission published a a further Communication entitled 'Preparing for Implementation of the Kyoto Protocol'[6] which recognised that, in light of increasing emissions, much more needed to be done at all levels if the Community was to meet its commitments under the Kyoto Protocol. The Communication focused on common and co-ordinated policies and identified potential priority areas for further Community action, including energy and fuel taxation.

The Communication also contained a discussion of how the Community should prepare for implementation of the flexible mechanisms under the Kyoto Protocol. It recognised that the flexible mechanisms were 'fundamentally different from the way the European Community and its Member States have organised their environmental policy over the last decades' and that therefore an informed debate on emissions trading was needed, drawing on trading initiatives already under way in some private companies and Member States.

To this end, the Communication proposed a wide consultation on options for a Community-wide trading scheme, focusing on: whether and, if so, how private entities could become involved in emissions trading; the point of regulation (i.e. whether to regulate emission sources themselves or energy producers, such as coal mines and oil and gas suppliers); and how to organise the initial allocation of allowances.

B. Green Paper on Greenhouse Gas Emissions Trading within the European Union, March 2000

In March 2000, building on the consultation exercise and a series of studies commissioned by the Commission into the options for developing an emissions trading scheme,[7] the Commission published a

---

[6] Communication from the Commission to the Council and the European Parliament, Preparing for Implementation of the Kyoto Protocol, COM(1999) 230 final.

[7] Center for Clean Air Policy (CCAP), (1999), *Design of a Practical Approach to Greenhouse Gas Emissions Trading Combined with Policies and Measures in the EC* (available at: http://

(continued...)

Green Paper on Greenhouse Gas Emissions Trading within the European Union.[8] The Green Paper acknowledged that by establishing a trading scheme by 2005 the EU would be not only be preparing for trading under the Kyoto Protocol but also moving ahead of the international process.[9] It was therefore important to design any scheme so that it was open to gradual extension in terms of the countries, sectors and gases covered.

The Green Paper made the case for action at Community level rather than a set of uncoordinated national schemes. The arguments for an EU–wide trading scheme included: the potential negative impact on the internal market of different approaches being taken in different Member States; the establishment of a single market with a single price for carbon across the EU; and cost savings of nearly one fifth compared to the costs of each Member State seeking to meet its commitment under the Burden Sharing Agreement with no cross-border trading.[10]

The Green Paper posed a set of questions for stakeholders on key aspects of the design of a trading scheme, namely, the coverage of sectors and of Member States, the respective roles of the Commission and Member States in determining the total quantity and distribution of allowances, and the mechanisms for ensuring compliance. These are considered in turn below.

1.   Coverage: Sectors and Gases

The proposal was for a scheme which focused initially at least on large fixed point sources of carbon dioxide. The sectors proposed were very similar to the sectors covered in the Directive as finally adopted. One significant difference was that the Green Paper proposed a threshold of 50MW rated thermal input for the heat and power sector (as opposed to the 20MW threshold in the subsequent formal Commission proposal

www.ccap.org/pdf/ECtrading.pdf); Foundation for International Environmental Law and Development (FIELD), (2000), *Designing Options for Implementing an Emissions Trading Regime for Greenhouse Gases in the EC* (available at: http://www.field.org.uk/PDF/ecet.pdf).
[8] Green Paper on greenhouse gas emissions trading within the European Union, COM(2000) 87 final.
[9] The paper was particularly prescient in considering the issues raised by the establishment of links between domestic schemes, acknowledging in particular the impact of linkage on the price of allowances (see further chapter 10).
[10] These savings (estimated at €1.7bn per year) are likely to be an underestimate because the baseline assumes that Member States would in the absence of a Community scheme carry out emissions trading effectively within their borders. See Annex I to the Green Paper and background analysis in: National Technical University of Athens, (2000), *The Economic Effects of EU-wide Industry-Level Emissions Trading to Reduce Greenhouse Gases – Results from the PRIMES model* and Institute for Protective Technological Studies (IPTS), (2000) *Preliminary Analysis of the Implementation of an EU-Wide Permit Trading Scheme on CO$_2$ Emissions Abatement Costs - Results from the POLES model* (both are available at: http://europa.eu.int/comm/environment/enveco/studies2.htm).

and ultimate Directive). The proposal estimated that the sectors covered by the proposal would be responsible for 45.1 per cent of the EU's carbon dioxide emissions.

2.   Coverage: Member States

While the Green Paper recognised that uniform coverage would be optimal, it acknowledged that not all Member States were likely to be ready to participate in an EU scheme at the same time. It therefore proposed two alternative approaches to securing full coverage over time: an opt-in approach whereby Member States could choose whether they wished to be part of the scheme; or an opt-out approach (which the Green Paper favoured) whereby Member States could opt out either by sector or completely for a limited period. Under either approach, sectors not covered by the scheme would need to be regulated by other policies and measures representing at least a similar economic effort in terms of emissions abatement.

3.   Total Quantity of Allowances and their Distribution: Roles of the Commission and Member States

The Green Paper acknowledged the central importance of defining the overall quantity of allowances for the trading sector and that this had to be based on a fair balance between the contribution of sectors covered by the trading scheme and other sectors. While it acknowledged that each Member State would have to decide how much reduction should take place through the scheme and how much through other policies and measures, it appeared to propose that, at least after 2008, Member States would have to agree collectively on how much would be allocated to the trading sectors in each Member State in order to provide a transparent framework within which the distribution of allowances could take place.

Notably, the Green Paper made a clear distinction between the setting of the total quantity of allowances and their distribution, and proposed limiting the role of the Commission to the latter. Although the distribution of allowances to companies would not affect the overall environmental outcome, the Green Paper acknowledged that negotiations on this issue would not be easy and that Member States may seek to favour some sectors more than others by setting unchallenging targets. The Commission would, therefore, have a key role to play in assessing proposed allocations under existing EU rules on state aid and the internal market. The specific question raised was whether these powers should be supplemented by specific powers and guidelines relating to the trading scheme.

The Green Paper also discussed the key options for the distribution of allowances (free allocation through grandfathering or benchmarking, or auctioning) and the need to take careful account of the situation of new entrants to avoid any discrimination against companies wishing to enter the market.

4.   Compliance and Enforcement

The Green Paper stressed the importance of strict compliance and enforcement provisions to ensure the environmental integrity of the trading scheme. A number of proposed compliance and enforcement mechanisms were suggested including the possible obligations relating to verification of emissions reports and the need for minimum monetary penalties fixed at EU level and exceeding the cost of complying with the basic obligations under the scheme.

Following the publication of the Green Paper, the central idea of an EU-wide trading scheme received a very high level of support from representatives of industry and NGOs and other European institutions. Both the Council and European Parliament welcomed the Green Paper and encouraged the Commission to move forward quickly in its work on establishing an EU emissions trading scheme.[11]

C.   Early Discussions in the Working Group on Flexible Mechanisms

Alongside the Green Paper the Commission also adopted a Communication entitled 'EU policies and measures to reduce greenhouse gas emissions: Towards a European Climate Change Programme (ECCP)'.[12] One of the working groups set up under the ECCP focussed on the flexible mechanisms under the Kyoto Protocol. The working group met from July 2000 and the Commission dedicated a large part of the discussions to an exchange of views and expertise on the shape of the proposed emissions trading scheme.[13] The final report of that working group,[14] published in July 2001, made the following recommendations:

---

[11] See Council Conclusions (Environment) 22 June 2000, Luxembourg; and European Parliament Resolution on the Commission Green Paper on greenhouse gas emissions trading within the European Union A5-0271/2000 OJ C 197, 12.7.2001, p.400. It is worth noting that the Parliament placed more emphasis on the need for an EU-wide scheme to be established quickly and made specific recommendations about its future shape.

[12] Supra Introduction, note 37.

[13] Details of these meetings are available at: http://europa.eu.int/comm/environment/climat/eccpl.htm.

[14] Available at http://europa.eu.int/comm/environment/climat/pdf/final_report.pdf.

a) The emissions trading scheme should start as soon as practicable and should not wait for progress in international fora. Rather it should be developed in the context of, and with a view to influencing the design of, an international scheme.

b) The EU framework should be established with Member States allowed to choose their own initial method of allocation including the use of auctioning, although the methodology chosen should encourage early action.

c) Absolute targets must be at the core of the scheme, but there could be a limited role for relative targets.[15]

d) The regulation of carbon dioxide should be at the core of the scheme, but it should be designed with a view to expansion to as many sectors, entities and gases as possible.

e) The use of credits from the Kyoto project mechanisms should be encouraged and consideration should be given to the establishment of mechanisms for generating credits from domestic projects which could be used to fulfill obligations under the scheme.

f) A high standard of monitoring, reporting and verification would be crucial for guaranteeing the environmental and financial credibility of emissions trading, both within the EU scheme and between linked national schemes.

g) Speedy and automatic financial penalties would be necessary, with minimum levels harmonised at the EU level.

III. The Negotiation of the Emissions Trading Directive

The formal legislative procedure for the Emissions Trading Directive began on 23 October 2001, when the Commission published its Proposal for a Framework Directive for Emissions Trading within the European Community,[16] alongside a draft Decision for the ratification by the EU of the Kyoto Protocol.[17] The timing of the publication was designed to

[15] The concept of relative targets was not included in the final Directive.
[16] Proposal for a framework Directive for Greenhouse Gas Emissions Trading within the European Community, COM(2001) 581 final, OJ C 75E, 26.3.2002, p33.
[17] Proposal for a Council Decision concerning the approval, on behalf of the European Community, of the Kyoto Protocol to the United Nations Framework Convention on Climate Change and the joint fulfilment of commitments thereunder, COM(2001) 579 final, OJ C 75E, 26.3.2002, p17.

reinforce the EU's position in the international climate change negotiations in Marrakech which took place shortly afterwards, and in particular to demonstrate that the European Commission had a clear vision of how greenhouse gases would be regulated in the future.

Negotiations on the Directive in the Council and the European Parliament under the co-decision procedure[18] proceeded rapidly over the next two years. The pace reflected the importance which the Member States and the Community institutions placed on ensuring that the scheme was in place well before the start of the first Kyoto Protocol commitment period in 2008. As a result of the shared political will to establish the scheme, the Council managed to agree its common position under the Danish Presidency on 9 December 2002.[19] Even more remarkably, it was then possible for the text of the Directive to be agreed between the European Parliament and the Council in its second reading in the European Parliament,[20] thus avoiding the need to convene the conciliation committee: a rare occurrence for a Directive with such wide-ranging implications.

The formal negotiations revolved around the same issues as arose during the Green Paper process described above.

---

[18] The co-decision procedure is the decision making procedure prescribed for measures to be adopted under Article 175(1) of the EC Treaty. Under the procedure, measures must be jointly agreed by the Council and the European Parliament before they can become Community law. If agreement is not reached after the second reading in each institution, a conciliation committee, made up of equal numbers of members from the Council and the European Parliament, is convened and tries to negotiate a compromise which can be agreed by both institutions. For further details of the co-decision procedure see Craig and de Burca, G, (2002), *EU Law: Text, Cases and Materials*, (3rd ed.), Oxford, Oxford University Press, p.144-7.

[19] Common Position adopted by the Council on 18 March 2003 with a view to the adoption of Directive of the European Parliament and of the Council establishing a scheme for greenhouse gas emission allowance trading within the Community and amending Council Directive 96/61/EC, 15792/1/02. Available at: http://register.consilium.eu.int/pdf/en/02/st15/st15792-re01en02.pdf See also Amended proposal for a directive of the European Parliament and of the Council establishing a scheme for greenhouse gas emission allowance trading within the Community and amending Council Directive 96/61/EC (presented by the Commission pursuant to Article 250 (2) of the EC Treaty), COM(2002) 680 final, and Commission Communication to the European Parliament concerning the Council's Common Position on the adoption of a Directive establishing a scheme for greenhouse gas emission allowance trading within the Community and amending Council Directive 96/61/EC, SEC(2003) 364 final.

[20] See Opinion of the Commission on the European Parliament's amendments to the Council's common position regarding the proposal for a Directive of the European Parliament and of the Council, COM(2003) 463 final. Available at: http://europa.eu.int/cgi-bin/eur-lex/udl.pl?REQUEST=Seek-Deliver&COLLECTION=com&SERVICE=eurlex&LANGUAGE=en&DOCID=503PC0463&FORMAT=pdf See also European Parliament first reading report, 13 September 2002, Rapporteur Moreira da Silva, PE 232.374 – the European Parliament proposed 73 amendments to the proposal at its first reading.

A.   Coverage: Sectors and Gases

Given the subsequent controversy over the definition of combustion installation and the continuing discussions about the impact of the scheme on smaller operators,[21] there was surprisingly little discussion during the negotiations of the detailed wording of Annex I to the Directive (which sets out the sectors covered and the applicable thresholds for their inclusion). This is particularly surprising given that the proposal included a threshold for combustion installations of 20 MW rated thermal input, significantly lower than the 50MW threshold proposed in the earlier Green Paper.

Instead, discussions focussed on whether to extend the coverage of the scheme at this stage, in particular in relation to gases other than carbon dioxide, and how to provide for future flexibility. The Commission Proposal envisaged that the scheme would be extended only by the adoption of further Directives adopted by the Council and the European Parliament through the full co-decision procedure. This cautious approach was founded on the concerns of the Commission and NGOs about uncertainties in the accuracy of methodologies for the monitoring of other gases.

In the end, following heavy industry lobbying and pressure from the European Parliament, Member States were allowed to provide for the unilateral inclusion of additional activities subject only to approval by a Commission Decision adopted through the comitology procedure.[22] The final Directive also mandated the Commission to consider the expansion of the scheme to other sectors both before the end of 2004 and again in a report on the scheme which it was required to produce by 30 June 2006.[23] This report was required to consider in particular expansion to the aluminium, chemicals and transport sectors.

B.   Coverage: Member States

While the Green Paper had proposed a number of possible approaches to the question of whether the scheme would be mandatory, the subsequent Commission Proposal was for a compulsory regime from 1 January 2005 onwards. Whether the scheme should be mandatory or voluntary prior to 2008 took up a large part of the discussions in the Council. Eventually, a compromise was brokered which consisted of three elements:

---

[21] See chapter 2.
[22] For a discussion of unilateral inclusion under article 24 and an account of the comitology procedure, see chapter 2, sections V and VII.
[23] See chapter 10.

a)   First, Member States may apply for installations to be temporarily excluded from the Directive in its first phase. This provision was advocated strongly by the United Kingdom which was concerned that it would be required to abandon the policies and measures which it had already put in place to tackle climate change.[24] It was only during the discussions with the European Parliament which led to the second reading deal on the Directive that the scope of this article was limited to specific installations rather than entire activities.

b)   Secondly, under pressure in particular from Spain which was concerned about the potential impact of a drought on its hydro-electric capacity, the Commission agreed to allow Member States to apply to the Commission for permission to issue additional allowances in the first phase in situations of *force majeure*.[25]

c)   Thirdly, under pressure from German industry and then the German government, provisions were introduced allowing pooling in the first and second phases of the scheme.[26] Obtaining language on pooling became a central issue for Germany during the negotiations, such that the Commission and other Member States had little choice but to agree to its inclusion in order to keep the Directive alive. The original wording which emerged during the negotiations would have allowed all EU ETS sectors in a particular Member State to have formed a single pool managed by a trustee (e.g. the Member State government). This would have effectively meant that private entities could have been completely excluded from the mechanisms under the Directive and that the Directive could have resulted in little more than trading as envisaged under article 17 of the Kyoto Protocol. However, the final compromise text was significantly revised to limit the scope of the pooling provision and to include for example

---

[24] In particular, the UK Emissions Trading Scheme and the Climate Change Agreements discussed in chapters 22 and 23 respectively. The UK was particularly concerned that its industry which had just adapted to the operation of these innovative measures should not be required to switch to the EU ETS without a transitional period. In practice, a number of Member States have also made successful applications for the use of this provision to exclude sources from the scheme. For a discussion of the Directive mechanisms for temporary exclusion, see section VI of chapter 2.

[25] For discussion of the provisions in the Directive relating to force majeure, see section III of chapter 6.

[26] For discussion of pooling, as it emerged in the final text of the Directive, see section III of chapter 7.

provisions on the residual liability of individual operators. These provisions made it unlikely that the provision would be used extensively, if at all.

C.   Total Quantity of Allowances and their Distribution: Roles of the Commission and Member States

One of the criticisms of the Directive from the perspective of economic theory is the failure to separate out the process for setting the total quantity of allowances to be allocated within the EU from the distribution of those allowances between and within Member States.[27] By combining these in NAPs, the level of ambition under the scheme risks being diluted as a result of industry lobbying over their allocations (rather than ensuring that negotiation over the distribution is a zero sum game with no impact on the environmental integrity of the scheme). The European Parliament introduced amendments calling for a fixed ceiling on the total quantity of allowances for each Member State but these were not accepted. The final compromise wording inserted additional sentences into criterion 1 of Annex III to the Directive, designed to ensure that the total quantity in each Member State in the first phase was at least consistent with a pathway to meeting the Member State's targets under the Burden Sharing Agreement and Kyoto Protocol.

The respective roles of the Member States and the European Commission were also hotly debated during the negotiation process. The Commission clearly had a role to play in ensuring that there was no distortion of competition resulting from over-allocation by any Member State. Some Member States argued that the Commission's existing powers under the EU state aids and internal market rules were sufficient to allow it to police this aspect of the Directive. In the final compromise text, Member States were given significant discretion in the allocation methodology contained in their NAPs but the Commission ended up with powers to assess and reject NAPs, as well as an obligation to produce guidance on the development of NAPs.[28]

---

[27] This approach can be compared with a number of other examples of emissions trading. For example, the national emissions cap in the US Acid Rain Programme was fixed relatively early on in the process and accepted by all, so that rather than debating the costs and benefits of the goal, efforts were dedicated to the most effective means of achieving it. In the New Zealand fisheries quota trading scheme, a change was introduced in 1990 whereby quotas were defined as proportions of the total allowable catch rather than absolute tonnages per year (OECD Annex 1 Expert Group, supra. Introduction note 84). Finally, in the Ozone Transport Commission Scheme in the Eastern States of the USA, the agreement of the overall cap on NO_x emissions was made entirely separately from the decision to use emissions trading. Consequently decisions by individual States on allocation could not affect the environmental integrity of the scheme (Farrell, see Introduction note 84).

[28] For a discussion of the substantive requirements for NAPs and the related Commission Guidance, see chapter 4. For a discussion of the procedure for NAPs, see chapter 5.

One issue relating to allocation in which the Directive adopted a more harmonised approach was auctioning. In part to address industry concerns, the Commission Proposal provided for a fully harmonised approach under which no auctioning would be permitted in the first phase and thereafter the Commission would specify through comitology the method to be used by all Member States. However, the European Parliament and environmental NGOs were strongly in favour of at least some allowances being auctioned and the final text of the Directive provides that at least 95 per cent of allowances must be issued free of charge in the first phase and at least 90 per cent in the second phase. The further harmonisation of the method of allocation of allowances, including auctioning from 2013, will be considered as part of the review of the scheme under article 30 of the Directive, discussed in chapter 10.

D.   Compliance and Enforcement

It was generally recognised that the standard approach to sanctions in environmental Directives, whereby Member States are given discretion to set effective, proportionate and dissuasive penalties, was not sufficient for an emissions trading scheme. Member States accepted early on the need for the Directive to set out the minimum penalties that Member States must apply in relation to emissions for which allowances are not surrendered. The negotiations therefore focused on the level of penalties. The main changes from the Commission Proposal were the reduction of the penalty for the first phase from €50 to €40, and the removal of any link between the penalty and the market price.[29] This change resulted from arguments made by those industries and Member States with experience of emissions trading that this would cause too much uncertainty and would be difficult to administer in practice.

The final Directive came into force on its publication on 25 October 2003, just over two years after the publication of the original Commission Proposal. It set a very tight deadline for the transposition and implementation of the Directive: Member States were required to transpose the Directive into their domestic law by 31 December 2003 and the scheme would become operational on 1 January 2005.

Chapters 2 to 9 contain a detailed discussion of the provisions of the Directive as ultimately adopted, while chapters 11 to 21 consider the implementation of the Directive in the UK.

---

[29] The Proposal envisaged that the penalty would be at least twice the average market price during the preceding year.

IV.   Negotiation of the Linking Directive

The Commission Proposal for the Emissions Trading Directive did not provide for the detail of linking the EU ETS with the flexible mechanisms under the Kyoto Protocol. In part this was an issue of timing: the detailed rules for the Kyoto regime were only agreed in Marrakech in November 2001, a month after the Commission Proposal for the Emissions Trading Directive was published. It also allowed the negotiations on the Directive to proceed more quickly than if they had been encumbered by the diverging views of NGOs, industry and Member States on the use of credits from the Kyoto flexible mechanisms. However, in the final text of the Emissions Trading Directive the Commission was given a clear mandate to make a proposal for a further Directive establishing such a link.[30]

Indeed, even before the Emissions Trading Directive had been adopted, the Commission had initiated the process to amend it in order to establish a link to the project mechanisms under the Kyoto Protocol. Following the pattern established by the negotiations on the Emissions Trading Directive, the discussions began under the auspices of a working group of the European Climate Change Programme.[31] This process provided the Commission with a forum through which to seek the views of expert stakeholders prior to drawing up a legislative proposal. The Commission then issued a formal Proposal for a Directive on 23 July 2003[32] taking the form of an amendment to the Emissions Trading Directive. Following negotiations in the Council and European Parliament, the Directive was adopted on 27 October 2004.

The key questions discussed during the negotiations on the use of project credits were how, what, and how much?[33]

The question 'how?' related to the technical question of how to reconcile the differing status of the various credits under the Kyoto Protocol trading regime (AAUs, CERs, ERUs and RMUs) with the fact that under the EU ETS all allowances enjoy equal status. Originally the Proposal

---

[30] See recitals 18 and 19, and article 30(3): 'Linking the project-based mechanisms ... with the Community scheme is desirable and important to achieve the goal of both reducing global greenhouse gas emissions and increasing the cost-effective functioning of the Community scheme'.

[31] Details of these meetings are available at http://europa.eu.int/comm/environment/climat/ji_cdm.htm.

[32] Commission Proposal for a Directive of the European Parliament and of the Council amending the Directive establishing a scheme for greenhouse gas emissions allowance trading within the Community, in respect of the Kyoto Protocol's project mechanisms, COM(2003) 403 final.

[33] For a detailed analysis of the issues raised during the Linking Directive negotiations, see Part II of Yamin (supra. Introduction, note 10, p.126-139).

had provided for CERs and ERUs to be converted[34] into EU allowances but the final version of the Directive gets around the potential difficulties which would be caused by converting project credits into another form of Kyoto currency, by providing for conversion at the point of use, effectively allowing for their use by operators to demonstrate compliance with the operator's obligations under the EU ETS.[35]

The second issue was what types of project credits would be eligible for use by operators under the EU ETS. Industry favoured the unlimited use of all types of project credits, while NGOs argued for restrictions above and beyond those set out in the Marrakech Accords in order to preserve the environmental integrity of the EU ETS. Among the issues discussed during the negotiations were the practical difficulties involved in setting up additional eligibility tests and the potential fracturing of the emerging carbon market which might result from such additional requirements. The final text of the Linking Directive only prohibits the use of project credits generated from nuclear facilities (to the extent that they are excluded from the Kyoto trading regime) and land use, land-use change and forestry projects (or sinks).[36]

The most heated issue discussed during the negotiations was the question of the amount of credits from Kyoto project mechanisms which would be permitted to enter the EU ETS. The context for these discussions were the so called 'supplementarity' obligations in the Kyoto Protocol and implementing rules, which state that the 'use of the Kyoto mechanisms shall be supplemental to domestic actions and domestic action shall thus constitute a significant element of the effort made'.[37] Given that the EU had taken a hard line in the international context to ensure that domestic action would be prioritised by pushing for a quantitative constraint or 'concrete ceiling' on the use of the Kyoto mechanisms, it was hardly surprising that efforts would be made to limit the number of credits which could be used within the EU ETS and thereby maintain the incentive for emission reductions within the EU.

During the negotiation of the Emissions Trading Directive, the European Parliament had inserted what became the final sentence in article 30,

---

[34] The bald concept of conversion concealed a number of questions.

[35] For a discussion of the final provisions of the Directive, see chapter 8.

[36] This approach accorded with the EU line during the international negotiations under the Kyoto Protocol. Given the progress made to date on the international rules on sinks projects, this issue is likely to be revisited in future reviews of the Emissions Trading Directive. No restrictions are placed on the use of credits from large dams projects (although there are additional obligations relating to the approval of such projects in article 11b of the Emissions Trading Directive, as amended by the Linking Directive), but the Commission was required to consider this further as part of the review due in June 2006 (see chapter 10).

[37] See in particular Articles 6(1)(d) and 17 of the Kyoto Protocol.

stating that 'use of the mechanisms shall be supplemental to domestic action, in accordance with the relevant provisions of the Kyoto Protocol and Marrakech Accords'. However, opposition by industry and Member States concerned about the potential impact of the Directive on their companies led to a watering down of this original language in the formal Commission Proposal for the Linking Directive, which provided instead for a review once use of the credits from project mechanisms under the Kyoto Protocol reached six per cent of allowances, triggering the possibility of a cap being proposed through comitology procedures.

All that remained by the time the Linking Directive was finally adopted was a requirement on Member States to state in their NAPs the amount of credits which may be used and to demonstrate that this was consistent with the Member State's supplementarity obligations. Following interventions during the negotiations by the UK, Member States are required to state this amount as a percentage limit for each installation. This has the advantage of both providing certainty to industry on how they may meet their obligations and setting in place a mechanism for sharing out the total number of credits available in any Member State.

The following chapters now turn to look in detail at provisions of the Emissions Trading Directive and the Linking Directive.

CHAPTER 2

WHAT ACTIVITIES ARE COVERED BY THE DIRECTIVE?

This chapter gives an overview of the activities covered by the Emissions Trading Directive and the rationale for their selection (section I). It then considers the interpretation of the category of activity which gave rise to the most discussion in the preparation for the first phase of the scheme, namely 'combustion installations' (section II). It considers briefly the exception for research and development (section III) and the rules of the Directive relating to thresholds, including when activities must be aggregated for the purposes of applying the thresholds (section IV). The rules allowing Member States to unilaterally include additional activities, installations and greenhouse gases are considered in section V, and the rules which allowed Member States to temporarily exclude installations for the first phase are considered in section VI. The final section (section VII) briefly explains the comitology procedure under which the Commission takes Decisions approving such unilateral inclusion and temporary inclusion, and certain other Decisions under the Directive.

I.   Overview and Rationale

Article 2(1) sets the scope of the Directive in the following terms: 'This Directive shall apply to emissions from the activities listed in Annex I and greenhouse gases listed in Annex II'.

The gases listed in Annex II are carbon dioxide (CO_2), methane (CH_4), nitrous oxide (N_2O), hydrofluorocarbons (HFCs), perfluorocarbons (PFCs) and sulphur hexafluoride (SF_6). However, significantly, installations are required to have a permit for activities listed in Annex I only if those activities result in emissions specified in relation to that activity in Annex I.[1] Similarly it is only those emissions which installations are required to monitor and report.[2] The only emissions

[1] Article 4.

[2] This flows from Article 14. There is a possible, wider, reading, flowing from Article 6(2)(b). This provides that greenhouse gas emissions permits shall contain 'a description of the activities and emissions from the installation'. 'Emissions' is defined in Article 3(b) as 'the release of greenhouse gases into the atmosphere from sources in an installation'. It is therefore possible to read Article 6(2)(c), which requires a permit to contain 'monitoring requirements', as requiring the permit to include provisions relating to the monitoring of all greenhouse gases, not only those specified in relation to the activity in annex I. However, Article 14(2) provides that Member States shall ensure that emissions are monitored in accordance with the guidelines. As guidelines will be adopted, under Article 14(1) only for emissions of the gases specified in annex I, the better reading is

(continued...)

which are specified in Annex I are emissions of carbon dioxide. Therefore until the Directive is amended to specify additional gases in relation to activities listed in Annex I (or Member States chose to unilaterally include additional gases in accordance with the Directive), activities are covered by the scheme only if they lead to emissions of carbon dioxide and it is only their emissions of carbon dioxide which must be monitored.

The list of activities in Annex I is a subset of the activities listed in Annex I to the IPPC Directive.[3] The activities listed in Annex I to the Emissions Trading Directive therefore correspond to the equivalent activities listed in Annex I to the IPPC Directive, subject to two exceptions, both of which relate to the category of 'combustion installations'. The first difference is that, whereas the IPPC Directive covers combustion installations with a rated thermal input exceeding 50MW, the Emissions Trading Directive adopts a lower threshold, and covers combustion installations with a rated thermal input exceeding 20MW. The second difference is that the Emissions Trading Directive excludes from the category of combustion installations 'hazardous or municipal waste installations'.

The reasons for the choice of gases and activities covered were set out in the Commission's Explanatory Memorandum to its original Proposal[4] as follows:

10. COVERAGE OF GASES

The Community scheme being proposed by this Directive covers, in principle, emissions of all greenhouse gases covered by the Kyoto Protocol – as listed in Annex II. However, only carbon dioxide emissions from the activities listed in Annex I will be included from the start. In 1999, carbon dioxide accounted for over 80% of the Community's greenhouse gas emissions. Emissions of carbon dioxide are widely recognised as capable of generating good quality monitoring data on a consistent basis.

Inclusion of the other greenhouse gases listed in the Kyoto Protocol is desirable but would be dependent on resolving monitoring, reporting and verification issues, possible local impacts as well as other Community policies and measures addressing emissions of these gases. In particular, emissions trading presupposes a sufficiently accurate monitoring of

therefore that the obligation under Articles 6(2)(b) and (c) is fulfilled by the permit describing the emissions of the gases specified in annex I, and containing provisions for the monitoring only of those gases. This reading is also clearly supported by recital 11, which provides that 'Member States should ensure that the operators of certain specified activities hold a greenhouse gas emissions permit and that they monitor and report their emissions of greenhouse gases specified in relation to those activities'.
[3] Council Directive 96/61/EC of 24 September 1996 concerning Integrated Pollution Prevention and Control, OJ L 257, 10.10.96, p26.
[4] Supra chapter 1, note 16.

emissions, but the monitoring uncertainties are still too great for greenhouse gases other than carbon dioxide. For these reasons, emissions of greenhouse gases other than carbon dioxide are not included in the first phase of the scheme.

It is proposed that the inclusion of greenhouse gases other than carbon dioxide in Annex I should be considered in the context of an amendment of the Directive, as their inclusion is not an appropriate matter for a Regulatory Committee to decide.

11. COVERAGE OF SECTORS

The sectoral coverage of this Directive builds upon the framework of regulation arising from the IPPC Directive.

Initially, only carbon dioxide emissions from the activities listed in Annex I will be covered by the scheme. Inclusion in the scheme of the 'core activities' listed in Annex I will result in coverage of approximately 46% of estimated EU carbon dioxide emissions in 2010[5] comprising some 4000 to 5000 installations. Significant carbon dioxide emitters currently covered by the IPPC Directive, such as power and heat generation installations between 20-50 MW, will also be included as these are also significant sources of carbon dioxide emissions, and their number is likely to increase in the future.

The chemical sector and waste incineration sectors would not be included, although carbon dioxide emissions from any on-site power and heat generating capacity would be included if it exceeds the threshold of 20MW. The decision not to include the chemical sector initially is taken for two reasons: first the chemical sector's direct emissions of carbon dioxide are not so significant (approximately 26 million tonnes of carbon dioxide in 1990, which is less than 1% of the EU's total emissions of carbon dioxide in the same year). Second, the number of chemical installations in the Community is high, in the order of 34000 plants, and their inclusion would substantially increase the administrative complexity of the scheme. Finally, the waste incineration sector is not included due to the complexities of measuring the carbon content of the waste material that is being burnt.

It is proposed that the inclusion of additional activities in Annex I should be considered in the context of an amendment of the Directive, as their inclusion is not an appropriate subject matter for a Regulatory Committee to decide.

---

[5] Equivalent to approximately 38% of the European Community's projected total greenhouse gas emissions covered by the Kyoto Protocol in 2010.

II.   Meaning of 'Combustion Installations'

The scope of the activities listed in Annex I caused, with one exception, relatively little debate during the implementation of the first phase of the scheme. The exception was the activity of 'combustion installations'.[6]

During the development of NAPs for the first phase of the scheme, at least three interpretations of the term 'combustion installations' emerged. A wide interpretation would cover any installation in which fuels are oxidised in order to use the heat generated.[7] A middle interpretation would cover any installation in which fuels are burned to produce an energy product.[8] A narrow interpretation would cover only installations producing power for the commercial electricity market.[9] There are good arguments in favour of both the middle and wide interpretations.

In favour of the wide interpretation it is argued that this is consistent with the interpretation given to similar terms used in other Community legislation. The fact that Annex I of the Emissions Trading Directive is a truncated version (with some amendments) of Annex I of the IPPC Directive, and the cross-references to the IPPC Directive in Articles 8 and 26 of the Emissions Trading Directive suggest that the meaning of the words 'combustion installations' where used in the IPPC Directive will be, at least, a relevant factor in interpreting those words in the

---

[6] It appears from paragraph 2 of annex 9 to Commission Communication 'Further guidance on allocation plans for the 2008 to 2012 trading period of the EU Emission Trading Scheme' (COM(2005)703 final)(the 2005 Commission Guidance) that there may have been debate in the first phase in some Member States about the scope of the activity of manufacture of ceramic products in annex I. The Communication advises Member States that the words 'and/or' in the description of that activity give Member States a discretion as to whether to require both of the subsequent specified criteria to be met before an installation falls within the scheme. This is consistent with Declaration of the Council and the Commission of 4 September 1996 on Directive 96/61/EC of the Council on Integrated Pollution Prevention and Control, 9388/96, Inter-institutional dossier No, 00/0526 (SYN) which interpreted the same activity under paragraph 3.5 of annex I to the IPPC Directive.

[7] This interpretation was adopted by the Commission in its Non-Paper on the installation coverage of the EU Emissions Trading Scheme and the interpretation of annex I, September 2003. In that Non-Paper the Commission concluded that 'This includes apparatus where the heat is used in another piece of apparatus, through a medium such as electricity or steam, and apparatus where the heat resulting from combustion is used directly within that apparatus, for example, for melting other substances.'

[8] This interpretation was adopted by the UK for the first phase of the scheme. See EU Emissions Trading Scheme Guidance Note 1, revised March 2006 (available at www.defra.gov.uk/environment/climatechange/trading/eu/pdf/eu-ets-guidance01.pdf).

[9] This interpretation was initially adopted by France for the first phase of the scheme but was not accepted by the Commission, and as a result France later amended its national allocation plan for the first phase of the scheme to adopt the middle interpretation. See recital 6 to the Commission Decision on the French NAP, C(2004) 3982 / 7 final of 20 October 2004.

Emission Trading Directive. Although the words 'combustion installations' in paragraph 1.1 of Annex I of the IPPC Directive are not further defined, it is commonly accepted that the term 'combustion installation' in the IPPC Directive covers not just the power generation industry but also other industries where fuels are burned. This would support the wide interpretation and suggest that the inclusion of the activity under the heading 'Energy Industries' should not be interpreted narrowly.

Further, it is well-established that industries can fall within more than one activity category of the IPPC Directive. For example, integrated steel works carry out several activities falling within Annex I of that Directive and refineries include combustion installations. This would also seem to apply to the Emissions Trading Directive.

A similar term 'combustion plant' is used in the Large Combustion Plants Directive.[10] The Directive defines 'combustion plant' widely as 'any technical apparatus in which fuels are oxidised in order to use the heat as generated'. The Directive then sets out a list of limitations to the types of combustion plant to which the Directive applies for example 'reactors used in the chemical industry'. The absence of similar exclusions in the Emissions Trading Directive suggests that the types of combustion excluded from the Large Combustion Plants Directive are included in the Emissions Trading Directive where they are carried out in a stationary technical unit above the 20MW threshold.

The express exclusion of 'hazardous or municipal waste installations' might even be taken as indicating that the words 'combustion installations' are to be taken as covering any installation in which material is burned, irrespective of whether the heat generated is put to a useful purpose. This is because the exclusion necessarily implies that hazardous or municipal waste installations could, otherwise, fall within the category of combustion installations. However this is not conclusive, as such installations could have the purpose not only of disposing of waste, but also of generating electricity or useful heat. It could be argued that the effect of the exclusion is therefore to exclude this subset of hazardous or municipal waste installations.

The wide approach is also supported by the broad purpose of the Directive. Where the meaning of the wording of a Directive is ambiguous

---

[10] Directive 2001/80/EC of the European Parliament and of the Council on the limitation of emissions of certain pollutants into the air from large combustion plants, OJ L 309, 27.11.2001, p.1. See also the Sulphur in Liquid Fuels Directive (Directive 1999/32/EC relating to a reduction in the sulphur content of certain liquid fuels, OJ L 121, 11.05.1999, p.13) which uses the same term.

the ECJ interprets the provision in question by reference to the purpose and general scheme of the rules of which it forms part. Thus in a case[11] on the interpretation of the Environmental Impact Assessment Directive,[12] the ECJ gave a wide interpretation to the category of 'canalization and flood-relief works' in Annex II to that Directive because the wording of the Directive indicated that it had 'a wide scope and a broad purpose'. Article 1 of the Emissions Trading Directive indicates that the purpose of the Directive is 'to promote reductions of greenhouse gases in a cost-effective and economically efficient manner'. Thus it might be argued that the Directive has the broad purpose of reducing greenhouse gas emissions, and that a wide interpretation would be more in line with this purpose, as greater coverage would provide more flexibility for installations covered by the EU ETS to decide how to reduce their emissions in the most cost-effective and economically efficient manner.

Arguments in favour of the middle approach focus on the differences between the Emissions Trading Directive and the IPPC and Large Combustion Plants Directives,[13] the immediate context in which the term 'combustion installations' is used in the Emissions Trading Directive[14] and the staggered approach to the introduction of the EU ETS.[15]

---

[11] Case C-72/95 *Kraaijeveld and Others* [1996] ECR I-I403.

[12] Council Directive 85/337/EC of 27 June 1985 on the assessment of the effects of certain public and private projects on the environment, OJ No. L 175, 05.07.1985, p 40.

[13] It is argued that neither the IPPC Directive nor the Large Combustion Plants Directive provide any decisive aids to the interpretation of the term 'combustion installations' in the Emissions Trading Directive. Firstly, the words 'combustion installations' in paragraph 1.1 of annex I of the IPPC Directive are not further defined. Secondly, and equally importantly, it is not necessary under the IPPC Directive to reach a view on the difficult questions of interpretation which arise under the Emissions Trading Directive. For example, under the Emissions Trading Directive it is unclear whether a chemical installation where fuels are burned as part of the production process falls within the term 'combustion installations'. Under the IPPC Directive, however, such an installation falls within annex I by virtue of the express inclusion of chemical installations in paragraph 4 of annex I. Nothing turns, under the IPPC Directive, on whether an installation falls within paragraph 4 of annex I, as a chemical installation, or under paragraph 1.1 of annex I, as a combustion installation. It is not, therefore, necessary, under the IPPC Directive to resolve that issue. As regards the Large Combustion Plants Directive, it is argued that the definition in that Directive has no relevance to the interpretation of the term 'combustion installation' in the Emissions Trading Directive. Firstly, the term used in the Large Combustion Plants Directive is 'combustion plant' rather than the term 'combustion installations' used in the Emissions Trading Directive. Furthermore the context and purpose of the Large Combustion Plants Directive is different from that of the Emissions Trading Directive, and finally there are no specific cross-references to the Large Combustion Plants Directive in the recitals or in the body of the Emissions Trading Directive which would require its terms to be taken into account.

[14] The category 'combustion installations' appears under the italicised heading 'Energy activities' and alongside 'mineral oil refineries' and 'coke ovens'. As both mineral oil

(continued...)

In the light of the different approaches adopted by Member States, the Commission approved divergent NAPs for the first phase. The United Kingdom NAP was based, for example, on the middle interpretation. The Netherlands NAP was based on the wide interpretation. It is only the narrow interpretation, on which the French NAP was initially based, which the Commission rejected.

However, it is clear that this approach could lead to distortions of competition with similar installations being treated differently depending on the Member State in which they are situated. The 2005 Commission Guidance, therefore, states that the Commission will require the wide interpretation to be applied by all Member States in the second phase of the scheme.[16] This would include in the scheme both installations where the heat is used in another piece of apparatus, and installations where the heat resulting from combustion is used directly

refineries and coke ovens produce energy products, this lends some support to the view that the words 'combustion installations' indicate another example of a class of installation which produce energy products. With the exception of the category of 'Other activities', all activities listed under each heading have as their purpose the production of the product referred to in that heading. For example, metal or roasting or sintering installations and installations for the production of pig-iron or steel under the heading 'Production of ferrous metals', all produce ferrous metals. Similarly, listed installations for the production of clinker in rotary kilns, installations for the manufacture of glass, and installations for the manufacture of ceramic products by firing, listed under the heading 'Mineral industry', all produce mineral products. This suggests that the heading defines activities by reference to the products which they generate. This lends some support to the argument that the category 'combustion installations' would cover only activities which produce energy as a product.

[15] It is clear from Article 2 of the Directive that its scope is to be limited to emissions from activities listed in annex I and to the greenhouse gases listed in annex II. Annex I of the Emissions Trading Directive includes a subset of the activities listed in annex I of the IPPC Directive. Article 30 of the Directive also shows that the Council and European Parliament, when adopting the Directive, envisaged a staggered approach to the introduction of greenhouse gas emissions trading, with the scheme being extended to other types of installation and other greenhouse gases at a later stage. This is supported by the Commission's Explanatory Memorandum to its original proposal for the Directive, quoted above, which explained that under the proposal 'the chemical and waste incineration sectors would not be included'. The two reasons given in support of this approach were the sectors' relatively low level of direct emissions of carbon dioxide and the high number of chemical installations in the EU, which would increase the administrative complexity of introducing the scheme. This approach is reflected in Article 30(2) which requires the Commission to draw up a report on the application of the Directive considering in particular 'how and whether annex I should be amended to include other relevant sectors, *inter alia*, the chemicals, aluminium and transport sectors, activities and emissions of other greenhouse gases listed in annex II, with a view to further improving the economic efficiency of the scheme'. It could therefore be argued that the wide interpretation of the term 'combustion installations' would pre-empt a significant element of this review by bringing the majority of chemical installations and a significant number of non-ferrous metals installations into the EU ETS from the outset.

[16] Paragraphs 34 to 36 of and annex 8 to the 2005 Commission Guidance, supra. note 6. This was further clarified by the agreement of 'co-ordinated definitions' of additional combustion installations, see minutes of the Climate Change Committee of 31 May 2006.

within that apparatus. This would bring within the scheme all combustion processes involving crackers, carbon black, flaring (including offshore flaring), furnaces (including rock wool) and integrated steel works (including rolling mills, re-heaters, annealing furnaces and pickling). In its assessment of the first batch of NAPs for the second phase of the scheme, the Commission has rejected NAPs which did not contain all installations falling within the agreed definitions.[17] It is, therefore, expected that all Member States which adopted the middle approach in the first phase of the scheme will follow the guidance and move to the wide approach for the second phase. If not, it will fall to the European Court of Justice to determine the correct interpretation in any ensuing litigation. In doing so, it is likely that the Court would examine closely the arguments in support of the wide interpretation of the term 'combustion installations' set out in the 2005 Commission Guidance.

III. Research and Development

Although the meaning of the term 'combustion installations' is the most significant area of doubt about the scope of the Directive, some questions do arise in the application of the two interpretative notes at the beginning of Annex I to the Directive.

The first note provides that:

> Installations or parts of installations used for research, development and testing of new products and processes are not covered by this Directive.

This note is exactly the same as the note in Annex I to the IPPC Directive. The terms 'research, development and testing' are not further defined, and their application will require the exercise of judgment by the national competent authorities. Although there is a circularity in the use of the term 'installation' in this interpretative note,[18] the clear effect of this note is that units used for research development and testing of new products and processes do not form part of an activity, or part of an installation. Emissions from these units do not fall within the scheme.

---

[17] See for example, Commission Decision of 29 November 2006 concerning the National Allocation Plan for the allocation of greenhouse gas emission allowances notified by Germany in accordance with Directive 2003/87/EC of the European Parliament and of the Council.

[18] This circularity appears elsewhere in the Directive. For a fuller discussion see below, section IV. B.

IV. Thresholds and Aggregation

The second interpretative note in Annex I provides:

> The threshold values given below generally refer to production capacities or outputs. Where one operator carries out several activities falling under the same sub-heading in the same installation or on the same site, the capacities of such activities are added together.

Again, this note directly reflects a provision in the IPPC Directive. This explains some of the difficulties of trying to apply this note in the context of the Emissions Trading Directive.

A. Do Thresholds Refer to Capacity or Actual Output?

Although the first sentence of the second interpretative note appears to suggest that some of the thresholds in the Annex are not based on production capacities, an examination of the thresholds prescribed for the activities in Annex I reveals that this is not the case: all the thresholds specified in Annex I to the Emissions Trading Directive refer to production capacities.[19] However, although the inappropriate export from the IPPC Directive of this first sentence renders it odd, it does not in practice have any effect on the interpretation of the Directive, or give rise to any difficulties in practice.

B. What Activities are to be Aggregated in Applying the Threshold?

The second sentence of the second interpretative note requires operators to aggregate the capacities of all activities carried out either in the same installation or on the same site in order to determine whether they exceed the threshold, and so are caught by the Directive. The sentence raises two issues: namely, what is meant by installations 'falling within the same sub-heading', and what activities are 'carried in the same installation or out the same site'?

---

[19] In annex I to the IPPC Directive it is true that the threshold values given do 'generally' refer to production capacities or outputs: there are some thresholds relating to inputs (e.g. section 6.4(c) of annex I to the IPPC Directive) or capacities for disposal (as opposed to production) of waste (e.g. section 5.4(c) of annex I to the IPPC Directive). There are, however, no equivalents to these in the sub-set of activities which has been carried over into annex I to the Emissions Trading Directive: wherever thresholds are specified in that Annex, they always refer to production capacities or outputs. A further quirk of this first sentence of the note is the reference to 'outputs' as an apparent alternative to 'production capacities' in defining threshold values. This might, at first sight, suggest that some thresholds are defined by reference to actual output, as opposed to production capacity. However, an examination of the thresholds specified in annex I to the Directive reveals no thresholds which relate to actual output: they all relate to production capacity.

Looking at these in turn, there appears at first to be a question as to whether the reference to 'sub-headings' is to be taken as a reference to the italicised headings in the table in Annex I, or rather to the specific activities listed under those headings. For example, whether it operates to aggregate all activities under the title 'Energy activities' or only those falling within one of the specific listed activities, such as 'combustion installations'. It seems clear that the better interpretation is that the reference to 'sub-headings' is a reference to each of the specific activities listed under the italicised headings: the thresholds are specified differently in relation to the different listed activities, making it impossible to aggregate capacities across these activities[20]. This interpretation is endorsed by the 2005 Commission Guidance.[21]

The second question arises from the reference to the 'same installation' in the aggregation rule. This reference gives rise to some circularity. An 'installation' is defined in Article 3(e) as 'a stationary technical unit where an activity listed in Annex I is carried out' but the aggregation rule appears to need the scope of the installation to be defined in order to decide whether, by operation of the aggregation rule, an activity meeting the relevant threshold in Annex I is carried out. In practice this circularity is avoided by asking whether, ignoring the thresholds, activities falling within the same sub-heading in Annex I are carried out in the same stationary technical unit or on the same site. If so, then the capacities of each activity are aggregated and if the aggregated capacity exceeds the threshold in Annex I for that activity the activities will be covered by the Directive.

In accordance with Article 3(e) the 'installation' then includes not just the activities listed in Annex I but also any directly associated activities which have a technical connection with the activities carried out on that site and which could have an effect on emissions and pollution.

Experience of the first phase of the EU ETS has demonstrated that the combined effect of the 20 MW threshold for combustion installations and the aggregation rule is to bring into the scheme a large number of small installations comprised of small combustion units spread across a site, such as universities or hospitals. The Commission has recognised that the costs and benefits of including such installations require further consideration, and will be considered as part of the review of the scheme

---

[20] This reading is also supported by the French text of the Directive. It is another example of the consequence of exporting this interpretative note unchanged from the IPPC Directive, where activities are listed not just under headings, but also under second level numbered paragraphs.

[21] Paragraph 1 of annex 9 to the 2005 Commission Guidance, supra. note 6.

under Article 30(2) of the Directive which is expected in 2007. If appropriate in the light of the findings of that review, the Commission will propose amendments to the scope of the Directive. Such amendments could, for example, allow for small installations to be removed from the scheme, or exclude units with a rated thermal input of 3MW or less from the application of the aggregation rule.[22]

V.   Unilateral Inclusion of Activities, Installations and Other Greenhouse Gases by Member States

Article 24 allows Member States to unilaterally include additional activities, installations and gases in the EU ETS.

In relation to the first phase, the second paragraph of Article 24(1) entitled Member States to include installations carrying out activities below the thresholds specified in Annex I of the Directive. Member States have a wider power for the second phase, from 2008. For this phase the first paragraph of Article 24(1) allows Member States to include not just installations carrying out activities in Annex I below the thresholds specified in Annex I, but also to include activities, installations and greenhouse gases not listed in Annex I.

Although the Directive expresses this power in terms of 'apply[ing] emissions allowance trading in accordance with this Directive to activities, installations and greenhouse gases', this necessarily pre-supposes applying all the requirements of the Directive to the relevant installations, including the grant of a permit, and the allocation of allowances. Installations included under this power would, therefore, be treated identically to installations carrying out activities above the thresholds in Annex I.

The Directive sets out the conditions for inclusion. The unilateral inclusion must be approved in advance by a Decision of the Commission, taken under the comitology procedure referred to in Article 23(2).[23] In deciding whether to approve the unilateral inclusion, the Commission must take into account 'all relevant criteria, in particular effects on the internal market, potential distortions of competition, the environmental integrity of the scheme and reliability of the planned monitoring and reporting system'.

---

[22] See paragraphs 37 to 40 of the 2005 Commission Guidance, supra note 6 and chapter 10 section II. B.

[23] For an explanation of the comitology procedure see section VII of this chapter.

Article 24(3) provides that the Commission may adopt monitoring and reporting guidelines for emissions from activities, installations and greenhouse gases not listed in Annex I if it is possible to monitor and report such emissions with sufficient accuracy. The Commission must adopt such guidelines if requested by a Member State. Such a request could be made at anytime. A Member State could therefore request the Commission to adopt guidelines prior to applying for approval to unilaterally include certain activities, installations or greenhouse gases, and the Commission would be required to do so.

Article 24(2) requires any installations unilaterally included in the scheme to be included in the list of installations contained within the relevant Member State's NAP.[24] However the Commission has in practice interpreted this requirement flexibly and accepted applications for unilateral inclusion after the deadline for submission of NAPs. Consistent with the approach for installations carrying out activities listed in Annex I, NAPs would need to make provision for new entrants in the unilaterally included categories if such entrants are to be included.

Where a Member State unilaterally includes additional activities installations or greenhouse gases, Article 24(4) requires the Commission to consider as part of its review under Article 30 of the Directive whether Annex I should be amended to include those activities. Unilateral inclusion may, in this way, be a first step towards harmonised expansion of the scheme.

In the first phase the Commission approved the unilateral inclusion of installations in Sweden (66 installations), Austria (1 installation), Finland (209 installations), Latvia (23 installations) and Slovenia (15 installations).[25]

The inclusion of a specific mechanism for Member States to 'apply greenhouse gas emission allowance trading' to other activities raises the question as to the extent to which Member States retain competence

---

[24] For an explanation of NAPs and the ways in which this requirement could be fulfilled, see chapter 4.

[25] See Commission Decision C(2004)4240-1 of 29 October 2004 (Sweden); Commission Decision C(2005) 481 final of 8 April 2005 (Austria); Commission Decision C(2005) 481 final of 8 April 2005 amended by a further Commission Decision to extend the inclusion to 12 further installations and any new installations in the same category (Finland); Commission Decision C(2005) 481 final of 8 April 2005 extended by Commission Decision C(2006)584 final of 1 March 2006 to extend the inclusion to 4 further installations and any new installations in the same category (Latvia); and Commission Decision C(2005) 481 final of 8 April 2005 (Slovenia). Decisions are available from: http://ec.europa.eu/environment/climat/first_phase_ep.htm.

to introduce alternative national trading measures. Article 24 could be interpreted as setting out the procedure that Member States must follow if they wish to engage in any form of greenhouse gas emissions trading involving activities and gases which fall within the scope of the Directive but which are not contained in Annex I. The purpose of Article 24 is to allow emissions trading on a broader scale but to ensure that it does not have any negative effects on the internal market, distort competition, or damage the environmental integrity of the scheme. These concerns could also apply to a parallel domestic trading scheme, and so it is arguable that Article 24 is intended to apply to these schemes as well. If that were the case, then a Member State that wished to establish a domestic greenhouse gas emissions trading scheme would need to apply to the Commission for approval to include additional activities in the EU ETS.

The counter-argument to this is that if a Member State were seeking to set up a new parallel trading scheme with different allocation methodology and different allowances which are not usable interchangeably with EU ETS allowances, then it is not seeking to 'apply emission allowance trading *in accordance with the Directive'*. Further, although the wording 'may apply greenhouse gas emission allowance trading' may sound broad, the Directive's title is '.establishing a scheme for greenhouse gas emission allowance trading....'. This reinforces the argument that it is only intended to govern the extension of the Directive to other installations or gases, rather than the use of other greenhouse gas emissions trading schemes. If a Member State wished to establish a separate national emissions trading scheme applying to activities and gases which do fall within the scope of the Directive, but do not yet fall within Annex I, then it would be open for that Member State to do so, independently of the Directive. Support for this interpretation is found in recital 16 which states that:

> This Directive should not prevent any Member State from maintaining or establishing national trading schemes regulating emissions of greenhouse gases from activities other than those listed in Annex I or included in the Community scheme, or from installations temporarily excluded from the Community scheme.

VI. Temporary Exclusion of Installations During the First Phase

Article 27 of the Directive provided for the temporary exclusion of installations from the scheme for the first phase only. The effect of temporary exclusion is that the installation does not require a permit under the Directive[26] and is not obliged to comply with the requirements

---

[26] See Article 4.

relating to the monitoring and reporting of emissions or the surrender of allowances which flow from the permit. Member States wishing to exclude installations had to apply to the Commission.[27] The application had to be published to give the public an opportunity to comment.

The Commission was required to decide in accordance with the comitology procedure referred to in Article 23[28] and having considered any comments made by the public, whether the relevant installations would:

- as a result of national policies, limit their emissions as much as would be the case if they were subject to the provisions of the Directive;

- be subject to monitoring, reporting and verification requirements equivalent to those provided for pursuant to Articles 14 and 15 of the Directive; and

- be subject to penalties at least equivalent to those referred to in Articles 16(1) and (4) in the case of non-fulfilment of national requirements.

If the Commission decided that these conditions were met, and that the exclusion would not lead to any distortion of the internal market, it was to adopt a Decision to 'provide for the temporary exclusion of [the relevant] installations from the Community scheme'. There was room for argument about the effect of a Commission Decision providing for the temporary exclusion of installations. Did it have the effect of excluding the relevant installations from the scheme[29] or merely 'provide for' their exclusion, leaving it to the applying Member State to decide whether ultimately to exclude the relevant installations? In the end no dispute arose over this possible difference of interpretation.

---

[27] No specific deadline was provided for making an application to temporarily exclude installations from the first phase of the scheme. However, as the NAP was required to contain a list of installations covered by the scheme it seemed necessary to know in advance of the deadline for submitting NAPs which installations were excluded. In practice, however, the Commission adopted a pragmatic approach and accepted applications made at a later stage.

[28] For a description of the comitology procedure see section VII of this chapter.

[29] For example, Article 1 of Commission Decision C(2004)4240-2 of 29 October 2004 relating to installations subject to the UKETS provided that 'The 63 installations listed in the United Kingdom's application are excluded from the Community emission allowance trading scheme from 1 January 2005 to 31 December 2006'.

Installations were excluded in accordance with Article 27 in the UK, the Netherlands and Belgium.[30]

VII.   Comitology Procedure

As seen earlier in this chapter, Commission Decisions either approving the unilateral inclusion of activities, installations and gases, or providing for the temporary exclusion of installations are to be taken in accordance with the comitology procedure referred to in Article 23(2) of the Directive. This procedure also applies to the Commission's Decision under Article 14 to adopt guidelines on monitoring and reporting of emissions;[31] the Commission's adoption under Article 19(3) of a Regulation for a standardised and secure system of registries;[32] any Commission Decision under Article 22 to amend the criteria for NAPs in Annex III;[33] and any Commission Decision under Article 25(2) providing for mutual recognition of allowances from other greenhouse gas emissions trading schemes with which the Community scheme is linked.[34]

Note that the Commission's Decisions under Article 9(3) to approve or reject NAPs are not subject to this procedure.[35] Although NAPs are required by Article 9(2) to be 'considered within the committee referred to in Article 23(1)', the comitology procedure referred to in Article 23(2)

---

[30] The Commission adopted a Decision on 29 October 2004 providing for the temporary exclusion from 1st January 2005 to 31 December 2006 of 63 installations in the United Kingdom which were direct participants in the UK Emissions Trading Scheme (Ibid. For a description of the UKETS see chapter 23). On 23 December 2005, a further Decision provided for the temporary exclusion of installations in the United Kingdom subject to climate change agreements. The Decision specifically backdated this exclusion to the beginning of the scheme on 1 January 2005 (See Commission Decision C(2005)5714 final of 23 December 2005. This was extended to a further installation by Commission Decision C(2006)582 final of 1 March 2006 and to a further 2 installations by Commission Decision C(2006)4765 final of 11 October 2006). On 29 October 2004, the Commission also took a Decision providing for the temporary exclusion from 1 January 2005 to 31 December 2007 of 93 installations in the Netherlands which were 'subject to commitments under long term agreements on energy efficiency, benchmarking covenants and, if these commitments [were] not met, obligations from the Netherlands Environmental Management Act' (Commission Decision C(2004)4240-3 of 29 October 2004. This was extended to a further 57 installations by Commission Decision C(2005) 866 final of 22 March 2005. On 14 July 2006, the Commission approved the temporary exclusion of 22 Belgian installations (Commission Decision C(2006) 3169 final of 14 July 2006). The Commission Decisions are available at: http://ec.europa.eu/environment/climat/ first_phase_ep.htm.
[31] See section II of chapter 3.
[32] See section II of chapter 7.
[33] See section VII of chapter 4.
[34] See section III of chapter 8.
[35] See chapter 5.

is not applied, with the result that the Commission is required merely to have regard to the views of the committee when taking its Decisions under Article 9(3).

Where Article 23(2) applies, it provides that 'Articles 5 and 7 of Decision 1999/468/EC[36] [the Comitology Decision] shall apply, having regard to the provisions of Article 8 thereof.'

The Comitology Decision sets out three different procedures to be followed when the Commission is given power to adopt secondary legislation implementing a Directive or Regulation adopted by the Council and European Parliament. These embody differing degrees of control exercisable by the Member States over the Commission's exercise of discretion. In increasing order of control, these are the advisory procedure, the management procedure, and the regulatory procedure. Article 23(2) of the Emissions Trading Directive applies the third of these, the regulatory procedure.

Under the regulatory procedure:

a)   the Commission is assisted by a regulatory committee comprising representatives of the Member States and chaired by the representative of the Commission;

b)   the Commission submits a draft of the measures to be taken to the committee and the committee delivers an opinion on the draft within the time limit which the chairman lays down according to the urgency of the matter. The Commission must persuade a qualified majority[37] of the committee to support its measure;

c)   if the measures are not supported by a qualified majority or no opinion is delivered within the time limit set, the Commission must submit a (possibly amended) proposal to the Council and inform the European Parliament;

d)   the Council then has a maximum of three months (as set in Article 23(2) of the Emissions Trading Directive) to adopt the proposal by qualified majority, or block the proposal

---

[36] Council Decision 1999/468/EC laying down the procedures for the exercise of implementing powers conferred on the Commission, OJ L 184, 17. 7. 1999, p.23.

[37] Qualified majority voting is a procedure for adopting decisions under which each Member State is given a fixed number of votes based on population but weighted in favour of smaller Member States the accession of Romania and Bulgaria the total number of votes is 345 and a qualified majority is constituted by 255 votes.

by qualified majority (and where this happens the Commission may submit an amended proposal, resubmit its proposal or present a proposal for legislation);

e) if the Council takes no action the Commission must adopt the proposed measure.

Article 7 of the Comitology Decision contains provisions about the rules of procedure for the committee, public access to documents and informing the European Parliament.

Article 8 of the Comitology Decision provides that the European Parliament may pass a Resolution indicating that it considers that a measure proposed by the Commission exceeds the powers provided for in the basic instrument (i.e. Articles 14, 19, 22, 24, 25 or 27 of the Emissions Trading Directive) and setting out the grounds on which this opinion is based. Where this happens, the Commission may submit a new draft measure to the committee, continue with the procedure, or present a proposal for legislation.

A revision of the Comitology Decision was adopted in July 2006.[38] The revision introduces a new procedure, the regulatory procedure with scrutiny, which gives more power to the European Parliament. The procedure applies to the adoption of measures of general scope designed to amend non-essential elements of a basic instrument, inter alia by deleting or adding non-essential elements.[39]

Under this new procedure

a) the Commission is assisted by a regulatory committee comprising representatives of the Member States and chaired by the representative of the Commission;

b) the Commission submits a draft of the measures to be taken to the committee and the committee delivers an opinion on the draft within the time limit which the chairman lays down according to the urgency of the matter. The Commission must persuade a qualified majority[40] of the committee to support its measures;

---

[38] Council Decision 2006/512/EC amending Decision 1999/468/EC laying down the procedures for the exercise of implementing powers conferred on the Commission, OJ L 2006, 22.7.2006, p.11.
[39] Article 2(2) of Decision 1999/468/EC as amended.
[40] Supra. note 37.

c)   if the measures are not supported by a qualified majority, the Commission must submit the draft measures for scrutiny bv the European Parliament and the Council. Within 3 months of the submission of the draft measures, a majority of members of the Parliament or a qualified majority in the Council may oppose the adoption of the draft measures on the basis that they exceed the implementing powers provided for in the basic instrument; they are not compatible with the aim or the content of the basic instrument; or do not respect the principles of subsidiarity or proportionality. If the measures are opposed, the Commission cannot adopt the draft measures and may submit an amended draft of the measures or a legislative proposal under the Treaty. If neither the Council nor the Parliament oppose the measures within 3 months of their submission, the Commission must adopt the proposed measure ;

d)   if the measures proposed by the Commission are not in accordance with the opinion of the committee, or if no opinion is delivered within the time limit set, the Commission must submit a proposal to the Council and forward it to the European Parliament. The Council then has a maximum of two months to act on the proposal by qualified majority. If the Council opposes the proposal, the proposed measures will not be adopted and the Commission may present an amended proposal or present a legislative proposal on the basis of the Treaty. If the Council envisages adopting the proposed measures, it shall submit them to the Parliament without delay. If the Council does not act within the two month period the Commission must without delay submit the measures to the Parliament for scrutiny. Within 4 months of the forwarding of the draft measures, a majority of members of the Parliament may oppose the adoption of the draft measures on the basis that they exceed the implementing powers provided for in the basic instrument; they are not compatible with the aim or the content of the basic instrument; or do not respect the principles of subsidiarity or proportionality. If the measures are opposed, the Commission cannot adopt the draft measures and may submit an amended draft of the measures or a legislative proposal under the Treaty. If the Parliament does not oppose the measures within 4 months, the Commission or the Council must adopt the proposed measure.

The procedures provided for in the Directive will eventually need to be reassessed and updated to provide for the new procedure to apply where appropriate. This is likely to take place when amendments are proposed to the Directive as part of the review foreseen in Article 30 of the Directive.[41]

---

[41] See chapter 10.

CHAPTER 3

PERMITS

Article 4 of the Directive requires any installation carrying out an activity listed in Annex I and resulting in emissions of the specified gases to be covered by a permit. It is through the permit that all other obligations under the Directive are imposed on the operator. The permit is issued to the operator of the installation by the competent authority of the Member State in which the activity is carried out.[1]

This chapter considers the procedure for the application for and grant of permits (section I). It then considers the requirements relating to monitoring, reporting and verification which are imposed via the permit (sections II and III). Permits also include an obligation on the operator to surrender each year allowances equivalent to the emissions made in the previous year: this obligation is considered in chapter 9.

Finally this chapter considers the relationship between permits under the Emissions Trading Directive and the IPPC Directive[2] (section IV).

I.   Application for and Grant of Permits

The Directive contains very little detail about the procedure for applications for permits and the determination of those applications.

A.   Overview of the Directive Requirements

Article 4 establishes the requirement for greenhouse gas emissions permits:

Article 4

Greenhouse Gas Emissions Permits

> Member States shall ensure that, from 1 January 2005, no installation undertakes any activity listed in Annex I resulting in emissions specified in relation to that activity unless its operator holds a permit issued by a competent authority in accordance with Articles 5 and 6 or the installation

---

[1] Article 18 of the Directive requires the Member State to designate the appropriate competent authority or authorities and requires that where more than one competent authority is designated, the work of these authorities undertaken pursuant to the Directive must be co-ordinated.

[2] Supra chapter 2, note 3.

is temporarily excluded from the Community scheme pursuant to Article 27.

Articles 5 sets out the requirements for applications:

Article 5

Applications for Greenhouse Gas Emissions Permits

An application to the competent authority for a greenhouse gas emissions permit shall include a description of:

    (a)   the installation and its activities including the technology used;

    (b)   the raw and auxiliary materials, the use of which is likely to lead to emissions of gases listed in Annex I;

    (c)   the sources of emissions of gases listed in Annex I from the installation; and

    (d)   the measures planned to monitor and report emissions in accordance with the guidelines adopted pursuant to Article 14.

The application shall also include a non-technical summary of the details referred to in the first sub-paragraph.

Article 6 sets out the requirements which must be met before a permit may be granted, and the conditions which a permit must contain:

Article 6

Conditions for and Contents of Greenhouse Gas Emissions Permit

1.   The competent authority shall issue a greenhouse gas emissions permit granting authorisation to emit greenhouse gases from all or part of the installation if it is satisfied that the operator is capable of monitoring and reporting emissions.

A greenhouse gas emissions permit may cover one or more installations on the same site operated by the same operator.

2.   A greenhouse gas emissions permit shall contain the following:

    (a)   the name and address of the operator;

    (b)   a description of the activities and emissions from the installation;

    (c)   monitoring requirements, specifying monitoring methodology and frequency;

    (d)   reporting requirements; and

    (e)   an obligation to surrender allowances equal to the total emissions of the installation in each calendar year, as verified in accordance with Article 15, within four months following the end of that year.

Article 7 sets out the obligation of the operator to notify the competent authority of proposed changes in the operation of the installation, and specifies the two types of situation in which the permit must be updated. Firstly, situations where there is a change in the nature or functioning of the installation, or an extension. Secondly, where the identity of the operator changes:

Article 7

Changes relating to installations

The operator shall inform the competent authority of any changes planned in the nature or functioning, or an extension, of the installation which may require updating of the greenhouse gas emissions permit. Where appropriate, the competent authority shall update the permit. Where there is a change in the identity of the installation's operator, the competent authority shall update the permit to include the name and address of the new operator.

These provisions together give rise to the following questions of interpretation, which are considered in turn below:

- what is an operator?
- what is an installation?;
- to what dgree of detail must the prmit define an installation and its activities?; and
- can one installation be covered by more than one permit?

B.   What is an Operator?

The person who is required to hold a permit is the operator of an installation. Article 3(f) of the Directive defines an operator as 'any person who operates or controls an installation or, where this is provided for in national legislation, to whom decisive economic power over the technical functioning of the installation has been delegated'. Member States therefore have a choice as to how they identify operators. There is a question as to whether there is much difference in practice between the two approaches. On the one hand, the first approach could be interpreted as identifying someone who physically operates an installation whereas the second approach identifies the person who controls the financing.[34]

---

[3] This raises the question of what is meant by the term 'decisive economic power over the technical functioning of the installation'. This would seem to mean that the person is able

(continued...)

However the term 'operates or controls' could also be interpreted more widely to take into account economic power determining investment decisions alongside physical control. This wide interpretation would seem to fit best with the objective of the Directive which is to impose the obligations on the person who has most control over the operation of the installation and therefore the level of emissions.

C.   What is an Installation?

As seen in chapter 2, the Directive applies to the carrying out of 'activities' specified in its Annex I. Although in practice the key concept remains that of an activity, when the Directive imposes obligations on individual plants, it does so by using the concept of an 'installation'.[5]

'Installation' is defined in Article 2(e) as:

> a stationary technical unit where one or more activities listed in Annex I are carried out and any other directly associated activities which have a technical connection with the activities carried out on that site which could have an effect on emissions and pollution.

The definition is therefore made up of two elements: the stationary technical unit where the activity listed in Annex I is carried out, and activities which are directly associated to those activities.

Looking at these in turn, there is no definition of a 'stationary technical unit' in the Directive. However, the words 'technical unit' have the implication that it must be something which is identifiable as a unit, and which is able to carry out an Annex I activity on its own.[6] It is the requirement in this definition that the technical unit must be stationary that prevents the Directive applying to mobile plants.

to control the expenditure of the installation such as to determine the equipment used in the installation and investment decisions for the future.

[4] There is also a question as to whether the two limbs of the definition are mutually exclusive or could be applied together. There are good arguments to suggest that either one or the other should be applied to avoid creating confusion about who is the operator. This ambiguity already exists within the first limb since the person who 'operates' the installation and the person who 'controls' it may be different.

[5] For example, Article 4 requires a Member State to ensure that an installation has a permit.

[6] The term 'stationary technical unit' is used in the IPPC Directive (supra. chapter 2, note 3). In the context of the UK's implementation of the IPPC Directive the term has been interpreted as 'something which is functionally self-contained in the sense that the unit - which may consist of one component or a number of components functioning together - can carry out the Schedule 1 activity or activities on its own': see Integrated Pollution Prevention and Control: A Practical Guide, edition 4, July 2005 (available at http://www.defra.gov.uk/environment/ppc/ippcguide/). This definition has been adopted in the context of the UK implementation of the Emissions Trading Directive: see EU Emissions Trading Scheme Guidance Note 1 (supra chapter 2, note 8).

The second limb of the definition of installation brings in directly associated activities. Directly associated activities are only part of the installation if they meet three tests, namely that:

– they are carried out on the same site as one of the activities listed in Annex I;[7]

– they have a technical connection with one of the activities carried out in the stationary technical unit; and

– they have an effect on emissions and pollution.[8]

Despite the fact that the application of these tests can clearly give rise to questions of interpretation (e.g. the meaning of a 'technical connection'), no obligations flow directly from the view taken. This is because, although the Directive requires the application and permit to define the installation (and hence to determine what are the 'associated activities'), all the substantive obligations are then imposed by reference to the concept of 'an activity'.

D. To What Degree of Detail Must the Permit Define an Installation and its Activities?

The second interpretational question arising from Articles 5 and 6 is the degree of detail in which an installation and the activities it carries out must out be described, both in the application and in the permit.

The relevant provision in relation to the application is Article 5(a), which requires the application to contain a 'description of the installation and its activities including the technology used'. Turning to the permit, Article 6 does not expressly require any description of the installation, but Article 6(2)(b) requires a 'description of the activities and emissions from the installation'.

---

[7] In *United Utilities Water v Environment Agency for England and Wales* [2006] EWCA Civ 633, the Court of Appeal in the UK considered the meaning of 'installation' under the IPPC Directive (ibid.), which is defined in identical terms as in the Emissions Trading Directive. Although the Court did not need to rule on the point definitively, it expressed the view (at paragraph 42) that a directly associated activity does not need to be carried out on the same site as the primary activity in order to form part of the installation.

[8] Although the term 'emissions' is limited to emissions of greenhouse gases by Article 3(b) of the Directive, the term 'pollution' which is not defined in the Directive is a much broader term. 'Pollution' is defined in the IPPC Directive (ibid.) as 'the direct or indirect introduction as a result of human activity, of substances, vibrations, heat or noise into the air, water or land which may be harmful to human health or the quality of the environment, result in damage to material property or impair or interfere with amenities and other legitimate uses of the environment'.

The degree of detail in which the installation and the activities are described has two consequences: in relation to updating of the permit (under Article 7) and in relation to the provision which can be made in a NAP for the allocation of allowances to new entrants (under Article 3(h) and criterion 6 in Annex III). Both sets of provisions turn on the question of whether the permit requires 'updating'. The extent to which a permit requires updating clearly depends in part on the level of detail in which the installation and activities are described in the permit. As the most significant consequence of whether a permit needs updating is whether an operator is eligible to receive allowances as a new entrant, this issue is discussed in section IV. D of chapter 4, as part of the discussion of the substantive requirements for NAPs, and in particular criterion 6 in Annex III of the Directive relating to new entrants.

E.   Can One Installation be Covered by More than One Permit?

A further question arising from Articles 5 and 6 is whether the Directive allows more than one permit to be issued in respect of two parts of the same installation.

The need for more than one permit may arise where there is one large installation carrying out an activity which has two distinct parts and two distinct operators. An example of this might be a plant for the production of paperboard with an associated combined heat and power plant. Articles 5 and 6 contain no express prohibition on having separate permits for the two distinct parts of the installation, but nor do they contain any express authorisation.   However, Article 6(1) does refer to a greenhouse gas emissions permit 'granting authorisation to emit greenhouse gases from all or part of an installation'. Whilst this would seem to support the contention that a permit may relate to part only of an installation it is not conclusive: it could be interpreted simply as a reflection of the fact that only emissions from the activities listed in Annex I are required to be authorised by the permit (and not, for example, emissions from directly associated activities which are nevertheless part of the installation).

However, a further indication that a permit may relate to part only of an installation is found in Article 8. This requires Member States to take the necessary measures to ensure that 'where installations carry out activities that are included in Annex I to [the IPPC Directive], the conditions of, and procedure for, the issue of a greenhouse gas emissions permit are co-ordinated with those for the permit provided for in that Directive'. The definition of permit in Article 2 of the IPPC Directive provides that 'a permit may cover one or more installations or parts of installations on the same site operated by the same operator'. It is therefore clear that a permit under the IPPC Directive may cover only part of an installation. Although the express co-ordination provision in Article 8 of the Emissions Trading Directive relates only to the conditions of and procedure for the issue of greenhouse gas emissions permits (as opposed to their scope), the value of this co-ordination would be reduced unless it was also possible to co-ordinate the scope of permits.

There are therefore good arguments that the Emissions Trading Directive allows separate permits to be issued in respect of two or more parts of the same installation. Where this happens, references to 'the installation' in the Directive and the secondary legislation adopted under it should be read as references to the part of the installation covered by the relevant permit.

II. Monitoring and Reporting

A. Monitoring of Emissions

As mentioned above, Article 6(2)(c) and (d) require a greenhouse gas emissions permit to contain monitoring requirements which are to specify both the monitoring methodology and frequency, and reporting requirements.

Article 14 of the Directive required the Commission to adopt guidelines for monitoring and reporting emissions of the specified gases from activities listed in Annex I. It required those guidelines to be based upon the principles for monitoring and reporting set out in Annex IV to the Directive. Paragraphs 2 and 3 of Article 14 require Member States to ensure that the emissions are monitored in accordance with those guidelines, and that at the end of each year the operator reports the emissions from the installation in accordance with the guidelines.

The Commission adopted a Decision (the Monitoring and Reporting Decision) establishing guidelines for the monitoring and reporting of greenhouse gas emissions on 29 January 2004.[9] The guidelines are, as required by Article 14 of the Directive, based on the framework set out in Annex IV of the Directive. The Commission has subsequently adopted a second Decision establishing modified guidelines, which will apply from 2008 onwards (the Modified Monitoring and Reporting Decision).[10]

[9] Commission Decision 2004/156/EC of 29 January 2004 establishing guidelines for the monitoring and reporting of greenhouse gas emissions pursuant to Directive 2003/87EC of the Parliament and the Council.

[10] The Climate Change Committee adopted unanimously the document on 31st July 2006. At the time of writing formal adoption by the Commission is still pending. However a draft of the decision is available at: http://ec.europa.eu/environment/climat/emission/mrg_en.htm.

The guidelines are contained in eleven annexes to the Monitoring and Reporting Decision. The first of these annexes sets out general guidelines for monitoring and reporting of emissions. The later annexes set out additional activity-specific guidelines for combustion[11] and process emissions.[12]

Annex IV of the Directive provides that emissions shall be monitored either by calculation or on the basis of measurement. The details of how these two methods are to be applied is set out in the guidelines.

Under the calculation-based methodology emissions are determined by multiplying the activity data[13], the emission factor[14] and the oxidation factor[15], or such other approach as is defined in the annex relating to the specific activity in question.

Under the measurement-based methodology, emissions are determined using continuous emission measurement systems from each source using standardised or accepted methods.[16] Such methods may only be used if approved by the competent authority. The competent authority can only approve the use of a measurement-based methodology if it achieves greater accuracy than the calculation of emissions using the tier approach (described below).[17] If this approach is used, emissions determined using the measurement-based methodology must be corroborated by a supporting calculation of emissions, with the rules for selection of tiers being the same as those applied for the calculation approach.

---

[11] Combustion emissions are defined in paragraph 2(e) of the Monitoring and Reporting Decision as 'greenhouse gas emissions occurring during the exothermic reaction of a fuel and oxygen'.

[12] Process emissions are defined in paragraph 2(o) of Annex I to the Monitoring and Reporting Decision as 'greenhouse gas emissions other than combustion emissions occurring as a result of intentional and unintentional reactions between substances or their transformation, including the chemical or electrolytic reduction of metal ores, the thermal decomposition of substances, and the formation of substances for use as product or feedstuff'.

[13] See paragraph 4.2.2.1.5 of Annex I to the Monitoring and Reporting Decision. Activity data is information on material flow, consumption of fuel, input material or production output expressed as energy content (TJ) determined as net calorific value for fuels and mass or volume for input or output materials (t or m3).

[14] See paragraph 4.2.2.1.6 of Annex I to the Monitoring and Reporting Decision. Emission factors are based on the carbon content of fuels or input materials and expressed as tCO_2 /TJ (combustion emissions) or tCO_2 /t or $tCO_2/m2$(process emissions).

[15] See paragraph 4.2.2.1.7 of Annex I to the Monitoring and Reporting Decision. An oxidation factor is used if an emission factor does not reflect the proportion of the carbon that is not oxidised.

[16] See section 4.2.3 of Annex I to the Monitoring and Reporting Decision.

[17] See paragraph 4.3.2 of Annex I to the Monitoring and Reporting Decision for an explanation of how the operator can justify the use of a measurement-based methodology to the competent authority.

The activity-specific guidelines in Annexes II-XI to the Monitoring and Reporting Decision provide for a tiered approach to determining activity data, emission factors and oxidation factors. The higher the tier the more accurate the method. The guidelines provide that the highest tier approach should be used by all operators for calculating all variables.[18] However, operators can use a lower tier if using the highest tier would either not be technically feasible or would lead to unreasonably high costs.

The Guidelines go on to provide for a less stringent approach to the monitoring of emissions from minor sources in an installation. Where the competent authority agrees, the operator may apply a lower tier to calculate emissions from minor sources than is used to calculate emissions from the major sources in the installation. The guidelines provide that 'major sources including major streams of fuels and materials are those which, if ranked in order of their decreasing magnitude, cumulatively contribute at least ninety-five per cent to the total annual emissions of the installation'. Minor sources are 'those emitting 2.5 kilo tonnes or less per year or that contribute five per cent or less to the total annual emissions of an installation, whichever is the highest in terms of absolute emissions'.

The guidelines make a further provision for minor sources which jointly emit 0.5 kilo tonnes or less per year or which contribute less than one per cent of the total annual emissions of an installation, whichever is the highest in terms of absolute emissions. In such circumstances, the operator may apply a *de minimis* approach to the monitoring and reporting of emissions from those sources 'using his own no-tier estimation method'. This is subject to the approval of the competent authority. This provision for the application of *de minimis* methods benefits operators of larger installations who have a small number of smaller sources. However, it does not assist installations which are made up of a large number of small sources since the *de minimis* methods can only be applied to sources representing up to one per cent of the emissions from the installation.

The guidelines set out in the Modified Monitoring and Reporting Decision will apply less stringent monitoring requirements to installations whose average verified reported emissions during the first phase were 25,000 tonnes of carbon dioxide per year or, subject to certain conditions, if the competent authority projects that it will emit less that 25,000 tonnes of fossil carbon dioxide for each year over the following 5 years.[19]

---

[18] Paragraph 4.2.2.1.4 of Annex I to the Monitoring and Reporting Decision.
[19] See paragraph 16 of Annex I to the Modified Monitoring and Reporting Decision for an explanation of which requirements are waived.

B.   Monitoring Emissions from Biomass Fuels

The Directive itself lays down a special regime for the monitoring of emissions from installations using biomass fuels. Annex IV to the Directive provides that 'the emissions factor for biomass shall be zero'. This has the result that installations burning exclusively biomass fuels do not have to surrender allowances to cover emissions (although they are still required to hold a permit and report emissions).

Biomass is defined in paragraph 2 of Annex I to the guidelines as:

> non-fossilised and biodegradable organic material originating from plants, animals and micro-organisms. This shall also include products, by-products, residues and waste from agriculture, forestry and related industries, as well as the non-fossilised and biodegradable organic factions of industrial and municipal waste. Biomass also includes gases and liquids recovered from the decomposition of non-fossilised and biodegradable organic material. When burned for energy purposes biomass is referred to as biomass fuel.

Paragraph 9 of Annex I to the Monitoring and Reporting Decision contains a non-exhaustive list of materials which are considered to be biomass and to which an emissions factor of zero is to be applied. Although, as noted above, operators are still required to monitor emissions for pure biomass fuels, the guidelines permit an operator in certain circumstances to apply the lower tier approaches to the variables used to calculate the emissions.[20] The guidelines set out in the Modified Monitoring and Reporting Decision do not repeat this definition of biomass, but instead contain a list of materials which are considered to be biomass for this purpose.[21]

C.   Carbon Capture and Storage

Increasingly there is an interest in carbon capture and storage techniques as a means of reducing the release of greenhouse gases into the atmosphere. Carbon dioxide captured and stored (for example in a disused oil well under the sea) would not be released into the atmosphere and therefore would not fall within the definition of emissions within the Directive.[22] Guidelines are therefore necessary to determine how

---

[20] Paragraph 4.2.2.1.4 of Annex I to the Monitoring and Reporting Decision. The lowest tiers can be used unless the calculated emissions are to be used for the subtraction of biomass carbon from carbon dioxide emissions derived by means of continuous emission measurement.

[21] Paragraph 12 of Annex I to the Modified Monitoring and Reporting Decision.

[22] Article 3(b) of the Directive defines emissions as 'the release of greenhouse gases into the atmosphere from sources in an installation'.

such carbon dioxide captured and stored would be taken into account when calculating the emissions from an installation. As discussed above, the Monitoring and Reporting Guidelines require that emissions be monitored by calculation in most cases. This involves calculating emissions based on (amongst other things) fuel inputs. Whether the carbon dioxide produced from an installation is emitted to the atmosphere or securely stored underground will not affect fuel inputs or activity data. Therefore in order for captured and stored emissions not to be included in the reportable emissions, these must be separately monitored and deducted from the fuel based calculation. The Monitoring and Reporting Decision only permits certain types of 'transferred CO_2' to be deducted, and this does not include captured and stored carbon dioxide.[23]

However, the guidelines envisage the eventual adoption of further guidelines.[24] Such guidelines would take into account methodologies developed by the UNFCCC. In the meantime the guidelines provide that Member States may submit to the Commission interim guidelines for the monitoring and reporting of the capture and storage of carbon dioxide where covered by the Directive. Whether proposed by the Commission or put forward by Member States such guidelines would need to be formally adopted as a decision using the procedure provided for in Article 23(2) of the Directive.[25] In the absence of such guidelines, it is strongly arguable that the existing guidelines require any emissions captured and stored to be reported as any other emissions.

In the guidelines set out in the Modified Monitoring and Reporting Decision which will apply from 2008 onwards, there are three notable changes which are relevant to carbon capture and storage. The first is that the paragraph about the development of supplementary monitoring and reporting guidelines for Carbon Capture and Storage, including interim guidelines submitted by Member States, has not been repeated. Secondly the new section on transferred CO_2[26] now states generically that carbon dioxide which is not emitted to the atmosphere but transferred away from the installation may be deducted. It gives a list of illustrative examples, but unlike in the current guidelines, these are not exhaustive.

The third is a recital to the new Decision which states:

---

[23] Paragraph 4.2.2.1.2 of Annex I to the Monitoring and Reporting Decision.
[24] Paragraph 4.2.2.1.3 of Annex I to the Monitoring and Reporting Decision.
[25] See section VII of chapter 2.
[26] Paragraph 5.7 of Annex I to the Modified Monitoring and Reporting Decision.

(24) Recognition of activities relating to carbon capture and storage is not provided for in this Decision, but will depend on an amendment of Directive 2003/87/EC or by the inclusion of those activities pursuant to Article 24 of that Directive.

Although not referred to in the operative part of the Decision, it would appear that recital 24 to the Modified Monitoring and reproting Decision means that paragraph 5.7 should not be read as allowing emissions which are captured and stored to be deducted, unless the capture and storage activity is included pursuant to Article 24 of the Directive, or the Directive is amended to specifically permit it.

Therefore in order for captured and stored emissions to be deductible from an installation's reportable emissions, the Member State must make an application to the Commission under Article 24 for the activity of carbon capture and storage to be included in the EU ETS. The Commission must assess the application through comitology taking into account all relevant criteria including those specifically referred to in Article 24.[27] This includes the environmental integrity of the scheme and the reliability of the planned monitoring and reporting system. It is likely that these requirements mean that the operator will have to satisfy the Commission that someone will remain responsible for surrendering allowances to account for any greenhouse gases which may leak into the atmosphere from the capture and storage activities.

D.   Emission Reports

Article 14(3) of the Directive requires Member States to 'ensure that each operator of an installation reports the emissions from that installation during each calendar year to the competent authority after the end of the year in accordance with the guidelines [developed under Article 14(1)]'. Section 5 of Annex I to the Monitoring and Reporting Decision requires the report to be submitted to the competent authority by 31 March each year.

The information to be included in the report is prescribed in Annex IV to the Directive as supplemented by section 5 of Annex I to the Monitoring and Reporting Decision. Section 5 also requires operators to report in emissions reports the amount of biomass combusted or employed in processes and the carbon dioxide emissions from biomass where measurement is used to determine emissions.

---

[27] See section V of chapter 2.

III. Verification

Article 15 of and Annex V to the Emissions Trading Directive set out the requirements for verification of emissions reports.

A. Verification Procedure

Paragraph (1) of Article 15 requires Member States to ensure that reports submitted by operators of the emissions from an installation in the previous year are verified in accordance with the criteria in Annex V and that the competent authority is informed of such verifications.

Annex V of the Directive sets out some general principles and methodology for verification. The requirements in Annex V are supplemented by section 5 of the Annex I to the Monitoring and Reporting Decision. It is clear from Paragraph 2 of Article 15 and section 5 of Annex I to the Monitoring and Reporting Decision that verification of emissions reports must happen before the report is submitted to the competent authority by 31 March each year. Paragraph 11 of Annex V provides that a statement that the report is satisfactory may only be made if in the opinion of the verifier the total emissions are not materially misstated.

If an operator's report has not been verified as satisfactory by 31 March then paragraph 2 of Article 15 requires Member States to ensure that the operator cannot make further transfers of allowances until the report has been verified as satisfactory. This is given effect by Article 27 of the Registries Regulation which provides that:

> 1. If on 1 April of each year starting in 2006, an installation's annual verified emissions for the preceding year have not been entered into the verified emissions table in accordance with the verified emissions entry process set out in Annex VIII, the registry administrator shall block the transfer of any allowances out of the operator holding account for that installation.
>
> 2. When the installation's annual verified emissions for the year referred to in paragraph 1 have been entered into the verified emissions table, the registry administrator shall unblock the account.
>
> 3. The registry administrator shall immediately notify the relevant account holder and the competent authority of the blocking and unblocking of each operator holding account.
>
> 4. Paragraph 1 shall not apply to the surrender of allowances pursuant to Article 52 or the cancellation and replacement of allowances pursuant to Articles 60 and 61.

There is a difference between the trigger point under Article 15 of the Directive and Article 27 of the Registries Regulation. Article 15 applies where a report has not been verified as satisfactory by 31 March each year, whereas Article 27 applies where on 1 April the verified emissions figure has not been entered into the verified emissions table in the registry.

The effect of the provision in the Registries Regulation is therefore to establish a common procedure for demonstrating that a report has been verified. A report of emissions will not be considered as satisfactory until the verifier has entered or approved the entry of the verified emissions figure in the verified emissions table.

B.   Who Can Be a Verifier?

The question of who can be a verifier is addressed by paragraph 12 of Annex V to the Directive. This sets out minimum competency requirements. It provides that:

> The verifier shall be independent of the operator, carry out his activities in a sound and objective professional manner, and understand:

> (a)   the provisions of this Directive as well as relevant standards and guidance adopted by the Commission pursuant to Article 14(1);

> (b)   the legislative, regulatory and administrative requirements relevant to the activities being verified; and

> (c)   the generation of all information related to each source of emissions in the installation, in particular relating to the collection, measurement, calculation and reporting of data.

Providing that these criteria are met, the Directive leaves Member States with a discretion as to who may be a verifier. Member States may provide for verification to be carried out either by a public body or by private entities.

If a Member State decides to allow verification to be carried out by private entities, there is nothing express in the Directive which requires those entities to be accredited: paragraph 12 of Annex V sets out objective competency requirements, and it would appear that anybody meeting these requirements can act as a verifier, without any authorisation or accreditation. However, paragraph 2(f) of Annex I of the Monitoring and Reporting Decision defines 'verifier' as 'a competent independent accredited verification body'. This use of the word 'accredited' implies that in order to carry out verification a verifier must be approved in advance.

Although Article 14(1) of the Directive does not expressly empower the Commission to create new requirements relating to verification, the power in Article 14(1) is phrased widely, as allowing the Commission to adopt rules relating not just to monitoring and reporting in a narrow sense, but also verification.It therefore follows that the definition of 'verifier' in paragraph 2(f) in Annex I of the Monitoring and Reporting Decision has supplemented the requirements of Article 15 of and Annex V to the Directive, so as to require verifiers to be accredited in advance.[28]

C.   Effect of Verification

Under the Registries Regulation, each emissions trading registry is required to contain a verified emissions table.[29] The table contains space to enter the verified emissions for each installation for each year.[30] Once an installation's emissions have been verified as satisfactory, Article 51 of the Registries Regulation requires the amount of emissions from the installation during the previous year to be entered into the verified emissions table designated for that installation for that year in the registry. The entry shall either be made by the verifier himself or approved by him.

Paragraph (2) of Article 51 provides that the competent authority may instruct the registry administrator to correct the annual verified emissions for an installation for a previous year to ensure compliance with the detailed requirements established by a Member State pursuant to Annex V to the Directive. This procedure could be used, for example, to correct an error in the annual verified emissions discovered subsequent to these being entered in the verified emissions table.

Although the use of the term 'correction' suggests that a figure has to have already been entered into the table prior to this instruction to the registry administrator, to interpret the Regulation in this way would leave a lacuna in the legislation. It therefore seems that Article 51(2) should be interpreted as including the situation where an operator is unable to or simply has not entered its annual verified emissions into the verified emissions table by the end of March.

When considering entering a correction, it is necessary to bear in mind Article 57 of the Registries Regulation, which prohibits a substitute figure

---

[28] The definition of 'verifier' in the Registries Regulation similarly contains a reference to verifiers being 'accredited' (Article 2(x)).
[29] Article 24(1) of the Registries Regulation.
[30] Annex II of the Registries Regulation.

from being entered into the verified emissions table unless it has been 'calculated in accordance with the detailed requirements established by a Member State pursuant to Annex V of [the Directive]'. This prohibition is strict and the Registries Regulation does not provide for any exceptions. This may cause difficulties where an operator is unable to comply with the requirements of the Monitoring and Reporting Decision because, for example, there has been a fire at the installation. In such circumstances, it may be difficult to determine emissions in a manner which meets the requirements of Annex V and therefore to enter a verified emissions figure. The operator's account would be blocked in accordance with Article 27, but if the emissions cannot be verified it would not be possible to unblock it. If no figure is entered in the verified emissions table then it would also be impossible to know how many allowances the operator should have surrendered. Furthermore, it would be impossible to assess compliance in future years as such compliance is assessed by comparing the total verified emissions from the installation in all years of the relevant phase with the total number of allowances surrendered for those years. In order to avoid this unworkable situation it appears necessary, in these circumstances, to interpret the requirements of Annex V flexibly.

D.   Public Availability of Emission Reports

Article 17 of the Directive and the guidelines in the Monitoring and Reporting Decision[31] provide for emission reports held by the competent authority to be made available to the public subject to the restrictions laid down in the Environmental Information Directive[32] and provide that 'with regard to the application of the exception laid down in Article 4(2)(d) of that Directive, operators may indicate in their report which information they consider commercially sensitive.' Article 4 of that Directive sets out the exceptions to the requirements to provide access to information. However, many of these exceptions, including notably the exception for commercial confidentiality, do not apply to information on emissions into the environment.[33] It remains to be seen, therefore, what if any scope there remains for refusing public access to any emissions reports under the Directive.

IV.   Relationship with Permits Required by the IPPC Directive

There is a substantial overlap between the installations covered by the Emissions Trading Directive and the IPPC Directive.[34]

---

[31] Paragraph 5 of Annex I to the Monitoring and Reporting Decision.
[32] Directive 2003/4/EC of the European Parliament and of the Council of 28 January 2003 on public access to environmental information, OJ L 41, 14.2.2003, p.26.
[33] Article 4(2), second sub-paragraph of the Environmental Information Directive (ibid.).
[34] See section I of chapter 2.

Before the adoption of the Emissions Trading Directive, the greenhouse gas emissions regulated under that Directive were regulated by the imposition of emission limit values under the IPPC Directive.[35] Those emission limit values are required to be based on the best available techniques.[36] That Directive also requires the permit to include measures relating to energy efficiency.[37]

A. Amendment to the IPPC Directive Requirements Relating to Greenhouse Gas Emissions

The scope for reductions in greenhouse gas emissions to be made in the most economically efficient location, as sought by the Emissions Trading Directive, would be compromised if the relevant installations were simultaneously subject under the IPPC Directive to inflexible emission limit values for those gases. Article 26 of the Emission Trading Directive therefore amended the requirements of Article 9(3) of the IPPC Directive in relation to those gases. That Article, as amended, provides that:

> Where emissions of a greenhouse gas from an installation are specified in Annex I of the [Emissions Trading Directive] in relation to an activity carried out in that installation, the permit shall not include an emission limit value for direct emissions of that gas unless it is necessary to ensure that no significant local pollution is caused.

---

[35] Article 9(3) of the IPPC Directive (supra. chapter 2, note 3) requires the IPPC permit to include 'emission limit values for pollutants, in particular, those listed in Annex III, likely to be emitted from the installation concerned in significant quantities'. Some gases listed in Annex II of the Emissions Trading Directive are not expressly listed in Annex III of the IPPC Directive e.g. carbon dioxide, and methane. This does not mean that those gases were not required to be regulated under the IPPC Directive: Article 9(3) of the IPPC Directive requires emission limit values for 'pollutants ... likely to be emitted from the installation in significant quantities' and, as noted in the heading to Annex III to the IPPC Directive, the gases listed in that annex are merely an 'indicative list'. Greenhouse gases such as carbon dioxide and methane cause 'pollution' as defined in Article 2(2) of the IPPC Directive ('the introduction ... of substances ... into the air which may be harmful to ... the quality of the environment...') and so are required by Article 9(3) to be made subject to emission limit values in the IPPC permit if they are likely to be emitted in significant quantities. 'Emission limit value' is defined in Article 2(6) of the IPPC Directive as 'the mass, expressed in terms of certain specific parameters, concentration and/or level of an emission, which may not be exceeded during one or more periods of time'.

[36] Articles 9(4), 3(a) and 2(11) of the IPPC Directive (ibid).

[37] See Articles 9(1) and 3(d) of the IPPC Directive (ibid). Although Article 9(1) appears to impose an onerous requirement that 'the permit includes all measures necessary for compliance with the requirements of Article 3', and Article 3 in turn requires competent authorities to 'ensure that installations are operated in such a way that ... energy is used efficiently', considerable flexibility is introduced by the final words of Article 3: 'for the purposes of compliance with this Article, it shall be sufficient if Member States ensure that the competent authorities take account of the general principles set out in this article when they determine the conditions of the permit.'

Note that emissions of greenhouse gases still remain a relevant consideration in the determination of emission limit values for other pollutants under the IPPC permit. This is because emission limit values for all pollutants are to be based on the best available techniques[38], and 'best' in this context means 'most effective in achieving a high level of protection of the environment as a whole'. Thus it remains relevant, for example, to consider whether tightening an emission limit value for emissions of one substance (e.g. sulphur dioxide) will lead to a disproportionate increase in emissions of a greenhouse gas (e.g. carbon dioxide, as result of the energy demands of the necessary abatement equipment), even though the IPPC permit itself will not regulate emissions of that greenhouse gas.

B. Amendment to the IPPC Directive Requirement Relating to Energy Efficiency

The requirements of Article 9(3) of the IPPC Directive relating to energy efficiency were also modified, to provide that:

> For activities listed in Annex I to [the Emissions Trading Directive] Member States may choose not to impose requirements relating to energy efficiency in respect of combustion units or other units emitting carbon dioxide on the site.

The discretion given to Member States to continue to apply such energy efficiency requirements reflects the fact that the IPPC Directive leaves Member States substantial flexibility in what energy efficiency requirements to impose. It is therefore possible for a Member State to continue to impose energy efficiency requirements without undermining the benefits of the trading scheme established by the Emissions Trading Directive.

Note that competent authorities remain under a duty, when determining the conditions of IPPC permits relating to other aspects of an installation such as efficiency in the use of electricity to 'take account of the general principle' that 'energy is used efficiently'.[39]

The amended Article 9(3) of the IPPC Directive also requires that 'where necessary, the competent authorities will amend the permit as appropriate'.

---

[38] Article 9(4) of the IPPC Directive (ibid).
[39] Article 3 of the IPPC Directive (ibid).

C. Co-ordination of Permits under the Emissions Trading and IPPC Directives

Notwithstanding these amendments to the IPPC Directive to make it complement the Emissions Trading Directive, the fact that the two Directives both require the grant of permits creates the possibility for unnecessary administrative burdens on business. Article 8 of the Emissions Trading Directive therefore requires Member States to:

> take the necessary measures to ensure that, where installations carry out activities that are included in Annex I to [the IPPC Directive] the conditions of, and procedure for, the issue of a greenhouse gas emissions permit are co-ordinated with those for the permit provided for in [the IPPC] Directive.

Article 8 further provides that the requirements in the Emissions Trading Directive relating to applications, grant and content of permits, and changes to installations, may be integrated into the procedures provided for in the IPPC Directive.

D. Effect on IPPC Permits Where Activities are Unilaterally Included under Article 24

As seen in chapter 2,[40] the Emissions Trading Directive allows Member States to unilaterally include during the first phase installations carrying out regulated activities below the thresholds in Annex I. During the second and subsequent phases it allows Member States to unilaterally include activities, installations and gases not listed in Annex I.

There is scope for argument about the effect on the conditions of IPPC permits for installations which are unilaterally included under either of these provisions. This is because the amendments made by the Emissions Trading Directive to Article 9(3) of the IPPC Directive do not disapply the requirement for emission limit values for *any* installation regulated under the Emissions Trading Directive. Rather, they disapply those requirements only where both the activity and the gas are specified in Annex I to the Emissions Trading Directive.

Where, in the second or subsequent phases of the schemes, a Member State includes additional activities or gases these would not fall within the scope of the disapplication provision because those activities would not be specified in Annex I of the Directive. Therefore it would appear that the Member State remains under a duty to impose emission limit values on emissions of those gases under the relevant installations' IPPC

---

[40] See section V.

permits. It is also doubtful whether the disapplication provision would apply where a Member State unilaterally includes during the first phase installations falling below the thresholds in Annex I of the Emissions Trading Directive. This is because the descriptions of the activities listed in Annex I include the thresholds, and therefore activities below that threshold would not be activities 'specified in Annex I'.

There is no obvious logic to this differential treatment, and this may lead to arguments that Article 9(3) should be read as disapplying the requirement for emission limit values under the IPPC Directive to any gases and installations regulated under the Emissions Trading Directive, whether or not by virtue of unilateral inclusion. This would, it has to be admitted, require a strained reading of Article 9(3). There may therefore be pressure for the Commission, in any Decision approving unilateral inclusion under Article 24, to purport to clarify the position, notwithstanding that it is given no express power to do so under that Article.

If, as the natural reading of Article 9(3) suggests, the requirement for an emission limit value does indeed remain for gases and installations unilaterally included in the second phase, there is limited scope for Member States to mitigate the restriction which this imposes on emissions trading and thus the efficient allocation of emission reductions among the affected installations. The limited scope which may exist is through the concept of general binding rules under Article 9(8) of the IPPC Directive. This allows Member States to 'prescribe certain requirements for certain categories of installations in general binding rules instead of including them in individual conditions, provided that an integrated approach and an equivalent high level of environmental protection as a whole are ensured.' It can be argued that this allows a Member State to aggregate the emissions which would have been allowed from the relevant installations, and establish through general binding rules a mechanism which ensures that, provided that that overall quantity is not exceeded, the emissions can be made from any of the relevant installations. The effect would be to establish a total cap of emissions which would have to be met as a group by those unilaterally included installations.

Whether the ECJ would uphold this reading of Article 9(8) of the IPPC Directive is open to argument. It is also not obvious that, even if it did, it would be worthwhile establishing such a structure, given the restrictions on trading which would nevertheless remain on the unilaterally included installations as a group.

CHAPTER 4

ALLOWANCES: SUBSTANTIVE REQUIREMENTS FOR NATIONAL ALLOCATION PLANS

As mentioned in chapter 3, the Directive requires permits to include an obligation for the operator of each installation, by 30 April each year, to surrender allowances equal to the total emissions from that installation during the preceding calendar year. This chapter and the following two chapters consider the mechanisms by which allowances are allocated and issued to installations.

An allowance is defined in Article 3(a) of the Directive as:

> an allowance to emit one tonne of carbon dioxide equivalent during a specified period, which shall be valid only for the purposes of meeting the requirements of this Directive and shall be transferable in accordance with the provisions of this Directive.

The number of allowances allocated to each installation is not determined directly by the Directive. Rather, it is determined by each Member State through a two-stage procedure. The first stage involves the development by each Member State of a National Allocation Plan (NAP). In this plan the Member State sets out the total quantity of allowances which it intends to allocate to installations within its territory in the relevant phase, and how many it intends to allocate to each installation. The Member State can only proceed to the second stage once that NAP has been approved by the European Commission. At the second stage the Member State determines definitively the total quantity of allowances to be issued within its territory and the allocation of those allowances among installations.

This chapter considers the substantive requirements with which Member States' NAPs must comply, including the criteria in Annex III of the Directive.[1] Chapter 5 considers the procedure under which Member States submit and the Commission approves NAPs, and chapter 6 discusses how allowances are allocated and issued to installations.

This chapter sets out an overview of substantive requirements for NAPs (section I). It then considers in detail the criteria in Annex III of the

---

[1] With the exception of criterion 9, relating to how a Member State will take into account public comments on the NAP which is discussed in chapter 5 in the context of procedural requirements for NAPs.

Directive with which Member States' NAPs must comply. It considers in turn the criteria relating to the total quantity of allowances (section II), the criteria relating to the distribution of allowances between sectors (section III), and then the criteria relating to the allocation of allowances among installations (section IV). Within section IV there is a detailed discussion of the rules of the Directive relating to the allocation of allowances to new entrants. The Directive also provides for 'other criteria' than those in Annex III, and these are discussed in section V.

Section VI discusses whether the allocation of allowances in the NAP must be definitive or whether they may be conditional or subject to subsequent adjustments (so-called *ex post* adjustments). Finally, section VII considers the Commission's power to amend the criteria using the comitology procedure for the second and subsequent phases of the scheme.

I.   Overview of Substantive Requirements for NAPs

The substantive requirements for NAPs are set out in Articles 9(1) and 10 and Annex III to the Directive. Article 9(1) requires the Member State to ensure that the NAP it submits to the Commission is 'based on objective and transparent criteria, including those listed in Annex III'. Article 10 prescribes the percentage of allowances which Member States must distribute to installations free of charge. It provides that 95 per cent of the allowances must be allocated free of charge in the first phase, falling to 90 per cent in the second phase. It is therefore open to Member States to sell or auction up to 5 per cent of allowances allocated in the first phase and up to 10 per cent of allowances allocated in the second phase. No specific limits are placed on the amount of allowances which can be auctioned or sold in subsequent phases and therefore, in the absence of a further amendment to the Directive, it would be open to Member States to propose anything up to 100 per cent auctioning or sale in the subsequent phases. To date Member States have made only limited use of auctioning in the scheme. However, the complexity of developing and implementing rules for the free allocation of allowances and concerns about windfall profits from the free allocation of allowances is likely to mean that greater use of the provision for auctioning is made in the future.[2] This is a key issue in the review of the EU ETS. Proceeds of an auction could be used, *inter alia*, to cover administrative costs or to purchase additional Kyoto units. The Commission has made clear that where a Member State proposes to auction allowances, participation in the auction should be open to all persons in the Community without restrictions.[3]

---

[2] For further discussion on the future developments in the scheme, see section I of chapter 10.
[3] Communication from the Commission on Guidance to assist Member States in the Implementation of the criteria listed in Annex III to Directive 2003/87/EC establishing a scheme for greenhouse gas emission allowance trading within the European Community

(continued...)

The purpose of a NAP is to set out the total quantity of allowances that the Member State intends to allocate for the period, and how it proposes to allocate them among installations. This has three aspects. Firstly, the NAP must set out the proposed total quantity of allowances. Secondly, it must include a list of existing installations, and the quantity of allowances the Member State intends to allocate to each installation. Thirdly, it must contain information on how the Member State has calculated the allocations, including how new entrants will be treated. The criteria in Annex III set the framework within which Member States are to approach these decisions.

Article 9(1) required the Commission to develop guidance on the implementation of the criteria listed in Annex III by 31 December 2003. This guidance was published by the Commission in the form of a Communication[4] (the '2003 Commission Guidance'). It was supplemented in December 2005 by further guidance (the '2005 Commission Guidance') for the second phase of the trading scheme taking into account lessons learnt from the first phase.[5]

The 2003 Commission Guidance contained a helpful categorisation of the criteria in Annex III. It distinguished between those criteria which are mandatory and those which are optional.[6] It also distinguished between those criteria which relate to the total quantity of allowances to be allocated, their allocation to activities/sectors, and their allocation to installations.[7]

The Directive does not specify the level of detail required to be included in a NAP. This will in part be determined by the procedures in each Member State for giving effect to the NAP. However clearly the level of detail must be sufficient to demonstrate that the total quantity of allowances to be allocated and the methodology for allocating those allowances to individual installations is consistent with the Directive. In order to ensure that Member States include sufficient information, to increase transparency, and to make it easier for the Commission to assess NAPs, the 2003 Commission Guidance set out a common format for NAPs.[8] The 2005 Commission Guidance supplements this with a set of common tables summarising key information.[9]

and amending Council Directive 96/61/EC, and on the circumstances which force majeure is demonstrated, COM(2003) 830 final, paragraph 57.

[4] Ibid.

[5] Supra chapter 2, note 6.

[6] The 2003 Commission Guidance defines elements as mandatory if the Member State has an obligation to apply them and optional if the Member State has the choice whether or not to take specific action in relation to that element. However, for the majority of the criteria (with the possible exception of criterion 11) the Member State has an obligation to state whether or not it is applying that criterion in the National Allocation Plan.

[7] See paragraphs 5-8 of the 2003 Commission Guidance.

[8] Annex to the 2003 Commission Guidance.

[9] Annex 10 to the 2005 Commission Guidance.

II.   Criteria Relating to the Total Quantity of Allowances

Criteria 1 to 4 in Annex III to the Directive relate to the determination of the total quantity of allowances to be allocated.

The second sentence of criterion 1 provides that 'the total quantity of allowances to be allocated shall not be more than is likely to be needed for the strict application of the criteria of this Annex'. The effect of this provision is that the total quantity of allowances to be allocated in accordance with a Member State's NAP will be limited by whichever of the criteria in Annex III requires the lowest number of allowances to be allocated: this may be fewer than would be allowed by the application of any one criterion alone. This is supported by the 2005 Commission Guidance[10] and the Commission's approach to the assessment of NAPs for the second phase of the scheme.[11]

It also demonstrates that the criteria relating to the total quantity of allowances impose a ceiling on allocations only. So, for example, if a Member State wishes to set a total quantity which is lower than would be required to meet its obligations under the Burden Sharing Agreement[12] and Kyoto Protocol as required by criterion 1, it is free to do so. Indeed, where the application of other criteria imply a lower total quantity of allowances than that implied by criterion 1, it is obliged to do so. The words 'consistent with' used in the first sentence of criterion 1 should be interpreted as requiring Member States to ensure that the total quantity of allowances allocated is 'no greater than' an amount which would enable it to meet its obligations under the Burden Sharing Agreement and Kyoto Protocol rather than 'equal to' that amount. [13]

The total quantity of allowances allocated is of fundamental importance to the scheme. In order to ensure that the scheme delivers its full environmental and economic benefits, it is necessary to ensure that there is sufficient scarcity of allowances in the EU ETS. This will ensure that the scheme delivers real emissions reductions and incentivises the development and deployment of clean technologies, and will facilitate the creation of a dynamic and liquid global carbon market. The first set

---

[10] Supra chapter 2, note 6, paragraphs 10 – 16.

[11] See Communication from the Commission to the Council and the Parliament on the assessment of national allocation plans for the allocation of greenhouse gas emission allowances in the second period of the EU Emissions Trading Scheme accompanying Commission Decisions of 29 November 2006 on the national allocation plans of Germany, Greece, Ireland, Latvia, Lithuania, Luxembourg, Malta, Slovakia, Sweden and the United Kingdom in accordance with Directive 2003/87/EC (COM(2006) 725 final).

[12] Supra Introduction, note 30.

[13] Such an interpretation is also consistent with the objective of the Directive 'to promote reductions of greenhouse gas emissions in a cost-effective and economically efficient manner' (Article 1).

of independently verified emissions reports for the year 2005 confirmed suspicions of an overallocation of allowances in the first phase of the scheme: aggregate verified emissions for 2005 amounted to just over 2 billion tonnes, significantly below the annual average allocation for the period of nearly 2.2 billion tonnes. In assessing the compatibility of NAPs for the second phase of the scheme with the criteria in Annex III, the Commission has, therefore, focused on ensuring that the total quantity of allowances which a Member State proposes to allocate is consistent with its Kyoto Protocol commitment, emissions development and reduction potential. In order to ensure that the plans are assessed in a consistent, fair and transparent way, the Commission has developed and applied a number of formulae to determine the maximum quantity of allowances which may be allocated by each Member State.

A.   Criterion 1: Consistency with Kyoto Protocol Obligations

Of the criteria relevant to the determination of the total quantity of allowances, the most important is criterion 1. This provides that the total quantity of allowances to be allocated shall be consistent with the Member State's obligation to limit its emissions pursuant to the Burden Sharing Agreement and the Kyoto Protocol. The Burden Sharing Agreement sets a reduction target for each Member State for the period 2008-2012 in order to meet the Community's target under the Kyoto Protocol to reduce emissions to 8 per cent below 1990 emissions by the end of 2012. The UK's target under the Agreement is a 12.5 per cent reduction.

For the first phase of the scheme (i.e. the 3-year period 2005 to 2007), criterion 1 required the quantity to be consistent with the path towards achieving or over-achieving each Member State's target under the Decision. In other words, Member States had an obligation to demonstrate that the total quantity allocated for the period 2005 to 2007 is consistent with meeting their targets for the period 2008 to 2012.

In relation to the second phase of the scheme (2008-2012), the 2005 Commission Guidance emphasises that some Member States are not sufficiently on track to meet their target and that more needs to be done including fully realising the potential of emissions trading.[14] In assessing the consistency of NAPs for the second phase of the scheme with criterion 1, the Commission has taken into account progress towards the Kyoto commitment and, if relevant, the remaining gap to be closed, the reliance on and state of preparation and implementation of measures for the government purchase of Kyoto units, the reliance on and state of preparation and implementation of measures in non-trading sectors,

---

[14] See paragraph 9 of the 2005 Commission Guidance.

and the robustness of projections for the transport sector underlying the NAP.[15]

To the extent that Member States intend to purchase Kyoto units or use other policies and measures to meet their reduction target, this must also be substantiated in the NAP.[16] In its first batch of decisions on NAPS for the second phase, the Commission rejected NAPs for failing to comply with this criterion where another national policy or purchase of Kyoto units which the Member State claimed to be relying upon to meet its Kyoto target was insufficiently substantiated.[17] For intended purchases of Kyoto units, the Commission only found them to be substantiated where the Member State had an operational programme in place, signed contracts or initiated tenders, and had committed more than a minor share of the necessary budgetary resources. For other policies, the Commission required evidence that the proposed policy will be put in place, that the emissions reductions attributed to the policy are realistic in the 2008-12 timeframe and that the measure would not have significant overlapping effects to reduce emissions in installations covered by the EU ETS. The Commission has particularly required strong substantiation of emissions reduction policies in the transport sector as, in its opinion, emissions in that sector have grown significantly and reduction measures take considerable time for the intended effects to be achieved.[18]

B.   Criterion 2: Consistency with Emissions Projections and Criterion 3: Potential for Reduction

The second criterion requires the total quantity of allowances to be allocated to be 'consistent with assessments of actual and projected progress towards fulfilling the Member States' contributions to the Community's commitments made pursuant to Decision 93/389/EEC'. That Decision[19] established a mechanism for monitoring the Community's greenhouse gas emissions and was replaced in February 2004 by a further Decision.[20] The

[15] Supra note. 11, section 2.1.

[16] See paragraphs 18-20 of and Annexes 5 and 6 to the 2005 Commission Guidance.

[17] Supra note 11, section 2.1.

[18] For this, the Commission compared the trend in transport sector carbon dioxide emissions presented in NAPs with the trend development in the 'European Energy and Transport Trends to 2030 – update 2005' (available at: http://ec.europa.eu/dgs/energy_transport /figures/ trends_2030_update_2005/energy_transport_trends_ 2030_update_2005_en.pdf). Where a Member State has a remaining gap to close between its actual emissions and allowed emissions under its Kyoto target and the trend assumed in its NAP is substantially below the one in the 'European Energy and Transport Trends to 2030 – update 2005' the Commission considered the plan to be inconsistent with criterion 1 (supra note 11, section 2.1).

[19] Decision 93/389/EEC of 24 June 1993 for a monitoring mechanism of Community CO2 and other greenhouse gas emissions, OJ L 167, 9.7.1993, p.31 as last amended by Regulation (EC) No1882/2003 OJ L 284, 31.10.2003, p. 1.

[20] Decision 280/2004/EC of the European Parliament and of the Council concerning a mechanism for monitoring Community greenhouse gas emissions and for implementing

(continued...)

2004 Decision establishes a procedure under which Member States report annually their actual emissions of greenhouse gases in the year before the preceding year.[21] Article 3(2)(b) requires Member States to report to the Commission by 15 March 2005, and every two years thereafter, projections of their greenhouse gas emissions as a minimum for the years 2005, 2010, 2015 and 2020. The decision then provides, in Article 5, for the Commission to assess annually the progress of the Community and its Member States towards fulfilling their commitments under the Kyoto Protocol, as set out in the Burden Sharing Agreement. It also provides for the Commission, every two years, to assess the projected progress of the Community and its Member States. The first report of projected emissions in accordance with Article 3(1), and the first Commission assessment of projected progress of the Community and its Member States towards fulfilling their commitments under the Kyoto Protocol were, in accordance with Article 5(1) of the Monitoring Mechanism Decision, required in 2005, and in 2006 the Commission published a further report on progress towards meeting Kyoto objectives as required by Article 5(1).[22] It is this assessment which is relevant for the assessment of NAPs for the second phase of the scheme.

The first sentence of criterion 3[23] requires the total quantity of allowances to be consistent with the potential, including the technological potential, of activities covered by the scheme to reduce emissions.[24] This requires Member States, before determining the total quantity, to consider the scope for reductions in emissions in the activities covered by the scheme. The broad reference to 'potential', which is expressly stated to include technological potential, makes clear that the economic feasibility of reductions is to be taken into account. Member States are not, therefore, required to base their total quantity on a technological potential for

---

the Kyoto Protocol, OJ L 49, 19.2.2004, p.1. Article 11 of the Decision provides that any references to the 1993 Monitoring Mechanism Decision are to be construed as references to the 2004 Decision.

[21] Article 3(1)(a) of the 2004 Decision (ibid) requires Member States to report to the Commission by 15 January each year their anthropogenic emissions of greenhouse gases during the year before last.

[22] See Communication from the Commission: Report on demonstrable progress under the Kyoto Protocol (required under Article 5(3) of Decision 280/2004/EC concerning a mechanism for monitoring Community greenhouse gas emissions and for implementing the Kyoto Protocol), COM(2005) 655 final and Report from the Commission 'Progress towards achieving the Kyoto objectives' required under Decision 280/2004/EC of the European Parliament and of the Council concerning a mechanism for monitoring community greenhouse gas emissions and for implementing the Kyoto Protocol,COM(2006) 658 final.

[23] The second sentence of criterion 3 is relevant to the distribution of allowances between different activities, and not to the determination of the total quantity.

[24] As in criterion 1, the words 'consistent with' used in this criterion (and in criterion 2) must be interpreted as meaning 'no greater than' rather than 'equal to' that amount.

reductions in emissions if that is not economically feasible. The 2005 Commission Guidance makes clear that, in its analysis of the economic and technical potential to reduce emissions, the Commission will consider the annual GDP growth and carbon intensity reduction rates. These two factors together give the rate for the annual potential to reduce emissions.[25]

In assessing of NAPs for the second phase of the scheme, the Commission has applied a standard formula to assess the compatibility of a plan with criteria 2 and 3. In accordance with this formula the maximum allowed annual average cap is calculated by multiplying the 2005 verified emissions[26] by economic growth[27] and carbon intensity trend developments for 2005-2010.[28] The Commission has thus focused on actual greenhouse gas emissions of the sectors covered by the scheme in 2005. These are the most reliable and accurate emissions figures available as they have been independently verified and relate precisely to the scope of the scheme.[29] As regards carbon intensity trends, the Commission adopted the view that future trends are likely to bring carbon intensity lower more quickly than in the past because of the economic incentives created by the EU ETS itself, and because operators will become increasingly encouraged by other Member State policies to reduce emissions in the light of the EU's strong political commitment to tackling climate change, and increasingly by public opinion. It rejected NAPs where, in its view, Member States had underestimated the rate at which carbon intensity will reduce and therefore also underestimated the potential to reduce emissions. In addition the Commission considered that where a Member State allows for the banking of allowances between the first and second phases of the scheme, the NAP for the second phase will only be considered to be compatible with criterion 3 if the banked allowances are deducted from the total quantity of allowances to be allocated in the second pahse.[30]

---

[25] Paragraph 13 of the 2005 Commission Guidance.

[26] Including emissions for installations which were temporarily excluded from the scheme in 2005 in accordance with Article 27 (see chapter 2, section VI) and the annual average allocation to installations without an independently verified emissions report.

[27] The economic growth trend developments for the first set of NAPs assessed by the Commission are listed in Annex II to the Communication (supra note 11) and reflect the latest Commission forecasts for the years 2006, 2007 and 2008.

[28] To this is added additional emissions from installations included in NAPs for the second phase as a result of the decision to apply a wide definition of the term 'combustion installations'. See section II of chapter 2.

[29] The Commission considered that 2005 verified emissions figures can as a rule be regarded as representative. However in the case of exceptional circumstances the Commission will closely examine to what extent it is justified to adjust independently verified emissions for 2005 by an appropriate correction factor (section 2.1 of the Commission Communication (supra note 11).

[30] Section 2.4 of the Commission Communication (supra note 11). For a discussion on banking, see section II of chapter 9.

C. Criterion 4: Consistency with Other EC Instruments

The final criterion relating to the total quantity, criterion 4, requires the NAP to be consistent with other Community legislative and policy instruments. In particular, Member States are required to take account of unavoidable increases in emissions which result from new legislative requirements.

III. Criteria Relating to the Distribution of Allowances Between Sectors

Criteria 3 to 5 relate to the distribution of allowances between activities or sectors.[31]

Nothing in the Directive requires a Member State to provide in its NAP for allocations at an activity or sector level. It would be possible for Member States to simply provide for a total quantity of allowances and its distribution between individual installations. However, if a Member State's NAP does provide for activity or sector level allocations, then these criteria need to be taken into account.

A. Criterion 3: Potential for Reduction

As mentioned above, criterion 3 requires the quantities of allowances allocated to be consistent with the potential of the activity to reduce emissions. To the extent that the potential for reduction varies between sectors or activities, this factor may require the allocation to each sector or activity to vary. The second sentence of criterion 3 provides that Member States may distribute allowances on the basis of average emissions of greenhouse gases by product in each activity and achievable progress in each activity. This criterion therefore leaves open to Member States the possibility of allocating allowances on the basis of benchmarks, and allows these benchmarks to relate to specific products. The reference to 'achievable progress' is sufficiently flexible to allow not only the potential for reduction of emissions, but also the likely growth in that activity, to be taken into account.

B. Criterion 4: Consistency with Other EU Instruments

As noted above, criterion 4 requires that the allocations be consistent with other Community legislation and policy instruments.

---

[31] Annex III uses both the terms 'sector' and 'activity'. Although it is arguable that these terms are used interchangeably, there is a logic in giving them slightly different meanings such that an 'activity' is an activity listed in Annex I, whereas a 'sector' is a group of installations sharing similar characteristics. A sector could therefore be either larger or smaller than an activity.

C.   Criterion 5: Non-discrimination

Criterion 5 provides that 'the plan shall not discriminate between companies or sectors in such a way as to unduly favour certain undertakings or activities in accordance with the requirements of the Treaty, in particular Articles 87 and 88 thereof': this requires Member States to comply with the provisions on state aids in Articles 87 and 88 of the Treaty. At first sight, the use of the words 'in particular' in this criterion seem odd, as the only Treaty Articles dealing with discrimination between undertakings and which would appear to be relevant in this context are Articles 87 and 88 of the Treaty. However, this criterion could also be taken as a reference to the general principle of non-discrimination developed by the European Court of Justice.[32] This principle requires that in exercising the discretion afforded by EC law, Member States and the Community institutions must neither treat like cases differently, nor treat dissimilar cases alike, unless that treatment is objectively justified. In this context, the principle does not preclude Member States from applying rules of general application in determining installation level allocations where those rules do not take into account the particular circumstances of each installation.[33]

The 2003 Commission Guidance simply states that normal state aids rules will apply. This left many unanswered questions about the circumstances in which state aids issues might arise and the interaction of the state aids procedure with the NAP approval process under the Emissions Trading Directive. Therefore, on 17 March 2004, the Directors General of the Environment and Competition Directorates General of the Commission wrote to Member States to explain how the Commission would interpret criterion 5 of Annex III.[34] The letter identifies 4 situations in which the Commission considers state aid issues may arise:

a)   Where a Member State allocates to undertakings more allowances than needed to cover the projected emissions of those installations;

---

[32] See, for example, the Skimmed Milk Powder Case (1962) E.C.R 501.

[33] In *Cemex UK Cement Limited v Department for Environment, Food and Rural Affairs*, [2006] EWHC 3207 (Admin), Cemex argued that the prohibition on treating dissimilar cases alike meant that if a Member State was calculating allocations by applying a standard allocation formula to each installation's historic emissions during a baseline period, the government had to look at the individual circumstances of each installation to ensure that the baseline period chosen was representative of the installation's 'normal' level of activity, and therefore that all installations were not similar cases. The English High Court held that the government could regard types of installation with common features as being similar cases and need not look at the particular circumstances of each installation and regard every distinction between installations as making them dissimilar cases which may not therefore be treated similarly except where objectively justified.

[34] Letter from Catherine Day (Director General for Environment) and Philip Lowe (Director General for Competition) to Member States dated 17 March 2004.

b) Where a Member State overestimates measures in sectors not covered by the Emissions Trading Scheme or intends to purchase additional credits under the Joint Implementation provision or Clean Development Mechanism;

c) Where a Member State does not make full use of the ability to auction or otherwise charge for allowances under Article 10 of the Directive; and

d) Where a Member State provides for allowances to be banked between the first and second phase.[35]

Whilst it is clear that state aids issues may arise where a Member State uses its resources or foregoes resources to give its industry a competitive advantage it is not clear that this would be the case in all the situations outlined above. However, where a Member State over-allocates to its installations, it is clear that the Commission would be entitled to reject the NAP as being incompatible with other criteria in Annex III, in particular criteria 1, 2 and 3, or as being incompatible with the general principle of non-discrimination.

Despite the indications in the letter that the Commission would take issues of state aid in NAPs seriously, the Commission took a fairly relaxed approach to state aids issues in relation to the NAPs for the first phase of the scheme and did not, for example, pursue Member States who proposed to auction less than 5 per cent of available allowances. Early indications are that the Commission will take a similar approach in the second phase: the 2005 Commission Guidance does not address the issue of state aid and merely reminds Member States of the possibility of auctioning up to 10 per cent of allowances in the second phase.[36]

---

[35] This issue is further addressed by the Commission in section 2.4 of its Communication relating to the decisions on the first set of NAPs for the second phase of the scheme (supra note 11). This provides that 'The Commission takes the view that any national provisions related to the intended use of discretionary banking between the first and second trading periods must be notified to the Commission pursuant to Article 88(3) of the Treaty. The Commission at this stage considers that any issuance of banked allowances in the second period which is not based on an environmental counterpart by beneficiaries in terms of proven real emission reductions during the first trading period could constitute state aid which would likely be found incompatible with the internal market should it be assessed in accordance with Article 87 and 88 of the Treaty'.

[36] Supra chapter 2, note 6, paragraph 32. However, in its decision on the UK NAP for the second phase of the scheme, the Commission reserved its position in relation to the question of whether the plan could involve state aid if less than 10 per cent of allowances are auctioned by stating that the plan 'could potentially imply State aid pursuant to Article 87(1) of the Treaty' (see recital (4) of the Commission Decision of 29 November 2006 concerning the national allocation plan for the allocation of allowances notified by the United Kingdom in accordance with Directive 2003/87/EC of the European Parliament and of the Council, available at: http://ec.europa.eu/environment/climat/2nd_phase_ep.htm).

In terms of procedure the letter indicates that the Commission will not require NAPs to be formally notified under Article 88(3) of the Treaty but that it will scrutinise notified to it in accordance with the Directive for state aids incompatible with the Treaty, and if it finds such aid will use the state aids procedures. The question of exactly how this procedure would fit with the timetables in the Emissions Trading Directive for the development of NAPs is left unanswered, as is the question of how this relates to the requirements of state aids procedure.[37]

D.   Criterion 11: Competition from Outside the EU

The final criterion relating to the determination of the quantity of allowances allocated to activities or sectors is criterion 11. This provides that the NAP may contain information on the manner in which the existence of competition from countries or entities outside the European Union will be taken into account. The EC's High Level Group on Competitiveness, Energy and the Environment has recommended that Member States consider differentiated allocation between sectors in the second phase of the EU ETS, with a view to taking into account the external aspects of competitiveness.[38]

The 2003 Commission Guidance expresses the view that this criterion is not relevant in determining the total quantity of allowances, nor should it be taken into account in determining the allocation to individual installations. The latter proposition appears to be correct, as to treat differentially installations within the same sector, which are subject to the same international competition, would be contrary to the general principle of non-discrimination.

However, in certain circumstances, criterion 11 would appear to be relevant to determining the total quantity of allowances. A Member State may, for example, produce a figure for the total quantity by calculating the number of allowances to be allocated sector by sector, and adding these figures together. In calculating the number of allowances required by sector the 2003 Commission Guidance indicates that criterion 11 is relevant. As there is nothing in the Directive to prevent this approach to calculating the total quantity, it would appear to follow that Article 11 can be relevant to the determination of the total quantity. However, as indicated in section II above, the total quantity of allowances will be

---

[37] Council Regulation (EC) No 659/1999 of 22 March 1999 laying down detailed rules for the application of Article 93 of the EC Treaty OJ L 83, 27/03/1999, p.1.

[38] See paragraph 19 of the First Report of the High Level Group on Competitiveness, Energy and the Environment (an advisory platform launched in February 2006 bringing together the Members of the Commission for Enterprise and Industry, Competition, Energy and the Environment as well as all relevant stakeholders). Available at: http:// ec.europa.eu/enterprise/environment/hlg/doc_06/first_report_02_06_06.pdf.

limited by whichever of the criteria in Annex III requires the lowest number of allowances to be allocated, and therefore criterion 11 could not be used to justify an allocation of allowances higher than that permitted by criteria 1 to 4 of Annex III.

IV. Criteria Relating to the Distribution of Allowances Among Installations

The criteria in Annex III which are relevant to the allocation of allowances at installation level are criteria 4 to 8, and 10. It is criterion 10 which requires the NAP to contain a list of the installations covered by the Directive and the quantity proposed to be allocated to each: criteria 4-8 set the framework for determining those quantities.

A. Criterion 10: List of Installations

The list of installations required by criterion 10 must cover all installations which have received a greenhouse gas emissions permit up to the date of notification to the Commission of the NAP: although this is not expressly stated in criterion 10, it is a logical consequence of the definition of new entrant in Article 3(h).

Although not expressly required by criterion 10, the Commission takes the view in its 2003 Guidance that the NAP should not only set out the quantity proposed to be allocated to each installation, but also the quantity intended to be allocated each year to each installation.[39] This reflects the requirement in Article 11(4) that: 'the competent authority shall issue a proportion of the total quantity of allowances each year of the [phase] by 28th February of that year'.

The Directive does not expressly require those proportions to be equal, nor does it necessarily require the same approach to be taken to each installation. Indeed, provided some allowances are issued to at least one installation in any given year, there is nothing express in the Directive to require a proportion of allowances to be issued to *each* installation. However, the 2003 Commission Guidance asserts that Article 11(4) constitutes an obligation to issue a share of the total quantity to each installation each year.[40] Furthermore, it indicates that the proportion issued to all operators should, in principle, be equivalent in order to avoid undue discrimination.[41] Whilst the 2003 Commission Guidance recognises that a Member State may decide to issue different proportions of the total quantity of allowances each year, it draws attention to the

[39] Paragraph 100 of the 2003 Commission Guidance.
[40] Paragraph 101 of the 2003 Commission Guidance.
[41] Paragraph 102 of the 2003 Commission Guidance.

impact this could have on the market and recommends that the proportion of the total quantity of allowances issued each year does not substantially deviate from equal proportions.[42]

Where a Member State proposed to provide for different proportions of the total quantity to be issued each year or for different proportions to be issued in respect of different installations, then the NAP would need to explain the approach in order to satisfy the Commission that the proposed approach is consistent with the criteria in Annex III, in particular criterion 5 (non-discrimination).

Although the Directive requires a proportion of allowances to be issued by 28 February each year, there is also nothing in the Directive to prevent some allowances from being issued after 28 February. The Registries Regulation expressly provides for this in cases where it is envisaged in the Member State's NAP.[43] A Member State wishing to take advantage of this opportunity would need to make provision in its NAP for allowances to be issued at a later date.

B.   Criterion 10 and Unilateral Inclusion

The 2003 Commission Guidance provides that the list of installations should include both 'installations to be temporarily excluded in the first period pursuant to Article 27 and installations to be unilaterally included in any period pursuant to Article 24'.[44] In contrast to the provisions on installations which are temporarily excluded,[45] Article 24(2) provides expressly that allocations for installations carrying out activities which are unilaterally included shall be specified in the NAP. A number of questions do arise, however, about the inter-relationship of the procedure for unilateral inclusion under Article 24, the procedure for approval of NAPs under Article 9, and final decisions on allocation under Article 11.[46]

In theory a Member State should secure a decision for unilateral inclusion under Article 24 prior to submitting to its NAP to Commission under Article 9(1) and at the latest prior to the final allocation decision pursuant to Article 11. However, this has not been the practice of Member States

---

[42] Paragraph 101 of the 2003 Commission Guidance.

[43] Second paragraph of Article 40 and second paragraph of Article 46 of the Registries Regulation.

[44] Paragraph 98 of the 2003 Commission Guidance.

[45] If a decision had been made under Article 27(2) providing for the temporary exclusion of an installation prior to the submission of a NAP, that installation would not require a permit under the Directive, and therefore it is not obvious that it should be treated as an installation covered by the Directive for the purposes of criterion 10 of Annex III. For further information on temporary exclusion, see section VI of chapter 2.

[46] See section V of chapter 2 for further information on unilateral inclusion and chapters 5 and 6 for an explanation of the procedure for allocation under Articles 9 and 11.

in relation to the first and second phases of the scheme. In practice, the Commission has been willing to accept opt-in applications even where the installations to be opted-in have not been included in the NAP in accordance with Article 24(2), including after the final allocation decision.

C. Criterion 4: Consistency with Other EU Instruments and Criterion 5: Non-discrimination

Criterion 4 requires the NAP to be consistent with other Community legislative and policy instruments. Although the 2003 Commission Guidance only provides that this criterion is relevant when determining the total quantity of allowances to be allocated and the allocation of allowances at activity or sector level, it would appear to be equally applicable to allocations at the installation level. The reference in criterion 4 to 'the plan' could be read as a reference to all elements of the NAP including the proposed allocations at installation level, and there are good policy reasons why it should be read in this way. There could be Community legislation affecting some but not all installations within a particular activity or sector, and in such circumstances it would be contrary to the Community law principle of non-discrimination to treat all installations within the activity or sector in the same way. The principle not only requires installations in a like position to be treated in a like manner but also, conversely, installations in a different position to be treated in an appropriately different manner.

Criterion 5 requires that the NAP does not discriminate between companies or sectors and has been discussed above.

D. Criterion 6: New Entrants

Criterion 6 provides that the NAP must contain information on the manner in which 'new entrants will be able to begin participating in the Community Scheme in the Member State concerned'.

1. Definition of New Entrant

'New entrant' is defined in Article 3(8) as:

> (h) 'new entrant' means any installation carrying out one or more of the activities indicated in Annex 1, which has obtained a Greenhouse Gas Emissions Permit or an update of its Greenhouse Gas Emissions Permit because of a change in the nature or functioning or an extension of the installation, subsequent to the notification to the Commission of the National Allocation Plan.

The factor which distinguishes an existing installation from a new entrant is therefore that a new entrant either receives its permit after the NAP is notified to the Commission, or has its permit updated after the NAP is notified to the Commission. For the first phase this produced a cut-off date of 31 March 2004. For the second phase the cut-off date was 30 June 2006.

This element of the definition is clear conceptually, but does give rise to some difficulties in practice: it requires Member States to treat as existing installations all installations which are granted a permit up to and including the date of notification of the NAP. As the NAP must contain a list of existing installations, with a proposed allocation, this appears to require Member States either to impose a moratorium on granting or updating permits in the run-up to the date for notification of the NAP, or, if they do not impose a moratorium, to risk having to amend the list of installations, and proposed allocations, up to the last minute before notifying the NAP. In practice Member States have sought to find practical solutions to this problem by using the flexibility to update the allocations in the NAP when the final decision on allocation is made.[47]

2.   Obtaining a New Greenhouse Gas Emissions Permit

The category of new entrants can be sub-divided into two sub-categories. The first category, which is by far the clearest, includes those installations which have obtained a new greenhouse gas emissions permit. The majority of these will be brand new installations which were either not in existence or not in operation on the date of notification of the NAP to the Commission. However, this sub-category would also include installations which were in existence and operating at that date, but whose capacity subsequently extended with the result that they now exceed a threshold in Annex I. It might also include installations which were both in existence and operating and also exceeded the threshold in Annex I at the date of notification of the NAP, but which, for whatever reason, had not received a greenhouse gas emissions permit by that date. This was in particular the case in the first phase of the scheme.

3.   Change in Nature or Functioning

It is, however, the second sub-category which gives rise to more questions of interpretation. This second sub-category covers installations which

---

[47] Article 11 provides for the final decision on allocation to be 'based upon the national allocation plan'. This arguably provides some flexibility to include additional installations which were not included in the original NAP. For further discussion of the scope for changes between the NAP and the final decision under Article 11, see chapter 6.

obtain an update of their greenhouse gas emissions permit because of a change in the nature or functioning or an extension of the installation. In order to decide whether an installation falls within this sub-category, it is necessary to ask two questions: firstly, is there a change in the nature or functioning, or an extension of the installation, in the sense envisaged by Article 3(h)? Secondly, does this change or extension require an update of the greenhouse gas emissions permit? These two questions will be examined in turn below.

The first question, of whether there is a change in the nature or functioning or an extension of the installation, is likely to filter out relatively few cases. It will, nevertheless, have some application. For example, a change in the operator of an installation will require an update of the permit,[48] but this update would not be a consequence of a change in the nature or functioning or an extension of the installation.

Turning to the second question, there are, in contrast, likely to be many changes in the nature or functioning of an installation which do not require an update of the greenhouse gas emissions permit, and do not therefore render the installation a new entrant.

At first sight, the question of whether a change in the installation requires an update of the permit appears to turn on the arbitrary issue of the extent of detail which a Member State has included in the particular permit for that installation. The greater the level of detail, the greater the number of changes which would require an update of the permit. Indeed, if a Member State had included the operating hours of an installation within the permit, this line of argument could lead to the conclusion that an increased utilisation of the same capacity in a plant, through longer operating hours, could render the installation a new entrant. This appears unlikely to have been the intention behind the Directive: it would lead to differential treatment of like situations between Member States, with the same situation being treated in some Member States as constituting a new entrant, and in others not.

It is therefore necessary to look more closely at what is meant by an 'update' of a greenhouse gas emissions permit. Article 7 deals with the question of when the competent authority is required to update the permit, but merely provides that this is necessary 'where appropriate'. In order to determine what changes require an update in a permit, it is, therefore, necessary to deduce from the terms of Article 6 of the Directive, in its context, the level of detail which the Directive requires to be contained within a permit.

---

[48] Article 7 of the Directive.

Article 6(2) requires the greenhouse gas emissions permit to contain a 'description of the activities and emissions from the installation'. There are good arguments why such descriptions should include a description of the equipment used in the installation and the capacities of that equipment. It is necessary, firstly, to identify the boundaries of the activity and, secondly, to determine whether the activity being carried out in the installation is above the threshold specified in Annex I of the Directive for the relevant activity and therefore whether the plant is an installation for the purposes of the Directive. As this information is required by the Directive, a change requiring an update to information on the equipment and capacities of equipment used in the installation would be an update within the autonomous Community law sense of that word. A change within an installation of this type requiring an update of the permit would therefore bring the installation within the definition of 'new entrant' in the Directive.

Conversely, it is possible to imagine certain types of information which relate to the installation but upon which nothing in the Directive turns. For example, information on the proportion of capacity being used within the installation or on the number of hours in which the installation is operational.

These are two fairly clear-cut examples of types of information which the Directive does not require to be included in the permit. A Member State may of course choose to require such information, and in that case an update of the permit might be required where this information changes. But the mere fact that an update of the permit is required under national legistlation would not have the effect of bringing the installation within the definition of new entrant.

However, whilst the principle illustrated by these two examples may be clear, its application will not always be so. For example, Article 6(2)(c) requires the permit to include monitoring requirements. Article 14 requires Member States to ensure that emissions are monitored in accordance with the detailed guidelines on monitoring reporting adopted by the Commission, which in turn require detailed monitoring and plans to be developed. If the resulting monitoring plan is included in the permit, then relatively small changes may require an update to the greenhouse gas emissions permit. However, if the plan is not directly included in the permit but is instead developed by the operator pursuant to a requirement in the permit, then no update would be required. It is, therefore, unclear whether all changes to monitoring and reporting plans requiring an update of the permit would bring the installation within the definition of new entrant in the Directive.

4. Member State Discretion

As is clear from the definition in Article 3(h), a necessary condition for being eligible for allowances as a new entrant is either the grant of a new permit or a change in the installation which requires an update of the permit. If a change in an installation does not bring that installation within the definition of new entrant, it follows automatically that the installation cannot receive allowances as a new entrant under the Directive. The converse is not, however, true: not all new entrants must be allocated allowances. Criterion 6 in Annex III merely requires a Member State to include in its NAP information on the manner in which new entrants will be able to begin participating in the scheme, and therefore leaves Member States a wide discretion. The 2003 Commission Guidance recognises that a Member State has at least three options to enable participation of new entrants: it may have any new entrants buy all allowances on the market, it may make use of the possibility to set aside some allowances for periodic auctioning, or it may foresee a reserve in the NAP to issue allowances to new entrants free of charge.[49] The Guidance notes, however, that the operation of a reserve for new entrants increases the complexity and administrative costs of the emissions trading scheme.[50]

In deciding whether to provide for allowances to be allocated to any particular category of new entrant, the Member State will have to comply with the other relevant criteria in Annex III[51] and the general principles of Community law, including in particular the principle of non-discrimination.

The 2003 Commission Guidance argues that requiring new entrants to buy allowances on the market or to buy allowances in an auction would be consistent with that principle for the following reasons:[52]

> – the size of the EU-wide allowance market sets the correct conditions for liquidity, which ensures that new entrants will have access to allowances;
> – new entrants can be distinguished from existing installations because existing installations have made their investments without having been able to take the cost of carbon into account, in contrast to new entrants, who can minimise their carbon costs through investment choices;

---

[49] Paragraph 55 of the 2003 Commission Guidance.
[50] Paragraphs 64 of the 2003 Commission Guidance.
[51] In particular criteria 4, 5, 7, 8 and possibly 11.
[52] Paragraphs 56 and 59 of the 2003 Commission Guidance.

- new installations only fulfil the definition of a new entrant for a limited period of time, i.e. part of a trading period, and the cost of allowances for this limited period can be taken into account in the investment and timing decisions. The Directive guarantees that in the next allocation period the new entrant will be allocated allowances in the same manner as all other existing installations.

Where a Member States decides to provide for a new entrant reserve, the 2003 Commission Guidance provides that the NAP should indicate the size of the reserve justified by reference to an informed estimate of the expected number of new entrants during the trading period.[53] The rules and procedures by which new entrants would be issued allowances free of charge must be set out in the NAP. The Commission recommends that in order to respect the principle of equal treatment, the methodology for allocating allowances to new entrants should as far as possible be the same as the one used for comparable existing installations and Member States should not create several reserves dedicated to separate activities, technologies or specific purposes.[54] A Member State should further specify what use will be made of any allowances remaining in the reserve at the end of the period.[55] A Member State may auction any remaining allowances (subject to the limitation in Article 10 of the Directive). NAPs should also state the procedure which will be followed if new entrants apply for allowances and the reserve set aside for the period is already exhausted. [56]

In the first phase of the scheme, all Member States provided for a new entrant reserve. The 2005 Commission Guidance recognised that the difference in the rules developed by Member States for dealing with new entrants increased the complexity of the EU ETS, reduced transparency in the market and may lead to distortions in competition.[57] It therefore urged all Member States to consider simplifying the rules.

E. Criterion 7: Accommodation of Early Action

Criterion 7 provides that 'the Plan may accommodate early action and shall contain information on the manner in which early action is to be taken into account.'

---

[53] Paragraphs 60 of the 2003 Commission Guidance.
[54] Paragraphs 61 of the 2003 Commission Guidance.
[55] Paragraph 62 of the 2003 Commission Guidance.
[56] Paragraph 63 of the 2003 Commission Guidance.
[57] Paragraph 5 of Annex 4 to the 2005 Commission Guidance.

The term 'early action' is not defined in Annex III, but the 2003 Commission Guidance provides that 'early action is to be understood as actions undertaken in covered installations to reduce covered emissions before the National Allocation Plan is published and notified to the Commission'.[58] The principle behind this criterion is that it would be unfair for installations which have taken action to reduce emissions prior to the development of the NAP to be disadvantaged vis-a-vis other installations which have not taken such action.

The 2003 Commission Guidance suggests two main ways in which Member States may accommodate early action in their NAPs.[59] Firstly, they may provide for allocations to be based on historical emissions during a relatively early base period or to allow an operator to include an early year when it had higher emissions than in its base year.[60] Secondly, a Member State may accommodate early action by allocating allowances on the basis of benchmarks. This option is further expanded upon by the second sentence of criterion 7 which provides that:

> benchmarks derived from reference documents concerning the best available technologies may be employed by Member States in developing their National Allocation Plans, and these benchmarks can incorporate an element of accommodating early action.

Where allocations are based on benchmarks, installations will be allocated a set number of allowances (for example, on the basis of each unit of production) regardless of their actual emissions. This means that an installation which has taken early action to reduce its emissions may well receive more allowances than it needs.

In relation to the second phase of the scheme, the 2005 Commission Guidance encourages Member States not to use individual installations' emissions in the first phase of the scheme as a basis for allocating allowances to those installations for the second phase. If Member States follow this approach this would adequately recognise early action and dispense with other means of rewarding early action.[61]

---

[58] Paragraph 69 of the 2003 Commission Guidance.

[59] The 2003 Commission Guidance (paragraph 74) also includes another possible method of taking into account early action involving a second round of allocations giving a bonus to those installations in which operators have undertaken early action. This would only appear to be relevant where the baseline chosen by a Member State failed to take sufficient account of early action. For example, where allocations are made on the basis of one baseline year close the to the beginning of the period covered by the NAP.

[60] For details of how this was done in the UK National Allocation Plan for the first phase of the scheme, see chapter 19.

[61] Paragraphs 27 –30 of the 2005 Commission Guidance.

F.   Criterion 8: Clean Technology

Criterion 8 provides that 'the Plan shall contain information on the manner in which clean technology, including energy efficient technologies, are taken into account.'

The Commission Guidance notes that there is a link between criterion 7 on early action and criterion 8, since early action will typically have been an investment in clean or energy efficient technology. Nevertheless, criterion 8 is not redundant. For example, were a Member State to use a baseline methodology to reward companies which had taken early action, this would not reward installations which were inherently efficient and had been throughout the baseline period. Criterion 8 could justify an additional element in a Member State's methodology to reward such installations.

V.   Other Criteria

A.   Criterion 12: Use of Project Credits

From the second phase of the scheme, in addition to specifying the total quantity of allowances to be allocated and how they will be allocated between installations covered by the scheme, criterion 12 of Annex III[62] to the Directive also requires the national allocation plan to specify whether installations will be permitted to use project credits under the Kyoto Protocol to comply with their obligations under the scheme and, if so, the limit on the use of such credits. The use of such credits must be supplementary to domestic action. For the period 2008-12, NAPs are also required to set aside allowances for the avoidance of double counting between the scheme and any project activities under the Kyoto Protocol hosted by Member States which result in a reduction or limitation of emissions from installations covered by the Directive. Criterion 12 and the double counting issue is discussed in chapter 8.

B.   Criteria Not Listed in Annex III

It is clear from the wording of Article 9(1) that the criteria in Annex III are not exhaustive: Article 9(1) requires a plan to be 'based on objective and transparent criteria, including those listed in Annex III'. It is for this reason that the annex to the 2003 Commission Guidance, in setting out a common format for the NAPs for 2006-2007, suggested that Member States identify any criteria other than those in Annex III to the Directive which have been applied for the establishment of the notified NAP.[63]

---

[62] Inserted by Article 1(9) of the Linking Directive.
[63] At paragraph 9 of the Annex to the 2003 Commission Guidance.

In practice, however, the fact that the criteria in Annex III are phrased in a very open-ended way makes it possible to brigade under those criteria most, if not all, factors which a Member State is likely to wish to have regard to when drawing up a NAP.

VI.   'Ex post adjustments' and Future Commitments

So far this chapter has considered the rules in the Directive which govern the allocation of allowances to existing installations and to new entrants. There remains a question as to whether those allocations are final or whether it is compatible with the Directive for a NAP to make adjustments at a later date.

The Commission considers that, in line with Article 11 and criterion 10 of Annex III the Directive requires Member States to decide up-front the absolute quantity of allowances allocated in total and to each installation's operator, and that that decision may not be re-visited and no allowances may be reallocated by means of adding to or subtracting from the quantity allocated to each operator on the basis of a Government decision or a pre-determined rule.[64] The Commission has therefore rejected aspects of NAPs which provide for ex post adjustments as contravening criterion 10.[65]

Apart from the allocation of allowances to new entrants, the only situation in which the Commission has accepted that Member States may wish to make adjustments is where an installation ceases operation. There is nothing in the Directive which expressly prevents Member States from providing for no further allowances to be issued to closed installations whose permits have been withdrawn and indeed it can be seen as a logical corollary of providing for allocation of allowances to installations which start up during the phase, as new entrants. Furthermore the fact, as seen above, that Article 11(4) requires the issue of allowances to be staggered, with a proportion being issued at the beginning of each year of a phase, makes it possible to end the issuing

---

[64] Section 2.2 of the Communication on NAPs for the second phase of the scheme (supra note 11).

[65] For example, in the first phase, the Commission rejected aspects of the German NAP which provided for ex post adjustments to allocations to new entrants and to the allocations under the NAP to existing installations in the event that other installations in Germany closed or where the installation experienced lower capacity utilisation than foreseen, its annual emissions were less than 40 per cent of its base period emissions or that it generated a lower amount of power production from combined heat and power than in the base period (Commission Decision of 7 July 2004 concerning the national allocation plan for the allocation of greenhouse gas emission allowances notified by Germany in accordance with Directive 2003/87/EC of the European Parliament and of the Council, C(2004) 2515/2 final). This decision is the subject of a legal challenge, Case T-374/04, *Germany v Commission*.

of allowances where an installation has closed. The Commission, accordingly, approved NAPs for the first phase which took this approach, including NAPs containing detailed provisions on transfers. Whilst the 2005 Commission Guidance recognises that it is still too early to draw conclusions on best practice, it notes that detailed provisions on closures and transfers increase the complexity of the scheme and reduce transparency in the market which may distort competition.

In the first phase of the scheme, some Member States sought to include in their NAP a guarantee that a certain level of allowances would be allocated to particular allowances in future phases.[66] When assessing NAPs, the Commission only considered the allocation for the phase covered by the NAP and reserved its position in relation to allocations in future phases. The implementation of any guarantees will be considered as part of the Commission's assessment of the NAP for the phase in which they will be issued. In considering the implementation in second phase NAPs of guarantees given in first phase NAPs, the Commission has indicated that such guarantees may result in differential treatment between installations and thereby be inconsistent with criterion 5 of Annex III.[67] They would also be inconsistent with the environmental objectives of the scheme as they could create perverse incentives for investment, and be consistant with any future amendments to the Directive providing for greater harmonisation of allocation rules.

VII. Amendment of Annex III for the Second Phase

Article 21 of the Directive required Member States to report to the Commission by 30 June 2005 on the application of the Directive. The reports were to be based on a questionnaire drafted by the Commission.[68] Article 22 of the Directive gives the Commission power, in the light of these reports and of the experience of the application of the Directive, to amend the criteria in Annex III, with the exception of criteria 1 (consistency with Kyoto obligations), 5 (non-discrimination and state

---

[66] For example, the German NAP sought to guarantee that a new entrant in the first phase would be given a full allocation of allowances for 14 years into the future.
[67] Section 2.4 of the Communication on NAPs for the second phase of the scheme (supra note 11). Such guarantees may also constitute state aid
[68] See Commission Decision 2005/381/EC establishing a questionnaire for reporting on the application of Directive 2003/87/EC of the European Parliament and of the Council establishing a scheme for greenhouse gas emission allowance trading within the Community and amending Council Directive 96/61/EC, OJ L 126, 19.5.2005, p. 43 as amended by Commission Decision 2006/803/EC amending Decision 2005/381/EC establishing a questionnaire for reporting on the application of Directive 2003/87/EC of the European Parliament and of the Council establishing a scheme for greenhouse gas emission allowance trading within the Community and amending Council Directive 96/61/EC, OJ L 329, 25.11.2006, p. 38.

aids), and 7 (early action and benchmarks). This decision is to be made through the comitology procedure referred to in Article 23(2).[69]

The Commission has not made a proposal to amend the criteria in Annex III for the second phase of the scheme. Instead the 2005 Commission Guidance summarises the experience gained from NAPs for the first phase and general lessons learned for the second phase.[70] In particular the guidance emphasises the need to avoid ex post adjustments in order to create stable incentives for installations to reduce emissions, and encourages Member States to simplify their allocation methodologies particularly in relation to new entrants and closures, in order to increase transparency. Amendments to the allocation criteria will be considered as part of the strategic review of the scheme under Article 30(2) of the Directive. This would include in particular consideration of further harmonisation of the rules on allocation to new entrants and closures, including the possibility of an EU-level new entrant reserve accompanied by EU-wide administrative rules on closure and cross-border transfers.[71]

---

[69] See section VII of chapter 2.
[70] See in particular Annex 4 to the 2005 Commission Guidance.
[71] For further information on the review of the scheme, see Section I of chapter 10.

CHAPTER 5

ALLOWANCES: PROCEDURE FOR
NATIONAL ALLOCATION PLANS

This chapter considers the procedure by which Member States develop, and the Commission approves, NAPs. Section I discusses the procedure leading up to the notification of NAPs to the Commission. Section II discusses the way in which the Commission considers and approves NAPs. Section III looks at the public consultation required by the Directive after the Member State has notified the NAP to the Commission. Section IV looks at the procedure for amending a NAP after it has been notified to the Commission.

The requirements relating to NAPs are set out in Article 9 of the Directive:

Article 9

National Allocation Plan

1.   For each period referred to in Article 11(1) and (2), each Member State shall develop a national allocation plan stating the total quantity of allowances that it intends to allocate for that period and how it proposes to allocate them. The plan shall be based on objective and transparent criteria, including those listed in Annex III, taking due account of comments from the public. The Commission shall, without prejudice to the Treaty, by 31 December 2003 at the latest develop guidance on the implementation of the criteria listed in Annex III.

For the period referred to in Article 11(1), the plan shall be published and notified to the Commission and to the other Member States by 31 March 2004 at the latest. For subsequent periods, the plan shall be published and notified to the Commission and to the other Member States at least 18 months before the beginning of the relevant period.

2.   National allocation plans shall be considered within the committee referred to in Article 23(1).

3.   Within three months of notification of a national allocation plan by a Member State under paragraph 1, the Commission may reject that plan, or any aspect thereof, on the basis that it is incompatible with the criteria listed in Annex III or with Article 10. The Member State shall only take a decision under Article 11(1) or (2) if proposed amendments are accepted by the Commission. Reasons shall be given for any rejection decision by the Commission.

As mentioned in chapter 4, the Commission, as required by Article 9(1), adopted guidance to assist Member States in developing NAPs in December 2003 (the 2003 Commission Guidance)[1] and December 2005 (the 2005 Commission Guidance).[2]

I.   Notification of the NAP

A.   Public Consultation before Notification

The requirement in Article 9(1) that a Member State's NAP be developed 'taking due account of comments from the public' implies some form of public consultation before the Member State publishes the NAP and notifies it to the Commission in accordance with the second paragraph of Article 9(1). The Directive does not prescribe any procedure for this first round of public consultation.

The 2003 Commission Guidance suggests that the public should be consulted on the basis of a draft text of the NAP, and given a reasonable timeframe for submitting comments.[3] The Commission also suggests that feedback is to be given, in a general form, to the public about the decision taken and the main considerations on which it is based.

Whilst it is undoubtedly the legally safest route, there is room for argument as to whether the first consultation must take place on the basis of a draft text of the NAP. It might be possible to 'take due account of comments from the public' by consulting the public on general issues, rather than a full draft NAP. The requirements of Article 9 of the Directive may be contrasted in this respect with the requirements of other EU instruments in the environmental field requiring public consultation. The strategic environmental assessment Directive,[4] for example, expressly requires draft plans to which it applies to be 'made available to the ... public', who are to be given an 'early and effective opportunity to comment' on the draft plan.[5] A slightly less onerous requirement is found in the public participation Directive[6], which requires Member

---

[1] Supra chapter 4, note 3.
[2] Supra chapter 2, note 6.
[3] Paragraph 96 of the 2003 Commission Guidance applying the guidance given in paragraphs 94 and 95 relating to the consultation envisaged by paragraph 9 of Annex III to the Directive.
[4] Directive 2001/42/EC of the European Parliament and of the Council of 27th June 2001 on the assessment of the effects of certain plans and programmes on the environment, OJ L 197, 21.7.2001, p.30.
[5] Articles 6(1) and (2) of the strategic environmental assessment Directive (ibid).
[6] Directive 2003/35/EC of the European Parliament and of the Council of 26th May 2003 providing for public participation in respect of the drawing up of certain plans and programmes relating to the environment and amending with regard to public participation and access to justice Council Directives 85/337/EEC and 96/61/EC, OJ L 156, 25.6.2003, p.17.

States to ensure that the public is 'informed ... about any *proposals* [emphasis added] for [relevant] plans or programmes', that 'relevant *information about such proposals* [emphasis added] is made available to the public', and that 'the public is entitled to express comments and opinions when all options are open before decisions on plans and programmes are made'. Given the existence of these more detailed and explicit procedures to be found in instruments adopted in the two years before the Emissions Trading Directive, it can be argued that it was a deliberate decision of the Council and European Parliament, in that Directive, to impose a more flexible requirement to draw up a plan 'taking due account of comments from the public'.

A further factor to consider in interpreting the scope of the consultation requirement before Member States notify a NAP to the Commission is the relationship with the consultation which the Directive requires *after* the NAP has been notified. This requirement, which flows from paragraph 9 of Annex III and Article 11(1) will be explored further below. As will be seen below, the scope of that second stage of consultation was considered by the Court of First Instance in *United Kingdom v Commission*.[7] The judgment appears to leave open the possibility, advanced by the Commission, that the second stage of consultation may be limited to the allocation of allowances to individual installations.[8] As will be seen below, there are good arguments from the structure of the Directive to support this view. If it is correct, it strengthens the argument for a fuller consultation of the public at the first stage, before notification of the NAP. Otherwise the public would be denied the opportunity to comment on the Member State's detailed proposals relating to the determination of the total quantity. However even this may not necessarily require consultation on a full draft NAP.

If, on the other hand, and contrary to the arguments advanced by the Commission in *United Kingdom v Commission*, the correct interpretation of the Directive is that the second stage of consultation provides an opportunity for the public to comment on any aspect of the notified NAP, then a full consultation on a draft prior to notification would appear to provide superfluous duplication of consultation.

---

[7] *United Kingdom v Commission of the European Communities* (Case T 178-05), 23 November 2005, paragraphs 57-9.
[8] The Court, in paragraphs 58 and 59 of the judgment (ibid), dismissed the argument of the Commission that the second stage of consultation cannot lead to an increase in the total quantity of allowances. However, in paragraph 59 it left open the possibility that the Commission may have been correct in arguing that the second public consultation concerns only the allocation to individual installations (which may nevertheless, as the Court points out, lead to increases in the total quantity).

It is worth noting that neither the strategic environmental assessment Directive nor the public participation Directive mentioned above apply to the development of NAPs under the Emissions Trading Directive. The strategic environmental assessment Directive does not apply because, notwithstanding the fact that the NAP constitutes a 'plan' under Article 2(a) of that Directive, it does not 'set the framework for future development consent for projects' (a condition for the application of that Directive, set out in its Articles 3(2)(a) or 3(4)). This is because although the NAP sets the framework for decisions on the allocation of allowances under Article 11 of the Directive, these decisions are not development consents, as they do not entitle developers to proceed with any project.[9] Rather, if any decision under the Emissions Trading Directive constitutes a development consent, it is the grant of a permit under Article 4.The public participation Directive does not apply to the development of a NAP because the public participation requirements in Article 2 of that Directive apply only in relation to the plans listed in Annex I of that Directive, which does not include NAPs.

B.   Procedure for Notification of NAPs to the Commission

The Directive requires Member States, after having carried out public consultation, to publish the NAP and notify it to the Commission. For the first phase of the scheme, Member States were required to publish and notify their NAPs to the Commission by 31 March 2004. For the second and subsequent phases, NAPs must be published and notified 18 months prior to the start of the phase. For the second phase of the scheme, which starts on 1 January 2008, the deadline was therefore 30 June 2006.

For both the first and second phases of the scheme several Member States missed the deadline for publishing and notifying NAPs to the Commission.[10] This leads to the question of what legal consequences flow from missing the deadline.

The Commission can initiate proceedings against Member States under Article 226 of the EC Treaty for infringing the requirements of the Directive. If the Member State does not remedy the breach by notifying a NAP by the end of the period laid down in the Commission's reasoned opinion under Article 226 EC, the Commission could refer the matter to the ECJ for a declaration that the Member State is in breach of its obligations. However, this is a weak enforcement mechanism since, at

---

[9] See the definition of 'development consent' in the EIA Directive, supra chapter 2, note 12.

[10] By 29 November 2006, only 19 Member States had notified a NAP for the second phase of the scheme to the Commission.

this stage, the ECJ has no power to impose any financial penalty[11] and therefore no direct financial consequences would flow.

A failure to notify a NAP could also be invoked directly by affected operators and possibly environmental groups to seek a remedy in the domestic courts. Such an action could be brought either on the basis of a provision of national law which transposes the Directive requirement to publish and notify the NAP[12] or on the basis of the direct effect of Article 9 of the Directive.[13] Whilst a national court might, depending on the constitutional structure of the Member State, be able to make an order compelling the executive to publish and notify a NAP, it is difficult to imagine scenarios in which any individual could show that the failure to notify a NAP had caused them financial loss and so claim compensation: no individual entitlements to allowances are created by the notified NAP, as these flow only from the later allocation decision under Article 11.

II.   The Consideration of NAPs by the Commission

Article 9(3) of the Directive provides that within three months of a Member State notifying its NAP, the Commission may reject that NAP or any aspect of it on the basis that it is incompatible with the criteria listed in Annex III or with Article 10. Before deciding whether or not to reject a NAP, the NAP must be considered by the Climate Change Committee.[14] The effect of a decision rejecting all or any aspect of a NAP is to require the Member State to propose amendments. Only once these are accepted by the Commission can the Member State proceed to take the definitive decision under Article 11 as to the total quantity of allowances to be allocated and their allocation to individual installations.

---

[11] Financial penalties may only be imposed under Article 228 EC. Proceedings under Article 228 EC may only be brought if the Member State fails to comply with a judgment under Article 226 EC. The procedure under Article 228 EC is similar to the procedure under Article 226 EC. If a Member State fails to comply with the Commission's reasoned opinion under Article 228 EC then the ECJ may impose a lump sum penalty payment or a daily fine for as long as the breach persists, or both. See, for example, judgment of the ECJ on 12 July 2005 in Case C-304/02 Commission of the *European Communities v French Republic*.

[12] This requirement in Article 9(1) appears to meet the requirement of creating individual rights, as interpreted by the ECJ in, for example, Case C-131/88 *Commission v Germany* ECR I-825 and so must be transposed into national law.

[13] The requirement in Article 9(1) appears to meet the tests of clarity, precision and unconditional nature developed by the Court for a provision to have direct effect (*Van Duyn v Home Office* [1974] ECR 1337).

[14] Article 9(2). This is the same committee which is used where measures are to be adopted under the comitology procedure, see section VII of chapter 2.

A.   Submission of an Incomplete NAP

The question arises of how this provision applies where the Commission considers that a notified NAP is incomplete.

In preparation for the first phase of the scheme, the Commission adopted the practice of writing to Member States asking for the additional information it considered necessary to remedy any omissions. The Commission purported by these letters to suspend the three month period prescribed in Article 9(3) in the following terms:

> This information is requested in order to facilitate the definition of the Commission's position on the proposed plan. The Commission reserves the right to define its position only after the additional information is received, and in any case no later than three months after the receipt of this information.

This practice was considered by the CFI in *United Kingdom v Commission*:

> 73. ... if the NAP is incomplete or 'provisional', the Commission is entitled to reject it either because it fails to comply with the criteria laid down by the Directive or because the Commission is unable to assess its conformity with those criteria. In that situation, the Commission is entitled, by rejecting the NAP, to oblige the Member State to notify a new and complete NAP before it takes its definitive decision under Article 11(1) of the Directive. Contrary to the Commission's submission, there is no need to suppose that the notification of an incomplete NAP has the effect of stopping the three month period laid down by Article 9(3) of the Directive from running.

This suggests that in circumstances where the Commission receives a NAP which it considers incomplete, the Commission should take a decision rejecting the NAP, either in its entirety or in part. However, this was only part of the *obiter dicta* of the court's judgment. In practice, the Commission has continued to take a pragmatic approach in relation to the consideration of the NAPs for the second phase of the scheme where a NAP contains insufficient information for the Commission to assess its compatibility with the criteria in Annex III of the Directive. In such circumstances, the Commission practice is to write to Member States seeking further information from Member States before concluding its assessment of a NAP.[15] Under this approach a NAP is not considered to have been submitted until it is complete. Until the further information is provided, the Commission does not therefore need to take a formal Decision under Article 9(3) rejecting the NAP as incomplete. Rather, the

---

[15] See paragraph 7 of the 2005 Commission Guidance which provides that 'the Commission considers that the three month period foreseen in Article 9(3) can only commence once a complete national allocation plan has been submitted.'

Member State remains under a duty under Article 9(1) to notify a complete NAP. On this analysis the Commission could in such situations initiate proceedings under Article 226 of the Directive for failure to provide the further information necessary to constitute a complete NAP.

Once the Commission takes a decision rejecting a NAP, the period of three months under Article 9(3) becomes irrelevant: the Member State must notify amendments to the NAP and can only proceed to take a definitive decision on allocation once the Commission has accepted them. The Directive does not prescribe any period within which the Commission is to consider proposed amendments. However, as a Member State may not proceed to make its final allocation decision under Article 11 until proposed amendments have been accepted by the Commission, the general duty of Member States and the Commission under Article 10 of the EC Treaty to co-operate in the fulfilment of the tasks of the Community[16] requires both parties to co-operate in proposing and considering amendments expeditiously, to enable the decisions under Article 11 to be made by the deadlines set in that article.

B. Effect of the Commission not Reacting within Three Months

The CFI also considered in *United Kingdom v Commission* the effect of the Commission failing to take any decision within the three month period prescribed in Article 9(3). The Court stated in its *obiter dicta*[17] that 'if the Commission does not react to the NAP in the three months following notification, the plan must be considered as approved by the Commission'. One consequence, if this is correct, is that a Member State would then be entitled to proceed to take its allocation decision under Article 11 without any further approval from the Commission. A further consequence, expressly mentioned by the Court, would be that the NAP 'may not then be amended unless the proposed amendments are accepted by the Commission in accordance with Article 9(3) of the Directive'.

The fact that the Commission would be considered in these circumstances as having approved the NAP may bar the Commission from invoking in any other proceedings any alleged incompatibility of the NAP with the Directive.[18] In particular, it would prejudice the legal certainty (which Article 9(3) is designed to promote) if, notwithstanding its deemed

---

[16] Although the wording of Article 10 the EC only requires Member States to facilitate the achievement of the Community's tasks, the Community courts have determined that Article 10 imposes on both the European institutions and the Member States mutual duties of loyal cooperation with a view to attaining the objectives of the Treaty. See, for example, case C-2/88 *Imm Zwartveld* [1990] ECR I-3365, 16.

[17] Paragraph 55 (supra note 7).

[18] The exception to this may be state aids actions. See section III.C of chapter 4.

approval of a NAP under Article 9(3), the Commission could initiate proceedings under Article 226 EC alleging that a Member State is in breach of its obligations by virtue of alleged incompatibilities between the NAP and the Directive.[19]

C.   Scope of the Commission's Powers in Considering a NAP

The Commission may only reject the notified NAP, or any aspect of it, on the basis that it is incompatible with the criteria listed in Annex III of the Directive or with Article 10.[20] As seen in chapter 4, the criteria in Annex III are drafted broadly, with the consequence that there is a wide range of possible NAPs which a Member State could draw up, any one of which would be compatible with the criteria in Annex III. The Commission's power is limited to rejecting a NAP which is not compatible with those criteria. It does not have power to require a Member State to adopt a particular form of NAP. However, Article 9(3) of the Directive requires the Commission to give reasons for any decision to reject a NAP, in whole or in part. In doing so the Commission could indicate the changes which, in its view, would be consistent with the Directive.

In practice, in its assessment of first and second phase NAPs, the Commission has taken a pragmatic approach. Decisions rejecting elements of a NAP have been framed as conditional approvals, whereby if a Member State amends its NAP in the ways specified in the Decision and notifies such amendments to the Commission, the NAP will be deemed to have been approved by the Commission.[21] This approach

---

[19] However, the same arguments may not apply in relation to other parties wishing to invoke incompatibilities between the NAP and the Directive. It would be surprising, if a NAP were incompatible with the substantive requirements of the Directive (see chapter 4), for the Commission inaction to prevent this from being brought before the courts, for example by adversely affected operators or by environmental groups. Such proceedings would have to be brought against a national government in their national courts. This is because there would by definition be no Commission Decision which could the subject of a direct action in the CFI under Article 230 EC (even if operators and environmental groups were able to meet the restrictive rules on standing for bringing such an action). Nor would there appear to be any merit in trying to bring an action against the Commission for failure to act under Article 232 EC, as the Commission's power to reject a NAP would, according the decision of the CFI in *United Kingdom v Commission* supra note 7, have expired on the expiry of the three month period.

[20] Both the criteria in Annex III and Article 10 are discussed further below. This was supported by the CFI in *United Kingdom v Commission* (ibid, paragraph 54) in which the CFI recognised that the Article 9(3) 'stipulates that any rejection is to be based on the NAP's incompatibility with the criteria of Annex III to or with Article 10 of the Directive.' The Court also noted that no other ground for rejection of a NAP is provided for in the Directive.

[21] For example, Article 2 of the Commission Decision in relation to the UK NAP (supra chapter 4, note 36) for the second phase of the scheme provides: 'No objections shall be

(continued...)

avoids the need for a further Commission decision unless the Commission considers that the amendments notified to it are not consistent with the Decision, or the Member State wishes to propose an amendment which is inconsistent with the approach suggested by the Commission.

This question of the extent of the Commission's powers arose in relation to the NAP proposed by the United Kingdom for the first phase, and was considered by the CFI in *United Kingdom v Commission*.[22] In the course of proceedings, the UK questioned whether the Commission had been within its power to impose the condition in Article 3(1) of its Decision of 7 July 2004 requiring the total quantity of allowances to be capped at 736 million tonnes, notwithstanding the United Kingdom's statement that this figure was provisional and would be updated to reflect updated energy projections. The United Kingdom argued that the Commission had no power under Article 9(3) of the Directive of its own motion to fix the total quantity of allowances that a Member State may allocate and that the Commission could not therefore fix the total quantity of allowances at the provisional level, as its Decision purported to do.[23] Whilst the Court pointed out that ' a Member State cannot, by submitting an incomplete NAP, postpone indefinitely the adoption of a decision by the Commission under Article 9(3) of the Directive,' it did not rule on the wider question raised in the UK's submission.[24] It stated, rather, that it was 'immaterial' given the Court's decision that the United Kingdom was entitled under Article 9(3) of the Directive to propose an amendment to increase the total quantity notwithstanding the terms of Article 3(1) of the Commission's Decision.

---

raised to the national allocation plan, provided that the following amendment to the national allocation plan is made in a non-discriminatory manner and notified to the Commission as soon as possible, taking into account the time-scale necessary to carry out the national procedures without undue delay: the installations situated within the territory of Gibraltar and the quantities of allowances intended to be allocated to them are listed; those quantities being determined taking into account independently verified emissions and changes in production and carbon intensity that are reasonably anticipated'.

[22] Paragraph 73 of the judgment supra note 7.

[23] Paragraphs 25-27 of the judgment (ibid).

[24] Although the Court did indicate, at paragraph 73 of the judgment (ibid), that the correct response of the Commission to a 'provisional' NAP was to reject it in whole or in part if it either considered it incompatible with the Directive or considered that its provisional nature prevented the Commission from assessing its compatibility with the Directive, it did not address the issue of whether such rejection decisions could set the conditions under which a NAP would be deemed to be approved.

III. Public Consultation After Notification of a NAP

Paragraph 9 of Annex I to the Directive requires a NAP to:

> ...include provisions for comments to be expressed by the public, and contain information on the arrangements by which due account will be taken of these comments before a decision on the allocation of allowances is taken.

Correspondingly, Article 11(2) (Article 11(1) for the first phase) requires a Member State's decision on the total quantity of allowances and the allocation of these to each installation to be made 'taking due account of comments from the public'.

A question arises as to the scope of the consultation required by paragraph 9 of Annex I. At one extreme, it could be argued that the consultation is to address all aspects of the NAP. At the other, it can be argued that it is to be limited to the level of allocations to individual installations.

A number of arguments can be advanced for the latter, narrower view. Firstly, as we have seen earlier in this chapter, Member States are already required by Article 9(1) to have taken into account comments from the public before notifying the NAP to the Commission. To have a full public consultation on all aspects of the NAP at this second stage could be said to entail superfluous duplication. Secondly, the NAP has already been notified to the Commission: to carry out a full public consultation in parallel to the Commission's consideration of the NAP increases the risk of Member States having to propose changes either whilst the Commission is considering the NAP, or after the Commission has reached a decision. Thirdly, it could be argued that the structure of Articles 9 and 11, in requiring some form of public consultation, then notification of a NAP, followed by a Commission approval or rejection, further consultation, and finally a decision on the allocation of allowances to individual installations, is designed to provide a narrowing down of issues, and hence increasing certainty for the emissions trading market, as the beginning of each phase approaches.

The CFI considered the interpretation of these provisions in *United Kingdom v Commission*. The Commission argued that comments obtained during this second round of public consultation are only intended to serve the purposes of data variation and possibly the reallocation of allowances within the total quantity, and not to serve as a means to increase the total quantity.[25] The Commission further argued that the second consultation is limited to how the Commission Decision on the

---

[25] Paragraph 58 of the judgment (ibid).

NAP should be implemented within the context of its scope and to aspects with regard to which the Member State may exercise discretion.[26]

The CFI rejected the Commission's argument that the second consultation could not serve as a means to increase the total quantity. It left open the broader question, however, of whether the scope of the second consultation is more limited than the first:

> Even if the Commission was correct in arguing that the second public consultation concerned only the allocations to individual installations, the Court considers that the Commission has not shown why consequential amendments to the individual allocations might not necessitate making changes to the total quantity of allowances to be allocated. If, for example, there was an underestimation of the allowances to be allocated to an individual installation, while a competing installation of equivalent size received the correct allocation, it cannot be excluded that the allocation to the first installation, and consequently, the total allocation of allowances, would have to be amended.[27]

IV. Amendments After Commission Approval or Rejection of a NAP

A. Amendments to Remedy Deficiencies Identified by a Commission Decision

As we have seen above, if the Commission does reject a NAP, or any aspect of a NAP, the Member State must propose amendments to meet the Commission objection. It is only if those proposed amendments are accepted by the Commission that the Member State may proceed to take its final allocation decision under Article 11.[28]

Although the Directive uses the term 'accepted by the Commission' in relation to proposed amendments, and does not set out any criteria by which the Commission is to determine whether or not to accept amendments, the Commission's approach to this decision must be the same as its approach when it first considers a notified NAP: namely, that it can refuse to accept proposed amendments only if they are incompatible with one or more of the criteria in Annex III or Article 10.[29]

B. Power of the Member State to Propose Other Amendments

The central question in *United Kingdom v Commission* was whether the power of a Member State to propose amendments under Article 9(3) is

---

[26] Paragraph 40 of the judgment (ibid).
[27] Paragraph 59 of the judgment (ibid).
[28] Article 9(3).
[29] See e.g. paragraph 60 of the judgment in *United Kingdom v. Commission*(supra note 7).

limited to amendments designed to overcome any ground for rejection of the NAP identified in the Commission Decision.

The facts which gave rise to that case were as follows. On 30 April 2004 the United Kingdom submitted a NAP which contained a provisional total quantity of allowances representing 736 million tonnes of carbon dioxide. The NAP stated that this figure was 'subject to further revision in the light of ongoing work' relating to the projected energy consumption of industry over the first phase. After some correspondence about the NAP, but before the United Kingdom had updated the intended total quantity to reflect that ongoing work, the Commission proceeded on 7 July 2004 to adopt a Decision under Article 9(3).[30] The Decision identified particular incompatibilities between the NAP and the requirements of the Directive, relating to the treatment of new entrants and the absence of proposed allocations for installations in Gibraltar. The Decision also included the following condition:

> The total quantity of allowances to be allocated by the United Kingdom according to its [NAP] to installations listed therein and to new entrants ... shall not be exceeded.[31]

On 10 November 2004 the United Kingdom proposed amendments to the NAP, including one to increase the total quantity of allowances to 756.1 million tonnes. The Commission initially refused to adopt any further Decision, leading the United Kingdom on 23 December 2004 to formally call upon the Commission under Article 232 EC[32] to define its position in relation to the proposed amendments. This the Commission did in a further Decision of 12 April 2005, declaring the amendment proposed by the United Kingdom inadmissible.[33] The reasons given in the recitals to the Decision were that the United Kingdom was entitled under Article 9(3) only to propose amendments in order to address the incompatibilities identified in the earlier Decision, namely those relating to new entrants and Gibraltar, and that it would be contrary to Article

---

[30] Commission Decision of 7 July 2004 concerning the national allocation plan for the allocation of allowances notified by the United Kingdom in accordance with Directive 2003/87/EC of the European Parliament and of the Council, C(2004)2515/4 final.

[31] Article 3(1).

[32] Article 232 EC provides a remedy in situations where a Community institution fails to act. In such a situation there is no decision which can be challenged, therefore Article 232 empowers Member States or other institutions to call upon one of the Community institutions to take action. If the institution does not define its position within two months of being called upon to act, after a further two months an action may be brought against that institution.

[33] Commission Decision of 12/4/2005 concerning the proposed amendment to the national allocation plan for the allocation of greenhouse gas emission allowances notified by the United Kingdom in accordance with Directive 2003/87/EC of the European Parliament and the Council, C(2005) 1681 final.

3(1) of the Commission Decision to propose an amendment which would exceed the total quantity originally specified in the NAP.

The CFI considered that the United Kingdom's proposed amendments were admissible, and quashed the Commission Decision of 12 April 2005.

The CFI rejected the Commission's argument that a Member State's power to propose amendments is limited to amendments in response to elements of the NAP rejected by the Commission. The Court decided that Article 9(3) does not impose any limit to the permissible amendments which a Member State may propose. The Court rejected the arguments for a narrower interpretation for two reasons.

Firstly, a narrow interpretation of the power to propose amendments would deprive the second stage of public consultation of its effectiveness, given that this could identify material errors in the calculations underlying the NAP or disclose new information requiring an increase in the total quantity.[34]

The Court's second reason related to the purpose of the Directive and merits quoting in full, as it may act as a guide to the Court's approach to future questions on the interpretation of the Directive:[35]

> It must also be pointed out that the purpose of the Directive is to establish an efficient European market in greenhouse gas emission allowances, with the least possible diminution of economic development and employment (see Article 1 of and the fifth recital in the preamble to the Directive). Therefore, even though the Directive aims to reduce greenhouse gas emissions in accordance with the commitments of the Community and its Member States under the Kyoto Protocol, that aim must be achieved, in so far as possible, while respecting the needs of the European economy. It follows that the NAPs developed under the Directive must take due account of accurate data and information relating to emission forecasts for the installations and sectors covered by the Directive. If a NAP was based in part on incorrect information or erroneous evaluations relating to the level of emissions in certain sectors or certain installations, the Member State in question would have to be entitled to propose amendments to the NAP, including increases in the total quantity of allowances to be allocated, in order to address those problems before the market began functioning. That notwithstanding, in order to ensure that the environmental objectives of the Directive are respected, the Commission must still address whether the amendments proposed by the Member State are compatible with the criteria listed in Annex III to or with Article 10 of the Directive.

---

[34] Paragraphs 57 and 58 of the judgment (supra note 7).
[35] Paragraph 60 of the judgment (ibid).

C. Cut-off Date for Proposing Amendments

The Court also considered the cut-off date, after which the Member State would not be entitled to propose any further amendments to the NAP. As the purpose of the NAP is to form the basis for the decision under Article 11 of the Directive on the total quantity of allowances and their allocation to installations, there is a strong argument that the Member State cannot propose amendments to the NAP after it has taken that decision. The Court stated at paragraph 63:

> It follows from the express terms of the Directive, as well as from the general structure and objectives of the system which it establishes, that the United Kingdom was entitled to propose amendments to its NAP after it had been notified to the Commission and until its adoption of its decision under Article 11(1), and that the Commission is not entitled, when adopting a rejection decision in accordance with Article 9(3) of the Directive, to constrain the Member State in the exercise of its right.

This seems to clearly indicate that a Member State can propose amendments before it takes its Article 11 allocation decision. However, the judgment also indicates that the Commission may refuse to consider amendments proposed by a Member State after the date when the Member State should have taken its decisions under Article 11, even if, contrary to the requirements of the Directive, it has not done so. That was the factual situation underlying *United Kingdom v Commission*, as the United Kingdom should have take its decision under Article 11 by 30 September 2004, yet proposed the contested amendments only on 10 November 2004. The Court held:[36]

> The Commission implicitly relies on the fact that the United Kingdom should have taken its decision under Article 11(1) of the Directive by 30 September 2004 and that it was not entitled to propose any amendments after that date. It is worth mentioning in that regard that, even though this fact was referred to in the sixth recital in the preamble to the contested decision, it was not the basis for that decision. The amendments were rejected as inadmissible because they exceeded the total quantity fixed by the Decision of 7 July 2004.

> Furthermore, it is not contested that the United Kingdom was acting in good faith in continuing its work on the NAP after notification and in continuing its efforts to obtain more exact data regarding emission forecasts by the sectors affected by the Directive. In a letter to the United Kingdom's Permanent Representative of 11 October 2004, the Commission, after noting that the United Kingdom had not respected the 30 September 2004 deadline, mentioned 'the progress that your authorities are making towards fulfilling the requirements of the decision'

---

[36] Paragraphs 71 and 72 of the judgment (ibid).

and urged those authorities to notify the necessary information to the Commission as soon as possible. In view of the approach of the Commission at that time, it is not now entitled to claim that the date of 30 September 2004 was of the essence with regard to the entitlement of the Member States to propose amendments to their NAPs under Article 9(3) of the Directive.

On 22 February 2006 the Commission adopted a further Decision rejecting the UK's proposed amendment on the basis that it was proposed after 30 September 2004.[37] Despite its view that the proposed amendment was compatible with the Emissions Trading Directive, the UK Government announced on 28 April 2006 that in the interests of establishing certainty and in view of the time required to make a further challenge, it would not challenge this further decision.[38]

Notwithstanding the specific facts of this case, the judgment has clarified the position for the preparation of NAPs in the future. The passage suggests that the Commission is entitled to refuse to consider any amendments proposed after the date for taking the decision under Article 11 which do not respond to aspects of the NAP rejected by the Commission. This is important for practical reasons, as otherwise there could potentially be open-ended consideration of the NAPs which would undermine the operation of the other provisions of the Directive. In the second phase, the consequence would be that the Commission could have refused to consider amendments proposed after 31 December 2006. In the 2005 Commission Guidance, the Commission signalled its intention not to accept amendments submitted after that date, other than those which were required as a result of a Commission Decision.[39] It further emphasised this point by including in its Decisions on NAPs submitted for the second phase of the scheme conditions requiring that any amendments had to be notified to the Commission by 31 December 2006.[40] This did not apply to amendments which were required in order

---

[37] Commission Decision of 22 February 2006 concerning the proposed amendment to the national allocation plan for the allocation of greenhouse gas emission allowances notified by the United Kingdom in accordance with Directive 2003/87/EC of the European Parliament and of the Council, C(2006) 426 final.

[38] However, a separate challenge has been brought by certain electricity producers: Case T-130/06 *Drax Power and others v Commission*. It is not, however, clear that such operators would be considered to have sufficient standing to challenge the Commission's Decision.

[39] Paragraph 7 of 2005 Commission Guidance.

[40] See for example, Commission Decision of 29 November 2006 concerning the national allocation plan for the allocation of greenhouse gas emission allowances notified by the United Kingdom in accordance with Directive 2003/87/EC of the European Parliament and of the Council. Article 3(3) of that decision provides that 'any other amendments of the national allocation plan, apart from those made to comply with Article 2 of this Decision, must be notified by the deadline of 31 December 2006 referred to in Article 11(2) of this Directive and require prior acceptance by the Commission pursuant to Article 9(3) of the Directive'. Recital (10) to the Decision provides that 'the interpretation of the

(continued...)

to rectify aspects of the NAP which the Commission found to be incompatible with the Directive, amendments which consisted of a redistribution of allowances within the total quantity between installations resulting from improvements to data quality or amendments which reduced the quantity of allowances to be allocated free of charge.

deadline of 31 December 2006 specified in Article 11(2) as a 'cut-off deadline' is proportionate in balancing the interest of a Member State to exert its discretion on substantive issues and the interest of the Community to ensure the functioning of the emissions trading scheme.'

CHAPTER 6

ALLOWANCES: ALLOCATION AND ISSUE

The introduction to chapter 4 explained that the number of allowances allocated to each installation is determined not directly by the Directive, but rather by each Member State through a two-stage procedure. The first stage of that procedure, namely the development by each Member State of a NAP, was discussed in chapters 4 and 5. This chapter considers the second stage, the decision of each Member State under Article 11 of the Directive as to the total quantity of allowances it will allocate for a period, and the allocation of those allowances to the operators of installations.

Section I of this chapter considers the nature and timing of decisions under Articles 11(1) (for the first phase) and Article 11(2) (for subsequent phases) of the Directive. The procedure for issuing allowances is considered in section II (though the requirements of Article 11(4) of the Directive, relating to the timing of the issue of allowances, are discussed in section IV A of chapter 4, in the context of the list of allocations required by the NAP). Finally section III considers the possibility during the first phase of allocating additional allowances because of *force majeure* incidents.

I.   Nature and Timing of the Allocation Decision under Article 11

Article 11 of the Directive provides:

Article 11

Allocation and issue of allowances

1. For the three-year period beginning 1 January 2005, each Member State shall decide upon the total quantity of allowances it will allocate for that period and the allocation of those allowances to the operator of each installation. This decision shall be taken at least three months before the beginning of the period and be based on its national allocation plan developed pursuant to Article 9 and in accordance with Article 10, taking due account of comments from the public.

2. For the five-year period beginning 1 January 2008, and for each subsequent five-year period, each Member State shall decide upon the total quantity of allowances it will allocate for that period and initiate

the process for the allocation of those allowances to the operator of each installation. This decision shall be taken at least 12 months before the beginning of the relevant period and be based on the Member State's national allocation plan developed pursuant to Article 9 and in accordance with Article 10, taking due account of comments from the public.

3. Decisions taken pursuant to paragraph 1 and 2 shall be in accordance with the requirements of the Treaty, in particular Articles 87 and 88 thereof. When deciding upon allocation, Member States shall take into account the need to provide access to allowances for new entrants.

4. The competent authority shall issue a proportion of the total quantity of allowances each year of the period referred to in paragraph 1 or 2, by 28 February of that year.

Article 11(1) required each Member State to take its decision on the allocation of allowances for the first phase of the scheme by 30 September 2004. Article 11(2) required each Member State to take its decision on the total quantity of allowances to be allocated in relation to the second phase of the scheme by 31 December 2006. For subsequent phases the decision must be made at least twelve months before the beginning of the phase. However, unlike for the first period where the Directive clearly requires a decision on the total quantity of allowances to be allocated and their distribution to operators at the same time, Article 11(2) simply requires Member States to 'initiate the process for the allocation of those allowances to the operator of each installation'. This difference in wording suggests that the Community legislator intended the process in the second and subsequent phases to be different from the first. However, in practice the first stage of the process for the allocation of allowances to operators would be to decide upon the allocations and therefore the difference in wording does not appear to have any material effect. This is the approach which has been implicitly adopted by the Commission.[1]

As explained above, the allocation process is a two-stage process. This raises the question of the precise relationship between a Member State's decision under Article 11 of the Directive and its NAP.

---

[1] The Commission Guidance does not expressly address this point. However, Article 44(1) of the Registries Regulation requires Member States to notify to the Commission its NAP table setting out, inter alia, the allocation of allowances to each installation by 1 January 12 months before the start of each period. The assumption underlying this provision is that the decision on the total quantity of allowances and the allocations to individual operators will be made at the same time.

A.   Relationship with the Intended Total Quantity in the NAP

Article 11(1) and (2) requires a Member State's decision on allocation to be 'based on' its NAP. The words 'based on' raise a question as to the extent of a Member State's discretion under Article 11. As a NAP must state both the total quantity of allowances that the Member State intends to allocate and how it proposes to allocate them,[2] must a decision under Article 11, in order to be 'based on' the NAP, adopt the same figures as were set out in the NAP or does the Member State retain some flexibility? The arguments apply slightly differently in relation to, on the one hand, the total quantity, and on the other the allocation among installations. These two aspects are considered in turn below.[3]

Looking first at the total quantity there are two possible interpretations. It is possible to argue that Article 11 gives Member States some flexibility to determine a total quantity of allowances which is different from the total quantity stated in its NAP developed under Article 9. Article 9 only requires NAPs to state the total quantity of allowances which the Member State 'intends' to allocate and there is no provision in Article 9 for that intended amount to become binding when the Commission accepts a NAP. To accept that the decision became binding at that point would seem to make a further decision on the total quantity under Article 11 unnecessary. This points to interpreting the requirements in Article 11 for decisions to be based on the NAP as permitting a Member State to provide in a NAP a methodology for determining the total quantity of allowances.[4]

However, this was not the approach adopted by the Commission in its Decisions on NAPs for the first phase of the scheme. All those Decisions included the following provision:

> The total quantity of allowances to be allocated by [Member State] according to its national allocation plan to installations listed therein and to new entrants, taking into account amendments referred to in Article 2, shall not be exceeded.[5]

---

[2] See Article 9(1) of the Directive.

[3] The question of the extent to which Member States are required to consult upon the NAP is considered in chapter 5.

[4] The extent to which a Member State could decide under Article 11(1) or (2) to allocate a different total quantity from that stated in its NAP would therefore depend upon how the particular NAP was drafted. A NAP purporting to allow flexibility for the Member State to decide upon a different total quantity of allowances would have to clearly identify a methodology by which the total quantity would be calculated. The extent to which such an approach would be possible is limited by the need for the NAP to contain sufficient detail to enable the Commission to assess its compatibility with the criteria in Annex III.

[5] Copies of the Commission Decisions in relation to NAPs for the first phase of the scheme are available at: http://ec.europa.eu/environment/climat/first_phase_ep.htm.

The effect of this was that unless a Member State proposed amendments to its NAP under Article 9(3) the total quantity of allowances provided for in the decision under Article 11 had to be the same as, or lower than, the intended quantity stated in the NAP as accepted by the Commission. Although the question was not specifically addressed in *United Kingdom v Commission*,[6] the assumption in the judgment was that an amendment to the NAP would be required if a Member State wished to change the total quantity of allowances to be issued (even if that amendment were in line with the principles by which the 'intended' total quantity in the NAP was determined).[7]

B.   Relationship with the Intended Allocation to Installations in the NAP

As noted above, the decision under Article 11 has two elements relating, firstly, to the total quantity of allowances, and secondly to the allocation of those allowances to the operator of each installation. Similar arguments can be made about the Member State's discretion in deciding upon the allocation under Article 11 as are made above in relation to the Member State's discretion in deciding upon the total quantity. Just as Article 9(1) requires a Member State to include in its NAP the total quantity of allowances which it intends to allocate, similarly criterion 10 of Annex III requires the Member State to include in its NAP 'the quantities of allowances intended to be allocated to each installation'. Article 9(3) does not distinguish the effect of the Commission's approval in relation to the total quantity, and in relation to the intended allocation to installations. This would appear to suggest that, if the intended total quantity set out in the NAP notified by a Member State were to become binding upon the Commission's acceptance of that NAP under Article 9(3), similarly the intended allocation to installations would become binding. It would follow that the Member State retained no discretion when deciding upon the allocation of allowances under Article 11(1) and is required simply to repeat in that decision the intended allocation set out in the NAP.

However, in practice, the Commission has allowed for the possibility of amendments to the intended allocations to installations after it has approved the NAP under Article 9(3) through its decisions on the NAPs. All Commission decisions under Article 9(3) in relation to the first phase contained a provision allowing Member States to alter the proposed

---

[6] Supra. chapter 5, note 7.

[7] See paragraph 58 of the judgment (ibid) in which the Court uses the case where there are material errors in calculations underlying a plan as an example of when a Member State could propose an amendment. It does not allude to any circumstances in which a change could be made without Commission approval of the amendment.

allocations among installations which were set out in the NAP notified to the Commission onthe following terms:

> The national allocation plan may be amended without prior acceptance by the Commission if the amendment consists in modifications of the allocation of allowances to individual installations within the total quantity resulting from improvements to data quality.[8]

Similar provisions were contained in the Commission Decisions in relation to the second phase. These stated:

> The national allocation plan may be amended without prior acceptance by the Commission if the amendment consists in modifications of the allocation of allowances to individual installations within the total quantity to be allocated to installations listed therein resulting in improvements to data quality or to reduce the share of the allocation of allowances free of charge within the limits set in Article 10 of the Directive.[9]

The reason for this difference between treatment of the total quantity of allowances to be issued and the installation level allocations is that in terms of achieving the objective of the Directive to reduce emissions the total number of allowances to be allocated is key. The distribution of those allowances amongst installations does not affect total emissions. It is also the total quantity of allowances available on the market which determines the price of allowances and therefore is most important for providing certainty to the market. However, the provisions for variation of the allowances to individual installations were drafted broadly and could potentially have allowed quite significant amendments provided that they were required as a result of improvements in data quality. In the second phase of the scheme, the Commission has taken an even broader approach by allowing the share of allowances to be auctioned or sold to be increased to the maximum of 10 per cent of the total quantity prescribed by the Directive. Whilst the fact that any such amendments are deemed to be made to the NAP avoids the issue of the extent to

---

[8] Although a provision similar to this was included in all Decisions, there was some variation in the wording: not all Decisions expressly provided that amendments could be made 'without prior acceptance of the Commission' (see for example the Commission Decision of 27 December 2004 concerning the national allocation plan for the allocation of greenhouse gas emission allowances notified by Malta in accordance with Directive 2003/87/EC of the European Parliament and of the Council). The absence of this wording suggests that an amendment could not be made without Commission approval. Further some Decisions rather than simply referring to the 'total quantity' referred to 'the total quantity to be allocated to installations listed therein' (see for example the Decision on Malta's NAP). This additional wording may have the implication that this provision could not be used to include allocations to installations not originally included in the NAP.
[9] Copies of Decisions in relation to NAPs for the second phase of the scheme are available at: http://ec.europa.eu/environment/climat/2nd_phase_ep.htm.

which a decision under Article 11 is 'based' upon the NAP, it raises the question of how broad a provision for amendment subsequent to the approval of the NAP could be, without being inconsistent with the Commission's obligation to consider and decide whether or not to approve amendments to the NAP.

II.   Issue of Allowances

Once a Member State has made its decision under Article 11, the Registries Regulation sets out the further procedure which is to be followed before allowances can be issued to operators.

The Directive itself does not prescribe any mechanism for the Commission to check that a Member State's decision under Article 11 is 'based upon its national allocation plan' (as required by Article 11) before allowances are issued. However this potential loophole is plugged by Article 44 of the Registries Regulation (Article 38 for the first phase). This requires Member States to submit to the Commission a 'NAP table' containing details of the total number of allowances allocated, the total number of allowances reserved for new entrants, and the allocation of allowances to each installation in each year of the phase.[10] Article 44(1) goes on to provide:

> ...If the national allocation plan table is based upon the national allocation plan notified to the Commission which was not rejected under Article 9(3) of Directive 2003/87/EC or on which the Commission has accepted proposed amendments, the Commission shall instruct the Central Administrator to enter the national allocation plan table into the Community independent transaction log...

This effectively gives the Commission a power to prevent a Member State from issuing allowances if it does not consider that the decision under Article 11, as expressed in the NAP table, is based upon NAP which it approved.[11]

---

[10] See paragraphs 4-8 of Annex XIV to the Registries Regulation.

[11] It is not possible for a Member State to proceed to issue allowances without having submitted a NAP table. If it sought to, the Community Independent Transaction Log (CITL) would return a response code meaning 'the national allocation plan table has not been submitted to the Commission and therefore it is not possible for the issuance of allowances for the specified period to take place' (Code 7203, Annex IX and XII of the Registries Regulation). Likewise if the registry administrator seeks to issue more allowances than are provided for in the national allocation plan table, the CITL will return a response code meaning 'the amount of allowances for the specified period exceeds the amount approved by the Commission in the national allocation plan' (code 7201, Annex IX and XII of the Registries Regulation). Where such a response code is issued, the Registry administrator is required to terminate the process and inform the CITL (Article 28(2) of the Registries Regulation). Further under Article 6(3) the Commission 'may instruct the central administrator to temporarily suspend the communication link between a registry

(continued...)

Once the NAP table has been entered into the Community Independent Transaction Log (CITL), the Registries Regulation provides for allowances to be issued in a two-stage process.

Firstly, the registry administrator is required to issue the total number of allowances to be issued for the phase into the party holding account.[12] In the first phase, an allowance was issued by assigning to it a unique allowance identification code.[13] In the second and subsequent phases of the scheme, allowances will be issued by converting an equal quantity of Assigned Amount Units held in the Party holding account into allowances.[14]

Secondly, the registry administrator is required each year to transfer from the party holding account to the relevant operator holding accounts the number of allowances specified in the NAP table for that operator for that year.[15] In accordance with Article 11(4) of the Directive[16] this transfer should take place by 28 February each year. However, this is subject to one exception. The Registries Regulation provides that 'where foreseen for an installation in the national allocation plan of the Member State, the registry administrator may transfer that proportion at a later date of each year'.[17]

As regards the issue of allowances to new entrants, Articles 42 and 46 of the Registries Regulation provide for a less stringent procedure than outlined above. The registry administrator is simply required to transfer allowances to a new entrant where instructed to do so by the competent authority. There is no requirement for the allocation to be included in the NAP table entered into the CITL and therefore no procedure for Commission approval of the allocation before it is issued. However, additional procedures may be included in subsequent amendments to the Registries Regulation.

---

and the CITL or to suspend all or some of the processes referred to in Annexes VIII and IX, if that registry is not operated...in accordance with the provisions of [the] Regulation.'

[12] For a description of the types of accounts required to be held in the registry, see section II of chapter 7.

[13] Second paragraph of Article 39 of the Registries Regulation. The unique identification code will comprise the elements set out in Annex VI to the Registries Regulation.

[14] Article 45 of the Registries Regulation. The conversion will take place by adding an allowance element to the unique unit identification code of the AAU, comprising the elements set out in Annex VI to the Registries Regulation.

[15] Articles 40 and 46 of the Registries Regulation.

[16] For a discussion of the requirements of Article 11(4) of the Directive see section IV A of chapter 4.

[17] Second paragraph of Article 40 and second paragraph of Article 46 of the Registries Regulation.

III.   Force Majeure

As we have seen, the Directive requires the allocation of allowances to be determined for the whole phase at the start of each phase, and a proportion to be issued before 28 February of each year of the phase.

During the first phase only, there is an additional, exceptional power for Member States to issue allowances to take account of unforeseen events. Article 29 of the Directive allows Member States to 'apply to the Commission for certain installations to be issued with additional allowances in cases of *force majeure.*' The article provides that 'the Commission shall determine whether *force majeure* is demonstrated, in which case it shall authorise the issue of additional and non-transferable allowances by that Member State to the operators of those installations'.

Article 29 also required the Commission by 31 December 2003, to develop guidance to describe the circumstances in which *force majeure* is demonstrated. This the Commission did in the 2003 Commission Guidance.[18]

The Guidance noted[19] that applications for *force majeure* allowances may cause uncertainty in the market, and if granted may give advantages to certain companies which affect trade between Member States. The Commission noted that Article 29 is therefore without prejudice to the Treaty, a reference to the requirements relating to state aids in Articles 87 to 89.

The 2003 Commission Guidance expresses the view that circumstances that could have been insured against do not constitute *force majeure*, and that these are limited to 'exceptional and unforeseeable circumstances which cause a substantial increase in annual direct emissions of greenhouse gases covered by [the Directive] at an installation, which could not have been avoided if all due care had been exercised.'[20] It further provides that 'the circumstances must have been beyond the control of the operator of the installation concerned and of the Member State submitting an application to the Commission.' As examples of circumstances which it considers may constitute *force majeure*, the Commission cites 'natural disasters, war, threats of war, terrorist acts, revolution, riot, sabotage or acts of vandalism'.[21]

---

[18] Supra chapter 4, note 3.
[19] Paragraphs 110 to 117 of the 2003 Commission Guidance.
[20] Paragraph 113 of the 2003 Commission Guidance.
[21] Paragraph 114 of the 2003 Commission Guidance.

The Directive itself does not prescribe any procedure for the applications envisaged under Article 29, nor their content. The Commission plugs this gap by its Guidance, requiring that 'the presence of *force majeure* has to be demonstrated at an installation level and on a case-by-case basis',[22] and requires the application to include 'in respect of each installation, the Member State's best estimate of the increase in emissions resulting from the circumstances for which *force majeure* is pleaded and a substantiation of that estimate.'[23] It requires Member States to submit applications under Article 29 to the Commission by 31 January in the year following the year of the trading period during which the circumstances occurred for which *force majeure* is pleaded.[24] This is to enable applications to be approved in time for the Member State to issue allowances to affected operators before the deadline of 30 April by which allowances for the preceding year must be surrendered. It also enables the Commission to consider the effect of all applications from Member States as a whole, before approving any one application.

Article 43 of the Registries Regulation sets out the procedure for issuing *force majeure* allowances. As allowances cannot be traded, Article 43(3) provides for the number of force majeure allowances issued to be entered directly into the list of allowances surrendered. Article 29 of the Directive provides that *force majeure* allowances are non-transferable. The effect of this is that such allowances cannot be traded on the emissions allowance market and can therefore only be used for compliance by the operator to whom they are issued.

---

[22] Paragraph 115 of the 2003 Commission Guidance.
[23] Paragraph 116 of the 2003 Commission Guidance.
[24] Paragraph 117 of the 2003 Commission Guidance.

CHAPTER 7

TRADING (INCLUDING REGISTRIES AND POOLING)

This chapter sets out the basic principle enshrined in the Directive, that allowances can be freely traded (section I). Section II explains the system of registries established by the Directive and the Registries Regulation to enable trading. Finally section III considers the possibility in the first and second phases of the scheme of creating a 'pool' of installations under which one person is responsible for trading and surrendering allowances for all the installations in the pool.

I.   The Basic Principle that Allowances May Be Freely Traded

The Directive allows allowances to be traded freely between persons within the EU. In order that these traded allowances can be used by any operator to meet its obligations under the Directive, Article 12(2) of the Directive also expressly requires Member States to ensure that 'allowances issued by a competent authority of another Member State are recognised for the purpose of meeting an operator's obligations to surrender allowances to cover emissions.[1]

Where the EU scheme has been linked with another scheme outside the EU, the Directive also requires Member States to allow allowances to be traded with persons outside the EU. This trading may be made subject only to the restrictions envisaged in the agreement providing for the linking itself, or in any Decision under Article 25(2) of the Directive setting out provisions on mutual recognition.[2]

II.   Registries

In order to facilitate the trading of allowances, Articles 19 and 20 of the Directive provide for the creation of an electronic registries system.

Each Member State is required to 'provide for the establishment and maintenance of a registry in order to ensure the accurate accounting of the issue, holding, transfer and cancellation of allowances.' The registry is incorporated into each Member State's national registry established

---

[1] For further information on the obligation to surrender allowances, see chapter 9.
[2] For a discussion of the procedure for linking the EU scheme with other schemes, see section III of chapter 8.

for the purposes of the Kyoto Protocol under Article 6 of the Monitoring Mechanism Decision.[3]

In order to assist Member States in establishing registries and to ensure the compatibility of the registries with each other, Article 19(3) of the Emissions Trading Directive required the Commission to 'adopt a Regulation...for a standardised and secured system of registries in the form of standardised electronic databases containing common data elements to track the issue, holding, transfer and cancellation of allowances, to provide for public access and confidentiality as appropriate and to ensure that there are no transfers incompatible with obligations resulting from the Kyoto Protocol.'[4]

The Commission adopted a Regulation for a standardised system of registries on 21 December 2004 (the 'Registries Regulation').[5]

The Registries Regulation required the Member States and the Commission to establish a registry in the form of a standardised electronic database by the day after entry into force of the Regulation, that is by 31 December 2004.[6] The registries must be distinct but may be established, operated and maintained in a consolidated manner with other Member States or the Commission.[7] The registries established by Member States are overseen by the central administrator designated by the Commission under Article 20(1) of the Directive. The central administrator operates and maintains the Community Independent Transaction Log ('CITL') which records transactions in national registries and looks for irregularities.[8]

Each registry is required to contain a number of accounts and tables, including at least one party holding account.[9] The party holding account

---

[3] Supra, chapter 4, note 20. Article 6 requires Member States to 'establish and maintain registries in order to ensure the accurate accounting of the issue, holding, transfer, acquisition, cancellation and withdrawal of assigned amount units, removal units, emission reduction units and certified emission reductions and the carry-over of assigned amount units, emission reduction units and certified emission reductions.' It expressly requires that such registries 'shall incorporate registries established pursuant to Article 19 of [the Emissions Trading Directive].'

[4] Article 19(3) provides for the Regulation to be adopted in accordance with the procedure referred to in Article 23(2) of the Directive. For a description of this procedure see section VII of chapter 2. Article 19 was amended by the Linking Directive to provide that 'this Regulation shall also include provisions concerning the use and identification of CERs and ERUs in the Community scheme and the monitoring of the level of use'. For further detail on the requirements of the Linking Directive, see section II of chapter 8.

[5] See chapter 1, note 3.

[6] Article 3(1) of the Registries Regulation.

[7] Article 19(1) of the Directive and Article 4 of the Registries Regulation.

[8] Article 20(2) and (3) of the Directive.

[9] Article 11(1) of the Registries Regulation.

is the account used for the functions carried out by the Member State. In particular it is into the party holding account that allowances will be issued and from which they will be transferred to individual installations.[10] A Member State may have more than one party holding account[11] and therefore Member States could set up separate accounts for specific purposes such as to hold a reserve of allowances for new entrants.

The registry must also contain a retirement account, a cancellation account for the period 2005-2007 and a cancellation account for the period 2008-2012.[12] These will be used for the cancellation and retirement of allowances and Kyoto units.[13] [14]

Each Member State registry must also contain one operator holding account for each installation granted a permit by the Member State.[15] Allowances allocated in respect of an installation will be issued into the operator holding account for that installation. Allowances can only be surrendered from operator holding accounts to cover the emissions from the installation to which the account relates.

In addition to operator holding accounts, each registry may contain a number of person holding accounts.[16] These may be opened by operators or other persons and can be used for trading allowances. A person may have up to 99 person holding accounts.[17]

---

[10] Articles 39 and 40 of the Registries Regulation in relation to the first phase and Articles 45 and 46 in relation to the second and subsequent phase.

[11] See Articles 12-14 of the Registries Regulation for the procedure for opening and closing party holding accounts.

[12] Article 11(3) of the Registries Regulation.

[13] The retirement account is, in relation to the first phase, the account into which allowances will be transferred once surrendered by operators (Article 58 of the Registries Regulation). In the second and subsequent phases, the retirement account will be the retirement account for the purposes of retiring units in accordance with the requirements of the Kyoto Protocol and its implementing decisions (Articles 59 and 63). Member States will be required to retire Kyoto units equal to the number of allowances and CERs and ERUs surrendered by operators (Articles 58(1) and 59). The cancellation account will be used for the voluntary cancellation of allowances in the first phase (Article 62(2)). CERs used in the first phase of the scheme will be cancelled in the cancellation account for the period 2008-2012 which will serve as the cancellation account for the purposes of the Kyoto Protocol and its implementing decisions (Article 62(3)).

[14] For more information on the use of Kyoto units in the scheme, see section II of chapter 8.

[15] Article 11(2) of the Registries Regulation. See also Articles 15-18 for the procedure for opening and closing operator holding accounts.

[16] Article 11(2) of the Registries Regulation. See Articles 19–22 for the procedure for opening and closing person holding accounts.

[17] Article 19(2) of the Registries Regulation.

Each registry will also contain at least the following three tables: a verified emissions table, a surrendered allowance table and a compliance status table.[18] The verified emissions table will record the verified emissions for each year for each installation.[19] The surrendered allowance table will set out the number of allowances, CERs, ERUs and force *majeure* allowances surrendered in each year in respect of each installation.[20] The compliance status table will indicate for each year whether or not sufficient allowances have been surrendered in respect of an installation.[21]

All transactions concerning allowances are carried out electronically in accordance with the Registries Regulation. The Regulation sets out procedures for all transactions concerning allowances and Kyoto units used in the scheme including the allocation and issue of allowances,[22] the transfer,[23] cancellation[24] and surrender of allowances,[25] and the use of CERs and ERUs.[26] Transactions in the registry will be carried out by a registry administrator appointed by the Member State to operate and maintain the registry.[27]

The registries only reflect the holding of allowances. They do not have capacity to recognise charges over allowances. However, it is possible for the holder of an account in the registry to appoint an additional authorised representative whose approval is required before any transactions may be carried out concerning that account.[28]

III. Pooling

Article 28 allows operators of installations carrying out the same activity listed in Annex I to form a pool. This provision applies to the first and second phases of the scheme. It applies to installations carrying out the same activity so, for example, two or more mineral oil refineries could form a pool but a mineral oil refinery could not form a pool with a

---

[18] Article 24 of the Registries Regulation.
[19] Paragraph 1 of annex II to the Registries Regulation.
[20] Paragraph 2 of annex II to the Registries Regulation.
[21] Paragraph 3 of annex II to the Registries Regulation. The compliance status will be calculated in accordance with Article 55 of the Registries Regulation and entered into the table in accordance with Article 56.
[22] For the procedures for the period 2005-2007, see Articles 38-43 of the Registries Regulation and for the period 2008-2012 and subsequent five year periods, see Articles 44-48 of the Registries Regulation.
[23] See Article 49 of the Registries Regulation.
[24] See Article 62 of the Registries Regulation.
[25] See Article 52 of the Registries Regulation.
[26] See Article 53 of the Registries Regulation.
[27] Articles 2(q) and 8 of the Registries Regulation.
[28] Article 23(2) of the Registries Regulation.

combustion installation.[29] In broad terms, the forming of a pool means that all trading aspects of the scheme are handled by one trustee for all the installations forming part of a pool. However, all installations in the pool must have their own permit and will receive a separate allocation in the NAP.

A.   Procedure for Establishing a Pool

Each Member State has a discretion as to whether to allow pooling at all, and if so, the conditions under which it is to be permitted. A Member State could, for example, allow pooling only in respect of certain activities or even in respect of a specified subset of an activity. In exercising their discretion Member States are, however, constrained by the principle of non-discrimination. A Member State which allowed pooling for one activity but not for another would have to satisfy one of two tests. They would have to show either that the installations carrying out the different activities were not in a like situation or, alternatively, that though in a like situation, there was an objective justification which justified their differential treatment. The same tests would have to be met if Member States allowed pooling for installations in a subset of one activity but not in a different subset.

The procedure for establishing a pool is initiated by the installations wishing to form a pool submitting an application to the competent authority in the Member State. This application must contain three elements:

a)   The application must specify the installations which will be within the pool;

b)   The application must specify whether the pooling is to take place for the first phase, the second phase, or both;[30]

---

[29] The use of the word 'activity' in Article 28(1) is consistent with the structure of Annex I which lists activities under italic headings. This can be compared with the slightly anomalous use of the phrase 'activities falling under the same sub-heading' in the second interpretational note in Annex I; see section IV. B of chapter 2.

[30] Article 28(2) refers to 'the period for which they want the pool'. The better reading, and the reading assumed in the text above, is that this is a reference back to the word 'period' in Article 28(1) which is clearly a reference to the periods specified in Article 11, i.e., the phases 2005-2007, 2008-2012, etc. It is possible to read Article 28(2) more widely, as allowing pooling for any period of time within a phase, e.g. just for the year 2006. There is no practical impediment to a Member State establishing pooling for such shorter periods of time.   A Member State could, for example, require the surrender of allowances corresponding to emissions made during the subsistence of a pool before the dissolution of a pool, and then provide for installations which were within the pool to be treated in the normal way for the remainder of the phase.   Such a structure would also not be

(continued...)

c) The application must demonstrate that the trustee is able to comply with the obligations set out in Article 28(3) and (4). The evidence that is necessary to demonstrate this becomes clear when one looks in turn at the requirements of paragraphs (3) and (4). Firstly, the trustee is to be issued with allowances and to be responsible for surrendering allowances.[31] Secondly, the application must demonstrate that the trustee is to be restricted from making further transfers in the event that an operator's report of emissions has not been verified as satisfactory by 31 March following the end of the year to which it relates. Article 27(1) of the Registries Regulation will ensure that no transfers are made from the operator holding account relating to the installation in respect of which a verified report has not been submitted. It does not, however, restrict transfers from the operator holding accounts relating to the other installations within the pool. This drafting of the Registries Regulation may be taken as an indication that the Commission regards Article 28(3)(c) as being satisfied by the blocking of the operator holding account relating to the particular installation which has failed to submit a verified report (i.e. it does not require the blocking of the operator holding accounts for all installations within the pool). If, contrary to this view, it is necessary for the accounts for all installations within the pool to be blocked, it is not immediately apparent how this can be achieved by the operators themselves through a pooling agreement. The agreement could, perhaps, seek to restrain the trustee in his role as appointed representative from transferring allowances from any of the relevant accounts. However, such a contractual restraint would usually have to be enforced by one of the parties to the contract. Even then there could be situations in which the operators may not have an interest in enforcing that restriction, and it would be doubtful whether such a contractual arrangement would therefore meet the requirement in Article 28(2) of

---

inconsistent with the purpose of the Directive. It therefore remains possible to argue that the word 'period' in Article 28(2) is to have its ordinary meaning, and not be treated as a reference back to the word 'period' as used in Article 28(1).

[31] Article 28(3)(a) and (b). As the Registries Regulation does not provide for accounts to be established in the name of a pool trustee, allowances will continue to be issued into, and surrendered from, the operator holding account in respect of each installation within a pool. In order to demonstrate that the trustee will be able to be issued with allowances, and surrender them, it therefore appears necessary for operators to demonstrate that they have nominated the Trustee as their appointed representative under the Registries Regulation, (Article 15 and Annex III).

demonstrating that the trustee will be restricted from making transfers. If this wider interpretation of the requirements of Article 28(3)(c) is correct, it therefore appears that a Member State would need to provide for such a restriction in its transposing legislation.

d) The application must demonstrate that the trustee will be in a position to pay any penalties due because the pool has failed to surrender allowances equivalent to the total emissions from the pool.[32] This would appear to require the application to demonstrate either that the trustee has and will retain sufficient resources to pay these penalties or, more likely, that the contractual arrangements between the trustee and the operators provide the trustee with access to these resources.

When a Member State receives an application, it has a discretion as to whether or not to allow the pool to proceed. If it wishes to allow the pool to proceed, it must submit the application to the Commission. The Commission then has a period of three months in which to reject an application, on the basis that it does not fulfil the requirements of the Directive. If the Commission has not rejected the application within that three month period it loses the power to do so, and the operators have the right to establish the pool set out in the application. If, on the other hand, the Commission does reject the application, the pool may not be established unless the Commission accepts proposed amendments to the way the pool is established. Any decision of the Commission to reject a pool must give reasons for the rejection. Although Article 28(5) expresses the power of the Commission to reject an application broadly where the application does not 'fulfil the requirements of the Directive', it appears clear from the context that only three types of reason would be permitted:

– the installations to be included in the proposed pool do not fall within the same activity;
– the application is for a period other than the first phase of the scheme, the second phase of the scheme, or both;
– the application is not accompanied by evidence to demonstrate that the trustee will be able to fulfil the obligations referred to in Article 28(3) and (4) discussed above.[33]

---

[32] Article 28(4) of the Directive.
[33] The procedure in Article 28(5) is phrased in identical terms to the procedure for notification and amendments of NAPs under Articles 9(1) and (3). For a fuller discussion of that procedure see chapter 5 above.

B.   Operation of a Pool

When a pool is formed operators are still required to have a separate permit for each installation and the monitoring, reporting and verification requirements of the permit will apply separately to each installation.   Allowances will still be allocated to each installation separately in the NAP and subsequent decision under Article 11, and each installation will have a separate operator holding account within the Registry. Allowances will be issued into the individual operator holding accounts and will be surrendered from those holding accounts.[34]

It is clear from Article 28 that the trustee is not a trustee in the normal English law sense of the term. Instead, the trustee is someone who carries out the functions in Article 28(3) and (4). A trustee will carry out these functions by being the appointed representative for the operator holding accounts of all the installations forming part of the pool and by entering into the contract with each operator governing the exercise of his functions as trustee. As discussed above, a contract is likely to specify how the trustee will have access to funding for any penalties incurred in respect of any of the installations in the pool but it is likely that any contract would go further and provide more detail about the relationship between operators.[35]

Article 28(a) provides that allowances are to be issued to the trustee 'by way of derogation from Article 11, and Article 28(3)(b) provides that the trustee is to be responsible for surrendering allowances 'by way of derogation from Articles 6(2)(e) and 12(3)'. However, this is slightly misleading as the effect of the Registries Regulation is that the trustee essentially stands in the shoes of the operator and carries out his functions by operating each of the operator holding accounts.

If the trustee fails to surrender sufficient allowances to cover the emissions from all of the installations in the pool, then Article 28(4) provides for the trustee to be subject to the financial penalties which apply under the scheme. This would at first suggest that there would be a mechanism in the Registries Regulation for calculating the total emissions from installations in the pool and the total number of allowances surrendered by the trustee in respect of the pool. However, the Registries Regulation does not make such provision. The effect of

[34] This is not apparent from the wording of Article 28(3) but is a necessary consequence of the structure of the Registries Regulation which does not provide any specific provisions for pooling. This means that the trustee will have to fulfil his function by being nominated as the appointed representative of the operator holding accounts for each installation covered by the pool in accordance with Article 15 and Annex III of the Registries Regulation.

[35] The terms of such an agreement are likely to raise a number of legal issues, including potentially competition law issues.

this is that whether the trustee has complied with his obligations to surrender allowances will be judged on an installation-by-installation basis and a penalty will be imposed on the trustee in respect of each installation for which insufficient allowances were surrendered. The trustee will also be responsible for surrendering an amount of allowances in the following calendar year to make up for the shortfall of allowances surrendered in respect of emissions in the previous calendar year.[36] [37]

Article 28(6) requires that where a trustee fails to comply with penalties 'each operator of an installation in the pool shall be responsible under Articles 12(3) and 16 in respect of emissions from its own installation'. The Directive does not specify at what point a trustee fails to comply with the penalties. That is to say, it does not specify the date by which an excess emission penalty must be paid. However, this is something that would be specified in Member State legislation and therefore a trustee would be taken to have failed to comply with the penalties if he failed to comply by the date set out in the Member State legislation. The effect of the reference to Article 16 in Article 28(6) is that the operator of each installation in the pool for which insufficient allowances were surrendered will be responsible for paying the excess emissions penalty in respect of that installation. The effect of the reference to Article 12(3) is less clear. This appears in the first instance to be an attempt to pass back to the operator the obligation to surrender allowances covering the emissions in the previous calendar year. However, that cannot be the case because operators are required to surrender allowances by the 30 April each year. Article 28(6) can only apply if that date has already passed, and there is no provision within the Directive for the extension of that period. The better interpretation would seem to be that the reference to Article 12(3) transfers back to the operator the obligation to surrender allowances in respect of subsequent calendar years. Given that this result is directly in contradiction with Article 28(3)(b) there is a strong implication that this puts an end to the pool.

---

[36] Final sentence of Article 16(3) and (4). For a description of the financial penalties for non-compliance with the obligation to surrender allowances, see section III of chapter 9.
[37] Article 28(4) is expressed to be 'by way of derogation from Article 16(3) and (4)'. Although the penalty itself will relate to individual installations, the derogation from Article 16(3) and (4) comes because the penalty is imposed upon the trustee rather than the operator of the installations concerned. It is also expressed to be a derogation from Article 16(2) which requires Member States to publish the names of operators who are in breach of the requirements to surrender sufficient allowances under Article 12(3). In the case of a pool it is the trustee who is responsible for surrendering allowances and therefore by way of derogation to Article 16(2) it would be the trustee's name that Member State is required to publish.

C. Benefits of Pooling

Having considered the procedure for setting up a pool and the operation of pools it is useful to consider what are the possible benefits for operators of forming a pool. From the discussion above, it is clear that the effect of forming a pool is simply to transfer the obligation to pay the excess emissions penalty from the operator to the trustee of the pool and to require the publication of the trustee's name where there is a failure to surrender sufficient allowances, rather than the name of the individual operators. In all other respects the installations continue to be treated as if they were not part of a pool. Although it may be a benefit to avoid the negative publicity associated with publication of operators' names for failure to comply with their obligation to surrender allowances, it is not clear that the transfer of the obligation to pay the penalty to the trustee provides any real benefit to operators. This is because the contractual arrangements between the trustee and operators is likely to provide where the liability for paying any penalty will fall. Therefore the benefits of forming a pool would seem to be very slight. In addition, there is the added risk that Article 28(3)(c) could require all operator accounts for installations covered by the pool to be blocked where one installation fails to submit a verified report of its emissions to the competent authority on time. In conclusion it seems questionable whether Member States and operators will consider that the minimal benefits of forming a pool justify the administrative effort in establishing it.

CHAPTER 8

THE USE OF KYOTO UNITS AND
UNITS FROM SCHEMES IN OTHER STATES

This chapter sets out the conditions under which operators can use either credits under the Kyoto Protocol, or units from other trading schemes linked to the EU ETS to fulfil their obligations under the EU ETS.

It sets out first the relevant requirements of the Kyoto Protocol (section I), then the rules of the EU ETS allowing the use of Kyoto credits to fulfil obligations under the EU ETS (section II). Finally it sets out the provisions of the EU ETS which could enable the EU ETS to be linked with other schemes, and so allow operators to use units from any such linked scheme to fulfil their obligations under the EU ETS (section III).

I.   Trading under the Kyoto Protocol

The legal framework for international trading is found in the UNFCCC[1] and Kyoto Protocol,[2] and in the subsequent Decisions of the Parties setting out the detailed rules for the application of the Protocol.[3] The Introduction to this book set out the historical development and broad requirements of the UNFCCC and the Kyoto Protocol. What follows is a more detailed explanation of their provisions relevant directly to the EU ETS.

A.   Emission Limitation and Reduction Obligations in the Kyoto Protocol

The Kyoto Protocol provides for a reduction of emissions of the six main greenhouse gases from countries listed in Annex I of the UNFCCC by at least 5 per cent below 1990 levels over the period 2008-2012 (the first Kyoto commitment period). It seeks to achieve this reduction by committing countries listed in Annex B of the Protocol to legally binding

---

[1] Supra Introduction, note 6.
[2] Supra Introduction, note 10.
[3] The main decisions were adopted in Marrakech in 2001 and are collectively known as the 'Marrakech Accords'. Prior to the entry into force of the Kyoto Protocol, these decisions were adopted by the Conference of the Parties to the UNFCCC. The majority of these decisions are in the form of draft decisions which were then formally adopted at the first conference of the parties serving as the meeting of the parties to the Kyoto Protocol in December 2005. The decisions are available on the UNFCCC website at: http://unfccc.int/.

targets, known as quantified emission limitation and reduction obligations (QUELROs).

Each of these countries with a target under the Kyoto Protocol is required to calculate the total amount of greenhouse gases that it is allowed to emit during the period 2008-2012 in accordance with its target under the Protocol. This is done by multiplying each country's emissions in 1990 by the reduction or limitation target and then by five to cover the five years of the commitment period (2008-2012)). The amount of emissions which a country is allowed to emit during the period 2008-2012 is known as the Party's assigned amount. [4]

The assigned amount is then divided into assigned amount units (AAUs), each worth one metric tonne of carbon dioxide equivalent. In addition each country may issue removal units (RMUs) to reflect its net removals of greenhouse gas emissions through land use and forestry. At the end of each commitment period each country must retire one AAU (or RMU) for each tonne of carbon dioxide (or its equivalent) emitted in that country during the commitment period. [5]

B.   The Kyoto Protocol Flexible Mechanisms

As explained in the Introduction, the Kyoto Protocol provides for three so-called flexible mechanisms. The purpose of these is to assist countries in meeting their targets by allowing reductions of emissions to be made at least cost and to include countries without a target under the Kyoto Protocol in efforts to reduce emissions. The mechanisms are:

---

[4] See Article 3(7) and (8) of the Kyoto Protocol and the Modalities for the accounting of assigned amount under Article 7, paragraph 4, of the Kyoto Protocol (Decision 13/CMP.1 (Decision 19/CP.7)). Paragraph 6-8 of the Annex to Decision 13/CMP.1 require Parties with a target under the Kyoto Protocol to submit a report containing the information necessary to determine their assigned amount by 1 January 2007. This is reflected in Article 7 of the Monitoring Mechanism Decision (supra chapter 4, note 20) which required the Commission and Member States to submit a report to the UNFCCC Secretariat determining assigned amounts by 31 December 2006. To ensure coordination of reports to the UN, Member States were required to submit a report to the Commission by 15 January 2006 (Article 8(1)(e) of the Monitoring Mechanism Decision and Article 23 of the Commission Decision of 10 February 2005 laying down rules implementing Decision 280/2004/EC OJ L55, 1.3.2005, p.57).

[5] AAUs for emissions in the period 2008-12 do not have to be retired until 2015. See paragraph 34 of the Modalities for the accounting of assigned amounts under Article 7, paragraph 4 of the Kyoto Protocol as read with section XIII of the Procedures and mechanisms relating to compliance under the Kyoto Protocol (Decision 27/CMP.1 (Decision 24/CP.7)) the Guidelines for review under Article 8 of the Kyoto Protocol (Decision 22/CMP.1 (Decision 23/CP.7)) and Guidelines for the preparation of the information requirement under Article 7 of the Kyoto Protocol (Decision 15/CMP.1 (Decision 22/CP.7)).

a) emissions trading,[6]
b) Joint Implementation (JI)[7] and
c) the Clean Development Mechanism (CDM).[8]

The latter two are referred to collectively as the 'project mechanisms'. They enable credits to be generated by reducing emissions in another country. These project credits can then be traded like AAUs and used to meet targets under the Kyoto Protocol.

The detailed rules for the operation of the flexible mechanisms are set out in the Marrakech Accords and are considered briefly below.

1.   Trading of Assigned Amount Units

The Kyoto Protocol provides for countries with a target in Annex B of the Protocol to meet their targets through international emissions trading. AAUs can be traded between countries and can then be retired by the purchasing country in order to meet its target under the Protocol. By providing for allowances in the second phase of the EU ETS to be attached to an AAU,[9] the EU ETS provides a mechanism for individual operators to participate in this trading of AAU between Member States. This may be extended to other Parties to the Kyoto Protocol who agree to link their own trading schemes to the EU scheme.[10]

2.   Clean Development Mechanism

The CDM provides a mechanism for Annex I countries to carry out projects to reduce emissions in developing countries (those not included in Annex I of the UNFCCC). Article 12(2) of the Kyoto Protocol states that the CDM is intended 'to assist Parties not included in Annex I in achieving sustainable development and in contributing to the ultimate objective of the Convention'. Credits known as certified emissions reductions (CERs) are issued for the reductions achieved by such projects and can be used by Annex I countries to meet their reduction or limitation obligations under the Kyoto Protocol.

The CDM benefited from a so-called 'prompt start' under international rules: credits could be issued for emission reductions achieved from the year 2000 onwards and can then be used against targets in the first Kyoto

---

[6] Article 17 of the Kyoto Protocol.
[7] Article 6 of the Kyoto Protocol.
[8] Article 12 of the Kyoto Protocol.
[9] See section II of chapter 6.
[10] See section III of this chapter.

Protocol commitment period of 2008 to 2012.[11] Assuming that the necessary registry links are established in time, projects can be approved and credits may be issued and forwarded to national registries prior to the start of the first commitment period in 2008 and the establishment of the assigned amount. Trading of CERs between registries is, however, only allowable after the establishment of the assigned amount.

3.   Joint Implementation

JI provides an alternative or supplemental mechanism to international emissions trading. It enables an Annex I country to obtain additional credits by carrying out projects to reduce emissions in another Annex I country. These credits are known as emission reduction units (ERUs).

As ERUs are issued by converting an AAU (or an RMU) into an ERU,[12] ERUs cannot be issued until the country's assigned amount has been established. ERUs will therefore not be available until 2008 at the earliest.

4.   Supplementarity

The use of the flexible mechanisms is subject to limitations. By imposing express emissions limits only on Annex I countries, the Kyoto Protocol sought to ensure that developed countries would take the lead in reducing emissions of greenhouse gases. This is reinforced by the requirement in the Kyoto Protocol and the Marrakech Accords that the use of the mechanisms 'shall be supplemental to domestic action' and that 'domestic action shall thus constitute a significant element of the effort made by each Party…'.[13]. The requirement that the use of the project mechanisms is supplemental to domestic action is known as the principle of supplementarity. Although this in principle requires Annex I countries to make real efforts to reduce their own emissions rather than simply buying credits from overseas to meet their obligations, the international rules do not expand upon the precise balance between domestic action and the use of other credits.

---

[11] Article 12(10) of the Kyoto Protocol. Unlike the majority of decisions in the Marrakech Accords, the Modalities and Procedures for the clean development mechanism as defined in Article 12 of the Kyoto Protocol annexed to Decision 17/CP.7 (now adopted as Decision 3/CMP.1) were adopted directly by the Conference of the Parties to the UNFCCC. They therefore had effect prior to their adoption at the first meeting of the Conference of the Parties serving as the Meeting of the Parties of the Kyoto Protocol in Montreal in 2005.

[12] Paragraph 29 of annex to Decision 13/CMP.1 (supra note 4).

[13] See in particular Articles 6(1)(d) and 17 of the Kyoto Protocol and paragraph 1 of the Decision on principles, nature and scope of the mechanisms pursuant to Articles 6, 12 and 17 of the Kyoto Protocol, Decision 2/CMP.1 (Decision 15/CP.7).

II.   Linking the EU ETS to the Kyoto Project Mechanisms

A.   The Emissions Trading Directive and its Amendment by the Linking Directive

Article 30(3) of the Emissions Trading Directive as originally adopted in 2003 did not itself provide for the use of credits from the Kyoto project mechanisms under the EU ETS. It did, however, contain a strong commitment to provide for this in a further Directive:

Article 30

Review and Further Developments

3. Linking the project-based mechanisms, including Joint implementation (JI) and the Clean Development Mechanism (CDM) with the Community scheme is desirable and important to achieve the goals of both reducing global greenhouse gas emissions and increasing the cost-effective functioning of the Community scheme. Therefore, the emission credits from the project-based mechanisms will be recognised for their use in this scheme subject to provisions adopted by the European Parliament and the Council on a proposal from the Commission, which should apply in parallel with the Community scheme in 2005. The use of the mechanisms shall be supplemental to domestic action, in accordance with the relevant provisions of the Kyoto Protocol and the Marrakesh Accords.

Accordingly, on the basis of a proposal from the European Commission,[14] the Council and the European Parliament adopted on 13 October 2004 a Directive amending the Emissions Trading Directive giving effect to the commitment in Article 30(3) of the Emissions Trading Directive to provide for the linking of the EU ETS with the project mechanisms in the Kyoto Protocol.[15] It is commonly referred to as the 'Linking Directive'. Member States were required to bring into force the laws, regulations and administrative provisions necessary to give effect to the Directive in their national law by 13 November 2005.[16]

The provisions of the Linking Directive can be divided into two main categories:

a)   provisions providing for Member States to allow operators to use credits from the project-based mechanisms (CERs and ERUs) to comply with their obligations under the EU ETS; and

---

[14] See chapter 1, note 32.
[15] See chapter 1, note 2.
[16] For an account of the negotiation of the Linking Directive, see section III of chapter 1.

b)   provisions specifying additional criteria to be applied by
Member States when approving and authorising projects
under the Kyoto Protocol project mechanisms.

This chapter addresses only the former of these, as it is only these
provisions of the Linking Directive which are directly relevant to the
use of credits from the Kyoto project mechanisms in the EU ETS.

B.   Limits on the Use of Kyoto Credits under the EU ETS

The Linking Directive introduced into the Emissions Trading Directive
a new Article 11a:

Article 11a

**Use of CERs and ERUs from project activities in the Community
scheme**

1. Subject to paragraph 3, during each period referred to in Article 11(2),
Member States may allow operators to use CERs and ERUs from project
activities in the Community scheme up to a percentage of the allocation
of allowances to each installation, to be specified by each Member State
in its national allocation plan for that period. This shall take place through
the issue and immediate surrender of one allowance by the Member
State in exchange for one CER or ERU held by the operator in the national
registry of its Member State.

2. Subject to paragraph 3, during the period referred to in Article 11(1),
Member states may allow operators to use CERs from project activities
in the Community scheme. This shall take place through the issue and
immediate surrender of one allowance by the Member State in exchange
for one CER. Member States shall cancel CERs that have been used by
operators during the period referred in Article 11(1).

3. All CERs and ERUs that are issued and may be used in accordance
with the UNFCCC and the Kyoto Protocol and subsequent decisions
adopted thereunder may be used in the Community scheme:

(a)   except that, in recognition of the fact that, in accordance with the
UNFCCC and the Kyoto Protocol and subsequent decisions
adopted thereunder, Member States are to refrain from using
CERs and ERUs generated from nuclear facilities to meet their
commitments pursuant to Article 3(1) of the Kyoto Protocol and
in accordance with Decision 2002/358/EC, operators are to refrain
from using CERs and ERUs generated from such facilities in the
Community scheme during the period referred to in article 11(1)
and the first five-year period referred to in Article 11(2), and

(b)   except for CERs and ERUs from land use, land use change and forestry activities.

These provisions give Member States a discretion as to whether or not to allow operators to use CERs and ERUs in order to comply with their obligations under the EU ETS to surrender allowances to cover emissions. In the first phase only CERs may be used, since ERUs will not be available until after 2008. In the second and subsequent phases both CERs and ERUs can be used.

Member States' discretion to allow the use of Kyoto credits by operators in their jurisdiction to fulfil their obligations under the EU ETS is subject to limitations.

1.   Nuclear Projects

Firstly, Article 11a(3) of the Emissions Trading Directive (as amended by the Linking Directive) imposes qualitative restrictions on the types of CERs and ERUs which can be used in the EU ETS. The first of these relates to credits from nuclear projects. This reflects the statement in the preambles to the decisions on JI and CDM in the Marrakech Accords which provide that 'Parties included in Annex I to the Convention are to refrain from using emission reduction units generated from nuclear facilities to meet their commitments...'.[17] This limitation expressly applies only to the first and second phases of the EU ETS. This reflects the fact that the commitments under the Kyoto Protocol only extend to 2012 and therefore the Marrakech Accords do not exclude their use thereafter.

2.   Land Use, Land Use Change and Forestry

Secondly, the Directive excludes credits from land use, land use change and forestry activities. This reflects the fact that at the time the proposal for the Directive was made, no agreement had been reached on the international rules for crediting land use, land use change and forestry projects. International rules were then agreed at the end of 2003 and provided for the issue of temporary credits for such projects.[18] Incorporating the use of such temporary credits into an emissions trading scheme which is based on permanent emissions reductions would be

---

[17] See Guidelines for the implementation of Article 6 of the Kyoto Protocol Decision 9/CMP.1 (Decision 16/CP.7) and Decision 3/CMP.1, supra note 11).

[18] See Modalities and procedures for afforestation and reforestation project activities under the clean development mechanism in the first commitment period of the Kyoto Protocol, Decision 5/CMP.1 (Decision 19/CP.9).

difficult. However the Commission is required to consider, as part of its review of the EU ETS under Article 30 of the Directive, 'technical provisions...to allow operators to use CERs and ERUs resulting from land use, land use change and forestry project activities in the Community scheme from 2008'.[19]

3.   Quantitative Limits

As regards quantitative limits, the Directive makes a distinction between the first phase of the scheme, on the one hand, and the second and subsequent phases of the scheme, on the other. In the first phase, no limit on the use of CERs in the EU ETS is required (although a Member State could decide to impose one as a condition for allowing operators to use CERs).

However in the second phase the use of CERs and ERUs must be subject to a quantitative limit. The Directive does not determine the level of this limit but rather requires each Member State to set the limit in the NAP which it submits to the Commission under Article 9. To this end, the Linking Directive amended Annex III to the Emissions Trading Directive to include the following additional criterion:

> 12. The plan shall specify the maximum amount of CERs and ERUs which may be used by operators in the Community scheme as a percentage of the allocation of the allowances to each installation. The percentage shall be consistent with the Member State's supplementarity obligations under the Kyoto Protocol and the decisions adopted pursuant to the UNFCCC or the Kyoto Protocol.

This is further expanded upon by Article 30(3) which requires each Member State to include in its NAP 'its intended use of ERUs and CERs and the percentage of the allocation to each installation up to which operators are allowed to use ERUs and CERs in the Community scheme for that period.' This would appear to extend the requirement in Annex III by requiring the NAPs to contain the limit for use of CERs and ERUs, but also to confirm the total use of ERUs and CERs which the Member State intends to make, across not only sectors subject to the EU ETS but also other sectors not subject to the EU ETS.[20]

The application of the quantitative limit in the second and subsequent phases raises a number of questions of interpretation:

---

[19] Article 30(2)(o) of the Directive.

[20] The number of Kyoto units to be purchased is relevant to determining the consistency of the total quantity of allowances which a Member State proposes to allocate in its NAP with criterion 1 of Annex III, see section II.A of chapter 4.

a)　　There is a question as to whether the limit applies to each year of the scheme or to each phase of the scheme. Articles 11a(1) and 30(3) and criterion 12 of Annex III require the limit to be specified as a percentage of the 'allocation to each installation'. When read with Article 11(2) of the Directive which requires Member States to decide on the allocation for each period, this suggests that the limit relates to each period. This is supported by the 2005 Commission Guidance which recommends that limits are applied to the entire trading period.[21] However, this is expressed as a recommendation and therefore does not exclude the possibility that limits could be applied on a year by year basis.

b)　　The Directive requires the limit to be set as 'a percentage of the allocation of allowances to each installation'. This raises some questions about the level at which the limit should be applied. The use of the phrase 'to each installation' in Articles 11a(1) and 30(3) and criterion 12 of Annex III suggests that the percentage should be applied individually to each installation. This is supported by the wording of Article 53 of the Registries Regulation which provides that the 'registry administrator shall only accept requests to use CERs and ERUs up to a percentage of the allocation made to each installation'. However, the 2005 Commission Guidance adopts a more flexible approach stating that in the Commission's view Member States are free to choose whether to apply the limit individually in respect of each installation or collectively to all installations.[22] The Guidance recommends that Member States apply the limit collectively to installations.

c)　　If the limit is applied at an installation level, there is a question as to whether the percentage specified must be the same for each installation or whether a different percentage could be specified for different installations. The Directive refers to 'a percentage' rather than 'percentages', suggesting that there will only be one percentage which will apply to all installations. In practice, whether a

---

[21] Paragraph 25 of the 2005 Commission Guidance (supra chapter 2, note 6).

[22] This more flexible approach may also be supported by paragraph 3(a) of Annex XIV of the Registries Regulation which simply requires Member States to notify the Commission of 'the total number of ERUs and CERs which operators are allowed to use for each period pursuant to Article 11a(1) of [the Directive].' rather than requiring a percentage to be specified or an amount to be specified for each installation.

> differential approach is possible may ultimately depend more on whether there could ever be an objective justification for such difference in treatment.

Article 30(3) requires the total use of CERs and ERUs to 'be consistent with the supplementarity obligations under the Kyoto Protocol and the decisions adopted pursuant to the UNFCCC or the Kyoto Protocol.' The Directive does not specify what level of use would be consistent with the principle of supplementarity and therefore allows for flexible interpretation of the principle consistent with the interpretation of the international rules.[23]

In considering the NAPs for the second phase of the EU ETS, the Commission had to consider amongst other things, whether the NAP complied with criterion 12 of Annex III to the Directive. The Commission assessed compliance with the supplementarity principle by comparing the Member State's proposed limit on the amount of CERs and ERUs which may be used with the overall 'effort' required of the Member State by the Kyoto Protocol.[24] The Commission calculated 'effort' by taking the highest of the Member State's base year emissions, its greenhouse gas emissions in 2004 and its projected emissions in 2010 and comparing it with the Member State's target under the Kyoto Protocol. The Commission then took the view that supplementarity required at least half of this effort to be achieved within the EU. Therefore where the permitted use of credits under a Member State's NAP plus any proposed government purchase of credits exceeded 50 per cent of that Member State's effort, the NAP was rejected as breaching criterion 12 of Annex III.[25] Where this calculation resulted in operators in Member States being able to use credits up to a level of less than 10 per cent of their allocations, the Commission considered that installations should be allowed to use credits up to a level of 10 per cent, so as not to exclude operators from participation in the international emissions market.

---

[23] The original proposal for the Directive (supra chapter 1, note 32) sought to fix a quantitative limit which in the Commission's view would meet the supplementarity requirements. Article 11(ter)(2) provided that: 'At such time as the number of CERs and ERUs from project activities converted for use in the Community scheme reaches 6 per cent of the total quantity of allowances allocated by the Member States for the period, the Commission shall undertake an immediate review. In the light of this review, the Commission may consider whether a maximum of for example 8 per cent of the total quantity of allowances allocated by the Member States for the period should be introduced in accordance with the procedure in Article 23(2)'. However this was not accepted by the Council during the course of negotiating the Linking Directive.

[24] Section 2.3 of the Communication relating to the decision on the first set of NAPs for the second phase of the scheme (supra chapter 4, note 11).

[25] See, for example, Commission Decision of 29 November 2006 concerning the national allocation plan for the allocation of greenhouse gas emission allowances notified by Ireland in accordance with Directive 2003/87/EC of the European Parliament and of the Council, available at: http://ec.europa.eu/environment/climat/2nd_phase_ep.htm.

4.   Additional Limits

In addition to the express limitations on the use of credits from project activities provided for in the Directive, Member States also have a discretion to impose additional qualitative or quantitative limits. This flows from the fact that Member States are given a choice not to allow the use of credits at all.

5.   Review

Article 30(2)(d) required the Commission to consider the use of credits from project activities, including the need for harmonisation of the allowed use of ERUs and CERs in the Community scheme as part of its report of the scheme due by June 2006. However, the timing of this requirement made it difficult for the Commission to review the question of limits since the practical constraints on the use of CERs mean that no CERs had yet been used in the scheme and 30 June 2006 was also the deadline for Member States to notify the Commission of their NAPs for the second phase of the scheme. The review was deferred to 2007.[26]

C.   How ERUs and CERs are Used in the EU ETS

Article 11a(1) and (2) provide that the use of CERs and ERUs 'shall take place through the issue and immediate surrender of one allowance by the Member State in exchange for one CER or ERU held by the operator in the national registry of its Member State'. The purpose of this provision is to ensure that only allowances are surrendered in the EU scheme. To this end, the original Commission proposal for the Linking Directive provided for operators to apply to convert CERs and ERUs into allowances which could then be traded and surrendered in the EU scheme.[27] However, as well as leading to additional administration for operators, this would have required Member States to issue allowances, which in the second phase would mean issuing an AAU in return for the ERU or CER. This could have implications for a Member State's management of its AAU and its compliance overall with the principle of supplementarity. Therefore the requirement for conversion was amended during the negotiation of the Directive to provide for conversion at the point of use.

The procedure for surrendering CERs and ERUs is set out in Article 53 of the Registries Regulation. This provides for the operator to request transfer and for the number of transferred CERs or ERUs to be added to the number of allowances surrendered, for the purpose of assessing

---

[26] For further information on the review of the scheme, see chapter 10.
[27] Article 11bis(1) (supra chapter 1, note 32).

compliance with the operator's obligations under the scheme. For the operator this is the same as if the Directive provided for the direct use of CERs and ERUs. Likewise for the Member State, the requirement for conversion has little practical effect since although the Directive technically requires conversion of the CER or ERU into an allowance, which in the second phase would include an AAU, this does not have an impact on the type of Kyoto unit which the Member State is required to retire under the Registries Regulation to cover emissions in the EU ETS sectors.[28]

D.   Prevention of Double Counting

The Directive also contains provisions to ensure that there is no double counting where project activities are carried out in Member States. Double counting could occur if a project activity carried out in a Member State also directly or indirectly reduced the emissions of an installation covered by the Directive.[29] In such case there would both be an award of ERUs or CERs, and the operator would need to surrender fewer allowances. The risk of double counting existed independently of the decision to link the EU ETS to the project mechanisms, but the decision to link creates a risk of double counting within the EU ETS if the CERs or ERUs are used in the scheme.

The basic principle set out in Article 11b(2) is that Member States hosting project activities must ensure that 'no ERUs or CERs are issued for reductions or limitations of greenhouse gas emissions from installations falling within the scope of [the] Directive'. As this provision applies to the issue of CERs or ERUs it not only prevents double counting within the EU ETS scheme but also prevents credits from being issued and used for compliance with Kyoto obligations outside the EU scheme.

Articles 11b(3) and (4) provide exemptions from the basic principle up to 31 December 2012. The exemptions make a distinction between projects which directly reduce the emissions of an identifiable installation under the Directive and those which indirectly reduce emissions from a group of installations within the scope of the Directive. Article 11b(3) allows for the possibility of issuing CERs and ERUs for reductions or limitations which directly reduce the emissions of installations falling

---

[28] Article 59(b) of the Registries Regulation allows the Member State to specify the types of Kyoto unit which it wishes to retire to cover the emissions of installations under the EU scheme.

[29] For example a project activity for fuel switching occurs in an installation falling under the ETS, a project activity in the municipal heat generation sector results in a lower production in another installation under the scheme, or a project activity for a wind- or hydropower plant feeds electricity into the electricity grid, thereby replacing fossil fuel-based electricity generation.

within the scope of the Directive if 'an equal number of allowances is cancelled by the operator of that installation'. Article 11b(4) allows for the possibility of issuing CERs or ERUs for reductions or limitations which indirectly reduce the emissions of installations falling within the scope of the Directive if 'an equal number of allowances is cancelled from the national registry of the Member State of the ERUs' or CERs' origin.'

The Directive provides for the detailed provisions for the implementation of these exemptions to be adopted by the Commission using the comitology procedure.[30] The Commission adopted a Decision on 13 November 2006.[31] Under this Decision, Member States are required to include two set asides in their NAP for the second phase for project activities which result in emissions reductions or limitations in installations in the EU ETS: one for project activities which it has approved as host country and one for project activities in relation to which it intends to issue a letter of approval (planned projects).[32] The format for the inclusion of such information is set out in the annexes to the Decision. If a Member State issues a letter of approval in relation to a project prior to the deadline for making the final decision on allocation under Article 11(2) and before it has made that decision, it may move the allowances from the set aside for planned projects to the set aside for approved projects.[33]

ERUs and CERs may be issued for projects activities listed in the NAP if an equivalent number of allowances is converted into AAU from one of the set asides.[34] In relation to planned projects, when a Member State issues a letter approving the project, it is required to assign allowances from the set aside for planned projects to that project. Once assigned the allowances cannot be assigned to any other projects.

If by the end of 2012 not all allowances in the set aside for approved projects have been used then they may be sold or, in the case of projects causing a direct emission reduction or limitation, be issued to the installation where the reduction was expected to occur.[35] Any allowances

---

[30] Article 11b(5). For an explanation of the comitology procedure, see section VII of chapter 2.
[31] Commission Decision of 13 November 2006 on avoiding double counting of greenhouse gas emission reductions under the Community emissions trading scheme for project activities under the Kyoto Protocol pursuant to Directive 2003/87/EC C(2006) 5362 final. Although this Decision was adopted after the deadline for Member States to notify their NAPs for the second phase, Member States would have been aware of the likely final form of the Decision as they had the opportunity to scrutinise it through the comitology procedure. It appears that in practice the Commission expected second phase NAPs to comply with the Decision.
[32] Article 3(1) and (2) of the Decision (ibid).
[33] Article 3(3) of the Decision (ibid).
[34] Article 5(1) of the Decision (ibid).
[35] Article 5(2) of the Decision (ibid).

remaining in the set aside for intended project activities must be cancelled.[36]

III. Linking the EU ETS with Schemes in Other States

The concept of linking the EU ETS to other schemes was discussed in the Green Paper on Emissions Trading in the EU.[37] It is generally accepted that the establishment of links between trading schemes will lead to increased economic efficiency through increased liquidity, greater opportunities for low cost abatement and the establishment of a single carbon price across the schemes.[38]

Article 25 of the Emissions Trading Directive provides for the possibility of linking the EU ETS with other greenhouse gas emissions trading schemes. Recital 18 to the Directive indicates that the principle behind this provision is that linking 'will increase the cost-effectiveness of achieving the Community emission reductions target as laid down in [the Burden Sharing Agreement]'.

In some respects, the EU ETS is itself a system of 25 linked national schemes with common design elements regulated through the Directive, Monitoring and Reporting Guidelines, Registries Regulation and other legislative instruments. The first expansion of the scheme was therefore the inclusion of Bulgaria and Romania with their accession to the EU on 1 January 2007. Both countries put in place legislation to transpose the Directive into domestic law before the end of 2006.

A. The Conditions for Linking the EU ETS to other National Schemes

Article 25 provides that:

Article 25

Links with other greenhouse gas emissions trading schemes

1.   Agreements should be concluded with third countries listed in Annex B to the Kyoto Protocol which have ratified the Protocol to provide for the mutual recognition of allowances between the trading scheme

---

[36] Article 5(3) of the Decision (ibid).

[37] Green Paper on Greenhouse Gas Emissions Trading within the European Union, supra chapter 1, note 8.

[38] Because linking two schemes will mean that the price of allowances in both will become the same linking is only likely to be politically acceptable where the price of allowances in two schemes is already roughly equivalent. Given that a number of emerging schemes envisage that CERs can be used for compliance purposes, this may mean that CERs become established as a common currency whose price becomes the standard price for allowances in a number of schemes, thereby assisting subsequent linking.

and other greenhouse gas emissions trading schemes in accordance with the rules set out in Article 300 of the Treaty.

2.   Where an agreement referred to in paragraph 1 has been concluded, the Commission shall draw up any necessary provisions relating to the mutual recognition of allowances under that agreement in accordance with the procedure referred to in Article 23(2).

Article 25 therefore envisages a two stage process for linking to the EU ETS: firstly an international agreement with the state wishing to link its scheme to the EU ETS and then, if necessary, a Commission Decision (taken through the comitology procedure)[39] making provision for how allowances will be recognised under the agreement.

Article 25 places two limitations on linking. First, it only provides for linking with countries listed in Annex B of the Kyoto Protocol, namely those which have taken on a quantified emission limitation or reduction commitment.

Secondly, linking can only take place with another 'greenhouse gas emissions trading scheme'. The linked scheme must therefore cover one or more of the gases listed in Annex II to the Directive.[40]

However the Directive does not require the scope the EU ETS and the third country trading scheme to be the same, either in terms of activities or gases. This is consistent with the fact that a Member State may (subject to approval by the Commission) unilaterally include activities, installations or greenhouse gases which are not listed in Annex I and the fact that during the first phase Member States apply to temporarily exclude specified installations from the scheme.[41]

---

[39] For a description of the comitology procedure, see section VII of chapter 2.

[40] As the linking is to provide for the mutual recognition of 'allowances', and 'allowance' is defined in the Directive as 'an allowance to emit one tonne of carbon dioxide equivalent during a specified period', a literal reading would therefore only allow linking to schemes in which the tradable units were to emit one tonne of CO_2 equivalent. However, as the second paragraph of Article 25 provides for the adoption of necessary provisions for the mutual recognition, there seems no good practical reason why there should not be linking to a scheme with allowances of a different denomination. A further oddity flows from the requirement in the defined term 'allowance' in that an allowance 'shall be valid only for the purposes of meeting the requirements of this Directive and shall be transferable in accordance with the provisions of this Directive'. However, the purpose of these words is most likely to be to ensure that allowances do not assist operators in meeting their obligations under other EU pollution control measures. To make Article 25 workable, the references to 'the requirements of this Directive' and 'the provisions of this Directive' must be read as references to the requirements of any scheme to which the EU scheme is linked in accordance with Article 25.

[41] For a description of the provisions relating to unilateral inclusion and temporary exclusion, see sections V and VI of chapter 2.

The Directive does not place further requirements on the type of scheme which could be linked to the EU ETS. However as a matter of policy it is likely that the EU would not conclude an agreement with a third country providing for linking to the EU scheme unless it was satisfied that the third country scheme contained an appropriate method for capping the number of allowances in the scheme (this would have an effect on the liquidity of the two linked markets), equivalent monitoring and reporting and verification requirements, and equivalent penalties for non-compliance. Any agreement under Article 25(1) would need to contain provisions governing matters such as these. For example, the agreement could specify minimum standards for monitoring and reporting emissions and a minimum level of penalties.

Where the linking would apply after 1 January 2008, it would also be necessary to consider arrangements for the transfer of assigned amount units.

Recital 10 of the Emissions Trading Directive provides that:

> Starting with the [period beginning 1 January 2008], transfers of allowances to another Member State will involve corresponding adjustments of assigned amount units under the Kyoto Protocol.

This has been interpreted in the Registries Regulation to mean that all allowances will be attached to an AAU.[42] Therefore a trade in allowances will also be a trade in AAUs.

In concluding any agreement to link the EU scheme to a scheme in a third country it would be necessary to provide for the transfer of AAU (or other Kyoto units). If the scheme to which it was proposed to link the EU ETS also allowed AAU (or other Kyoto units) to be traded by operators directly, allowances could be traded directly across the schemes. If not, then any agreement would need to make provision for the transfer of AAU (or other Kyoto units) either at the point an allowance moves between schemes or at the end of each scheme year or phase, representing the net sale or purchase of allowances between the third country and the EU Member States in that period.

---

[42] Article 45 of the Registries Regulation provides that allowances will be issued by 'converting' AAU into allowances and that the conversion will take place 'through adding the allowance element to the unique unit identification code of each such AAU'.

B. The Procedure for Linking the EU ETS to other National Schemes

In order to provide for the EU ETS to be linked to another scheme, Article 25 requires an agreement to be concluded using the procedure under Article 300 of the EC Treaty.

This reference in Article 25 of the Directive to the procedure in Article 300 of the Treaty reflects the respective allocation between the Member States, on the one hand, and the European Community, on the other, of the power to conclude international agreements. Under the AETR principle,[43] the Community alone has competence to negotiate agreements which affect 'common binding rules' which exist within the Community. In the present context, an international agreement providing for the recognition within the Community scheme of allowances emanating from a scheme in a third country would have an effect on the 'common binding rules' in the Directive. It follows that it is the Community, applying the procedure in Article 300 of the Treaty, that has competence to conclude such agreements, and not any Member State acting unilaterally.[44] Under the procedure in Article 300, it is the European Commission which conducts the negotiations on behalf of the European Community, having first received a negotiating mandate from the Council. It is then the Council which takes the ultimate decision as to whether to conclude the agreement which the Commission has negotiated. The Council, both in granting the Commission a negotiating mandate, and in deciding whether to conclude the agreement, acts by qualified majority.[45] Throughout the negotiations the Commission will

---

[43] Under the AETR principle, once the Community passes secondary legislation in a particular field laying down common internal rules within the Community, it acquires exclusive competence in relation to international agreements which could affect those common rules. Otherwise the Community measure could be frustrated by a Member State entering into an incompatible international agreement. Case C-22/70 *Commission v Council* [1971] ECR 263. See also judgment in the 'Open Skies' case, C-467-469/98 and C-475-476/98 *Commission v Denmark, Sweden, Finland, Belgium, Luxemburg, Austria and Germany.*

[44] There may, nevertheless, be additional elements of any agreement which fall outside the scope of that AETR principle and so fall within Member State competence. The resulting so called 'mixed' agreement would be concluded both by the European Community and the Member States in relation to their respective competences. The allocation of competence is often a contentious issue, and has to be considered case by case.

[45] Qualified majority voting (QMV) is a procedure for adopting decisions under which each Member State is given a fixed number of votes based on population but weighted in favour of smaller Member States. From 1 November 2004, when the Treaty of Nice came into force, the total number of votes was 321. To pass a vote by QMV, the proposal had to be supported until 31 December 2006 by at least 232 votes and a majority of Member States. Any Member State could ask for verification that the countries supporting the proposal represented at least 62 per cent of the total EU population. Declarations No 20 and 21 adopted when the Treaty of Nice was signed set out the position for the negotiations on the accession of Romania and Bulgaria. They provided for 14 votes for Romania and 10 votes for Bulgaria. The total number of votes since 1 January 2007 has thus been 345 and a qualified majority constituted by 255 votes.

consult with a committee of representatives of the Member States. This consultation is the mechanism by which the Member States influence the course of the negotiations and by which the Commission satisfies itself that the agreement that it is negotiating will ultimately command qualified majority support in the Council.

As noted above, the Directive envisages a two-stage process for linking to another trading scheme: after the conclusion of an agreement with a third country, the Commission is to adopt, through the comitology procedure under Article 23(2) 'any necessary provisions relating to the mutual recognition of allowances under the agreement'. Despite the word 'any', it is difficult to imagine a situation in which it would not be necessary to adopt provisions for mutual recognition.[46]

What form of 'provisions' would be necessary would depend on a number of factors, such as the type of allowances to be recognised, the procedure for transferring allowances between the two schemes, and any limitations on the use of recognised allowances. For example, in order to allow Member States to set a cap on the number of recognised allowances which can be retired, it would be necessary to amend the Emissions Trading Directive itself. On the other hand, a mechanism to control the transfer of allowances into the Community scheme might be given effect through amendments to the Registries Regulation alone.

C.   Possible Future Developments

1.   Links with Countries that have not Ratified the Kyoto Protocol or are not Listed in Annex B to the Protocol

The Directive currently only refers to concluding agreements with third countries which are listed in Annex B to the Kyoto Protocol and have ratified that Protocol. It would not therefore be possible, without an amendment to the Directive, to link the Community scheme to schemes

---

[46] Under Article 300(7) EC, the agreement with the third country would itself be binding on the institutions of the Community and on the Member States. Furthermore, if the agreement met the tests of precision and unconditionality laid down in the case law of the Court, it could have direct effect so as to confer rights on individuals. (For a discussion of the precise scope of the obligations on Member States, and the extent of direct effect of international agreements concluded by the Community see MacLeod, Hendry and Hyett, (1996), *The External Relations of the European Communities*, Oxford, Claerendon Press, pp 125-128 and 135-137). It may therefore at least theoretically be possible for some of the changes in Community law necessary to provide mutual recognition to be effected through the conclusion of the international agreement itself. But in practice it would almost certainly be necessary to make at least some changes to the Registries Regulation and possibly the Emissions Trading Directive itself. Further, it is likely that all the necessary adjustments would, for reasons of legal certainty, in practice be effected by such changes to the existing legislation.

in countries which are either not listed in Annex B or are listed in Annex B but have not ratified the Protocol.

It would appear possible, at least theoretically, for the third country to develop a unilateral link by recognising the cancellation of an allowance in the EU ETS as a method of compliance with the obligations under its scheme. However, there would appear to be practical limitations to a one-sided linkage such as this. It would appear difficult, for example, to establish in this way a structure which would enable industry within the Community and the third country to benefit equally from opportunities for lower abatement costs.

However, there are indications that there may be some political desire to amend the Directive in this respect. In response to an amendment tabled by the European Parliament in the course of the negotiations, the Linking Directive contains the following recital:[47]

> Following entry into force of the Kyoto Protocol, the Commission should examine whether it could be possible to conclude agreements with countries listed in Annex B to the Kyoto Protocol which have yet to ratify the Protocol, to provide for the recognition of allowances between the Community Scheme and mandatory greenhouse gas emissions trading schemes capping absolute emissions established within those countries.

This could be achieved by an amendment to the Directive to provide for linking agreements under Article 25 not only with Annex B parties which have ratified the Kyoto Protocol but also with Annex B Parties which have not ratified the Protocol. Any such amendment could also allow for linking agreements to be made with parties not listed in Annex B to the Protocol.

If such an amendment were made, there would, in relation to the first phase of the scheme, seem to be no greater difficulty linking to such a scheme than linking to a scheme in an Annex B party which has ratified the Kyoto Protocol since, other than the use of CERs, there is no interaction with the Kyoto regime. However, there could be practical

---

[47] Recital 18. This was included as part of negotiations to reach a 'first reading deal' with the European Parliament. The Parliament had proposed the following amendment to Article 25 of the Directive: 'Prior to third countries listed in Annex B to the Kyoto Protocol ratifying the Protocol, agreements may be concluded with regional authorities in those countries to provide for the mutual recognition of allowances between the Community Scheme and mandatory greenhouse gas emissions trading schemes capping absolute emissions established by those authorities.' The justification given for this proposal was the political situation in Australia and the US where there is interest in emissions trading at a regional level despite the decisions of the national governments not to ratify the Kyoto Protocol. The proposal was not accepted by the Council but it was agreed that an additional recital should be included.

difficulties in linking to such schemes in the second phase of the EU ETS, when it is envisaged that allowances in the EU ETS will be backed by AAU. If the third country is not a party to the Kyoto Protocol then it will not have AAU or be able to take part in international emissions trading under Article 17 of the Kyoto Protocol. If allowances are allowed to be transferred freely (even if the AAU is removed from the allowance before it is transferred to the third country scheme) and there is a net flow of allowances into the EU scheme, then EU Member States could find that they have insufficient AAU to cover the emissions of greenhouse gases from the EU.[48]

Possible ways of dealing with these difficulties might include introducing some form of gateway controlling the transfer of allowances between the linked schemes,[49] or a system involving the transfer of CERs instead of AAU.

2.   Linking with Schemes at a Sub-national Level

Further, Article 25 would not allow the Community Scheme to be linked to a scheme at sub-national level, for example to schemes at state level in the USA or Australia.

Unlike in relation to Annex B countries which have not ratified the Kyoto Protocol (or parties which are not listed in Annex B), there is a question as to whether it would be legally possible to conclude an agreement to link to schemes at a sub-national level. Such a link would only be possible if the sub-national entity itself has the necessary legal competence to enter into international agreements[50] and if so, only if the EU itself has competence to enter into such agreements.[51]

---

[48] This issue currently arises in any case under the EU ETS because Cyprus and Malta are not Annex B parties. However, as the amounts of AAU involved are small, provision will be made for the borrowing of AAU to cover allowances issued by Malta and Cyprus.

[49] Such as the gateway that operates in the UK Emissions Trading Scheme to delineate between allowances obtained through achievement of an absolute target and those relating to a relative target. For further information on the UK ETS, see chapter 22.

[50] It is not clear that such states would have the necessary legal competence to enter into international agreements. Some federal constitutions give states limited powers to enter into international agreements. For example, Article I(10)(3) of the US Constitution permits US States to enter into agreements or compacts with other states but only with the consent of Congress.

[51] It is not clear whether the EU would have competence to enter into such an agreement. Article 300 EC allows the Community to conclude agreements with one or more states or international organisations where the Treaty provides for the conclusion of agreements. In the field of environment, Article 174(4) provides for cooperation with third countries and competent international organisations. Whether the EU could conclude an agreement with an individual state in a federation would depend on the interpretation of the term 'state' and 'countries' in these articles.

If it is possible to conclude such agreements, then the Directive could be amended to enable the Community scheme to be linked with sub-national schemes. This would raise similar issues and practical difficulties to those raised if the Community scheme were linked to schemes in countries not listed in Annex B of the Kyoto Protocol, or Annex B countries which have not ratified the Kyoto Protocol.

D. Snapshot of the Current State of other National Schemes

The remainder of this chapter considers the current state of development of other domestic trading schemes around the world (including in countries which have not yet ratified the Kyoto Protocol), and so provides the political context within which discussions about linking the EU ETS will take place.

1. Norway

Norway is a member of the European Economic Area.[52] The EEA comprises the EU Member States plus Norway, Iceland and Liechtenstein. Under the EEA Agreement most, but not all, EU environmental law is binding on these three EEA states that are not EU Member States. Individual Directives are extended (sometimes with modifications) to these three states by a decision of the EEA Joint Committee under the EEA Agreement. It is likely that the Emissions Trading Directive will be extended. In preparation for this, Norway has put in place a trading scheme modelled on the EU ETS which began operation in January 2006. Rather than link directly into the EU ETS now, Norway is intending to apply for a temporary exclusion until the end of 2007 when it will join the EU ETS formally.

2. Switzerland

The Swiss Government is intending to introduce a tax on process and heating fuels. In an approach similar to the UK's Climate Change Agreements,[53] companies will be exempted from the tax if they sign up to binding reduction targets via a negotiated agreement. From 2008, it is envisaged that these companies will be allowed to meet their targets through emissions trading, including the use of CERs and ERUs. The scheme will cover around 10-15 per cent of the total Swiss carbon dioxide emissions, with similar sectors to the EU ETS but smaller coverage

---

[52] The European Economic Area (EEA) enables Norway, Iceland and Liechtenstein to participate in the internal market while not assuming the full responsibilities and rights of EU membership.
[53] For information on Climate Change Agreements, see chapter 24.

because hydro-electricity and nuclear power dominate the Swiss electricity generation industry.

The Swiss Government have indicated that they intend to establish a link to the EU ETS as soon as possible and have already started discussions with the Commission.

3.   Canada

In July 2005, the Government of Canada published a Notice of Intent to Regulate setting out its plans to establish a mandatory greenhouse gas emissions trading scheme for its large final emitters. The Notice stated that Large Final Emitters Regulations would be in place by 1 January 2008. The scheme would cover around 700 companies in the energy-intensive mining and manufacturing, oil and gas, and thermal electricity sectors which are responsible for approximately 50 per cent of total Canadian greenhouse gas emissions.

The scheme would operate on a baseline and credit basis with relative targets based emissions intensity. Regulated companies would be given substantial flexibility on how to meet their targets, including use of JI and CDM credits and credits from domestic offset projects. The Government also intended to introduce a price cap of fifteen Canadian dollars (approximately ten euros) per tonne at least until 2012, although it did not indicate how such a cap would operate. The Notice envisaged a potential link to the EU ETS at some stage in the future but issues relating to the operation of the price cap and relative targets present difficulties. However, following the change in government in January 2006, there is some uncertainty as to whether the trading scheme will proceed.

4.   Japan

As part of its plans to meet its Kyoto Protocol target, the Japanese Ministry of Environment launched a voluntary cap and trade scheme in March 2006 for large final emitters. After an open call for applicants, 34 companies were selected to participate in the scheme, which started in April 2006 and ran to March 2007, operating alongside subsidies for investment in abatement equipment. Companies which cannot meet the target of a 21 per cent reduction on average emissions from 2002 to 2004 will be required to refund their subsidy. Companies will be allowed to use CERs and ERUs for compliance.

The government will review the scheme, as well as its reduction target plan, before deciding whether or not to roll it out more widely, or turn it into a mandatory scheme.

5.   Australia

Following the Australian Government's decision not to ratify the Kyoto Protocol, state and territory level governments announced an agreement in April 2005 to investigate the design of a potential national Australian ETS covering energy sector emissions. In August 2006, the State and Territory governments released a discussion paper for a potential scheme design. The scheme would operate on a cap and trade basis covering the Kyoto Protocol basket of six greenhouse gases, with offsets possible through use of CERs and ERUs.[54] There is currently no commitment by the state and territory governments to implement a scheme.

6.   United States

The decision by President George W Bush not to ratify the Kyoto Protocol in 2002 nearly proved fatal to its coming into force. Despite the United States' role in including emissions trading as one of the flexible mechanisms in the Protocol and the success of its domestic trading scheme to address acid rain, the US Federal Government continues to oppose the concept of mandatory cap and trade emissions trading in relation to greenhouse gases.

However, the negative attitude of the Federal Government has not stopped work on developing trading schemes at the state level. States including California, Oregon and New Mexico have set up inventories of emissions by companies of greenhouse gases and are considering establishing trading schemes. The most advanced proposals are the Regional Greenhouse Gas Initiative (RGGI) on the east coast but California is catching up fast.

Following the adoption of a memorandum of understanding issued in December 2005, the RGGI programme will cover the carbon dioxide emissions from power plants in seven Northeast and Mid-Atlantic states including New York, New Jersey and Connecticut.[55] In August 2006, the participating states issued a so-called 'model rule'[56] setting out the details of the proposed programme which will form the basis of

---

[54] See www.emissionstrading.net.au.
[55] See http://www.rggi.org/docs/mou_12_20_05.pdf. Massachusetts and Rhode Island dropped out of RGGI in December 2005 following the rejection by the other states of their proposals for a price cap.
[56] Available at: http://www.rggi.org/docs/model_rule_8_15_06.pdf.

implementation by individual states. The scheme will cover all fossil fuel power plants with a capacity of 25 megawatts or more (approximately 730 installations) and will start in 2009. Emissions will be capped at current levels until 2015 and will then be reduced annually so that the cap in 2019 will represent a 10 per cent reduction compared to current levels or a 35 per cent reduction from 'business as usual' levels. At least 25 per cent of allowances will be allocated to a set-aside account from which allowances will be sold or distributed in order to fund energy efficiency or clean energy technology programmes. Plants may utilise domestic offsets (such as landfill gas recovery or reforestation) from anywhere in the United States for up to 3.3 per cent of their overall emissions. A safety valve mechanism applies whereby the amount of offsets which may be used rises if the average annual price of allowances passes certain thresholds.

Meanwhile on the west coast, the California State Legislature passed the Global Warming Solutions Act in August 2006 setting a mandatory emissions cap on California's emissions (reduction to 1990 levels by 2020) and providing for enabling powers to introduce mitigation measures. Provisions requiring the establishment of an emissions trading scheme were dropped in the final negotiations for the Act but market mechanisms such as emissions trading are envisaged to play a key role in achieving these targets. In July 2006, UK Prime Minister Blair and Calafornian Governor Schwarzenegger announced a UK-California collaboration on efforts to address climate change. The collaboration will include sharing best practice on emissions trading and lessons learned in Europe and exploring the potential for linkages between market-based mechanisms. In October 2006, Governor Schwarzenegger and New York Governor Pataki announced that they had agreed to explore ways to link the California carbon market with RGGI.

CHAPTER 9

SURRENDER OF ALLOWANCES AND PENALTIES

This chapter considers the obligation of operators to surrender allowances to cover their emissions, and the penalties if they fail to do so.

I.   Obligation to Surrender Allowances

As we have already seen, an operator of an installation is required to have their reported emissions from an installation for a year verified by 31 March in the following year.[1] Article 12(3) of the Directive then requires the operator to surrender the number of allowances equal to those total verified emissions, by 30 April. The procedure by which this is done is set out in Article 52 of the Registries Regulation. Surrendered allowances are transferred back into the party holding account and cancelled on 30 June each year. In the first phase allowances are cancelled by transferring them to the retirement account.[2] For the second and subsequent periods, the cancellation of allowances is a two-stage process.[3] Firstly, the surrendered allowances are converted to AAUs by removing the allowance element from the identification code. Secondly Kyoto units equal to the number of allowances and Kyoto units surrendered are transferred to the retirement account. Member States can specify the types of Kyoto units to be retired.

A record of the number of allowances and Kyoto units surrendered by each operator is kept in the surrendered allowance table.[4] The number of allowances and other Kyoto units surrendered is then compared with the verified emissions for the relevant installation to determine whether the operator has complied with the requirements of the Directive and is recorded in a compliance status table.[5]

II.   Banking of Allowances Between Phases

Article 13(1) provides that allowances are valid for surrender only in respect of the period for which they are issued. In the absence of this

---

[1] For further information on the obligations to report emissions, see section II of chapter 3.

[2] Article 58 of the Registries Regulation.

[3] Article 59 of the Registries Regulation

[4] See Articles 52 and 53 of the Registries Regulation.

[5] Article 55 of the Registries Regulation.

provision operators would have been able to surrender allowances in respect of one phase to meet their obligations relating to emissions in a different phase. For example, the Directive requires a proportion of allowances in respect of the second phase to be issued by 28 February 2008, which is before the date by which allowances must be surrendered in respect of emissions made during the first phase.

The Directive does, however, provide for a form of banking between phases. This is achieved by the cancellation of allowances which have not already been surrendered or cancelled, and the issue of new allowances to replace them in the following phase. For the transition from the first to second phase, Member States have a discretion as to whether or not to provide for such banking.[6] Between the second and third and subsequent phases, Member States are required to provide for banking.[7] The procedure for cancellation and replacement of allowances is set out in Articles 60 and 61 of the Registries Regulation. This requires the Member State on 1 May following the end of each phase to cancel any allowances which have not been surrendered and replacing them with allowances for the subsequent phase.

III.   Penalties

The Directive requires Member States to impose penalties for breach of the national measures implementing the Directive which are effective, proportionate and dissuasive. It prescribes specific penalties in two circumstances.

Firstly where an operator's annual emissions report is not verified as satisfactory by 31 March the following year, the second paragraph of Article 15 requires Member States to ensure that the operator cannot make further transfers of allowances until a report from that operator has been verified as satisfactory.[8]

Secondly the Directive prescribes the sanction to be imposed on an operator who fails to surrender sufficient allowances. In the first phase this is a penalty of 40 euros per tonne of carbon dioxide equivalent emitted by the installation for which the operator has not surrendered allowances. For the second and subsequent phases the penalty rises to 100 euros. The payment of these penalties does not absolve the operator of the obligation to surrender in the following year allowances

---

[6] Article 13(2) of the Directive. The majority of Member States (23) decided not to allow the banking of allowances between the first and second phases. Where Member States do allow for banking there could be state aids implications, see chapter 4, note 35.
[7] Article 13(3) of the Directive.
[8] This is given effect by Article 27 of the Registries Regulation.

corresponding to the shortfall.[9] Member States are also required to publish the names of operators who are in breach of the requirement to surrender allowances.[10]

IV. Voluntary Cancellation

In addition to these provisions relating to compulsory surrender of allowances, the Directive provides for voluntary cancellation of allowances. Article 12(4) of the Directive requires Member States to take the necessary steps to ensure that allowances will be cancelled at any time at the request of the person holding them. The effect of this is to enable any operator or other person to acquire allowances and cancel them, thereby reducing the total quantity of allowances available in the trading scheme and therefore the total quantity of emissions permitted by the scheme.

---

[9] Article 16(3) and (4) of the Directive.
[10] Article 16(2) of the Directive.

CHAPTER 10

FUTURE DEVELOPMENTS

This chapter considers the continuation of the ETS after 2013 (section I) and the review of the scheme required by Article 30 of the Directive (section II). It also considers another key issue in the future development of the scheme, namely the inclusion of emissions from aviation (section III).

I.   Continuation of the Emissions Trading Scheme

The EU ETS Directive contains no 'sunset clause' providing for the expiry of the scheme after a certain time period. The scheme will therefore continue to operate after the second phase ends in 2012 on the same basis as beforehand. In the absence of any revision to the Directive, the only design feature of the scheme which would change post 2012 would be the method of allocation.[1] However, broader uncertainty about the future of the international climate change regime following the end of the first Kyoto Protocol commitment period[2] also creates uncertainty about the future of the ETS. In order to respond to these concerns, the Environment Council adopted conclusions in October 2005 recognising that the EU ETS will remain essential in the EU's medium and long-term strategy to tackle climate change.[3]

II.   Review of the Emissions Trading Directive

Under Article 30(1) of the Directive, the Commission was given a mandate to consider by 31 December 2004 whether to bring forward an amendment to include other activities and emissions of greenhouse gases in future phases of the scheme. No such legislative proposal was brought forward nor was any decision published confirming that the Commission had decided not to extend the scheme at that time. However, as the

---

[1] As discussed in section I of chapter 4, Article 10 of the Directive does not place any specific limits on the amount of allowances which can be auctioned after the second phase of the scheme and therefore, in the absence of a further amendment to the Directive, it would be open to Member States to propose 100 per cent auctioning from 2013.
[2] See section III.A of the Introduction.
[3] Available at http://ue.eu.int/ueDocs/cms_Data/docs/pressData/en/envir/86627.pdf. The Council conclusions recalled and emphasised the Council's commitment to the conclusions of the 2005 Spring European Council and the (Environment) Council which inter alia asserted that reduction pathways by the group of developed countries in the order of 15-30 per cent by 2020 and 60-80 per cent by 2050 compared to the baseline envisaged in the Kyoto Protocol should be considered.

Commission does not require authority from the Emissions Trading Directive to make a proposal for amendment to extend the activities or gases covered by the scheme, the passing of this date has no practical significance.

Article 30(2) of the Directive required the Commission, on the basis of experience of the application of the Directive and of progress achieved in the monitoring of emissions of greenhouse gases, and in the light of developments in the international context, to draw up a report on the application of the Directive.

The Directive contains the following list of the issues to be included in the report:

(a) how and whether Annex I should be amended to include other relevant sectors, inter alia the chemicals, aluminium and transport sectors, activities and emissions of other greenhouse gases listed in Annex II, with a view to further improving the economic efficiency of the scheme;

(b) the relationship of Community emission allowance trading with the international emissions trading that will start in 2008;

(c) further harmonisation of the method of allocation (including auctioning for the time after 2012) and of the criteria for national allocation plans referred to in Annex III;

(d) the use of credits from project mechanisms, including the need for harmonisation of the allowed use of ERUs and CERs in the Community scheme;

(e) the relationship of emissions trading with other policies and measures implemented at Member State and Community level, including taxation, that pursue the same objectives;

(f) whether it is appropriate for there to be a single Community registry;

(g) the level of excess emissions penalties, taking into account, inter alia, inflation;

(h) the functioning of the allowance market, covering in particular any possible market disturbances;

(i) how to adapt the Community scheme to an enlarged European Union;

(j) pooling;

(k) the practicality of developing Community-wide benchmarks as a basis for allocation, taking into account the best available techniques and cost-benefit analysis; ...

(o)   whether to allow operators to use CERs and ERUs resulting from land use, land-use change and forestry project activities in the Community scheme from 2008.

Although the list is fairly extensive, the scope of the review is not limited to the issues expressly listed in the Directive. The issues raised in Article 30(2) mostly represent issues raised during the negotiation of the Directive and the Linking Directive which were not addressed in the final compromise texts (such as expansion to other sectors, activities and gases and harmonised use of auctioning). However, the list continues to represent the key issues which the review and any subsequent amendment to the Directive seem likely to address.

A.   Process

Article 30(2) required the Commission to submit the report on the application of the Directive to the European Parliament and the Council by 30 June 2006, accompanied by proposals as appropriate. Although the Commission had commenced the review process, it did not meet this deadline.

In some respects, the timetable envisaged in Article 30(2) was out of step with the implementation of the scheme. If the Commission had published a legislative proposal in June 2006, it would have been too early to take account of issues arising during the development and assessment of NAPs for the second phase which were required to be submitted to the Commission by exactly the same deadline. It would also have been too early to take into account reports on the operation and enforcement of the scheme submitted by Member States under Article 21 of the Directive, also due by the end of June 2006.[4]

The Commission began the review process by undertaking a survey of Member States, regulated industry, environmental NGOs and other stakeholders in late summer 2005 and published the preliminary results in November 2005.[5] The Commission adopted a Communication entitled

---

[4] Article 21 requires Member States to submit to the Commission each year by 30 June a report on the application of the Directive paying particular attention to the arrangements for the allocation of allowances, the operation of registries, the application of the monitoring and reporting guidelines, verification and issues relating to compliance with the Directive and on fiscal treatment of allowances, if any. The report must be drafted on the basis of a questionnaire prescribed by the Commission (see Commission Decision, supra. chapter 4, note 68). Within 3 months of receiving such reports from Member States, the Commission is required to publish a report on the application of the Directive. See EEA Technical Report No 2/2006 'Application of the emissions trading Directive by EU Member States' available at: http://reports.eea.europa.eu/technical_report_2006_2/en/technicalreport_2_2006.pdf.

[5] See 'Review of EU Emissions Trading Scheme: Survey Highlights' available at: http://ec.europa.eu/environment/climat/pdf/highlights_ets_en.pdf.

'Building a Global Carbon Market – Report pursuant to Article 30 of Directive 2003/87/EC' on 13 November 2006.[6] The Communication did not set out concrete proposals and was not accompanied by a legislative proposal. Instead, it recognised that while there was a growing consensus on the key strategic issues for review, more experience and evaluation were needed for addressing these issues. The Communication, therefore, sought to draw a number of general conclusions about the operation of the scheme so far and its future direction and outlined a number of issues to be considered further by a stakeholder working group on the review of the ETS.

The Communication identified the following categories of issues for further consideration as part of the review:[7]

- the scope of the Directive
- further harmonisation and increased predictability
- robust compliance and enforcement; and
- links to emissions trading schemes in third countries.

B. Scope of the Scheme

As mentioned above, an issue of particular concern has been the question of the scope of the scheme both in relation to the interpretation of the term 'combustion installations' in Annex I of the Directive and to the cost-effectiveness of including small installations in the EU ETS. Following on from the 2005 Commission Guidance, the interpretation of the term 'combustion installations' will be considered further in the review with a view to facilitating harmonised application by Member States through the development of more specific technical descriptions of types of combustion installation. As regards the small installations, the review will consider whether the participation of small installations in the scheme is cost-effective or whether there is sufficient justification for removing small installations from the scheme. The latter approach would require a workable threshold to be agreed and for emissions from excluded installations to be addressed by other policies and measures achieving the same environmental results.

---

[6] COM(2006) 676 final.

[7] The review will also consider the relationship between the EU ETS and other market-based instruments, in particular the interplay between emissions trading and energy taxation (which is governed by Council Directive 2003/96/EC). It will also cover wider issues related to the involvement of developing countries and countries in economic transition in the emissions abatement efforts through the Clean Development Mechanism and Joint Implementation.

The review will also assess the scope for expanding the scheme to other sectors and gases. The Communication identified the following sectors and gases which could considered for a possible expansion of the scheme: N_2O from the production of nitric acid; CO_2 from the production of petrochemicals; CO_2 and N_2O from the production of ammonia, other fertilisers than nitric acid and adipic acid; CO_2 and PFCs from the production of aluminium and CH_4 from coal mines. This work is likely to be informed by work undertaken by some Member States on the possibility of 'opting-in' additional installations into the scheme for the second phase under Article 28 of the Directive. The review will also consider whether the procedures for unilateral inclusion continue to be appropriate.

One sector expressly mentioned in the recitals and Article 30 of the Directive is transport. There has already been significant interest in the potential to incorporate emissions from aviation into the scheme in the future (see section III of this chapter) and it seems likely that the review will need to consider whether emissions from surface transport might also be included in the scheme. However, the Communication recognised that the extension of the scheme to other sectors and gases should be part of a comprehensive and coherent policy mix. In this context, it noted that the inclusion of direct emissions from the road and maritime transport would involve much greater administrative costs while a number of Member States have other policies and measures in place.

The review will also assess appropriate methodologies for the recognition of CO_2 capture and geological storage activities in the scheme. This will take into account the need for comparable treatment of low or non-CO_2 emitting activities and for a level playing field between various options for capture and storage.

C.   Further Harmonisation and Increased Predictability

The review will consider ways in which to reduce differences in the allocation rules across the EU and to give sufficient predictability to operators when taking investment decisions.

As explained in chapters 4 and 5 the Emissions Trading Directive leaves a large measure of discretion to Member States to determine the total quantity of allowances that they will allocate and how to allocate them between installations, including new entrants. This results in a wide variation between the rules applied in each Member State. This increases the complexity of the scheme for operators operating in more than one Member State and can result in distortions in competition. For this reason there is an increasing recognition that further harmonisation may be desirable, particularly in relation to the rules on new entrants and closures.[8] The administrative complexity of developing the first round of NAPs and concern about windfall profits made by some sectors as a result of free allocation[9] has increased the attractiveness of auctioning as an allocation option.[10]

The Communication confirms that the review will consider the possibilities for greater harmonisation in the procedure for the allocation of allowances. In particular, it will consider whether Member States should continue to determine the total quantity of allowances to be allocated separately or whether a total quantity should be set EU-wide, or the amount to be allocated by each Member State specified in the Directive. As regards the distribution of allowances between sectors and operators, the review will consider the scope for greater harmonisation. In this context, it will asses the methodologies and in particular the use of auctioning and specific issues related to benchmarking. A key issue for consideration in the review is the treatment of new entrants and installations which close. In the first phase all Member States chose to set aside some allowances for new entrants but the size and exact rules governing access to and allocation from the reserve differed considerably. The review will therefore consider whether the creation of a new entrant, reserve is necessary, did if so whether for internal market reasons, the allocation from such reserve should be harmonised, either by adopting common rules or by constituting and administering the reserve at EU level.

In this context, the review will also assess the balance between the tasks performed at Community level and those at Member State level and consider how existing institutional arrangements could be adapted and administrative costs reduced.

A further issue concerns the extent to which the allocation process gives sufficient certainty to enable participants in the scheme to make informed

---

[8] See the First Report of the High Level Group on Competitiveness, Energy and the Environment (supra. chapter 4, note 38) which recommended that 'the Commission investigates how rules, notably for new entrants and closure can be more harmonised, including the possibility for using a benchmarking approach' (paragraph 23).

[9] The competitive situation in the electricity sector enabled electricity producers to pass-through costs to electricity prices. This resulted in windfall profits for the electricity producers and increased base costs for the energy intensive industries which affects their competitive position vis-à-vis international competitors not subject to similar legislation.

[10] Footnote (3) to the first Report of the High Level Group on Competitiveness, Energy and Environment (supra chapter 4, note 38) notes that 'a significant proportion of HLG members considered that a more extensive use of auctioning should be explored for the post-2012 period.' Some Member States are also proposing to make greater use of the possibility for auctioning allowances in the second phase of the scheme. See chapter 21 for details of the use of auctioning proposed in the UK's NAP for the second phase of the scheme.

investment decisions. The current design of the scheme makes it very difficult to predict the price of allowances or the way in which individual installations will be treated in future phases. Under Article 11, final decisions on the total quantity of allowances and the way in which they will be distributed are made just twelve months before the beginning of any phase. This means that operators and others with an interest in the carbon market will have a clear picture of the number of allowances available for the following six years. However, as the period progresses, this period of foresight will gradually reduce back to a single year until the allocation decisions for the following phase are made.

This uncertainty has two negative impacts, both of which undermine the potential of the scheme to incentivise investment in the development and deployment of low-carbon technology.

Firstly, the lack of clarity about future limits on emissions means that it is very difficult to predict the likely price of allowances in future phases and therefore the extent to which any investment in abatement technology will provide a return. While it is true that the price of allowances is driven by a number of factors about which it is impossible to be certain (such as weather, fuel prices and the cost abatement curve not only of installations covered by the scheme but also those which might take part in the Kyoto project mechanisms), lack of clarity about the demand for allowances generates an additional element of regulatory risk which may discourage investment.

Secondly and likely to be of equal significance is the lack of certainty about the allocation methodology which will apply to an installation in future years. As a result, anyone investing in abatement technology now has no certainty that he will continue to receive his current level of allocation in future phases. The risk that in some way his allocation will reduce as a result of this early action cannot be eliminated. The fact that from the second phase onwards installations will be able to bank allowances into subsequent phases means that there will be some additional incentives for early investment but these could be reinforced by greater certainty about the extent to which allowances will be sold rather than allocated for free in future phases and how any free allocation will be carried out.

The review will therefore also consider whether the allocation period should be extended in order to give certainty to operators further in advance. In order to further increase predictability, the review will further consider whether the design of the scheme should be reviewed periodically in the future and if so, at what intervals. It will also look at how emissions information is provided to the market.

D.   Robust Compliance and Enforcement

The Communication noted that compliance in first year of the first phase of the scheme had been good.[11] However, the review will consider if there is a need for further harmonisation, particularly in relation to the verification of emissions reports and the accreditation of verifiers. It will also consider enforcement provisions in particular relating to non-submission of verified emissions reports, the late submission of verified emission reports, errors in verified emission reports and inaccuracies in data reported for the purposes of allocation.

E.   Links with Emissions Trading Schemes in Third Countries

The review will assess the design of third country emissions trading schemes which are in operation or planned and the possibility of such schemes linking to the EU ETS. It will consider in particular whether Article 25 of the Directive which provides for links with third countries listed in Annex B to the Kyoto Protocol which have ratified the Protocol could be extended to allow linking with other national or regional schemes.

F.   Outcome of the Review

The Communication provided for the Working Group set up to consider these issues to report by 31 March 2007 with a view to enabling the Commission to make a legislative proposal later in 2007. Any proposal would then go through the co-decision procedure involving consideration by both Council and European Parliament, with any amendment to the Directive unlikely to be finalised earlier than 2009. The Communication stated that 'the Commission takes the firm view that for reasons of regulatory stability and predictability, any changes emanating from [the] review should take effect at the start of the third trading period in 2013'.

III.   Aviation

On 27 September 2005, the Commission adopted a Communication on 'reducing the climate change impact of aviation'.[12] The Communication

---

[11] By the 31 March 2006 (the first deadlines for reporting emissions under the scheme), emissions reports had been received for 8,980 installations, accounting for more than 99 per cent of the allowances allocated to installations in the 21 Member States with functioning registries on 30 April 2006.

[12] Communication from the Commission to the Council, European Parliament, the European Economic and Social Committee and the Committee of the Regions 'Reducing the Climate Impact of Aviation' COM(2005)459 final.

proposed that the aviation sector should be included in the emissions trading scheme[13] and indicated that the Commission aimed to put forward a legislative proposal to this effect by end 2006. The approach of including aviation in the EU ETS was a top priority for the UK Presidency of the EU in the second half of 2005 and was supported in conclusions adopted by the Environment Council on 2 December 2005[14] and the European Council on 15/16 December 2005,[15] and by a resolution of the European Parliament.[16]

Although the Communication indicated that the Commission had decided in principle that extending the emissions trading scheme to the aviation sector was preferable to alternative policy instruments, such as emissions taxes or charges, it recognised that the inclusion of the aviation sector in the emissions trading scheme created a number of practical challenges and that further consideration of the key design elements for applying the scheme to the aviation sector would be necessary before a legislative proposal could be made.

The design of the scheme needs to take into account a number of, sometimes competing, objectives. The extension of the scheme should be designed so as to incorporate aviation into the scheme in a way which maximises the environmental effectiveness and economic efficiency of the scheme and without damaging the competitiveness of the EU aviation industry and dependent industries. It must also be consistent with existing legislation: this means that as well as being complementary to the existing EU ETS and international climate change commitments, the manner in which aviation is incorporated into the EU ETS must also be consistent with the international regulation of aviation.[17] Given the global nature of the aviation industry, it is also important that the scheme is capable of expansion to other countries.

---

[13] This was part of a comprehensive package of measures including research into cleaner air transport, better air traffic management and the removal of legal barriers to taxing aircraft fuel.

[14] Available at: http://europa-eu-un.org/articles/fr/article_5400_fr.htm.

[15] Available at: http://www.consilium.europa.eu/ueDocs/cms_Data/docs/pressData/en/ec/87642.pdf.

[16] On 4 July 2006 the European Parliament adopted a resolution welcoming the Commission's Communication and recognising that emissions trading has the potential to play a role in reducing the climate change impact of aviation, available at: http://www.europarl.europa.eu/sides/getDoc.do?Type=TA&Reference=P6 TA 2006 0296&language=EN. The European Economic and Social Committee also adopted an opinion in response to the Commission's Communication. Available at: http://eescopinions.esc.eu.int/EESCopinionDocument.aspx?identifier=ces\nat\nat299\ces598-2006_ac.doc&language=EN.

[17] The principal instrument regulating international civil aviation is the Convention on International Civil Aviation adopted at Chicago on 7 December 1944. This is supplemented by a plethora of bilateral air service agreements. (Member States have concluded around 1,600 such agreements).

The Commission's Communication identified four central issues requiring further consideration before a legislative proposal could be made.

The first related to the entity which should be responsible for complying with the obligations under the scheme. The Communication indicates that the Commission considered that aircraft operators should be the entities responsible as they have the most direct control over the type of aircraft in operation and the way in which they are flown.

The second issue related to the fact that the impact aviation is not limited to its emissions of CO_2. Instead aviation affects the climate through nitrogen oxides, water vapour and sulphur and soot particles. The total impact of aviation on the climate may be between two and four times higher than the effect of its CO_2 alone. The Communication indicated that both the CO_2 and the non-CO_2 impacts of aviation should be addressed to the extent possible. However, there is considerable scientific uncertainty surrounding the non-CO_2 impacts of aviation on the climate. The Communication therefore suggested that pending scientific progress a pragmatic approach would be needed. In the short term it identified the following two possibilities:

– a requirement for aviation to surrender a number of allowances corresponding to its CO_2 emissions multiplied by a precautionary average factor reflecting other impacts; or

– an approach where initially only CO^2 is included, but ancillary instruments are implemented in parallel such as differentiation of airport charges according to NOx emissions.

The third issue concerned the scope of the scheme, that is, which types of flights should be covered by the scheme. The Communication concluded that 'in environmental terms, the preferred option would be to cover all flights departing EU airports, as limiting the scope to intra-EU flights, which both depart and land in the EU, would address less than 40 per cent of the emissions from all flights departing from the EU.'

The final issue identified in the Communication related to the allocation of allowances to the aviation sector. The Communication concluded that 'given the level of integration in the Community's air transport market, a harmonised allocation methodology should be agreed.' Also as international aviation emissions are not covered by the targets under the Kyoto Protocol and therefore are not covered by Member States' assigned amount, it would be necessary to ensure that the inclusion of the aviation sector did not affect the accounting system linking the EU ETS to the international system under the Kyoto Protocol.

These and other technical issues were initially considered in the feasibility study carried out for the Commission prior to the adoption of the Communication.[18] Following the adoption of the Communication, the Commission set up a stakeholder working group[19] to advise the Commission further on these issues. The deliberations of the group are set out in its final report which was published at the end of April 2006.[20]

On the 20 December 2006, the Commission adopted a formal proposal for the inclusion of aviation in the EU ETS. Under the proposal, emissions from flights between EU airports would be subject to a cap and aircraft operators would be required to surrender allowances to cover their emissions from 2011. From 2012 the scope of the scheme would be expanded to cover all flights to and from EU airports. In order to avoid any double counting, the proposal provides that where a third country adopts measures to address the climate change impact of aviation, the EU ETS will not apply to flights arriving from that country.

Key elements of the proposal are:

-   aircraft operators would be the entities responsible for complying with the obligations imposed by the scheme;
-   flights by state aircraft, flights under visual flight rules, circular flights, flights for testing navigation equipment or for training purposes, rescue flights and flights by aircraft with a maximum take-off weight of less than 5 700 kg would be excluded from the scheme;
-   to address other gases, by the end of 2008 the Commission would put forward a proposal to address the nitrogen oxide emissions from aviation after a thorough impact assessment;
-   in order to avoid duplication and an excessive administrative burden on aircraft operators, each aircraft operator, including operators from third countries, would be administered by one Member State only;

---

[18] 'Giving wings to emissions trading – Inclusion of aviation under the European emission trading scheme (ETS): design and impacts', CE Delft, July 2005, available at: http://ec.europa.eu/environment/climat/pdf/aviation_et_study.pdf.
[19] This was part of the second European Climate Change Programme which was launched in October 2005 (see section III.B of the Introduction). The Group was made up of representatives from the aviation industry, sectors covered by the existing scheme and NGOs.
[20] Available at: http://ec.europa.eu/environment/climat/pdf/eccp_aviation_final.pdf. For a further discussion of the design issues, see also Barton, J , 'Tackling Aviation Emissions: the challenges ahead' (2006) *JEEPL* 4, p.316.

- in contrast to the existing scheme, the method of allocating allowances would be harmonised across the Community;
- the total number of allowances to be allocated to the aviation sector would be determined at Community level by reference to average emissions from aviation in the years 2004-2006;
- a fixed percentage of the total quantity of allowances would be allocated free of charge on the basis of a benchmark to aircraft operators which submit an application (the earliest application relating to 2008 data). For the period 2011-2012 this percentage would correspond to the average percentage proposed by the Member States including auctioning in their NAPs. Thereafter this would be reviewed in the light of the results of the general review of the emissions trading scheme;
- the details of how auctioning would work such as appropriate design and timing would be set out in a Commission Regulation. Auctioning proceeds should be used to mitigate and adapt to the impacts of climate change and to cover administrative costs;
- like other participants in the Community scheme, aircraft operators would have to monitor their emissions of carbon dioxide and report them to the competent authority of its administering Member State by 31 March each year. The reports must be verified to make sure that they are accurate. The basic principles for monitoring, reporting and verifying emissions set out in the proposal would be elaborated by guidelines;
- aircraft operators would be able to buy allowances from other sectors in the Community scheme for use to cover their emissions;
- aircraft operators would also be able to use ERUs and CERs up to a harmonised limit equivalent to the average of the limits prescribed by Member States in their national allocation plans for other sectors in the Community scheme;
- domestic aviation would be included in the scheme and treated in the same way as international aviation;
- no specific provisions are made for remote or isolated regions. Special consideration to the treatment of air services to remote or isolated regions, which are particularly dependent on air transport services, can best be given within the framework of existing measures such as public service obligations and aid having a social character under Article 87(2) of the Treaty.

The proposal will now be discussed by the Council and the European Parliament and must be adopted under the co-decision procedure.[21] This process generally takes two to three years.

---

[21] For a brief explanation of the co-decision procedure, see chapter 1, note 18.

PART II

IMPLEMENTATION OF THE EU ETS IN THE UK

CHAPTER 11

EVOLUTION OF THE UK REGULATIONS AND DIVISION OF RESPONSIBILITIES

This chapter provides a brief account of the evolution of the UK Regulations (section I) and an explanation of how the Regulations allocate different functions under the scheme to different public bodies within the UK (section II).

I.   Evolution of the UK Regulations

The Greenhouse Gas Emissions Trading Scheme Regulations 2003[1] (the 2003 ETS Regulations) entered into force on 31 December 2003. These Regulations transposed the Emissions Trading Directive into UK law and provided the framework for the operation of the EU ETS in Great Britain and Northern Ireland.[2] The UK was the only Member State to transpose the Directive by the deadline of 31 December 2003.

The Regulations were amended on 13 January 2005 by the Greenhouse Gas Emissions Trading Scheme (Amendment) Regulations 2004[3] in order to   allow regulators to recover the costs involved in the first year of administering the scheme. The operation of the scheme in the UK is now governed by the Greenhouse Gas Emissions Trading Scheme Regulations 2005 (the ETS Regulations).[4] These entered into force on 21 April 2005 and consolidated and largely replaced the 2003 ETS Regulations and the 2004 amending Regulations (although elements of the 2003 ETS Regulations imposing obligations relating to the first phase allocation process remained in force until those obligations were discharged).[5] The ETS Regulations were further amended by the Greenhouse Gas Emissions Trading Scheme (Amendment) and National Emissions Inventory Regulations 2005[6] which entered into force on 13 November 2005, the Greenhouse Gas Emissions Trading Scheme (Amendment) Regulations 2006 (which entered into force on 6th April 2006)[7] and the Greenhouse Gas Emissions Trading Scheme (Amendment) Regulations 2007 (which entered into force on 16th March 2007).[8] The

---

[1] S.I. 2003/3311.
[2] Separate legislation was subsequently made to give effect to the Directive in Gibraltar.
[3] S.I. 2004/3390.
[4] S.I. 2005/925.
[5] See Regulation 47 of the ETS Regulations.
[6] S.I. 2005/2903.
[7] S.I. 2006/737.
[8] S.I. 2007/465.

2005 amendments transposed the requirements of the Linking Directive[9] permitting the use of Kyoto project credits within the EU ETS. The 2006 amendments put in place a procedure for late entrants to apply for allowances and made a number of other miscellaneous amendments including some relating to permitting and charging.[10] The 2007 amendments enable the Secretary of State to enter into agreements to allocate allowances for the first phase of the scheme to registry account holders in exchange for payment where allowed for by the approved national allocation plan. This gives the government the power to sell any surplus of allowances in the new entrant reserve by auction or sale as envisaged in the NAP for the first phase of the scheme.[11]

The process of transposing and implementing the Emissions Trading Directive included extensive consultation with interested parties.[12]

II.   Division of Responsibilities

The ETS Regulations provide for a number of bodies to be involved in the operation of the EU ETS in the UK. The Regulations make a distinction between the function of setting the framework for the scheme by, for example, developing the UK NAP and the function of managing the day-to-day administration of the scheme.

The function of setting the framework for the scheme is reserved to the Secretary of State. It is the Secretary of State who is required to develop the NAP and to take the final decision on the allocation of allowances.[13] However, to reflect the fact that competence for environmental issues is generally devolved, this must be done in agreement with the administrations for Scotland, Wales and Northern Ireland in so far as it relates to installations in those parts of the UK.[14]

---

[9] Chapter 1, note 2. For a description of the requirements of the Linking Directive, see chapter 8.

[10] See chapter 12.

[11] See section IV of chapter 20.

[12] Consultation on draft implementing regulations took place in September 2003 with consultations on amendments in October and November 2004. In June 2005 a consultation was carried out on draft Regulations to implement the Linking Directive. A consultation on amendments relating to the allocation of allowances to late installations, i.e. those installations identified after allocations for the first phase of the scheme were finalised, was carried out in October 2005. No specific consultation was carried out on the 2007 regulations but consultation took place on the principle of auctioning or sale in August 2003 and July 2005 and on the method of auctioning or sale in April 2005. Consultations are available on the Defra website: http://www.defra.gov.uk.

[13] See Regulations 20 and 21 discussed in section I of chapter 15.

[14] Regulation 46(1). However, if no agreement is reached and the Secretary of State considers that it is necessary to act to ensure that the UK complies with its obligations under the Emissions Trading Directive, the Secretary of State may act without the agreement of the devolved administration concerned (Regulation 46(2) – (8)).

In contrast, the day to day responsibility for administering the scheme, including the issuing and administering of permits and enforcing the requirements of the scheme, is assigned to one of four regulators. The identity of the regulator varies according to the location of the installation:

a)   for installations in England and Wales, the regulator is the Environment Agency;[15]

b)   for installations in Scotland, the regulator is Scottish Environment Protection Agency;[16]

c)   for installations in Northern Ireland, the regulator is the Chief Inspector;[17] and

d)   for offshore installations, the regulator is the Secretary of State.[18]

The regulators are also responsible for allocating allowances to new entrants in accordance with the provisions of the NAP.[19] The functions of the regulators are overseen by the Secretary of State and the devolved administrations by virtue of the fact that appeals against decisions of the regulators are determined by the Secretary of State and those administrations.[20] In addition, the Secretary of State and the devolved administrations may issue directions and guidance to the regulators.[21] A regulator must comply with any direction given to it[22] and must have regard to any guidance issued in carrying out functions under the ETS Regulations.[23] A direction may direct the regulator to exercise any of its powers or to do so in specified circumstances or in a particular manner, or direct a regulator not to exercise a power in particular circumstances or in a particular manner.[24] Directions may be revoked or varied by

---

[15] The Environment Agency is a statutory body created under chapter I of part I of the Environment Act 1995 (c.25).

[16] The Scottish Environment Protection Agency is a statutory body created under chapter II of part I of the Environment Act 1995 (ibid).

[17] The Chief Inspector is constituted under Regulation 8(3) of the Pollution Prevention and Control Regulations (Northern Ireland) 2003, S.R. (NI) 2003 No 46, amended by S.I. 2003/496 (section 2 of the ETS Regulations). Regulation 6 of the ETS Regulations permits the Chief Inspector to delegate any functions conferred or imposed on him by the ETS Regulations to any inspector appointed by him under Regulation 8 of the Northern Ireland Regulations.

[18] Regulation 2.

[19] Regulation 22. For further details see chapter 15.

[20] Regulation 32. For further details see chapter 18.

[21] Regulations 42 and 43.

[22] Regulation 42(4).

[23] Regulation 43(2).

[24] Regulation 42(2).

further directions.[25] These mechanisms give the Secretary of State and the devolved administrations ultimate control over the operation of the scheme in the UK.

Finally the Registries Regulation requires Member States to appoint a registry administrator who will be responsible for operating and maintaining the emissions allowance registry. The Environment Agency is registry administrator for the UK.[26] However, this function is overseen by the Secretary of State who may issue directions and guidance to the registry administrator.[27]

---

[25] Regulation 42(3).

[26] Regulation 26(2). Subject to the exceptions referred to in Regulation 26(4), the regulators will be the 'competent authority' for the purposes of the Registries Regulation (Regulation 26(3)). The Secretary of State will also act as the relevant body as provided in Regulation 26(5).

[27] Regulations 44 and 45. It is also the Secretary of State who is responsible for establishing a registry consistent with article 19 of the Directive and the Registries Regulation (Regulation 26(1)). These powers are the same as the Secretary of States powers to give directions and guidance to regulators described above.

CHAPTER 12

PERMITTING: APPLICATIONS AND GRANT OF PERMITS

This chapter considers the provisions of the ETS Regulations which govern applications for permits, the grant of permits, and the conditions to be included in permits. Chapter 13 considers the variation and transfer of permits and chapter 14 considers the surrender and revocation of permits.

The provisions of the ETS Regulations relating to the permitting of installations closely reflect the provisions for permitting under the UK implementation of the IPPC Directive.[1]

I.   Obligation to Hold a Permit

The ETS Regulations impose a general prohibition on carrying out an activity listed in schedule 1 to the Regulations and resulting in specified emissions, except under and to the extent authorised by a greenhouse gas emissions permit.[2]

In order to determine whether a permit is required an operator must consider, firstly, whether the activity which it carries out is an activity listed in schedule 1 to the ETS Regulations. If so and if that activity results in emissions specified in respect of that activity, then a permit will be required.

The activities listed in schedule 1 exactly reflect the activities[3] and thresholds listed in Annex I of the Emissions Trading Directive and, like

---

[1] This was implemented in respect of England and Wales by the Pollution Prevention and Control Regulations 2000 (S.I. 2000/1973), in respect of Scotland by the Pollution Prevention and Control (Scotland) Regulations (SSI 2000/323), in respect of Northern Ireland by the Pollution Prevention and Control (Northern Ireland) Regulations 2003 (supra, chapter 11, note 17), and in respect of offshore installations by the Offshore Combustion Installation (Prevention and Control of Pollution) Regulations 2001 (SI 2001/1091). These Regulations were amended by Regulation 38 of and schedule 5 to the 2003 Regulations as required by Article 26 of the Emissions Trading Directive. For a description of the requirements of the Directive, see section IV of chapter 3.

[2] Regulation 7.

[3] As explained in chapter 2 above, in the first phase of the scheme there has been a divergence of views about the scope of the term 'combustion installation'. The interpretation applied by the United Kingdom in the first phase of the scheme is set out in the EU Emissions Trading Scheme: Guidance Note 1(see chapter 2, note 8). This provides that 'the term 'combustion installation' refers to a stationary technical unit which burns

(continued...)

the Directive, only carbon dioxide emissions are currently specified.[4] In relation to the first phase of the scheme, therefore, the UK did not take advantage of the possibility under Article 24 of the Directive for Member States to lower the thresholds for any of the activities included in Annex I of the Directive in the first phase of the scheme. If, in relation to the second scheme phase or a later scheme phase, the Directive is amended to extend the activities listed in Annex I or the UK decides to exercise the discretion in Article 24 to allow for the inclusion of additional activities, installations or gases, an amendment to schedule 1 of the ETS Regulations would be required.

In order to determine whether a particular activity is a schedule 1 activity, two specific rules are set out in the ETS Regulations:

a)   First, activities 'carried out for research, development or testing of new products or processes' are not considered to fall within schedule 1. It follows that no permit is required to carry out these activities.[5]

b)   Secondly, where more than one activity is carried out in the same stationary technical unit or in different stationary technical units on the same site,[6] an aggregation rule is

---

fuel for the production of an energy product. The energy product could be electricity, heat or mechanical power. Where energy is produced as heat it may be transferred using different media such as steam, hot oil, hot water and hot air. If the energy is produced and used within the same technical unit such that the main product of the unit is not an energy product (electricity or heat) then the technical unit is not considered to fall within the definition of 'combustion installation'. The Guidance note was amended in March 2006 to provide that the definition of 'combustion installation' will be broadened in the light of the 2005 Commission Guidance. From the second phase, the scheme will include additional sources of CO_2 emissions under the definition of combustion installation (namely, emissions from glass, mineral wool, integrated steel works, petrochemical, crackers, flaring from offshore oil and gas production, gypsum and carbon black).

[4] Neither the Regulations nor the Directive specify how the capacity is to be assessed for the purpose of determining whether an activity is over the threshold, however there are good reasons for concluding that this should be judged by reference to the potential capacity rather than using an historical average or current operating level. This is the approach taken in the EU Emissions Trading Scheme: Guidance Note 1 (supra Chapter 2, note 8) which concludes that stand-by generation or boiler capacity should be included in aggregation for a combustion installation 'provided that it is technically feasible for stand-by generators or boilers to be run concurrently with the main generators or boilers on site'.

[5] Paragraph 2 of part 2 of schedule 1.

[6] There is no definition of 'site' in the Emissions Trading Directive or the ETS Regulations. EU Emissions Trading Scheme: Guidance Note 1 (supra chapter 2, note 8) provides that 'same site' means 'the same location or situation and is a question of judgement for each installation. A site does not necessarily become two sites merely because two parcels of land are separated by a physical barrier such as a stream. Two parcels of land do not need to touch physically to form part of the same site provided they are technically connected.'

applied. This provides that the capacities of each part or unit will be added together and used for the purposes of determining whether each activity falls within a description in schedule 1.[7]

II. Applications for a Greenhouse Gas Emissions Permit

An application for a greenhouse gas emissions permit must be made in respect of an 'installation'. 'Installation' is defined in the ETS Regulations as:

(i) a stationary technical unit where one or more schedule 1 activities are carried out; and

(ii) any other location on the same site where any other directly associated activities are carried out which have a technical connection with the activities carried out in the stationary technical unit and which could have an effect on greenhouse gases and pollution.

The ETS regulations provide for an application for a permit to be made by 'the operator' of the installation. 'Operator' is defined as the person who has control over the operation of the installation.[8] Whether a person is the operator is a question of fact to be assessed in each case. However, the following factors are identified in Government Guidance[9] as being potentially relevant to such assessment:

a) who has day to day control of plant operation including the manner and rate of operation;

b) who will have the ability to ensure that permit conditions will be complied with;

c) who hires and fires key staff;

d) who makes investment decisions;

e) who ensures that operations are shut down in an emergency; and

---

[7] Paragraph 3 of part 2 of schedule 1.

[8] Regulation 2(1) and (2)(a). This definition is phrased differently from the definition in Article 3(f) of the Directive (which refers to the person who 'operates or controls' the installation) to provide greater certainty in the definition since the person who operates and the person who controls the installation could be different. The ETS Regulations do not make express use of the possibility under Article 3(f) of the Directive to define 'operator' as a person 'to whom decisive economic power over the technical functioning of the installation has been delegated'. However, the question of who has economic power over the installation will be a relevant consideration when determining who is the operator of the installation.

[9] EU Emissions Trading Scheme: Guidance Note 1(supra chapter 2, note 8).

f) who is the operator for the purposes of other relevant environmental permits or agreements in respect of the installation e.g. pollution prevention and control permits or climate change agreements.

It is possible that there could be more than one operator of an installation, each running distinct parts of the installation. For example, an installation producing paper may consist of a paper mill and a combined heat and power (CHP) plant supplying that paper mill with steam. The operator of the paper mill may be different from the operator of the CHP plant. In such circumstances, a separate permit application may be made in respect of each part of the installation.[10] It is arguable whether a part of an installation which carries out a schedule 1 activity but does not result in any greenhouse gas emissions requires a separate permit. There are arguments that in such circumstances no permit would be required because the operator of that part of the installation would not be carrying out an activity resulting in specified emissions.

An application for a greenhouse gas emissions permit must be made to the regulator and must include the information required by Regulation 8(2) and (3). This includes the information required to be included in applications by Article 5 of the Emissions Trading Directive as well as some additional domestic requirements. In particular the application must include a description of the installation. This includes not only the schedule 1 activities carried out in the installation but also any directly associated activities.[11] In practice the regulators have developed a standard application form for permits which must be used by persons wishing to apply for permits.[12] Except in the case of offshore installations,[13] the application must be accompanied by the fee prescribed in relation to applications.[14]

---

[10] The definition of 'installation' in Regulation 2 of the ETS Regulations provides that 'references to an installation include references to part of an installation'.

[11] Guidance on how to identify directly associated activities can be found in EU Emissions Trading Scheme: Guidance Note 1 (supra chapter 2, note 8).

[12] See, for example, the application for a permit in respect of installations in England and Wales available on the Environment Agency website http://www.environment-agency.gov.uk. Regulation 5(1) permits the regulator to require an application to be made on a form developed by the regulator. The development of specific forms for applications is not compulsory but any form made available by the regulator for that purpose must specify the information required by the regulator to determine the application, including the information required to be included by the specific provision of the ETS Regulations (in this case Regulation 8) (Regulation 5(2)). Where a regulator makes available a form, Regulation 5(3) requires applications to be made on that form.

[13] Where an application relates to an offshore installation, no fee is required to be paid at the time of the application. Instead liability to pay the fee is incurred when the permit is granted and will be payable within 28 days of the regulator serving a notice on the operator requesting payment (Regulation 9(5) and paragraph 1(1)(b) of schedule 5). Regulation 9(5) also contains a transitional provision relating to payment of fees on grant of a permit where the application was made prior to the ETS Regulations entering into force.

[14] Regulation 8(1). Paragraph 1(1)(a) of schedule 5 sets out the fees prescribed in relation

(continued...)

Applications may be made electronically.[15] Where an application is made electronically, the fee may be sent separately but the application will not be treated as having been received until the fee has been received.[16]

An application may be withdrawn by the applicant at any time before it is determined.[17]

III.   Requests for Further Information

Once an application has been submitted to the regulator, the first step is for the regulator to consider whether the application has been 'duly made'.[18] The ETS Regulations do not contain an express definition of the term 'duly made'. However, it follows from the context that it means an application that has been made in accordance with the procedural and substantive requirements of the Regulations. A regulator could reject an application as not 'duly made' which does not contain the information required by Regulation 8(2), is not made on the form made available by the regulator for making applications or (except where an application relates to an offshore installation) is not accompanied by the fee prescribed in relation to application for permits.

Alternatively, where the missing information is capable of being identified in a notice, the regulator could treat the application as having been duly made and serve a notice on the applicant requesting him to furnish the further information needed to determine the application.[19] A request for further information must specify the additional information which is required and the period within which the information must be furnished. If the applicant does not furnish the information within the specified period, the regulator may repeat the request or notify the applicant that it is treating the application as having been withdrawn.

---

to an application for a permit. These are: for an installation emitting less than 50kt per year, £1230; for an installation emitting between 50 and 500kt per year, £2300 and in relation to an installation emitting more than 500kt per year, £5490. These fees may be superseded by charging schemes developed by the regulators (see Regulation 18(3) to (5)). Charging schemes may be developed in relation to installations in England and Scotland under the Environment Act 1995 (see sections 41, 41A, 42 and 56 as amended by schedule 6 to the ETS Regulations) and in relation to offshore installations under Regulation 19 of the ETS Regulations (see also amendments to the Pollution Prevention and Control Act 1999 in schedule 6).

[15] Regulation 5(4).
[16] Regulation 5(5).
[17] Regulation 5(6).
[18] Regulation 9(4).
[19] Regulation 8(5).

IV.   Determination of Applications

A.   Determination period

Once an application has been made, the regulator has two months to determine the application, although this period can be extended if agreed in writing by the regulator and the applicant.[20] Where a formal request for further information is made the two month period does not include the period from the date of the request to receipt of the requested information.[21] If the regulator fails to determine the application within the two month period, or such longer period as may have been agreed with the applicant, the applicant may at any point notify the regulator that it is treating the application as having been refused.[22] This would enable the applicant to appeal against the refusal.[23] If the applicant does not expressly notify the regulator that it wishes to treat the application as having been refused, the determination period simply continues until the regulator determines the application.

Once the regulator has all the information necessary to consider an application, it must determine the application. It may either grant the permit subject to conditions or refuse the application.[24]

B.   Refusal of a Permit

The ETS Regulations set out two express circumstances in which the regulator is required to refuse an application. Both are aimed at ensuring that any person to whom a permit is granted is able to comply with the obligations in the permit.

Firstly, the regulator must refuse an application where it considers that the applicant will not be the operator of the installation after the grant of the permit.[25] This requirement seeks to ensure that the permit holder is the person who has control over the operation of the installation and is therefore best placed to ensure compliance with the conditions of the permit and has the ability to make decisions affecting the emissions from the installation.

---

[20] Regulation 9(1). It should be noted that this period runs from the date on which the application is submitted and not from the date on which the regulator determines that the application has been 'duly made'.

[21] Regulation 9(2). This suspension only relates to the calculation of the period of two months provided for in Regulation 9(1) of the ETS Regulations. If a longer period were agreed, it would be necessary for a suspension where a request for further information is made to be agreed as part of the extended period.

[22] Regulation 9(3).

[23] See chapter 18 below.

[24] Regulation 9(4).

[25] Regulation 9(6)(a).

Secondly, the regulator must refuse an application where it is not satisfied that the applicant will ensure that the installation is operated so as to comply with the monitoring and reporting conditions in the permit.[26] This requires an assessment not only of whether the operator is capable of monitoring emissions in accordance with the permit but also the likelihood of compliance.[27] Examples of situations where an application might be refused on this ground include where the regulator is not satisfied that the operator has the technological capability to monitor emissions as required by the Monitoring and Reporting Decision or where the applicant has a poor compliance history.

There is, at least, a theoretical possibility that a regulator could reject an application in other circumstances. However, in practice, the majority of situations in which a regulator would not consider it appropriate to grant a permit is likely to either be sufficient to justify a rejection of the application on the grounds that the application is not duly made or fall within one of the express grounds for refusal.

C.   Grant of a Permit

If a regulator does not refuse an application, it must grant a permit to the operator. The permit must contain the information required by Regulation 9(8) as well as the conditions required by Regulation 10. Although the permit is required to contain a description of the installation, the conditions of the permit will only relate to the schedule 1 activities carried out in the installation resulting in specified emissions. This reflects the conceptual structure of the Emissions Trading Directive described in chapter 3 above.

D.   Conditions in Permits

The ETS Regulations set the framework for the conditions to be included in the permit but leave the detail of the conditions to be determined by the regulator.[28]

First the permit must contain conditions relating to the monitoring and reporting of emissions including, in particular conditions to ensure that:

---

[26] Regulation 9(6)(b).
[27] See *R v Secretary of State for the Environment and NC Compton, ex parte West Wiltshire District Council* [1996] Env LR 312 (QBD) which concerned the interpretation of a similar provision in section 6(4) of the Environmental Protection Act 1990.
[28] Regulation 10(1) provides for a greenhouse gas emissions permit to include 'such conditions as the regulator considers appropriate and in particular such conditions as the regulator considers appropriate to comply with paragraphs (2) to (6)'.

a) specified emissions from the installation are monitored in accordance with the Monitoring and Reporting Decision;[29]

b) the total specified emissions which arise from the installation each year (the 'annual reportable emissions') are recorded in a report to the regulator in accordance with the Monitoring and Reporting Decision;[30] and

c) all reports are verified in accordance with the requirements of Annex V of the Emissions Trading Directive.[31]

The permit is also required to include a condition requiring the operator to notify the regulator as soon as possible if he becomes aware of any factor which might prevent him from complying with any of the monitoring and reporting conditions included in the permit.[32] This requirement is intended to ensure that regulators are made aware at the earliest opportunity if an operator is facing difficulties with the requirement to monitor and report emissions and can consider the appropriate enforcement action to take to bring the operator into compliance.

In practice the regulators tend to issue permits which require an operator to monitor and report emissions in accordance with a 'monitoring and reporting plan'. This is a plan prepared by the operator setting out the detailed requirements for the monitoring and reporting of emissions in accordance the Monitoring and Reporting Decision. The plan must either be included in the application and included in the permit or, where an application relates to an installation which will not be put into operation for some time, be submitted to the regulator at least two months prior to the installation being brought into operation.[33] In the latter case, the plan must be approved by the regulator before the installation is put into operation.

Permits also normally include additional conditions to give effect to other requirements of the Monitoring and Reporting Decision. These require operators to retain information relating to emissions and to submit it to the regulator by 30 June each year a report justifying the continued use of anything less than the highest tier set out in the Monitoring and

---

[29] Regulation 10(2)(a)(i). For further information on the Monitoring and Reporting Decision (supra chapter 3, note 9), see section II of chapter 3. The Regulations will be amended in relation to the second phase of the scheme to refer to the Modified Monitoring and Reporting Decision (supra chapter 3, note 10).
[30] Regulation 10(2)(a)(ii).
[31] Regulation 10(2)(b).
[32] Regulation 10(2)(c).
[33] A similar procedure applied for permits issued prior to the start of the scheme.

Reporting Decision and identifying any potential improvements to the monitoring and reporting requirements.

Secondly, the permit must contain conditions to ensure that the operator surrenders allowances equal to his annual reportable emissions from the installation within 4 months of the end of the scheme year during which the emissions arose (i.e. by end of April in the year following the year in which the emissions arose).[34] This requirement to surrender allowances is discussed further in chapter 14.

In addition to these key provisions, regulators have a discretion to include in permits 'such conditions as [they] consider appropriate'.[35]

E.   Review of Permit Conditions

Regulators are required to review the conditions of permits periodically and are given the power to carry out a review at any time.[36] The ETS Regulations do not prescribe the procedure for or substance of such review. However, to comply with the obligation, the regulator would need to consider, from time to time, whether the requirements of the conditions of the permit are still appropriate to comply with the requirements of the ETS Regulations and the reasons for including any additional requirements still stand.

The majority of the conditions required to be included in a permit are standard for all permits and could be reviewed on a universal level and any changes incorporated into the permits for all installations. However, some, such as the monitoring and reporting requirements, will require a review at the level of the individual installation.

The extent of the review required for the regulator to be satisfied that the conditions of the permit remain appropriate will vary depending on the installation in question (taking into account such matters as its size and the complexity of monitoring requirements included in the permit, its regulatory history and the length of time since the conditions of the permit were previously reviewed).

F.   Special Provisions for Operators of More than One Installation

As discussed in chapter 3 above, Article 6(1) of the Emissions Trading Directive provides that 'a greenhouse gas emissions trading permit may

---

[34] Regulation 10(3).
[35] Regulation 10(1).
[36] Regulation 10(9).

cover one or more installations on the same site operated by the same operator'. Therefore the ETS Regulations allow for the possibility of a permit covering more than one installation.[37] There are two ways in which an operator could obtain a permit covering more than one installation.

Firstly, it could apply for such a joint permit. Where an operator makes a joint application, the application must contain the information required by Regulation 8(2) for each installation covered by the application.[38]

Secondly, the regulator may replace the separate permits applying to installations on the same site operated by the same operator with a consolidated permit applying to the same activities and subject to the same conditions.[39]

In practice, it is likely that an operator operating more than one installation on the same site would wish to have a joint permit since the operator would then only be liable to pay one subsistence charge.[40]

G.   Special Provisions for Temporarily Excluded Installations

As described in chapter 2 above, Article 27 of the Emissions Trading Directive provides for the possibility of certain installations being temporarily excluded from the first phase of the Emissions Trading Scheme.[41] The Commission has approved two types of application for the temporary exclusion of installations in the UK, relating to installations covered by direct participant agreements for the UK Emissions Trading Scheme and Climate Change Agreements respectively.[42]

Where the Commission makes a Decision providing for the temporary exclusion of an installation under Article 27 of the Directive, the Secretary of State is required to publish the Decision.[43] An operator of an installation covered by the Decision then has 2 months[44] to apply for a certificate of temporary exclusion.[45]

---

[37] Regulation 9(7).
[38] Regulation 8(4).
[39] Regulation 9(9).
[40] For a description of subsistence charges, see subsection H below.
[41] For a description of the Directive provisions relating to temporary exclusion, see section VI of chapter 2.
[42] See Commission Decisions, supra chapter 2, note 30. For an account of the UK ETS, see chapter 22 and for an account of CCAs, see chapter 23.
[43] Regulation 11(2).
[44] The Secretary of State or the relevant devolved administration may accept applications after this date (Regulation 11(4)). In considering whether to exercise this discretion, the Secretary of State will need to consider whether the delay in applying for a temporary

(continued...)

A certificate of temporary exclusion identifies the installation, its operator and regulator, states the date from which the installation is excluded and the duration of the exclusion[46] and specifies any conditions applying to the exclusion. The conditions included in the certificate include a requirement that the installation continues to be covered by the national policy which justifies its exclusion. Further conditions may be determined by the terms of the Commission decision providing for the exclusion of the installation.[47]

Holding a certificate of temporary exclusion does not exempt the operator from the requirement under the ETS Regulations to hold a permit.[48] However, permits are required to include two specific provisions for installations which are covered by a certificate. The first of these provisions deems the operator to be in compliance with the conditions of the permit requiring the operator to monitor and report emissions and surrender allowances to cover emissions for the duration of the period of temporary exclusion.[49] The second of the provisions deems the permit to authorise any changes in operation made to the installation during the period for which it is temporarily excluded.[50] However, this deemed authorisation only applies for the period of temporary exclusion and therefore operators will need to apply for a variation of the permit before the end of the period of temporary exclusion. In any event, the operator is required to notify the regulator of any changes in operation during the period of temporary exclusion at least 2 months before the end of the exclusion period.[51]

---

exclusion certificate has led to allowances being issued which would not have been issued if the application had been made on time.

[45] Regulation 11(1) and (3). An application must contain the information required by Regulation 11(5). An application must be made to the 'responsible authority' i.e. the person who is responsible for the national policy by virtue of which the European Commission has provided for the temporary exclusion of the installation (Regulation 11(12) and (13)). For policies covering the whole of the UK, such as the UK Emissions Trading Scheme and Climate Change Agreements, this is the Secretary of State. If an application for temporary exclusion were to be made in relation to a policy developed by one of the devolved administrations, the relevant devolved administration would be the responsible authority and would be responsible for issuing certificates.

[46] Regulation 11(6). Where, as was the case in relation to the exclusion of installations covered by climate change agreements, the certificate of temporary exclusion is not issued until after the start of the scheme phase, the certificate may, if the Commission decision so provides, provide for the exclusion to be backdated (Regulation 11(7)). This avoided operators being required to monitor and report emissions and surrender allowances for emissions in the period prior to the issue of the certificate.

[47] A condition of the temporary exclusion of installations covered by climate change agreements was that no allowances were issued to excluded installations, and that the total quantity of allowances for the first phase was reduced accordingly. Regulation 27(8) requires the Secretary of State to take such steps as are necessary to ensure that this happens.

[48] Regulation 7.

[49] Regulation 10(6).

[50] Regulation 10(7).

[51] Regulation 10(6)(b).

If an operator fails to comply with the conditions specified in the certificate of temporary exclusion, the certificate may be revoked.[52] Where a certificate is revoked the operator will have to monitor and report emissions and surrender allowances in accordance with the conditions of its permit from the date of the revocation. He will also be required, within 10 days of the date of the revocation, to notify the regulator of any changes in operation during the period of temporary exclusion.[53]

H.   Subsistence Charges

Operators holding a permit are required to pay an annual subsistence charge in respect of the permit.[54] The fee covers the costs of administering the scheme including enforcement and compliance costs. It also contains an element to cover the registry administrator's costs relating to the subsistence of the operator's holding account in the registry.[55]

However, the following installations are exempt from the subsistence charge:[56]

a)   installations which are covered by a temporary exclusion certificate;

b)   installations in respect of which a retention notice has been served;[57] and

c)   planned installations, that is to say installations which have not been put into operation. In order to qualify for this exemption the intended operator of the installation must notify the regulator by 1 April each year that he does not intend to put the installation into operation in the following financial year.[58] If the installation is brought into operation during the financial year the operator is required to notify the regulator within 14 days of the date on which the installation is put into operation and the exemption will no longer apply.[59]

---

[52] Regulation 11(9).
[53] Regulation 10(6)(b).
[54] Regulation 18(1).
[55] Paragraph 14 of schedule 5 to the ETS Regulations requires the registry administrator to notify the regulator of its charges and for the regulator to pass on to the registry administrator any operator charges it receives.
[56] Paragraph 6 of schedule 5.
[57] For a description of retention notices, see section I.A of chapter 16.
[58] Regulation 13(1). 'Planned installation' is defined in paragraph 15 of schedule 5 as 'an installation in respect of which an operator has notified the regulator under Regulation 13(1)'. In the year in which the permit is issued, this notice must be given within 14 days of the date on which the permit is granted (Regulation 13(1)(b)). In relation to the financial year 2005-6 the notice was required to be given within 14 days of the entry into force of the ETS Regulations (Regulation 13(1)(c)).
[59] Regulation 13(2).

The amount of the subsistence charge is set out in paragraphs 4 and 5 of schedule 5 to the ETS Regulations.[60] The amount of the subsistence charge depends on the total number of installations covered by the scheme on 1 April each year[61] and the estimated annual emissions of the installation.[62] Each year the Secretary of State is required to calculate and publish the total number of installations covered by the scheme and the subsistence charges which apply.[63] The regulator will then serve a notice on each operator specifying the charge applying to the installation and requesting payment.

As a general rule, payment of the full charge must be made within 28 days of that notice.[64] However, operators may opt to pay the subsistence fee in quarterly instalments.[65] Payment will then be due on the first day of each quarter or within 28 days of the notice from the regulator specifying the charge applying to the installation in that financial year.

Where a permit is granted during the financial year, an installation ceases to be covered by one of the exemptions described above or is the subject of a partial transfer, the subsistence charge will be payable calculated pro-rata to the nearest day.[66] Likewise where a permit is surrendered or revoked during the financial year or one of the exemptions starts to apply, the operator will receive a refund of the amount paid pro-rata to the nearest day.[67]

---

[60] In the longer term, the regulators have the power to develop charging schemes which will supersede the charges provided for in the Regulations (supra note 14).

[61] 'Total number of installations' is defined in paragraph 15 of schedule 5 to exclude planned installations, installations which are temporarily excluded from the scheme and installations covered by a retention notice in accordance with Regulation 24.

[62] See paragraphs 4, 5 and 6 of schedule 5 to the ETS Regulations. The Regulations adopted this approach because at the time of making the Regulations it was not known whether the UK's application under Article 27 of the Directive for the temporary exclusion of installations covered by Climate Change Agreements would be accepted by the European Commission. The acceptance of the application substantially affected the number of installations in the scheme and therefore affected the costs per installation of administering the scheme. The approach adopted in the Regulations will also allow for the amount of the subsistence charge to reflect the entry into the scheme of temporarily excluded installations in 2007 and 2008.

[63] Paragraph 7 of schedule 5.

[64] Paragraph 10 of schedule 5.

[65] Paragraphs 11 and 12 of schedule 5.

[66] Paragraph 9 of schedule 5.

[67] Paragraph 13 of schedule 5.

CHAPTER 13

VARIATION AND TRANSFER OF PERMITS

This chapter considers the provisions in the ETS Regulations relating to the variation of greenhouse gas emission permits (section I) and the transfer of greenhouse gas emission permits (section II).

I.   Variation of Greenhouse Gas Emissions Permits

The most common type of change requiring a variation to a permit is a change in the activities carried out in an installation. Before making a change to the operation of the installation, an operator will need to consider whether the proposed change must be notified to the regulator or whether the change would require a variation of the permit.

A.   Notifying Changes in Operation

Unless the installation is an excluded installation, an operator is required to notify the regulator at least 14 days before making a 'change in operation'. A 'change in operation' is defined as:

> a change in the nature, functioning or scope of the installation which-
> (i)   affects any information included in the greenhouse gas emissions permit pursuant to Regulation 9(8)(d) [that is to say, the description of the installation, the Schedule 1 activities to be carried out in the installation and the specified emissions from those activities]; or
> (ii)   might, in the opinion of the regulator, require any monitoring and reporting condition to be amended[1]

Where an operator proposes to make a change to the installation which does not fall within this definition, the change may be made without notifying the regulator. An example of such a change might be the employment of additional staff or an increase in the operating hours of the installation.

A notification of a proposed change must be sent in writing to the regulator at least 14 days before the change in operation is effected and must describe the proposed change including whether and, if so, why the change falls within the definition of a change of operation.[2] The regulator is required to acknowledge receipt of the application.[3]

---

[1] Regulation 2(1).
[2] Regulation 12(1) and (2).
[3] Regulation 12(3).

In most cases a change falling within the definition of 'change in operation' will also be a change necessitating a variation of the provisions of the permit. An example of such a change might be the extension of the capacity of the installation beyond the capacity authorised in the permit. In such circumstances, the operator may apply to the regulator for a variation of the provisions of the permit.[4] Where an application for a variation of the permit is made, the operator does not have to make a separate notification of the change in operation.[5] The requirement to notify a change in operation does not apply to changes to installations covered by a temporary exclusion certificate where the change is made more than 2 months before the end of the period for which the installation is excluded from the scheme.[6]

B.   Applications for Variation

If a change in operation is likely to require a variation to the provisions of a permit, the operator may apply directly for a variation of its permit. A variation can be made to any provision of the permit, including the description of the installation.

An application must contain the information required by Regulation 14(5)[7] which includes:

a)   a description of the proposed change in operation and a statement of any changes to the information provided as part of the application for the permit under Regulation 8(2)(c) to (f);[8] and

b)   an indication of the variations to the provisions of the permit which the operator would like.

Where the regulator has developed a standard form for applications for variation, the application must be made on that form.[9] The application must, except in the case of an application relating to an offshore installation,[10] be accompanied by the prescribed fee.[11] An application

---

[4] Regulation 14(2).
[5] Regulation 12(4)(b) disapplies the requirement to notify a change in operation where an application for a variation of the permit is made in respect of the change.
[6] Regulation 12(4)(a).
[7] Regulation 14(3).
[8] Namely, the description of the installation and the activities to be carried out, the raw and auxiliary materials used for carrying out those activities, the sources of specified emissions and the measures planned to monitor and report emissions.
[9] Regulation 5(1) – (3). See section II of chapter 12 above.
[10] Offshore installations will be required to pay the fee within 28 days of a notice requesting payment (see Regulation 14(4)).
[11] Regulation 14(3). The fee prescribed in respect of an application for the variation of the provisions of a permit is £240 (see paragraph 1(d) of part 1 of schedule 5 to the ETS
(continued...)

may be made electronically.[12] Where the application is made electronically, the fee may be sent separately but the application will not be treated as having been received until the fee is received.[13]

Where the regulator needs further information to determine an application for a variation, the regulator may serve a notice on the operator requesting further information.[14] The notice must specify the further information required and the time period for furnishing the information. If the operator fails to supply the information within the time specified in the notice, the regulator may serve a notice on the applicant indicating that the application is being treated as having been withdrawn.

Although changes in the operation of installations which are the subject of a temporary exclusion certificate are deemed to be authorised by the permit for the duration of the exclusion period,[15] an operator may still apply for a variation of its permit.[16] This will be necessary if the operator wishes to apply for an allocation of allowances from a new entrant reserve.[17]

Where an application is duly made,[18] the regulator must determine the application and give notice to the operator within two months or such longer period as is agreed with the applicant.[19] If the regulator fails to give notice of its determination of an application within two months (or any longer period agreed between the operator and the regulator), the operator may notify the regulator that it is treating the application as refused.[20] This will enable the operator to appeal against the refusal.[21]

C.   Variations Initiated by the Regulator

In addition to providing for permits to be varied on application by the operator, the ETS Regulations also provide for the possibility of variations being initiated by the regulator.[22] Such variations may be made

---

Regulations). This fee may be superseded by a charging scheme adopted by the regulators (supra. chapter 12, note 14).

[12] Regulation 5(4).

[13] Regulation 5(5).

[14] Regulation 14(6).

[15] Regulation 10(7).

[16] Regulation 10(8).

[17] Regulation 22(2). For a description of the procedure relating to the allocation of allowances from the new entrant reserve, see section II of chapter 15.

[18] For a discussion of the term 'duly made', see section III of chapter 12 above.

[19] Regulation 14(7). For the purposes of calculating the two month period, no account is taken of any period in which there was an outstanding request for further information under Regulation 14(6) (Regulation 14(8)).

[20] Regulation 14(12).

[21] For further information on appeals, see chapter 18.

[22] Regulation 14(1).

at any time and may include a variation of the extent to which a permit authorises a schedule 1 activity.

The regulator is required to vary the permit where it becomes apparent to it that the ETS Regulations require an amendment to the description of the installation in the permit or the conditions of the permit.[23] A regulator may become aware that different provisions are required through the periodic reviews of the conditions included in permits; a notification of a proposed change in operation; or enforcement steps taken by the regulator.

D.   Notice of Variation

Where a regulator decides to vary the provisions of a permit (whether as result of an application or on its own initiative), it must serve a notice on the operator specifying the variations of the provisions of the permit and the date or dates on which the variations will take effect.[24] Where the variation is initiated by the regulator, the notice will require the operator to pay, within the period specified in the notice, the fee prescribed in respect of the variation notice.[25]

Where a regulator refuses an application for variation of a permit, it must notify the operator.[26]

Variations to a permit will take effect on the date specified in the notice unless the variation is withdrawn.[27]

E.   Minor Changes

The fees applicable to variations do not apply where the regulator considers that the variation relates to a minor change.[28] It may be difficult to identify in advance whether a change will be considered to be minor as this requires a judgment in each case based on the regulatory effort involved in making the change. As the question of whether a change is minor is to be judged by the regulator, a operator would be advised to contact the regulator before submitting an application for variation without a fee to avoid the risk that the regulator would reject the application as not having been 'duly made'.

---

[23] i.e. where Regulations 9(8) or 10 require provisions to be included in the permit which are different from the subsisting provisions.
[24] Regulation 14(9).
[25] Regulation 14(10). The fee prescribed for this purpose is £240 (paragraph 1(1)(e) of part 1 of schedule 5). This fee may be superseded by a charging scheme adopted by the regulators (supra chapter 12, note 14).
[26] Regulation 14(11).
[27] Regulation 14(9).
[28] Paragraph 1(1)(d) and (e) of part 1 of schedule 5 to the ETS Regulations.

F. Variations of a Purely Administrative nature

The procedure for variation of the permit can also be used to vary administrative details of the permit such as the contact details of the permit holder (although not to transfer the permit to another operator, on which see section II below). The fees applicable to variations do not apply where the regulator considers that the variation relates to a change of a purely administrative nature.[29]

II. Transfer of Permits

Where control of an installation passes to another operator the new operator will be required to hold a permit in respect of the activities carried out in the installation.

Except where a permit relates to an installation which no longer carries out a schedule 1 activity,[30] an operator may transfer all or part of his permit to another person. This enables a purchaser to take over the operation of an installation, or part of an installation, without having to apply for a new greenhouse gas emissions permit.

There are two types of transfer:

a) A full transfer takes place where the whole installation is transferred to another person. After the transfer only the new owner needs to hold a permit in respect of the installation and therefore the whole permit is transferred to the new operator.

b) A partial transfer takes place where an operator transfers only part of his installation to another person. In this case, both the original operator and the owner of the part transferred will be carrying out a schedule 1 activity resulting in specified emissions. Each will require a greenhouse gas emissions permit, and so only part of the permit is transferred.

A. Procedure for Transfer

Applications to transfer a permit must be made to the regulator and must be made jointly by the operator and the person to whom the permit, or part of the permit, is to be transferred.[31] Where the regulator has

---

[29] Paragraph 1(1)(d) and (e) of part 1 of schedule 5 to the ETS Regulations.
[30] Regulation 15(2).
[31] Regulation 15(1).

developed a standard form for applications for transfers, the application must be made on that form.[32] The applications must contain the information specified in Regulation 15(3) or, in the case of a partial transfer, Regulation 15(4). Except where the application relates to an offshore installation,[33] the application must be accompanied by the transfer fee.[34] An application may be made electronically.[35] Where the application is made electronically, the fee may be sent separately but the application will not be treated as having been received until the fee is received.[36]

If an application contains insufficient information to enable the regulator to determine it, the regulator may serve a notice on the operator requesting further information.[37] The notice must specify the information required and the time limit for furnishing the information. If the operator does not furnish the additional information within the time limit specified in the notice, the regulator may treat the application as having been withdrawn.[38]

The regulator is not required to determine applications for transfer within any specified timeframe. However, if after a period of two months (which does not include the time during which a request for further information is outstanding)[39] or such longer period as agreed by the regulator and the operator, the regulator has not either effected the transfer or refused the application, the applicants may notify the regulator that they are treating the application as having been refused,[40] giving them a right to appeal.[41]

The regulator must effect a transfer unless one of the two grounds for refusal specified in the ETS Regulations applies.[42] The regulator must reject an application if it considers that the person to whom the permit is to be transferred will not:

---

[32] Regulation 5(1) – (3). See section II of chapter 12 above.

[33] Regulation 15(5) provides that the fee in respect of an application relating to an offshore installation will be payable within 28 days of the regulator serving a notice on the operator requesting payment.

[34] The fee prescribed in paragraph 1(1)(f) of part 1 of schedule 5 in respect of a transfer is £240. This fee may be superseded by a charging scheme adopted by the regulators (supra. chapter 12, note 14).

[35] Regulation 5(4).

[36] Regulation 5(5).

[37] Regulation 15(10).

[38] Regulation 15(11)(b).

[39] Regulation 15(11)(a).

[40] Regulation 15(9).

[41] For further information on appeals, see chapter 18.

[42] Regulation 15(6).

a) be the operator of the installation, or the part of the installation to which the transfer relates, after the transfer is effected; or

b) ensure that the installation is operated so as to comply with any monitoring and reporting condition.[43]

Where the regulator accepts an application for transfer, the procedure for effecting the transfer varies depending on whether the transfer is full or partial. To effect a full transfer, the regulator will simply update the permit to record the transfer and reissue the permit.[44]

Effecting a partial transfer is a two-stage process.[45] The first stage is for the regulator to issue a new permit to the proposed transferee applying to the part of the installation transferred. The second stage is to update and vary the existing permit to apply only to the remainder of the installation which is not the subject of the transfer. The revised permit will then be reissued to the operator. Where a partial transfer is effected, the conditions of each permit should be the same as the conditions in the original permit but may be varied to ensure that they are relevant to the part of the installation covered by the permit.[46]

The transfer will take effect on the date agreed with the applicants and specified in the permit.[47]

B. Relationship with Allocation of Allowances

The transfer of a permit in whole or in part will have implications for the allocation and issue of allowances. These are described in section II of chapter 16 below.

C. Transfer of Excluded Installations

If an excluded installation is transferred to another operator, the regulator is required to notify the authority which issued the certificate of temporary exclusion.[48] If the whole installation is being transferred and

---

[43] The grounds for refusal reflect the grounds set out in Regulation 9(6) for which an application for a permit may be refused. For discussion of these grounds see section IV.B of chapter 12.

[44] Regulation 15(7)(b).

[45] Regulation 15(7)(a).

[46] Regulation 15(8).

[47] Regulation 15(7).

[48] Regulation 11(10). As noted in section II of chapter 11, the 'responsible authority' is the Secretary of State in relation to installations excluded on the basis of the UK Emissions Trading Scheme or Climate Change Agreements.

the authority is satisfied that the installation will continue to be covered by the national policy for which it was excluded, the responsible authority will serve a notice on the operator and the regulator specifying the change of operator.[49] If only part of the installation is being transferred, then the authority will revoke the certificate and issue new certificates in respect of each part of the installation which it is satisfied will continue to be covered by the national policy for which the installation was originally excluded.[50] If the authority is not satisfied that the installation or any part of the installation will continue to be covered by the national policy for which it was excluded, it will revoke the certificate of temporary exclusion.[51]

---

[49] Regulation 11(11)(a).
[50] Regulation 11(11)(b).
[51] Regulation 11(11)(c).

CHAPTER 14

SURRENDER AND REVOCATION OF PERMITS

This chapter considers the provisions in the ETS Regulations relating to the surrender and revocation of greenhouse gas emissions permits. It first outlines the procedure for surrender (section I) and revocation (section II) and then considers the implications of the notice of surrender or revocation on the operation of the permit (section III).

I. Procedure for Surrender of a Permit

An operator is required to surrender its permit where it 'has ceased carrying out in an installation all of the schedule 1 activities authorised by a greenhouse gas emissions permit in relation to that installation'.[1] Such application must be made within one month of ceasing to carry out the activities.[2]

Where an operator ceases to carry out only some of the schedule 1 activities carried out in the installation, or ceases to carry out a schedule 1 activity in only some of the installations covered by the same permit, there is no obligation to surrender the permit.[3] An operator would simply be required to notify the regulator of the change in operation or apply for a variation of the permit.[4]

There are two exceptions to the requirement to surrender.[5] These exceptions are necessary to allow for the possibility of allowances continuing to be issued following the closure of an installation if required by the NAP and are discussed in more detail in chapter 16.

A. Applications for Surrender

An application to surrender a permit must be made to the regulator[6] and must, except where the application relates to an offshore installation,[7] be accompanied by the fee prescribed in relation to

---

[1] Regulation 16(1).
[2] Regulation 16(2).
[3] Regulation 16(3)(c).
[4] For a description of the rules relating to changes in operation and variations of the permit, see section I of chapter 13.
[5] Regulation 16(3).
[6] Where the regulator has developed a standard application form, the application must be made on that form (Regulation 5(3)).
[7] Where an application relates to an offshore installation, the fee will be payable within 28 days of the regulator serving a notice requesting payment (Regulation 16(5)).

applications for surrender. [8] Applications may be made electronically.[9] Where an application is made electronically, the fee may be sent separately but the application will not be treated as having been received until the fee has been received.[10]

B.   Determination of Applications for Surrender

Once an application has been duly made, the regulator is required to serve a notice of its determination of the application on the operator within two months of the date on which the application is made.[11] Where the regulator approves the application, this notice is known as a 'notice of surrender'. Where a regulator requires further information in order to determine an application, it may serve a notice on the operator requesting him to furnish the information within the period specified by the notice.[12] It is an offence to fail to furnish such further information when requested.[13]

II.   Revocation

The regulator may revoke a greenhouse gas emissions permit at any time.[14] A revocation is effected by serving a notice (known as a 'revocation notice') on the operator and, where an installation is included in a pool,[15] the Secretary of State or the devolved administration which authorised the pool. The revocation notice must specify the date on which the revocation will take effect.[16] This must be at least 28 days after the date on which the notice is served.

The serving of a revocation notice may serve two purposes.

---

[8] Regulation 16(4). The fee prescribed in paragraph 1(1)(g) of Schedule 5 is £620. This fee may be superseded by a charging scheme adopted by the regulators (supra chapter 12, note 14).

[9] Regulation 5(4).

[10] Regulation 5(5).

[11] Regulation 16(6). Where the application for surrender relates to an installation included in a pool in accordance with Regulation 27, the regulator is also required to serve the notice on the appropriate authority. For a discussion of Regulation 27, see section III of chapter 16.

[12] Regulation 16(16). Any period during which a request for further information is outstanding will not be included when calculating the two-month period in which the regulator must determine an application (Regulation 16(17)).

[13] Regulation 38(1)(e).

[14] Regulation 17(1). This power does not apply where the national allocation plan provides for allowances to continue to be issued to a closed installation. For further details see section I of chapter 16.

[15] For further detail on pooling, see section III of chapter 16.

[16] Regulation 17(4).

Firstly, it is a used as an enforcement tool to ensure compliance with the conditions of the permit.[17] Although the Regulations require regulators to revoke a permit where an operator has failed to comply with a requirement to surrender the permit[18] and expressly permit revocation where an operator fails to pay a subsistence charge,[19] the power to revoke is not limited to such circumstances. Revocation could therefore be used as a sanction of last resort where an operator fails to comply with the requirements of the permit or to pay the subsistence charge. Further, the fact that a power to revoke exists is an incentive for compliance and the threat of exercising the power to revoke a permit can be used as an enforcement tool. The ETS Regulations expressly provide that a notice revoking a permit may be withdrawn before the notice takes effect.[20] A regulator may therefore serve a notice revoking a permit and then withdraw the notice if the operator complies before the notice takes effect.

Secondly, the service of a revocation notice enables the regulator to ensure that an operator who fails to comply with the requirement to surrender his permit does not benefit from that failure by receiving additional allowances after the installation has ceased operation. The ETS Regulations therefore require the regulator to revoke a permit where an operator has failed to comply with an obligation to surrender the permit or with the conditions of a retention notice. They also require the regulator to revoke a permit where an application to retain allowances is withdrawn or refused, or was accepted on the basis of false or misleading information.[21] In such circumstances, the revocation notice will require an operator to pay the fee prescribed in respect of a revocation notice.[22] This is to ensure that operators are not able to avoid the surrender fee by waiting for the regulator to revoke the permit. The fee is payable within the period specified in the notice.

III.   Effect of a Notice of Surrender or Revocation Notice

A notice of surrender or revocation notice serves two main purposes:

Firstly, it specifies the date on which the surrender or revocation of the permit will take effect. From that date, the permit will cease to have

---

[17] For further detail on its use as an enforcement mechanism, see section I of chapter 17 below.

[18] Regulation 17(2).

[19] Regulation 18(2).

[20] Regulation 17(14).

[21] Regulations 17(2), 24(11) and 25(3). For more information on allowances and retention notices, see section I of chapter 16.

[22] Regulation 17(6). The fee prescribed in paragraph 1(1)(h) of Schedule 5 is £620. This fee is the same as the fee in respect of an application to surrender a greenhouse gas emissions permit under paragraph 1(1)(g) of schedule 5. This fee may be superseded by a charging scheme adopted by the regulators (supra chapter 12, note 14).

effect to authorise the carrying out of a schedule 1 activity or to require the monitoring of emissions.[23] If the application is made prior to the date on which the applicant ceases to carry out all the schedule 1 activities authorised by the permit, the date will be determined by reference to the date on which the operator indicates that it will cease carrying out the activity. Where the operator has already ceased carrying out a schedule 1 activity, the notice may take effect after the operator has been given sufficient period to appeal.

Secondly, it provides for the date for reporting emissions and surrendering of allowances to be brought forward. The notice will require the operator to submit to the regulator a verified report of its emissions in the year in which the notice takes effect and to surrender allowances to cover those emissions by the dates specified in the notice.[24] The emissions report must be prepared and verified in accordance with the monitoring and reporting conditions in the permit.[25] The requirements relating to surrender of allowances are described further in chapter 15 below.

The permit will cease to authorise any activities and to require the monitoring of emissions from the date on which the surrender or revocation takes effect. However, it does not follow that all the provisions of the permit cease to have effect from that date. Instead, conditions of the permit will continue to have effect so far as they are not superseded by the requirements of the notice.[26] For example, if the surrender or revocation takes effect before 30 April in any year, the operator's obligation to report its emissions for the previous year and surrender allowances to account for those emissions will not be removed by the notice. The detailed requirements of the permit in relation to the content and verification of emissions reports would not be superseded and would, therefore, continue to apply.

---

[23] Regulations 16(9) and 17(7).
[24] Regulations 16(7) and 17(5).
[25] Regulations 16(8) and 17(5)(a).
[26] Regulations 16(9) and 17(7).

CHAPTER 15

ALLOCATION, ISSUE AND SURRENDER OF ALLOWANCES

This chapter and chapter 16 consider the provisions in the ETS Regulations prescribing the process for the allocation, issue and surrender of allowances in the UK. This chapter considers the provisions governing the development of the NAP and decisions on allocations to existing installations (section I) and the allocation to new entrants (section II). The allocation of allowances in the first phase of the scheme to so called 'late entrants' is considered in section III. Rules relating to transactions concerning allowances are considered in section IV. Finally section V considers the requirement to surrender allowances.

Chapter 16 considers the rules governing changes, such as closure, after the initial allocation which can affect the allocation and issue of allowances.

The provisions discussed in this chapter and chapter 16 are intended to remain largely unchanged through successive phases.

The method of calculating the actual number of allowances to which any particular installation is allowed is, in contrast, regulated not by the ETS Regulations themselves but by the NAP for the relevant phase. The NAP is, by its nature, relevant only to one phase, and its provisions are, in that sense, transient. The content of the UK NAP for the first phase is considered separately in chapters 19 (development and allocations to existing installations) and 20 (new entrants, closures and auctioning). Chapter 21 considers the UK NAP for the second phase of the scheme.

I. Development of the NAP and Allocation to Existing Installations

As discussed in chapters 5 and 6 above the Emissions Trading Directive provides for allowances to be allocated in a two stage process. This two stage process is reflected in the ETS Regulations.

Firstly, for each scheme phase, the Secretary of State[1] is required to develop a NAP for the United Kingdom.[2] The NAP must be developed

---

[1] For information on the role of the devolved Administrations in the allocation process, see section II of chapter 11.

(continued...)

in accordance with Articles 9 and 10 of and Annex III to the Emissions Trading Directive[3] and must be published[4] at least 18 months before the start of the scheme phase to which it relates.[5]

Once the NAP has been developed the Secretary of State will notify the NAP to the Commission in accordance with the requirements of Article 9 of the Directive. The Commission may reject the plan or any aspect of the plan within three months of the notification of the NAP. Amendments to the NAP must be approved by the Commission. The Secretary of State is required to publish, as soon as possible, information on whether the Commission has accepted or rejected the NAP, or any aspect of it, and any amendments proposed to the NAP.[6]

Secondly, the Secretary of State is required to decide upon the allocation of allowances for the phase (the 'final allocation decision'). This includes deciding on the total quantity of allowances to be allocated for the phase and the allocation of allowances in respect of each installation, including the number of those allowances to be issued in each scheme year in that phase.[7] Where there is more than one greenhouse gas permit relating to an installation, the decision on allocation will also specify the allocation of allowances to each part of the installation covered by a separate permit.[8]

As required by Article 11 of the Directive, the final allocation decision must be based on the NAP which has been approved by the Commission[9]

---

[2] Regulation 20(1). The NAP must cover Great Britain and Northern Ireland. Although the United Kingdom is responsible as an EU Member State for Gibraltar, the ETS Regulations extend only to Great Britain and Northern Ireland.

[3] Regulation 2(1) defines 'national allocation plan' as 'a plan developed in accordance with Articles 9 and 10 of and Annex III to the [Emissions Trading] Directive'. For an explanation of the procedural and substantive requirements for NAPs set out in the Directive, see chapters 4 and 5 above.

[4] The requirement to publish is subject to Regulation 37(1) which excludes information from publication if it would be contrary to the interests of national security.

[5] Regulation 20(3). In relation to the first phase the national allocation plan was required to be published by 31st March 2004 (Regulation 18 of the 2003 Regulations). The NAP for the first phase of the scheme was prepared under Regulation 18 of the 2003 Regulations which continued to apply until it was fulfilled (Regulation 47(1) and (2)).

[6] Regulation 20(4) and (5). Information relating to the plan must be sent to the devolved administrations who are required to publish it (Regulation 20(2) and (6)–(8)).

[7] Regulation 21(1). The decision must be notified to the devolved administrations (Regulation 21(4)). The final allocation decision for the first phase of the scheme was prepared under Regulation 19 of the 2003 Regulations which continued to apply until it was fulfilled (Regulation 47(1) and (3)).

[8] Regulation 21(1)(c). This reflects the possibility that there may be more than one operator of a single installation and that in such circumstances each operator will hold a permit. See section II of chapter 12 above.

[9] Regulation 21(2)(a). In accordance with the Emissions Trading Directive a NAP can be taken to be 'approved' if the Commission has not rejected the plan or any aspect of it

(continued...)

and must take into account comments from the public.[10] The final allocation decision must be made and published[11] at least 12 months before the start of the scheme phase to which it relates.[12]

In parallel to the making of the final allocation decision, the NAP, as approved by the Commission, will be designated in Regulations as the 'approved NAP' for that phase.[13] This is necessary because, in addition to setting out the total number of allowances to be allocated and the intended allocation to each installation, the NAP will almost certainly also contain additional rules and procedures relating to the allocation of allowances, such as, the rules relating to allocations to new entrants.[14] Instead of amending the ETS Regulations to reflect the detailed provisions of each NAP, the ETS Regulations are designed to set out general procedures and to cross-refer to the provisions of the approved NAP. The fact that the approved NAP is designated in Regulations gives Parliament the opportunity to scrutinise the provisions of the NAP given effect through this mechanism.[15]

II. Allocation to New Entrants

Annex III to the Emissions Trading Directive requires Member States to provide in their national allocation plans information on how new entrants[16] will be able to begin participating in the scheme. There are a number of ways in which a Member State could fulfil this obligation. It could simply provide that no additional allowances will be allocated and new entrants will have to buy on the market or it could provide for one or more new entrant reserves from which allocations will be made

---

within three months of the Commission being notified of the plan or if the Commission has accepted amendments to the plan (Article 9). For a further discussion of when a NAP will be deemed to be approved, see chapter 5 above.

[10] Regulation 21(2)(b). See chapter 5 above for a discussion of these requirements.

[11] The requirement to publish is subject to Regulation 37(1) which excludes information from publication if it would be contrary to the interests of national security.

[12] Regulation 21(3). In relation to the first scheme phase, a decision on allocations was required to be made at least 3 months before the start of the scheme phase (Regulation 19 of the 2003 Regulations).

[13] "See definition of 'approved national allocation plan' and 'approved NAP regulations' in Regulation 2 of the ETS Regulations.

[14] Other rules set out in the NAP may relate to the treatment of allowances where an installation closes or the circumstances in which allowances may be issued later than 28th February in accordance with Articles 40 and 46 of the Registries Regulation.

[15] See in relation to the first phase of the Scheme, the Greenhouse Gas Emissions Trading Scheme (Approved National Allocation Plan) Regulations 2005, S.I. 2005/1387.

[16] 'New entrant' has the same meaning in the ETS Regulations as in the Emissions Trading Directive (Regulation 2). It is defined in Article 3(h) of the Directive as 'any installation...which has obtained a greenhouse gas emissions permit or an update of its greenhouse gas emissions permit because of a change in the nature or functioning or an extension of the installation subsequent to the notification to the Commission of the national allocation plan'. For a discussion of this definition, see section IV.D of chapter 4.

to new entrants (or specific types of new entrants). Where a NAP provides for a new entrant reserve, it will need to identify the types of new entrant which will be eligible for an allocation of allowances from the reserve and to set out the methodology for determining the number of allowances and the procedure for the operation of the reserve.

A. Applications for an Allocation from a New Entrant Reserve

Where the approved NAP in respect of a particular phase provides for a new entrant reserve, an allocation of allowances may be obtained by applying to the regulator.[17] As the ability to make an application depends on the provisions of the approved NAP, an application cannot be made before Regulations designating the relevant NAP as the 'approved NAP' have entered into force.[18] As an installation will only fall within the definition of new entrant if it is the subject of a new permit or a variation of an existing permit, an application for an allocation from a new entrant reserve must be combined with an application for a new permit or an application for a variation of the existing permit.[19] An operator must, therefore, consider at the relatively early stage of obtaining a permit or a variation of the permit whether it wishes to apply for an allocation from the new entrant reserve. If it does not apply at that stage, it will not be able to apply later.

An application for an allocation from the new entrant reserve must contain the information required by the regulator for the purpose of determining the application in accordance with the provisions of the NAP.[20] Where the regulator has made available a standard application form, the application must be made on that form.[21] An application may be made electronically.[22]

---

[17] Regulation 22(1).
[18] In relation to the first phase, the Regulations made special provision allowing applications to be made before Regulations specifying a NAP as the 'approved NAP' for the first phase entered into forced (see Regulation 22(5) and (6)).
[19] Regulation 22(2).The requirement for applications to be combined does not apply if the application for an allocation from the new entrant reserve relates to an installation in respect of which an application for a new permit or variation of the permit was made prior to the Regulations specifying the approved NAP for that phase entered into force (Regulation 22(3)).This exception will apply in each phase because there will always be a gap between the date on which installations are considered to be new entrants on the one hand, and the approval of the NAP by the Commission and specification as 'the approved NAP' in domestic regulations on the other.
[20] Regulation 22(4).
[21] Regulation 5(3).
[22] Regulation 5(4).

Except in the case of an application relating to an offshore installation, the application must be accompanied by the prescribed fee.[23] Where an application is made electronically, the fee may be sent separately but the application will not be treated as having been received until the fee has been received.[24] An operator of an offshore installation will be required to pay the fee within 28 days of the date on which the regulator serves a notice requesting the fee.[25] If the operator of an installation refuses to comply with this obligation then the regulator may reject the application.[26] In order to reduce the risk of non-payment, it is likely that the regulator of offshore installations will request the fee promptly on receipt of the application and wait for receipt of the fee before determining the application.

B.   Determination of Applications, Conditional Allocations, and Reservation of Allowances

Once an application is submitted the regulator will consider whether the application is duly made[27] and may request further information from the applicant.[28] If the operator fails to supply the information within the time specified in the notice, the regulator may serve a notice on the applicant indicating that the application is being treated as having been withdrawn.

The regulator is required to determine an application for an allocation from the new entrant reserve within two months of the date on which the application was received or within such longer period as may be agreed in writing with the applicant.[29] If the regulator fails to give notice of its determination of an application within two months (or any longer period agreed between the operator and the regulator), the operator may notify the regulator that it is treating the application as refused.[30] The operator could then appeal against the refusal.[31]

---

[23] The fee prescribed in respect of an application for an allocation from the new entrant reserve is £1030 (see paragraph 2(1)(a) of part 2 of schedule 2 to the ETS Regulations). This fee may be superseded by a charging scheme adopted by the regulators (see chapter 12, note 14).

[24] Regulation 5(5).

[25] Regulation 22(7).

[26] Regulation 22(8).

[27] See section III of chapter 12 for a discussion of what is required for an application to be duly made.

[28] Regulation 22(9).

[29] Regulation 22(10). The two-month period does not include any period during which a request for further information under Regulation 22(9) is outstanding (Regulation 22(11)(a)) or, where an application for an allocation is combined with an application for a permit or variation of a permit, any period during which a request for further information is made in relation to the application with which it is combined 22(11)(b)).

[30] Regulation 22(12).

[31] For further information on appeals, see chapter 18.

The regulator may determine an application in one of three ways:

a)   determine the eligible allocation subject to conditions;
b)   determine the eligible allocation and allocate the allowances; or
c)   reject the application.[32]

The regulator must determine the application by applying the rules set out in the NAP. If the applicant is not entitled to an allocation from the new entrant reserve in accordance with the NAP, the regulator will reject the application. If the operator is eligible for allowances under the NAP, then the regulator will determine the amount of allowances which may be allocated in respect of the installation subject to sufficient allowances being available in the new entrant reserve. This amount is known as the 'eligible allocation'.[33]

The methodology for determining the eligible allocation will be set out in the NAP. In order to apply the methodology, it will be necessary to establish certain key facts about the new entrant. For example, when it started or will start operating, its capacity and the type of technology used. Where a new entrant has already started operating[34] or is about to start operating imminently, it may be possible to establish these facts with some certainty. In such circumstances the NAP may provide for the regulator to determine the eligible allocation and allocate the allowances to the relevant operator.

However, operators may apply for a permit or variation of the permit some time in advance of the start of operation. In such situations it is likely that certain of the key facts will change prior to the start of operation. In particular the date for the start of operation may be delayed. This is likely to affect the number of allowances to which the operator is eligible.

In such circumstances it is likely that the NAP will provide for the making of conditional allocations. The conditions of the allocation must be set out in the notice determining the application and must be determined in accordance with the approved NAP.[35]

---

[32] Regulation 22(13). The determination is effected by serving a notice on the operator.
[33] See Regulation 22(23).
[34] This is most likely to be the case where a new entrant begins operating in the period between the notification of the NAP to the Commission and the start of the scheme phase. The requirement to combine an application for an allocation from the new entrant reserve with an application for a permit or variation of the permit does not apply where an application for the permit or variation of the permit is made prior to the date on which Regulations specifying the approved NAP for a particular scheme phase enter into force (Regulation 22(2) and (3)).
[35] Regulation 22(13)(a) and (15)(a).

Where a conditional allocation is made, the operator will be required to notify the regulator of any change to the information provided in respect of the application or provided to the regulator in response to an application for further information.[36]

Where the regulator becomes aware, either through notification by the operator or otherwise, of information affecting the provisions of a notice of conditional allocation, it is required to amend the provisions of the notice to ensure that the conditional allocation is consistent with the provisions of the NAP.[37] An example of where this might happen include where a new entrant will begin operating later than previously planned and therefore the operator will be eligible for fewer allowances than previously provided. If the regulator becomes aware that the operator is no longer eligible for an allocation at all, it must serve a notice rejecting the application.[38]

A NAP which provides for conditional allocations may also provide a mechanism for reserving allowances which are the subject of a conditional allocation. Such a mechanism could be used to ensure that allowances from a new entrant reserve are allocated on a first come, first served basis regardless of the date on which the new entrant will actually start operating. Where the NAP makes provision for the reservation of allowances, a notice of conditional allocation will specify whether and if so, how many, allowances have been reserved.[39]

Once the regulator is satisfied that all the conditions specified in a notice of conditional allocation are met, it will allocate the allowances to the operator.[40] From that point the allocation is no longer conditional.

C. What Happens where there are Insufficient Allowances Available in a New Entrant Reserve?

Where a NAP provides for a new entrant reserve, the size of the reserve will be fixed in the NAP and set out in the NAP table required for the registry.[41] Where the eligible allocation is greater than the number of

---

[36] Regulation 22(15)(b).
[37] Regulation 22(16).
[38] Regulation 22(17).
[39] Regulation 22(15)(c).
[40] Regulation 22(18).
[41] The size of the reserve may simply be the difference between the total quantity of allowances allocated for the scheme phase and the total number of allowances allocated in the final allocation decision, or where there is more than one reserve, the NAP may specify the size of each one. The total size of the reserve(s) is required to be expressly specified in the NAP table notified to the Commission in accordance with Articles 38(1) and 44(1) of the Registries Regulation.

available allowances, the regulator may only allocate the available allowances.[42] The NAP may provide that unissued allowances allocated to installations which close during the scheme phase may be added to the new entrant reserve. If that is the case, then the NAP will set out the rules for how any further allowances which become available should be allocated. The regulator may then make additional allocations of allowances in accordance with these rules.[43]

A notice allocating allowances will identify the operator and the installation and will specify the allocation of allowances including the number of allowances to be issued in each remaining year or part year of the scheme phase to which the allocation relates.[44]

III. Allocation of Allowances to Late Installations

In relation to the first phase of the scheme, the ETS Regulations set out the procedure for allocating allowances to late installations in accordance with the approved NAP. An element of the new entrant reserve was set aside for this purpose.[45]

The allocation procedure is set out in Regulation 22A. Operators wishing to receive an allocation were required to make an application to the Secretary of State within 30 working days of either the grant of its greenhouse gas emissions permit or 6 April 2006, [46] whichever was the later.[47] The Secretary of State was required to determine the application within 20 days or such longer period as is agreed with the applicant.[48] If the Secretary of State approved an application, he was required to

---

[42] Regulation 22(14) When allowances are 'available' will be determined in accordance with the provisions of the NAP (Regulation 22(23)). For example, if the NAP provides for a queuing system, allowances may not be available even though they have not yet firmly been allocated to an operator.

[43] Regulation 22(14) and (19).

[44] Regulation 22(20).

[45] 'Late installations' for the purpose of the UK national allocation plan for the first phase were installations which received or applied for their greenhouse gas emissions permits after 1 January 2005 and were not included in the final allocation of allowances for the first phase of the scheme under Regulation 19 of the 2003 Regulations.

[46] This is the date on which Greenhouse Gas Emissions Trading Scheme (Amendment) Regulations 2006 (supra. Chapter 11, note 7) which introduced this provision entered into force.

[47] Regulation 22A(1) and 22A(5)(e). This period does not include any period in which applicant has an outstanding request to the Secretary of State regarding the manner in which he should compile and submit information (see Regulation 22A(7)).

[48] Regulation 22A(8). This period would not include any period during which the Secretary of State has requested further information from the operator (Regulation 22A(9)). If the Secretary of State does not determine the application within that time the operator may treat the application as having been refused and may appeal against the refusal (Regulation 22A(10) and Regulation 33(3)).

allocate allowances to the operator within 15 working days by serving a notice on the operator and the registry administrator.[49]

Regulation 22B and schedule 7 set out how the allocations should be calculated. The allocations were based on the same methodology as that applied to existing installations who received their allocations in the final allocation decision for the first phase of the scheme: that is, based on verified emissions from the installation for the period 1998 to 2003, or where not available, benchmarked data.[50] However, a reduction factor of 10 per cent was applied to ensure that operators received fewer allowances than they would have done if the installation had been included in the final allocation decision. The reduction factor was increased to 25 per cent from 1 September 2005. No further applications could be made after 28 February 2007.[51]

If the Secretary of State became aware that there were insufficient allowances in the element of the new entrant reserve set aside for late entrants to satisfy all the applications which had been received, allocations would be made in the order in which the operators applied for their greenhouse gas emissions permits.[52]

IV.   Allocation by Auction or Sale

Regulation 21A permits the Secretary of State to enter into an agreement with a registry account holder to allocate allowances to that person in exchange for payment.[53]

This power only applies to allowances for the first phase which are held in the new entrant reserve and qualify for auction or sale in accordance with the approved national allocation plan.[54]

The effect of this provision is to give effect to the statement in appendix C to the NAP for the first phase of the scheme[55] that the government intended to auction or sell any surplus allowances in the new entrant reserve.

---

[49] Regulations 22A(11) and (16).
[50] This was essentially the same methodology as set out for existing installations in the approved NAP for the first phase of the scheme. However, late installations are not able to take advantage of the baseline changes, rationalisation and commissioning rules. These are discussed in chapter 19.
[51] Regulation 22A(1).
[52] Regulation 22A(12)-(15).
[53] Regulation 22A(1).
[54] Regulation 22A(2).
[55] See section IV of chapter 20.

Where the Secretary of State enters into such agreement, whether as a result of the account holder being the winning bidder in an auction or through a straightforward sale arrangement, the purchaser must pay the agreed sum by the date agreed. If he fails to do so, then the Secretary of State may decide not to allocate the allowances to him and seek to sell them to another purchaser or, if he prefers, pursue the original purchaser for payment as a debtor.[56] If the Secretary of State chooses to pursue the payment and manages to recover it before 10th April 2008, then he must allocate the allowances within the timescales as if he were paid on time.[57] if, on the other hand, the money is received after that date, the Secretary of State will not be required to allocate allowances to the purchaser. This reflects the fact that the registry administrator is required to cancel all remaining first phase allowances on 1st May 2008[58] and the prescribed timescales mean that it may not be possible to transfer the allowances to the purchaser before that date. This does not preclude the Secretary of State from recovering the money after this date as it does not change the fact that the purchaser has defaulted on an agreed payment.

If the purchaser does pay on time[59] the Secretary of State must allocate the allowances to him within 7 days of receiving such payment, which takes the form of serving a notice on the purchaser and the registry administrator.[60] This notice will instruct the registry administrator to transfer the allowances into the purchaser's account.[61]

The Secretary of State's decision to allocate allowances by way of auction or sale may only be taken with the consent of the devolved administrations.[62] This consent is to be to the principle and manner of allocating by auction or sale (for example whether to sell through an auction or not and, if so, the design of the auction) rather than to each individual sale agreement entered into.[63]

If he wishes to do so, the Secretary of State, with the consent of the devolved administrations, may appoint the Environment Agency or the Scottish Environment Protection Agency to carry out an auction or sale.[64]

---

[56] Regulation 21A(5)-(6).
[57] Regulation 21A(7).
[58] See section II of chapter 9.
[59] Or where the defaulting purchaser subsequently pays by 10th April 2008 (except where the Secretary of State has made a decision to not allocate the allowances to that defaulting purchaser).
[60] Regulation 21A(4).
[61] Regulation 22A(7)-(11).
[62] Regulation 46(5).
[63] Regulation 46(7).
[64] Regulations 22A(12) and 46(6)

V.   Transactions Concerning Allowances

The procedures for all transactions concerning allowances are set out in the Registries Regulation.[65] The Registries Regulation is directly applicable and therefore does not require transposition into UK law. However the ETS Regulations do include some supplementary provisions including provision for registry fees. [66]

Fees relating to operator holding accounts are included as part of the subsistence charge for the greenhouse gas emissions permit.[67] Applications for person holding accounts are required to be accompanied by the prescribed fee.[68] Fees are also prescribed for a change of authorised representatives or appointment of additional authorised representatives.[69]

A.   Issue of Allowances

Once allowances have been allocated they will be issued in accordance with the Registries Regulation. Allowances are issued by transferring the allowances from the party holding account to the relevant operator holding account in the registry.[70]

For installations included in the final allocation decision, the number of allowances issued each year will be set out in the NAP table and will be issued by the registry administrator in accordance with Articles 40 or 46 of the Registries Regulation.

Where the regulator makes an allocation from the new entrant reserve, it must also serve the notice on the registry administrator who will issue allowances to the operator in accordance with the notice.[71] A notice allocating allowances will therefore identify the operator and the

---

[65] For further information on the registries system, see section II of chapter 7. At the time of writing the Registries Regulation is being amended. The changes are likely to include changes to the procedure for issuing allowances to new entrants.

[66] Regulation 26.

[67] Regulation 18(1) and paragraph 14 of schedule 5.

[68] Regulation 26(12). Paragraph 3(a) of part 3 of schedule 5 provides for the fee in respect of an application for the creation of a person holding account to be £175. This fee may be superseded by a charging scheme adopted by the regulators (supra. chapter 12, note 14).

[69] Regulation 26(13) and paragraph 3 of schedule 5 which provides for verifiers and persons wishing to be appointed as authorised representatives to pay a fee of £175 to be included in the registry and £50 to appoint additional users to use the registry on their behalf (after the first two appointments).

[70] For further information on the issue of allowances, see section II of chapter 6.

[71] Regulation 22(21) provides that a notice specifying an allocation served on the registry administrator will be treated as an instruction to the registry administrator for the purposes of Article 42 (in relation to the first scheme phase) or Article 48 (in relation to the second and subsequent phases) of the Registries Regulation which empower the registry administrator to issue allowances to new entrants.

installation and permit identification code for the installation and will specify the allocation, how the allocation will be divided between the remaining scheme years in the scheme phase and the date on which allowances will be issued in the year in which the notice is served.[72]

B.   Delaying the Issue of Allowances

As discussed in chapter 6 above, the Emissions Trading Directive requires a proportion of the total quantity of allowances to be allocated by 28 February each year. The Registries Regulation requires all allowances allocated for a specific scheme year to be issued by 28 February in that year. However as an exception to this rule, 'where foreseen for an installation in the NAP of the Member State, the registry administrator may transfer that proportion at a later date of each year'.[73]

Where the NAP specifies circumstances in which the issue of allowances will be postponed[74] and these circumstances apply, the registry administrator and the operator will be notified.[75] The notice will specify either the precise date on which the allowances will be issued or the conditions which must be met before the allowances will be issued.[76] Once the conditions are met a further notice will be served on the operator and the registry administrator who will issue the allowances.[77] A similar procedure applies where the initial allocation was made from the new entrant reserve, including the late entrant reserve.[78]

C.   Force Majeure Allowances

As discussed in section III of chapter 6 above, the Emissions Trading Directive allows for the possibility during the first phase of the scheme of additional, non-transferable allowances being issued in cases of force majeure. The issue of such allowances must be authorised by the Commission who will determine whether force majeure is demonstrated. Where the Commission authorises force majeure allowances to be issued, the Secretary of State will instruct the registry administrator to issue the additional allowances in accordance with Article 43(1) of the Registries Regulation.[79]

---

[72] Regulation 21(20).

[73] Article 40 and 46 of the Registries Regulation. For a discussion of this provision, see section II of chapter 6.

[74] The approved NAP for the first phase of the scheme identified a number circumstances in which the issue of allowances would be delayed. For a description of these, see chapter 19.

[75] Regulation 21(6). Notice will be given by different authorities depending on which conditions for delay apply (Regulation 21(10).

[76] Regulation 21(7).

[77] Regulation 21(8) and (9).

[78] Regulation 22C.

[79] Regulation 21(5).

D. Transfers and other Transactions

Once allowances have been issued to the operator, allowances may be transferred, cancelled or surrendered in accordance with the provisions of the Registries Regulation. As discussed in chapter 7 above, Article 23(2) of the Registries Regulation creates the concept of an additional authorised representative whose agreement is required for any transactions carried out from the account which it represents. The ETS Regulations allow holders of accounts in the UK Registry to appoint authorised representatives.[80]

VI. Surrender of Allowances

As indicated in chapter 12 above, each operator is required by its permit to surrenders allowances equal to its annual reportable emissions from the installation by the end of April in the year following the year in which the emissions arose.[81]

The number of allowances which an operator is required to surrender is determined by the amount of its annual reportable emissions for the previous scheme year. 'Annual reportable emissions' are the emissions over the year arising from schedule 1 activities carried out in the installation of gases specified in schedule 1 in relation to those activities.[82] Each allowance is equal to one tonne of carbon dioxide equivalent[83] and therefore so long as the greenhouse gas emissions covered by the emissions trading scheme are limited to carbon dioxide only, the number of allowances which an operator will be required to surrender will usually be equal to the number of tonnes of carbon dioxide emitted from the schedule 1 activities carried out in the installation.

This is subject to one exception. Where an operator has failed to comply with the obligation to surrender allowances in relation to the previous scheme year, the annual reportable emissions for the following scheme year will be deemed to be increased by the amount of the earlier failure.[84] This ensures that the operator is required to surrender the allowances which it failed to surrender in the previous year. For example, if an installation emits 100,000 tonnes of carbon dioxide in 2006 and by 30 April 2007 the operator has surrendered only 80,000 allowances, the operator would have failed to comply with its obligation and would

---

[80] Regulation 26(7).
[81] Regulation 10(3).
[82] Regulation 2(1).
[83] See Regulation 2 of the ETS Regulations and Article 3(a) of the Directive to which it refers.
[84] Regulation 10(4)-(5).

have to surrender the additional 20,000 allowances by 30 April 2008 along with allowances to cover its emissions in 2007.[85]

Where a permit is surrendered or revoked the requirement to surrender allowances will be brought forward. The notice of surrender or revocation notice will specify a date by which allowances must be surrendered.[86] In addition to requiring allowances to be surrendered to cover the reportable emissions specified in the report required by the notice,[87] the notice of surrender must also require allowances to be surrendered to cover:

a) any emissions in respect of which the operator failed to surrender allowances in the previous year (although this does not cover emissions in respect of which the deadline for surrendering allowances has not yet occurred);[88]

b) where an error has been discovered in the amount of emissions from the installation reported in a previous year, any emissions for which allowances were not surrendered as a result of the error;[89]

c) where a supplementary decision has been made,[90] the number of allowances which have been issued in respect of the installation but which should not have been issued;[91]

d) where an operator did not apply to surrender its permit within one month of ceasing to carry out all schedule 1 activities, any allowances which would not have been issued if the operator had applied to surrender its permit within one month of ceasing to carry out the schedule 1 activities.[92]

As discussed in chapter 8, the Emissions Trading Directive was amended by the Linking Directive to allow for the possibility of using CERs and ERUs in the scheme. The ETS Regulations were therefore similarly amended to allow UK operators to use CERs and ERUs to comply with their obligations to surrender allowances under the greenhouse gas emissions permit or to comply with an obligation under a notice of

---

[85] In addition, the operator will also be required to pay a financial penalty. For further information on penalties, see chapter 17.
[86] Regulations 16(7) and 17(5).
[87] Regulations 16(7)(a) and 17(5)(a).
[88] Regulations 16(7)(b)(ii) and 17(5)(b)(ii). For example, if a notice of surrender took effect in March 2007, this will not affect the operator's obligations to report its 2006 emissions by 31 March 2007 and to surrender allowances equal to those emissions by 30 April 2007.
[89] Regulations 16(7)(b)(iii) and 17(5)(b)(iii).
[90] For more information on supplementary decisions made under Regulation 25(2) or (7), see section I.E of chapter 17.
[91] Regulations 16(7)(b)(iv) and 17(5)(b)(iv).
[92] Regulations 16(7)(b)(v) and 17(5)(b)(v).

surrender or revocation notice.[93] In the first phase operators may use any combination of CERs and allowances without limit.[94] In the second and subsequent phases of the scheme operators will be able to use any combination of allowances, CERs and ERUs provided that the amount of CERs and ERUs used during the phase does not exceed the limit provided for in the NAP.[95] As required by the Directive, operators may not use CERs or ERUs generated from nuclear facilities or land use, land use change and forestry activities.[96]

---

[93] Regulation 27A(1) inserted into the ETS Regulations by the Greenhouse Gas Emissions Trading Scheme (Amendment) and National Emissions Inventory Regulations 2005, S.I. 2005/2903.
[94] Regulation 27A(2) prevents operators from using ERUs in the first phase.
[95] Regulation 27A(4). This limit must be specified as a percentage of the allocation of allowances to each installation (see chapter 8 for a discussion of Article 11a(1) of and criterion 12 of Annex III to the Emissions Trading Directive).
[96] Regulation 27A(3).

CHAPTER 16

CHANGES AFFECTING THE ALLOCATION AND
ISSUE OF ALLOWANCES

This chapter considers changes which affect the allocation and issue of allowances, namely the closure of an installation (section I) and the transfer of an installation (section II). Section III considers the effect of a pool on the allocation and issue of allowances.

The allocation and issue of allowances may also be affected by the making of supplementary decisions where an allocation was based on false or misleading information. This situation is considered in chapter 17 on enforcement and penalties.

I.   Closure of an Installation

As discussed in chapter 14, an operator is required to surrender its permit within one month of ceasing to carry out in the installation covered by the permit all of the schedule 1 activities authorised by a greenhouse gas emissions permit. If an operator fails to comply with this obligation then the regulator is required to revoke the permit.[1] If a permit covers more than one installation, and the operator continues to carry out a schedule 1 activity in at least one of those installations but ceases to carry out all of the schedule 1 activities in any one of the other installations, then the operator is required to vary its permit so that it no longer applies to those installations in which the schedule 1 activities have ceased.[2]

Where a permit is surrendered, revoked or varied so that it no longer applies to an installation, the regulator must notify the Secretary of State. The regulator and the Secretary of State are required to 'take such steps as [are considered] necessary to ensure that no further allowances are issued in respect of the installation from the date on which the notice of surrender, the revocation notice or the notice of variation takes effect.'[3]

Where the surrender, revocation or variation relates to an installation in respect of which allowances were allocated in the final allocation decision, this will require the Secretary of State to notify a correction to the NAP table held in the registry.[4] In the case of an installation which

---

[1] Regulation 17(2).
[2] Regulation 16(1) and (3).
[3] Regulation 23.
[4] In accordance with the procedure set out in articles 38(2) and 44(2) of the Registries Regulation.

received an allocation from the new entrant reserve, the regulator would need to notify the registry administrator not to issue any further allowances in respect of the new entrant.

A.   Retention of Allocation

Although the general rule is that a permit must be surrendered or revoked where an installation no longer carries out a schedule 1 activity, there are two exceptions to this.

Firstly, there is an automatic exception where 'the approved NAP provides for allowances allocated in respect of an installation in which a schedule 1 activity is no longer carried out to continue to be issued during the same scheme phase to which the approved NAP relates.' In such circumstances the obligation on the operator to surrender the permit does not apply[5] and the regulator may only revoke the permit 'after 28 February in the last scheme year in that scheme phase.'[6]

Secondly, there is a possible exception where the NAP provides that if certain conditions are met allowances allocated in respect of an installation or a proportion of those allowances may be retained.[7] In such circumstances, an operator may make an application to retain the allowances.[8] Such an application must be made within one month of the date on which activities ceased in order for the obligation to surrender the permit to be disapplied.[9] If an application to retain an allocation of allowances is withdrawn by the operator or refused by the regulator or any conditions on the retention are no longer met, the permit must be revoked.[10]

B.   Application to Retain Allowances

An application to retain allowances must be made to the regulator and contain such information as the regulator may reasonably require to determine the application in accordance with the approved NAP.[11] An application may be made electronically.[12] If the regulator makes available a standard form for such an application then all applications must be made on the standard form.[13] The purpose of the application process is

[5] Regulation 16(3)(a).
[6] Regulation 17(3).
[7] Regulation 16(3)(b).
[8] Regulation 24(1).
[9] Regulation 16(3).
[10] Regulation 24(11).
[11] Regulation 24(1) and (2).
[12] Regulation 5(4).
[13] Regulation 5(3).

to determine whether the conditions set out in the NAP for retention of allocation are met.

There is no requirement for an application to retain an allocation of allowances to be accompanied by a fee. Instead, a fee is payable for the determination of the application based on the number of hours which the regulator required to determine the application.[14] The fee will be payable by the date specified in the notice requesting payment.[15]

C. Determination of an Application to Retain Allocation

Once an application is submitted, the regulator will consider whether the application is duly made[16] and may request further information from the operator.[17] If the operator fails to supply the information within the time specified in the notice, the regulator may serve a notice on the applicant indicating that the application is being treated as having been withdrawn.

The regulator is required to determine an application within two months of the date it received the application or such longer period as is agreed with the applicant.[18] Any period during which an application for further information is outstanding does not count for calculating the period of two months for determining the application.[19] If the regulator fails to give notice of its determination of an application within two months (or any longer period agreed between the operator and the regulator), the operator may notify the regulator that it is treating the application as refused,[20] generating a right to appeal.[21]

Where an application is duly made, the regulator is required to determine the application in accordance with the NAP. If the conditions provided for in the NAP are not met, then the regulator must refuse the application. If the conditions provided for in the NAP are met then the regulator must accept the application and provide for the allocation of allowances or a proportion of the allocation to be retained.[22]

---

[14] Regulation 24(8)(b). The hourly rate currently prescribed in the regulations is £115 (paragraph 2(1)(b) of part 2 of schedule 5). This fee may be superseded by a charging scheme adopted by the regulators (supra. chapter 12, note 14).

[15] Paragraph 2(2) of part 2 of schedule 5.

[16] See chapter 12 for a discussion of when an application will be considered to be duly made.

[17] Regulation 24(6).

[18] Regulation 24(3).

[19] Regulation 24(4).

[20] Regulation 24(5).

[21] For further information on appeals, see chapter 18.

[22] Regulation 24(7).

The NAP may provide for an application to be accepted subject to conditions. Where the NAP requires conditions to be imposed the notice providing for the retention of the allocation will specify the conditions. If the conditions are not met, the regulator must revoke the permit.[23]

The ETS Regulations provide for the possibility that a NAP may provide for an operator to retain only a proportion of the allowances allocated to it. Where the determination is that only a proportion of the allocation may be retained, the regulator and the Secretary of State are required to take 'such steps as [they] consider necessary to ensure that, from the date on which the retention notice takes effect, only such proportion of the allowances as is specified in the notice are issued in respect of the installation'.[24] Where allowances were allocated in respect of the installation in the final allocation decision, this will require the Secretary of State to notify a correction to the NAP table held in the registry.[25] In the case of an installation which received an allocation from the new entrant reserve, the regulator would need to notify the registry administrator of an amendment to the instruction to issue allowances.

Where an application to retain an allocation of allowances is successful, the permit may not be revoked until after 28 February in the last year of the phase for which the allowances were allocated.[26] However, it will be necessary to vary the content of the permit to reflect the fact that a schedule 1 activity will no longer be carried out in the installation. Therefore, a notice providing for the retention of allowances will also specify the variations to the provisions of the greenhouse gas emissions permit which the regulator considers appropriate and the date on which the provisions will take effect.[27]

Where an application to retain allowances is withdrawn or refused, the permit will be revoked and no further allowances will be issued.[28]

II.   Transfer of an Installation

Chapter 13 explained the procedure for transferring a greenhouse gas emissions permit where an installation is transferred in whole or in part to another operator. Whilst the effect of transferring the permit is that the new operator will have the obligation to surrender allowances, there

---

[23] Regulation 24(11)(b).

[24] Regulation 24(9). Regulation 24(10) provides that the decision will not take effect for 15 days from the date of the decision.

[25] In accordance with the procedure set out in articles 38(2) and 44(2) of the Registries Regulation.

[26] Regulation 17(3).

[27] Regulation 24(8)(a).

[28] Regulations 23 and 24(11)(a).

is also a need to make clear what the effect of a transfer is on the allocation of allowances.

Where a permit is wholly transferred, the position is relatively simple. The regulator will notify the registry administrator of the transfer[29] and the operator will be required to notify the registry administrator of the changes to the details of the operator holding account.[30] Allowances allocated in respect of the installation will continue to be issued into the operator holding account for the installation and therefore will be issued to the new operator.

The position is, however, more complicated in relation to partial transfers. Where a permit is only partially transferred, a new permit will be issued in respect of the transferred part and a new operator holding account will be opened in the registry.[31]

Any allowances already allocated in respect of the installation will continue to be issued to the operator holding account relating to the original permit. Operators may therefore wish to make provision for the transfer of some of these allowances as part of the transfer agreement.

Any allowances allocated after the transfer will be allocated to the specific part of the installation to which they relate. Where an application for allowances from the new entrant reserve or late entrant reserve is outstanding at the time of an application for a partial transfer, the application must specify whether the application for allowances relates to the transferred part.[32] If so and the transfer is effected prior to the allocation of allowances from the reserve, the allowances will be allocated and issued to the operator of the transferred part.

A partial transfer also has implications for the use of CERs and ERUs by the operator. As explained in chapter 8, in the second and subsequent phases of the scheme the use of CERs and ERUs will be subject to a limit set in the NAP as a percentage of the allocation to the installation for the phase. This raises the question of how the limit would be applied when an installation is the subject of a partial transfer. As the limit is set as a percentage of the allocation to installations in relation to each phase of the scheme, operators would be able to vary the number of project credits which they use to meet their obligations under the scheme each year. This makes it difficult to develop a system to apply the limit pro rata where an installation is transferred. Therefore, the ETS Regulations

---

[29] Regulation 15(12).
[30] Article 15(3) of the Registries Regulation (read with regulation 26(6) of the ETS Regulations).
[31] Article 15(1) of the Registries Regulation.
[32] Regulation 15(4)(c) and (e) and 22(22).

provide that during a phase in which a partial transfer takes place, the operator of the transferred part may not use CERs or ERUs to meet its obligation to surrender allowances.[33] This would need to be taken into account in the context of the transfer agreement between the parties.

III. Installations in a Pool

As discussed in chapter 7, Article 28 of the Emissions Trading Directive provides for the possibility of allowing operators of installations carrying out the same activity to form a pool in the first and second phases of the scheme. In broad terms, the forming of a pool means that a trustee is appointed to handle all trading aspects of the scheme for all the installations forming part of a pool. The UK has made limited use of the possibility for pooling to operators of installations which:

a)   fall within the same description in schedule 1 to the ETS Regulations (e.g. all combustion installations); and

b)   do not fall within any description in Annex I of the IPPC Directive.[34]

These criteria effectively limit the types of installation which have the option of forming a pool to combustion installations below 50MW. An application to form a pool must be made jointly to the Secretary of State or relevant devolved administration[35] at least 6 months before the start of the phase.[36] Applications for the second scheme phase must therefore be made by 30 June 2007. Applications for the first scheme phase can no longer be made.

An application must:

a)   identify the installations to be included in the pool;

b)   contain the names and postal addresses of the operators of the installations and, if different, the addresses to which correspondence should be sent;

---

[33] Regulation 27A(5).
[34] Regulation 27(2) and supra. chapter 2, note 3.
[35] The appropriate authority for making an application will be the Secretary of State in relation to installations in England and offshore installations, and the relevant devolved administrations for installations in the devolved regions (see definition of appropriate authority in regulation 2). If an application relates to installations in more than one country, the Secretary of State will be the appropriate authority but will act with the agreement of the relevant devolved administrations (regulation 27(15) and regulation 46). Regulation 27(18) and (19) set out specific provisions for where no agreement is reached.
[36] Regulation 27(1) and (3).

c)   contain a copy of the greenhouse gas emissions permit relating to each installation and identify the regulator which issued the permit;

d)   nominate a person to act as pool administrator[37] and contain a declaration from that person that he is willing to act as pool administrator; and

e)   contain evidence that the pool administrator will be able to fulfil the obligations of the pool administrator.[38]

A.   Determination of Applications

Once an application has been duly made, the Secretary of State or relevant devolved administration will decide whether it considers it appropriate to allow the pool. If it does then the Secretary of State or relevant devolved administration will notify the operator of each installation, the relevant regulator(s) and the proposed pool administrator[39] and submit the application to the Commission for approval.[40]

Under the Directive, the Commission may within three months of an application reject the application if it does not meet the requirements of the Directive, or is inconsistent with the EC Treaty. The Commission must give reasons for any decision rejecting an application. Where the Commission rejects an application, the Secretary of State or the relevant devolved administration[41] will notify the operator of each installation, the regulator(s), and the proposed pool administrator. The operators may resubmit an amended application within four weeks of being notified that the application has been rejected by the Commission.[42] If the Secretary of State or relevant devolved administration considers that the amended application addresses the reasons given by the Commission for rejection of the application, it will notify the operator, regulator and proposed pool administrator and submit the amended application to the Commission.[43]

---

[37] This is the name given to the role of trustee referred to in article 28 of the Directive to avoid confusion with the existing term 'trustee' under English law.

[38] Regulation 27(3).

[39] Regulation 27(5).

[40] Regulation 27(4). Where the application is made to a devolved administration, the application will be submitted to the Commission via the Secretary of State.

[41] Where an application to form a pool was made to a devolved administration, the Secretary of State must notify the relevant devolved administration if the Commission rejects the application (regulation 27(6)).

[42] Regulation 27(7).

[43] Regulation 27(8) and (9). Where the application is made to a devolved administration, the application will be submitted to the Commission via the Secretary of State.

If the Commission does not reject the application within three months of its submission, or where the Commission accepts an amended application, then the application will be taken to have been approved. In such cases the Secretary of State or relevant devolved administration will serve a notice authorising the pool on the operators of each installation in the pool, the regulator, the pool administrator and the registry administrator.[44] The notice will identify the installations included in the pool and the pool administrator and the phase for which the pool is approved and will specify any conditions applying to the pool.[45]

B.   Effect of a Pool

Where a pool is authorised, the pool administrator will be primarily responsible for managing and trading allowances.

The Registries Regulation does not make special provision for issuing allowances to installations covered by a pool. The allowances will therefore be issued to the operator holding accounts relating to the installations covered by the pool rather than into a single account managed by the pool administrator. In order to ensure that pool administrator has access to the allowances and the power to carry out transfers on behalf of the operator, the operator of each of the installations included in the pool are required to appoint the pool administrator as their primary authorised representative.[46] It would, however, be open to a pool administrator to open a separate person holding account and to transfer into that account all allowances issued in respect of installations covered by the pool.

For the duration of the pool arrangement, the pool administrator will be responsible for surrendering allowances to cover the emissions of installations included in the pool.[47] Unless the notice authorising the pool is revoked, the operator will be deemed to have complied with its permit obligation to surrender allowances.[48] As the Registries Regulations makes no specific provisions for the installations in a pool, the allowances to cover emissions from each installation in the pool will need to be surrendered from the operator holding account for that installation.[49]

---

[44] Regulation 27(10). Where an application to form a pool was made to a devolved administration, the Secretary of State must notify the relevant devolved administration if the Commission accepts the application.
[45] Regulation 27(11).
[46] Regulation 27(12)(a).
[47] Regulation 27(12)(b).
[48] Regulation 27(13).
[49] In accordance with article 52 of the Registries Regulation.

The ETS Regulations do not prescribe the relationship between operators in the pool and between operators and the pool administrator. It is anticipated that these matters would be regulated in an agreement between the operators in the pool. This should take into account the rules provided for in the ETS Regulations.

There are two circumstances in which the pool may be amended mid-way through a phase: on the revocation or surrender of the greenhouse gas emissions permit relating to an installation included in the pool, and on the transfer or partial transfer of the installation.

Where an installation's permit is surrendered or revoked, the Secretary of State will amend the notice authorising the pool to remove that installation from the notice authorising the pool.[50]

Where an installation included in the pool is transferred part-way through a phase, the application for transfer must indicate whether the installation, or in the case of a partial transfer each part of the installation, should remain in the pool.[51] The notice authorising the pool will be amended to take account of the transfer.[52]

---

[50] Regulation 27(16).
[51] Regulation 15(4)(d).
[52] Regulation 27(17).

CHAPTER 17

ENFORCEMENT AND PENALTIES FOR NON-COMPLIANCE

In order to ensure the proper functioning of the scheme and to maintain its environmental integrity, it is necessary to ensure that operators comply fully with their obligations and that there are measures in place to remedy any failure to comply.

This chapter describes the enforcement mechanisms (including financial penalties) under the ETS Regulations. The ETS Regulations employ a combination of penalties designed to penalise non-compliance and act as a deterrent to non-compliance and active measures to ensure that any breach of the requirements of the scheme are remedied in the shortest possible time. These are considered in section I. The ETS Regulations also create a number of criminal offences for failure to comply with the requirements under the ETS Regulations. These are considered in section II. Section III considers briefly the application of the ETS Regulations to the Crown.

I.   Enforcement Mechanisms

The enforcement mechanisms provided for under the ETS Regulations vary according to the type of non-compliance.

A.   Enforcement of Monitoring, Reporting and Verification Requirements

As discussed above, one of the essential components of an emissions trading scheme is the requirement for operators to monitor emissions and submit a verified report of emissions to the regulator.[1] Effective enforcement of these requirements is essential both to maintain the environmental integrity of the scheme and to underpin confidence in the financial market for allowances.

Regulators have a specific duty under the ETS Regulations to ensure that the monitoring and reporting conditions of the permit are complied with.[2] In carrying out this duty, a regulator will be assisted by its general powers to request information and to enter land:

---

[1] Such conditions are included in the permit pursuant to Regulation 10(2)(a) and (b). For information on the requirements to monitor and report emissions, see sections II and III of chapter 3 and section IV.D of chapter 12.

[2] Regulation 28.

a) For the purposes of discharging its functions under the ETS Regulations, the regulator may serve a notice on any person requiring them to furnish information.[3] The information required and the deadline for providing the information must be specified in the notice. A notice may require a person to furnish information which it is reasonable to require him to compile as well as information which is in his possession.[4]

b) Regulators also have power to enter onto land. This can be used for carrying out inspections to check operators are complying with their permit conditions.[5]

Where as a result of a request for information or an inspection or otherwise, the regulator becomes aware of an actual or potential breach of the monitoring and reporting conditions in the permit, the regulator may serve an enforcement notice on an operator.[6] The notice, which may be withdrawn at any time,[7] must:

a) state that the regulator is of the opinion that an operator has contravened, is contravening or is likely to contravene any monitoring or reporting condition;

b) specify the matters constituting the contravention or the matters making it likely that the contravention will arise;

c) specify the steps which must be taken to comply with the monitoring and reporting condition or to the extent possible to remedy any failure to comply; and

d) specify the period within which those steps must be taken.[8]

---

[3] Regulation 35(2).
[4] Regulation 35(3).
[5] Subject to Regulation 3(4) (Crown premises), the Environment Agency and Scottish Environment Protection Agency are given powers under section 108 of the Environment Act 1995 and the Department of the Environment in Northern Ireland has powers of entry set out in Regulation 27 of the Pollution Prevention and Control Regulations (Northern Ireland) 2003 (supra. chapter 11, note 17). Regulation 31 of the ETS Regulations sets out the powers of entry for the Secretary of State as regulator of offshore installations. These powers of entry are backed up by offences in the relevant legislation including obstructing an authorised person in the exercise of his powers or duties, falsely pretending to be an authorised person, failing to provide facilities or assistance, and preventing a person from answering questions asked by an authorised person.
[6] Regulation 29(1).
[7] Regulation 29(3).
[8] Regulation 29(2).

In addition to the power to issue an enforcement notice, the ETS Regulations provide for the regulator to determine the reportable emissions from an installation where:

a) an operator notifies the regulator of factors which might prevent it from complying with the monitoring and reporting conditions of the permit and requests the regulator to determine all or part of its annual emissions;[9] or

b) an operator fails to submit a verified report of its emissions in accordance with the permit or in accordance with a notice of surrender or revocation notice.[10]

In determining emissions the regulator must take into account the requirements in Annex V of the Directive[11] and notify the emissions figure to the operator[12] and the registry administrator. This figure will then be used to determine the number of allowances which an operator is required to surrender.[13] This procedure ensures that even where the operator of an installation does not comply with the requirements of the scheme, it is possible to determine the emissions from the installation and therefore the number of allowances which he is obliged to surrender. The regulator may recover the cost of determining reportable emissions from the operator.[14]

The ETS Regulations do not set a time limit for the determination of emissions by the regulator, as the time needed to determine emissions will depend on the number of operators requiring determinations and the data available to assist the regulator. There is, however, no extension of the deadline for surrendering allowances. Where the regulator does not determine the reportable emissions from an installation prior to 30 April, the operator will need to estimate the number of allowances to surrender. If the allowances surrendered by 30 April deadline turn out to be insufficient to cover the emissions determined by the regulator, the operator will be treated as any other operator which fails to fully comply with the obligation to surrender sufficient allowances to cover its emissions in the previous calendar year.

---

[9] Regulation 30(1)(a).
[10] Regulation 30(1)(b) and (c).
[11] Regulation 30(2).
[12] Regulation 30(3).
[13] Regulation 30(3) and (4). The notice will be treated as an instruction to the registry administrator for the purposes of Article 51(2) of the Registries Regulation.
[14] Regulation 30(5).

In addition to the enforcement mechanisms for identifying and remedying breaches of the monitoring and reporting requirements, there are two penalties designed to penalise operators who fail to comply and to deter non-compliance.

The Emissions Trading Directive itself[15] provides for operator accounts to be blocked where an operator fails to have its emissions report verified as satisfactory by 31 March each year. This requirement is given effect by Article 27 of the Registries Regulation. The ETS Regulations extend this to where an operator fails to comply with the requirement to submit a report of emissions under a notice of surrender or a revocation notice.[16]

As an ultimate sanction a regulator may revoke the greenhouse gas emissions permit.[17]

The ETS Regulations also make it an offence to fail to comply with the monitoring and reporting conditions in the permit[18] or to fail to comply with the requirements of an enforcement notice.[19] A failure to comply with an obligation to monitor and report emissions contained in a notice of surrender or revocation notice will be treated in the same way as a failure to comply with a condition of the permit.[20]

B.   Penalties for Failure to Surrender Sufficient Allowances

Another key requirement of an emissions trading scheme, as discussed above, is the requirement to surrender allowances to cover emissions in the previous scheme year.[21]

Where an operator fails to comply with this requirement, it will be liable to a civil penalty known as the 'excess emissions penalty'.[22] The amount of the excess emissions penalty is prescribed by the Emissions Trading Directive: 40€ in the first phase and 100€ in the second and subsequent

---

[15] Second paragraph of Article 15.
[16] Regulation 26(14). Where the registry administrator blocks an account it must notify the operator of why the account has been blocked and the period for which it will be blocked (Regulation 26(16)).
[17] Regulation 17(1). For further information on the revocation of permits, see section II of chapter 14.
[18] Regulation 38(1)(b).
[19] Regulation 38(1)(d).
[20] Regulation 16(13) and (14) and Regulation 17(11) and (12) provide that a requirement in a notice of surrender or revocation notice to prepare a report will be treated as if it were a monitoring and reporting condition, that is a condition of the permit relating to the monitoring and reporting of emissions from the installation.
[21] For information on the obligation to surrender allowances, see section I of chapter 9 and section VI of chapter 15.
[22] Regulation 39(1).

phases.[23] The amount of the penalty is converted into sterling using the rate of conversion in the C series of the Official Journal of the European Communities in September of the year preceding that in which liability to the penalty arose.[24] The total amount of the penalty which an operator is required to pay is calculated by multiplying the excess emissions of the installation (i.e. the amount of tonnes of CO_2 equivalent by which the annual reportable emissions from the installation exceeded the number of allowances surrendered) by the excess emissions penalty.[25]

The ETS Regulations also provide for the excess emissions penalty to apply where an operator understates its reportable emissions in a report required to be prepared in relation to the surrender or revocation of a permit.[26] However, in this case the operator may avoid liability to the penalty if, before 30 April, he surrenders or cancels allowances equal to the amount of the understatement and satisfies the appropriate authority that he did not knowingly or recklessly understate the emissions from the installation.[27]

Civil penalties are administered by the regulator. Where a person is liable to a penalty, the regulator will calculate the amount of the penalty and notify the person liable.[28] The penalty will be due two months from the date of notification and will be paid to the regulator[29] who will pass the penalties on to the Secretary of State a relevant devolved administration.[30] If it is paid after that date it will carry interest from the due date until the date it is paid,[31] charged at a rate of one percent above LIBOR on a day-to-day basis.[32] The amount and any interest is recoverable on demand.[33]

As soon as possible after the passing of the 30 April deadline for surrendering allowances, the regulator will publish a list of the operators who are liable to the excess emissions penalty.[34]

The application of an excess emissions penalty does not relieve the operator from the continuing obligation to surrender allowances to cover

[23] Article 16(3) and (4) of the Directive and Regulation 39(3).
[24] Regulation 39(4).
[25] Regulation 39(2).
[26] Regulation 40.
[27] Regulation 40(2).
[28] Regulation 41(2).
[29] Regulation 41(3).
[30] Regulation 41(8).
[31] Regulation 41(4).
[32] Regulation 41(5) and (6).
[33] Regulation 41(7).
[34] Regulation 36. The requirement to publish is subject to Regulation 37(2) which excludes information from publication if it would be contrary to the interests of national security.

all its emissions in the previous year.[35] The operator is required to surrender allowances by 30 April in the following year to cover the shortfall. If he does not do so a further excess emissions penalty will be applied.

In addition, where an operator fails to comply with the obligation to surrender allowances, the operator holding account for the installation will be blocked until sufficient allowances have been surrendered.[36] When the account is blocked, no allowances may be transferred out of the account except for the purposes of surrendering them or for the cancellation and replacement of the allowances at the end of a scheme phase.[37]

This penalty is in addition to the penalties required by the Directive and, so far as allowances are held in the operator holding account, provides an incentive for operators to surrender any outstanding allowances as soon as possible in order to resume trading through its operator holding account.

Finally, as an ultimate sanction, a regulator may revoke the permit of an operator who fails to comply with the obligation to surrender allowances.[38]

C. Failure to Surrender Allowances as Required by a Notice of Surrender or Revocation Notice

A failure to surrender allowances by the date specified in a notice of surrender or revocation notice will be treated as a failure to comply with an obligation in a permit to surrender allowances.[39] An operator which fails to surrender sufficient allowances to comply with a notice of surrender or revocation notice would therefore be liable to a civil penalty and the registry administrator will be informed so that its account in the registry can be blocked.[40] The obligation to surrender the outstanding allowances will continue and a further civil penalty imposed each year[41] until sufficient allowances have been surrendered or the regulator certifies that there is no reasonable prospect of recovering

---

[35] See Article 16(3) and (4) and Regulation 10(4).
[36] Regulation 26(8). Where an account is blocked, the registry administrator will notify the operator of the reason and the period during which the account will be blocked (Regulation 26(16)).
[37] Regulation 26(15). For more details on how allowances are surrendered and cancellation and replacement see sections I and II of chapter 9.
[38] Regulation 17(1). For more detail on revocation see chapter 14.
[39] Regulations 16(14) and 17(12).
[40] Regulations 16(15), 17(13) and 26(8).
[41] Regulations 16(10) and 17(8).

further allowances. In the latter case, a further civil penalty will be applied at the point of certification.[42]

D. Failure by Pool Administrator to Surrender Allowances

Where an installation is included in a pool[43] and the pool administrator fails to surrender sufficient allowances in respect of one or more of the installations, the pool administrator will be liable to pay a civil penalty.[44] However, if the pool administrator fails to pay the civil penalty on time (that is, within two months of the date on which he is notified of the penalty) the notice authorising the pool will be revoked.[45] Each operator included in the pool will then become responsible for the emissions from his own installation.

E. Over-allocation as a Result of False or Misleading Information

As described in chapter 15 above, allowances are allocated for each scheme phase in advance. Whilst certainty about allocations is necessary to establish a market for trading allowances, allocation in advance creates a risk that operators may provide false or misleading information at the time that the allocation of allowances is determined in order to receive more allowances. It was therefore necessary to develop a mechanism to prevent operators from benefiting from providing false or misleading information.

In developing such a mechanism, the distinction between the allocation and issue of allowances is important. Allocation is the process of earmarking particular allowances to particular operators. However, it is only at the point of issue that the allowances move into the operator's possession and can be physically traded in the emissions trading market.

The characteristics of an allowance are defined in the legislation and therefore it is, theoretically at least, possible to provide for an allowance to be issued subject to a right to revoke if it is discovered that there has been an over-allocation. However, such an approach would have negative impacts for the market in allowances as it would require buyers to consider whether allowances had been properly allocated.

The ETS Regulations do not therefore make any provision for allowances to be revoked once issued. Instead they allow for the Secretary of State

---

[42] Regulations 16(11) and (12) and 17(9) and (10).
[43] For a discussion of the possibility of pooling see section III of chapter 7 and section III of chapter 16.
[44] Regulations 27(12)(c) and 39.
[45] Regulation 27(14).

or the regulator to make a supplementary decision adjusting the number of allocated allowances to be issued to the operator in the future in order to compensate for the earlier over-allocation.

Regulators and the Secretary of State have a discretion whether or not to make a supplementary decision.[46] However, once a decision is made there is no discretion in relation to the amount of the reduction.

A supplementary decision may be made in two situations:

a)   First, where following an allocation of allowances it is discovered that that operator provided false or misleading information and as a result was allocated a higher allocation of allowances than it is entitled to.

b)   Secondly, where an operator provides false or misleading information in relation to an application to retain allowances and as a result the number of allowances which he is allowed to retain is higher than the number of allowances which he should have been allowed to retain in accordance with the approved NAP.[47]

Supplementary decisions will be made by the regulator or the Secretary of State depending on who made the decision on the allocation of allowances which is affected by the false or misleading statement. This means that a supplementary decision will be made by the Secretary of State where the false or misleading information was provided in relation to the Secretary of State's decision on allocation prior to the start of the phase or for an allocation from the late entrant reserve, and by the regulator where the false or misleading statement was made in connection with an application for an allocation from a new entrant reserve or an application to retain an allocation of allowances.

A supplementary decision is made by serving a notice on the operator.[48] The decision must identify the false or misleading statement and the

---

[46] On 21 December, the Department for Environment, Food and Rural Affairs announced a decision not to make supplementary decisions in the first phase of the scheme (although it reserved the right to make a supplementary decision in respect of any blatant and very significant over-allocation that come to light in the remainder of the first phase of the scheme). It indicated that Regulations 25(7) will be applied rigorously in the second phase of the scheme and that criteria for the application of the regulation will be published. The criteria will take into consideration the materiality of the error and the significance of the over-allocation.

[47] Regulation 25(1),(7) and (16).

[48] Regulation 25(1) and (7).

amount of the over-allocation resulting from that statement and set out the steps which will be taken by the regulator or Secretary of State.[49]

Where a regulator or the Secretary of State makes a supplementary decision, it is required to take such steps as it considers necessary to ensure that the amount of allowances issued to the operator in the future is reduced by the amount of the over-allocation.[50] Where allowances were allocated in respect of the installation in the final allocation decision, this will require the Secretary of State to notify a correction to the national allocation plan table held in the registry.[51] In the case of an installation which received an allocation from the new entrant reserve or late entrant reserve, the regulator would need to notify the registry administrator of an amendment to the instruction to issue allowances.

Where an over-allocation is made from the new entrant reserve and there are insufficient unissued allowances to cover the over-allocation, the regulator must notify the Secretary of State who may make a supplementary decision to reduce the number of allowances allocated by the Secretary of State.[52] Conversely a supplementary decision by the Secretary of State may include directions to the regulator to reduce allowances issued from the new entrant reserve. [53]

If there are insufficient unissued allowances in the scheme phase in which the error is discovered to cover the over-allocation, the number of allowances to be issued to the operator in the subsequent scheme phase may be reduced.[54] Such a decision will be made by the Secretary State. This may represent an additional penalty since the cost of allowances may be higher in subsequent phases to reflect the fact that the total quantity of allowances allocated will be lower.

A supplementary decision by the Secretary of State is subject to similar obligations to publish as decisions on allocation.[55] A notice of a

---

[49] Regulation 25(4), (8) and (16).
[50] Regulation 25(5) and (9).
[51] In accordance with the procedure set out in Articles 38(2) and 44(2) of the Registries Regulation.
[52] Regulation 25(6) and 25(8)(b).
[53] Regulation 25(10). Directions are given directly to the regulator where it relates to an installation in England. Where the supplementary decision relates to an installation in Scotland, Wales or Northern Ireland, the Secretary of State may arrange with the devolved administrations for a direction to be given to the regulator under Regulation 42 (Regulation 25(10)(b)).
[54] Regulation 25(9)(b) and (16).
[55] Regulation 25(11) to (14). The requirement to publish is subject to Regulation 37(1) which excludes information from publication if it would be contrary to the interests of national security.

supplementary decision will take effect two months from the date it is served.[56] Up to this point, it may be withdrawn.[57]

F.   Penalties for Failure to Comply with the Terms and Conditions of the Registry

Where an operator or other person holding an account in the registry fails to comply with any terms or conditions for the use of the account, the registry administrator may block accounts of that type held in the registry in the name of that person until the operator or other person complies with the terms and conditions.[58] The blocking of the account would prevent the transfer of any allowances out of the account other than for the purposes of surrender or cancellation and replacement.[59]

Where, following the blocking of a person holding account, an account holder continues to fail to comply with the terms or conditions, the registry administrator may close the account. Before closing the account it must serve a notice on the operator indicating that the account will be closed on the expiry of a specified period.[60] If after the notice has been served, the account holder starts to comply with the terms and conditions, the registry administrator may withdraw the notice.[61]

II.   Offences

The ETS Regulations create a number of criminal offences for failure to comply with the requirements placed on operators under the scheme. These include, for example, failing to hold a permit for the activity being carried out; failing to comply with or contravening a condition of a permit (except where the breach leads to a civil penalty); failing to notify the regulator of a change in operation; failing to comply with an enforcement notice; making a statement which the person knows to be false or misleading in a material particular, or recklessly making a statement which is false or misleading in a material particular, where the statement is made for the purposes of obtaining the grant, variation, transfer or surrender of a permit, or as part of verifying a report required under a monitoring or reporting condition of a permit.[62]

---

[56] Regulation 25(4) and (8).
[57] Regulation 25(15).
[58] Regulation 26(11) (operator holding accounts) and 26(9)(a) (person holding accounts). Where the registry administrator blocks an account it must notify the account holder of the reason and the period during which the account will be blocked (Regulation 26(16)).
[59] Regulation 26(15).
[60] Regulation 26(9)(b).
[61] Regulation 26(10).
[62] Regulation 38(1). Additional offences in relation to the powers of entry are set out in section 110 of the Environment Act 1995, and Regulation 18(1)(f) of the Offshore Combustion Installations (Prevention and Control of Pollution) Regulations 2001, S.I. 2001/1091.

These offences may incur, on summary conviction, a fine of up to the statutory maximum and up to 3 months imprisonment or, on conviction in the Crown Court, an unlimited fine and up to two years imprisonment.[63]

Where an offence is committed by a body corporate or a partnership, the director, manager or similar officer may be held liable for the offence.[64] Such an offence may be committed not only where a director had direct involvement in the commission of offence but also where he or she turns a blind eye to the commission of the offence or negligently fails to perform his duty.[65]

Further where a person commits an offence as a result of the act or default of another person, that other person may be charged or convicted of the offence regardless of whether proceedings are brought against the person who committed the offence.[66] It would, therefore be possible, for an employee to be prosecuted for an offence committed by a corporate operator.

III. Application to the Crown

The ETS Regulations apply to the Crown and people who work for the Crown.[67] However, the Crown cannot be held criminally liable for a contravention of the Regulations.[68] Instead the regulator can apply to the High Court, or in relation to Scotland the Court of Session, to have an act or omission of the Crown contravening the ETS Regulations declared unlawful.

---

[63] Regulation 38(2).
[64] Regulation 38(3) and (4).
[65] *Huckerby v Eliott* [1970] 1 All ER 189; *Re Hughes* [1943] 2 All ER 269.
[66] Regulation 38(5).
[67] Regulation 3(1) and (3).
[68] Regulation 3(2).

CHAPTER 18

APPEALS

The ETS Regulations allow operators to appeal against decisions made under the Regulations. Appeals against decisions of the regulator or registry administrator are made to the appropriate authority for the area in which the installation is situated.[1] Other types of appeal are made to the original decision maker for reconsideration of its decision.[2]

The table in section I below sets out the decisions against which an appeal can be made, the time-limits for appeal (and any possibility of extension) and the effect of making an appeal. The procedure for appeals is set out in section II.

---

[1] Regulation 32. The appropriate authority for installations in England and offshore installations is the Secretary of State. For installations in a devolved region it is the relevant devolved administration (see definition of 'appropriate authority' in regulation 2).
[2] Regulation 33.

I.   Rights of Appeal

Decision	Appeals	Time for making an appeal	Suspension while appeal determined?
Refusal to grant a permit	Appeal against refusal – regulation 32(1)(a)	Within 6 months of refusal – paragraph 2(1)(a) schedule 2 – although a later application may be accepted paragraph 2(2)	N/A
Grant of a permit	Appeal against provisions of permit – regulation 32(1)(c)	Within 6 months of the grant of the permit – paragraph 2(1)(a) schedule 2 – although a later application may be accepted – paragraph 2(2)	No suspension – regulation 32(9)(a)
Refusal of an application for variation of a permit	Appeal against refusal – regulation 32(1)(b)	Within 6 months of the refusal – paragraph 2(1)(a) schedule 2 – although a later application may be accepted – paragraph 2(2)	N/A
Variation of a permit following an application for variation	Appeal against provisions of the permit following variation-regulation 32(1)(c)	Within 6 months of the variation – paragraph 2(1)(a) of schedule 2 although a later application may be accepted – paragraph 2(2)	No suspension – regulation 32(9)(a)
Refusal of an application to transfer a permit	Appeal against refusal – regulation 32(1)(d)	Within 6 months of the refusal – paragraph 2(1)(a) schedule 2 although a later application may be accepted – paragraph 2(2)	N/A
Variation of a permit to take account of a transfer	Appeal against provisions of permit to take account of the transfer – regulation 32(1)(d)	Within 6 months of variation – paragraph 2(1)(a) although a later application may be accepted – paragraph 2(2)	No suspension – regulation 32(9)(a)
Refusal of an application for surrender	Appeal against refusal – regulation 32(1)(e)	Within 6 months of refusal – paragraph 2(1)(a) schedule 2 although a later application may be accepted – paragraph 2(2)	N/A
Service of a notice of surrender	Appeal against provisions of a notice of surrender – regulation 32(1)(e)	Before the date on which the notice of surrender takes effect – paragraph 2(1)(b) schedule 2	Suspension - regulation 32(10)(a)

Decision	Appeals	Time for making an appeal	Suspension while appeal determined?
Determination of reportable emissions	Appeal against the determination - regulation 32(1)(f)	Within 2 months of the date of notice of determination – paragraph 2(1)(d) schedule 2	No suspension – regulation 32(9)(b) but regulation 32(12) provides that the determination will not be used to assess compliance with the requirement to surrender allowances until the appeal has been determined or withdrawn
Variation by regulator under regulation 14(9)	Appeal against provisions of the permit following variation – regulation 32(2)	Within 2 months of the date of variation (paragraph 2(1)(c) schedule 2) although a later application may be accepted (paragraph 2(2))	No suspension – regulation 32(9)(c)
Service of a revocation notice	Appeal against revocation notice – regulation 32(2)	Before date on which the revocation notice takes effect -Paragraph 2(1)(b) schedule 2	Suspension – regulation 32(10)(a)
Service of an enforcement notice	Appeal against the enforcement notice – regulation 32(2)	Within 2 months of the date of the notice – paragraph 2(1)(c) of schedule 2 although a later application may be accepted (paragraph 2(2))	No suspension – regulation 32(9)(c)
Delaying allowances in the final decision on allocation	Appeal against instruction to registry administrator not to issue allowances Where instruction is given by: (a) the regulator – regulation 32(3)(a) (b) the responsible authority – regulation 33(2)(b); or (c) the Secretary of State - regulation 33(3)(b)	Within 2 months of the date of the instruction - paragraph 2(1)(c) schedule 2 although a later application may be accepted (paragraph 2(2))	No suspension – regulations 32(9)(d) and (e) and 33(5)(b)
Refusal of an application for an allocation from the new entrant reserve	Appeal against refusal – regulation 32(3)(b)	Within 15 working days of refusal – paragraph 2(1)(e) schedule 2	N/A
Refusal of an application for an allocation from the late entrant reserve	Appeal against refusal of an allocation or against the amount of an allocation – regulation 33(3)(b)	Within 10 working days of notice of determination – paragraph 2(1)(h) schedule 2	N/A

Decision	Appeals	Time for making an appeal	Suspension while appeal determined?
Allocation from the new entrant reserve	Appeal against terms of a allocation or conditional allocation – regulation 32(3)(b)	Within 15 working days of refusal – paragraph 2(1)(e) schedule 2	No suspension – regulation 32(9)(d)
Refusal of an application to retain allowances after closure	Appeal against refusal - regulation 32(3)(c)	Within 15 working days of refusal – paragraph 2(1)(e) schedule 2	Para 32(11) provides that where a refusal is appealed the application will not be considered to have been refused for the purposes of regulation 24(11) unless the refusal is affirmed on appeal or the appeal is withdrawn
Issue of a retention notice	Appeal against the terms of a retention notice – regulation 32(3)(c)	Within 15 working days of refusal – paragraph 2(1)(e) schedule 2	Suspension – regulation 32(10)(b)
Supplementary decision by regulator	Appeal against notice of supplementary decision – regulation 32(3)(d)	Within 2 months of the supplementary decision – paragraph 2(1)(f) schedule 2	Suspension – regulation 32(10)(b)
Decision to impose civil penalty – regulation 41(2)(b)	Against notice – 32(3)(e)	Within 2 months of the notice – paragraph 2 (1)(f) schedule 2	Suspension – regulation 32(10)(b)
Decision to block account in the registry	Appeal against the blocking of the account – regulation 32(4)(a)	Within 2 months of the decision – paragraph 2(1)(f) schedule 2	No suspension – regulation 32(9)(f)
Decision to close a person holding account	Appeal against closing account - regulation 32(4)(b)	Within 2 months from decision – paragraph 2(1)(f) schedule 2	Suspension – regulation 32(10)(c)
Refusal to forward pool application to the Commission	Against decision not to allow pool or not to accept amendment – regulation 33(1)	Within 2 months of the decision – paragraph 2(1)(c) schedule 2 – although a later application may be accepted –paragraph 2(2)	N/A
Notice revoking pool	Against decision to revoke authorisation of the pool – regulation 33(1)	Within 2 months of decision – Paragraph 2(1)(c) of schedule 2 although a later application may be accepted – paragraph 2(2)	Suspension – regulation 33(6)(b)
Notice amending the pool to take account of the surrender, revocation or transfer of a permit relating to an installation in the pool	Appeal against notice – regulation 33(1)	Within 2 months of decision – Paragraph 2(1)(c) of schedule 2 although a later application may be accepted – paragraph 2(2)	No suspension – regulation 33(5)(a)

Decision	Appeals	Time for making an appeal	Suspension while appeal determined?
Issue of temporary exclusion certificate under regulation 11(6)	Appeal against certificate of temporary exclusion – regulation 33(2)(a)	Within 2 months from date of certificate or notice – paragraph 2(1)(g) schedule 2	No suspension – regulation 33(5)(b)
Revocation of temporary exclusion certificate under regulation 11(9) or 11(11)(c)	Appeal against revocation notice – regulation 33(2)(a)	Within 2 months of notice revoking certificate or notice – paragraph 2(1)(g) of schedule 2	No suspension – regulation 33(5)(b)
Issue of certificate of temporary exclusion to take account of a transfer under regulation 11(11)(a) or (b)	Appeal against notice or certificate recognising the transfer – regulation 33(2)(a)	Within 2 months of issue of certificate – paragraph 2(1)(g) of schedule 2	No suspension – regulation 33(5)(b)
Supplementary decision by Secretary of State under regulation 25(7)	Appeal against supplementary decision – regulation 33(3)(a)	Within 2 months of the decision – paragraph 2(1)(f) schedule 2	Suspension -regulation 33(6)(a)

II.   Procedure and Powers

A.   Making and Determination of Appeals

Schedule 2 of the ETS Regulations sets out the procedure for appeals.[3]

An appeal must be made in writing and attach relevant documents.[4] Where an appeal is made, persons having a 'particular interest' in the appeal must be notified of the appeal and given opportunity to comment.[5]

The appeal may be dealt with through written representations[6] or a hearing may be held. The Secretary of State or relevant devolved administration may in any case choose to appoint a person to hold a hearing.[7] However, a hearing must be held if the operator or, where appropriate, the regulator or registry administrator request a hearing. A hearing may be held wholly or partially in private.[8]

An appeal may be withdrawn at any time.[9]

B.   Outcome of Appeals

Where an appeal is made against the decision of the regulator or registry administrator, the Secretary of State or relevant devolved administration may affirm the decision; quash all or part of the decision; vary the decision; or give directions to the regulator or registry administrator.[10]

Where appeal is made to the original decision-maker, the decision-maker must reconsider its original decision and may affirm, reverse or vary that decision.[11]

The Secretary of State or relevant devolved administration must give notice of his determination of the appeal to the parties to the appeal and

---

[3] The procedure for appeals under regulation 32 relating to installations in Northern Ireland is set out in schedule 4.

[4] Paragraph 1(1) and (2) of schedule 2. Where an appeal is made against the decision of the registry administrator, regulation 32(5) provides that references in schedule 2 to 'regulator'
shall be read as references to the 'registry administrator '.

[5] Paragraph 3 of schedule 2.

[6] Paragraph 5 of schedule 2.

[7] Paragraph 4(1) and (2) of schedule 2. Paragraph 4(3) to (9) set out the procedure for hearings.

[8] Paragraph 4(2) of schedule 2.

[9] Paragraph 1(3) of schedule 2.

[10] Regulation 32(7).

[11] Regulation 33(4). See also regulation 32(8) in relation to an appeal against a decision of the Secretary of State in his capacity as regulator of offshore installations.

to any person who made representations in writing or at a hearing and, where a hearing was held, provide a copy of the report of the hearing.[12]

If the determination of an appeal is successfully challenged by way of judicial review, the Secretary of State (or devolved administration) is required to notify everyone who was notified of his original determination and invite further representations for the purposes of re-determining the appeal.[13] He may hold or reopen a hearing.[14] The Secretary of State must give notice of his re-determination of the appeal to the parties to the appeal and any person who made representations.[15]

C. Delegation of Powers to Determine Appeals

The Secretary of State (or devolved administration) has two options for delegating its powers in relation to appeals:

a)   he may appoint a person to determine an appeal (or any element of an appeal) on her behalf;[16] or

b)   he may refer any matter involved in an appeal to a person (with or without payment), for that person to report on that matter him.[17]

An appointment to determine appeals must be in writing[18] and may relate to one or more specific appeals (or specific elements of the appeals) or to a wider category of appeals[19] and may be subject to conditions.[20] It may be repealed at any time by notice in writing.[21]

The decisions of the appointed person will be treated as acts of the Secretary of State or devolved administration[22] and the appointed person will have the same powers and duties in relation to the appeals covered by the appointment.[23]

---

[12] Paragraph 6(1) and (2) of schedule 2.
[13] Paragraph 7(a) and (b) of schedule 2.
[14] Paragraph 7(c). The same rules apply to a hearing held following a court order quashing the determination of an appeal as apply to a hearing held prior to the determination.
[15] Paragraph 6 of schedule 2 will apply to the re-determination of appeals as it applies to the determination of an appeal (paragraph 7).
[16] Regulation 34(2)(a).
[17] Regulation 34(2)(b).
[18] Paragraph 2 of schedule 3.
[19] Paragraph 2(a) of schedule 3.
[20] Paragraph 2(b) of schedule 3.
[21] Paragraphs 2(c) and 5 of schedule 3.
[22] Paragraph 6 of schedule 3. This does not include any acts in relation to the contract of appointment and any criminal offences.
[23] Paragraph 3 of schedule 3. This does not include the power to appoint a person to hold a hearing or to refer a matter or question to that person.

The ETS Regulations permit the appointed person power to hold a local inquiry or other hearing.[24] He must hold a hearing if requested to by either party to the appeal[25] and he must hold a local inquiry if directed to do so by the Secretary State or devolved administration.[26] Where an appointed person holds a local inquiry or hearing, the Secretary of State or devolved administration may appoint an assessor to sit with the appointed person at the inquiry or hearing and advise him on any matters arising (although the appointed person is to determine the appeal).[27] Where an inquiry or other hearing is held the Secretary of State or devolved administration will pay for the hearing.[28] However, the Secretary of State or devolved administration may give directions requiring a party to the hearing to pay the costs of the hearing and the costs of other parties.[29]

In exercise of the power to delegate function of determining appeals in England, the Secretary of State has appointed the Planning Inspectorate to hear appeals against decisions taken by the regulator relating to greenhouse gas emissions permits.

---

[24] Paragraph 4(2) of schedule 3. Paragraph 4(5) of schedule 3 provides for section 250(3) and (4) of the Local Government Act 1972 (c.70) to apply – these allow for the possibility of issuing a summons to require a person to attend to give evidence or to produce documents relating to the inquiry or hearing. It is an offence to fail to comply with such summons. Paragraph 4(6) of schedule 3 provides for section 210 of the Local Government (Scotland) Act 1973 (c.65) to apply in relation to an appeal to Scottish Ministers.
[25] Paragraph 4(1) of schedule 3.
[26] Paragraph 4(2) of schedule 3.
[27] Paragraph 4(3) of schedule 3.
[28] Paragraph 4(4) of schedule 3.
[29] Paragraph 4(5) of schedule 3 and section 250(4) and (5) of the Local Government Act 1972.

CHAPTER 19

UK NAP FOR THE FIRST PHASE: DEVELOPMENT AND ALLOCATIONS TO EXISTING INSTALLATIONS

This chapter contains in section I an account of the development of the UK NAP for the first phase of the scheme. Section II explains how the finally approved NAP for the first phase allocated allowances to existing installations. As such, this chapter is largely of historic interest.

The NAP for the first phase of the scheme continues to be relevant, however, for the allocation of allowances to any new installations put into operation until the end of 2007 or installations which close before that date. These aspects are considered separately in chapter 20.

The UK NAP for the second phase of the scheme is considered in chapter 21.

I.   Development of the UK NAP for the First Phase

Work on the development of the UK NAP for the first phase of the scheme began in 2003. The finalised installation-level allocations were published in May 2005 alongside a consolidated version of the NAP incorporating the methodology as approved by the European Commission.

The first formal consultation on allocation arrangements was published in August 2003. Given that Article 10 of the Directive requires 95 per cent of allowances to be distributed for free in the first phase and that generally allocations must be fixed well in advance of the start of the relevant phase,[1] the consultation documents focused on determining how to calculate allocations on the basis of historical or current information. The consultation also sought views on related issues including the treatment of new entrants and closures, whether or not a restriction should be imposed on the banking of allowances between the first and second phases, the use of auctioning, the use of the provisions in the Directive on temporary exclusion and the interpretation of the term 'combustion installations' proposed by the UK.[2] At the same time as this consultation was being carried out and the results analysed,

---

[1] With the exception of allocations to new entrants discussed in chapter 20.
[2] Full details of this consultation including the consultation document and summary of responses received can be found at http://www.defra.gov.uk/corporate/consult/eu-emissions/index.htm.

Defra and the UK regulators continued to identify installations falling within the scope of the scheme and to collect historic data from those installations. This ensured that Defra was in a position to publish provisional installation level allocations alongside the proposed methodology in the draft NAP.

A draft NAP was published in January 2004 together with a consultation paper and draft regulatory impact assessment.[3] The plan followed the common format provided for in the annex to the 2003 Commission Guidance.[4] It explained how the total quantity of allowances had been determined and how allowances would be distributed to installations and drew extensively on the views expressed in the earlier consultation. Following strong support during the earlier consultation, a two-stage allocation methodology was proposed. Sector totals were calculated on the basis of the total 'with measures' projected emissions of the installations in each sector.[5] The sector totals were then to be distributed on the basis of each installation's share of the sector's historic emissions during the baseline period of 1998-2002.[6] The draft plan also proposed that new entrants during the first phase would receive allowances for free from a 'new entrant reserve' set aside from the total for this purpose and that plants which closed during the phase would lose their entitlement to allowances not yet issued.

In May 2004, the UK notified a NAP to the Commission.[7] The methodology proposed was closely based on that set out in the draft NAP but had been refined to take account of the principal concerns raised by stakeholders in the January consultation. In particular, it proposed that account should be taken of significant changes in the activities undertaken at installations during the baseline period (e.g. commissioning periods, the opening and closure of units within an installation and the transfer of production from closed sites). It also proposed that the baseline period should be extended to 2003. The consultation document published alongside the provisional NAP sought views on these proposed refinements, and decisions on the application

---

[3] The consultation documents are available at http://www.defra.gov.uk/corporate/consult/eu-etsnap/index.htm.

[4] Supra chapter 4, note 3.

[5] 'With measures' projections already take account of the projected effect of policies and measures for which there were already firm plans in place e.g. the Climate Change Agreements and UK Emissions Trading Scheme (see chapters 23 and 24) and are therefore significantly lower than 'business as usual' emissions.

[6] It was also proposed that in order to ensure that there was no incentive to over-report emissions during the baseline period, reports would need to be verified by third party verifiers.

[7] Available at: http://www.defra.gov.uk/corporate/consult/euetsnap-stagethree/index.htm.

of additional rules were published in July 2004.[8] The consultation document also sought views on further details of the proposed treatment of new entrants and closures.

The plan notified to the Commission indicated that it was provisional in that the figures it contained (including the total figure of 736 million allowances and the sector totals) would change once underlying data was finalised. This included the updating by the Department of Trade and Industry of the UK's energy and emissions projections,[9] the review of Climate Change Agreement targets and the submission of verified emissions data by operators of UK installations.

In July 2004, the Commission adopted a Decision stating that it had no objection to the UK plan, provided that the UK adopted amendments (a) providing further information on the way in which new entrants would begin participating in the scheme and (b) listing the installations situated in Gibraltar and the quantities of allowances allocated to them.[10] The Decision also contained provisions relating to: the restrictions on increases in the total quantity of allowances to be issued by the UK; modifications of allocations to individual installations arising from improvements in data quality; and the procedure for agreeing other amendments.

Following the completion of the outstanding work referred to in the provisional NAP, it emerged that estimated UK emissions for the period 2005-7 were considerably higher than the provisional projections used to calculate the allocations in the NAP submitted to the Commission. Therefore, in November 2004, the UK Government submitted a proposal to the Commission to increase the total quantity of allowances by around 20 million (which represented less than half the amount by which the projections had increased) to 756 million. In February 2005, the Government announced how the 756 million allowances would be allocated at installation level. To provide as much regulatory certainty to industry as possible the Government also confirmed that any allocation below 756 million would be achieved by reducing the number of allowances given to the electricity generation sector.

During February 2005, the Commission indicated that it was not prepared to consider the substance of the UK amendment on the grounds

---

[8] Available at: http://www.defra.gov.uk/environment/climatechange/trading/eu/nap/pdf/nap-decisions-0704.pdf .

[9] To support the analysis underlying the introduction of the EU ETS, DTI undertook an update of its energy and emissions projections which were last published in Energy Paper 68 in November 2000. Details of the documents published during this process are available at: http://www.dti.gov.uk/energy/sepn/uep.

[10] Supra chapter 5, note 30.

that it did not consider that the provisional total of 736 million allowances notified to them in April 2004 could be exceeded. The formal Decision rejecting the amendment was adopted by the Commission in April 2005.[11]

Following the annulment of the April Decision by the Court of First Instance in November 2005,[12] the Commission was obliged to take a new Decision on the UK amendment. In February 2006, the Commission rejected the UK amendment again, this time on the grounds that it was submitted after 30 September 2004, the date by which the Directive required Member States to take their final allocation decision.[13] In April 2006, the UK government announced that, although it still considered its NAP amendment to be compatible with the Directive, it would not launch further court action against the February 2006 Decision in the interests of establishing certainty for UK installations for the remainder of the first phase of the scheme. Given the time that it would take for a further legal challenge to be determined, there was a strong possibility that the UK would not be able to allocate the allowances within the first phase even if the challenge were successful.

II. UK NAP for the First Phase: Allocations to Existing Installations

As explained above, the UK published and submitted a provisional NAP to the Commission in May 2004. Further submissions were made to the Commission in June, September, November and December 2004 and February and March 2005, providing further, updated information requested by the Commission.

In May 2005, the UK published a consolidated version of the plan (the approved NAP) drawing together the substance of the amendments notified to and approved by the Commission.[14] This was published alongside the final list of installation level allocations. Following the Commission's rejection in April 2005 of the UK amendment to the total quantity of allowances to be allocated, these documents were prepared on the basis of the provisional total of $736MtCO_2$ approved by the Commission, with the additional 19.8 million allowances to be allocated later in the event that the UK's challenge was successful and the Commission accepted the proposed amendment.

---

[11] Supra chapter 5, note 25

[12] The UK's successful application to the Court of First Instance to annul the Commission Decision rejecting the UK's proposed amendment is considered in section IV.B of chapter 5.

[13] Supra chapter 5, note 37.

[14] Available at: http://www.defra.gov.uk/environment/climatechange/trading/eu/nap/approved.htm.

The approved NAP follows the structure set out in the 2003 Commission Guidance on the preparation of NAPs. In particular, it addressed the following issues:

a) Determination of the total quantity of allowances
b) Sector totals
c) Installation level allocations
d) New entrants and closures
e) Other issues

A. Determination of the Total Quantity of Allowances

The first section of the NAP outlined the broad approach taken by the UK to the calculation of the total quantity of allowances to be issued under the scheme, in the context of commitments made under the Kyoto Protocol and Burden Sharing Agreement. It explained how the UK had used a combination of projected emissions figures and its ongoing position as set out in the National Climate Change Programme[15] and Third National Communication[16] to inform the level of allowances to be allocated. It then discussed how policies and measures designed to address emissions from sources not covered by the scheme (such as energy efficiency and renewable energy sources) would enable a continual reduction of UK emissions, in line with domestic commitments to move towards a reduction in CO_2 emissions of 20 per cent by 2010 and 60 per cent by 2050. Further detail was set out in supporting appendices. Appendix D set out estimates of the emissions reductions which the policies in the UK Climate Change Programme are projected to achieve. Appendix F set out details of the updated energy projections from November 2004.

B. Determination of the Quantity of Allowances at Activity Level

The second section of the NAP explained that the UK had decided to adopt a two-stage approach to allocation and set out how the sector totals had been calculated. It explained that all sectors other than power stations would receive allocations equivalent to their projected emissions during the first phase of the scheme: the power station sector would receive the remainder of the total quantity of allowances. The technical detail of how individual sector totals had been calculated and in

---

[15] For further information on the UK Climate Change Programme, see section III.C of the Introduction

[16] The UK's Third National Communication under the United Nations Framework Convention on Climate Change (2001), available at: http://www.defra.gov.uk/environment/climatechange/pubs/3nc/pdf/climate_3nc.pdf.

particular the way in which the updated energy projections and revised Climate Change Agreement targets had been used was set out in Appendix B to the NAP.

C.   Determination of the Quantity of Allowances at Installation Level

Section three of the NAP discussed individual installation level allocations.

The starting point for calculating an allocation for an installation was to calculate its 'relevant emissions' during the baseline period from 1998 to 2003. In most cases, an installation's 'relevant emissions' would be its average annual emissions for the years in the baseline period when the installation was in operation, after excluding the lowest year's emissions. The sector total was then divided between installations in the sector according to each installation's share of the sector's aggregate relevant emissions.

It was decided to use historic emissions as the basis for allocation mainly for reasons of transparency and consistency and due to the infeasibility of projecting installation level emissions consistently and accurately across installations in all sectors. The choice of baseline was intended to use the widest range of available verifiable data so that both the most recent data (which represented the most likely picture for the installation in the first phase), and the earliest available data (rewarding early action) could be taken into account. The exclusion of the lowest year's emissions was intended to minimise the impact where an installation had had an anomalous year with unusually low emissions.

As explained above, consultation on the NAP raised a number of generic instances where the standard methodology disadvantaged large groups of installations under particular circumstances, e.g. where an installation had installed a new technical unit on site during the baseline period. The UK Government therefore adopted additional rules to address a number of generic situations where it appeared that the methodology led to significant anomalies. These included rules in relation to commissioning, rationalisation and changes at an installation during the baseline period. Detailed explanations of how these rules operated can be found in Appendix F to the plan.

Following publication of the decisions on these rules in July 2004, operators whose installation had undergone commissioning, rationalisation or other changes in the baseline period were asked to notify the Government and to provide any necessary supporting evidence within a specified period. Applications relating to these changes during the baseline period were analysed and operators were notified of the Government's decision.

D. Reserve for Late Installations

A small reserve of 1.5 million allowances was set aside for installations which were in operation on or before 31 December 2003 but did not receive an allocation in the final allocation decision because they did not supply information to Government in time. In order to qualify for access to this reserve, the installation had to have obtained its permit on or after 1 January 2005.

The procedure for accessing these allowances is set out in Regulation 22A of the 2005 regulations.[17]

Allocations were calculated on the same basis as incumbent installations, although a reduction factor of 10 or 25 per cent was applied. This was to reflect the fact that the installations had failed to supply the relevant information in time to receive an allocation in the final allocation decision and represented a balance between encouraging installations covered by the Directive to come forward and the expectations of other installations who supplied information to Government within tight deadlines, under the threat that they may not receive an allocation if they did not. Late installations were also not permitted to take advantage of the rules to take account of commissioning, rationalisation or other changes in the baseline period, as a strict deadline had been imposed on other installations who did not apply for these rules within a specified period.

---

[17] For a further explanation of this provision, see section III of Chapter 15.

CHAPTER 20

UK NAP FOR THE FIRST PHASE: NEW ENTRANTS, CLOSURE AND AUCTIONING

This chapter describes those aspects of the UK NAP for the first phase of the scheme which remain relevant in determining changes in allocations to installations over the remainder of the first phase up to the end of 2007. These aspects are the rules governing allocations to new entrants (section I), closure (section II), rationalisation (III) and auctioning (section IV).

I.   Allocations to New Entrants

Appendix C of the NAP sets out the manner in which new entrants are to be treated. The approach taken by the UK in the first phase was to set aside a reserve of allowances from within its total quantity of allowances, to be allocated free of charge to eligible new entrants. The reserve contains 46.8 million allowances, representing 6.3 per cent of the total quantity of allowances allocated by the UK.[1]

A.   Types of Eligible New Entrant

In order to qualify for an allocation from the new entrant reserve, the installation must fall within one of the categories of eligible new entrant set out below and apply for and obtain either a new permit or a variation of an existing permit,[2] as a result of falling within one of these categories.[3]

1.   New Installation Commencing an Activity Covered by the Scheme

This category encompasses installations which begin to carry out an activity listed in Annex I to the Directive for the first time after 31 December 2003, including installations which increase their capacity so that it exceeds one of the thresholds set out in Annex I to the Directive.

---

[1] Paragraph 2 of Appendix C to the approved NAP (supra chapter 19, note 14).
[2] Paragraph 15 of Appendix C to the approved NAP.
[3] For a discussion of the provisions of the Emissions Trading Directive relating to the allocation to new entrants see section IV.D of chapter 4 and for a discussion of the relevant provisions of the ETS Regulations, see Section II of chapter 15.

2.   Extension in Capacity

This category applies where a physical change[4] takes place at an installation after 31 December 2003, which results in an increase in net output capacity of an activity listed in Annex I to the Directive.[5]

3.   Offshore Tiebacks

This category encompasses drilling centres and oil and gas wells which serve new fields or new areas of a reservoir that are connected to existing installations which are physically offshore. It also covers modifications to existing installations which are physically offshore and which are directly related to the production, processing and delivery of the offshore oil and gas reserves. In addition, for both of these categories, the tieback or modification must result in both a quantified enhanced recovery of reserves and a quantified increased additional power demand that will generate additional emissions from the existing power plant.[6] Drilling programmes related to the enhancement projects are not eligible.[7]

4.   Increase in Good Quality Combined Heat and Power (CHP) Capacity

This category encompasses an installation whose good quality[8] CHP capacity is increased after 31 December 2003 by a minimum of 5 per cent as a proportion of total power capacity.

5.   Recommencement of Operations

This category encompasses an installation which ceased carrying out an activity covered by the scheme on or before 31 December 2003 and which subsequently recommenced the activity after 31 December 2003.[9]

B.   Allocation Process

In order to obtain allowances from the new entrant reserve, eligible new entrants must make an application to the regulator. As the reserve contains a finite number of allowances, allowances are allocated based

---

[4] In most circumstances, the physical change must involve a piece of equipment which directly produces emissions which must be accounted for under the scheme. See paragraphs 19-21 of Appendix C to the approved NAP.
[5] Paragraphs 18-25 of Appendix C to the approved NAP.
[6] Paragraphs 26 and 29 of Appendix C to the approved NAP.
[7] Paragraph 28 of Appendix C to the approved NAP.
[8] The definition of 'good quality' is set out in paragraphs 36-37 of Appendix C to the approved NAP.
[9] Paragraph 15 of Appendix C to the approved NAP.

upon a queuing system. An operator's place in the queue is determined by the date upon which the application was received by the Environment Agency.[10] In order to take its place in the queue, the application must be 'duly made', that is it must contain all of the information that the regulator requires in order to determine the application.

If the reserve is exhausted, it is likely that some eligible new entrants will not be able to receive allowances free of charge. If that were to happen, such an installation would have to obtain any allowances which it might require to comply with its obligations on the open market.

If, at the time that the reserve runs out, there are still applications which have not yet been determined or which have been approved but in respect of which allowances have not been allocated, those applications will remain in the queue. If the reserve is subsequently replenished,[11] those allowances will be distributed to the next in the queue.

An operator may make an application to the reserve before the change which forms the basis of the eligibility takes place (for example an application may be made, and determined, before a new installation commences operation). In such circumstances, the application must contain evidence that the application is not speculative, contingent, unrealistic or false.

If an application is rejected, the operator has a right to appeal to the Secretary of State within 15 working days of the rejection.[12] The operator will retain his place in the queue until the period for appeal expires and the allowances for which he has applied will not be allocated to another operator until that period has expired or, if the operator decides to appeal, until that appeal is determined.[13]

C.   Issue of Allowances

On approval of an application, the allowances to which the operator is entitled for the remainder of the phase are 'tagged'. This means that they are no longer available for allocation to other installations. Once the regulator has received confirmation that the new entrant activity has commenced, the regulator will issue the allowances tagged in respect

---

[10] For installations which are not situated in England or Wales, an application must be submitted to both the regulator who will determine the application and the Environment Agency, who manage the queue on a UK-wide basis. See paragraph 93 of Appendix C to the approved NAP.

[11] See Subsection C on the issue of allowances and section II on closures.

[12] For further information on appeals under the ETS Regulations, see chapter 18.

[13] Paragraphs 97-98 of Appendix C to the approved NAP.

of the first year of operation, within 10 working days. At the same time, the regulator will serve a notice on the registry administrator setting out the number of allowances to be issued to the installation in subsequent years of the phase.

If allowances are allocated before the change which forms the basis of the eligibility takes place and the expected start date is subsequently delayed, the allocation will be adjusted downwards to reflect the fact that the installation will start operation on a later date. This will result in a number of allowances being 'untagged'.

D.   Calculation of Allocation

The number of allowances to be allocated is determined through a standardised benchmarking methodology reflecting technology, load and fuel specific factors and the date upon which the relevant change takes place. A reduction factor is also applied to reflect each installation's contribution to the CHP reserve and an additional reduction factor is applied to installations who fall within the power station sector. This latter adjustment was to reflect the reduced allocation to be made to that sector, as a result of the Commission's refusal of the UK's proposed amendment to the total quantity of allowances to be allocated for the first phase of the scheme.[14]

An operator can calculate his eligible allocation by inserting the relevant factors into a spreadsheet[15] and must submit a copy of a completed spreadsheet with his application.[16] Certain factors will require third party verification.[17]

E.   Ring-fenced CHP Reserve

Part of the new entrant reserve has been ring-fenced for use by fully qualified good quality CHP eligible new entrants.

Eligible new entrants which comprise of fully qualified good quality CHP will make an application and be granted a place in the queue in the usual way but will receive their allocation from this element of the reserve.[18] Should this element run out they will receive their allocation from the main new entrant reserve.

---

[14] For further information, see section I of Chapter 19.

[15] Available at http://www.dti.gov.uk/energy/environment/euets/phase1/new-entrants/page26999.html.

[16] See subsection B on applications.

[17] Paragraphs 30 and 31 of Appendix C to the approved NAP,

[18] Applications which consist of partial good quality CHP will receive part of their allocation from the ring-fenced element and the other part of their allocation from the main reserve.

If the main new entrant reserve should run out but allowances still remain in the ring-fenced element, applications for fully qualified good quality CHP new entrants will be able to receive allowances from this element of the reserve even if this means jumping ahead of other applications further ahead in the queue which are not entitled to access the ring-fenced element.

Where a good quality CHP installation closes, the allowances which had been allocated to that installation and would have been issued to it in subsequent years of the first phase are transferred to the ring-fenced element of the reserve. Should this ring-fenced part of the reserve be under-utilised, the government may transfer the remaining allowances to the main new entrant reserve.

II.   Closure

Installations which close – that is, permanently cease carrying out an activity covered by the scheme (whether by ceasing the activity or by reducing it so that if falls below the threshold in Annex I of the Directive) - will retain allowances issued in the year in which the activity ceased. However, subject to the rationalisation rule,[19] no allowances will be issued to the installation for subsequent years in the first phase.

Allowances which had been allocated to that installation and were to be issued to it in subsequent years of the phase are transferred to the new entrant reserve.

III.   Rationalisation

The rationalisation rule operates as an exception to the rule that a closed installation will not receive any allowances in years after the year in which closure occurs. Rationalisation occurs where the operator ceases carrying out an activity covered by the scheme at one installation and transfers production to another existing installation. In such a case, the operator may apply to the regulator to continue to have allowances issued to the closed installation in subsequent years of the phase.

Rationalisation applies only where at least 50 per cent of the production[20] from the closing site is transferred to another site which has the same permit holder and is classified within the same sector. In addition, the production transferred must be classified as the same product according to the Standard Industrial Classification code, taken to a three-digit level.

---

[19] See section III on rationalisation.
[20] Calculated as the final output of goods and services as an average of the previous three years.

The rationalisation rule does not apply to installations in the power stations sector.[21]

Where the regulator is satisfied that an application for rationalisation meets these conditions, it will approve the application, which will entitle the operator to continue to receive a percentage of the closed installation's allocation, equal to the percentage of the production transferred, in subsequent years of the first phase.

If, subsequent to a rationalisation application being granted, the installation or to which the production was transferred itself closes, then subject to a further rationalisation application being made and approved, no further allowances will be issued to either of the closed installations.

IV.   Auctioning

Should there be a sufficient surplus of allowances remaining in the new entrant reserve, the government intends to distribute them by way of auction or sale.[22]

---

[21] Paragraph 63 of Appendix C to the approved NAP.
[22] Paragraphs 112-114 of Appendix C to the approved NAP.

CHAPTER 21

UK NAP FOR THE SECOND PHASE: DEVELOPMENT AND SUBSTANTIVE PROVISIONS

This chapter contains in section I an account of the development of the UK NAP for the second phase of the EU ETS. Section II explains how the NAP proposes to allocate allowances in the second phase.

I. Development of the NAP for the Second Phase of the Scheme

In contrast to the development of the NAP for the first phase of the scheme, the UK government did not need to devise its allocation rules from scratch, but was able to draw on lessons from the first phase by assessing each of the rules applied in the first phase and determining their appropriateness for the second phase.

As in the first phase, a significant amount of public consultation was undertaken during the course of developing the NAP.[1]

On 31 March 2005, the Government published an informal communication paper outlining its general approach to the second phase of the scheme, which sought comments on a range of issues on the operation of the scheme including on its scope and the allocation methodology. This set out the UK government's main objectives for the second phase which included consistency with national emissions reduction goals, expanding the scheme where appropriate, maintaining the competitiveness of UK industry, consistency with energy policy and facilitating the development of an efficient trading market which incentivises emissions reductions and provides appropriate signals for long term investment.

On 19 July 2005 the Government published a consultation document setting out its intended approach to the second phase of the scheme and seeking to gather information and stakeholder views on a number of issues, including on methodologies for installation-level allocations, limits on the use of credits generated from the Kyoto project mechanisms, use of auctioning, allocations to new entrants and treatment of small installations. It also expressed indications of the types of activity to which

---

[1] Documents relating to the development of the NAP for the second phase of the scheme are available at: http://www.defra.gov.uk/ENVIRONMENT/climatechange/trading/eu/phase2/index.htm.

the UK was considering expanding the scheme, seeking views on the appropriateness of including these activities and on the importance of harmonisation across the EU. Following publication of this consultation, the Government invited representatives from individual sectors to comment on more detailed papers and held a number of meetings with sectors.

On 24 November 2005 some interim decisions on the allocation methodology for the second phase of the scheme were published.

The Department for Trade and Industry launched a consultation on the emissions projections that would inform sector allocations in February 2006 and in March 2006 the Government published a detailed guide to its allocation methodology for the second phase. This set out how, in general terms, allocations would be calculated for the second phase and invited applications from operators to apply for the special allocation methodology rules. This was followed by the publication for consultation later in March 2006 of a draft NAP for the second phase. This and sought responses on a number of more specific questions relating to how the final total cap should be determined, the allocation methodology for good quality CHP plants and large electricity suppliers, the level of auctioning and use of project credits and how the new entrant reserve should function. It also sought views on what other better regulation and simplification measures should be considered. It stated that the total cap would be set within a range of 3 and 8Mt of carbon below total 'business as usual' emissions projections.

On 29 June 2006, the day before the NAP was due to be notified to the Commission, the Secretary of State for Environment, Food and Rural Affairs announced in Parliament that the cap would be set at 8Mt carbon below projections, that 7 per cent of allowances would be auctioned, and the limit on the use of project credits for operators in the UK would be set at 8 per cent of their allocation, which amounted to around two thirds of the 'effort'[2] required in the second phase.

A further short consultation began on 17 July 2006 on whether certain small emitters should be excluded from the scheme for the second phase. This sought views on whether a *de minimis* threshold should be introduced and whether the UK should adopt a narrower interpretation of the provision in annex I of the Directive on ceramic producing activities. Following this consultation decisions were made to exclude some smaller emitters from the scheme through the introduction of a *de minimis* threshold of 3MW in the application of the aggregation rule[3]

---

[2] The difference between the total cap and the 'business as usual' projections.
[3] See section IV.B of chapter 2.

and the narrowing of the UK's interpretation of the activity of ceramic producing activities.

On 21 August 2006, the NAP was published and sent to the Commission.[4] Further information was notified to the Commission in October and November. Also on 21 August, the Government commenced an eight-week consultation on the draft allocations to individual installations to ensure that there were no errors in the allocations. Applications were also invited for operators to apply to take advantage of the *de minimis* rule.

On 29 November, the Commission published a Decision on the UK NAP. It found only one inconsistency with the Directive, namely that it did not list installations in Gibraltar and the proposed allocations to those installations.[5]

At the time of writing the government had announced that it would hold one further short consultation on a new list of installation level allocations before making its final allocation decision.[6]

II. Substance of the UK NAP for the Second Phase of the EU ETS

A. Determination of the Total Quantity of Allowances

The first section of the NAP outlines how the total quantity of allowances would be calculated. It stated that the UK would allocate 246,175,995 allowances per annum, a total of 1230,879,990 allowances over the five year period (2008-12). In order to compare this with allocations in the first phase of the scheme, this figure can be broken down into 236,676,739 allowances per annum for installations that were covered by the first phase of the scheme, thus implying a reduction compared to the first phase of approximately 9 million allowances per year. An additional 9,499,259 allowances would be allocated to cover emissions from types of activity which the UK had not considered to be covered by the scheme in the first phase but are now to be brought in for the second phase.[7]

This total number of allowances will be further reduced when the UK comes to take its final allocation decision so as not to allocate allowances to installations which it will exclude from the scheme as a result of the

---

[4] Available at: http://www.defra.gov.uk/ENVIRONMENT/climatechange/trading/eu/phase2/phase2nap.htm.
It was registered by the Commission on 30 August 2006.
[5] Supra chapter 4, note 36.
[6] The final allocation decision was published on 16 March 2007, and can be found at the site referred to in note 4 above.
[7] For a discussion of the issues relating to the interpretation of the term combustion installation, see section II of chapter 2.

implementation of the voluntary *de minimis* threshold and the change in interpretation of the category of ceramic producing activities.[8]

This total quantity of allowances was arrived at by deciding that the annual allocation for the new types of activities included in the second phase should be their estimated 'business as usual' projections, and the other activities (which were included in the first phase of the scheme) should be 8MtC (29.3 Mt CO_2) below 'business as usual' projections which were stated to be 266.01 $MtCO_2$ per year.

B. Determination of the Quantity of Allowances at Activity Level

The NAP goes on to explain that the 8MtC reduction in allowances against 'business as usual' projections would be borne entirely by the Large Electricity Producers ('LEP') sector. Although this corresponds largely to the 'power stations' sector used in the NAP for the first phase, there are some differences.[9] All other sectors will be allocated allowances equivalent to their projected 'business as usual' emissions taking account of contributions to the new entrant reserve (NER).[10] This follows the model used in the first phase of the scheme, where the entire burden was borne by the power stations sector. The NAP explains this difference in treatment on the basis that the LEP sector is relatively insulated from international competition and is able to pass on the cost of allowances to its customers.

The UK adopted a different way of classifying activities in the second phase: installations were classified into 19 sectors rather than the 51 used in relation to the first phase. This was an attempt to simplify the allocation procedures for the second phase but also to reflect the fact that, unlike in the first phase, Climate Change Agreement targets have not been used directly to determine projections for the second phase.

Another notable change from the first phase was the introduction of a separate sector for Good Quality Combined Heat and Power ('GQ CHP') installations.

---

[8] See sub-section I.

[9] Large Electricity Producers are defined in section 2.6 of the proposed NAP as any operator of a combustion installation (except a hazardous or municipal waste installation) (a) which has a thermal rated input of above 20MW; and (b) which generates electricity and is normally capable of exporting more than 100 MW of electrical power to either the total system in Great Britain or the total system in Northern Ireland and (c) for which the operator is not exempted under section 5 of the Electricity Act 1989 or, as the case may be, Article 9(1) of the Electricity (Northern Ireland) Order 1992 from the requirement to hold a generation licence. For detailed reasoning behind this definition, see Appendix B to the proposed NAP.

[10] The details of how the sector totals were calculated is set out in Appendix B to the NAP.

C.   Determination of the Quantity of Allowances at Installation Level

As in the first phase of the scheme, the starting point for calculating installation level allocations is the 'standard allocation methodology' which used each installation's 'relevant emissions' as its starting point. For most installations, 'relevant emissions' is the average of the installation's annual emissions for the baseline period 2000-2003 inclusive after excluding the lowest operational[11] year's emissions. The sector total (minus a portion which had been deducted for the new entrant reserve) would then be divided up between the installations in the sector according to each installation's share of the sector's relevant emissions.

For installations which began operation in 2003, the standard methodology is applied to a baseline of 2003 to 2004 meaning that relevant emissions are the installation's annual emissions for 2003 or 2004, whichever is the higher.

As in the first phase, the UK has modified this methodology slightly to cater for a number of specific cases by providing special rules to take account of rationalisation (both during and after the baseline period), changes at an installation (both during and after the baseline period), low levels of emissions during an installation's first year of operation and temporary closure during the baseline period. Operators were able to make an application for these rules to be used in calculating their allocations. Applications had to be made within a specific time limit and contain supporting evidence.

For installations which began operation in 2004, 2005 or the first half of 2006 (before 1 July), the standard allocation methodology would be applied and 'relevant emissions' for these installations would be the output of a benchmark.

A modified methodology will be adopted for the GQ CHP sector, which will consist of installations which are fully qualified under the UK's Combined Heat and Power Quality Assurance (CHPQA) programme.[12] In this sector, a baseline of 2001 to 2003 will be used as this data is available from the CHPQA database. Adjustments will be made for CHP schemes which are only partially qualified under the CHPQA programme.

---

[11] i.e. the years before the installation began emitting carbon dioxide are omitted.
[12] More information about the CHPQA programme is available at: http://www.defra.gov.uk/environment/energy/chp/index.htm.

A significant change from the first phase was the move away from using historical emissions towards allocating on the basis of benchmarks. For the Large Electricity Producers sector, the share of the sector total will be based upon a benchmark formula instead of the standard methodology. The benchmark takes into account a number of factors including load factor, capacity, emissions factor, fuel used and whether the installation has opted out of the Large Combustion Plants Directive.[13]

D.   New entrants

As in the first phase, a New Entrants Reserve (NER) will be set aside for new entrants. This will contain a finite number of allowances in order to provide free allowances for installations which begin operation (or in some cases extend) on or after 1 July 2006. The methodology will depend on whether the installation begins operation during the first or second phase of the scheme.

For those which begin operation during the first phase, on or after 1July 2006 (referred to as 'Later Phase I New Entrants'), allocations will be calculated on the basis of the new entrant benchmark for the second phase.

The reserve is created by taking a share of allowances from each sector total. This share would be calculated on the basis of whether there was expected to be a high level of new entry within that sector. With one exception,[14] no parts of the NER are ring-fenced for the use of particular sectors. The whole reserve is accessible on a first come first served basis.

1.   Eligibility

In order to be eligible for an allocation from the NER, and installation must have varied its permit after the UK NAP for the second phase was notified to the Commission and commence or extend an activity listed in Annex I to the Directive between 1 January 2008 and 31 December 2012.

New installations commencing an activity listed in Annex I to the Directive for the first time are eligible for an allocation from the new entrant reserve so long as they fulfill these criteria.

In addition to the criteria listed above, extensions must involve the introduction of a new piece of equipment that increases the net output capacity of the activity listed in Annex I to the Directive and the increase

---

[13] Supra chapter 2, note 10.
[14] The GQ CHP ring-fence. This operates in the same way as the ring-fence operated in relation to the first phase of the scheme described in section I.E of chapter 20.

must apply to a technology listed in the new entrant calculation spreadsheet in order to be eligible for an allocation from the NER. An extension will only be eligible for allowances from the NER in respect of a piece of equipment which directly produces emissions which must be accounted for under the scheme. As in the first phase of the scheme, offshore tiebacks and increase in good quality CHP will also be eligible.

2. Allocation Process

The allocation process including the queuing procedure remains the same as in the first phase. The opening date for applications is 1 May 2007 for Later Phase I New Entrants and 1 August 2007 for other new entrants for the second phase.

3. Rate of Allocation

As in the first phase, the allocation will be calculated through the output of a benchmark spreadsheet. These spreadsheets have been updated for the second phase to increase standardisation. Unlike the first phase, as a general principle new entrants will be allocated 95 per cent of the amount of allowances as calculated by the spreadsheets in order to show some movement towards the government's long term goal of moving away from free allocation. The exceptions to this rule are that GQ CHP new entrants will be allocated at 100 per cent of the amount of allowances as calculated by the spreadsheet and boilers (which are a direct alternative to GQ CHP) will be allocated at 90 per cent. New entrants in the Large Electricity Producers sector will be allocated at 70 per cent, which is the same cut in allocation as for incumbent installations in the Large Electricity Producers sector.

4. Later Phase I New Entrants

Later Phase I New Entrants will be eligible for allowances from the NER for the second phase if they were eligible for allowances from the NER for the first phase. They will have priority access to the NER, being able to apply three months earlier than those beginning operation after the beginning of the second phase. They will also not be subject to a reduced rate of allocation like most other new entrants for the second phase.

E. Closure and rationalisation

The rules on closure are the same as in the first phase.[15] Installations that permanently cease operation will retain their allocation for the year

---

[15] See section II of chapter 20.

in which they cease operation but, subject to the application of the rationalisation rule,[16] will be not be issued with further allowances. The one change from the first phase is that the allowances which would otherwise have been issued to the closed installation will be auctioned or sold rather than returned to the new entrant reserve. If this number of allowances exceeds 3 per cent of the total quantity of allowances, then the remainder will be cancelled. This is because the UK is already planning to auction or sell 7 per cent of its total quantity of allowances and Article 10 of the Directive prohibits more than 10 per cent of the total number of allocated allowances from being auctioned or sold in the second phase.

The rationalisation rule is continued as it applied in the first phase with one change. An operator who makes a successful application for rationalisation will continue to receive the complete allocation for the closed installation rather than only a percentage of the allocation corresponding to the amount of production transferred.

F. Use of CERs and ERUs

Following the coming into force of the Linking Directive, Member States are required to state the limit that that they will impose on the use of CERs and ERUs for compliance purposes. This limit must be expressed as a percentage of the allocation to each installation.[17] This limit has been set at 8 per cent of each installation's allocation except for installations in the Large Electricity Producers sector where the limit is set at 9.3 per cent.[18] The 8 per cent limit represents approximately two thirds of the 'effort' required by the NAP (that is the difference between the total quantity of allowances and 'business as usual' projections for all activities covered by the scheme).

G. Auctioning

The UK will auction 7 per cent of the total quantity of allowances in the second phase of the scheme, currently equating to 17,232,320 allowances. These allowances will be deducted from the allocation to the Large Electricity Producers sector.

---

[16] See section III of chapter 20.

[17] For further information on the Linking Directive and this requirement, see section II of chapter 8.

[18] This is stated to be equivalent to 8 per cent of the allocation to these installations before allowances being deducted from that sector for the auctioning reserve.

H. Contingency Fund

Within the new entrant reserve 2.4 million allowances has been ring fenced as a contingency fund. This fund will be used to grant allocations to installations who enter the scheme late and to correct any allocation mistakes which are uncovered after the final allocation decision has been taken. As in the first phase, installations which join the scheme late will receive a reduced allocation depending on how late they join, so as to incentivise operators to identify themselves as falling within the scope of the Directive at an early stage. If there are still allowances in the fund at the end of 2009, a decision will be taken at that stage as to whether they should be returned to the main new entrant reserve.

I. Better Regulation and Simplification Measures

The NAP states the UK's intention to adopt a narrower interpretation of ceramic producing activities than that adopted in the first phase. Annex I to the Directive states that all ceramic producing activities with:

> production capacity exceeding 75 tonnes per day, and/or with a kiln capacity exceeding 4m³ and with a setting density per kiln exceeding 300kg/m³

should be included. For the second phase, the words 'and/or' will be interpreted so as to require the simultaneous presence of all of these sub-elements in order for the activity to be covered.

It also signals an intention to adopt a voluntary *de minimis* interpretation in applying the aggregation rule to the definition of 'combustion installation' in Annex I to the Directive.[19] When aggregating combustion units on the same site to see whether they collectively cross the 20MW threshold for combustion installations, operators may choose to take no account of units which have a capacity of less than 3MW. If they do collectively amount to 20MW or more then all units, including those below 3MW, will be included in the scheme. The purpose of this new interpretation is to exclude small installations whose emissions are very low but because of the number of small units would incur disproportionately high monitoring, reporting and verification costs.

Allowances which would have been allocated to installations which choose to drop out of the scheme as a result of applying this new rule will be removed from the UK's total quantity of allowances before the final allocation decision is taken. Allocations to other installations will be unaffected.

---

[19] See section IV.B of chapter 2.

PART III

RELATED UK POLICY INSTRUMENTS: UK EMISSIONS TRADING SCHEME AND CLIMATE CHANGE AGREEMENTS

CHAPTER 22

THE UK GREENHOUSE GAS EMISSIONS
TRADING SCHEME

I.   Policy Evolution

The UK Greenhouse Gas Emissions Trading Scheme began operating in 2002. It was a voluntary scheme open to nearly all organisations in the UK and was the first economy-wide emissions trading scheme anywhere in the world.

Following the signature of the Kyoto Protocol in 1997, which paved the way for international caps on emissions of greenhouse gases and trading in emissions allowances by 2008, the Chancellor of the Exchequer, Gordon Brown, commissioned a task force of senior civil servants from a broad cross section of government departments and led by Lord Marshall to assess whether and if so, how best to use new economic instruments to improve commercial and industrial commercial use of energy and help reduce emissions of greenhouse gases. This task force consulted industry extensively and reported in November 1998.[1] One of its conclusions was that as a system of international greenhouse gas emissions trading was likely to emerge in 2008, the UK government should seriously consider a pilot with interested players as soon as possible as a means of UK industry learning lessons for participation in the international scheme.

The UK scheme was subsequently introduced in 2002 with several aims. The first was to incentivise the reduction of greenhouse gas emissions within the UK. The second was to give UK industry and government early experience of emissions trading in order to give them an advantage ahead of international trading. The third was to establish the UK and the City of London in particular as an international centre in emissions trading services, through the establishment of services such as emissions allowance brokers and verification.

II.   Overview of the Scheme

A.   Gases

The scheme was a voluntary cap and trade[2] scheme, with the aim of producing absolute reductions in emissions of greenhouse gases by its

---

[1] Supra Introduction, note 49.
[2] Although it has elements which operate on a baseline and credit basis. See sections II.B.2, IV and III.J and section IV.A of the Introduction.

participants. The scheme covered all six of the greenhouse gases identified by the Kyoto Protocol, namely carbon dioxide (CO_2), methane (CH_4), nitrous oxide (N_2O), hydrofluorocarbons (HFCs), perfluorocarbons (PFCs) and sulphur hexafluoride (SF_6).

As well as direct emissions of each of these gases, the scheme also included indirect emissions of carbon dioxide – that is the emissions which would have been emitted into the atmosphere in order to generate the electricity, heat and steam used.[3]

The different gases covered by the scheme have varying impacts on climate change and it is therefore more valuable to reduce emissions of one gas by one tonne than it is to reduce emissions of another by one tonne. The impact of one tonne of each of the gases was therefore compared to the impact of one tonne of carbon dioxide, and given a weighted value measured in units of tonnes of carbon dioxide equivalent (tCO_2e). Their relative values are set out in the table below.

B.   Participation in the Scheme

Gas	1 tonne measured in tonnes of carbon dioxide equivalent (tCO_2e)
Carbon Dioxide	1
Methane	21
Nitrous Oxide	310
Hydrofluorocarbons	140 - 11,700[4]
Perfluorocarbons	6,500 - 9,200[5]
Sulphur Hexafluoride	23,900

There are three ways in which a natural or legal person could take part in the scheme. These are as a direct participant, a climate change agreement participant or as a trading participant. At the outset of the scheme, it was also envisaged that at some point in the future it would be possible to participate as a project participant. However the scheme

---

[3] Details of how indirect emissions are calculated is contained in section III.G.1.

[4] The difference in value depends upon the precise compound used. The relative values of each of these compound are set out in paragraph 3.4 of the Guidance for Measurement and Reporting of Emissions by direct participants in the UK Emissions Trading Scheme, available at: http://www.defra.gov.uk/environment/climatechange/trading/uk/reporting.htm.

[5] The difference in value depends upon the precise compound used. The relative values of each of these compound are set out in paragraph 3.4 of the Guidance for Measurement and Reporting of Emissions by Direct Participants in the UK Emissions Trading Scheme (ibid).

was never extended to project participants. This chapter looks at all forms of participation in the scheme, but focuses mainly on direct participation.

1.   Direct Participants[6]

The direct participant element of the scheme ran from 2002 to 2006 and 31 direct participants took part in the scheme. They came from a wide variety of backgrounds, and included multinational oil companies, other manufacturers, supermarkets, an airline, universities, a local council and a museum. Over a period of 5 years (2002 - 2006) each of these participants received an annual financial incentive payment from government in return for achieving an annual emissions reduction target, which was measured against their baseline emissions. The targets themselves were notional in that, as would be expected in a trading scheme, participants were free to choose whether to reduce their emissions to the target level, to reduce their emissions below the target level and bank or sell their surplus allowances, or to exceed their target level and make up the shortfall by obtaining and using additional allowances, which may have been banked from previous years or purchased from other participants.

2.   Climate Change Agreement Participants

Persons who are already party to a Climate Change Agreement (CCA) with the Secretary of State[7] to reduce their emissions of carbon dioxide and/or to improve their energy efficiency may apply separately to become a CCA participant in the scheme.[8] Once a party to a CCA has become a CCA participant, they may convert any overachievement against their CCA targets into allowances, which are freely tradable within the scheme.[9] Equally, if a CCA participant fails to meet its CCA target, it may make up the shortfall by buying and retiring[10] scheme allowances.[11] This is therefore an example of how a cap and trade scheme can be combined with a baseline and credit system.[12]

---

[6] See section III.

[7] Climate Change Agreements are discussed in more depth in chapter 23.

[8] Section 3 of the Framework for the UK Emissions Trading Scheme, available at: http://www.defra.gov.uk/environment/climatechange/trading/uk/documents.htm.

[9] Subject to the operation of the gateway – this is discussed further at section VII.B.

[10] See section VII.

[11] The retirement of allowances by a CCA participant technically has the effect of making the CCA target easier to meet. This is discussed in more detail in section VIII of chapter 23.

[12] For further details on cap and trade and baseline and credit systems see section IV.A of the Introduction.

3.   Trading Participants

Persons who do not wish to take on an emissions reductions target as a direct participant or a CCA participant, may still buy and sell allowances as a trading participant. A trading participant is entitled to a trading account in the scheme registry, through which it can freely buy and sell allowances from and to other participants.

C.   Legal Framework of the Scheme

There is no legislation setting out the rules of the UK Emissions Trading scheme. The primary legal framework for the scheme is the UK Emissions Trading Scheme 2002 (the Scheme Rules)[13] which is made up of a number of rules. This scheme was made by the Secretary of State for Environment, Food and Rural Affairs in September 2002 and has been subsequently amended several times.[14] The Scheme Rules are incorporated into a series of separate standard form agreements between the Secretary of State on the one hand and each of the scheme participants on the other hand.[15] For this reason it does not confer any rights or obligations on any party other than the participants themselves and the Secretary of State.[16]

In addition to these agreements and the Scheme Rules, some detail of how the scheme operates is delegated to guidance. For example, because of the lack of a legal framework for the scheme before participants entered into their agreements, much of the detail of how the scheme would operate were set out in the UK Emissions Trading Scheme Framework Document (the Framework Document).[17] Following the adoption of agreements and the Scheme Rules, the Framework Document continued to have legal effect as, in some cases, the rules

---

[13] Available at: http://www.defra.gov.uk/environment/climatechange/trading/uk/documents.htm.

[14] As amended by the UK Greenhouse Gas Emissions Trading (Amendment) Scheme 2004, the UK Greenhouse Gas Emissions Trading (Amendment) Scheme 2005, the UK Greenhouse Gas Emissions Trading (Amendment) (No. 2) Scheme 2005 and the UK Greenhouse Gas Emissions Trading (Amendment) Scheme 2006 and the UK Greenhouse Gas Emissions Trading (Amendment) Scheme 2007.

[15] It is not clear whether the agreements should rightly be interpreted according to the law of contract or by the Secretary of State within the rationality constraints imposed by public law. On the one hand, they are bilateral agreements with binding rights and duties, which are binding through the law of contract. On the other hand they are regulating behaviour in what is clearly a public law rather than a private law field. For an example of where an agreement has been interpreted in the latter way, see *R* v *Director General of Electricity Supply ex parte First Hydro Company*, 2 March 2000 (unreported).

[16] Although in accordance with the Contracts (Rights of Third Parties) Act 1999, the Participation Agreements specifically state that they are not intended to give rights to third parties, the Secretary of State also undertakes to apply the scheme to other participants as necessary to promote the purposes of the scheme.

[17] Supra note 8.

specifically require things to be done in accordance with the Framework Document. Many of the rules on the operation of the registry are contained in the Registry User Manual[18] and details of how direct participants are to monitor and report their emissions are contained in the Monitoring, Reporting and Verification Guidelines (the MRV Guidelines).[19]

There are two elements of the scheme which arguably needed statutory authority. The first was the imposition of financial penalties on participants who fail to comply with their emissions limitation commitment for the final two years of the scheme. Such penalties are given statutory authority by section 39 of the Waste and Emissions Trading Act 2003 which provides that:

> if incorporated in a participation agreement, the penalty provisions of the scheme shall have statutory effect between the parties to the agreement.

The second was the power to incur expenditure on incentive money over a number of years. This is authorised by section 153(1)(aa) of the Environmental Protection Act 1990 which authorises the Secretary of State, with the consent of the Treasury, to 'give financial assistance....for the purposes of....the United Nations Framework Convention on Climate Change.'

III.   Direct Participants

A.   Key Obligation of the Participants: Emissions Limitation Commitment

Direct participants are natural or legal persons who entered into a direct participant agreement with the Secretary of State. A direct participant's central obligation was to comply with its emissions limitation commitment[20] each year. This means that each year on 31 March ('the reconciliation deadline') direct participants had to ensure that they had sufficient allowances equal to or greater than their actual verified emissions[21] in the previous commitment year.[22] Those allowances had to subsequently be retired. For these purposes, one allowance was equal

---

[18] Emissions Trading Registry User Manual, available at: http://etr.defra.gov.uk/pdf/ETR_User_Manual.pdf.

[19] Guidelines for the Measurement and Reporting of Emissions by Direct Participants in the UK Emissions Trading Scheme, available at: http://defraweb/environment/climatechange/trading/uk/pdf/trading-reporting.pdf.

[20] Rule C1(2)(a) of the Scheme Rules.

[21] See section III.G.1

[22] Rule C6(1) of the Scheme Rules. Commitment year is a calendar year.

to one tonne of carbon dioxide equivalent (tCO_2e). The number of allowances in the compliance account on this date was referred to as the 'reconciliation balance'.[23]

B. Key Obligations of the Secretary of State

The Secretary of State had two central obligations: to pay each direct participant its incentive payment each year and to allocate allowances to direct participants each year.[24]

1. Incentive Money

The amount of incentive money payable to each participant each year was directly proportionate to the total emissions reduction target that it had taken on and was calculated by multiplying its annual target[25] by £53.37, which was the final auction clearing price.[26] The Secretary of State was required to pay this sum every year following reconciliation, and it had to be paid within 28 days of the participant submitting its verification opinion[27] or on the first banking day after 31st March, whichever was the later.[28]

2. Allocation of Allowances

In the first year of the scheme, the Secretary of State was obliged to allocate allowances by 1st April 2002 or, if later, within 15 days of the direct participant submitting a verification opinion in respect of its baseline.[29] In subsequent years, the obligation to allocate allowances occurred 15 days after the direct participant had submitted a verification opinion in respect of its annual emissions[30]. Allowances were allocated by transferring them to the direct participant's compliance account in the registry.[31]

No allowances were to be allocated following receipt of the verification opinion in respect of the final year's emissions. This is not immediately apparent from rule C5(1)(b) but that rule only provides for the allocation of allowances following the submission of a verification opinion 'in

---

[23] As defined in schedule 1of the Scheme Rules. See also section III.G.2-3.
[24] Rule C1(1) of the Scheme Rules.
[25] 'Annual target' means the total amount by which each participant's allocation of allowances (representing its notional emissions) are to reduce each year. This is calculated by dividing the final emissions reduction target by 5.
[26] Rule C2(1) of the Scheme Rules. See also section III.F.
[27] See also section III.G.
[28] Rule C2(2) of the Scheme Rules.
[29] See Section III.E.9.
[30] Rule C5(1) of the Scheme Rules. See also section III.G.1.
[31] Rule C5(4) of the Scheme Rules. See section VII.A

subsequent commitment years'. The verification opinion in respect of 2006 cannot be received in a 'subsequent commitment year', as a commitment year is defined in schedule 1 to the Scheme Rules as any calendar year during the period 1st January 2002 to 31st December 2006. It would have been impossible to verify the emissions which occurred in 2006 before the end of that calendar year.

The number of allowances allocated to each participant was calculated according to its baseline and its target and reduced each year.[32]

C.   Eligibility

The Scheme Rules do not contain any conditions which had to be fulfilled in order to become a participant in the scheme. The reason for this is that, as the agreement of the Secretary of State was necessary in order to participate in the scheme, he had discretion at common law to decide whether or not to enter into an agreement with each individual applicant, and if so, on what terms. Even if conditions were contained in the Scheme Rules to stipulate how he should exercise that discretion, they could not be enforced by an aggrieved applicant who had been refused participation in the scheme as the rules were only binding between parties once they have been incorporated into an agreement between them. It is possible, however that a clearly worded statment could have given rise to a legitimate expectation in public law.

The Secretary of State published in the Framework Document[33] the criteria upon which he would decide whether to permit an applicant to join the scheme as a direct participant. This stated that participation as a direct participant was open to any individual or organisation on the basis of the greenhouse gases it emits both directly and indirectly within the UK. There were some emissions which were ineligible to be included:[34] namely emissions from land and water transport, emissions from landfill sites covered by the Landfill Directive,[35] emissions from households and emissions from facilities within a target unit covered by a CCA.[36] The purpose of this final exclusion is to ensure that the same source cannot be entered into both schemes, leading to a person receiving a financial incentive as a result of direct participation in the scheme, and a Climate Change Levy discount,[37] in respect of emissions reductions from the same source.[38]

---

[32] See section III.E.
[33] Section 2 of the Framework Document.
[34] Paragraph 2.4 of the Framework Document.
[35] Council Directive 1999/31/EC on the landfill of waste, L 182, 16.7.1999, p.1.
[36] For more information on Climate Change Agreements, see chapter 23.
[37] In accordance with schedule 6 to the Finance Act 2000.
[38] Paragraph 2.6 of the Framework Document.

As indirect emissions are included in the scheme,[39] it was important to ensure that the same emissions reductions were not counted and paid for both at the generation stage and the energy use stage. For this reason, direct emissions from electricity or heat generation were excluded, except where the electricity and heat was both generated and used on the same site.[40] Where that was the case, only the direct emissions of that energy generation were counted; indirect emissions attributable to that energy use were not.[41]

In order to ensure the environmental integrity of the scheme, the Framework Document also stated that a discretion would be retained to refuse to enter into direct participant agreements with applicants if they could not demonstrate to the government's satisfaction that they intended to and were capable of complying with the Scheme Rules.[42] In particular, individual consideration would be given where the emissions to be entered into the scheme were already subject to a restriction imposed by other forms of regulation.[43] Such restrictions might be found in permit conditions issued under part I of the Environmental Protection Act 1999 or the Pollution Prevention and Control Regulations 2000[44] which require certain techniques to be used or certain emission limit values to be complied with when carrying out a particular process. In general, the approach taken by the government was that these applicants were permitted to join the scheme and this problem was resolved by adapting the way in which their baselines were set.[45]

Despite these exceptions and discretion, the scheme was still open to the vast majority of businesses and organisations in the UK, making it the first economy-wide emissions trading scheme anywhere in the world.

Apart from one limited circumstance,[46] any person wishing to join the scheme had to do so by entering into a direct participant agreement and taking part in the auction which took place in April 2002.

D. Direct Participant Agreements

Direct Participant Agreements took a standard form. It stated that it would come into force on the date upon which it was signed[47] and

---

[39] Indirect emissions are the emissions made in order to generate energy (such as electricity, heat or steam) which is subsequently used by the participant.
[40] Paragraph 2.4(i) of the Framework Document.
[41] Paragraph 2.5 of the Framework Document.
[42] Paragraph 2.2 of the Framework Document.
[43] Paragraph 2.8 of the Framework Document.
[44] Supra chapter 12, note 1. The PPC Regulations implement the IPPC Directive (supra chapter 2, note 3).
[45] See section III.E.
[46] Namely, where a person acquired a source already in the scheme.
[47] Clause 2.1 of the Direct Participant Agreement as read with clause 1.1.

terminate on 31 March 2007. This duration clause was expressly made subject to the Secretary of State's power to amend the scheme at any time.[48]

The Scheme Rules were incorporated into and form part of the Agreement, and both parties undertook to comply with the obligations imposed upon them by the Scheme Rules. The Secretary of State also undertook to apply the Scheme Rules to other participants as necessary to promote the purposes of the scheme.

The Agreement set out each party's main obligations under the scheme, in particular the Secretary of State undertook to open and maintain the direct participant's accounts at the registry, to allocate allowances to the direct participant's account and to pay annual incentive payments.

The direct participant undertook to abide by the rules of the scheme, taking on an emissions reduction target, monitoring emissions, submitting emissions statements and verification opinions in respect of its baseline and its annual emissions, retiring allowances each year equal to its annual emissions, and in certain circumstances repaying incentive money received.

The Direct Participant Agreement also set out the consequences of the direct participant withdrawing from the scheme or becoming insolvent.

E.   Source Lists, Baselines and Targets

The absolute reductions in greenhouse gas emissions to be achieved by the scheme were measured by way of a series of stepped targets which were measured against baselines. Targets represented the level of absolute emissions that each direct participant committed itself to in an auction and were directly proportionate to the amount of incentive money that it received. The source list listed emissions sources[49] which the participant entered into the scheme. The baseline was an absolute emissions figure measured in tonnes of carbon dioxide equivalent (tCO_2e) and represented the level of historic emissions attributable to those sources and against which absolute emissions reductions were measured. The combination of the baseline and target determined the number of allowances to be allocated to each direct participant.

---

[48] Clause 3.1 of the Direct Participant Agreement as read with clause 1.1. The Secretary of State's power to amend the scheme is discussed further at section VIII.A.
[49] For discussion on what constitutes a source, see section III.E.1

1. Sources and the Source List

Given the voluntary nature of the scheme, the direct participant had a certain amount of flexibility in deciding which of its emissions sources to enter into the scheme. In order to ascertain the direct participant's sources included in the scheme, it was necessary to refer to its source list annexed to its Direct Participant Agreement. However, this flexibility in deciding what sources of emissions to include in the source list was limited as the list had to be prepared in accordance with methodology set out in the Framework Document and the Reporting Guidelines, any guidance issued by the Secretary of State and, where there were any changes of operation[50] during the baseline period, schedule 3 to the Scheme Rules.[51]

2. What is a Source?

A 'source' meant a point source or a collection of point sources of the same type on the same site. A point source meant any separately identifiable point from which greenhouse gases are emitted.[52] As an example, all indirect emissions from electricity from a site represented a single source.[53]

3. The Source List

Each direct participant had a source list, which was a definitive list of the emissions sources which it had entered into the scheme. In drawing up its source list, the participant was required to go through several steps, some of which dictated whether certain sources would be included in the scheme and others of which gave him a limited choice as to whether to include or exclude certain sources.

Firstly, it had to identify all sources over which it had management control.[54]

Secondly it had to separate all of those sources into subsets based upon the industrial sector to which those sources belong. It is at this second stage that the participant had arguably its most significant discretion, in that it could decide to include some industrial subsets and exclude others, but could not pick and choose sources from within each subset. One reason for this is that it would have made it possible for a direct

---

[50] Changes in operation are discussed further in section III.H.
[51] Paragraph 1 of schedule 2 to the Scheme Rules.
[52] Paragraph 3 of schedule 2 to the Scheme Rules and paragraph 2.3 of the Framework Document.
[53] Paragraph 2.3 of the Framework Document.
[54] Management control is discussed further at section III.H.1.

participant to reduce its emissions from one source, which it elected to include in the scheme by reducing activity at that site, and transferring that activity to another of its sites in the same industrial sector. This would have to an increase in its emissions from that other source, which it could have elected to exclude from the scheme. The net effect of these actions would have been that there would be no overall reduction in the direct participant's emissions. However, as the source from which emissions had decreased would be included in the scheme and the source from which emissions had increased would be excluded, it would be credited with having achieved a reduction in emissions. This potential for 'leakage' from the scheme would have left it open to abuse.

Thirdly, from these subsets, it had to identify those subsets for which there was verifiable emissions data for the years 1998, 1999 and 2000, which are collectively known as 'the baseline period'.[55] This is because without this data it would not have been possible to determine the baseline.[56]

Fourthly, it had to check that the subsets of sources in question were eligible for the scheme.[57]

Fifthly, it could choose to include either the subset of the sources which had not yet been excluded which emit carbon dioxide and exclude those which emit other greenhouse gases or, alternatively, decide to enter sources of all greenhouse gases.

Finally, it had to check which of the subset of sources are covered by the protocols to the Monitoring Reporting and Verification Guidelines.[58] If any were not covered by these protocols, it could choose not to include them.

4.   Approval of the Source List

The source list had to be approved by the Secretary of State. The direct participant was required to submit its source list to the Secretary of State for approval and, in doing so, had to include a written record of how the list was arrived at including decisions taken in the process, such as which sources had been excluded from the list.[59] The Secretary of State could also request other information from the direct participant if

---

[55] See definition of 'baseline period' in schedule 1 to the Scheme Rules. The exact nature of the baseline period is explored further at section III.E.5-9.

[56] Calculation of the baseline is explored in section III.E.5-9.

[57] See section III.C on eligibility for the scheme.

[58] At the beginning of the scheme the protocols were in draft. The Guidelines and the protocols to them are discussed at section III.G.2.

[59] Paragraph 2 of schedule 2 to the Scheme Rules.

required in order to determine whether to approve the source list. The Secretary of State had to approve the source list in writing.[60]

5. The Baseline

Each direct participant had a baseline which was a figure measured in tonnes of carbon dioxide equivalent ('tCO_2e') and which represented the historic emissions level of all of the sources in the source list. It was the responsibility of the direct participant to calculate the baseline and it had to do so in accordance with several factors. These are the methodology set out in the Framework Document and the Monitoring, Reproting and Verfication Guidelines, instructions or comments from the Secretary of State, made in the source list approval, any other guidance given by the Secretary of State and, where there have been any changes of operation during the baseline period, schedule 3 to the Scheme Rules.[61] Once calculated, the baseline could not be amended, except where there was a change in operation or an error in compiling the source list. In such a case, any amendment to the baseline had to be made in accordance with schedule 3 to the Scheme Rules.[62]

The Framework Document sets out the rules for calculating the baseline. In most cases, it was the average of annual emissions of all of the sources in the approved source list for the period 1998 to 2000 inclusive.[63]

6. The Size Threshold

If at this stage it became apparent that during that period any one of its sources had average emissions of less than 10,000 tCO_2e or 1 per cent of the source list total (whichever was less), then the direct participant could choose to exclude these emissions from the baseline, although they would remain in the source list.[64]

7. Evolution of Baseline Policy

During the run-up to the commencement of the scheme, it became apparent that the proposed standard policy of setting baselines as the average of 1998-2000 emissions, as set out in the Framework Document was not always practical, could affect the environmental integrity of the scheme and, in some cases, could lead to a risk of elements of the incentive payments being unlawful State Aids.[65]

---

[60] Paragraph 4(1) of schedule 2 to the Scheme Rules.
[61] Paragraph 5(1) of schedule 2 to the Scheme Rules.
[62] Paragraph 5(3) of schedule 2 to the Scheme Rules.
[63] section III.E.7-9.
[64] Paragraphs 2.18 and 2.19 of the Framework Document.
[65] The applicability of the law on State Aids to the scheme is explored further in section III.L.

The government initially proposed that the baseline should be based on an average of three years' emissions as this would help to mitigate any unfairness by picking just one previous year if, for example, that year was unrepresentative for the prospective direct participant or if a source only became operational partway through the period. However, not all companies who wished to take part had verifiable data which covered all three years. In view of this, it was permitted to only use an average of data for 1999 and 2000, or just data for 2000 where the participant could satisfy an independent verifier that emissions data for the earlier years was not available. This is reflected in the definition of 'baseline period' in schedule 2 to the Scheme Rules.

8.   Use of Regulatory Requirements in Baseline Calculation

It had always been clear that it would be problematic for the scheme to include sources where the emissions of greenhouse gases were already regulated under other legislation. This was particularly the case where a regulatory restriction on emissions was imposed or tightened during the baseline period or between the end of the baseline period and the commencement of the scheme. Because actual emissions data from the period before the new regulatory restriction was imposed was used to calculate the baseline, this could have led to the situation where the level of emissions allowed under the regulatory restriction was actually lower than the direct participant's baseline or, in some cases, lower than baseline minus its overall target.[66] This would mean that incentive money would have been paid in respect of emissions reductions which the direct participant would be required by law to make anyway. Given that the purpose of the scheme was to give a financial incentive to firms who are prepared to go beyond their legal obligations under other regimes, it would not have been appropriate for this situation to have prevailed.

It was for these reasons that the Framework Document made clear that prospective direct participants must inform the government if they wished to include such sources in the scheme and that the government would consider their eligibility on a case-by-case basis. Applications were received in respect of such sources and the government decided that some could be included in the scheme, so long as the regulatory restrictions were taken into account in calculating the baseline. Further guidance was accordingly given on how the Secretary of State would exercise his discretion in these circumstances in April 2002.[67]

---

[66] See section III.E.10. This possibility is demonstrated in the National Audit Office Report, 'The UK Emissions Trading scheme - A New Way to Combat Climate Change', 21 April 2004, available at: http://www.nao.org.uk/publications/nao_reports/03-04/0304517.pdf. See in particular paragraph 2.12 and Figure 4. This is discussed in more detail in section X.
[67] Treatment of Regulatory Requirements in the UK Emissions Trading Scheme, April 2002.

This guidance defined a relevant regulatory requirement as 'including mass limits, concentration limits, improvement conditions and any other measures or conditions specified by a regulatory body which set direct or indirect controls on the emissions of greenhouse gases from a source'.[68] Where, in any one or more of the baseline years (1998 to 2000), the actual emissions from a source were greater than the level permitted under regulatory restriction for that same source which was in force at any time from 1998 to the date of the auction, then for the purpose of taking the average of those years, the emissions for that year were taken to be at the level of the lowest regulatory limit. This ensured that the baseline could not be higher than the lowest of any regulatory limit which was in place at any time from the beginning of the baseline period to the auction.

Where the regulatory restriction was not a limit on the absolute mass of emissions emitted, such as improvement conditions or concentration limits, the Secretary of State would, in consultation with the proposed participant and the relevant regulator, determine a deemed emissions figure equivalent to the limit.

Participation in the scheme did not affect participants' obligations under other regulatory regimes and the purchase and retirement[69] of allowances could not assist them in meeting those other obligations.

9.   Verification of the Baseline

Each direct participant was required to calculate its baseline figure itself in accordance with the methodology set out in the Framework Document, the Monitoring, Reporting and Verfication Guidelines and other guidance issued by the Secretary of State. This figure then had to be verified by an accredited third party verifier.

Each direct participant was required to submit a report of its baseline emissions, along with a signed statement from the verifier endorsing the report ('a verification opinion') at the beginning of the scheme. The Secretary of State would not allocate any allowances to the participant until it complied with this obligation.

10.   Targets

The target was an absolute emissions figure measured in tonnes of carbon dioxide equivalent (tCO_2e) and was determined by the amount that the direct participant bid into the auction.[70] The figure which was bid is

---

[68] Ibid, paragraph 3.
[69] See section III.G.2-3.
[70] See section III.F.

referred to as the 'overall target' and represented the total emissions reduction target for the final year of the scheme i.e. 2006.[71] The effect of this is that in 2006, the number of allowances allocated to the participant was the baseline minus the overall target. In the other years of the scheme, the number of allowances allocated decreased on an annual basis in a stepped manner so as to require a steady increase in effort throughout the scheme by one fifth of the overall target ('the annual target')[72] per year. So the number of allowances allocated was the baseline level minus the annual target in the first year of the scheme, baseline minus twice the annual target in the second year, baseline minus three times the annual target in the third year and baseline minus four times the annual target in the fourth year. This is illustrated in the diagram below.

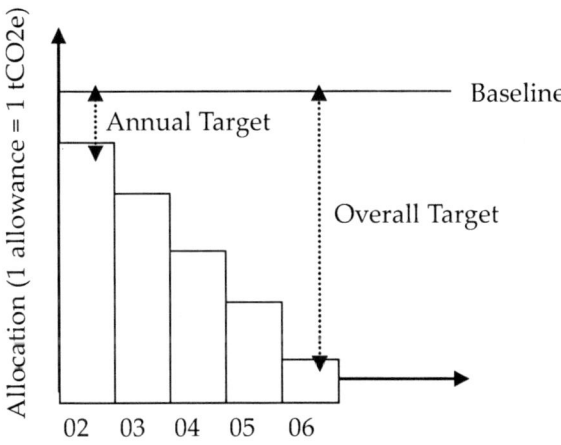

Targets were fixed for the duration of the scheme, except that they could be varied downwards along with baselines to reflect structural re-organisations by the participant. Most commonly, this would occur if the direct participant sold or otherwise disposed of one ofitsemissions sources.[73]

---

[71] See definition of 'overall target' in schedule 1 and paragraph 6(1) of schedule 2 to the Scheme Rules.

[72] See definition of 'annual target' in schedule 1 and paragraph 6(2) of schedule 2 to the Scheme Rules.

[73] See generally, schedule 3 to the Scheme Rules. See also section IIII.H.

F.   The Auction[74]

Direct participants' participation in the scheme began with them taking part in an auction, which took place on 11 and 12 March 2002. It took place electronically and took the form of a 'descending clock auction', which operated as follows.

The auction began by the auctioneer announcing a price per tonne of CO2e emissions reduction, against which all bidders would bid the amount of emissions reductions that they were willing to make, in 2006, at that price.[75] The government then assessed the total cost of accepting all of these bids at the opening price. If the total cost was less than the total amount of money which had been set aside for incentive payments, then the government had the choice of either accepting the bids, in which case the auction would close, or abandoning the auction. It was deemed necessary for the government to have this option at this stage, in case very few participants bid in, which would lead to an unviable market. If the total cost was more than the total amount set aside for incentive payments, then the auctioneer would begin a new round by announcing a second, lower price per tonne, against which direct participants would bid again. This process would continue until the total cost of accepting all bids was equal to, or less than, the total amount of available incentive money. When that point was reached, the Secretary of State would announce the clearing price and each participant's overall target.[76]

A bidder would be able to withdraw from the auction at any time, by putting in a bid of zero, but once it had withdrawn, it would not be able to bid again in subsequent rounds. However, if a bidder did make a bid, and that round turned out to be the final round in which the clearing price was reached, the bidder would then be bound to participate in the scheme on the basis of that bid. In assessing the bids, the government also agreed not to allow any one direct participant to receive more than 20 per cent of the total incentive money.[77]

The auctioneer opened the bidding at £100 per tonne, which was based upon analysis of the social cost of carbon. After nine rounds of bidding, the auction closed at just over 4.03 million tCO_2e at a price of £53.37 per tonne. As has been discussed,[78] the amount bid in to the auction represents the direct participants' targets for the final year of the scheme. Over the lifetime of the scheme, this worked out at a total reduction in

---

[74] See generally, schedule 5 to the Scheme Rules.
[75] See section III.E.10 on the overall target.
[76] See section III.B.1
[77] See section III.L.
[78] See section III.E.10

emissions of 12.1 million tonnes at a price of £17.79 per tonne. In 2004 six direct participants to waive their entitlement to a total of approximately 8.9 million allowances (partly in response to the National Audit Office report). As a result the scheme can now be said to be achieving a total of 21 million tonnes of emissions reductions at a price of just over £10 per tonne.

The government considered other methods of auction, including a 'sealed bid' method, which the National Audit Office considered could have secured a lower price per tonne for emissions reductions. This would have involved each bidder submitting their bids (in terms of quantity of emissions reductions) at a variety of prices and allowing the government to assess these together to decide how much of the incentive money to spend. However, this method was considered at the time by government and rejected on the grounds that many businesses may have been deterred from bidding by this method.[79]

It was initially envisaged that there would be three auctions, but in the event there has only been the one and at the time of writing, the government has not announced any plans to hold another. In its consultation of September 2004[80] on options to improve the scheme, the government proposed holding a further auction, in which original targets would be scrapped, and direct participants as well as non-direct participants could bid in new emissions reductions targets for the final two years of the scheme. However, following consultation, this option was not taken any further.

G. Complying with the Emissions Limitation Commitment

1. Monitoring and Reporting Emissions

In order to protect the environmental integrity of the scheme, it was necessary to have robust systems in place to ensure that direct participants' greenhouse gas emissions were measured consistently and accurately. This was necessary for both for the purposes of calculating the participant's baseline (which would partially determine the number of allowances allocated)[81] and in subsequently monitoring their annual emissions. The latter was needed to establish the number of allowances which the participant would be required to retire.

---

[79] See National Audit Office Report (e note 65) in particular paragraphs 2.35–2.43 and Appendix 6.
[80] Available at: http://defraweb/environment/climatechange/trading/uk/pdf/ukets-defra260804.pdf.
[81] See section III.E.

A direct participant was required to monitor and calculate its annual emissions for each commitment year in accordance with the detailed standards set out in the Monitoring, Reporting and Verification Guidelines. These guidelines contain detailed rules on how emissions were to be monitored and reported based on the principles of faithful representation, completeness, consistency, reliability and transparency. Each year, the direct participant was required to compile a statement of emissions, which was then submitted to an accredited third party verifier to assess its accuracy. The participant was then required to submit the statement of annual emissions, together with a signed statement from the verifier endorsing the statement (known as a 'verification opinion'), to the Secretary of State. These documents had to be submitted by the reconciliation deadline of 31 March following each commitment year.[82]

Amongst other things, the guidelines set out how the use of energy, such as electricity was to be accounted for in terms of indirect emissions measured in tCO_2e. The guidelines make clear that this was to be done by using a common emission factor for all electricity used from the grid. The one exception to this was that the use of renewable energy was calculated as giving rise to zero emissions.

2. Reconciliation

Reconciliation was the annual process by which the Secretary of State retired allowances from each direct participant's compliance account and assessed whether each direct participant had complied with its emissions limitation commitment. Retirement of allowances was the process of transferring allowances from a participant's compliance account into the national retirement account[83], following which no further use could be made of them.

The emissions limitation commitment was assessed by reference to each direct participant's reconciliation balance, which was the total quantity of valid allowances held in the direct participant's compliance account on the 31 March following the end of each commitment year (known as the 'reconciliation deadline')[84]. The Secretary of State calculated each direct participant's reconciliation balance as soon as practicable after the reconciliation deadline or the date upon which the direct participant delivers its verification opinion, whichever is the earlier[85] and then notified the direct participant of the balance[86].

---

[82] Rule C8 of the Scheme Rules.
[83] Section VII.A
[84] Rule C7(2) of the Scheme Rules. See also interpretation of 'reconciliation deadline' in schedule 1 to the Scheme Rules.
[85] Rule C7(2) of the Scheme Rules.
[86] Rule C7(4) of the Scheme Rules.

As soon as practicable after he had both calculated the reconciliation balance and received a verification opinion from the direct participant, the Secretary of State was required to retire allowances from the direct participant's compliance account. He had to retire the number of allowances equal to the total verified emissions in the preceding year, as set out in the verification opinion, or if there were not enough valid allowances in the reconciliation balance, all valid allowances in the compliance account.[87] If there were sufficient allowances in the reconciliation balance, then the direct participant had complied with its emissions limitation commitment and the Secretary of State was required to notify it of the fact.[88] If there were not sufficient allowances in the reconciliation balance, then the direct participant had failed to comply with its emissions limitation commitment and the Secretary of State was required to notify the participant of that fact.[89]

3. Method of Retirement

The Secretary of State retired allowances by transferring them from the direct participant's compliance account to the national retirement account.[90] Allowances also had to be retired in a particular order. Allowances from the earliest vintage had to be retired first and, within vintages, they had to be retired in the order in which they were allocated or transferred into the compliance account.

H. Changes in Operation

Businesses and other organisations do not stand still. During the five years' duration of the scheme, it would be surprising if no direct participants sold or closed any of its sources or transferred activity from one source to another. The Scheme Rules seek to deal with future changes to the structure of direct participants' activities in schedule 3. Underlying the rules in this schedule were several presumptions. The first was that a direct participant could not count emissions reductions towards achieving its target where those reductions were caused by it closing or otherwise disposing of a source or by substantially reducing the activities that it carried out. Secondly, if it closed or transferred a source to another person, its absolute emissions reduction target should become proportionately less onerous, as it had fewer sources from which to make those reductions. However, as the incentive payment was calculated by reference to the overall target, its incentive payment would be reduced by the same proportion.[91]

---

[87] Rule C7(5)-(8) of the Scheme Rules.
[88] Rule C7(9) of the Scheme Rules.
[89] Rule C7(10) of the Scheme Rules.
[90] For more on the national retirement account, see section VII.
[91] See section III.B.1.

This section sets out the different types of change in operation envisaged by the rules, and the consequences (if any) of each of those changes. Each direct participant was required to notify the Secretary of State of all of its changes in operation by 31 December each year.[92]

1.   Management Control

The application of many of the rules contained in schedule 3 depended upon an assessment of who had management control over a source. This was a question of fact. A person was regarded as having management control over a source if he exercised dominant influence over its emissions, through having the ability to direct the financial and operating policies governing its emissions.[93] A transfer of management control was therefore likely to include a sale or lease of a source or the outsourcing of the activity carried out by the source.

Where such a transfer of management control took place, it was deemed to take place on the contractual date of completion. If there was no such date, it was deemed to take place on the date on which the transfer took legal effect, and if that date could not be ascertained, on the date on which the transfer took practical effect.[94]

2.   Triggering the Change Threshold

Another key concept in schedule 3 was the 'change threshold'. In order to avoid administrative burdens when direct participants made only minor changes in operation, a change in operation would not have any consequences on baselines, targets and incentive payments unless the change, when taken in conjunction with previous changes in operation by the same direct participant, was substantial enough to trigger the change threshold.

The change threshold was triggered when the source in question, plus any of the direct participant's other sources which had previously been the subject of a change in operation, had verified baseline emissions totalling either 25,000 tCO_2e or 2.5 per cent of the direct participant's verified original baseline.

---

[92] Paragraph 7(1) of schedule 2 to the Scheme Rules.
[93] Paragraph 1(2) of schedule 3 to the Scheme Rules and paragraph A.2 of Annex A to the Framework Document.
[94] Paragraph 1(5) of schedule 3 to the Scheme Rules.

3. Types of Change of Operation

a. *Transfer of Management Control of a Source to another Direct Participant*

The first situation occurred where a direct participant transferred management control of a source to another direct participant. If the transferring direct participant's change threshold was triggered in relation to this transfer, then the source would be removed from its source list and its baseline and target would both be reduced downwards.[95] As a consequence, its incentive payment would be reduced proportionately.[96]

If the recipient direct participant's change threshold was triggered by the transfer, then that source would be added to its source list and its baseline and target would be adjusted upwards.[97] As a consequence its incentive payment would be increased proportionately.

If allowances had already been allocated in the year in question, and the change threshold was triggered in relation to both participants, then the Secretary of State was required to transfer the number of allowances attributable to the transferred source[98] from the transferring participant to the recipient participant. If the change threshold was only triggered in relation to the transferring participant then the Secretary of State was just required to cancel those allowances from the transferring participant's account. If the change threshold was only triggered in relation to the recipient participant then the Secretary of State was required to allocate an additional number of allowances to the recipient participant. The number of allowances to be transferred, cancelled or allocated was the amount which would not have been allocated to the transferring participant had its change threshold been triggered and the adjustment to its baseline and target had been made before allocation.

b. *Transfer to a Non-direct Participant*

Where a direct participant transferred management control of a source to a non-direct participant and its change threshold was triggered in relation to this transfer, the source would be removed from its source list and its baseline and target would both be reduced downwards.[99] As

---

[95] Paragraphs 2 and 17 of schedule 3 to the Scheme Rules.
[96] This is because the incentive payment directly relates to the target (rule C2(1) of the Scheme Rules).
[97] Part 3 of schedule 3 to the Scheme Rules.
[98] Part 3 of schedule 3 to the Scheme Rules.
[99] Paragraph 2 and 17 of schedule 3 to the Scheme Rules.

a consequence, its incentive payment would be reduced proportionately.[100]

On acquiring management control over the source, the recipient could apply to the Secretary of State to join the scheme as a direct participant. This was the only way in which a person who did not take part in the auction in 2002 could join the scheme as a direct participant at a later date. If it joined the scheme, then its source list was comprised of the transferred source only and its baseline and target was the amount by which the transferring participant's baseline and target was reduced.

If allowances had already been allocated in the year in question and the change threshold had been triggered then the Secretary of State was required to cancel a number of allowances from the transferring participant's account so that the annual allocation was equal to the amount which would have been allocated had the baseline and target been adjusted before allowances were allocated.

c. Closure of a Source

Where a direct participant closed a source, this was treated as if it were a divestment of management control of the source.[101] This prevented the direct participant from being able to count emissions reductions directly attributable to a source's closure towards its emissions reductions target.

d. Withdrawal of the Consent of a Joint-venture Partner

Where a source was entered into the scheme by two or more joint-venture partners and one joint-venture partner withdrew its consent to that source's inclusion in the scheme, this was treated as if it had divested management control over the source.

e. Substantial Closure or Divestment of a Source

Where a direct participant did not close or divest management control of a source but a substantial part of that source was closed or divested, the Secretary of State had a discretion to treat this as if that part of the source were a source itself and therefore subject to the rules which applied when management control over a source is divested. This enabled the rules to be applied equitably between direct participants,

---

[100] This is because the incentive payment directly relates to the target (rule C2(1) of the Scheme Rules).

[101] Paragraph 3(1)(b) and (4) of schedule 3 to the Scheme Rules.

such as in a case where one direct participant reduced its emissions substantially by ceasing production and thereby closing a source, where another direct participant substantially reduced production by the same amount as the first but the source remained open, emitting on a much reduced scale.

Because of this need to ensure equity, the Secretary of State had a wide discretion in deciding how and when to apply this rule. However, in doing so, he was required to have regard to the purposes of the scheme as set out in rule A1 of the Scheme Rules, namely achieving reductions in emissions of greenhouse gases in a cost-effective manner, facilitating compliance with the UK's obligations under the UN Framework Convention on Climate Change and the Kyoto Protocol and implementing the UK's climate change programme.[102]

One criticism of wide discretionary powers such as these is that in planning changes to its business, the direct participant could not be sure of how this discretion would be exercised and was therefore required to make such decisions in an uncertain environment. To reduce this risk, the Secretary of State could issue guidance on how he would apply this paragraph[103] and a direct participant had the right to seek the Secretary of State's opinion on how this rule would be applied to any particular case.[104]

The Secretary of State issued guidance on this point in June 2005[105] and made clear that action would be considered where operation of discrete parts of the source (or separate point sources comprised within the source) ceased, and in cases where there was a general scaling back of the relevant production processes or activities within the source as a whole.

Where this rule applied, the direct participant was required to notify its verifier and the Secretary of State as part of its verification and reporting obligations.[106] This notification had to identify the relevant source, the activity performed by the source, attach a copy of any opinion received from the Secretary of State (that is, an opinion received under paragraph 7(4) of schedule 3 to the Scheme Rules), and if there had been a *force majeure* event, a description of effect of that event on the source.[107]

---

[102] Rule A1(2) to the Scheme Rules.
[103] Paragraph 7(3) of Schedule 3 to the Scheme Rules.
[104] Paragraph 7(4) of schedule 3 to the Scheme Rules.
[105] UK Greenhouse Gas Emissions Trading scheme 2002 - Guidance on the application of schedule 3 paragraph 7 (substantial Closure or divestment of a source), available at: http://defraweb/environment/climatechange/trading/uk/pdf/substantial-closure.pdf.
[106] Paragraph 7(5) of schedule 3 to the Scheme Rules.
[107] Paragraph 7(6) of schedule 3 to the Scheme Rules.

f.   Substantial Increase in Emissions from a Source

If there had been a *force majeure* event which caused the emissions of a particular source to significantly increase, the Secretary of State could allow the direct participant to treat the source as if it were closed, if it would have been unreasonable to require it to retain the source within the scheme. This would have the result that the emissions from that source would no longer be counted towards its emissions limitation commitment and its baseline, target and incentive payment would be consequently reduced.

Clearly this would have a negative impact on the environmental integrity of the scheme, and so given the purposes of the scheme, it was expected that the Secretary of State would set a very high threshold on the application of this rule, particularly given that it was open to participants to buy additional allowances or use banked allowances in such an eventuality.

As this was another rule which grants the Secretary of State a fairly wide discretion in its application, again the Secretary of State could issue guidance on its application[108] and a direct participant could seek the Secretary of State's opinion on how this rule would be applied in any particular case.[109]

Where this rule applied, as in the case of a substantial closure or divestment, the direct participant was required to notify its verifier and the Secretary of State as part of its verification and reporting obligations.[110] This notification had to identify the relevant source, the activity performed by the source, attach a copy of any opinion received from the Secretary of State (that is, an opinion received under paragraph 8(3) of schedule 3 to the Scheme Rules), and if there had been a *force majeure* event, a description of the effect of that event on the source.[111]

g.   Acquisition of a Substitute Source, Including Restructuring of Operations

The rules on closures and divestments had two effects. Firstly, by removing the source from the direct participant's source list and adjusting the baseline accordingly, it prevented a direct participant from claiming credit for the emissions reductions obtained from the closure or divestment of a source. Secondly, by adjusting its target it reduced

---

[108] Paragraph 8(2) of schedule 3 to the Scheme Rules.
[109] Paragraph 8(3) of schedule 3 to the Scheme Rules.
[110] Paragraph 8(4) of schedule 3 to the Scheme Rules.
[111] Paragraph 8(5) of schedule 3 to the Scheme Rules.

the burden on the direct participant to make emissions reductions to recognise the fact that it no longer had as many sources from which to make those reductions. Its incentive payment was also reduced proportionately.

The rule on substitute sources provided an exception to these rules, which applied where the direct participant continued the activity which was performed by the closed or divested source. For example, this allowed the direct participant to reach its emissions reduction target by transferring its activities from one highly polluting or energy inefficient source to another new low polluting or more energy efficient source. Looking at the purpose of the scheme, action such as this is just as worthy of credit as the installation of a piece of equipment within a source which would have the effect of reducing the emissions directly from that particular source.

In order to qualify for this exception to the closure or divestment rules, the direct participant was required to continue the activity performed by the divested or closed source by either acquiring a new source, re-opening a closed source, or resuming normal activities at a source which was deemed to be closed because of a substantial increase in emissions[112] or by any similar transaction or restructuring of operations. This final category was loosely defined but could include the situation where a direct participant closed a source and, rather than acquiring a direct substitute, transferred the activities of the closed source to one or more of its existing sources, thus making a more efficient use of its existing capacity.

4.   Effect of Changes in Operation

When the scheme first began, the effect of a change in operation was backdated to the beginning of the commitment year in which it occurred. This meant that the source was deemed to have been removed from the source list and the baselines and targets were deemed to have been adjusted at the beginning of the year. The new annual target therefore applied to the whole year and emissions from the source in that year before the change in operation were not counted. If allowances have already been allocated for that year then allowances were cancelled so that the allocation was reduced to reflect the new baseline and target.

However, experience of the scheme began to show that, although relatively simple to operate from an administrative point of view, where a direct participant closed or divested management control of a source towards the end of a year, it received no credit for any abatement action

---

[112] I.e. in accordance paragraph 8 of schedule 3 to the Scheme Rules.

taken earlier in that year. This meant that it had no incentive to carry out that abatement.

As a result of this, the Secretary of State used her power to amend the scheme to adapt this rule so that if the change in operation took place in the second 6 months of a year, the baseline, targets and incentive payment remained unchanged for the first 6 months of that year and emissions from that source were counted for that period.[113]

Targets, baselines and incentive payments for previous years of the scheme remained unaffected.

I.   Penalties for Non-compliance

The Scheme Rules set out penalties for failing to comply with various parts of the scheme.

1.   Failure to Comply with the Emissions Limitation Commitment

The direct participant's primary obligation under the scheme was to comply with its emissions limitation commitment each year, i.e. to retire a number of allowances equal to its verified annual emissions, calculated in tonnes of carbon dioxide equivalent.[114] If a direct participant failed to comply with this obligation in any one year, it would not receive its annual incentive payment, and would also be liable to further penalties.

For the first three years of the scheme (2002-2004), a direct participant who failed to comply with its emissions limitation commitment would also have its allocation for the following year reduced by the number of allowances by which it failed to comply, multiplied by a factor of 1.3.[115] For the final two years of the scheme, a direct participant who failed to comply with its emissions limitation commitment would have its allocation for the following year reduced by the number of allowances by which it failed to comply, and was required to pay a financial penalty of £30 for each allowance that it failed to retire.[116]

This financial element of the penalties did not initially apply because of concerns that if a court were to regard Direct Participant Agreements as contracts, then the financial penalty provision may be regarded as a penalty clause, which would be unenforceable under the law of

---

[113] Part 3 of schedule 3 to the Scheme Rules.

[114] See section III.A and III.G on the emissions limitation commitment.

[115] See rule C6(3)(a), (4) and (6) of the Scheme Rules.

[116] See rule C6(3)(b), (5) and (6) of the Scheme Rules.

contract.[117] It was therefore explicitly stated in the Scheme Rules that financial penalties would only become payable once legislation came into force giving them a statutory basis. This statutory basis is now provided by section 39 of the Waste and Emissions Trading Act 2003, which gives statutory effect to the penalty provisions of all scheme participation agreements. This came into force on 20 April 2004 and so the financial penalty provisions applied to direct participants from the 2005 commitment year onwards.[118]

The penalty provisions ensured the environmental integrity of the scheme by providing very strong incentives for direct participants to comply with their emissions limitation commitment. As each direct participant could ensure that it complied with its emissions limitation commitment by purchasing allowances on the market, given that at any one time the cost of buying enough allowances to comply[119] was likely to be below the cost of failing to comply (i.e. £30 per allowance plus loss of annual incentive payment plus reduction in allocation for following year), participants were unlikely to make a conscious decision to fail to comply and face the penalties.

A further penalty provision was provided for direct participants who failed to comply with their emissions limitation commitment for 2006, the final year of the scheme for direct participants. This is because one of the usual penalties - a reduction in the allocation of allowances for the following year - was not available in relation to the final year, as allowances are not allocated following the final reconciliation process. Therefore, any direct participant who failed to have sufficient allowances in its compliance account on the final reconciliation deadline was required to repay all incentive payments received.[120] The incentive payments were repayable forthwith, and interest was payable for the period from the date on which the incentive payment was made until the date upon which it is repaid.[121]

2. Failure to Provide a Verification Opinion or Submission of a Qualified Verification Opinion

If a direct participant failed to submit a verification opinion on or before the reconciliation deadline, it would not be allocated any allowances

---

[117] See *Dunlop Pneumatic Tyre Co v New Garage and Motor Co* [1915] AC 79.
[118] See rule C6(3)(b) and (5) of the Scheme Rules, Section 39 of the Waste and Emissions Trading Act 2003 and Article 2 of the Waste and Emissions Trading Act 2003 (Commencement No 1) Order 2004, S.I. 2004/1163.
[119] At the time of writing, the market price per allowance has not exceeded £12, and has rarely been above £4.
[120] Rule C3(1) of the Scheme Rules.
[121] Rule C3(6) of the Scheme Rules.

for the following commitment year. In addition, the reconciliation process for the commitment year in question would be carried out as if the verified emissions for that year were equal to the baseline - that is, as if no emissions reductions had been made.[122]

If the direct participant submitted the verification opinion after the reconciliation deadline, then it could still be treated as if it had complied with this obligation but only if it notified the Secretary of State before the reconciliation deadline that it would not be able to submit its verification opinion on time. That notification was also required to demonstrate that the delay was due to a *force majeure* event or some other good explanation and to provide the Secretary of State with satisfactory evidence in support.[123]

If a direct participant was only able to provide a qualified verification opinion due to inadequate or incomplete emissions data for any of its sources, whether due to a *force majeure* event or for any other reason, then the Secretary of State had discretion to treat it as if it had complied with the obligation to provide a verification opinion.[124] Notwithstanding this, the Secretary of State could still reduce its allocation by an amount up to and including the baseline emissions of the source in question and reduce its incentive payment by an amount up to and including the proportion of its baseline which was made up by the source in question.[125]

3.   Failure to Comply with the Monitoring, Reporting and Verification Rules and the Changes in Operation Rules

If the direct participant failed to comply with any of the monitoring, reporting or verification rules, or failed to comply with the requirements relating to changes in operation, then the Secretary of State had a choice of sanctions at its disposal. These were: declaring that the direct participant's statement of emissions or verification opinion for that year is invalid (in whole or in part); withholding or delaying any incentive payment (in whole or in part); withholding or delaying the allocation of allowances (in whole or in part); and requiring the direct participant to amend its source list, baseline and targets and to make any associated transfer or cancellation of allowances.[126] The Secretary of State had a wide discretion in deciding which of these penalties to apply. He could decide to apply none, all or any combination of these penalties.

---

[122] Rule C9(3) of the Scheme Rules.
[123] Rule C9(1) of the Scheme Rules.
[124] Rule C9(4)(a) of the Scheme Rules.
[125] Rule C9(4)(b) and (c) of the Scheme Rules.
[126] In accordance with schedule 3 and rule C10(1) of the Scheme Rules.

However, in making that decision, he was required to have regard to the nature and seriousness of the failure to comply, taking into account the purposes of the scheme.[127] In addition, before applying any of the penalties, he was required to notify the direct participant that he considered it to have failed to comply with its obligations, inform it of the obligation that it had not complied with, his reasons and the penalty or penalties that he was minded to apply.[128] If the direct participant was aggrieved, it could invoke the dispute resolution process.[129]

4. Failure to Cancel Allowances

In a number of circumstances, the Secretary of State had an obligation to cancel allowances from a direct participant's compliance account. This occurred, for example, where a change in operation occurred or where, following its temporary exclusion from the EU ETS,[130] a direct participant received an overallocation of allowances. In such cases, if there were not sufficient allowances in the participant's compliance account, then on receipt of a notification from the Secretary of State to this effect, the participant had an obligation to obtain the shortfall of allowances within 7 days and to cancel them. If it failed to do so, then it was deemed to have failed to comply with its emissions limitation commitment. The penalties applicable in those circumstances therefore also applied here.

5. Withdrawal from the Scheme

A direct participant could not withdraw from the scheme before the end of 2006 without the consent of the Secretary of State.[131] If a direct participant withdrew from the scheme, it was required to repay all of its incentive payments received to date.[132] Interest was payable for the period beginning on the date on which the incentive payment was made until the date upon which it was repaid.[133] As soon as the direct participant notified the Secretary of State of its intention to withdraw from the scheme, all allowances held by the direct participant in trading accounts were automatically transferred to its compliance account, its trading accounts closed and its compliance account suspended.[134]

---

[127] Rule C10(2) of the Scheme Rules.
[128] Rule C10(3) of the Scheme Rules.
[129] Rule C10(4) of the Scheme Rules. For more on the dispute resolution process, see section IX.
[130] See section III.K.1.
[131] Clauses 9.1(a)(vi), (b)(ii) and 9.2 of the Direct Participant Agreement.
[132] Rule C3(3) of the Scheme Rules and clauses 9.1(a)(ii), (b)(i) and 9.2 of the Direct Participant Agreement.
[133] Rule C3(6) of the Scheme Rules.
[134] Clause 9.3 of the Direct Participant Agreement. See section VII.A.

If the direct participant intended to withdraw from the scheme at the end of a commitment year, then it was required to submit a verification opinion and retire allowances through the usual reconciliation process.[135] If it had any allowances left in its compliance account following the reconciliation process, the Secretary of State could allow it to transfer these remaining allowances to a third party or to enter into a trading participant agreement and transfer the remaining allowances to a trading account.[136] The compliance account was then closed[137] and it was not allocated any allowances for the following commitment year.

If the direct participant intended to withdraw from the scheme partway through a commitment year, then the Secretary of State would cancel the number of allowances allocated to the direct participant for that commitment year.[138] As with direct participants who withdrew at the end of a commitment year, if it had any allowances left in its compliance account following this cancellation, the Secretary of State could allow it to transfer these remaining allowances to a third party or to enter into a trading participant agreement and transfer the remaining allowances to a trading account.[139] The compliance account would then be closed.[140]

J.   Group Participation

The scheme provided a mechanism whereby two or more organisations could pool their resources to participate in the scheme together in more or less the same manner as single direct participants. The purpose of this was to make it easier for smaller enterprises to take part in the scheme by allowing them to spread out the costs of administering the scheme.

The members of the group were represented by the group participant, who could be a member of the group or a third party agent. The group participant entered into a Group Participant Agreement with the Secretary of State. The other members of the group would not have a direct legal relationship with the Secretary of State but the group participant would be required to enter into agreements with each of them. For the purposes of the scheme, the group participant would have one baseline, one target, one annual allocation and one annual incentive payment. It was for the group members to decide how these should be divided up between them.

---

[135] Clause 9.1(a)(iii) and (v) of the Direct Participant Agreement.
[136] Clause 9.4 of the Direct Participant Agreement.
[137] Clause 9.5 of the Direct Participant Agreement.
[138] Clause 9.1(b)(ii) of the Direct Participant Agreement.
[139] Clause 9.4 of the Direct Participant Agreement.
[140] Clause 9.5 of the Direct Participant Agreement.

The Group Participant Agreement was set out in a standard form very similar to the Direct Participant Agreement and incorporated the Scheme Rules into the Agreement, subject to some modifications.

The major modification was the way in which participants could trade their allowances. Like direct participants, allowances were allocated to them at the beginning of the scheme and shortly after reconciliation each year. However, unlike direct participants they were not free to begin selling those allowances straightaway because of perceived favourable market conditions or in anticipation of overachieving their target in that year. Like Climate Change Agreement Participants they could only trade on a 'baseline and credit' basis, that is they could only sell allowances once it was clear that they had overachieved their targets and only to the extent that they had overachieved.

Their compliance account was suspended immediately on opening, and allowances could only be transferred into and out of the account by the Secretary of State on the group participant's application.[141] The Secretary of State permitted allowances to be transferred into the account without restriction[142] but transfers out of the account were only permitted following reconciliation each year and, even then, only to the extent of its overachievement.[143]

The group participant was generally liable for any act or omission of a group member which would have amounted to non-compliance had the group member been a direct participant.[144] The group participant was required to be able to demonstrate to the Secretary of State's satisfaction that it could pursue the group members directly in the event that it was unable to carry out its obligations under the scheme.

If a group member withdrew from a group, this was treated as a closure of the sources controlled by that group member.[145] In this respect, the group member gained an advantage over a direct participant in a similar position in that the direct participant would have been required to repay all incentive payments received to date if it withdrew before the end of the scheme.[146] If the group participant withdrew from the scheme, then it was required to repay all incentive payments received. The one exception to this, was if a member of the group decided to become a member of another group or to become a direct participant in its own right. In this case, the incentive payment attributable to that group member did not become repayable.[147]

---

[141] Paragraph 1.2 and 1.3 of schedule 1 to the Group Participant Agreement.
[142] Paragraph 1.4 of schedule 1 to the Group Participant Agreement.
[143] Paragraph 1.5 of schedule 1 to the Group Participant Agreement.
[144] Paragraph 2.1 of schedule 1 to the Group Participant Agreement.
[145] Rule F1(4)(i) of the Scheme Rules.
[146] See section III.I above on penalties.
[147] Rule F1(4)(j) of the Scheme Rules.

K.   Relationship with EU Emissions Trading Scheme

The EU Emissions Trading scheme came into operation in January 2005, 3 years into the 5 year duration of the Direct Participant Agreements. The EU scheme is mandatory but covers only direct emissions of carbon dioxide from specified industrial activities.[148] Despite these differences, there are a significant number of sources covered by the UK scheme which were also covered by the EU scheme.

In order to mitigate the problem of a company having the same emissions subject to two similar, but different trading schemes, the Directive establishes the possibility of some installations covered by the Directive being temporarily excluded from the EU scheme for all or part of the first phase.

1.   Temporary Exclusion

As discussed above,[149] Article 27 of the Emissions Trading Directive permits the European Commission to provide for the temporary exclusion of certain installations from the EU scheme, if they meet a three-pronged equivalence test. The Commission can only make such provision if the installation would be subject to a national policy for the duration of the period of exclusion which would require the installations to limit their emissions as much as would be the case if they were not excluded, which imposed requirements as to the monitoring, reporting and verification of emissions which are equivalent to those required by the Directive, and which imposed penalties equivalent to those required by the Directive for failing to comply with those obligations.

The Commission may only consider excluding an installation on receipt of an application from a Member State and may not exclude an installation on its own initiative. Installations may not apply directly to the Commission to be excluded and Member States are under no obligation to apply on their behalf even if, in their opinion, a suitably equivalent national policy is in place.

In April 2004, the UK applied to the Commission to provide for a number of installations which fell within the scope of the Directive and belonging to direct participants in the UK scheme to be temporarily excluded from the EU scheme. This exclusion was to last from 2004 to the end of 2006, when the Direct Participant Agreements ended. Before the application was made, the government asked all of the affected direct participants

---

[148] For a discussion of the activities covered by the EU ETS, see chapter 2.

[149] For a discussion of the provisions on temporary exclusion in the Directive, see section VI of chapter 2.

whether they would like an application to be made on their behalf. An application was subsequently submitted to the Commission seeking the temporary exclusion of 64 installations belonging to 11 direct participants. This comprised the vast majority of installations which would potentially have been covered by both schemes.

This application was approved unanimously by all 25 Member States at the Climate Change Committee in September 2004, and subsequently agreed by Commission Decision on 29 October 2004.[150] This means that participants were entitled to apply for a certificate of temporary exclusion which would entitle them to not take part in the EU scheme until 2007,[151] the final year of its first phase.

In order to demonstrate that the direct participants met the equivalence criteria, the application to the Commission stated that in addition to the existing Scheme Rules, the applicants would also be required to monitor and report their emissions which would have been subject to the EU scheme as if they were not excluded and that their allocations would be adjusted they were more generous than what they would have received under the EU scheme. Amendments to the scheme have been made to give effect to these additional obligations.[152]

2.   Non-excluded Sources

Where a source which falls within the EU scheme was not temporarily excluded, it was treated as if the direct participant divested management control over it on 1 January 2005 (the date upon which the main obligations under the EU scheme became operational) or, if later, the date upon which it began carrying out one of the industrial activities covered by the scheme. It therefore ceased to form part of the direct participant's source list from that date.[153]

From 1 January 2005, a source which was covered by the EU scheme could not be introduced into the UK scheme as a substitute source for another source which had been closed or divested.[154]

L.   State Aid

The incentive payments made to direct participants had the potential to be an unlawful state aid, and as such required approval from the

---

[150] See chapter 2, note 30.
[151] Although they still had to hold a greenhouse gas emissions permit, see section IV.G of chapter 12.
[152] Rule C5(2A)-(2B) of and paragraph 15(3) of Schedule 2 to the Scheme Rules.
[153] Paragraphs 1(5)(aa) and 3(1)(d) of schedule 3 to the Scheme Rules.
[154] Paragraph 10(3) of schedule 3 to the Scheme Rules.

European Commission.[155] The incentive payments received under the scheme received such approval subject to certain conditions, one of which was that participants in the scheme would not be able to claim incentive payments in respect of emissions reductions resulting from European Community law.[156]

IV.   Climate Change Agreement Participants

Operators or sector associations who are party to a Climate Change Agreement with the Secretary of State may subsequently enter into a separate CCA Participant Agreement with the Secretary of State. This enables the operator to generate and use allowances on a baseline and credit basis. A CCA participant may convert an overachievement against their CCA targets into allowances, which are freely tradable within the scheme.[157] Equally, if a CCA participant fails to meet its CCA target, it may make up the shortfall by buying and retiring scheme allowances.[158]

When an operator who is party to a CCA becomes a CCA participant, a compliance account is opened in relation to each target unit in relation to which it wishes to participate in the scheme.[159] If the target unit in question has an absolute target, then the compliance account will be located in the absolute sector of the registry, and if the target unit has a relative target, then the compliance account will be located in the relative sector of the registry.[160]

Where a sector association applies to be a CCA participant, a single compliance account will be opened in the relative sector on behalf of each trading group on whose behalf the sector association is participating.[161]

V.   Trading Participants

Any person may apply to enter into a trading participant agreement with the Secretary of State. A person who is party to such an agreement may open and operate a trading account in the registry. They are not subject to any binding emissions commitments themselves but may buy and sell allowances from other participants in accordance with the rules of the scheme, in particular the registry provisions.

---

[155] Articles 87 and 88 of the EC Treaty.
[156] See section III.E.8.
[157] Subject to the operation of the gateway. This is discussed further in section VII.B.
[158] The retirement of allowances by a CCA Participant technically has the effect of making the CCA target easier to meet. This is discussed in more detail in section VIII of chapter 23.
[159] Rules D3(1) and (2) and D2(2) of the Scheme Rules.
[160] See section VIII.B.
[161] Rule D3(8) and (9) of the Scheme Rules.

VI.   Project Participation

At the outset of the scheme it was envisaged that allowances would be able to be earned by persons carrying out a project which reduces greenhouse gas emissions by a quantifiable amount.[162] These allowances could then be sold to direct participants or CCA participants in order for them to use to meet their obligations.

However, to date the government has not introduced rules governing this form of participation.

VII.   Registry

The registry is crucial to the operation of the scheme, as it is the registry which records allocations, transfers and use of allowances, in the manner of an online bank account. Allowances in the scheme can be freely traded between participants, but must be held electronically in an account in the scheme registry. If allowances are to be regarded as assets they only exist in tangible form within the confines of the registry.

The registry is in electronic form and is accessible through the scheme website.[163] Allowances can only be transferred between accounts in the registry and so they can only be transferred to, or held by, a person who holds an account in the registry.[164]

The registry contains a national cancellation account, a national retirement account and participants' compliance and trading accounts.

A.   Types of accounts

An account can only be opened by a person who has entered into an agreement with the Secretary of State to be a participant in the scheme, whether that be as a direct participant, CCA participant or a trading participant.[165] If a person simply wants to be able to hold allowances without taking on any emissions targets, they may apply to become a trading participant and hold a trading account.[166]

Direct participants and CCA participants must hold a compliance account. Allowances may only be retired for compliance purposes from this account.

All participants may open as many trading accounts as they wish.

---

[162] Section 6 of the Framework Document.
[163] Rule B2 of the Scheme Rules. See www.defra.gov.uk/etr
[164] Rule B3 of the Scheme Rules.
[165] Rule B4 and B5 of the Scheme Rules.
[166] See section V.

B.   The Gateway

A gateway was introduced into the scheme to take account of the fact that some CCA participants have relative rather than absolute targets.[167] Because of the potential for absolute emissions to rise even where relative targets are met, a policy decision was taken that allowances generated by CCA participants through the overachievement of relative targets should not be used for meeting absolute targets, whether for CCA participants or direct participants unless there was an equal or greater flow of allowances in the opposite direction.[168] The registry therefore contains an absolute sector and a relative sector, with a gateway in between. Each compliance account is located in one or other of these sectors and the gateway operates so as to ensure that there will be no net transfer of allowances out of the relative sector into the absolute sector. No transfers of allowances will be permitted from the relative to the absolute sector of the registry unless the net number of allowances from the absolute to the relative sector exceeds the number of allowances which have transferred the other way.

C.   Transfer of Allowances between Accounts

If a participant wishes to transfer allowances from one account to another, it must submit a transfer request in accordance with the registry User Manual. A request to transfer allowances made in accordance with the Manual will be accepted if the transferring account contains sufficient allowances and neither the transferring account nor the recipient account are suspended.[169] The exception to this rule is that the transfer will not be permitted if it is a transfer from a relative sector account to an absolute sector account and the gateway is closed.[170] Once allowances have been transferred into the national retirement account or cancellation account, they may not be subject to any further transactions unless they were transferred there in error.

D.   Suspension of Compliance Accounts

The Secretary of State may block the transfer of allowances in and out of a compliance account (save for the retirement of allowances) in certain circumstances by suspending the account for example if he has reason to believe that an unauthorised person may attempt to access the account, or where the participant (or any of its directors, partners or controllers)

---

[167] See section VI.B of chapter 23.
[168] See paragraphs 5.21-5.23 and figure 5.1 of the Framework Document.
[169] Rule B13(2) of the Scheme Rules.
[170] Rule B13(3) of the Scheme Rules. See section VII.B.

is subject to an investigation relating to certain types of fraud, dishonesty or professional misconduct.[171]

E.   Continuation of the Registry Post-2006

Direct Participant Agreements end in March 2007. However, Climate Change Agreement Participants will continue to be able to take part in the scheme, use allowances for compliance purposes and convert overachievements into allowances. The registry will continue to operate for these purposes after 2006, and any allowances held by direct participants shall continue to be valid.[172]

F.   Further Development of the Registry

The UK government has subsequently developed and modified the registry software so that it is compatible with the requirements of the EU Emissions Trading Scheme. This modified software is also compatible with other international emissions trading as envisaged by the Kyoto Protocol and so far the UK has licensed its software to fourteen countries.

VIII. Additional Powers of the Secretary of State

A.   Power to Amend the Scheme

Despite the fact that the scheme operates by way of agreement between the Secretary of State and the participants, the Secretary of State may amend the Scheme Rules unilaterally at any time. This is a particularly wide power, and is not limited to particular types of amendment or to amendments to achieve a particular purpose. The power is specifically stated in the rules to encompass amendments to correct errors in the scheme,[173] to improve the functioning of the scheme in the light of experience,[174] to extend the scheme to Climate Change Agreements, to additional direct participants following another auction, to projects or to allow international trading,[175] to transfer the Secretary of State's responsibilities to another body,[176] to comply with any state aid decision taken by the European Commission,[177] in connection with future

---

[171] Rules D7 and B20 of the Scheme Rules.
[172] Rules B17 and D6 of the Scheme Rules. See also the document entitled "UK Greenhouse Gas Emissions Trading Scheme 2002, Application of rule B17 (Emissions Trading Registry) and D6 (CCA Participants) (Banking)" available at: http://defraweb/environment/climatechange/trading/uk/pdf/restrictguide051130.pdf.
[173] Rule G1(2)(a) of the Scheme Rules.
[174] Rule G1(2)(b) of the Scheme Rules.
[175] Rule G1(2)(c) of the Scheme Rules.
[176] Rule G1(2)(d) of the Scheme Rules.
[177] Rule G1(2)(e) of the Scheme Rules. For further information on State aids see section III.L.

legislation dealing with greenhouse gases[178] or for making any incidental, supplemental or transitional provisions.[179]

However, these circumstances are not exhaustive and these examples are specifically stated to be without prejudice to the general nature of the power of amendment.[180]

This power extends to giving the Secretary of State the power to adjust the level of incentive payments and targets. This is made clear by the fact that rule C1(1)[181] - the rule which obliges the Secretary of State to make incentive payments at a specified level and to allocate a specified number of allowances in specified circumstances - is expressly stated to be subject to the power of amendment (rule G1).

This power of amendment may not be exercised unless and until the Secretary of State has consulted the participants whom he considers may be affected by proposed amendments.[182] There are three limited exceptions to this obligation to consult. These are where the proposed amendment is in relation to the provisions related to auctions, it is for the purpose of complying with a State aid decision taken by the European Commission or where the Secretary of State considers that the matter is so urgent that it is inappropriate to consult.[183]

The government exercised the power to make amendments in December 2004, March and May 2005, January 2006 and January 2006to make some administrative changes to the scheme.[184]

B.   Power to Issue Guidance

Guidance issued by the Secretary of State in order to assist in the interpretation of the scheme has special status. Where guidance has been issued, everyone acting under the scheme is obliged to do so in accordance with the guidance.[185]

IX.   Dispute Resolution

The scheme contains a procedure for resolving disputes arising out of, or relating to, a Direct Participant Agreement or Trading Participant

---

[178] Rule G1(2)(f) of the Scheme Rules.
[179] Rule G1(2)(g) of the Scheme Rules.
[180] Rule G1(1) and (2) of the Scheme Rules.
[181] Rule C1(1) of the Scheme Rules.
[182] Rule G1(3) of the Scheme Rules.
[183] Rule G1(4) of the Scheme Rules.
[184] Supra note 14.
[185] Rule G5 of the Scheme Rules.

Agreement. This process has two stages, namely mediation and adjudication.

A.   Mediation

Before resorting to other means, the parties to a dispute (likely to be a participant and the Secretary of State) must use their best efforts to discuss and settle the matter in good faith.[186]

B.   Adjudication

If a dispute cannot be resolved through discussion and agreement, either party may refer the matter to an adjudicator.[187] The adjudicator must be someone who is appropriately qualified to consider the dispute in question[188] and should be chosen by agreement between the parties.[189] If agreement cannot be reached, then the Secretary of State may appoint someone who he considers to be appropriately qualified.[190]

The procedure to be followed will be determined by the adjudicator, but there are certain minimum standards to which he must comply. He must take account of the wishes of the parties and the scheme timetable, consult the parties before taking expert advice and ensure that each party has adequate opportunity to respond to any representations made by the other party and to any evidence that the adjudicator proposes to take into account.[191] If a party receives a notice to make representations to the adjudicator, it will have 20 working days to make those representations and must copy them to the other party.[192]

The adjudicator may also request further information from either party (which must be copied to the other party), may hold an oral hearing and may impose a time limit for a party to take any particular action. He may extend such a time limit and, if he considers it appropriate, may proceed to adjudicate if a time limit is not complied with.[193]

The adjudicator must make a report giving his findings of fact, his reasons and his recommendation[194] and send a copy of the report to each party.[195]

---

[186] Paragraph 2 of schedule 4 to the Scheme Rules.
[187] Paragraph 3(1) of schedule 4 to the Scheme Rules.
[188] Paragraph 3(1) of schedule 4 to the Scheme Rules.
[189] Paragraph 3(2) of schedule 4 to the Scheme Rules.
[190] Paragraph 3(2) of schedule 4 to the Scheme Rules.
[191] Paragraph 5 of schedule 4 to the Scheme Rules.
[192] Paragraph 4 of schedule 4 to the Scheme Rules.
[193] Paragraph 5 of schedule 4 to the Scheme Rules.
[194] Paragraph 6(1) of schedule 4 to the Scheme Rules.
[195] Paragraph 6(2) of schedule 4 to the Scheme Rules.

If the dispute involves a challenge to any decision, act or omission of the Secretary of State, then he must reconsider in the light of the adjudicator's report.[196] He must then notify any affected participant of the action that he proposes to take as a result of that reconsideration, as well as his reasons for doing so.[197] However, the adjudicator's report and recommendation are not binding on either party, and both are free to accept or reject his recommendations.[198] They are also free at this point to commence any legal proceedings relating to the matters in dispute.[199] Unless it has been agreed otherwise, the adjudicator's report may be produced as evidence in any legal proceedings.[200]

However, if both parties accept the adjudicator's recommendation or otherwise settle the dispute, then that agreement must be recorded in writing and, once signed, will be binding on the parties.[201]

The costs of appointing the adjudicator and the costs of the adjudication proceedings are to be determined by the adjudicator and this determination will be binding upon the parties. However, neither party is able to recover their legal or other professional costs.[202]

The scheme specifically states that this procedure shall not restrict either party's freedom to commence legal proceedings if it is necessary to preserve confidentiality or any proprietary right or remedy.[203]

X.   National Audit Office Report

In April 2004, the National Audit Office (NAO) published a report into the scheme, entitled, 'The UK Emissions Trading scheme: A new way to combat Climate Change'.[204] The NAO is an independent public body whose function is to scrutinise public spending, and report to Parliament on the economy, efficiency and effectiveness with which taxpayers' resources have been used.

The NAO found that the scheme had had significant achievements in that a novel and well functioning emissions trading scheme had been established, which had the potential to benefit the UK economy. It found that companies have managed to establish services related to emissions

---

[196] Paragraph 8(1) of schedule 4 to the Scheme Rules.
[197] Paragraph 8(2) of schedule 4 to the Scheme Rules.
[198] Paragraph 9 of schedule 4 to the Scheme Rules.
[199] Paragraph 9 of schedule 4 to the Scheme Rules.
[200] Paragraph 11 of schedule 4 to the Scheme Rules as read with paragraph 7.
[201] Paragraph 10 of schedule 4 to the Scheme Rules.
[202] Paragraph 3 of schedule 4 to the Scheme Rules.
[203] Paragraph 12 of schedule 4 to the Scheme Rules.
[204] See note 65.

trading in the UK, such as brokerage and verification, which will give them a head start in competing for business as emissions trading develops at an international level. It also found that the UK scheme had helped develop the EU scheme and helped both government and industry prepare for the launch of the EU scheme, in particular in having a functioning registry which other Member States had expressed an interest in adopting.

On a more negative note, they questioned the method of setting baselines, noting that a number of direct participants had significantly overachieved against their targets in the first year of the scheme and that some of these emissions reductions may have occurred anyway and that some targets were undemanding. They also questioned whether an alternative method of auction might have resulted in companies bidding in more emissions reductions at a lower price.

It was partly in response to this report that the government announced in October 2004 that six direct participants had volunteered to cancel an additional 9 million allowances over the course of the remainder of the scheme.

XI.   Future Developments – Energy Performance Commitment

As the direct participant element of the UK Emissions Trading Scheme was about to expire, the government published a consultation paper in November 2006 on a new proposal for a new mandatory cap and trade scheme called the Energy Performance Commitment.[205]

Under the proposal, the new scheme would be centred around organisations rather than the earlier site-based schemes and around the emissions indirectly attributable to electricity consumption, rather than direct emissions of greenhouse gases themselves. Organisations would be covered by the new scheme if their annual total consumption of electricity exceeded 30,000 MW hours, although they would be exempt if 25 per cent or more of the organisation's emissions were covered by a Climate Change Agreement.

Affected organisations would not be allocated allowances free of charge – they would have to purchase a capped number of allowance from government via an auction, following which they would be freely tradable with each other. The revenue raised by the auction would be recycled to participants, so the proposal would be broadly revenue neutral to government.

---

[205] Available at: http://www.defra.gov.uk/environment/climatechange/trading/epc/index.htm.

The consultation also highlighted other options such as a system of voluntary benchmarking and changes to building regulations so it is not clear at this stage whether the proposal is likely to go ahead.

CHAPTER 23

CLIMATE CHANGE AGREEMENTS

I.   Introduction

In 1997, the Government published a Statement of Intent on environmental taxation which set out key principles for using the UK tax system as part of a policy package for improving the protection of the environment.[1] One principle was to use environmental taxation to shift the burden of tax from 'goods' to 'bads', to encourage polluters to face the true costs of their environmental impact. The Statement of Intent also recognised that environmental taxation must be well designed to address its implications for international competitiveness.

The Statement of Intent was followed in 1998 by a Budget announcement of a review into the feasibility of introducing a tax on the use of energy.[2] Lord Marshall's subsequent report recommended that energy taxation could contribute to a reduction in greenhouse gas emissions in the UK.[3]

The report set out reasons why a 'downstream' tax on the final use of energy, with tax rates reflecting the carbon content of different fuels, was the preferred option. It recommended that revenues should be recycled in full to business, with some of the revenues channelled into schemes aimed directly at promoting energy efficiency. The report also emphasised that any energy taxes adopted 'must be subject to careful design in order to protect the competitiveness of British industry and maximise their environmental benefit'.[4]

Taking into account the recommendations in Lord Marshall's report, the Chancellor announced the introduction of a climate change levy in the March 1999 Budget,[5] although the levy only took effect in April 2001 in order to allow for consultation and negotiation with industry.

---

[1] 'Statement of intent on environmental taxation' HM Treasury, 2 July 1997.
[2] Budget Speech 1998, Section 6: Protecting the environment.
[3] Supra Introduction, note 49.
[4] Paragraph 41 of Lord Marshall's report (ibid.).
[5] See speech, available at: http://archive.treasury.gov.uk/budget/1999/speech.html.

II.   Climate Change Levy

A.   What is it?

The climate change levy is a tax on the non-domestic use of certain types of energy. The main provisions of the climate change levy are contained in section 30 and schedules 6 and 7 of the Finance Act 2000. The Act received Royal Assent on 28 July 2000 and came into effect on 1 April 2001. The Act also provides for secondary legislation relating to the implementation of the climate change levy, such as registration and accounting procedures.

B.   What are its Specific Policy Aims?

The climate change levy is intended to promote energy efficiency to help meet the UK's international commitment to reduce greenhouse gas emissions under the Kyoto Protocol. The levy also aims to encourage the use of renewable energy and environmentally-friendly energy technologies.

The climate change levy is intended to be a revenue-neutral fiscal measure on business as a whole. Revenue raised from the climate change levy is returned to the non-domestic sector through a reduction in the rate of employers' national insurance contributions, exclusions from and reduced rate payments of the climate change levy, and a programme of assistance to industry through the Carbon Trust, including enhanced capital allowances.

C.   What is its scope?

The climate change levy is chargeable on the supply of taxable commodities including electricity, natural gas as supplied by a gas utility, liquefied petroleum gas, coal, and coke.[6] It is not chargeable on the supply of hydrocarbon oil, road fuel gas, heat, steam or waste (as defined).[7] The levy does not apply to diesel, petrol and road fuel gases as these are already subject to road fuel duties.

The climate change levy is chargeable on the following private and public sectors: industry, commerce, agriculture, public administration and other services. The climate change levy is not chargeable on the supply of energy to the domestic sector or on energy used by registered charities other than in the course of a business.[8] There is also a de minimis rule

---

[6] Paragraph 3(1) of schedule 6 to the Finance Act 2000.

[7] Paragraph 3(2) of schedule 6 to the Finance Act 2000.

[8] Paragraph 8(1) of schedule 6 to the Finance Act 2000.

that allows energy supplies to be made, free of climate change levy, up to a specified threshold.[9]

There are a number of other exemptions from the climate change levy set out in schedule 6 to the Finance Act 2000 including where electricity is derived from renewable sources (where certain conditions are met)[10] and where fuel is used by good quality combined heat and power systems.[11]

D.   Rates of Levy and Responsibility for Collection

The levy is applied as a specific rate per unit of energy. There is a different rate for each type of taxable commodity. The different tax rates are intended to reflect the energy content of each taxable commodity. The different measurements for gas and electricity in kilowatt hours and kilograms for other taxable commodities recognise the different billing conventions for those commodities.

The climate change levy rate for electricity is higher than for other forms of energy which takes account of the fact that a considerable proportion of the energy content of fossil fuels used in electricity generation is lost in combustion, transmission and distribution.

Taxable commodity supplied	Full-rate of climate change levy[12]
Electricity	£0.00441 per kilowatt hour
Gas supplied by a gas utility or any gas supplied in a gaseous state that is of a kind supplied by a gas utility	£0.00154 per kilowatt hour
Any petroleum gas, or other gaseous hydrocarbon supplied in a liquid state	£0.00985 per kilogram
Coal and any other taxable commodity	£0.01201 per kilogram

The levy is added to bills from an energy supplier before Value Added Tax is calculated and usually appears as a separate item on energy bills.

The climate change levy is collected by Her Majesty's Revenue and Customs (HMRC).[13] The responsibility for administering the levy falls on suppliers of taxable commodities (mainly electricity or gas suppliers) who are required to register and to pay to HMRC the levy that is due.

---

[9] Paragraph 9 of schedule 6 to the Finance Act 2000.
[10] Paragraph 19 of schedule 6 to the Finance Act 2000.
[11] Paragraph 15(1) of schedule 6 to the Finance Act 2000.
[12] Paragraph 42 of schedule 6 to the Finance Act 2000. Rates as at 1 April 2007.
[13] HM Revenue and Customs was formed in 2005 by the merger of the Inland Revenue and HM Customs and Excise. Prior to the merger, the climate change levy was collected by HM Customs and Excise.

III. Climate Change Agreements

A. Introduction

In line with Lord Marshall's recommendations, schedule 6 of the Finance Act 2000 made provision to protect the international competitiveness of the most energy-intensive sectors of UK industry from the full consequences of the introduction of the climate change levy by providing for climate change agreements.

Climate change agreements are agreements entered into between energy intensive sectors of industry and the Secretary of State for Environment, Food and Rural Affairs.[14] These agreements contain negotiated energy efficiency targets in return for an 80 per cent discount of the climate change levy.

Paragraph 42(1)(c) of schedule 6 to the Finance Act 2000 provides that the amount payable by way of the climate change levy on the supply of taxable commodity is 20 per cent of the full rate if the supply qualifies as a reduced-rate supply. Paragraphs 44 to 52 of schedule 6 set out the circumstances in which a supply is a reduced-rate supply.

To be reduced-rate supply, a supply has to be supplied to a facility which is certified by the Secretary of State as being covered by a climate change agreement and the facility must be listed in a notice published by the Commissioners for Revenue and Customs.[15] Once published, the facilities covered by agreements are eligible to receive supplies of taxable commodities at the reduced rate.

The notice published by the Commissioners for Revenue and Customs must contain certain information,[16] including the day on which it is published, the facility or facilities which are eligible for the reduced rate and the relevant certification periods for each facility listed in the notice.

---

[14] Formerly the Secretary of State for Environment, Transport and the Regions. In this chapter, unless otherwise indicated, any reference to the Secretary of State is a reference to the Secretary of State for Environment, Food and Rural Affairs.

[15] Paragraph 44(3) of schedule 6 to the Finance Act 2000.

[16] Para 44(2) of schedule 6 to the Finance Act 2000.

Example of extract of a notice published by HMRC[17]

PARAGRAPH 44 OF SCHEDULE 6 TO THE FINANCE ACT 2000
CLIMATE CHANGE LEVY: REDUCED RATE CERTIFICATE
The Secretary of State certifies that the following facilities in the Aluminium Federation sector are to be taken as being covered by a climate change agreement:

Facility Number	Facility Address	Date of Publication
AF/AGFA/00503	Agfa-Gevaert UK Manufacturing, Agfa-Gevaert UK Manufacturing, Coal Road, Leeds, West Yorkshire	01/04/2003
AF/ALCA2/00102	Alcan Aluminium UK Limited, Alcan Smelting & Power UK, Lynemouth Smelter, Ashington, Northumberland	01/04/2003

The first day of the period to which these certificates relate is 1 April 2003 and the last day is 31 March 2007. This notice may be varied by later notices.

Once a notice has been published, the operator of a facility should notify its energy supplier of its eligibility for the climate change levy discount. In the same document, they will also notify the energy supplier of a whole range of other exemptions from the climate change levy. This is done through a document known as a PP11 Supplier Certificate issued by HMRC, which sets out the effect of all the discounts and exemptions into one percentage figure which represents the percentage of the supply of taxable commodities eligible for relief from the climate change levy.[18] The energy supplier takes this percentage reduction and applies it to the energy bills it supplies to that operator.

B. Umbrella and Underlying Climate Change Agreements

Climate change agreements are made up of a combination of umbrella agreements and underlying agreements, which operate in tandem.[19] Each industrial sector (e.g. aluminium sector, chemical sector) eligible for a climate change agreement will have one umbrella agreement and a number of underlying agreements. Together, these two agreements must:

---

[17] Extract from Reduced Rate Certificate published on HMRC website at www.hmrc.gov.uk.

[18] Available at: http://www.defra.gov.uk/environment/ccl/pdf/pp11.pdf.

[19] Provision is made in paragraph 47 of schedule 6 to the Finance Act 2000 for a different type of climate change agreement, where one sole agreement is entered into by all the eligible facilities within a sector. This is known as an 'Option 1' agreement. There have been only two such climate change agreements to date.

(a)   set, or provide for the setting of, targets for the facilities to which the underlying agreement applies,

(b)   specify certification periods for the facilities to which the underlying agreement applies, and

(c)   provide for five-yearly (or more frequent) reviews by the Secretary of State of targets set by or under the agreements for those facilities and for giving effect to outcomes of such reviews.[20]

An umbrella agreement is an agreement:

(a) entered into with the Secretary of State,
(b) expressed to be entered into for the purposes of the reduced rate of climate change levy,
(c) identifying the facilities to which it applies, and
(d) to which a representative of each facility to which it applies is a party.[21]

In practice, the requirement for an umbrella agreement to be entered into by a representative of each facility in paragraph (d) above has resulted in the Secretary of State for Environment, Food and Rural Affairs entering into umbrella agreements with a relevant trade association (e.g. the British Cement Association) which represents a particular sector. Under the umbrella agreements, trade associations are known as sector associations. Umbrella agreements are public documents that can be found on Defra's website.[22] Defra has entered into approximately 52 umbrella agreements.

An underlying agreement is an agreement:

(a)   expressed to be entered into for the purposes of the umbrella agreement,

(b)   entered into-
    (i)   with the Secretary of State, or
    (ii)   with a party to the umbrella agreement other than the Secretary of State,

(c)   approved by the Secretary of State if he is not a party to it,

(d)   identifying which of the facilities to which the umbrella agreement applies are the facilities to which it applies, and

(e)   to which a representative of each facility to which it applies is a party.[23]

---

[20] Paragraph 48(3) of schedule 6 to the Finance Act 2000.
[21] Paragraph 48(4) of schedule 6 to the Finance Act 2000.
[22] Available at: http://www.defra.gov.uk/environment/ccl/agreements.htm.
[23] Paragraph 48(5) of schedule 6 to the Finance Act 2000.

There are two types of underlying agreements: (i) those entered into between the Secretary of State and the operator of an eligible facility (known as 'Option 2 agreements') and (ii) those entered into between the sector association and the operator of an eligible facility with the approval of the Secretary of State (known as 'Option 3 agreements'). Most sector associations chose Option 2 agreements. There are approximately 6,000 underlying agreements covering about 15,000 facilities.

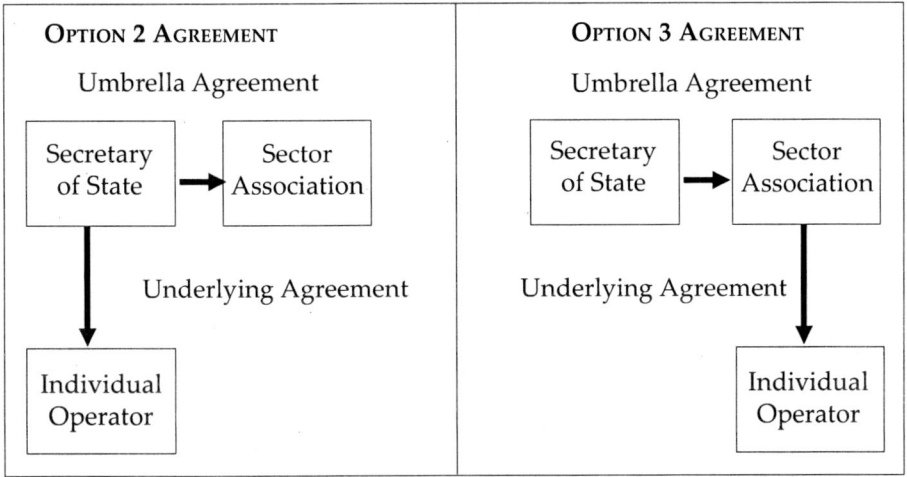

The provisions in the umbrella and underlying agreements are similar across all sectors, but they have been modified where necessary to reflect specific issues relating to individual sectors.[24] Specific terms in the agreements such as targets are of course different as between sectors.

The reason behind this two tier structure of umbrella and underlying agreements is to allow the sector association to act as the conduit between the Secretary of State and numerous operators of individual facilities belonging to that industrial sector.

C. Relationship between Sector Association and Individual Operators

The Secretary of State has only negotiated climate change agreements with sector associations that have agreed to include non-members within its umbrella agreements and not to disadvantage non-members in comparison with their members. Clause 6.2 of the umbrella agreements

---

[24] For example, in the aerospace sector, if the Secretary of State intends to disclose information relating to the umbrella agreement without the sector association's consent, he must notify the Secretary of State for Defence before disclosing the information where he considers it appropriate to do so.

expressly provides that 'The sector association shall not impose unreasonable requirements on non-members who operate facilities within the sector and wish to enter into underlying agreements with the Secretary of State'.

Sector associations usually charge operators for the services they provide in relation to the management of climate change agreements, to cover costs such as negotiating targets and collating data to send to the Secretary of State. Clause 6.3 of the umbrella agreements provides that 'The sector association shall not...impose unreasonable charges on operators or potential operators (whether members or non-members) in respect of the negotiation of this agreement or underlying agreements or the carrying out of its obligations under this agreement'. Any charges imposed by the sector association on the operator must therefore be objectively justifiable and should not discriminate unreasonably against non-members.

In addition, sectors should be careful not to reveal commercially sensitive information to competitor companies within the same sector, as this may constitute anti-competitive behaviour.

Defra produced a model template agreement for sector associations to enter into with individual operators which sets out the relationship between sector associations and individual operators in respect of climate change agreements. This agreement is known as a 'participation agreement'. Sector associations are not obliged to use the participation agreement produced by Defra and are free to enter into participation agreements on different terms, subject to not discriminating against non-members.

D.   State Aid Approval

The provision for an 80 per cent discount in the climate change levy in return for entering into a climate change agreement is subject to state aid approval by the European Commission. The EC Guidelines on state aid for environmental protection 2001 set out the framework under which state aid approval for climate change agreements was determined.[25]

In respect of a reduction in the rate of environmental taxes, such as the climate change levy, the Commission distinguishes between schemes which are national in their origin and those with Community origins. In cases involving purely national taxation schemes, such as climate change agreements, temporary reductions may be permitted. Even in these cases, the Commission must assess whether a reduction in the

---

[25] Community guidelines on State aid for environmental protection, OJ C 37, 3.2.2001, p.3.

rate of the levy would undermine the reason for the climate change levy. Such reductions may be permitted, where agreements are concluded 'between the Member State concerned and the recipient firms whereby the firms or associations of firms undertake to achieve environmental protection objectives during the period for which the exemptions apply'.

In respect of climate change agreements, the Commission concluded that for a ten year period 'the levy reduction for companies covered by climate change agreements is compatible with the requirements of the environmental aid guidelines and Article 87(3)(c) of the EC Treaty'.[26]

In terms of timing, the first batch of climate change agreements were entered into prior to the receipt of the state aid approval. Provision was made for the termination of the agreement if state aid approval was not granted. Similarly, if state aid approval ceases to apply, the agreements may be terminated.[27] In addition, the negotiated targets in the agreements may be varied to take account of any significant changes to the terms of state aid approval at any time during the agreement.[28]

IV. Eligibility

A. General Rule

A facility is eligible to enter into a climate change agreement if it is: (a) either an energy intensive installation, (b) a site on which there is one or more such installations[29] or (c)it meets the extended eligibility criteria set out in the Climate Change Agreements (Eligible Facilities) Regulations 2006.[30]

Energy intensive installations are defined by reference to part A installations listed in part 1 of schedule 1 to the Pollution Prevention and Control (England and Wales) Regulations 2000 (the PPC Regulations).[31]

An energy intensive installation[32] means:

---

[26] Invitation to submit comments pursuant to Article 88(2) of the EC Treaty, concerning measure C 18/2001 (ex N 123/2000) – Climate change levy, OJ C 185, 30.6.2001, p22.
[27] Clause 9.3 of the umbrella agreement and clause 9.6 of the underlying agreement.
[28] Paragraph 7 of schedule 6 of the umbrella agreement and paragraph 7 of schedule 5 of the underlying agreement.
[29] Paragraphs 50-52 of schedule 6 to the Finance Act 2000, as amended by SI 2001/1139 and SI 2006/1848.
[30] SI 2006/60. For further information, see section IV.D.
[31] Supra chapter 12, note 1.
[32] Further guidance on what constitutes an installation and a directly associated activity can be found in 'IPPC: A Practical Guide' (supra chapter 3, note 6). See also *United Utilities Water v Environment Agency for England and Wales [2006] EWCA Civ 633* , supra chapter 3, note 7.

(i) a stationary technical unit where one or more activities listed in part A of part 1 of schedule 1 to the PPC Regulations are carried out; and
(ii) any other location on the same site where any other directly associated activities are carried out which have a technical connection with the activities carried out in the stationary technical unit and which could have an effect on pollution.

A stationary technical unit is taken to mean a unit which is functionally self-contained that can carry out the schedule 1 activity on its own. Where there are two or more units on the same site, those units should be regarded as a single technical unit if:

(a)   they carry out successive steps in one integrated activity;
(b)   one of the listed activities is a directly associated activity of the other; or
(c)   both units are served by the same directly associated activity.

For the purpose of the definition of an energy intensive installation, most of the capacity thresholds for part A activities under part 1 of schedule 1 do not apply, with the exception of the 50MW threshold for combustion plants and the 3MW limit for burning waste oil. Similarly, certain exemptions in part 1 of schedule 1 also do not apply for the purpose of the definition of an energy intensive installation.[33] This means that eligible facilities not only include installations subject to a Pollution Prevention and Control permit, but all those installations carrying out part A activities that would have been subject to a permit had it not been for the size of the installation or some other relevant exception. Eligibility is not therefore dependent on whether or not a facility has a Pollution Prevention and Control permit.

For facilities in Scotland and Northern Ireland, eligibility is also defined by reference to the PPC Regulations.[34] An installation situated in Scotland or Northern Ireland will be an eligible facility if that installation would be a part A installation under the PPC Regulations were it situated in England or Wales.[35] Defra manages climate change agreements in

---

[33] For example, electric arc furnaces with a designed holding capacity of less than 7 tonnes are excluded from part A(1) activities under section 2.1 of part 1 of schedule 1 to the PPC Regulations but are eligible for a climate change agreement.
[34] Theoretically, where there are differences between the schedule 1 activities listed in the PPC Regulation and the Scottish or Northern Irish regulations, an installation may be subject to a Pollution Prevention and Control permit in Scotland/Northern Ireland, but not eligible for the Climate Change Levy discount.
[35] By virtue of entry 4 in the table in paragraph 51 of schedule 6 to the Finance Act 2000.

relation to facilities in Scotland and Northern Ireland, as well as those in England and Wales.

B.   Directly Associated Activities

To qualify as a directly associated activity, the activity must satisfy[36] three criteria:

1.   the activity must be directly associated with the stationary technical unit. It must be on the same site as the stationary technical unit and must serve the unit (and not the other way round);

2.   the activity must have a technical connection with the listed activities carried out in or by the stationary technical unit. This would include, for example, input activities concerned with the storage and treatment of inputs into the stationary technical unit and output activities concerned with the treatment of waste (or other emissions) from the stationary technical unit; and

3.   the activity must be capable of having an effect on emissions. This criterion covers both activities which have an effect on emissions and pollution from the listed activities with which they are directly associated and activities which have such an effect in their own right.

An example of a directly associated activity is an effluent treatment works which serves a chemical plant on the same site. The chemical plant is the stationary technical unit and the effluent treatment plant is a directly associated activity with a technical connection to the stationary technical unit.

C.   The 90/10 Rule

This rule is provided for under the Climate Change Agreements (Eligible Facilities) Regulations 2001.[37]

The 90/10 rule provides that a whole site[38] on which there is one or more installations may be an eligible facility, if at least 90 per cent of the total energy supplied to the site is used in the energy intensive installations

---

[36] Paragraph 51(2B) of schedule 6 to the Finance Act 2000 as inserted by the Climate Change Agreements (Energy-intensive Installations) Regulations 2001, SI 2001/1139.
[37] SI 2001/662.
[38] In this context, a site is taken to be an area of land falling within a continuous boundary which encloses the land used in connection with the operation of a Pollution Prevention and Control installation. See Defra guidance paper PP8.

on the site. Where the whole site is an eligible facility under this rule, this avoids the need for such a site to sub-meter small amounts of energy supplied to parts of the site which do not carry out part A activities (e.g. an office block or canteen).

If the energy intensive installation uses less than 90 per cent of the site's entire energy use, the energy intensive installation will be an eligible facility, but the energy used in the rest of the site will not be eligible for the climate change levy discount. The energy used in the eligible facility would need to be metered separately from the rest of the site.

D.   Extension of Eligibility for Climate Change Agreements

In the 2004 Budget, the Chancellor of the Exchequer announced 'new eligibility criteria for climate change agreements, which will increase the number of businesses that can participate in the scheme'.[39]

1.   Reasons for the Extension

The reason for this extension was to allow other energy-intensive sectors of industry to benefit from the climate change levy discount, where they were not eligible under the Pollution Prevention and Control criteria. Pollution Prevention and Control criteria are based on how much direct pollution an industry causes, rather than how much energy it uses, although there are overlaps between the two. This has meant that some industrial sectors did not previously qualify as energy intensive under the Pollution Prevention and Control criteria, although they use a large amount of energy in their processes. Previous eligibility criteria also meant that whilst some sectors were eligible for the discount, competing sectors were excluded.

2.   Criteria

The additional qualifying criteria for CCAs are set out in the Climate Change Agreements (Eligible Facilities) Regulations 2006.[40] These Regulations apply to installations described in the schedule to the Climate Change Agreements (Energy-intensive Installations) Regulations 2006[41] that are not otherwise eligible to enter into climate change agreements under the Pollution Prevention and Control criteria.

---

[39] Chapter 7 of the 2004 Budget, available at: http://www.hm-treasury.gov.uk/media/E69/6C/ACF1541.pdf.
[40] Supra note 31.
[41] S.I. 2006/59.

To qualify under the new eligibility criteria, the installation or site must meet the energy intensity criteria set out in the Regulations[42] or it must belong to a business sector which meets those criteria. The energy intensity criteria will be met where the predicted energy costs of the installation or site or the business sector to which the installation or site belongs amount to (a) 10 per cent or more of production value; or, (b) if less than 10 per cent but at least 3 per cent, where there is an import penetration ratio[43] of at least 50 per cent.[44] In addition, the energy supplied to the installation or site must be used for the purposes of carrying out an eligible process. An eligible process means a process described in the schedule to the Regulations which meets the energy intensity criteria set out above.

The rule relating to directly associated activities and the 90/10 rule continue to apply to facilities that qualify under the extended criteria.

A facility that is eligible under Pollution Prevention and Control criteria is not permitted to exit their current agreement and join another agreement relying on the extended eligibility criteria. Pollution Prevention and Control criteria will take precedence. However, to facilitate good managment, where eligible processes under the two different criteria are carried out on the same site, it is possible that the two undelying aggreements can be combined and the targets merged, provided that the energy intensity criteria are fully met and targets for that process agreed here the merge takes place.

Facilities that are eligible to enter into climate change agreements under the extended criteria will either be members of a new sector association or be able to join existing sectors and be included in an existing umbrella agreement. Where this occurs, amendments to the elgibility clauses of the umbrella agreement will be neccessary.

E. Application Process to Determine Eligibility for a Climate Change Agreement

Prospective applicants will need to identify any eligible facilities that they wish to include in a climate change agreement. They will need to assess:

---

[42] The eligibility criteria set out in the Regulations draw from the definitions of energy intensity set out in the Taxation of Energy Products Directive (Council Directive 2003/96/EC restructuring the Community framework for the taxation of energy products and electricity, OJ L 283, 31.10.2003, p.51).

[43] This is the percentage ratio of imports to home demand (where home demand is defined as total manufacturers' sales plus imports minus exports).

[44] These tests will be applied once only, at the beginning of negotiations, and at sector level based on measuring a four year period of data.

–   the extent of part A activity
–   the extent of the directly associated activity
–   the applicability of the 90/10 rule
–   whether the facility meets the extended energy intensity
    criteria.

Applicants are required to complete a PP4 eligibility form and send it to Defra for an assessment of their eligibility.[45]

Even where an installation is eligible to enter into a climate change agreement, an operator ought to consider whether the benefits of eligibility for the climate change levy discount outweigh the additional costs associated with entering into such an agreement, such as those associated with sub-metering and fees payable to sector associations. For small companies, with low energy use, it may be more cost-effective to reduce the amount of climate change levy payable by improving energy efficiency, rather than by joining a climate change agreement.

F.   **Changes to a Facility's Eligibility for a Climate Change Agreement**

If the extent of the facility changes, an operator should submit a new PP4 form to Defra showing the new extent of the eligible facility.[46] This may occur, for example, if part of a Pollution Prevention and Control installation is mothballed and the 90/10 rule no longer applies. These changes may also affect the negotiated targets under the umbrella and underlying agreements which may need to be renegotiated.

V.   **How Climate Change Agreements Operate: Certification, Targets and Target Periods**

A.   **Certification**

In order for a facility to be eligible for the climate change levy discount, the Secretary of State must give a certificate to HMRC stating that for the period specified in the certificate, the facility is to be taken as being covered by a climate change agreement. This process is known as 'certification' and the period of time specified in the certificate is known as a 'certification period'.

---

[45] Available at: http://www.defra.gov.uk/environment/ccl/papers.htm.
[46] Defra guidance paper CCA 01: Climate Change Agreements: Administration Procedures for varying umbrella and underlying agreements, available at: http://www.defra.gov.uk/environment/ccl/pdf/cca01.pdf.

Certification periods run consecutively and are set out in part 2 of schedule 1 of the umbrella agreement. Certification periods last for two years generally beginning on 1 April.[47]

Operators who have entered climate change agreements are certified prospectively for eligibility for the following certification period. The first certification period is essentially a 'free' period for eligibility for the levy discount, in that whatever the energy performance of a facility over that period, an operator will be entitled to the discount during that period.

However, for subsequent certification periods, an operator will only be recertified and hence continue to remain eligible for the discount if it has met its negotiated target in the preceding target period.[48] To recertify a facility, the Secretary of State certifies that a facility is covered by a climate change agreement for the next certification period. If a facility continues to meet its target, this process will occur once every two years.

B.   Target and Target Periods

Target setting is at the heart of achieving the environmental benefit from climate change agreements. Climate change agreements must 'set, or provide for the setting of, targets for the facilities to which the underlying agreement applies'.[49] Every umbrella and underlying agreement contains targets which must be met in order to remain eligible for the climate change levy discount.[50]

A target means a target relating to: '(a) energy, or energy derived from a source of any description, used in the facility or an identifiable group of facilities within which the facility falls or (b) emissions, or emissions of any description, from the facility or such a group of facilities'.[51]

---

[47] Paragraph 49(1) of schedule 6 to the Finance Act 2000 provides that 'The first certification period specified by a climate change agreement for a facility to which it applies shall begin with the later of-(a) the date on which the agreement, so far as relating to the facility, is expressed to take effect, and (b) 1st April 2001; and each subsequent certification period so specified shall begin immediately after the end of a previous certification period.'.

[48] For more information on meeting targets, see Section VII.

[49] Paragraph 48(3)(a) of schedule 6 to the Finance Act 2000.

[50] When a new entrant on a green field site enters into an underlying agreement, its targets may be set at a later date. This is because it will not have the baseline energy data to calculate a realistic target. Defra will require a new entrant to collect baseline data (usually within 6 months of the date of the underlying agreement) and then propose targets for Defra to consider. Once targets have been agreed, they will be inserted into the underlying agreement.

[51] Paragraph 49(7) of schedule 6 to the Finance Act 2000 defines a 'target' for the purposes of paragraphs 47-49 of schedule 6 to that Act.

Targets are set in relation to energy use or carbon emissions over a twelve month period ending between October and December, known as a target period. For sectors that entered into agreements in 2001, the first of these target periods ended between 30 September and 31 December 2002. Each sector association was required to decide on its target periods. Facilities may have a different target period from those of the sector association to which it belongs.

Target periods only occur once every two years. This means that the energy use or emissions of a facility are only relevant for meeting climate change agreement targets one year out of two. Theoretically, a facility could have a significant increase in energy use or emissions during the intervening period which is not a target period and still meet its target if it brings its energy use or emissions down during the relevant target period.

The following table illustrates which targets are relevant in order to remain eligible for the climate change levy discount in respect of particular certification periods.

Relevant target period in which target must be met to remain eligible for the climate change levy discount	Period of eligibility for the climate change levy discount, (certification period), if met target in proceeding target period.
None	1 April 2001 – 31 March 2003 ('Free' certification period)
1 October 2001 – 30 September 2002	1 April 2003 – 31 March 2005
1 October 2003 – 30 September 2004	1 April 2005 – 31 March 2007
1 October 2005 – 30 September 2006	1 April 2007 – 31 March 2009
1 October 2007 – 30 September 2008	1 April 2009 – 31 March 2011
1 October 2009 – 30 September 2010	1 April 2011 – 31 March 2013

VI.   Targets

A.   Setting Targets

1.   Sector Targets

At the start of the climate change agreement process in 2001, sector associations negotiated sector-level targets with Defra. Defra based negotiations of sector targets on energy efficiency trends to 2010, as well as data from questionnaires, site surveys and audits. The targets were derived at sector level by negotiating what percentage improvements could be made, in a cost effective way, compared with a base year. The base year was chosen by the sector and was any year between 1990 and 2000.The base year data is required to be verifiable by audit.

The sector targets can be derived by using a 'bottom-up' approach, based on an assessment of the energy efficiency improvements available to each potential climate change agreement participant in that sector or by using a 'top-down' approach which assesses the potential energy efficiency improvements of the sector as a whole.

The Government's objective in the target review negotiations was 'to agree demanding but achievable energy efficiency or carbon saving targets with each sector, requiring the implementation of all cost-effective energy efficient measures by 2010'.[52]

2.   Sub-sector Targets

Certain sectors opted to break down the sector target into sub-sector targets. For example, within the aluminium sector, sub-sector targets exist for a number of different processes, including aluminium rolling and aluminium extrusion.

3.   Individual Targets

Sectors allocate individual targets to their participant facilities based on whether they used a bottom-up or top down approach in negotiating the sector targets.[53] Each individual operator may choose whether to have a single composite target which applies to all or some of its facilities or whether to have a separate target for each facility. In either case, this target is known as a target unit target.

The advantage of a composite target unit target covering all the facilities run by an operator is one of simplicity. It also gives the operator flexibility in how it can meet its target across its different facilities. If one facility has an unexpected increase in energy use, this could be offset against a decrease in energy used by another facility. Whereas if each facility has individual underlying agreements, the facility that had an unexpected increase in energy use would fail to meet its target unless it bought carbon allowances to make up the shortfall.[54] On the other hand, a very poor performance by one facility could theoretically risk all facilities in the target unit failing to meet its composite target, unless carbon allowances were bought.

---

[52] Defra guidance paper PP5: CCL Background.
[53] The method of allocation has to be fair in order to avoid anti-competitive behaviour that may contravene the requirements of the Competition Act 1998.
[54] For further information on the use of allowances to offset a shortfall, see section VIII.

4.   Tolerance Bands and Product Mix Adjustments

In order to increase the flexibility of climate change agreements, operators are given the option of choosing between the risk management options of (i) tolerance bands and (ii) product mix adjustments. Tolerance bands give a facility extra leeway in meeting its targets during a target period.[55] Tolerance bands are set out in schedule 2 of the underlying agreement but do not apply to the final two target periods.

Product mix adjustments allow operators the option of adjusting relative targets to account for changes in product mix (PMO) of a facility during the relevant target period.[56] Whether product mix changes have occurred is assessed in relation to a baseline situation. For example, if relative targets were agreed on the basis of a facility producing 40 widgets and 60 gadgets during a target period and in the relevant target period 50 widgets were made and 100 gadgets, the target may be adjusted to reflect the different proportions of widgets to gadgets made in that target period.

In order to adjust a target, the operator must agree a procedure for the adjustment with the Secretary of State[57] and must serve a notice on the Secretary of State no more than 10 working days after a target period ending before 1 January 2007 stating that it wishes the target to be adjusted to take account of product mix and/or throughput.[58] An operator cannot use this procedure if it has an absolute target, if the relative target already has a tolerance band or in relation to the final two target periods.

B.   **Target Currencies**

Targets must be quantitative. Participants may choose one of four 'currencies'[59] in which to express their targets:

(a)   absolute carbon target
(b)   absolute energy target
(c)   relative carbon target
(d)   relative energy target

---

[55] For more information on meeting facility targets, see section VII.A 2.
[56] Paragraph 1.2 of schedule 2 of the underlying agreement.
[57] To be acceptable to Defra, the PMO procedure must show in quantitative terms the relationship between energy use and product mix/throughput; i.e. be a mathematical relationship.
[58] Some climate change agreements allow for adjustments in throughput to continue after 2007, even though product mix adjustments must not be made to targets in target periods ending after 1 January 2007.
[59] Paragraph 5.1 of schedule 2 to the umbrella agreement.

An operator is permitted to have a different target currency for its target unit target from the sector target currency. However, provision is made in the umbrella agreements to co-ordinate the currencies of targets in umbrella and underlying agreements.[60] Where more than 50 per cent of underlying agreement targets are either absolute or relative targets, the umbrella target must correspond to the majority. Likewise, where more than 50 per cent are carbon or energy targets, the sector target must also reflect the majority.

1. Absolute or Relative Targets

Targets can either be absolute targets or relative targets. Absolute targets set an absolute limit on energy use by a facility during a target period. Relative targets are targets set per unit of throughput in a given target period and provide no limit on the amount of energy used by a facility so long as relative energy use remains within the target. Facilities which predict a fall in production are more likely to benefit from choosing absolute targets[61] whereas relative targets may be more attractive to facilities with rising production, as it is likely that per unit energy consumption will decrease.

2. Energy or Carbon Targets

As well as being able to choose between absolute and relative targets, sectors and operators have the option of targets based either on energy used or carbon emitted during the target period.

C. Changes to the Currency of Targets

An operator may vary the currency of a target by agreement with the Secretary of State. Variations to the currency of a target may be made at any time before the beginning of the target period. The underlying agreements set out limits on the number of times an operator may vary the currency of its targets.[62] An operator may make only:

    (a)   two changes between absolute and relative targets, in the case of a target unit whose targets were initially set as absolute targets;

    (b)   one change between absolute and relative targets, in the case of a target unit whose targets were initially set as relative targets; and

---

[60] Paragraph 7 of schedule 2 to the umbrella agreement.
[61] Note, however, provisions on falls in throughput (see section IX.B).
[62] Paragraph 8 of schedule 5 to the underlying agreement.

(c)   one change between absolute energy targets and absolute carbon targets or between relative energy targets and relative carbon targets.

Any changes to the currency of the targets must follow the conversion conventions set out in the umbrella agreement, which provide various formulae for the conversion.[63] For example, to convert an absolute target into a relative target, the absolute target should be divided by the assumed throughput during the target period.

Where a change of the currency in an underlying agreement results in the currency of the sector target no longer being co-ordinated with the currency of the underlying agreements, the sector target currency must change.

D.   Target Reviews

Umbrella and underlying agreements must 'provide for five-yearly (or more frequent) reviews by the Secretary of State of targets set by or under the agreements for those facilities and for giving effect to outcomes of such reviews'.[64]

To give effect to this requirement, the umbrella agreements provide that 'the Secretary of State shall carry out a review at the end of 2004 of the sector targets for the final three target periods and shall carry out a further review at the end of 2008 of the sector target for the final target period.[65']

The purpose of the review is to ensure that the sector targets being reviewed continue to represent the potential for cost effective energy savings taking account of any changes in technical or market circumstances. In carrying out such a review, the Secretary of State must consult with the relevant sector association and take account of its representations on the review.[66]

Where the Secretary of State and the sector association agree the variations to the sector targets, the sector association must then notify the Secretary of State about the variations it proposes should be made to the targets in the underlying agreements to take account of the sector variations.[67]

---

[63] Paragraph 6 of schedule 2 to the umbrella agreement.
[64] Paragraph 48(3)(c) of schedule 6 to the Finance Act 2000.
[65] Clause 5.4 of the umbrella agreement.
[66] Clauses 5.5 and 5.6 of the umbrella agreement.
[67] Paragraph 6.9 of schedule 6 to the umbrella agreement.

Each operator must have an opportunity to comment on the sector association's proposed variations. The sector association and Secretary of State must consider any representations made by the operators, but if the Secretary of State and sector association agree the variations then the targets will be varied.[68]

Where the Secretary of State and sector association fail to agree on the variation of the sector targets or the targets in the underlying agreements, the Secretary for State may terminate the umbrella agreement, and hence the underlying agreement which cannot exist as a stand alone agreement.[69]

VII. Meeting Targets

A. Eligibility for Levy Discount

Apart from the first certification period, in order to remain eligible for the levy discount, the Secretary of State will only recertify a facility if:

> it appears to the Secretary of State that progress made in the immediately preceding certification period towards meeting targets for the facility by the agreement...is, or is likely to be, such as under the provision of the agreement in question is to be taken as being satisfactory.[70]

The legislation gives some flexibility in determining how targets may be met. The legislation allows a climate change agreement to provide that progress towards meeting any targets for a facility is to be taken as being satisfactory if, in the absence (or partial absence) of any such progress required under the agreement, alternative requirements provided for by the agreement are satisfied.[71]

These provisions allow climate change agreements to provide for a hierarchy of requirements for facilities to fulfil in order to remain eligible for the levy discount.

1. Meeting Sector Targets

Under clause 7.3 of the umbrella agreement, progress made towards meeting the targets set for a facility is to be taken as being satisfactory if

---

[68] Paragraphs 6.10 and 6.11 of schedule 6 to the umbrella agreement.
[69] Paragraph 6.12 of schedule 6 to the umbrella agreement and clause 9.6 of the underlying agreement.
[70] Paragraph 49(3)(b) of schedule 6 to the Finance Act 2000.
[71] Paragraph 49(4) of schedule 6 to the Finance Act 2000.

the sector target for the target period falling within the immediately preceding certification period has been met. This means that regardless of the individual performance of a facility, if the sector to which it belongs has met its target, the facility will remain eligible for the levy discount for the next certification period.

This concept is key to understanding how climate change agreements are designed to work. It is possible that over achievement by some facilities (where they have reduced energy use to below their target) can contribute to helping the sector meet its overall target, thus carrying through other companies that have failed to meet their individual targets.[72]

Similarly, progress made towards meeting the targets will also be satisfactory for facilities belonging to a sub-sector, where the sub-sector target is met, regardless of individual performance of those facilities.

2.   Meeting Facility Targets

Under clause 7.3 of the umbrella agreement, even where the sector target is not met, progress made towards meeting the targets set for a facility is also taken as being satisfactory if:

(a)   the target set for the facility in the relevant underlying agreement has been met by the facility and there is no tolerance band in the underlying agreement for that facility;

(b)   the target set for the facility in the relevant underlying agreement has been met by the facility and there is a tolerance band, but there is no need to take account of the tolerance band; or

(c)   (i) the facility has only met its target taking account of the tolerance band; and (ii) qualitative requirements set for the facility have been met.[73]

3.   Recertification Even where Facility Target not Met

Even where a facility is not part of a sector that has met its target or the facility has not met its individual target, progress made towards meeting the targets set for a facility will be taken as being satisfactory where:

---

[72] Note however that UK ETS trading arrangements may restrict the extent to which over-achievement by some facilities within a sector may be relied on by those facilities that fail to meet their target (see section VIII.B.1).
[73] For the meaning of 'qualitative requirements', see paragraph 4 below.

(a)   qualitative requirements set for the facility have been met; and

(b)   the facility target has not been met because of a relevant constraint or requirement which had a major impact on the performance of the operator of the facility and prevented the target from being achieved.[74]

4.   Meaning of Qualitative Requirements and Relevant Constraint

Qualitative requirements set for a facility are set out in schedule 3 to the underlying agreements and include the preparation and implementation of an energy plan, setting up a system for monitoring and controlling progress in the implementation of the energy plan and identifying improvement actions and making progress reports on energy use and management. These requirements are intended to improve the energy management process within a facility.

'Relevant constraint or requirement' is defined in clause 7.5 of the umbrella agreement and means, in relation to a target –

(a)   a constraint or requirement imposed by or under town and country planning, environmental, health and safety or food hygiene legislation;

(b)   a constraint or requirement imposed on the construction or operation of a combined heat and power plant under section 14 of the Energy Act 1976 or section 36 of the Electricity Act 1989; or

(c)   a constraint imposed by the gas or electricity network,

where the constraint or requirement is inconsistent with an assumption used for setting or reviewing the target.

For example, where it was assumed in setting a target that a new piece of energy saving equipment would be built by a certain date and, after the target was set, the operator failed to obtain planning permission for that piece of equipment, an operator could claim it had suffered from a relevant constraint that prevented it from meeting its target. A relevant constraint may also include a situation where there was an assumption

---

[74] For the meaning of 'relative constraint or requirement', see paragraph 4 below.

about a relevant constraint used for setting a target, but the assumption proved inaccurate as to the extent of the constraint.

B.   Procedure for Assessing Energy Performance

1.   Information Required at the End of the Target Period

In order to assess whether a facility should remain eligible for the levy discount, the Secretary of State requires the sector association to supply certain information regarding energy use and emissions.

Clause 6.4 of the umbrella agreements requires the sector association to supply to the Secretary of State, by the end of January in alternative years starting with 2003, information in relation to the most recently completed target period. This information must include:

(a)   the total number of units of energy used by relevant facilities with a sufficient breakdown of that information to determine whether the currencies of the sector targets need to be changed;

(b)   if the sector target is a carbon target, the total number of units of carbon emitted from the relevant facilities;

(c)   the total throughput for the relevant facilities in the sector;

(d)   the adjustment to be made to the sector target in accordance with paragraph 1.2 (adjustments for emissions trading) of schedule 2;[75] and

(e)   for each target unit with an absolute target, the throughput of that target unit during that period.

Where the sector target has not been met, the sector association must also supply to the Secretary of State the following information:

(a)   for each target unit, the total number of units of energy used by the target unit in relation to each type of fuel;

(b)   for each target unit with a carbon target, the total number of units of carbon emitted from the target unit;

(c)   for each target unit with a relative target, the throughput of that target unit;

---

[75] For further information on the use of emissions trading, see section VIII.

(d)   for each target unit where the target is to be adjusted under paragraph 1.2 (product mix adjustment) or paragraph 1.3 (emissions trading adjustment) of schedule 2 to the relevant underlying agreement, the information needed to calculate the adjustment;

(e)   for each operator that must rely on having met the qualitative requirements set for the facility in order to remain eligible for the levy discount, a copy of its energy plan and a description of the steps taken to implement the plan; and

(f)   for each operator that must rely on having met the relevant constraints provision to remain eligible for the levy discount, details of the relevant constraint or requirement and its impact on the performance of the operator.

2.   Calculation of Performance

Umbrella and underlying agreements set out rules for how the information that is required to be provided to the Secretary of State should be calculated.

2.1   Calculation of Units of Energy Used

Units of energy used must be measured in kilowatt hours. The total number of units of energy used by the sector during a target period is the sum of the units of energy used by the facilities within that sector. The method for calculating the units of energy used by a facility is set out in paragraph 2 of schedule 2 of the underlying agreements. These provisions breakdown the calculation according to the type of energy used, depending on whether the energy was derived from fossil fuels, general electricity imports, energy from a combined heat and power plant, renewable energy and energy from waste.

In order to encourage the use of renewable energy, certain types of renewable energy do not have to be counted as part of a facility's energy use. For example, electricity which is certified by the supplier as being from a new renewable energy supply and as being electricity which the supplier is not relying on for the purpose of fulfilling any obligation imposed on it by any legislation in relation to the generation of such electricity. New renewable energy supplies are defined in the climate change agreements and include supplies from wind energy, photovoltaics, landfill gas and municipal industrial wastes. They do not include hydropower from plants exceeding 10MW.

2.2   Calculation of Carbon Emissions

Carbon emissions from a facility are calculated by multiplying the units of energy from each fuel used in the facility by the relevant carbon emission factors for that fuel.[76] Carbon emission factors are designed to reflect the carbon content of each type of fuel. The total number of units of carbon emitted by the sector during a target period is the sum of the units of carbon emitted from the facilities within that sector.

Carbon emissions from industrial processes must not be counted as part of a facility's carbon emissions unless they result from combustion or oxidation of fossil fuels.[77]

2.3   Calculation of Throughput

The total throughput of the sector during a target period is the sum of the throughput of each facility within that sector. The method for calculating throughput is set out in paragraph 4 of schedule 2 of the underlying agreements. This method can vary between different facilities within a sector. If it varies between facilities within a sector, a formula needs to be used to aggregate throughput of each facility into the total throughput of the sector.

VIII.   Adjustments of Targets to Account for Emissions Trading

A.   Climate Change Agreement Participants in the UK ETS

Emissions trading under the UK ETS has an important role to play in how climate change agreement targets can be met. Operators of facilities that have not met their targets for the preceding target period may buy and retire UK ETS allowances to make up any shortfall in their target and therefore remain eligible for the climate change levy discount for the next certification period. Operators of facilities that overachieve their target (i.e. they use less energy or emit less carbon during a target period than permitted by their CCA target) are able to convert their surplus into UK ETS allowances and either bank their allowances or sell them on the UK ETS. Both where there is a surplus or a shortfall in the target, facilities are subject to the rules of the UK Greenhouse Gas Emissions Trading Scheme 2002 (as amended).[78]

---

[76] Carbon emissions factors are set out in paragraph 3 of schedule 2 of the underlying agreement.
[77] This is a key difference between climate change agreements and EU Emissions Trading, where under EU ETS process emissions do count towards the total emissions of an installation.
[78] Supra chapter 22, note 13 and 14.

Operators and sector associations covered by climate change agreements may join the UK ETS as a Climate Change Agreement Participant. Rule D of the Scheme Rules sets out the way in which operators and sector associations may participate in emissions trading for the purpose of their umbrella and underlying agreements. These rules are explored in greater detail in sub-section D.

B.   Emissions Trading Arrangements

Sector associations and operators are given a choice on how to approach emissions trading within their climate change agreements. Defra has set out a number of alternative potential management options.[79] Paragraph 8.2.1 and 8.2.2 describe the two most common management options.

1.   Independent Emissions Trading by Individual Target Holders (Model 1 Trading)

Under Model 1 Trading, each individual target unit within an underlying agreement may participate in the UK ETS if it chooses to do so. Each target unit needs to meet its own climate change agreement target without being able to rely on other target units' overachievement. Any target unit that overachieved its target would not automatically contribute this overachievement to the sector association. It could choose to sell or bank those allowances.

Therefore, an operator that fails to meet its target would have to buy sufficient UK ETS allowances to make up the shortfall. It cannot rely on the sector passing in order to remain eligible for the levy discount. If this option is selected, there is no mutually supportive action at sector level.

This model is likely to generate the least amount of work for sector associations. Sector associations would need to consolidate data from individual target units to pass to the Secretary of State,[80] but generally have no wider role in terms of trading on the UK ETS.

2.   Sector Emissions Trading (Model 2 Trading)

In contrast to Model 1 trading, sector associations have a much greater role if Model 2 trading is chosen. Under Model 2 trading, individual operators surrender their right to trade on the UK ETS. Instead, the sector

---

[79] See annex to climate change agreement paper 006 dated April 2002, available at: http://www.defra.gov.uk/environment/ccl/papers.htm.
[80] Schedule 3 of umbrella agreement sets out the information the sector association must supply to the Secretary of State.

association trades on the UK ETS to balance any overall surplus or deficit arising from the consolidated performance of all its climate change agreement members.

If there is an overall shortfall, sector associations must purchase necessary allowances to meet the sector target so that operators within that sector may remain eligible for the climate change levy discount. If there is an overall overachievement, the sector association is responsible for the sale or banking of the UK ETS allowances generated from the overachievement.

Under this model, the sector association and operators of facilities need to agree on the allocation of costs[81] and proceeds associated with trading on the UK ETS within the sector.

Target units are free to stay outside the model 2 trading system and trade independently using another model. If they do this within a sector that has opted for a Model 2 trading arrangement, the sector association does not have any UK ETS responsibilities in relation to any target units which have chosen to opt out. This means targets in the umbrella agreement would be split with an aggregate target for all participants opting for Model 2 trading and individual targets for those opting out of Model 2 trading.

Given its greater burden on sector associations, to date only a limited number of sectors have opted for Model 2 trading arrangements. Model 2 trading does, however, allow those target units that fail to meet their target to benefit from the overachievement of other target units within the sector.

C.   Registration and Compliance Accounts

Climate change agreement operators (or sector associations if Model 2 trading applies) can only trade on the UK ETS if they have complied with the registration procedure under the Scheme Rules[82] and have become climate change agreement participants. In order to register, a climate change agreement operator or sector association must submit a registration form to the Secretary of State together with other information including a list of the relevant target units that wish to participate in the UK ETS and consent to the publication of certain information.[83]

---

[81] For example, administration costs and costs of buying allowances.
[82] Rule D2 of the Scheme Rules (see chapter 22, note 13).
[83] Rule D2(1)(b) and D2(3)(c) of the Scheme Rules.

Once an operator or a sector association has submitted a registration form to the Secretary of State, the Secretary of State will open a compliance account for that operator or sector association.[84] Where Model 1 trading has been chosen, a separate compliance account will be opened for each target unit.[85] Where Model 2 trading has been chosen, a single compliance account will be opened for each trading group on whose behalf the sector association is participating.[86]

Climate change agreement operators and sector associations may also open trading accounts under the UK ETS.[87] However, the allowances in these accounts cannot be used to meet CCA targets, although allowances may be transferred from a trading account into a CCA participant's compliance account in order to meet CCA targets.

D. Mechanism for Adjusting Target Unit Targets for Emissions Trading

Where Model 1 trading is chosen, rules D4 and D5 of the Scheme Rules and schedule 2 of the underlying agreements set out the relevant provisions for how targets of a target unit must be adjusted to take account of any emission trading by that target unit on the UK ETS.

The underlying agreement provides that the target for a target unit will be adjusted in relation to a target period 'where carbon emissions allowances have been transferred to or from the target unit in accordance with a trading scheme established or approved by the Secretary of State which applies to this agreement and there is a positive or negative balance in relation to that target unit for that period'.[88] This provision means that where an operator has either failed to meet its target or has exceeded its target and it has traded on the UK ETS as a climate change agreement participant, the operator's target must be adjusted.

1. UK ETS Trading Where an Operator has Failed to Meet its Target

Where an operator has failed to meet its target, it may increase its target by retiring allowances under the UK ETS. Retirement of carbon emissions allowances will: 'for the purpose of the emissions trading provisions of the underlying agreement which applies to the facilities making up that target unit, be deemed to be the transfer of carbon emission allowances

---

[84] Rule D3 of the Scheme Rules.
[85] Rule D3(2) of the Scheme Rules.
[86] Rule D3(9) of the Scheme Rules.
[87] Rule B4 of the Scheme Rules.
[88] Paragraph 1.3 of schedule 2 of the underlying agreement.

to that target unit in accordance with a trading scheme established by the Secretary of State which applies to that agreement and the amount of allowances retired shall be deemed to be a positive balance in relation to that target unit'.[89]

There is a positive balance where there have only been transfers to the target unit or the number of carbon emission allowances transferred to the target unit exceeds the number of carbon emission allowances transferred from it.[90] So where an operator only retires allowances under the UK ETS or retires more allowances than it sells, a positive balance will exist. Where a positive balance exists, the target in relation to the target unit must be increased (i.e. relaxed) by the amount of the positive balance for the relevant target period.[91]

If sufficient allowances have been retired, the target will be increased so that it is equal to the energy used during the target period. If this occurs, there is no longer a shortfall between a target unit's target and the amount of energy used or carbon emitted during the relevant target period and the operator will be deemed to have met its target and remain eligible for the climate change levy discount.[92]

---

[89] Rule D5(2) of the Scheme Rules.
[90] Paragraph 1.4 of schedule 2 of the underlying agreement.
[91] Paragraph 1.5 schedule 2 of the underlying agreement.
[92] An operator may rely on a combination of emissions trading and a relevant constraint, where the relevant constraint only accounts for part of the shortfall in meeting its target unit target.

Flowchart showing effect of retiring UK ETS allowances on a CCA target

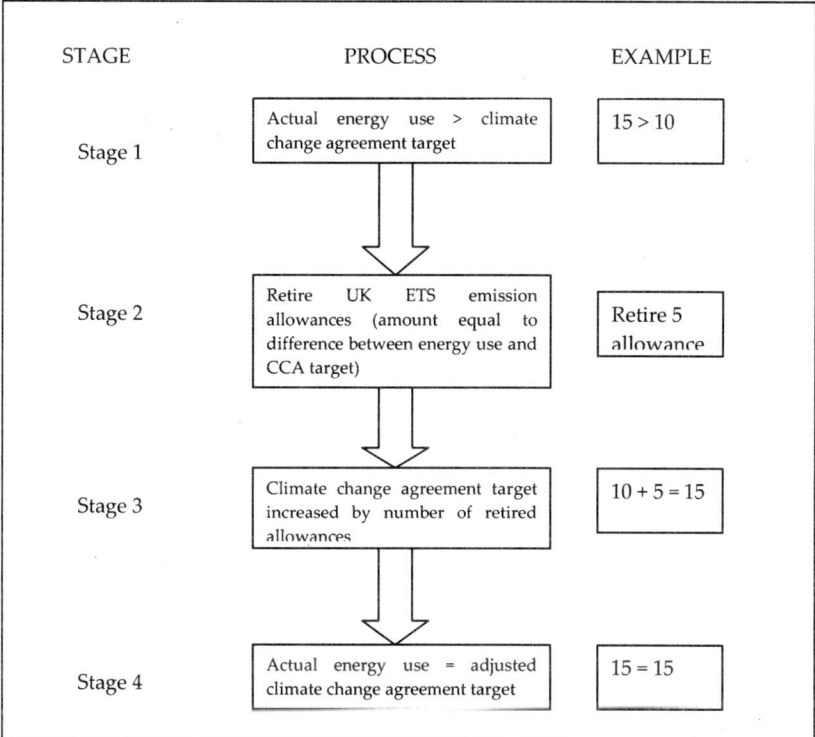

The Secretary of State will only retire allowances where the total quantity of allowances held in the target unit's compliance account on the climate change agreement reconciliation deadline (or an earlier date agreed between the sector association and the Secretary of State) is equal to or greater than the number of allowances needed to increase the target to a level where the target would be met.[93]

An operator seeking to increase its climate change agreement target must provide the Secretary of State with certain information (such as the amount by which the target is to be increased and the number of allowances to be retired) on or before the climate change agreement reconciliation deadline.[94] This deadline is, in respect of any target period, 12 noon on the next business day after 6 February following the end of the target period,[95] although this deadline can be extended in certain circumstances.[96]

---

[93] Rule D5(6) and (7) of the Scheme Rules.
[94] Rule D5(3) of the Scheme Rules.
[95] Rule D5(15) of the Scheme Rules.
[96] See rules D5(16), and (17) of the Scheme Rules. The Secretary of State may also decide to extend this deadline for other reasons, by amending the Scheme Rules, relying on his powers to do so under rule G1.

2.   UK ETS Trading Where an Operator has Met its Target

Where an operator has met its target by emitting less carbon or using less energy than permitted by its climate change agreement target during a target period, it may convert that overachievement into carbon emission allowances which it can then sell in the UK ETS. Alternatively, it can preserve the overachievement for conversion to allowances at a later date (for example, to use against meeting a future target).This is known as 'ring-fencing'.[97]

Where the operator notifies an overachievement to the Secretary of State, the notification will:

> for the purpose of the emissions trading provisions of the underlying agreement which applies to the facilities making up that target unit, be deemed to be the transfer of carbon emission allowances from that target unit in accordance with a trading scheme established by the Secretary of State which applies to that agreement and the number of allowances so reported shall be deemed to be a negative balance in relation to that target unit.[98]

There is a negative balance where there have only been transfers from the target unit or the number of carbon emission allowances transferred from the target unit exceeds the number of carbon emission allowances transferred to it.[99] So where an operator only notifies an overachievement under the UK ETS or the number of allowances in the notification exceeds the number of allowances it retires, a negative balance will exist. Where a negative balance exists, the target in relation to the target unit must be decreased (i.e. tightened) by the amount of the negative balance for the relevant target period.[100]

Where an operator fails to meet its target, there is a clear need to adjust the targets so that a target unit may remain eligible for the discount. By contrast, it may be less clear why targets need be adjusted where there has been an overachievement. The reason for this is to prevent the operator being able to sell or ring-fence the same overachievement more than once in the UK ETS. This is prevented because any notification of overachievement will reduce the amount of the difference between the climate change agreement target and the energy used/carbon emitted, thus reducing the number of allowances an operator may sell or ring-fence.

---

[97] Where an operator meets its target exactly by emitting the exact amount of carbon or using the exact amount of energy permitted by its target, it will not generate any overachievement so will not be able to ring-fence or sell allowances under the UK ETS.
[98] Under rule D4(3) of the Scheme Rules.
[99] Paragraph 1.4 of schedule 2 of the underlying agreement.
[100] Paragraph 1.5 schedule 2 of the underlying agreement.

Flowchart showing effect of a CCA overachievement on a CCA target
In order to convert a CCA overachievement into allowances or to ring-fence the overachievement, an operator must notify the sector association and Secretary of State of the overachievement by the reporting deadline set by its sector association and provide the following information:

—   Facility numbers of the target unit that overachieved its target;

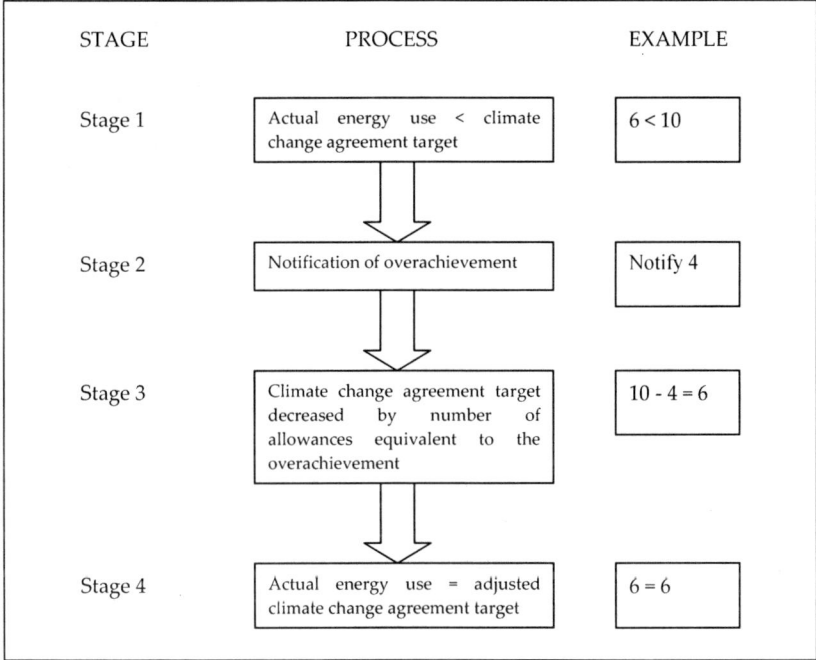

—   The amount by which the target has been over-achieved and the number of allowances to which the over-achievement is equivalent;
—   Any other information needed to calculate the adjustments to be made to the climate change agreement target. [101]

An operator wishing to convert a climate change agreement overachievement into allowances must submit its application to the Secretary of State by 31 December 2012. The application must also include:[102]

---

[101] Rule D4(5) and (6) of the Scheme Rules.
[102] Rule D4(7) of the Scheme Rules.

- a statement of over achievement prepared by the operator. The operator is required to certify that it has not submitted a previous claim in respect of the same target unit over the same target period and that it has complied with the notification requirements set out in rule D4(5) in respect of the over achievement; and

- an unqualified verification opinion from a verifier, who must be accredited by the UK Accreditation Service to assess the accuracy of emissions and energy use. To obtain such an opinion, the operator must supply the verifier with certain information to enable the verifier to assess whether there has been, and if so the extent of, any overachievement.[103]

3.   Adjusting Sector Targets to Account for Emissions Trading under Model 1 Trading

Where Model 1 trading is chosen, sector targets for a target period must be adjusted to reflect any adjustments made to targets in underlying agreements as a result of emissions trading.[104]

4.   Adjusting Sector Targets to Account for Emissions Trading under Model 2 Trading

Where Model 2 trading is chosen, sector targets will be adjusted to take account of any retirement of allowances or overachievement at a sector level. The same basic mechanism for adjusting targets applies to Model 2 trading, as applies to Model 1 trading.[105]

E.   **Suspension of Compliance Accounts of CCA Participants under the UK ETS**

The Secretary of State may suspend the compliance accounts of a CCA participant[106] if:

(a)   the participant (or any of its directors, partners or controllers) is subject to an ongoing investigation by any authority which may lead to conviction, or a legal or

---

[103] Paragraph 4 of schedule 6 to the Scheme Rules.
[104] Paragraph 1.2 of schedule 2 of the umbrella agreement.
[105] Rules for adjusting sector targets under Model 2 trading can be found under rules D4(9)-(13) and D5(8)-(13) of the Scheme Rules and schedule 2 of the umbrella agreements.
[106] Rule D7 of the Scheme Rules.

regulatory sanction, for an offence or other act involving fraud, dishonesty or professional misconduct relating to finance and markets or the formation and activities of a body corporate or other business or professional entity;

(b)   the Secretary of State has reason to believe that the participant has failed to comply with any of the rules relating to the operation of the emissions trading registry; or

(c)   the Secretary of State has reason to believe that an unauthorised person may attempt to access the account.

The effect of suspending a compliance account is that no allowances can be transferred into or out of the account until the suspension is lifted, except in exceptional circumstances. However, allowances may still be retired from the account.[107]

F.   Ceasing to be a CCA Participant under the UK ETS

A CCA participant may cease to be a participant under the UK ETS in the following circumstances:

(a)   where the Secretary of State serves a termination notice on a CCA participant;

(b)   where a climate change agreement to which a CCA participant is a party is terminated; or

(c)   where the CCA participant serves a withdrawal notice on the Secretary of State.[108]

Where the circumstances in paragraphs (a) or (b) above apply, all compliance accounts and trading accounts held by the participant will be suspended. Where there are allowances remaining in any compliance or trading account held in the participant's name, the Secretary of State will either transfer the remaining allowances into an account held in the name of a third party or cancel the allowances depending on the circumstances under which those allowances were acquired. Following the transfer or cancellation of the remaining allowances, the Secretary

---

[107] Rule D7(3) of the Scheme Rules.
[108] See rule D8(2), (4) and (5) of the Scheme Rules for further information on when each of these circumstances may apply.

of State will close all compliance accounts or trading accounts held in the participant's name.

IX.   Variation of Targets

As well as being able to adjust targets to take account of emissions trading and product mix procedures, sector and underlying targets may also be varied.[109] The provisions for the variation of sector targets are set out in part 2 of schedule 6 of the umbrella agreement and the corresponding provisions for underlying targets are set out in part 2 of schedule 5 of the underlying agreement. Sub-sections A and B set out two important examples of when targets must be varied under these provisions.

A.   Inclusion and Exclusion of Facilities

Sector targets must be varied where:

(a)   in the case of relative and absolute targets, a facility enters into an underlying agreement and joins a sector with an existing umbrella agreement;

(b)   in the case of relative targets, a facility no longer has an underlying agreement in relation to that sector.[110]

Any variation of sector targets must be carried out at the end of the first target period which ends after the inclusion or exclusion of the facilities. Sector targets for the relevant target period and any subsequent target periods must be varied to reflect such inclusions and exclusions.

Sector targets must also be varied where a facility (or part of it) has been excluded from an underlying agreement on the basis that it was not eligible for inclusion in that agreement.[111]

Targets in underlying agreements may also be varied where a facility is added to the list of facilities covered by the underlying agreement. Similarly, targets must be varied if facilities are excluded from an underlying agreement.[112]

---

[109] Clause 10.2 of the umbrella agreement and clause 10.2 of the underlying agreement.
[110] Paragraph 2.1 of schedule 6 to the umbrella agreement.
[111] Paragraph 3 of schedule 6 to the umbrella agreement.
[112] Paragraphs 2 and 3 of schedule 5 to the underlying agreement.

B.   Falls in Throughput

For sectors with absolute targets, to ensure that they do not benefit from a windfall of UK ETS allowances when throughput drops significantly, provision is made to tighten targets.[113]

Where the annual level of throughput of the sector during the target period is less than 90 per cent of the annual throughput of the sector when the sector targets were set, the sector targets must be varied to take account of the fall in throughput.

A reference year is used to compare throughput during the most recent targets period to assess whether throughput has dropped by 10 per cent or more.[114]

Similar provisions are contained in underlying agreements with absolute targets, so that where the annual level of throughput of a target unit during a target period is less than 90 per cent of the annual throughput of that target unit when the targets applying to the unit were set, those targets must be varied to take account of that fall of throughput.[115]

X.   Decertification

Decertification means the facility is not certified as eligible for the climate change levy discount for the next certification period. A facility may however remain in the climate change agreement and if it meets its target for next target period, it can be recertified for the following certification period. For example, where a facility did not meet its target during a target period covering January – December 2004, it would be decertified and therefore not eligible for the climate change levy discount for the certification period from 1 April 2005 to 31 March 2007. However, if it then met its target for the next target period from January – December 2006, it would become eligible for the climate change levy discount again for the next certification period from 1 April 2007 to 31 March 2009.

---

[113] Paragraph 4 of schedule 6 to the umbrella agreement.

[114] In 2004, the Secretary of State sent notices to each sector association proposing variations to the provisions which specify the reference year for the final three target periods. The notices proposed that the reference year should be the level of throughput agreed between the Secretary of State and the sector association during the 2004 target review. Where the sector associations agreed to the proposals, umbrella agreements have been amended to effect this change.

[115] In line with proposed changes to the umbrella agreements in 2004, the Secretary of State proposed to amend the underlying agreements to change the reference year. Where operators agree to such a change, the underlying agreements have been amended.

Target period	Target met	Certification period	Eligibility for climate change levy discount during certification period
1/1/04-31/12/04	×	1/4/05-31/3/07	×
1/1/06-31/12/06	✓	1/4/07-31/3/09	✓

A. Circumstances where the Secretary of State must decertify a facility

There are three circumstances where the Secretary of State must decertify a facility:[116]

(a) where a facility fails to make satisfactory progress in an immediately preceding certification period towards meeting its target. For example, where the sector has failed to meet its target and the facility has failed to meet its target and there was no relevant constraint the facility could rely on;

(b) where a facility is not eligible for inclusion in an umbrella agreement;[117] or

(c) where the facility is excluded from a climate change agreement either where the Secretary of State and the sector association agree to vary the agreement to exclude the facility from the climate change agreement or the facility has been excluded from an underlying agreement on the basis that it was not eligible for inclusion in that agreement.

B. Circumstances Where the Secretary of State May Decertify a Facility

Clause 7.9 of the umbrella agreements sets out a number of circumstances where the Secretary of State has the discretion to decertify a facility. These occur if:

(a) the sector association has failed to supply the Secretary of State with the information required in relation to that facility or the information supplied is incomplete or inaccurate;

---

[116] Clause 7.8 of the umbrella agreement.
[117] Clause 3.2 of the umbrella agreement sets out the criteria for eligibility for inclusion of a facility under an umbrella agreement.

(b)   the sector association or the operator of the facility has failed:

   (i)   to co-operate with a person appointed to carry out an audit of information provided in relation to that facility; or

   (ii)   to keep proper records for that purpose or make them available for inspection by the auditor;

(c)   the operator of the facility has failed to comply with its obligations under clause 6 of the underlying agreement; or

(d)   the sector association or the operator fails to meet its share of the costs of an adjudication in relation to which it is a party.

C.   Procedure for Decertification

Where the Secretary of State is minded not to certify that a facility is covered by a climate change agreement (or where a certificate has already been issued, he is minded to issue a variation certificate),[118] he must serve a notice on the sector association and the operator of the facility stating that he is so minded and set out his reasons.[119] He does not have to do this where a termination notice has already been served or where the facility concerned is no longer a facility to which an underlying agreement applies.[120]

Where the sector association or the operator receives such a notice, either party may within 10 working days of its receipt, serve a notice on the Secretary of State setting out why it considers that the Secretary of State should certify that the facility is covered by a climate change agreement (or not issue a variation certificate).

Where such a notice is served by either the sector association or the operator, the certification dispute resolution procedure set out in schedule 4 of the umbrella agreement applies. This procedure only allows for adjudication if there is a dispute on the facts.

---

[118] If a facility is covered by a climate change agreement for a specified target period and a variation certificate is given by the Secretary of State during that target period, the effect of the variation certificate is either (i) that a facility is to be taken as not being covered by a climate change agreement for the whole of that target period; or (ii) that a facility is to be taken as no longer being covered by a climate change agreement from a date specified in the variation certificate. See paragraph 45 of schedule 6 to the Finance Act 2000.
[119] Clause 7.10 of the umbrella agreement.
[120] Clause 7.12 of the umbrella agreement.

Once an adjudicator has been nominated, both parties may make representations. Any representations must be copied to the other parties to the dispute. An adjudicator may request in writing further information from any of the parties to the dispute.

The adjudicator will then, on the basis of the representations provided to him, make a finding on the disputed questions of fact and notify the parties of that finding. The adjudicator's finding on the disputed questions of fact is binding on the parties.[121] No time limit is set out in the climate change agreements for making such a finding.

Once the adjudicator has made a finding, the Secretary of State must serve a notice on the sector association and operator stating whether or not he intends to change her proposed decision and, if not, setting out her reasons.

XI.   Termination of Climate Change Agreements

In contrast to decertification, where an agreement is terminated, the facility cannot remain in a climate change agreement and attempt to regain eligibility for the climate change levy discount for the subsequent certification period by meeting its target. It could only regain eligibility by entering into a new climate change agreement.

A.   Duration

Unless climate change agreement agreements are terminated by either party, they continue in force from the date on which they are made until 31 March 2013.[122] Sector associations and individual operators may drop out of the agreements at any time.

B.   Grounds for Terminating Umbrella Agreements

Umbrella agreements may be terminated before 31 March 2013 in the following circumstances:[123]

(a)   by notice served by the sector association on the Secretary of State. For example, for sectors with low energy use, a sector association may find that the costs of entering into climate change agreements outweigh the benefits of the climate change levy reduction;

---

[121] Paragraph 5 of schedule 4 of the umbrella agreement.
[122] Clause 9.1 of the umbrella agreement and 9.1 of the underlying agreement. The reason for this date is that state aid approval only allows for a 10-year period of approval for exemptions from paying the full rate of the climate change levy.
[123] Clause 9.2 of the umbrella agreement.

(b)   by termination notice served by the Secretary of State on the sector association where:

     (i)   the Secretary of State considers the sector association has failed to comply with an obligation set out in clause 6 of the umbrella agreement (Obligations of the Sector Association), the Secretary of State has served a notice setting out the steps he considers should be taken to comply with the obligation and the sector association fails to carry out those steps;

     (ii)   state aid approval for the reduction in climate change levy is not granted by the European Commission or ceases to apply;

     (iii)   where, despite any finding by an adjudicator, the Secretary of State and the sector association fail to agree the variation of sector targets resulting from inclusion or exclusion of facilities within or from a sector;

     (iv)   where, despite any finding by an adjudicator, the Secretary of State and the sector association fail to agree the variation of the sector targets resulting from a fall in throughput;

     (v)   where the Secretary of State and the sector association fail to agree the variation of the sector targets or the targets in the underlying agreements following a target review in 2004 or 2008 or where the sector association fails to serve a notice on the Secretary of State setting out the variations that it proposes should be made to targets in the underlying agreements following such a review.

The underlying agreements must terminate if the umbrella agreement is terminated.

C.   Grounds for Terminating Underlying Agreements

Underlying agreements may be terminated before 31 March 2013 in the following circumstances:[124]

(a)   by notice served by the operator on the Secretary of State;

(b)   by termination notice served by the Secretary of State on the operator where:

---

[124] Clause 9.2 of the underlying agreement.

(i)   the Secretary of State considers the operator has failed to ensure that an appraisal is carried out of the potential for use, or additional use, of combined heat and power in the facilities;

(ii)   where the operator has not supplied the sector association with the information specified in schedule 4 in relation to the most recently completed target period and the sector association has served a notice on the Secretary of State informing him of this;

(iii) where, despite any finding by an adjudicator, the Secretary of State and the operator fail to agree on the exclusion of a facility from an underlying agreement or any variation of the target following the exclusion of a facility;

(iv)   where, despite any finding by an adjudicator, the Secretary of State and the operator fail to agree the variation of the underlying targets resulting from a review of targets following a combined heat and power appraisal; or

(v)   where, despite any finding by an adjudicator, the Secretary of State and the operator fail to agree the variation of the underlying targets following a fall in throughput.

D.   Procedure for Terminating Climate Change Agreements

The procedure for terminating climate change agreements depends on the reason for the termination. In general, however, a termination notice served by the Secretary of State should state the date on which the climate change agreement ceases to have effect (which should be at least 10 working days after the date on which the notice is served), unless the notice is withdrawn before that date. In some circumstances, the climate change agreements require the Secretary of State to give reasons for his decision to issue a termination notice.

XII.   Adjudication

The climate change agreements provide for a number of circumstances where disputes on the facts between the Secretary of State and either the sector association or operator may be referred to an adjudicator.

The nominated adjudicator must notify the parties of his appointment and inform them of the address to which representations must be sent. Except where there are express provisions in the climate change agreements on procedural matters relating to the adjudication, it is up to the nominated adjudicator to determine the procedure for the adjudication. In determining the procedure, the adjudicator should take account of the wishes of the parties. He must ensure that each party has an adequate opportunity to respond to any representations made by the other party. The adjudicator may impose or extend time limits and take action if the time limit is not complied with.[125]

The parties to the adjudication must bear their own costs and the cost of the adjudication must be shared equally between the parties.[126]

XIII. Audits

The Secretary of State may appoint a person to undertake an independent audit of information provided by the sector association (and information provided by the operators to the sector association).[127] The sector association must co-operate with any auditor and is required to keep proper auditable records and make them available for inspection when required by the auditor. Similar requirements are imposed on operators.[128]

To help ensure operators measure data accurately and to help prepare for a possible independent audit, Defra has produced guidance[129] which recommends that certain information is kept by each operator including:

(a) written procedures for data collection, handling, transfer and error checking;

(b) a map or equivalent document indicating eligible areas, input and output, energy flows and production and metering equipment;

(c) production records; and

(d) energy records - copies of each invoice for energy streams, e.g. electricity, gas, and fuel oil, or if own meters are used instead of utilities metering, then verifiable records of readings should be kept.

---

[125] Paragraph 4 of schedule 5 of the umbrella agreement.
[126] Paragraphs 5 and 6 of schedule 5 of the umbrella agreement.
[127] Clause 6.11 of the umbrella agreement.
[128] Clause 6.7 of the underlying agreement.
[129] Defra guidance document: Climate Change Agreement 03, available at: http://www.defra.gov.uk/environment/ccl/papers.htm.

Where an operator or sector association fails to co-operate with an auditor or keep proper records or make them available to the auditor, the Secretary of State has the discretion to decertify a facility.

XIV. Confidentiality of the Climate Change Agreements

Clause 8 of the umbrella and underlying agreements sets out three rules for governing the extent to which the Secretary of State may disclose information relating to the umbrella and underlying agreements.

First, the Secretary of State is entitled to publish, without the sector association or operator's consent, the umbrella agreement and a list of the facilities certified as being covered by a climate change agreement.

Secondly, the Secretary of State is also entitled to disclose, without the sector association or operator's consent, information relating to the umbrella and underlying agreements in the following circumstances:

(a)   where the disclosure is made under and in accordance with the terms of any legislation. This includes both primary and secondary legislation and would include circumstances where disclosure is required under the Freedom of Information Act 2000[130] and the Environmental Information Regulations 2004.[131]

(b)   where the disclosure is made to a relevant authority[132] for the purposes of:
(i)   the Secretary of State's functions under schedule 6 to the Finance Act 2000; or
(ii)   the authority's functions; or

(c) where the disclosure is made in the course of legal proceedings.

In these circumstances, the Secretary of State must consult the sector association before making any disclosure under clause 8.2 where he considers it appropriate in the circumstances to do so.

---

[130] 2000, c. 36.

[131] S.I. 2004/3391.

[132] The relevant authorities referred to in clause 8.2(b) of the umbrella and underlying agreements are (a) either House of Parliament (including any committee);(b) the European Commission; (c) the Commissioners of Customs and Excise (now the Commissioners of Revenue and Customs); (d) the relevant environmental regulator for a facility under part I of the Environmental Protection Act 1990 or regulations made under Section 2 of the Pollution Prevention and Control Act 1999 or corresponding legislation for Northern Ireland; (e) an auditor of climate change agreement information; and (f) the authorities charged with regulating under the Competition Act 1998.

Thirdly, in all other circumstances, the Secretary of State may only disclose information relating to the umbrella or underlying agreement with the consent of the sector association or operator respectively.[133]

The Secretary of State must take steps to prevent any auditor of climate change agreement data from disclosing information obtained in carrying out the audit to anyone other than the Secretary of State, except to the extent needed to carry out the audit.[134]

XV. Variation of the Terms of the Climate Change Agreements

A. General Variations

Save as discussed below, any provision of the umbrella agreement may be varied by agreement of the Secretary of State and the sector association. Part 1 of schedule 6 sets out the procedure for making the variation. The party that wishes to vary any provision must serve a notice on the other party. Such a notice must be in writing and can be in paper or electronic form.[135] The party that receives the notice has 20 working days after its receipt to respond in writing. Where the variation is agreed, the agreement must be varied accordingly. Where the variation is not agreed, the agreements cannot be varied.

As an exception to the general rule, no variation of the sector targets or the currency of the targets is permitted under the umbrella agreements, unless the variation is in accordance with specific provisions in part 2 of schedule 6 of the umbrella agreement[136]. Part 2 of schedule 6 sets out a finite number of circumstances where targets and currencies may be varied, for example due to a fall in throughput or exclusion of facilities. The underlying agreements contain similar restrictions on the variation of the list of facilities covered by the underlying agreement, the target unit targets and the target currencies.[137]

B. Change of Ownership

Where there is a change of ownership of the facilities that are covered by a climate change agreement, the new operator of the facility is required to sign a form[138] which is a declaration stating that the processes undertaken by the facility have not changed.

---

[133] Clause 8.5 of the umbrella agreement and clause 8.5 of the underlying agreement.
[134] Clause 8.7 of the umbrella agreement and clause 8.7 of the underlying agreement.
[135] Clause 2.1 of the umbrella agreement defines a 'notice' as including any document whether in paper or electronic form.
[136] Clause 10.2 of the umbrella agreement.
[137] Clause 10.2 of the underlying agreement.
[138] Form PP4B, available at: http://www.defra.gov.uk/environment/ccl/pdf/pp4b.pdf.

Where the facility is the only facility in a current underlying agreement, the underlying agreement with the vendor will be terminated and the buyer will enter into a new agreement. Targets will remain the same. If the buyer already has an underlying agreement, it may choose either to vary an existing agreement to include the new facility in its current agreement or enter into a new agreement for that facility.

Where the facility is one of several facilities in a current underlying agreement and only that facility is sold to the buyer, the vendor must also vary its underlying agreement.

XVI. Interpretation of Climate Change Agreements

The umbrella and underlying agreements are not intended to give rise to contractual obligations between the parties.[139] The framework for climate change agreements is provided for in the Finance Act 2000 and their purpose is to provide a mechanism to allow a tax reduction. This framework does not sit easily with the notion that climate change agreements are contracts. It follows that in interpreting climate change agreements, public law principles are more likely to be applicable.[140]

Apart from the adjudication procedures set out in the climate change agreements where there is a dispute of the facts, judicial review would be the most appropriate means for challenging allegedly unlawful actions by the Secretary of State in relation to climate change agreements.

XVII. Relationship with the EU Emissions Trading Scheme

A. Overlap Between Climate Change Agreements and the EU Emissions Trading Scheme

Some facilities that are eligible for the climate change levy discount, also fall within the definition of an 'installation' for the purpose of the Emissions Trading Directive. The reason for this is that the scope for inclusion of operating facilities within both climate change agreements and the EU ETS is based on Pollution Prevention and Control criteria, although there are some differences in scope between the two schemes. For example, in respect of climate change agreements, there is a 50 MW threshold for combustion installations, whereas this threshold has been lowered to 20MW for inclusion of combustion installations within the EU ETS.

---

[139] Clause 1.4 of the umbrella agreement and clause 1.4 of the underlying agreement.
[140] Defra guidance paper NA (00) 13, available at: http://www.defra.gov.uk/environment/ ccl/papers.htm. See also *R v DG Electricity Supply ex parte First Hydro* (Moses J, 2 March 2000).

The two schemes are also both intended to control emissions of carbon dioxide, although not all emissions covered by the climate change agreements are covered by the EU ETS and vice versa. For example, climate change agreements cover indirect emissions,[141] whereas the EU ETS does not and the EU ETS covers process emissions,[142] whereas in general climate change agreements do not.

As a result of the overlap between climate change agreements and the EU ETS, Defra has had to develop policy to manage the overlap. One policy option was to reduce the scope of the climate change agreement target so that direct emissions of carbon dioxide covered by the EU ETS should fall outside climate change agreements and be governed solely by EU ETS, leaving other direct and indirect emissions within climate change agreements.

After consultation with sector associations, Defra decided not to pursue this option for the EU ETS. The climate change agreement targets remain the same in scope as they always have been, covering both direct and indirect emissions. The result of this is that where operating facilities are covered by both schemes, the direct emissions produced by those facilities are regulated by both the EU ETS and climate change agreements. This gives rise to the possibility of 'double trading'.

B.   Double Trading

Where an installation falls under the EU ETS and is also a party to a climate change agreement, it will be able to trade allowances on both the EU ETS and the UK ETS. This raises the prospect of double trading, where operators could trade on both schemes in respect of the same emissions.

For example, where an installation reduces direct emissions to a level below its EU ETS allocation it will be able to sell surplus allowances on the EU ETS. Where the same reduction in emissions results in the installation emitting less carbon (or using less energy) than permitted by its climate change agreement target, it will also be able to convert such overachievement into UK ETS allowances for sale on the UK ETS. This ability to 'double sell' would result in a double benefit to the installation because the same emissions reductions are covered by both schemes.

---

[141] For example, emissions generated from grid electricity.
[142] These are emissions generated from industrial processes, excluding emissions generated from combustion of oxidation of fossil fuels. For example, CO2 is produced as a by-product of making cement. For a full definition of process emissions, see paragraph 2(o) of Annex I of the Monitoring and Reporting Decision, supra chapter 3, note 9.

Similarly, where an installation increases emissions to a level above its EU ETS allocation it will have to buy allowances on the EU ETS to comply with the rules of the scheme. Where the same increase in emissions results in the installation emitting more carbon (or using more energy) than permitted by its climate change agreement target, it will also have to buy and retire allowances on the UK ETS to remain eligible for the climate change levy discount. This requirement to 'double buy' results in a double penalty to the installation because the same emissions are controlled by both schemes.

C.   Amendments to the Scheme Rules to Address Double Trading

To avoid double trading, the Scheme Rules have been amended[143] so that any difference between an installation's allocation and actual emissions under the EU ETS is stripped out of an operator's ability to sell or need to buy allowances on the UK ETS.

In order to do this, it will be necessary to calculate the amount of the difference between an installation's allocation and actual emissions on the EU ETS. Where EU ETS emissions cover emissions which are not covered by the climate change agreement, such as process emissions, it will be necessary to deduct these emissions from any calculation of an EU ETS overachievement or underachievement.

D.   Temporary Exclusion

As discussed in section VI of chapter 2, Article 27 of the Emissions Trading Directive provides for the temporary exclusion of certain installations from the EU ETS until 31 December 2007 (i.e. for the first phase of the EU ETS), if the conditions for temporary exclusion set out in Article 27(2) of the Directive are satisfied. Member States wishing to exclude installations from the EU ETS had to apply to the Commission. Operators were required to notify Defra if they wished to be temporarily excluded from the EU ETS. On 23 December 2005, the Commission adopted a formal Decision[144] giving its approval to the UK's application to temporarily exclude 330 installation from the EU ETS on the basis of existing Climate Change Agreements. Operators whose installations were covered by the Commission's Decision had to apply to Defra for a certificate of temporary exclusion.[145]

---

[143] The UK Greenhouse Gas Emissions Trading (Amendment) Scheme 2007.
[144] Supra chapter 2, note 30.
[145] See section IV.G of chapter 12.

Glossary

2003 Commission Guidance
Communication from the commission on 'Guidance to assist Member States in the implementation of the criteria listed in Annex III to Directive 2003/87/EC establishing a scheme for greenhouse gas emission allowance trading within the European Community and amending Council Directive 96/61/EC, and on the circumstances which force majeure is demonstrated', COM(2003) 830 final.

2003 Regulations
The Greenhouse Gas Emissions Trading Scheme Regulations 2003, S.I. 2003/3311. These Regulations transposed the requirements of the Emissions Trading Directive into UK law.

2005 Commission Guidance
Commission Communication on 'Further guidance on allocation plans for the 2008 to 2012 trading period of the EU Emission Trading Scheme', COM(2005) 703 final

Allowance
Tradable unit equal to one tonne of carbon dioxide equivalent (see Article 2(a) of the Emmissions Trading Directive).

Annex I country
The countries listed in annex I to the UNFCCC. They include the 24 original OECD members, the European Union, and 14 countries with economies in transition.

Assigned Amount
The amount of greenhouse gases which each Annex I country under the Kyoto Protocol is allowed to emit during the commitment period.

Assigned Amount Unit (AAU)
A unit of assigned amount under the Kyoto Protocol equivalent to one tonne of carbon dioxide.

Burden Sharing Agreement
Council Decision 2002/358/EC concerning the approval, on behalf of the European Community, of the Kyoto Protocol to the UNFCCC and the

joint fulfilment of commitments thereunder, OJ L 130, 15.5.2002, p.1. This decision set out how the Community's target under the Kyoto Protocol would be divided between Member States.

Climate Change Agreement (CCA)
Agreements entered into between energy intensive sectors of industry and the Secretary of State for Environment, Food and Rural Affairs containing negotiated energy efficiency targets in return for an 80 per cent discount of the Climate Change Levy (paragraph 42(1)(c) of schedule 6 to the Finance Act 2000).

Certified Emission Reduction (CER)
Kyoto credits resulting from CDM projects.

CFI
Court of First Instance of the European Communities.

Clean Development Mechanism (CDM)
The project mechanism provided for under Article 12 of the Kyoto Protocol. These are projects in developing countries which reduce emissions of greenhouse gases or enhance sinks.

Codecision Procedure
The codecision procedure under Article 251 of the EC Treaty was introduced by the Treaty of Maastricht. It gives the European Parliament the power to adopt instruments jointly with the Council of the European Union. The procedure comprises one, two or three readings.

Comitology Procedure
Procedure for the supervision of the Commission's exercise of delegated powers to adopt measures implementing a Directive or Regulation. The exercise of delegated law making powers is supervised by a committee made up of representatives of the Member States chaired by a Commission representative. Procedure is governed by the Council Decision laying down the procedure for the exercise of implementing powers conferred on the Commission, Decision 1999/468/EC, OJ L 184, 17.7.1999, p.23 as amended by Council Decision 2006/512/EC amending Decision 1999/468/EC laying down the procedures for the exercise of implementing powers conferred on the Commission, OJ L 2006, 22.7.2006, p.11. There are 4 different procedures entailing increasing levels of control: the advisory procedure, the management procedure, the regulatory procedure and the regulatory procedure with scrutiny.

Common Position
Position adopted by the Council on a Commission proposal for legislation. Common position is required if it does not agree with the position taken by the European Parliament on the first reading of the proposed legislation.

Commitment Period
The period in which the targets for reduction of greenhouse gas emissions set out in the Kyoto Protocol apply. The first commitment period runs from 1 January 2008 to 31 December 2012.

Community Independent Transaction Log (CITL)
The independent transaction log provided for in Article 20(1) of the Emissions Trading Directive 2003/87/EC for the purpose of recording the issue, transfer and cancellation of allowances.

Environment Agency (EA)
The Environment Agency of England and Wales, a government agency created by the Environment Act 1995 to take over the roles of National Rivers Authority, Her Majesty's Inspectorate of Pollution (HMIP) and the Waste Regulation Authorities in England and Wales including the London Waste Regulation Authority. It acts as regulator for installations in England and Wales and registry administrator for the UK under the Emissions Trading Directive

EC Treaty
Treaty establishing the European Community (as amended).

Emissions Reduction Unit (ERU)
Kyoto unit resulting from JI projects.

Emissions Trading Directive
Directive 2003/87/EC of the European Parliament and of the Council establishing a scheme for greenhouse gas emission allowance trading within the Community and amending Council Directive 96/61/EC, OJ L 275, 25.10.2003, p.32. The Directive sets the framework for the EU ETS. It was amended in 2004 by the Linking Directive.

ETS Regulations
The Greenhouse Gas Emissions Trading Scheme Regulations 2005 (S.I. 2005/925, as amended by the Greenhouse Gas Emissions Trading Scheme (Amendment), the National Emissions Inventory Regulations 2005, S.I. 2005/2903, the Greenhouse Gas Emissions Trading Scheme (Amendment) Regulations 2006, S.I. 2006/737 and Greenhouse Gas Emissions Trading

Scheme (Amendment) Regulations 2007, S.I. 2007/465. The ETS Regulations govern the application of the EU ETS in the UK.

EU ETS
European Union Emissions Trading Scheme.

Flexible Mechanisms
Collective term for JI, CDM, and emissions trading under the Kyoto Protocol.

Greenhouse Gases
Atmospheric gases that contribute to the greenhouse effect. Greenhouse gases covered by the Kyoto Protocol are carbon dioxide (CO_2), methane (CH_4), nitrous oxide (N_2O), hydrofluorocarbons (HFCs), perfluorocarbons (PFCs) and sulphur hexafluoride (SF_6).

Integrated Pollution Prevention and Control (or IPPC) Directive
Council Directive 96/61/EC concerning integrated pollution prevention and control, OJ L 257, 10.10.1996, p.26. The Directive sets common rules for permitting and controlling industrial installations. Operators of industrial installations covered by annex I of the Directive are required to obtain a permit from Member State authorities.

Joint Implementation (JI)
The project mechanism provided for under Article 6 of the Kyoto Protocol. These are projects which reduce emissions of greenhouse gases or enhance sinks in developed countries with targets under the Kyoto Protocol.

Kyoto Protocol
The Kyoto Protocol to the UNFCCC adopted in Kyoto on 11 December 1997. Among other things, this Protocol sets binding targets for the reduction of greenhouse-gas emissions by industrialised countries.

Large Combustion Plants Directive
Directive 2001/80/EC of the European Parliament and of the Council on the limitation of emissions of certain pollutants into the air from large combustion plants, OJ L 309, 27.11.2001, p.1. The Directive aims to reduce acidification, ground level ozone, and particles throughout Europe by controlling emissions of sulphur dioxide, nitrogen oxides and particulate matter from large combustion plants.

Linking Directive
Directive 2004/101/EC of the European Parliament and of the Council of 27 October 2004 amending Directive 2003/87/EC establishing a scheme

for greenhouse gas emission allowance trading within the Community, in respect of the Kyoto Protocol's project mechanisms, OJ L 338, 13.11.2004, p.18. The Directive amended the Emissions Trading Directive to allow for the use of CERs and ERUs in the EU ETS.

Marrakech Accords

Decisions setting out the detailed rules for the implementation of the provisions of the Kyoto Protocol. The Accords include modalities and procedures for a clean development mechanism as defined in Article 12 of the Kyoto Protocol, and guidelines for the implementation of Article 6 of the Kyoto Protocol. These decisions were formally adopted at the first meeting of the parties to the Kyoto Protocol in December 2005.

Modified Monitoring and Reporting Decision

The Climate Change Committee adopted on 31st July 2006 a revised Decision for the monitoring and reporting of emmissions under the Emmissions Trading Directive. The Decision will apply from 2008.

Monitoring and Reporting Decision

Commission Decision 2004/156/EC of 29 January 2004 establishing guidelines for the monitoring and reporting of greenhouse gas emissions pursuant to Directive 2003/87/EC of the European Parliament and of the Council, OJ L 59, 26.02.2004 p.1.

Monitoring Mechanism Decision

Decision 280/2004/EC of the European Parliament and of the Council concerning a mechanism for monitoring Community greenhouse gas emissions and for implementing the Kyoto Protocol, OJ L 49, 19.2.2004, p.1. The Decision replaces Decision 93/389/EEC of 24 June 1993 for a monitoring mechanism of Community CO_2 and other greenhouse gas emissions, OJ L 167, 9.7.1993, p.31 as last amended by Regulation (EC) No1882/2003 OJ L 284, 31.10.2003, p. 1. The Decision sets out the detailed rules for the monitoring and reporting of emmissions under the Directive for the period 2005-7. From 2008, it will be replaced by the Modified Monitoring and Reporting Decision.

National Allocation Plan (NAP)

Plan developed by each Member State for each phase of the EU ETS as required by Article 9 of the Emissions Trading Directive. The plan sets out the total quantity of allowances which a Member State intends to allocate in the phase and how it proposes to allocate them

NAP table

Table setting out the allocations for a particular phase in accordance with a Member States decision on allocation under Article 11 of the Emissions Trading Directive.

OECD

The Organisation for Economic Co-operation and Development which groups 30 member countries sharing a commitment to democratic government and the market economy.

Operator Holding Account

Account in an Emissions Trading Registry held by an operator of an installation covered by the EU ETS. The account can be used for trading and compliance.

Party Holding Account

Government account in an Emissions Trading Registry through which allowances are issued and surrendered

Person Holding Account

Trading account in an Emissions Trading Registry

Phase

The allocation and trading periods laid down in Article 11 of the Emissions Trading Directive. The first phase runs from 1 January 2005 to 31 December 2007. The second and subsequent phases run for five years, the first being from 1 January 2008 to 31 December 2012.

Pooling

Arrangement permitted by Article 28 of the Emissions Trading Directive under which operators appoint a trustee to take on the responsibility of trading allowances and surrendering allowances to cover the emissions of all the installations in the pool.

PPC Regulations

The Pollution Prevention and Control (England and Wales) Regulations, S.I. 2000/1973. These Regulations transposed the requirements of the IPPC Directive into UK law.

Project Mechanisms

Collective term for CDM and JI.

QUELRO

Quantified Emissions Limitation and Reduction Obligation under the Kyoto Protocol.

Registries Regulation

Commission Regulation No 2216/2004 for a standardised and secured system of registries pursuant to Directive 2003/87/EC of the European

Parliament and of the Council and Decision 280/2004/EC of the European Parliament and of the Council, OJ L 386, 29.12.2004, p.1. This Regulation governs the establishment and operation of electronic registries to track emission allowances in the EU ETS and Kyoto units.

Registry Administrator
Person designated by a Member State or the Commission to operate and maintain a registry in accordance with the requirements of the Emissions Trading Directive and the Registries Regulation.

Scottish Environment Protection Agency (SEPA)
Government agency created alongside the Environment Agency by the Environment Act 1995 in respect of Scotland. It acts as regulator for installations in Scotland in the UK implementation of the Emissions Trading Directive.

Supplementarity
The principle that the use of the project mechanisms should be supplemental to domestic action to reduce greenhouse gas emissions.

UK ETS
UK Emissions Trading Scheme – the world's first economy-wide greenhouse gas emissions trading scheme launched in 2002. The scheme for direct participants runs until December 2006. Under the scheme, 33 organisations ('direct participants' in the scheme) have voluntarily taken on emission reduction targets to reduce their emissions against 1998-2000 levels. They have committed to reducing their emissions by 3.96m tonnes of carbon dioxide equivalent (CO_2e) by the end of the scheme. The scheme is also open to the 6000 companies with Climate Change Agreements. These companies can use the scheme either to buy allowances to meet their targets under their Agreements, or to sell any over-achievement of these targets.

UNFCCC
United Nations Framework Convention on Climate Change approved by Council Decision 94/69/EC of 15 December 1993 concerning the conclusion of the UNFCCC (OJ L 33, 7.2.1994, p.11). 189 countries around the world have joined this international treaty that sets general goals and rules for confronting climate change. The Convention sets an ultimate objective of stabilising greenhouse gas emissions 'at a level that would prevent dangerous anthropogenic (human induced) interference with the climate system.' As a 'framework' document it is something to be amended or augmented over time.

ANNEX 1

THE EMISSIONS TRADING DIRECTIVE

Directive 2003/87/EC of the European Parliament and of the Council of October 2003 establishing a scheme for greenhouse gas emission allowance trading within the Community and amending Council Directive 96/61/EC

(Text with EEA relevance)

THE EUROPEAN PARLIAMENT AND THE COUNCIL OF THE EUROPEAN UNION,

Having regard to the Treaty establishing the European Community, and in particular Article 175(1) thereof,

Having regard to the proposal from the Commission,[1]

Having regard to the opinion of the European Economic and Social Committee,[2]

Having regard to the opinion of the Committee of the Regions,[3]

Acting in accordance with the procedure laid down in Article 251 of the Treaty,[4]

Whereas:

(1) The Green Paper on greenhouse gas emissions trading within the European Union launched a debate across Europe on the suitability and possible functioning of greenhouse gas emissions trading within the European Union. The European Climate Change Programme has considered Community policies and measures through a multi-stakeholder process, including a scheme for greenhouse gas emission allowance trading within the Community (the Community scheme) based on the Green Paper. In its Conclusions of 8 March 2001, the Council

---

[1] OJ C 75 E, 26.3.2002, p. 33.
[2] OJ C 221, 17.9.2002, p. 27.
[3] OJ C 192, 12.8.2002, p. 59.
[4] Opinion of the European Parliament of 10 October 2002 (not yet published in the Official Journal), Council Common Position of 18 March 2003 (OJ C 125 E, 27.5.2003, p. 72), Decision of the European Parliament of 2 July 2003 (not yet published in the Official Journal) and Council Decision of 22 July 2003.

recognised the particular importance of the European Climate Change Programme and of work based on the Green Paper, and underlined the urgent need for concrete action at Community level.

(2) The Sixth Community Environment Action Programme established by Decision No 1600/2002/EC of the European Parliament and of the Council[5] identifies climate change as a priority for action and provides for the establishment of a Community-wide emissions trading scheme by 2005. That Programme recognises that the Community is committed to achieving an 8 % reduction in emissions of greenhouse gases by 2008 to 2012 compared to 1990 levels, and that, in the longer-term, global emissions of greenhouse gases will need to be reduced by approximately 70 % compared to 1990 levels.

(3) The ultimate objective of the United Nations Framework Convention on Climate Change, which was approved by Council Decision 94/69/EC of 15 December 1993 concerning the conclusion of the United Nations Framework Convention on Climate Change,[6] is to achieve stabilisation of greenhouse gas concentrations in the atmosphere at a level which prevents dangerous anthropogenic interference with the climate system.

(4) Once it enters into force, the Kyoto Protocol, which was approved by Council Decision 2002/358/EC of 25 April 2002 concerning the approval, on behalf of the European Community, of the Kyoto Protocol to the United Nations Framework Convention on Climate Change and the joint fulfilment of commitments thereunder,[7] will commit the Community and its Member States to reducing their aggregate anthropogenic emissions of greenhouse gases listed in Annex A to the Protocol by 8 % compared to 1990 levels in the period 2008 to 2012.

(5) The Community and its Member States have agreed to fulfil their commitments to reduce anthropogenic greenhouse gas emissions under the Kyoto Protocol jointly, in accordance with Decision 2002/358/EC. This Directive aims to contribute to fulfilling the commitments of the European Community and its Member States more effectively, through an efficient European market in greenhouse gas emission allowances, with the least possible diminution of economic development and employment.

(6) Council Decision 93/389/EEC of 24 June 1993 for a monitoring mechanism of Community CO_2 and other greenhouse gas emissions,[8]

---

[5] OJ L 242, 10.9.2002, p. 1.

[6] OJ L 33, 7.2.1994, p. 11.

[7] OJ L 130, 15.5.2002, p. 1.

[8] OJ L 167, 9.7.1993, p. 31. Decision as amended by Decision 1999/296/EC (OJ L 117, 5.5.1999, p. 35).

established a mechanism for monitoring greenhouse gas emissions and evaluating progress towards meeting commitments in respect of these emissions. This mechanism will assist Member States in determining the total quantity of allowances to allocate.

(7) Community provisions relating to allocation of allowances by the Member States are necessary to contribute to preserving the integrity of the internal market and to avoid distortions of competition.

(8) Member States should have regard when allocating allowances to the potential for industrial process activities to reduce emissions.

(9) Member States may provide that they only issue allowances valid for a five-year period beginning in 2008 to persons in respect of allowances cancelled, corresponding to emission reductions made by those persons on their national territory during a three-year period beginning in 2005.

(10) Starting with the said five-year period, transfers of allowances to another Member State will involve corresponding adjustments of assigned amount units under the Kyoto Protocol.

(11) Member States should ensure that the operators of certain specified activities hold a greenhouse gas emissions permit and that they monitor and report their emissions of greenhouse gases specified in relation to those activities.

(12) Member States should lay down rules on penalties applicable to infringements of this Directive and ensure that they are implemented. Those penalties must be effective, proportionate and dissuasive.

(13) In order to ensure transparency, the public should have access to information relating to the allocation of allowances and to the results of monitoring of emissions, subject only to restrictions provided for in Directive 2003/4/EC of the European Parliament and of the Council of 28 January 2003 on public access to environmental information.[9]

(14) Member States should submit a report on the implementation of this Directive drawn up on the basis of Council Directive 91/692/EEC of 23 December 1991 standardising and rationalising reports on the implementation of certain Directives relating to the environment.[10]

---

[9] OJ L 41, 14.2.2003, p. 26.
[10] OJ L 377, 31.12.1991, p. 48.

(15) The inclusion of additional installations in the Community scheme should be in accordance with the provisions laid down in this Directive, and the coverage of the Community scheme may thereby be extended to emissions of greenhouse gases other than carbon dioxide, inter alia from aluminium and chemicals activities.

(16) This Directive should not prevent any Member State from maintaining or establishing national trading schemes regulating emissions of greenhouse gases from activities other than those listed in Annex I or included in the Community scheme, or from installations temporarily excluded from the Community scheme.

(17) Member States may participate in international emissions trading as Parties to the Kyoto Protocol with any other Party included in Annex B thereto.

(18) Linking the Community scheme to greenhouse gas emission trading schemes in third countries will increase the cost-effectiveness of achieving the Community emission reductions target as laid down in Decision 2002/358/EC on the joint fulfilment of commitments.

(19) Project-based mechanisms including Joint Implementation (JI) and the Clean Development Mechanism (CDM) are important to achieve the goals of both reducing global greenhouse gas emissions and increasing the cost-effective functioning of the Community scheme. In accordance with the relevant provisions of the Kyoto Protocol and Marrakech Accords, the use of the mechanisms should be supplemental to domestic action and domestic action will thus constitute a significant element of the effort made.

(20) This Directive will encourage the use of more energy-efficient technologies, including combined heat and power technology, producing less emissions per unit of output, while the future directive of the European Parliament and of the Council on the promotion of cogeneration based on useful heat demand in the internal energy market will specifically promote combined heat and power technology.

(21) Council Directive 96/61/EC of 24 September 1996 concerning integrated pollution prevention and control[11] establishes a general framework for pollution prevention and control, through which greenhouse gas emissions permits may be issued. Directive 96/61/EC should be amended to ensure that emission limit values are not set for direct emissions of greenhouse gases from an installation subject to this Directive and that Member States may choose not to impose

[11] OJ L 257, 10.10.1996, p. 26.

requirements relating to energy efficiency in respect of combustion units or other units emitting carbon dioxide on the site, without prejudice to any other requirements pursuant to Directive 96/61/EC.

(22) This Directive is compatible with the United Nations Framework Convention on Climate Change and the Kyoto Protocol. It should be reviewed in the light of developments in that context and to take into account experience in its implementation and progress achieved in monitoring of emissions of greenhouse gases.

(23) Emission allowance trading should form part of a comprehensive and coherent package of policies and measures implemented at Member State and Community level. Without prejudice to the application of Articles 87 and 88 of the Treaty, where activities are covered by the Community scheme, Member States may consider the implications of regulatory, fiscal or other policies that pursue the same objectives. The review of the Directive should consider the extent to which these objectives have been attained.

(24) The instrument of taxation can be a national policy to limit emissions from installations temporarily excluded.

(25) Policies and measures should be implemented at Member State and Community level across all sectors of the European Union economy, and not only within the industry and energy sectors, in order to generate substantial emissions reductions. The Commission should, in particular, consider policies and measures at Community level in order that the transport sector makes a substantial contribution to the Community and its Member States meeting their climate change obligations under the Kyoto Protocol.

(26) Notwithstanding the multifaceted potential of market-based mechanisms, the European Union strategy for climate change mitigation should be built on a balance between the Community scheme and other types of Community, domestic and international action.

(27) This Directive respects the fundamental rights and observes the principles recognised in particular by the Charter of Fundamental Rights of the European Union.

(28) The measures necessary for the implementation of this Directive should be adopted in accordance with Council Decision 1999/468/EC of 28 June 1999 laying down the procedures for the exercise of implementing powers conferred on the Commission.[12]

---

[12] OJ L 184, 17.7.1999, p. 23.

(29) As the criteria (1), (5) and (7) of Annex III cannot be amended through comitology, amendments in respect of periods after 2012 should only be made through codecision.

(30) Since the objective of the proposed action, the establishment of a Community scheme, cannot be sufficiently achieved by the Member States acting individually, and can therefore by reason of the scale and effects of the proposed action be better achieved at Community level, the Community may adopt measures, in accordance with the principle of subsidiarity as set out in Article 5 of the Treaty. In accordance with the principle of proportionality, as set out in that Article, this Directive does not go beyond what is necessary in order to achieve that objective, HAVE ADOPTED THIS DIRECTIVE:

Article 1
Subject matter
This Directive establishes a scheme for greenhouse gas emission allowance trading within the Community (hereinafter referred to as the "Community scheme") in order to promote reductions of greenhouse gas emissions in a cost-effective and economically efficient manner.

Article 2
Scope
1. This Directive shall apply to emissions from the activities listed in Annex I and greenhouse gases listed in Annex II.

2. This Directive shall apply without prejudice to any requirements pursuant to Directive 96/61/EC.

Article 3
Definitions
For the purposes of this Directive the following definitions shall apply:

(a) "allowance" means an allowance to emit one tonne of carbon dioxide equivalent during a specified period, which shall be valid only for the purposes of meeting the requirements of this Directive and shall be transferable in accordance with the provisions of this Directive;

(b) "emissions" means the release of greenhouse gases into the atmosphere from sources in an installation;

(c) "greenhouse gases" means the gases listed in Annex II;

(d) "greenhouse gas emissions permit" means the permit issued in accordance with Articles 5 and 6;

(e) "installation" means a stationary technical unit where one or more activities listed in Annex I are carried out and any other directly associated activities which have a technical connection with the activities carried out on that site and which could have an effect on emissions and pollution;

(f) "operator" means any person who operates or controls an installation or, where this is provided for in national legislation, to whom decisive economic power over the technical functioning of the installation has been delegated;

(g) "person" means any natural or legal person;

(h) "new entrant" means any installation carrying out one or more of the activities indicated in Annex I, which has obtained a greenhouse gas emissions permit or an update of its greenhouse gas emissions permit because of a change in the nature or functioning or an extension of the installation, subsequent to the notification to the Commission of the national allocation plan;

(i) "the public" means one or more persons and, in accordance with national legislation or practice, associations, organisations or groups of persons;

(j) "tonne of carbon dioxide equivalent" means one metric tonne of carbon dioxide (CO_2) or an amount of any other greenhouse gas listed in Annex II with an equivalent global-warming potential;

(k) "Annex I Party" means a Party listed in Annex I to the United Nations Framework Convention on Climate Change (UNFCCC) that has ratified the Kyoto Protocol as specified in Article 1(7) of the Kyoto Protocol;

(l) "project activity" means a project activity approved by one or more Annex I Parties in accordance with Article 6 or Article 12 of the Kyoto Protocol and the decisions adopted pursuant to the UNFCCC or the Kyoto Protocol;

(m) "emission reduction unit" or "ERU" means a unit issued pursuant to Article 6 of the Kyoto Protocol and the decisions adopted pursuant to the UNFCCC or the Kyoto Protocol;

(n) "certified emission reduction" or "CER" means a unit issued pursuant to Article 12 of the Kyoto Protocol and the decisions adopted pursuant to the UNFCCC or the Kyoto Protocol.

Article 4
Greenhouse gas emissions permits
Member States shall ensure that, from 1 January 2005, no installation undertakes any activity listed in Annex I resulting in emissions specified in relation to that activity unless its operator holds a permit issued by a competent authority in accordance with Articles 5 and 6, or the installation is temporarily excluded from the Community scheme pursuant to Article 27.

Article 5
Applications for greenhouse gas emissions permits
An application to the competent authority for a greenhouse gas emissions permit shall include a description of:

(a) the installation and its activities including the technology used;

(b) the raw and auxiliary materials, the use of which is likely to lead to emissions of gases listed in Annex I;

(c) the sources of emissions of gases listed in Annex I from the installation; and

(d) the measures planned to monitor and report emissions in accordance with the guidelines adopted pursuant to Article 14.

The application shall also include a non-technical summary of the details referred to in the first subparagraph.

Article 6
Conditions for and contents of the greenhouse gas emissions permit

1. The competent authority shall issue a greenhouse gas emissions permit granting authorisation to emit greenhouse gases from all or part of an installation if it is satisfied that the operator is capable of monitoring and reporting emissions.
A greenhouse gas emissions permit may cover one or more installations on the same site operated by the same operator.

2. Greenhouse gas emissions permits shall contain the following:

(a) the name and address of the operator;

(b) a description of the activities and emissions from the installation;

(c) monitoring requirements, specifying monitoring methodology and frequency;

(d) reporting requirements; and

(e) an obligation to surrender allowances equal to the total emissions of the installation in each calendar year, as verified in accordance with Article 15, within four months following the end of that year.

Article 7
Changes relating to installations
The operator shall inform the competent authority of any changes planned in the nature or functioning, or an extension, of the installation which may require updating of the greenhouse gas emissions permit. Where appropriate, the competent authority shall update the permit. Where there is a change in the identity of the installation's operator, the competent authority shall update the permit to include the name and address of the new operator.

Article 8
Coordination with Directive 96/61/EC
Member States shall take the necessary measures to ensure that, where installations carry out activities that are included in Annex I to Directive 96/61/EC, the conditions of, and procedure for, the issue of a greenhouse gas emissions permit are coordinated with those for the permit provided for in that Directive. The requirements of Articles 5, 6 and 7 of this Directive may be integrated into the procedures provided for in Directive 96/61/EC.

Article 9
National allocation plan
1. For each period referred to in Article 11(1) and (2), each Member State shall develop a national plan stating the total quantity of allowances that it intends to allocate for that period and how it proposes to allocate them. The plan shall be based on objective and transparent criteria, including those listed in Annex III, taking due account of comments from the public. The Commission shall, without prejudice to the Treaty, by 31 December 2003 at the latest develop guidance on the implementation of the criteria listed in Annex III.

For the period referred to in Article 11(1), the plan shall be published and notified to the Commission and to the other Member States by 31 March 2004 at the latest. For subsequent periods, the plan shall be published and notified to the Commission and to the other Member States at least 18 months before the beginning of the relevant period.

2. National allocation plans shall be considered within the committee referred to in Article 23(1).

3. Within three months of notification of a national allocation plan by a Member State under paragraph 1, the Commission may reject that plan, or any aspect thereof, on the basis that it is incompatible with the criteria listed in Annex III or with Article 10. The Member State shall only take a decision under Article 11(1) or (2) if proposed amendments are accepted by the Commission. Reasons shall be given for any rejection decision by the Commission.

Article 10
Method of allocation
For the three-year period beginning 1 January 2005 Member States shall allocate at least 95 % of the allowances free of charge. For the five-year period beginning 1 January 2008, Member States shall allocate at least 90 % of the allowances free of charge.

Article 11
Allocation and issue of allowances
1. For the three-year period beginning 1 January 2005, each Member State shall decide upon the total quantity of allowances it will allocate for that period and the allocation of those allowances to the operator of each installation. This decision shall be taken at least three months before the beginning of the period and be based on its national allocation plan developed pursuant to Article 9 and in accordance with Article 10, taking due account of comments from the public.

2. For the five-year period beginning 1 January 2008, and for each subsequent five-year period, each Member State shall decide upon the total quantity of allowances it will allocate for that period and initiate the process for the allocation of those allowances to the operator of each installation. This decision shall be taken at least 12 months before the beginning of the relevant period and be based on the Member State's national allocation plan developed pursuant to Article 9 and in accordance with Article 10, taking due account of comments from the public.

3. Decisions taken pursuant to paragraph 1 or 2 shall be in accordance with the requirements of the Treaty, in particular Articles 87 and 88 thereof. When deciding upon allocation, Member States shall take into account the need to provide access to allowances for new entrants.

4. The competent authority shall issue a proportion of the total quantity of allowances each year of the period referred to in paragraph 1 or 2, by 28 February of that year.

Article 11a
Use of CERs and ERUs from project activities in the Community scheme
1. Subject to paragraph 3, during each period referred to in Article 11(2), Member States may allow operators to use CERs and ERUs from project activities in the Community scheme up to a percentage of the allocation of allowances to each installation, to be specified by each Member State in its national allocation plan for that period. This shall take place through the issue and immediate surrender of one allowance by the Member State in exchange for one CER or ERU held by the operator in the national registry of its Member State.

2. Subject to paragraph 3, during the period referred to in Article 11(1), Member States may allow operators to use CERs from project activities in the Community scheme. This shall take place through the issue and immediate surrender of one allowance by the Member State in exchange for one CER. Member States shall cancel CERs that have been used by operators during the period referred to in Article 11(1).

3. All CERs and ERUs that are issued and may be used in accordance with the UNFCCC and the Kyoto Protocol and subsequent decisions adopted thereunder may be used in the Community scheme:

(a) except that, in recognition of the fact that, in accordance with the UNFCCC and the Kyoto Protocol and subsequent decisions adopted thereunder, Member States are to refrain from using CERs and ERUs generated from nuclear facilities to meet their commitments pursuant to Article 3(1) of the Kyoto Protocol and in accordance with Decision 2002/358/EC, operators are to refrain from using CERs and ERUs generated from such facilities in the Community scheme during the period referred to in Article 11(1) and the first five-year period referred to in Article 11(2); and

(b) except for CERs and ERUs from land use, land use change and forestry activities.

Article 11b
Project activities
1. Member States shall take all necessary measures to ensure that baselines for project activities, as defined by subsequent decisions adopted under the UNFCCC or the Kyoto Protocol, undertaken in

countries having signed a Treaty of Accession with the Union fully comply with the acquis communautaire, including the temporary derogations set out in that Treaty of Accession.

2. Except as provided for in paragraphs 3 and 4, Member States hosting project activities shall ensure that no ERUs or CERs are issued for reductions or limitations of greenhouse gas emissions from installations falling within the scope of this Directive.

3. Until 31 December 2012, for JI and CDM project activities which reduce or limit directly the emissions of an installation falling within the scope of this Directive, ERUs and CERs may be issued only if an equal number of allowances is cancelled by the operator of that installation.

4. Until 31 December 2012, for JI and CDM project activities which reduce or limit indirectly the emission level of installations falling within the scope of this Directive, ERUs and CERs may be issued only if an equal number of allowances is cancelled from the national registry of the Member State of the ERUs' or CERs' origin.

5. A Member State that authorises private or public entities to participate in project activities shall remain responsible for the fulfilment of its obligations under the UNFCCC and the Kyoto Protocol and shall ensure that such participation is consistent with the relevant guidelines, modalities and procedures adopted pursuant to the UNFCCC or the Kyoto Protocol.

6. In the case of hydroelectric power production project activities with a generating capacity exceeding 20 MW, Member States shall, when approving such project activities, ensure that relevant international criteria and guidelines, including those contained in the World Commission on Dams November 2000 Report "Dams and Development A New Framework for Decision-Making", will be respected during the development of such project activities.

7. Provisions for the implementation of paragraphs 3 and 4, particularly in respect of the avoidance of double counting, and any provisions necessary for the implementation of paragraph 5 where the host party meets all eligibility requirements for JI project activities shall be adopted in accordance with Article 23(2).

Article 12
Transfer, surrender and cancellation of allowances
1. Member States shall ensure that allowances can be transferred between:

(a) persons within the Community;

(b) persons within the Community and persons in third countries, where such allowances are recognised in accordance with the procedure referred to in Article 25 without restrictions other than those contained in, or adopted pursuant to, this Directive.

2. Member States shall ensure that allowances issued by a competent authority of another Member State are recognised for the purpose of meeting an operator's obligations under paragraph 3.

3. Member States shall ensure that, by 30 April each year at the latest, the operator of each installation surrenders a number of allowances equal to the total emissions from that installation during the preceding calendar year as verified in accordance with Article 15, and that these are subsequently cancelled.

4. Member States shall take the necessary steps to ensure that allowances will be cancelled at any time at the request of the person holding them.

Article 13
Validity of allowances
1. Allowances shall be valid for emissions during the period referred to in Article 11(1) or (2) for which they are issued.

2. Four months after the beginning of the first five-year period referred to in Article 11(2), allowances which are no longer valid and have not been surrendered and cancelled in accordance with Article 12(3) shall be cancelled by the competent authority.

Member States may issue allowances to persons for the current period to replace any allowances held by them which are cancelled in accordance with the first subparagraph.

3. Four months after the beginning of each subsequent five-year period referred to in Article 11(2), allowances which are no longer valid and have not been surrendered and cancelled in accordance with Article 12(3) shall be cancelled by the competent authority.

Member States shall issue allowances to persons for the current period to replace any allowances held by them which are cancelled in accordance with the first subparagraph.

Article 14
Guidelines for monitoring and reporting of emissions

1. The Commission shall adopt guidelines for monitoring and reporting of emissions resulting from the activities listed in Annex I of greenhouse gases specified in relation to those activities, in accordance with the procedure referred to in Article 23(2), by 30 September 2003. The guidelines shall be based on the principles for monitoring and reporting set out in Annex IV.

2. Member States shall ensure that emissions are monitored in accordance with the guidelines.

3. Member States shall ensure that each operator of an installation reports the emissions from that installation during each calendar year to the competent authority after the end of that year in accordance with the guidelines.

Article 15
Verification

Member States shall ensure that the reports submitted by operators pursuant to Article 14(3) are verified in accordance with the criteria set out in Annex V, and that the competent authority is informed thereof. Member States shall ensure that an operator whose report has not been verified as satisfactory in accordance with the criteria set out in Annex V by 31 March each year for emissions during the preceding year cannot make further transfers of allowances until a report from that operator has been verified as satisfactory.

Article 16
Penalties

1. Member States shall lay down the rules on penalties applicable to infringements of the national provisions adopted pursuant to this Directive and shall take all measures necessary to ensure that such rules are implemented. The penalties provided for must be effective, proportionate and dissuasive. Member States shall notify these provisions to the Commission by 31 December 2003 at the latest, and shall notify it without delay of any subsequent amendment affecting them.

2. Member States shall ensure publication of the names of operators who are in breach of requirements to surrender sufficient allowances under Article 12(3).

3. Member States shall ensure that any operator who does not surrender sufficient allowances by 30 April of each year to cover its emissions during the preceding year shall be held liable for the payment of an excess emissions penalty. The excess emissions penalty shall be EUR 100 for each tonne of carbon dioxide equivalent emitted by that installation for which the operator has not surrendered allowances. Payment of the excess emissions penalty shall not release the operator from the obligation to surrender an amount of allowances equal to those excess emissions when surrendering allowances in relation to the following calendar year.

4. During the three-year period beginning 1 January 2005, Member States shall apply a lower excess emissions penalty of EUR 40 for each tonne of carbon dioxide equivalent emitted by that installation for which the operator has not surrendered allowances. Payment of the excess emissions penalty shall not release the operator from the obligation to surrender an amount of allowances equal to those excess emissions when surrendering allowances in relation to the following calendar year.

Article 17
Access to information
Decisions relating to the allocation of allowances, information on project activities in which a Member State participates or authorises private or public entities to participate, and the reports of emissions required under the greenhouse gas emissions permit and held by the competent authority, shall be made available to the public in accordance with Directive 2003/4/EC.

Article 18
Competent authority
Member States shall make the appropriate administrative arrangements, including the designation of the appropriate competent authority or authorities, for the implementation of the rules of this Directive. Where more than one competent authority is designated, the work of these authorities undertaken pursuant to this Directive must be coordinated. Member States shall in particular ensure coordination between their designated focal point for approving project activities pursuant to Article 6 (1)(a) of the Kyoto Protocol and their designated national authority for the implementation of Article 12 of the Kyoto Protocol respectively designated in accordance with subsequent decisions adopted under the UNFCCC or the Kyoto Protocol.

Article 19
Registries

1. Member States shall provide for the establishment and maintenance of a registry in order to ensure the accurate accounting of the issue, holding, transfer and cancellation of allowances. Member States may maintain their registries in a consolidated system, together with one or more other Member States.

2. Any person may hold allowances. The registry shall be accessible to the public and shall contain separate accounts to record the allowances held by each person to whom and from whom allowances are issued or transferred.

3. In order to implement this Directive, the Commission shall adopt a Regulation in accordance with the procedure referred to in Article 23(2) for a standardised and secured system of registries in the form of standardised electronic databases containing common data elements to track the issue, holding, transfer and cancellation of allowances, to provide for public access and confidentiality as appropriate and to ensure that there are no transfers incompatible with obligations resulting from the Kyoto Protocol. That Regulation shall also include provisions concerning the use and identification of CERs and ERUs in the Community scheme and the monitoring of the level of such use.

Article 20
Central Administrator

1. The Commission shall designate a Central Administrator to maintain an independent transaction log recording the issue, transfer and cancellation of allowances.

2. The Central Administrator shall conduct an automated check on each transaction in registries through the independent transaction log to ensure there are no irregularities in the issue, transfer and cancellation of allowances.

3. If irregularities are identified through the automated check, the Central Administrator shall inform the Member State or Member States concerned who shall not register the transactions in question or any further transactions relating to the allowances concerned until the irregularities have been resolved.

Article 21
Reporting by Member States

1. Each year the Member States shall submit to the Commission a report on the application of this Directive. This report shall pay particular attention to the arrangements for the allocation of allowances, the use of ERUs and CERs in the Community scheme, the operation of registries, the application of the monitoring and reporting guidelines, verification and issues relating to compliance with the Directive and the fiscal treatment of allowances, if any. The first report shall be sent to the Commission by 30 June 2005. The report shall be drawn up on the basis of a questionnaire or outline drafted by the Commission in accordance with the procedure laid down in Article 6 of Directive 91/692/EEC. The questionnaire or outline shall be sent to Member States at least six months before the deadline for the submission of the first report.

2. On the basis of the reports referred to in paragraph 1, the Commission shall publish a report on the application of this Directive within three months of receiving the reports from the Member States.

3. The Commission shall organise an exchange of information between the competent authorities of the Member States concerning developments relating to issues of allocation, the use of ERUs and CERs in the Community scheme, the operation of registries, monitoring, reporting, verification and compliance with this Directive.

Article 21a
Support of capacity-building activities
In accordance with the UNFCCC, the Kyoto Protocol and any subsequent decision adopted for their implementation, the Commission and the Member States shall endeavour to support capacity-building activities in developing countries and countries with economies in transition in order to help them take full advantage of JI and the CDM in a manner that supports their sustainable development strategies and to facilitate the engagement of entities in JI and CDM project development and implementation.

Article 22
Amendments to Annex III
The Commission may amend Annex III, with the exception of criteria (1), (5) and (7), for the period from 2008 to 2012 in the light of the reports provided for in Article 21 and of the experience of the application of this Directive, in accordance with the procedure referred to in Article 23(2).

Article 23
Committee

1. The Commission shall be assisted by the committee instituted by Article 8 of Decision 93/389/EEC.

2. Where reference is made to this paragraph, Articles 5 and 7 of Decision 1999/468/EC shall apply, having regard to the provisions of Article 8 thereof.
The period laid down in Article 5(6) of Decision 1999/468/EC shall be set at three months.

3. The Committee shall adopt its rules of procedure.

Article 24
Procedures for unilateral inclusion of additional activities and gases

1. From 2008, Member States may apply emission allowance trading in accordance with this Directive to activities, installations and greenhouse gases which are not listed in Annex I, provided that inclusion of such activities, installations and greenhouse gases is approved by the Commission in accordance with the procedure referred to in Article 23(2), taking into account all relevant criteria, in particular effects on the internal market, potential distortions of competition, the environmental integrity of the scheme and reliability of the planned monitoring and reporting system.
From 2005 Member States may under the same conditions apply emissions allowance trading to installations carrying out activities listed in Annex I below the capacity limits referred to in that Annex.

2. Allocations made to installations carrying out such activities shall be specified in the national allocation plan referred to in Article 9.

3. The Commission may, on its own initiative, or shall, on request by a Member State, adopt monitoring and reporting guidelines for emissions from activities, installations and greenhouse gases which are not listed in Annex I in accordance with the procedure referred to in Article 23(2), if monitoring and reporting of these emissions can be carried out with sufficient accuracy.

4. In the event that such measures are introduced, reviews carried out pursuant to Article 30 shall also consider whether Annex I should be amended to include emissions from these activities in a harmonised way throughout the Community.

Article 25
Links with other greenhouse gas emissions trading schemes

1. Agreements should be concluded with third countries listed in Annex B to the Kyoto Protocol which have ratified the Protocol to provide for the mutual recognition of allowances between the Community scheme and other greenhouse gas emissions trading schemes in accordance with the rules set out in Article 300 of the Treaty.

2. Where an agreement referred to in paragraph 1 has been concluded, the Commission shall draw up any necessary provisions relating to the mutual recognition of allowances under that agreement in accordance with the procedure referred to in Article 23(2).

Article 26
Amendment of Directive 96/61/EC

In Article 9(3) of Directive 96/61/EC the following subparagraphs shall be added:"Where emissions of a greenhouse gas from an installation are specified in Annex I to Directive 2003/87/EC of the European Parliament and of the Council of 13 October 2003 establishing a scheme for greenhouse gas emission allowance trading within the Community and amending Council Directive 96/61/EC[13] in relation to an activity carried out in that installation, the permit shall not include an emission limit value for direct emissions of that gas unless it is necessary to ensure that no significant local pollution is caused.

For activities listed in Annex I to Directive 2003/87/EC, Member States may choose not to impose requirements relating to energy efficiency in respect of combustion units or other units emitting carbon dioxide on the site.

Where necessary, the competent authorities shall amend the permit as appropriate.

The three preceding subparagraphs shall not apply to installations temporarily excluded from the scheme for greenhouse gas emission allowance trading within the Community in accordance with Article 27 of Directive 2003/87/EC."

Article 27
Temporary exclusion of certain installations

1. Member States may apply to the Commission for installations to be temporarily excluded until 31 December 2007 at the latest from the Community scheme. Any such application shall list each such installation and shall be published.

---

[13] OJ L 275, 25.10.2003, p. 32.

2. If, having considered any comments made by the public on that application, the Commission decides, in accordance with the procedure referred to in Article 23(2), that the installations will:

(a) as a result of national policies, limit their emissions as much as would be the case if they were subject to the provisions of this Directive;

(b) be subject to monitoring, reporting and verification requirements which are equivalent to those provided for pursuant to Articles 14 and 15; and

(c) be subject to penalties at least equivalent to those referred to in Article 16(1) and (4) in the case of non-fulfilment of national requirements; it shall provide for the temporary exclusion of those installations from the Community scheme.
It must be ensured that there will be no distortion of the internal market.

Article 28
Pooling

1. Member States may allow operators of installations carrying out one of the activities listed in Annex I to form a pool of installations from the same activity for the period referred to in Article 11(1) and/or the first five-year period referred to in Article 11(2) in accordance with paragraphs 2 to 6 of this Article.

2. Operators carrying out an activity listed in Annex I who wish to form a pool shall apply to the competent authority, specifying the installations and the period for which they want the pool and supplying evidence that a trustee will be able to fulfil the obligations referred to in paragraphs 3 and 4.

3. Operators wishing to form a pool shall nominate a trustee:

(a) to be issued with the total quantity of allowances calculated by installation of the operators, by way of derogation from Article 11;

(b) to be responsible for surrendering allowances equal to the total emissions from installations in the pool, by way of derogation from Articles 6(2)(e) and 12(3); and

(c) to be restricted from making further transfers in the event that an operator's report has not been verified as satisfactory in accordance with the second paragraph of Article 15.

4. The trustee shall be subject to the penalties applicable for breaches of requirements to surrender sufficient allowances to cover the total emissions from installations in the pool, by way of derogation from Article 16(2), (3) and (4).

5. A Member State that wishes to allow one or more pools to be formed shall submit the application referred to in paragraph 2 to the Commission. Without prejudice to the Treaty, the Commission may within three months of receipt reject an application that does not fulfil the requirements of this Directive. Reasons shall be given for any such decision. In the case of rejection the Member State may only allow the pool to be formed if proposed amendments are accepted by the Commission.

6. In the event that the trustee fails to comply with penalties referred to in paragraph 4, each operator of an installation in the pool shall be responsible under Articles 12(3) and 16 in respect of emissions from its own installation.

Article 29
Force majeure

1. During the period referred to in Article 11(1), Member States may apply to the Commission for certain installations to be issued with additional allowances in cases of force majeure. The Commission shall determine whether force majeure is demonstrated, in which case it shall authorise the issue of additional and non-transferable allowances by that Member State to the operators of those installations.

2. The Commission shall, without prejudice to the Treaty, develop guidance to describe the circumstances under which force majeure is demonstrated, by 31 December 2003 at the latest.

Article 30
Review and further development

1. On the basis of progress achieved in the monitoring of emissions of greenhouse gases, the Commission may make a proposal to the European Parliament and the Council by 31 December 2004 to amend Annex I to include other activities and emissions of other greenhouse gases listed in Annex II.

2. On the basis of experience of the application of this Directive and of progress achieved in the monitoring of emissions of greenhouse gases and in the light of developments in the international context, the

Commission shall draw up a report on the application of this Directive, considering:

(a) how and whether Annex I should be amended to include other relevant sectors, inter alia the chemicals, aluminium and transport sectors, activities and emissions of other greenhouse gases listed in Annex II, with a view to further improving the economic efficiency of the scheme;

(b) the relationship of Community emission allowance trading with the international emissions trading that will start in 2008;

(c) further harmonisation of the method of allocation (including auctioning for the time after 2012) and of the criteria for national allocation plans referred to in Annex III;

(d) the use of credits from project activities, including the need for harmonisation of the allowed use of ERUs and CERs in the Community scheme;

(e) the relationship of emissions trading with other policies and measures implemented at Member State and Community level, including taxation, that pursue the same objectives;

(f) whether it is appropriate for there to be a single Community registry;

(g) the level of excess emissions penalties, taking into account, inter alia, inflation;

(h) the functioning of the allowance market, covering in particular any possible market disturbances;

(i) how to adapt the Community scheme to an enlarged European Union;

(j) pooling;

(k) the practicality of developing Community-wide benchmarks as a basis for allocation, taking into account the best available techniques and cost-benefit analysis.

(l) the impact of project mechanisms on host countries, particularly on their development objectives, whether JI and CDM hydroelectric power production project activities with a generating capacity exceeding 500 MW and having negative environmental or social impacts have been approved, and the future use of CERs or ERUs resulting from any such hydroelectric power production project activities in the Community scheme;

(m) the support for capacity-building efforts in developing countries and countries with economies in transition;

(n) the modalities and procedures for Member States' approval of domestic project activities and for the issuing of allowances in respect of emission reductions or limitations resulting from such activities from 2008;

(o) technical provisions relating to the temporary nature of credits and the limit of 1 % for eligibility for land use, land-use change and forestry project activities as established in Decision 17/CP.7, and provisions relating to the outcome of the evaluation of potential risks associated with the use of genetically modified organisms and potentially invasive alien species by afforestation and reforestation project activities, to allow operators to use CERs and ERUs resulting from land use, land-use change and forestry project activities in the Community scheme from 2008, in accordance with the decisions adopted pursuant to the UNFCCC or the Kyoto Protocol.

The Commission shall submit this report to the European Parliament and the Council by 30 June 2006, accompanied by proposals as appropriate.

3. In advance of each period referred to in Article 11(2), each Member State shall publish in its national allocation plan its intended use of ERUs and CERs and the percentage of the allocation to each installation up to which operators are allowed to use ERUs and CERs in the Community scheme for that period. The total use of ERUs and CERs shall be consistent with the relevant supplementarity obligations under the Kyoto Protocol and the UNFCCC and the decisions adopted thereunder.

Member States shall, in accordance with Article 3 of Decision No 280/ 2004/EC of the European Parliament and of the Council of 11 February 2004 concerning a mechanism for monitoring Community greenhouse gas emissions and for implementing the Kyoto Protocol, report to the Commission every two years on the extent to which domestic action actually constitutes a significant element of the efforts undertaken at national level, as well as the extent to which use of the project mechanisms is actually supplemental to domestic action, and the ratio between them, in accordance with the relevant provisions of the Kyoto Protocol and the decisions adopted thereunder. The Commission shall report on this in accordance with Article 5 of the said Decision. In the light of this report, the Commission shall, if appropriate, make legislative or other proposals to complement provisions adopted by Member States to ensure that use of the mechanisms is supplemental to domestic action within the Community.

Article 31
Implementation

1. Member States shall bring into force the laws, regulations and administrative provisions necessary to comply with this Directive by 31 December 2003 at the latest. They shall forthwith inform the Commission thereof. The Commissio shall notify the other Member State these laws, regulations and adminstrative provisions.

Where Memeber States adopt these measures, they shall contain a reference to this Directive or be accompanied by such a reference on to the occasion of their oficial publication. The methods of making such reference shall be laid don by Member States.

2. Member States shall communicate to the Commission the text of the provisions of national law which they adopt in the field covered by this Directive. the Commission shall inform the other Member States thereof.

Article 32
Entry into force

This Directive shall enter into force on the day of its publication in the Official Journal of the European Union.

Article 33
Addressees

This Directive is addressed to the Member States.
Done at Luxembourg, 13 October 2003.
For the European Parliament
The President
P. Cox
For the Council
The President
G. Alemanno

ANNEX I
CATEGORIES OF ACTIVITIES REFERRED TO IN ARTICLES 2(1), 3, 4, 14(1), 28 AND 30

1. Installations or parts of installations used for research, development and testing of new products and processes are not covered by this Directive.

2. The threshold values given below generally refer to production capacities or outputs. Where one operator carries out several activities falling under the same subheading in the same installation or on the same site, the capacities of such activities are added together.

Activities	Greenhouse Gases
Energy Activities	
Combustion Installations with rated thermal input exceeding 20 MW (except hazardous or municipal waste installations)	Carbon dioxide
Mineral oil refineries	Carbon dioxide
Coke ovens	Carbon dioxide
Production and processing ferrous metals	
Metal ore (including sulphide ore) roasting or sintering installations	Carbon dioxide
Installations for the production of pig iron or steel (primary or secondary fusion) including continuous casting, with capacity exceeding 2,5 tonnes per hour	Carbon dioxide
Mineral industry	
Installations for the production of cement clinker in rotary kilns with a production capacity exceeding 500 tones per day or lime in rotary kilns with production capacity exceeding 50 tonnes per day in other furnaces with a production capacity exceeding 50 tonnes per day	Carbon dioxide
Installations for the manufacture of glass including glass fibre with a melting capacity city exceeding 20 tonnes per day	Carbon dioxide
Installations for the manufacture of ceramic products by firing, in particular roofing tiles, bricks, refractory bricks, tiles stoneware or porcelain, with a production capacity exceeding 75 tonnes per day, and/or with kiln capacity exceeding 4 m^3 and with a setting density per kiln exceeding 300 kg/ m^3	Carbon dioxide
Other activities	
Industrial plants for the production of	Carbon dioxide
(a) pulp from timber or other fibrous materials	
(b) paper and board with production capacity exceeding 20 tonnes per day	Carbon dioxide

ANNEX II
GREENHOUSE GASES REFERRED TO IN ARTICLES 3 AND 30

Carbon dioxide (CO2)
Methane (CH4)
Nitrous Oxide (N2O)
Hydrofluorocarbons (HFCs)
Perfluorocarbons (PFCs)
Sulphur Hexafluoride (SF6)

ANNEX III
CRITERIA FOR NATIONAL ALLOCATION PLANS REFERRED
TO IN ARTICLES 9, 22 AND 30

1. The total quantity of allowances to be allocated for the relevant period shall be consistent with the Member State's obligation to limit its emissions pursuant to Decision 2002/358/EC and the Kyoto Protocol, taking into account, on the one hand, the proportion of overall emissions that these allowances represent in comparison with emissions from sources not covered by this Directive and, on the other hand, national energy policies, and should be consistent with the national climate change programme. The total quantity of allowances to be allocated shall not be more than is likely to be needed for the strict application of the criteria of this Annex. Prior to 2008, the quantity shall be consistent with a path towards achieving or over-achieving each Member State's target under Decision 2002/358/EC and the Kyoto Protocol.

2. The total quantity of allowances to be allocated shall be consistent with assessments of actual and projected progress towards fulfilling the Member States' contributions to the Community's commitments made pursuant to Decision 93/389/EEC.

3. Quantities of allowances to be allocated shall be consistent with the potential, including the technological potential, of activities covered by this scheme to reduce emissions. Member States may base their distribution of allowances on average emissions of greenhouse gases by product in each activity and achievable progress in each activity.

4. The plan shall be consistent with other Community legislative and policy instruments. Account should be taken of unavoidable increases in emissions resulting from new legislative requirements.

5. The plan shall not discriminate between companies or sectors in such a way as to unduly favour certain undertakings or activities in

accordance with the requirements of the Treaty, in particular Articles 87 and 88 thereof.

6. The plan shall contain information on the manner in which new entrants will be able to begin participating in the Community scheme in the Member State concerned.

7. The plan may accommodate early action and shall contain information on the manner in which early action is taken into account. Benchmarks derived from reference documents concerning the best available technologies may be employed by Member States in developing their National Allocation Plans, and these benchmarks can incorporate an element of accommodating early action.

8. The plan shall contain information on the manner in which clean technology, including energy efficient technologies, are taken into account.

9. The plan shall include provisions for comments to be expressed by the public, and contain information on the arrangements by which due account will be taken of these comments before a decision on the allocation of allowances is taken.

10. The plan shall contain a list of the installations covered by this Directive with the quantities of allowances intended to be allocated to each.

11. The plan may contain information on the manner in which the existence of competition from countries or entities outside the Union will be taken into account.

12. The plan shall specify the maximum amount of CERs and ERUs which may be used by operators in the Community scheme as a percentage of the allocation of the allowances to each installation. The percentage shall be consistent with the Member State's supplementarity obligations under the Kyoto Protocol and decisions adopted pursuant to the UNFCCC or the Kyoto Protocol.

ANNEX IV
PRINCIPLES FOR MONITORING AND REPORTING REFERRED
TO IN ARTICLE 14(1)

Monitoring of carbon dioxide emissions

Emissions shall be monitored either by calculation or on the basis of measurement.

Calculation

Calculations of emissions shall be performed using the formula:

Activity data × Emission factor × Oxidation factor

Activity data (fuel used, production rate etc.) shall be monitored on the basis of supply data or measurement.

Accepted emission factors shall be used. Activity-specific emission factors are acceptable for all fuels. Default factors are acceptable for all fuels except non-commercial ones (waste fuels such as tyres and industrial process gases). Seam-specific defaults for coal, and EU-specific or producer country-specific defaults for natural gas shall be further elaborated. IPCC default values are acceptable for refinery products. The emission factor for biomass shall be zero.

If the emission factor does not take account of the fact that some of the carbon is not oxidised, then an additional oxidation factor shall be used. If activity-specific emission factors have been calculated and already take oxidation into account, then an oxidation factor need not be applied. Default oxidation factors developed pursuant to Directive 96/61/EC shall be used, unless the operator can demonstrate that activity-specific factors are more accurate.

A separate calculation shall be made for each activity, installation and for each fuel.

Measurement

Measurement of emissions shall use standardised or accepted methods, and shall be corroborated by a supporting calculation of emissions.

Monitoring of emissions of other greenhouse gases

Standardised or accepted methods shall be used, developed by the Commission in collaboration with all relevant stakeholders and adopted in accordance with the procedure referred to in Article 23(2).

Reporting of emissions

Each operator shall include the following information in the report for an installation:

A. Data identifying the installation, including:
- Name of the installation;
- Its address, including postcode and country;

- Type and number of Annex I activities carried out in the installation;
- Address, telephone, fax and email details for a contact person; and
- Name of the owner of the installation, and of any parent company.

B. For each Annex I activity carried out on the site for which emissions are calculated:

- Activity data;
- Emission factors;
- Oxidation factors;
- Total emissions; and
- Uncertainty.

C. For each Annex I activity carried out on the site for which emissions are measured:

- Total emissions;
- Information on the reliability of measurement methods; and
- Uncertainty.

D. For emissions from combustion, the report shall also include the oxidation factor, unless oxidation has already been taken into account in the development of an activity-specific emission factor.

Member States shall take measures to coordinate reporting requirements with any existing reporting requirements in order to minimise the reporting burden on businesses.

ANNEX V
CRITERIA FOR VERIFICATION REFERRED TO IN ARTICLE 15

General Principles

1. Emissions from each activity listed in Annex I shall be subject to verification.

2. The verification process shall include consideration of the report pursuant to Article 14(3) and of monitoring during the preceding year. It shall address the reliability, credibility and accuracy of monitoring systems and the reported data and information relating to emissions, in particular:

(a) the reported activity data and related measurements and calculations;

(b) the choice and the employment of emission factors;

(c) the calculations leading to the determination of the overall emissions; and

(d) if measurement is used, the appropriateness of the choice and the employment of measuring methods.

3. Reported emissions may only be validated if reliable and credible data and information allow the emissions to be determined with a high degree of certainty. A high degree of certainty requires the operator to show that:

(a) the reported data is free of inconsistencies;

(b) the collection of the data has been carried out in accordance with the applicable scientific standards; and

(c) the relevant records of the installation are complete and consistent.

4. The verifier shall be given access to all sites and information in relation to the subject of the verification.

5. The verifier shall take into account whether the installation is registered under the Community eco-management and audit scheme (EMAS).

Methodology

Strategic analysis
6. The verification shall be based on a strategic analysis of all the activities carried out in the installation. This requires the verifier to have an overview of all the activities and their significance for emissions.

Process analysis
7. The verification of the information submitted shall, where appropriate, be carried out on the site of the installation. The verifier shall use spot-checks to determine the reliability of the reported data and information.

Risk analysis
8. The verifier shall submit all the sources of emissions in the installation to an evaluation with regard to the reliability of the data of each source contributing to the overall emissions of the installation.

9. On the basis of this analysis the verifier shall explicitly identify those sources with a high risk of error and other aspects of the monitoring and reporting procedure which are likely to contribute to errors in the determination of the overall emissions. This especially involves the choice of the emission factors and the calculations necessary to determine the level of the emissions from individual sources. Particular attention shall be given to those sources with a high risk of error and the abovementioned aspects of the monitoring procedure.

10. The verifier shall take into consideration any effective risk control methods applied by the operator with a view to minimising the degree of uncertainty.

Report
11. The verifier shall prepare a report on the validation process stating whether the report pursuant to Article 14(3) is satisfactory. This report shall specify all issues relevant to the work carried out. A statement that the report pursuant to Article 14(3) is satisfactory may be made if, in the opinion of the verifier, the total emissions are not materially misstated. Minimum competency requirements for the verifier

12. The verifier shall be independent of the operator, carry out his activities in a sound and objective professional manner, and understand: (a) the provisions of this Directive, as well as relevant standards and guidance adopted by the Commission pursuant to Article 14(1);

(b) the legislative, regulatory, and administrative requirements relevant to the activities being verified; and

(c) the generation of all information related to each source of emissions in the installation, in particular, relating to the collection, measurement, calculation and reporting of data.

ANNEX 2

RECITALS TO THE LINKING DIRECTIVE

Directive 2004/101/EC of the European Parliament and of the Council
of 27 October 2004

amending Directive 2003/87/EC establishing a scheme for greenhouse
gas emission allowance trading within the Community, in respect of
the Kyoto Protocol's project mechanisms

(Text with EEA relevance)

THE EUROPEAN PARLIAMENT AND THE COUNCIL OF THE
EUROPEAN UNION,

Having regard to the Treaty establishing the European Community, and
in particular Article 175(1) thereof,

Having regard to the proposal from the Commission,

Having regard to the opinion of the European Economic and Social
Committee,[1]

After consulting the Committee of the Regions,

Acting in accordance with the procedure laid down in Article 251 of the
Treaty,[2]

Whereas:

(1) Directive 2003/87/EC[3] establishes a scheme for greenhouse gas
emission allowance trading within the Community (the Community
scheme) in order to promote reductions of greenhouse gas emissions in
a cost-effective and economically efficient manner, recognising that, in
the longer-term, global emissions of greenhouse gases will need to be
reduced by approximately 70 % compared to 1990 levels. It aims at
contributing towards fulfilling the commitments of the Community and

---

[1] OJ C 80, 30.3.2004, p. 61.
[2] Opinion of the European Parliament of 20 April 2004 (not yet published in the Official
Journal) and Council Decision of 13 September 2004 (not yet published in the Official
Journal).
[3] OJ L 275, 25.10.2003, p. 32.

its Member States to reduce anthropogenic greenhouse gas emissions under the Kyoto Protocol which was approved by Council Decision 2002/358/EC of 25 April 2002 concerning the approval, on behalf of the European Community, of the Kyoto Protocol to the United Nations Framework Convention on Climate Change and the joint fulfilment of commitments thereunder .[4]

(2) Directive 2003/87/EC states that the recognition of credits from project-based mechanisms for fulfilling obligations as from 2005 will increase the cost-effectiveness of achieving reductions of global greenhouse gas emissions and shall be provided for by provisions for linking the Kyoto project-based mechanisms, including joint implementation (JI) and the clean development mechanism (CDM), with the Community scheme.

(3) Linking the Kyoto project-based mechanisms to the Community scheme, while safeguarding the latter's environmental integrity, gives the opportunity to use emission credits generated through project activities eligible pursuant to Articles 6 and 12 of the Kyoto Protocol in order to fulfil Member States' obligations in accordance with Article 12(3) of Directive 2003/87/EC. As a result, this will increase the diversity of low-cost compliance options within the Community scheme leading to a reduction of the overall costs of compliance with the Kyoto Protocol while improving the liquidity of the Community market in greenhouse gas emission allowances. By stimulating demand for JI credits, Community companies will invest in the development and transfer of advanced environmentally sound technologies and know-how. The demand for CDM credits will also be stimulated and thus developing countries hosting CDM projects will be assisted in achieving their sustainable development goals.

(4) In addition to the use of the Kyoto project-based mechanisms by the Community and its Member States, and by companies and individuals outside the Community scheme, those mechanisms should be linked to the Community scheme in such a way as to ensure consistency with the United Nations Framework Convention on Climate Change (UNFCCC) and the Kyoto Protocol and subsequent decisions adopted thereunder as well as with the objectives and architecture of the Community scheme and provisions laid down by Directive 2003/87/EC.

(5) Member States may allow operators to use, in the Community scheme, certified emission reductions (CERs) from 2005 and emission reduction units (ERUs) from 2008. The use of CERs and ERUs by operators from 2008 may be allowed up to a percentage of the allocation to each

---

[4] OJ L 130, 15.5.2002, p. 1.

installation, to be specified by each Member State in its national allocation plan. The use will take place through the issue and immediate surrender of one allowance in exchange for one CER or ERU. An allowance issued in exchange for a CER or ERU will correspond to that CER or ERU.

(6) The Commission Regulation for a standardised and secured system of registries, to be adopted pursuant to Article 19(3) of Directive 2003/87/EC and Article 6(1) of Decision No 280/2004/EC of the European Parliament and of the Council of 11 February 2004 concerning a mechanism for monitoring Community greenhouse gas emissions and for implementing the Kyoto Protocol,[5] will provide for the relevant processes and procedures in the registries system for the use of CERs during the period 2005 to 2007 and subsequent periods, and for the use of ERUs during the period 2008 to 2012 and subsequent periods.

(7) Each Member State will decide on the limit for the use of CERs and ERUs from project activities, having due regard to the relevant provisions of the Kyoto Protocol and the Marrakesh Accords, to meet the requirements therein that the use of the mechanisms should be supplemental to domestic action. Domestic action will thus constitute a significant element of the effort made.

(8) In accordance with the UNFCCC and the Kyoto Protocol and subsequent decisions adopted thereunder, Member States are to refrain from using CERs and ERUs generated from nuclear facilities to meet their commitments pursuant to Article 3(1) of the Kyoto Protocol and pursuant to Decision 2002/358/EC.

(9) Decisions 15/CP.7 and 19/CP.7 adopted pursuant to the UNFCCC and the Kyoto Protocol emphasise that environmental integrity is to be achieved, inter alia, through sound modalities, rules and guidelines for the mechanisms, and through sound and strong principles and rules governing land use, land-use change and forestry activities, and that the issues of non-permanence, additionality, leakage, uncertainties and socioeconomic and environmental impacts, including impacts on biodiversity and natural ecosystems, associated with afforestation and reforestation project activities are to be taken into account. The Commission should consider, in its review of Directive 2003/87/EC in 2006, technical provisions relating to the temporary nature of credits and the limit of 1 % for eligibility for land use, land-use change and forestry project activities as established in Decision 17/CP.7, and also provisions relating to the outcome of the evaluation of potential risks associated with the use of genetically modified organisms and potentially invasive alien species in afforestation and reforestation project activities,

---

[5] OJ L 49, 19.2.2004, p. 1.

to allow operators to use CERs and ERUs resulting from land use, land use change and forestry project activities in the Community scheme from 2008, in accordance with the decisions adopted pursuant to the UNFCCC or the Kyoto Protocol.

(10) In order to avoid double counting, CERs and ERUs should not be issued as a result of project activities undertaken within the Community that also lead to a reduction in, or limitation of, emissions from installations covered by Directive 2003/87/EC, unless an equal number of allowances is cancelled from the registry of the Member State of the CERs' or ERUs' origin.

(11) In accordance with the relevant treaties of accession, the acquis communautaire should be taken into account in the establishment of baselines for project activities undertaken in countries acceding to the Union.

(12) Any Member State that authorises private or public entities to participate in project activities remains responsible for the fulfilment of its obligations under the UNFCCC and the Kyoto Protocol and should therefore ensure that such participation is consistent with the relevant guidelines, modalities and procedures adopted pursuant to the UNFCCC or the Kyoto Protocol.

(13) In accordance with the UNFCCC, the Kyoto Protocol and subsequent decisions adopted for their implementation, the Commission and the Member States should support capacity building activities in developing countries and countries with economies in transition in order to help them take full advantage of JI and the CDM in a manner that supports their sustainable development strategies. The Commission should review and report on efforts in this regard.

(14) Criteria and guidelines that are relevant to considering whether hydroelectric power production projects have negative environmental or social impacts have been identified by the World Commission on Dams in its November 2000 Report "Dams and Development A New Framework for Decision-Making", by the OECD and by the World Bank.

(15) Since participation in JI and CDM project activities is voluntary, corporate environmental and social responsibility and accountability should be enhanced in accordance with paragraph 17 of the Plan of implementation of the world summit on sustainable development. In this connection, companies should be encouraged to improve the social and environmental performance of JI and CDM activities in which they participate.

(16) Information on project activities in which a Member State participates or authorises private or public entities to participate should be made available to the public in accordance with Directive 2003/4/EC of the European Parliament and of the Council of 28 January 2003 on public access to environmental information.[6]

(17) The Commission may mention impacts on the electricity market in its reports on emission allowance trading and the use of credits from project activities.

(18) Following entry into force of the Kyoto Protocol, the Commission should examine whether it could be possible to conclude agreements with countries listed in Annex B to the Kyoto Protocol which have yet to ratify the Protocol, to provide for the recognition of allowances between the Community scheme and mandatory greenhouse gas emissions trading schemes capping absolute emissions established within those countries.

(19) Since the objective of the proposed action, namely the establishment of a link between the Kyoto project-based mechanisms and the Community scheme, cannot be sufficiently achieved by the Member States acting individually, and can therefore by reason of the scale and effects of this action be better achieved at Community level, the Community may adopt measures, in accordance with the principle of subsidiarity as set out in Article 5 of the Treaty. In accordance with the principle of proportionality, as set out in that Article, this Directive does not go beyond what is necessary in order to achieve that objective.

(20) Directive 2003/87/EC should therefore be amended accordingly

---

[6] OJ L 41, 14.2.2003, p. 26.

ANNEX 3

THE REGISTRIES REGULATION

Commission Regulation (EC) No 2216/2004
of 21 December 2004 for a standardised and secured system of
registries pursuant to Directive 2003/87/EC of the European
Parliament and of the Council and Decision No 280/2004/EC of the
European Parliament and of the Council

(Text with EEA relevance)

TABLE OF CONTENTS

THE COMMISSION OF THE EUROPEAN COMMUNITIES,

Having regard to the Treaty establishing the European Community,

Having regard to Directive 2003/87/EC of the European Parliament and of the Council of 13 October 2003 establishing a scheme for greenhouse gas emission allowance trading within the Community and amending Council Directive 96/61/EC ('), and in particular Article 19(3) thereof,

Having regard to Decision 280/2004/EC of the European Parliament and of the Council of 11 February 2004 concerning a mechanism for monitoring Community greenhouse gas emissions and for implementing the Kyoto Protocol (²), and in particular the first subparagraph, second sentence, Article 6(1) thereof,

Whereas:

(1)   An integrated Community system of registries, consisting of the registries of the Community and its Member States established pursuant to Article 6 of Decision No 280/2004/EC that incorporate the registries established pursuant to Article 19 of Directive 2003/87/EC and the Community independent transaction log established pursuant to Article 20 of that Directive, is necessary to ensure that the issue, transfer and cancellation of allowances does not involve irregularities and that transactions are compatible with the obligations resulting from the United Nations Framework Convention on

Climate Change (UNFCCC) and the Kyoto Protocol.

(2)   In accordance with Directive 2003/4/EC of 28 January 2003 on public access to environmental information (') and Decision 19/CP.7 of the Conference of the Parties to the UNFCCC, specific reports should be made public on a regular basis to ensure that the public has access to information held within the integrated system of registries, subject to certain confidentiality requirements.

(3)   Community legislation concerning the protection of individuals with regard to the processing of personal data and on the free movement of such data, in particular Directive 95/46/EC on the protection of individuals with regard to the processing of personal data and on the free movement of such data (²), Directive 2002/58/EC concerning processing of personal data and the protection of privacy in the electronic communications sector (³) a   n   d Regulation (EC) No 45/2001 on the protection of individuals with regard to the processing of personal data by the Community institutions and bodies and on the free movement of such data (⁴), should be respected where these are applicable to information held and processed pursuant to this Regulation.

(4)   Each registry should contain one Party holding account, one retirement account, and the cancellation and replacement accounts required pursuant to Decision 19/CP.7 of the Conference of the Parties to the UNFCCC for each commitment period, and each registry established

('), OJ L 41, 14.2.2003, p. 26.
(²) OJ L 281, 23.11.1995, p. 31.
(³) OJ L 201, 31.7.2002, p. 37.
(⁴) OJ L 8, 12.1.2001, p. 1.

('), OJ L 275, 25.10.2003, p. 32.
(²) OJ L 49, 19.2.2004, p. 1.

pursuant to Article 19 of Directive 2003/87/EC should contain holding accounts required to implement the requirements of that Directive for operators and for other persons. Each such account should be created in accordance with standardised procedures to ensure the integrity of the registries system and public access to information held in this system.

(5) Article 6 of Decision No 280/2004/EC requires the Community and its Member States to apply the functional and technical specifications for data exchange standards for registry systems under the Kyoto Protocol, elaborated pursuant to Decision 24/CP.8 of the Conference of the Parties to the UNFCCC, for the establishment and operation of registries and the Community independent transaction log. The application and elaboration of these specifications in relation to the integrated Community registries system allows the incorporation of the registries established pursuant to Article 19 of Directive 2003/87/EC into the registries established pursuant to Article 6 of Decision No 280/2004/EC.

(6) The Community independent transaction log will perform automated checks on all processes in the Community registries system concerning allowances, verified emissions, accounts and Kyoto units, and the UNFCCC independent transaction log will perform automated checks on processes concerning Kyoto units to ensure that there are no irregularities. Processes that fail these checks will be terminated to ensure that transactions in the Community registries system comply with the requirements of Directive 2003/87/EC and the requirements elaborated pursuant to the UNFCCC and the Kyoto Protocol.

(7) All transactions in the Community registries system should be executed in accordance with standardised procedures and, where necessary, on a harmonised timetable, in order to ensure compliance with the requirements of Directive 2003/87/EC and with the requirements elaborated pursuant to the UNFCCC and the Kyoto Protocol, and to protect the integrity of that system.

(8) Minimum security standards and harmonised requirements on authentication and access rights should be applied to protect the security of information held in the integrated Community registries system.

(9) The Central Administrator and each registry administrator should ensure that interruptions to the operation of the integrated Community registries system are kept to a minimum by taking all reasonable steps to ensure the availability of the registries and the Community independent transaction log and by providing for robust systems and procedures for the safeguarding of all information.

(10) Records concerning all processes, operators and persons in the Community registries system should be stored in accordance with the data logging standards set out in the functional and technical specifications for data exchange standards for registry systems under the Kyoto Protocol elaborated pursuant to Decision24/CP.8 of the Conference of the Parties to the UNFCCC.

(11) A transparent system of fees and a prohibition to charge account holders for specific transactions in the Community registries system will help ensure the integrity of that system.

(12) The measures provided for in this Regulation are in accordance with the opinion of the Committee referred to in Article 23(1) of Directive 2003/87/EC and Article 9(2) of Decision No 280/2004/EC.

HAS ADOPTED THIS REGULATION:

CHAPTER I

SUBJECT MATTER AND DEFINITIONS

Article 1

Subject matter

This Regulation lays down general provisions, functional and technical specifications and operational and maintenance requirements concerning the standardised and secured registries system consisting of registries, in the form of standardised electronic databases containing common data elements, and the Community independent transaction log. It also provides for an efficient communication system between the Community independent transaction log and the UNFCCC independent transaction log.

Article 2

Definitions

For the purposes of this Regulation, the definitions laid down in Article 3 of Directive 2003/87/EC shall apply. The following definitions shall also apply:

(a) '2005-2007 period' means the period from 1 January 2005 to 31 December 2007 as referred to in Article 11(1) of Directive 2003/87/EC;

(b) '2008-2012 period and subsequent five-year periods' means the period from 1 January 2008 to 31 December 2012 plus consecutive five-year periods as referred to in Article 11(2) of Directive 2003/87/EC;

(c) 'account holder' means a person who holds an account in the registries system;

(d) 'assigned amount' means the amount of greenhouse gas emissions in tonnes of carbon dioxide equivalent calculated in accordance with the emission levels determined pursuant to Article 7 of Decision No 280/2004/EC;

(e) 'assigned amount unit'(AAU) means a unit issued pursuant to Article 7(3) of Decision No 280/2004/EC;

(f) 'authorised representative' means a natural person authorised to represent the Central Administrator, a registry administrator, an account holder or a verifier pursuant to Article 23;

(g) 'CDM registry' means the clean development mechanism registry established, operated and maintained by the executive board of the clean development mechanism pursuant to Article 12 of the Kyoto Protocol and the decisions adopted pursuant to the UNFCCC or the Kyoto Protocol;

(h) 'Central Administrator' means the person designated by the Commission pursuant to Article 20 of Directive 2003/87/EC to operate and maintain the Community independent transaction log;

(i) 'Community independent transaction log' means the independent transaction log provided for in Article 20(1) of Directive 2003/87/EC for the purpose of recording the issue, transfer and cancellation of allowances, and established, operated and maintained in accordance with Article 5;

(j) 'competent authority' means the authority or authorities designated by a Member State pursuant to Article 18 of Directive 2003/87/EC;

(k) 'discrepancy' means an irregularity detected by the Community independent transaction log or UNFCCC independent transaction log whereby the proposed process does not conform to the requirements specified under Directive 2003/87/EC as elaborated in this Regulation and the requirements elaborated pursuant to the UNFCCC or the Kyoto Protocol;

(l) 'force majeure allowance' means a force majeure allowance issued pursuant to Article 29 of Directive 2003/87/EC;

(m) 'inconsistency' means an irregularity detected by the Community independent transaction log or UNFCCC independent transaction log whereby the information regarding allowances, accounts or Kyoto units provided by a registry as part of the periodic reconciliation process differs from the information contained in either independent transaction log;

(n) 'Kyoto unit' means an AAU, RMU, ERU or CER;

(o) 'process' means any one of the processes referred to in Article 32;

(p) 'registry' means a registry established, operated and maintained pursuant to Article 6 of Decision No 280/2004/EC, incorporating a registry established pursuant to Article 19 of Directive 2003/87/EC;

(q) 'registry administrator' means the competent authority, persons or person, designated by the Member State or the Commission, that operates and maintains a registry in accordance with the requirements of Directive 2003/87/EC, Decision No 280/2004/EC and this Regulation;

(r) 'removal unit' (RMU) means a unit issued pursuant to Article 3 of the Kyoto Protocol;

(s) 'temporary CER' (tCER) is a CER issued for an afforestation or reforestation project activity under the CDM which, subject to the decisions adopted pursuant to the UNFCCC or the Kyoto Protocol, expires at the end of the commitment period following the one during which it was issued;

(t) 'long-term CER' (lCER) is a CER issued for an afforestation or reforestation project activity under the CDM which, subject to the decisions adopted pursuant to the UNFCCC or the Kyoto Protocol, expires at the end of the crediting period of the afforestation or reforestation project activity under the CDM for which it was issued;

(u) 'third country registry' means a registry established, operated and maintained by a country listed in Annex B to the Kyoto Protocol which has ratified the Kyoto Protocol and is not a Member State;

461

(v) 'transaction' means the issue, transfer, acquisition, surrender, cancellation and replacement of allowances and the issue, transfer, acquisition, cancellation and retirement of ERUs, CERs, AAUs and RMUs and carry-over of ERUs, CERs and AAUs;

(w) 'UNFCCC independent transaction log' means the independent transaction log established, operated and maintained by the Secretariat of the United Nations Framework Convention on Climate Change;

(x) 'verifier' means a competent, independent, accredited verification body with responsibility for performing and reporting on the verification process, in accordance with the detailed requirements established by the Member State pursuant to Annex V of Directive 2003/87/EC;

(y) 'year' means a calendar year, defined according to Greenwich Mean Time.

CHAPTER II

REGISTRIES AND TRANSACTION LOGS

Article 3

Registries

1. A registry in the form of a standardised electronic database shall be established by each Member State and the Commission by the day after the entry into force of this Regulation.

2. Each registry shall incorporate the hardware and software set out in Annex I, be accessible via the Internet, and conform to the functional and technical specifications required by this Regulation.

3. Each registry shall be capable of executing correctly all the processes concerning verified emissions and accounts set out in Annex VIII, the reconciliation process set out in Annex X and all the administrative processes set out in Annex XI by the day after the entry into force of this Regulation.

Each registry shall be capable of executing correctly all the processes concerning allowances and Kyoto units set out in Annex IX by the day after the entry into force of this Regulation, with the exception of the processes with process types 04-00, 06-00, 07-00 and 08-00.

Each registry shall be capable of executing correctly the processes concerning allowances and Kyoto units with process types 04-00, 06-00, 07-00 and 08-00 set out in Annex IX by 31 March 2005.

Article 4

Consolidated registries

A Member State or the Commission may establish, operate and maintain their registry in a consolidated manner together with one or more other Member States or the Community, provided that its registry remains distinct.

Article 5

The Community independent transaction log

1. The Community independent transaction log shall be established by the Commission in the form of a standardised electronic database by the day after the entry into force of this Regulation.

2. The Community independent transaction log shall incorporate the hardware and software set out in Annex I, be accessible via the Internet, and conform to the functional and technical specifications required by this Regulation.

3. The Central Administrator designated pursuant to Article 20 of Directive 2003/87/EC shall operate and maintain the Community independent transaction log in accordance with the provisions of this Regulation.

4. The Central Administrator shall provide the administrative processes referred to in Annex XI in order to facilitate the integrity of the data within the registries system.

5. The Central Administrator shall only perform processes concerning allowances, verified emissions, accounts or Kyoto units where necessary to carry out its functions as Central Administrator.

6. The Community independent transaction log shall be capable of executing correctly all the processes concerning allowances, verified emissions, accounts or Kyoto units set out in Annex VIII and Annex IX by the day after the entry into force of this Regulation.

The Community independent transaction log shall be capable of executing correctly the reconciliation process set out in Annex X and the administrative processes set out in Annex XI by the day after the entry into force of this Regulation.

Article 6

Communication link between registries and the Community independent transaction log

1. A communication link between each registry and the Community independent transaction log shall be established by 31 December 2004. The Central Administrator shall activate the communication link after the testing procedures set out in Annex XIII and the initialisation procedures set out in Annex XIV have been completed successfully and notify the relevant registry administrator thereof.

2. From 1 January 2005 until the communication link referred to in Article 7 has been established, all processes concerning allowances, verified emissions and accounts shall be completed through the exchange of data via the Community independent transaction log.

3. The Commission may instruct the Central Administrator to temporarily suspend a process referred to in Annexes VIII and IX initiated by a registry if that process is not being executed in accordance with Articles 32 to 37.

The Commission may instruct the Central Administrator to temporarily suspend the communication link between a registry and the Community independent transaction log or to suspend all or some of the processes referred to in Annexes VIII and IX, if that registry is not operated and maintained in accordance with the provisions of this Regulation.

Article 7

Communication link between the independent transaction logs

A communication link between the Community independent transaction log and the UNFCCC independent transaction log shall be established promptly after the UNFCCC independent transaction log has been established.

After such a link is established, all processes concerning allowances, verified emissions, accounts and Kyoto units shall be completed through the exchange of data via the UNFCCC independent transaction log and thereon to the Community independent transaction log.

Article 8

Registry administrators

1. Each Member State and the Commission shall designate a registry administrator to operate and maintain its registry in accordance with the provisions of this Regulation.

Member States and the Commission shall ensure that there is no conflict of interest between the registry administrator and its account holders or between the registry administrator and the Central Administrator.

2. Each Member State shall notify the Commission of the identity and contact details of the registry administrator for its registry by 1 September 2004 in accordance with the initialisation procedures set out in Annex XIV.

3. The Member States and the Commission shall retain ultimate responsibility and authority for the operation and maintenance of their registries.

4. The Commission shall coordinate the implementation of the requirements of this Regulation with the registry administrators of each Member State and the Central Administrator.

CHAPTER III

CONTENTS OF THE REGISTRIES

SECTION 1

Reporting and Confidentiality

Article 9

Reporting

1. Each registry administrator shall make available the information listed in Annex XVI at the frequencies and to the recipients set out in Annex XVI in a transparent and organised manner via his registry web site. Registry administrators shall not release additional information held in the registry.

2. The Central Administrator shall make available the information listed in Annex XVI at the frequencies and to the recipients set out in Annex XVI in a transparent and organised manner via the Community independent transaction log web site. The Central Administrator shall not release additional information held in the Community independent transaction log.

3. Each web site shall allow the recipients of the reports listed in Annex XVI to query those reports using search facilities.

4. Each registry administrator is responsible for the accuracy of the information that originates from his registry and is made available via the Community independent transaction log website.

5. Neither the Community independent transaction log nor registries shall require account holders to submit price information concerning allowances or Kyoto units.

Article 10

Confidentiality

1. All information, including the holdings of all accounts and all transactions made, held in the registries and the Community independent transaction log shall be considered confidential for any purpose other than the implementation of the requirements of this Regulation, Directive 2003/87/EC or national law.

2. Information held in the registries may not be used without the prior consent of the relevant account holder except to operate and maintain those registries in accordance with the provisions of this Regulation.

3. Each competent authority and registry administrator shall only perform processes concerning allowances, verified emissions, accounts or Kyoto units where necessary to carry out their functions as competent authority or registry administrator.

SECTION 2

Accounts

Article 11

Accounts

1. From 1 January 2005 onwards, each registry shall contain at least one Party holding account created in accordance with Article 12.

2. From 1 January 2005 onwards, each Member State registry shall contain one operator holding account for each installation created in accordance with Article 15 and each registry shall contain at least one person holding account for each person created in accordance with Article 19.

3. From 1 January 2005 onwards, each registry shall contain one retirement account and one cancellation account for the 2005-2007 period and one cancellation account for the 2008-2012 period, created in accordance with Article 12.

4. From 1 January 2008 and from 1 January of the first year of each subsequent five-year period, each registry shall contain one retirement account and the cancellation and replacement accounts required by the relevant decisions adopted pursuant to the UNFCCC or the Kyoto Protocol for the 2008-2012 period and for each subsequent five-year period, created in accordance with Article 12.

5. Unless otherwise provided, all accounts shall be capable of holding allowances and Kyoto units.

SECTION 3

Party accounts

Article 12

Creation of Party accounts

1. The relevant body of the Member State and the Commission shall submit an application to their respective registry administrator for the creation in their registries of the accounts referred to in Article 11(1), (3) and (4).

The applicant shall provide the registry administrator with the information reasonably required by the registry administrator. That information shall include the information set out in Annex IV.

2. Within 10 days of the receipt of an application in accordance with paragraph 1 or the activation of the communication link between the registry and the Community independent transaction log, whichever is the later, the registry administrator shall create the account in the registry in accordance with the account creation process set out in Annex VIII.

3. The applicant referred to in paragraph 1 shall notify its registry administrator within 10 days of any changes in the information provided to its registry administrator pursuant to paragraph 1. Within 10 days of the receipt of such a notification the registry administrator shall update that information in accordance with the account update process set out in Annex VIII.

4. The registry administrator may require the applicants referred to in paragraph 1 to agree to comply with reasonable terms and conditions addressing the issues set out in Annex V.

Article 13

Closure of Party accounts

Within 10 days of the receipt of an application from the relevant body of a Member State or from the Commission to close a Party holding account, its registry administrator shall close the account in accordance with the account closure process set out in Annex VIII.

Article 14

Notification

The registry administrator shall immediately notify the account holder of the creation or update of his Party accounts and of the closure of his Party holding accounts.

SECTION 4

Operator holding accounts

Article 15

Creation of operator holding accounts

1. Within 14 days of the issue of each greenhouse gas emissions permit to the operator of an installation where the installation has not previously been covered by such a permit or the activation of the communication link between the registry and the Community independent transaction log, whichever is the later, the competent authority, or the operator where the competent authority so requires, shall provide the registry administrator of the Member State registry with the information set out in Annex III.

2.   Within 10 days of the receipt of the information in paragraph 1 or the activation of the communication link between the registry and the Community independent transaction log, whichever is the later, the registry administrator shall create an operator holding account referred to in Article 11(2) for each installation in its registry in accordance with the account creation process set out in Annex VIII.

3.   The competent authority, or the operator where the competent authority so requires, shall notify the registry administrator within 10 days of any changes in the information provided to the registry administrator pursuant to paragraph 1. Within 10 days of the receipt of such a notification the registry administrator shall update the operator's details in accordance with the account update process set out in Annex VIII.

4.   The registry administrator may require operators to agree to comply with reasonable terms and conditions addressing the issues set out in Annex V.

Article 16

Holding of Kyoto units in operator holding accounts

An operator holding account shall be capable of holding Kyoto units where authorised by Member State or Community legislation.

Closure of operator holding accounts

1.   The competent authority shall notify the registry administrator within 10 days of a greenhouse gas emissions permit being revoked or surrendered for an installation that is, as a result, not covered by any such permit. Without prejudice to paragraph 2, the registry administrator shall close all operator holding accounts relating to that revocation or surrender in accordance with the account closure process set out in Annex VIII on 30 June the year after the revocation or surrender took place if the relevant installation's entry in the latest year of the compliance status table is greater than or equal to zero. If the relevant installation's entry in the latest year of the compliance status table is less than zero, the registry administrator shall close its account the day after the entry is greater than or equal to zero or the day after the competent authority has instructed the registry administrator to close the account because there is no reasonable prospect of further allowances being surrendered by the installation's operator.

2.   If there is a positive balance of allowances or Kyoto units in an operator holding account which the registry administrator is to close in accordance with paragraph 1, the registry administrator shall first request the operator to specify another account within the registry system to which such allowances or Kyoto units shall then be transferred. If the operator has not responded to the registry administrator's request within 60 days, the registry administrator shall transfer the balance to the Party holding account.

Article 18

Notification

The registry administrator shall immediately notify the account holder of the creation, update or closure of his operator holding account.

Person holding accounts

Article 19

Creation of person holding accounts

1. An application for the creation of a person holding account shall be submitted to the registry administrator of the registry concerned.

The applicant shall provide the registry administrator with the information reasonably required by the registry administrator. That information shall include the information set out in Annex IV.

2. Within 10 days of the receipt of an application in accordance with paragraph 1 or the activation of the communication link between the registry and the Community independent transaction log, whichever is the later, the registry administrator shall create a person holding account in its registry in accordance with the account creation process set out in Annex VIII.

The registry administrator shall not establish more than 99 person holding accounts in any one person's name in its registry.

3. The applicant shall notify the registry administrator within 10 days of any changes in the information provided to the registry administrator pursuant to paragraph 1. Within 10 days of the receipt of such a notification the registry administrator shall update the person's details in accordance with the account update process set out in Annex VIII.

4. The registry administrator may require the applicants referred to in paragraph 1 to agree to comply with reasonable terms and conditions addressing the issues set out in Annex V.

Article 20

Holding of Kyoto units in person holding accounts

A person holding account shall be capable of holding Kyoto units where authorised by Member State or Community legislation.

Article 21

Closure of person holding accounts

1. Within 10 days of the receipt of an application from a person to close a person holding account, the registry administrator shall close the account in accordance with the account closure process set out in Annex VIII.

2. If a person holding account has a zero balance and no transactions have been recorded during a period of 12 months, the registry administrator shall notify the account holder that the person holding account shall be closed within 60 days unless the registry administrator receives within that period a request from the account holder that the person holding account be maintained. If the registry administrator does not receive any such request from the account holder, it shall close the account in accordance with the account closure process set out in Annex VIII.

Article 22

Notification

The registry administrator shall immediately notify each account holder of the creation, update or closure of his person holding account.

Article 23

Authorised representatives

1.   Each account holder shall appoint a primary and a secondary authorised representative for each account created in accordance with Articles 12, 15 and 19. Requests to the registry administrator to carry out processes shall be submitted by an authorised representative on behalf of the account holder.

2.   Each Member State and the Commission may allow account holders in its registry to nominate an additional authorised representative whose agreement is required in addition to the agreement of the primary or secondary authorised representative to submit a request to their registry administrator to carry out one or more of the processes pursuant to Articles 49(1), 52, 53 and 62.

3.   Each verifier shall appoint at least one authorised representative to enter or approve the entry of the annual verified emissions for an installation into the verified emissions table in accordance with Article 51(1).

4.   Each registry administrator and the Central Administrator shall appoint at least one authorised representative to operate and maintain their registry and the Community independent transaction log on behalf of that administrator.

SECTION 6

Tables

Article 24

Tables

1.   From 1 January 2005 onwards, each Member State registry shall contain one verified emissions table, one surrendered allowances table, and one compliance status table. Each registry may contain additional tables for other purposes.

2.   The Community independent transaction log shall contain one national allocation plan table for each Member State for the 2005-2007 period, the 2008-2012 period and for each subsequent five-year period.

The Community independent transaction log may contain additional tables for other purposes.

3.   The tables in each Member State registry shall contain the information set out in Annex II. The operator holding accounts and person holding accounts shall contain the information set out in Annex XVI.

The national allocation plan table in the Community independent transaction log shall contain the information set out in Annex XIV.

SECTION 7

Codes and identifiers

Article 25

Codes

Each registry shall contain the input codes set out in Annex VII and the response codes set out in Annex XII in order to ensure the correct interpretation of information exchanged during each process.

Article 26

Account identification codes and alphanumeric identifiers

Before creating an account the registry administrator shall assign to each account a unique account identification code and the alphanumeric identifier specified by the account holder as part of the information given under Annexes III and IV respectively. Before creating an account, the registry administrator shall also assign to the account holder a unique account holder identification code comprising the elements set out in Annex VI.

CHAPTER IV

CHECKS AND PROCESSES

SECTION 1

Blocking of accounts

Article 27

Blocking of operator holding accounts

1. If, on 1 April of each year starting in 2006, an installation's annual verified emissions for the preceding year have not been entered into the verified emissions table in accordance with the verified emissions entry process set out in Annex VIII, the registry administrator shall block the transfer of any allowances out of the operator holding account for that installation.

2. When the installation's annual verified emissions for the year referred to in paragraph 1 have been entered into the verified emissions table, the registry administrator shall unblock the account.

3. The registry administrator shall immediately notify the relevant account holder and the competent authority of the blocking and unblocking of each operator holding account.

4. Paragraph 1 shall not apply to the surrender of allowances pursuant to Article 52 or the cancellation and replacement of allowances pursuant to Articles 60 and 61.

SECTION 2

Automated checks and the data reconciliation process

Article 28

Detection of discrepancies by the Community independent transaction log

1. The Central Administrator shall ensure that the Community independent transaction log conducts the automated checks set out in Annex VIII, Annex IX and Annex XI for all processes concerning allowances, verified emissions, accounts and Kyoto units to ensure that there are no discrepancies.

2. If the automated checks referred to in paragraph 1 identify a discrepancy in a process under Annex VIII, Annex IX and Annex XI, the Central Administrator shall immediately inform the registry administrator or administrators concerned by returning an automated response detailing the exact nature of

the discrepancy using the response codes set out in Annex VIII, Annex IX and Annex XI. Upon receiving such a response code for a process under Annex VIII or Annex IX, the registry administrator of the initiating registry shall terminate that process and inform the Community independent transaction log. The Central Administrator shall not update the information contained in the Community independent transaction log. The registry administrator or administrators concerned shall immediately inform the relevant account holders that the process has been terminated.

Article 29

Detection of inconsistencies by the Community independent transaction log

1.   The Central Administrator shall ensure that the Community independent transaction log periodically initiates the data reconciliation process set out in Annex X. For that purpose the Community independent transaction log shall record all processes concerning allowances, accounts and Kyoto units.

Through that process, the Community independent transaction log shall check that the holdings of Kyoto units and allowances in each account in a registry are identical to the records held in the Community independent transaction log.

2.   If an inconsistency is detected during the data reconciliation process, the Central Administrator shall immediately inform the registry administrator or administrators concerned. If the inconsistency is not resolved, the Central Administrator shall ensure that the Community

independent transaction log does not allow any further process under Annex VIII and Annex IX concerning any of the allowances, accounts or Kyoto units which are the subject of the earlier inconsistency to proceed.

Article 30

Detection of discrepancies and inconsistencies by the UNFCCC independent transaction log

1.   If the UNFCCC independent transaction log, following an automated check, informs the registry administrator or administrators concerned of a discrepancy in a process, the registry administrator of the initiating registry shall terminate the process and inform the UNFCCC independent transaction log thereof. The registry administrator or administrators concerned shall immediately inform the relevant account holders that the process has been terminated.

2.   If the UNFCCC independent transaction log has detected an inconsistency, the Central Administrator shall ensure that the Community independent transaction log does not allow any further process under Annex VIII and Annex IX concerning any of the Kyoto units which are the subject of the earlier inconsistency, and which is not subject to the UNFCCC independent transaction log's automated checks, to proceed.

Article 31

Registry automated checks

Prior to and during the execution of all processes the registry administrator shall ensure that appropriate automated checks are conducted within the registry, in order to detect discrepancies and thereby terminate processes in advance of automated checks being conducted by the Community independent transaction log or UNFCCC independent transaction log.

SECTION 3

Execution and finalisation of processes

Article 32

Processes

Each process shall follow the complete sequence for message exchanges for that type of process as set out in Annex VIII, Annex IX, Annex X and Annex XI. Each message shall conform to the format and informational requirements described using web services description language as elaborated pursuant to the UNFCCC or the Kyoto Protocol.

Article 33

Identification codes

The registry administrator shall assign to each process referred to in Annex VIII a unique correlation identification code and to each process referred to in Annex IX a unique transaction identification code. Each such identification code shall comprise the elements set out in Annex VI.

Article 34

Finalisation of processes concerning accounts and verified emissions

All processes referred to in Annex VIII shall be final when both independent transaction logs successfully inform the initiating registry that they have not detected any discrepancies in the proposal sent by the initiating registry.

However, prior to the communication link between the Community independent transaction log and the UNFCCC independent transaction log being established, all processes referred to in Annex VIII shall be final when the Community independent transaction log successfully informs the initiating registry that it has not detected any discrepancies in the proposal sent by the initiating registry.

Article 35

Finalisation of processes concerning transactions within registries

All processes referred to in Annex IX, except the external transfer process, shall be final when both independent transaction logs inform the initiating registry that they have not detected any discrepancies in the proposal sent by the initiating registry and the initiating registry has successfully sent confirmation to both independent transaction logs that it has updated its records in accordance with its proposal.

However, prior to the communication link between the Community independent transaction log and the UNFCCC independent transaction log being established, all processes referred to in Annex IX, except the external transfer process, shall be final when the Community independent transaction log informs the initiating registry that it has not detected any discrepancies in the proposal sent by the initiating registry and the initiating registry has successfully sent confirmation to the Community independent transaction log that it has updated its records in accordance with its proposal.

Article 36

Finalisation of the external transfer process

The external transfer process shall be final when both independent transaction logs inform the acquiring registry that they have not detected any discrepancies in the proposal sent by the initiating registry and the acquiring registry has successfully sent confirmation to both independent transaction logs that it has updated its records in accordance with the initiating registry's proposal.

However, prior to the communication link between the Community independent transaction log and the UNFCCC independent transaction log being established, the external transfer process shall be final when the Community independent transaction log informs the acquiring registry that it has not detected any discrepancies in the proposal sent by the initiating registry and the acquiring registry has successfully sent confirmation to the Community independent transaction log that it has updated its records in accordance with the initiating registry's proposal.

Article 37

Finalisation of the reconciliation process

The reconciliation process referred to in Annex X shall be final when all inconsistencies between the information contained in a registry and the information contained in the Community independent transaction log for a specific time and date have been resolved, and the reconciliation process has been successfully re-initiated and completed for that registry.

CHAPTER V

TRANSACTIONS

SECTION 1

Allocation and issue of allowances for the 2005-2007 period

Article 38

National allocation plan table for the 2005-2007 period

1.   By 1 October 2004, each Member State shall notify to the Commission its national allocation plan table, corresponding to the decision taken under Article 11 of Directive 2003/87/EC. If the national allocation plan table is based upon the national allocation plan notified to the Commission which was not rejected under Article 9(3) of Directive 2003/87/EC or on which the Commission has accepted proposed amendments, the Commission shall instruct the Central Administrator to enter the national allocation plan table into the Community independent transaction log in accordance with the initialisation procedures set out in Annex XIV.

2.   A Member State shall notify each correction to its national allocation plan together with each corresponding correction in its national allocation plan table to the Commission. If the correction to the national allocation plan table is based upon the national allocation plan notified to the Commission which was not rejected under Article 9(3) of Directive 2003/87/EC or on which the Commission has accepted amendments and that correction is in accordance with methodologies set out in that national allocation plan or results from improvements in data,

the Commission shall instruct the Central Administrator to enter the corresponding correction into the national allocation plan table held in the Community independent transaction log in accordance with the initialisation procedures set out in Annex XIV. In all other cases, the Member State shall notify the correction to its national allocation plan to the Commission and if the Commission does not reject this correction in accordance with the procedure in Article 9(3) of Directive 2003/87/EC, the Commission shall instruct the Central Administrator to enter the corresponding correction into the national allocation plan table held in the Community independent transaction log in accordance with the initialisation procedures set out in Annex XIV.

3. The registry administrator shall, subsequent to any correction made pursuant to paragraph 2 which occurs after allowances have been issued under Article 39 and which reduces the total quantity of allowances issued under Article 39 for the 2005-2007 period, transfer the number of allowances specified by the competent authority from the holding accounts referred to in Article 11(1) and (2) in which the allowances are held to the cancellation account for the 2005-2007 period.

The correction shall take place in accordance with the correction to allowances process set out in Annex IX.

Article 39

Issue of allowances

After the national allocation plan table has been entered into the Community independent transaction log and,

subject to Article 38(2), by 28 February 2005, the registry administrator shall issue the total quantity of allowances set out in the national allocation plan table into the Party holding account.

When issuing such allowances the registry administrator shall assign a unique unit identification code to each allowance comprising the elements set out in Annex VI.

Allowances shall be issued in accordance with the allowance issue (2005-2007) process set out in Annex IX.

Article 40

Allocation of allowances to operators

Without prejudice to Articles 38(2) and 41, by 28 February 2005 and by 28 February of each year thereafter for the 2005-2007 period, the registry administrator shall transfer from the Party holding account to the relevant operator holding account the proportion of the total quantity of allowances issued under Article 39 which has been allocated to the corresponding installation for that year in accordance with the relevant section of the national allocation plan table.

Where foreseen for an installation in the national allocation plan of the Member State, the registry administrator may transfer that proportion at a later date of each year.

Allowances shall be allocated in accordance with the allowance allocation process set out in Annex IX.

Article 41

Surrender of allowances on instruction of the competent authority

If instructed to do so by the competent authority pursuant to Article 16(1) of Directive 2003/87/EC, the registry administrator shall surrender part or all of the proportion of the total quantity of allowances issued under Article 39 which has been allocated to an installation for a specific year, by entering the number of surrendered allowances into the section of the surrendered allowance table designated for that installation for that year. These surrendered allowances shall remain in the Party holding account.

Allowances surrendered on instruction of the competent authority shall be surrendered in accordance with the allowance allocation process set out in Annex IX.

Article 42

Allocation of allowances to new entrants

If instructed to do so by the competent authority, the registry administrator shall transfer a proportion of the total quantity of allowances issued under Article 39 that are remaining in the Party holding account to the operator holding account of a new entrant.

Allowances shall be transferred in accordance with the internal transfer process set out in Annex IX.

Article 43

Issue of force majeure allowances

1.   If instructed to do so by the competent authority, the registry administrator shall issue into the Party holding account the number of force majeure allowances authorised by the Commission for the 2005-2007 period pursuant to Article 29 of Directive 2003/87/EC.

Force majeure allowances shall be issued in accordance with the force majeure allowance issue process set out in Annex IX.

2.   The registry administrator shall enter the number of issued force majeure allowances into the sections of the surrendered allowance table designated for those installations and years for which authorisation was given.

3.   When issuing force majeure allowances the registry administrator shall assign a unique unit identification code to each such force majeure allowance comprising the elements set out in Annex VI.

SECTION 2

Allocation and issue of allowances for the 2008-2012 period and each subsequent five year period

Article 44

National allocation plan table for the 2008-2012 period and each subsequent five year period

1.   By 1 January 2007 and by 1 January 12 months before the start of each subsequent five year period, each Member State shall notify to the Commission its national allocation

plan table, corresponding to the decision taken under Article 11 of Directive 2003/87/EC. If the national allocation plan table is based upon the national allocation plan notified to the Commission which was not rejected under Article 9(3) of Directive 2003/87/EC or on which the Commission has accepted proposed amendments, the Commission shall instruct the Central Administrator to enter the national allocation plan table into the Community independent transaction log in accordance with the initialisation procedures set out in Annex XIV.

2.   A Member State shall notify each correction to its national allocation plan together with each corresponding correction in its national allocation plan table to the Commission. If the correction to the national allocation plan table is based upon the national allocation plan notified to the Commission which was not rejected under Article 9(3) of Directive 2003/87/EC or on which the Commission has accepted amendments and that correction results from improvements in data, the Commission shall instruct the Central Administrator to enter the corresponding correction into the national allocation plan table held in the Community independent transaction log in accordance with the initialisation procedures as set out in Annex XIV. In all other cases, the Member State shall notify the correction to its national allocation plan to the Commission and if the Commission does not reject this correction in accordance with the procedure in Article 9(3) of Directive 2003/87/EC, the Commission shall instruct the Central Administrator to enter the corresponding correction into the national allocation plan table held in the Community independent

transaction log in accordance with the initialisation procedures set out in Annex XIV.

3.   The registry administrator shall, subsequent to any correction made pursuant to paragraph 2 which occurs after allowances have been issued under Article 45 and which reduces the total quantity of allowances issued under Article 45 for the 2008-2012 period or subsequent five-year periods, convert the number of allowances specified by the competent authority into AAUs by removing the allowance element from the unique unit identification code of each such AAU comprising the elements set out in Annex VI.

The correction shall take place in accordance with the correction to allowances process set out in Annex IX.

Article 45

Issue of allowances

After the national allocation plan table has been entered into the Community independent transaction log and, subject to Article 44(2), by 28 February of the first year of the 2008-2012 period and by 28 February of the first year of each subsequent five-year period, the registry administrator shall issue the total quantity of allowances set out in the national allocation plan table into the Party holding account by converting an equal quantity of AAUs held in that holding account into allowances.

This conversion shall take place through adding the allowance element to the unique unit identification code of each such AAU, comprising the elements set out in Annex VI.

The issue of allowances for the 2008-2012 period and each subsequent five-year period shall take place in accordance with the allowance issue (2008-2012 onwards) process set out in Annex IX.

Article 46

Allocation of allowances to operators

Without prejudice to Articles 44(2) and 47, by 28 February 2008 and by 28 February in each year thereafter, the registry administrator shall transfer from the Party holding account to the relevant operator holding account the proportion of the total quantity of allowances issued under Article 45 which has been allocated to the corresponding installation for that year in accordance with the relevant section of the national allocation plan table.

Where foreseen for an installation in the national allocation plan of the Member State, the registry administrator may transfer that proportion at a later date of each year.

Allowances shall be allocated in accordance with the allowance allocation process set out in Annex IX.

Article 47

Surrender of allowances on instruction of the competent authority

If instructed to do so by the competent authority pursuant to Article 16(1) of Directive 2003/87/EC, the registry administrator shall surrender part or all of the proportion of the total quantity of allowances issued under Article 45 which has been allocated to an installation for a specific year, by entering the number of surrendered allowances into the section of the surrendered allowance table designated for that installation for that year. These surrendered allowances shall remain in the Party holding account.

Allowances surrendered on instruction of the competent authority shall be surrendered in accordance with the allowance allocation process set out in Annex IX.

Article 48

Allocation of allowances to new entrants

If instructed to do so by the competent authority, the registry administrator shall transfer a proportion of the total quantity of allowances issued under Article 45 that are remaining in the Party holding account to the operator holding account of a new entrant.

Allowances shall be transferred in accordance with the internal transfer process set out in Annex IX.

SECTION 3

Transfers and eligibility

Article 49

Transfers of allowances and Kyoto units by account holders

1.   The registry administrator shall carry out any transfer between holding accounts referred to in Article 11(1) and (2):

(a) within its registry as requested by an account holder in accordance with the internal transfer process set out in Annex IX;

(b) between registries as requested by an account holder for allowances issued for the 2005-2007 period in accordance with the external transfer (2005-2007) process set out in Annex IX; and

(c) between registries as requested by an account holder for allowances issued for the 2008-2012 period and subsequent five-year periods and Kyoto units in accordance with the external transfer (2008-2012 onwards) process set out in Annex IX.

2.   Allowances may only be transferred from an account in a registry to an account in a third country registry or the CDM Registry, or acquired from an account in a third country registry or the CDM Registry by an account in a registry, where an agreement has been concluded in accordance with Article 25(1) of Directive 2003/87/EC and such transfers are in accordance with any provisions relating to the mutual recognition of allowances under that agreement drawn up by the Commission pursuant to Article 25(2) of Directive 2003/87/EC.

Article 50

Eligibility and the commitment period reserve

1.   A Member State may not transfer or acquire ERUs or AAUs, or use CERs, until 16 months have elapsed since the submission of its report in accordance with Article 7(1) of Decision 280/2004/EC, unless the Secretariat to the UNFCCC has informed that Member State that compliance procedures will not be commenced.

Pursuant to Article 8 of Decision No 280/2004/EC, if the Secretariat to the UNFCCC informs a Member State that it does not meet the requirements allowing it to transfer or acquire ERUs or AAUs, or use CERs, the relevant body of the Member State shall instruct the registry administrator not to initiate those transactions requiring such eligibility.

2.   When, from 1 January 2008 onwards, the holdings of ERUs, CERs, AAUs and RMUs valid for the relevant five-year period in the Party holding accounts, operator holding accounts, person holding accounts and retirement accounts in a Member State approach a breach of the commitment period reserve, calculated as 90 % of the Member State's assigned amount or 100 % of five times its most recently reviewed inventory, whichever is the lowest, the Commission shall notify that Member State.

SECTION 4

Verified emissions

Article 51

Verified emissions of an installation

1.   Upon the verification as satisfactory, in accordance with the detailed requirements established by the Member State pursuant to Annex V of Directive 2003/87/EC, of an operator's report on the emissions from an installation during a previous year, the verifier shall enter or approve the entry of the annual verified emissions for that installation for that year into the section of the verified emissions table designated for that installation for that year in accordance with the verified emissions update process set out under Annex VIII.

2. The competent authority may instruct the registry administrator to correct the annual verified emissions for an installation for a previous year, to ensure compliance with the detailed requirements established by the Member State pursuant to Annex V of Directive 2003/87/EC, by entering the corrected annual verified emissions for that installation for that year into the section of the verified emissions table designated for that installation for that year in accordance with the verified emissions update process set out under Annex VIII.

SECTION 5

Surrender of allowances

Article 52

Surrender of allowances

An operator shall surrender allowances for an installation by requesting or, where provided in Member State legislation, be deemed to have requested, the registry administrator to:

(a) transfer a specified number of allowances for a specified year from the relevant operator holding account into the Party holding account of that registry;

(b) enter the number of transferred allowances into the section of the surrendered allowance table designated for that installation for that year.

The transfer and entry shall take place in accordance with the allowance surrender process set out under Annex IX.

Article 53

The use of CERs and ERUs

The use of CERs and ERUs by an operator in accordance with Article 11a of Directive 2003/87/EC in respect of an installation shall take place through an operator requesting the registry administrator to:

(a) transfer a specified number of CERs or ERUs for a specified year from the relevant operator holding account into the Party holding account of that registry;

(b) enter the number of transferred CERs and ERUs into the section of the surrendered allowance table designated for that installation for that year.

From 1 January 2008 onwards, the registry administrator shall only accept requests to use CERs and ERUs up to a percentage of the allocation made to each installation, as specified by that administrator's Member State in its national allocation plan for that period.

The transfer and entry shall take place in accordance with the allowance surrender process set out under Annex IX.

Article 54

Surrender of force majeure allowances

The issue of force majeure allowances in accordance with Article 43 shall constitute the surrender of those force majeure allowances.

Article 55

Calculation of compliance status figures

Upon an entry being made into the section of the surrendered allowance table or verified emissions table designated for an installation, the registry administrator shall:

(a) during the years 2005, 2006 and 2007 determine the compliance status figure for that installation and for each year by calculating the sum of all allowances surrendered from the year 2005 up to and including the current year minus the sum of all verified emissions from the year 2005 up to and including the current year;

(b) during the year 2008 and each year thereafter determine the compliance status figure for that installation and for each year by calculating the sum of all allowances surrendered from the year 2008 up to and including the current year minus the sum of all verified emissions from the year 2008 up to and including the current year, plus a correction factor. The correction factor shall be zero if the 2007 figure was greater than zero, but shall remain as the 2007 figure if the 2007 figure is less than or equal to zero.

Article 56

Entries into the compliance status table

1.   The registry administrator shall enter the installation's compliance status figure calculated in accordance with Article 55 for each year into the section of the compliance status table designated for that installation.

2.   On 1 May 2006 and on 1 May of each year thereafter the registry administrator shall notify the compliance status table to the competent authority. In addition, the registry administrator shall notify any changes to the entries for previous years of the compliance status table to the competent authority.

Article 57

Entries into the verified emissions table

Where, on 1 May 2006 and on 1 May of each year thereafter, no verified emissions figure has been entered into the verified emissions table for an installation for a previous year, any substitute emissions figure determined pursuant to Article 16(1) of Directive 2003/87/EC which has not been calculated in accordance with the detailed requirements established by the Member State pursuant to Annex V of Directive 2003/87/EC shall not be entered into the verified emissions table.

SECTION 6

Cancellation and retirement

Article 58

Cancellation and retirement of surrendered allowances and force majeure allowances for the 2005-2007 period

On 30 June 2006, 2007 and 2008 the registry administrator shall cancel a number of allowances, CERs, and force majeure allowances held in the Party holding account pursuant to Articles 52, 53 and 54. The number of allowances, CERs, and force majeure allowances to be cancelled shall be equal to the total number of surrendered allowances entered in the surrendered allowance table for the periods 1 January 2005 to 30 June 2006, 30 June 2006 to 30 June 2007, and 30 June 2007 to 30 June 2008.

Cancellation shall take place by transferring CERs, with the exception of CERs resulting from projects referred to in Article 11a(3) of Directive 2003/87/EC, from the Party holding account into the cancellation account for the 2008-2012 period, and by transferring allowances and force majeure allowances from the Party holding account to the retirement account for the 2005-2007 period, in accordance with the retirement (2005-2007) process set out in Annex IX.

Article 59

Cancellation and retirement of surrendered allowances for the 2008-2012 period and subsequent periods

On 30 June 2009 and on 30 June of each year thereafter, the registry administrator shall cancel allowances surrendered for the 2008-2012 period and each subsequent five year period, in accordance with the retirement (2008-2012 onwards) process set out in Annex IX, by:

(a) converting a number of allowances issued for that five-year period and held in the Party holding account, equal to the total number of allowances surrendered pursuant to Article 52 as entered in the surrendered allowance table since 1 January 2008 on 30 June 2009 and since 30 June of the preceding year on 30 June of the subsequent years, into AAUs by removing the allowance element from the unique unit identification code of each such AAU comprising the elements set out in Annex VI; and

(b) transferring a number of Kyoto units of the type specified by the competent authority, with the exception of Kyoto units resulting from projects referred to in Article 11a(3) of Directive 2003/87/EC, equal to the total number of allowances surrendered pursuant to Articles 52 and 53 as entered in the surrendered allowance table since 1 January 2008 on 30 June 2009 and since 30 June of the preceding year on 30 June of the subsequent years, from the Party holding account to the retirement account for the relevant period.

SECTION 7

Cancellation and replacement

Article 60

Cancellation and replacement of allowances issued for the 2005-2007 period

On 1 May 2008, each registry administrator shall cancel and, if instructed to do so by the competent authority, replace allowances held in his registry in accordance with the allowance cancellation and replacement process set out in Annex IX by:

(a) transferring a number of allowances, equal to the number of allowances issued for the 2005-2007 period minus the number of allowances surrendered pursuant to Articles 52 and 54 since 30 June of the preceding year, from their holding accounts referred to in Article 11(1) and (2) to the cancellation account for the 2005-2007 period;

(b) if instructed to do so by the competent authority, issuing a number of replacement allowances specified by the competent authority by converting an equal number of AAUs issued for the 2008-2012 period held in the Party holding account into allowances by adding the allowance element to the unique unit identification code of each such AAU comprising the elements set out in Annex VI;

(c) transferring any such replacement allowances referred to in (b) from the Party holding account into the operator and person holding accounts specified by the competent authority from which allowances were transferred under point (a).

Article 61

Cancellation and replacement of allowances issued for the 2008-2012 period and subsequent periods

On 1 May in 2013 and on 1 May in the first year of each subsequent five year period, each registry administrator shall cancel and replace allowances held in its registry in accordance with the allowance cancellation and replacement process set out in Annex IX by:

(a) transferring all allowances issued for the preceding five-year period from their operator and person holding accounts to the Party holding account;

(b) converting a number of allowances, equal to the number of allowances issued for the preceding five-year period minus the number of allowances surrendered pursuant to Article 52 since 30 June of the preceding year, into AAUs by removing the allowance element from the unique unit identification code of each such AAU comprising the elements set out in Annex VI;

(c) issuing an equal number of replacement allowances by converting AAUs issued for the current period held in the Party holding account into allowances by adding the allowance element to the unique unit identification code of each such AAU comprising the elements set out in Annex VI;

(d) transferring a number of those allowances issued under point (c) for the current period from the Party holding account into each operator and person holding account from which allowances were transferred under point (a), equal to the number of allowances that were transferred from those accounts under point (a).

SECTION 8

Voluntary cancellation and retirement

Article 62

Voluntary cancellation of allowances and Kyoto units

1. The registry administrator shall carry out any request from an account holder pursuant to Article 12(4) of Directive 2003/87/EC to voluntarily cancel allowances or Kyoto units held in any of his holding accounts. The voluntary cancellation of allowances and Kyoto units shall take place in accordance with paragraphs 2 and 3.

2. For allowances issued for the 2005-2007 period the registry administrator shall transfer the number of allowances specified by the account holder from his account to the cancellation account for the 2005-2007 period in accordance with the allowance cancellation (2005-2007) process set out in Annex IX.

3. For Kyoto units and allowances issued for the 2008-2012 period and subsequent five-year periods the registry administrator shall transfer the number of Kyoto units or allowances specified by the account holder from his account to the appropriate cancellation account for the 2008-2012 period and subsequent five-year periods in accordance with the cancellation (2008-2012 onwards) process set out Annex IX.

4. Allowances or Kyoto units held in a cancellation account may not be transferred to any other account in the registries system or to any account in the CDM registry or in a third country registry.

Article 63

Retirement of Kyoto units

1. If instructed by the relevant body of the Member State, the registry administrator shall transfer any quantity and types of Kyoto units specified by that body which have not already been retired pursuant to Article 59 from the Party holding account to the appropriate retirement account in his registry in accordance with the retirement (2008-2012 onwards) process set out in Annex IX.

2. An operator or person shall not be able to transfer allowances from his operator or person holding account into a retirement account.

3. Kyoto units held in a retirement account may not be transferred to any other account in the registries system or to any account in the CDM registry or in a third country registry.

CHAPTER VI

SECURITY STANDARDS, AUTHENTICATION AND ACCESS RIGHTS

Article 64

Security standards

1. Each registry shall comply with the security standards set out in Annex XV.

2. The Community independent transaction log shall comply with the security standards set out in Annex XV.

Article 65

Authentication

The Member States and the Community shall use the digital certificates issued by the Secretariat to the UNFCCC, or an entity designated by it, to authenticate their registries and the Community independent transaction log to the UNFCCC independent transaction log.

However, from 1 January 2005 until the communication link between the Community independent transaction log and UNFCCC independent transaction log is established, the identity of each registry and the Community independent transaction log shall be authenticated using digital certificates and usernames and passwords as specified in Annex XV. The Commission, or an entity designated by it, shall act as the certification authority for all digital certificates and shall distribute the usernames and passwords.

Article 66

Access to registries

1.   An authorised representative shall only have access to the accounts within a registry which he is authorised to access or be able to request the initiation of processes which he is authorised to request pursuant to Article 23. This access or these requests shall take place through a secure area of the website for that registry.

The   registry administrator shall issue each authorised representative with a username and password to permit the level of access to accounts or processes to which he is authorised. Registry administrators may apply additional security requirements at their discretion if they are compatible with the provisions of this Regulation.

2.   The registry administrator may assume that a user who has entered a matching username and password is the authorised representative registered under that username and password, until such point that the authorised representative informs the registry administrator that the security of his password has been compromised and requests a replacement. The registry administrator shall promptly issue such replacement passwords.

3.   The registry administrator shall ensure that the secure area of the registry website is accessible to any computer using a widely available Internet browser. Communications between the authorised representatives and the secure area of the registry website shall be encrypted in accordance with the security standards in Annex XV.

4.   The registry administrator shall take all necessary steps to ensure that unauthorised access to the secure area of the registry website does not occur.

Article 67

Suspension of access to accounts

1.   The Central Administrator and each registry administrator may only suspend an authorised representative's password to any accounts or processes to which he would otherwise have access if the authorised representative has, or that administrator has reasonable grounds to believe the authorised represent ti ve has:

(a) attempted to access accounts or processes which he is not authorised to access;

(b) repeatedly attempted to access an account or a process using a non-matching username and password; or

(c) attempted, or is attempting, to undermine the security of the registry or the registries system.

2.   Where access to an operator holding account has been suspended pursuant to paragraph 1 or pursuant to Article 69 between 28 April and 30 April in any year from 2006 onwards, the registry administrator shall, if so requested by the account holder and following submission of his authorised representative's identity by means of supporting evidence, surrender the number of allowances and use the number of CERs and ERUs specified by the account holder in accordance with the allowance surrender process set out in Article 52 and 53 and Annex IX.

CHAPTER VII

AVAILABILITY AND RELIABILITY OF INFORMATION

Article 68

Availability and reliability of registries and the Community independent transaction log

The Central Administrator and each registry administrator shall take all reasonable steps to ensure that:

(a) the registry is available for access by account holders 24 hours a day, 7 days a week, and that the communication link between the registry and the Community independent transaction log is maintained 24 hours a day, 7 days a week, thereby providing backup hardware and software in the event of a breakdown in operations of the primary hardware and software;

(b) the registry and the Community independent transaction log respond promptly to requests made by account holders.

They shall ensure that the registry and Community independent transaction log incorporate robust systems and procedures for the safeguarding of all data and the prompt recovery of all data and operations in the event of a disaster.

They shall keep interruptions to the operation of the registry and Community independent transaction log to a minimum.

Article 69

Suspension of access

The Central Administrator may suspend access to the Community independent transaction log and a registry administrator may suspend access to his registry if there is a breach of security of the Community independent transaction log or of a registry which threatens the integrity of the Community independent transaction log or of a registry or the integrity of the registries system and the back-up facilities under Article 68 are similarly affected.

Article 70

Notification of suspension of access

1. In the event of a breach of security of the Community independent transaction log that may lead to suspension of access, the Central Administrator shall promptly inform registry administrators of any risks posed to registries.

2. In the event of a breach of security of a registry that may lead to suspension of access, the relevant registry administrator shall promptly inform the Central Administrator who shall, in turn, promptly inform other registry administrators of any risks posed to registries.

3. If the registry administrator becomes aware that it is necessary to suspend either access to accounts or other operations of the registry, he shall give all relevant account holders and verifiers, the Central Administrator and other registry administrators such prior notice of the suspension as is reasonably practicable.

4.   If the Central Administrator becomes aware that it is necessary to suspend access to operations of the Community independent transaction log, it shall give all registry administrators such prior notice of the suspension as is reasonably practicable.

5.   The notices referred to in paragraphs 3 and 4 shall include the likely duration of the suspension and shall be clearly displayed on the public area of that registry's web site or on the public area of the Community independent transaction log's web site.

Article 71

Testing area of each registry and the Community independent transaction log

1.   Each registry administrator shall establish a testing area within which any new version or release of a registry can be tested in accordance with the testing procedures set out in Annex XIII so as to ensure that:

(a) any testing procedures on a new version or release of a registry are completed without reducing the availability to account holders of the version or release of the registry which currently has a communication link with the Community independent transaction log or UNFCCC independent transaction log; and

(b) any communication link between a new version or release of a registry and the Community independent transaction log or UNFCCC independent transaction log is established and activated with minimum disruption to its account holders.

2.   The Central Administrator shall establish a testing area so as to facilitate testing procedures referred to in paragraph 1.

3.   The registry administrators and the Central Administrator shall ensure that the hardware and software of their testing area shall perform in a manner that is representative of the performance of the primary hardware and software referred to in Article 68.

Article 72

Change management

1.   The Central Administrator shall coordinate with registry administrators and the Secretariat to the UNFCCC the preparation and implementation of any amendments to this Regulation resulting in changes in the functional and technical specifications of the registry system before their implementation.

2.   If, as a result of a these amendments, a new version or release of a registry is required, each registry administrator shall successfully complete the testing procedures set out in Annex XIII before a communication link is established and activated between the new version or release of that registry and the Community independent transaction log or UNFCCC independent transaction log.

3.   Each registry administrator shall continuously monitor the availability, reliability and efficiency of his registry in order to ensure a level of performance which meets the requirements of this Regulation. If, as a result of this monitoring or the suspension of the communication link pursuant to Article 6(3), a new version or release of a registry is required, each

registry administrator shall successfully complete the testing procedures set out in Annex XIII before a communication link is established and activated between the new version or release of that registry and the Community independent transaction log or UNFCCC independent transaction log.

CHAPTER VIII

RECORDS AND FEES

Article 73

Records

1. The Central Administrator and each registry administrator shall store records concerning all processes and account holders set out in Annex III, Annex IV, Annex VIII, Annex IX, Annex X and Annex XI for 15 years or until any questions of implementation relating to them have been resolved, whichever is the later.

2. Records shall be stored in accordance with the data logging standards elaborated pursuant to the UNFCCC or the Kyoto Protocol.

Article 74

Fees

Any fees charged by the registry administrator to account holders shall be reasonable and shall be clearly displayed on the public area of that registry's web site. Registry administrators shall not differentiate any such fees on the basis of the location of an account holder within the Community. Registry administrators shall not charge account holders for transactions of allowances pursuant to Article 49, Articles 52 to 54 and Articles 58 to 63.

CHAPTER IX

FINAL PROVISIONS

Article 75

Entry into force

This Regulation shall enter into force on the day following that of its publication in the *Official Journal of the European Union*.

This Regulation shall be binding in its entirety and directly applicable in all Member States.

Done at Brussels, 21 December 2004.

> *For the Commission*
> Stavros DIMAS
> *Member of the Commission*

ANNEX I

Hardware and software requirements of registries and the Community independent transaction log

Architecture requirements

1. Each registry and the Community independent transaction log shall include the following hardware and software in their architecture:

   (a) web server;

   (b) application server;

   (c) database server installed on a separate machine to that or those used for the web server and application server;

   (d) firewalls.

Communication requirements

2. From 1 January 2005 until the communication link between the Community independent transaction log and the UNFCCC independent transaction log is established:

   (a) the record of the time in the Community independent transaction log and each registry shall be synchronised to Greenwich Mean Time;

   (b) all processes concerning allowances, verified emissions and accounts shall be completed by the exchange of data written in extensible markup language (XML) using the simple object access protocol (SOAP) version 1.1 over the hypertext transfer protocol (HTTP) version 1.1 (remote procedure call (RPC) encoded style).

3. After the communication link between the Community independent transaction log and the UNFCCC independent transaction log is established:

   (a) the record of the time in the UNFCCC independent transaction log, Community independent transaction log and each registry shall be synchronised, and

   (b) all processes concerning allowances, verified emissions, accounts and Kyoto units shall be completed by the exchange of data, using the hardware and software requirements set out in the functional and technical specifications for data exchange standards for registry systems under the Kyoto Protocol, elaborated pursuant to Decision 24/CP.8 of the Conference of the Parties to the UNFCCC.

ANNEX II

Tables to be contained in Member State registries

1. Each Member State registry shall be capable of tabulating the following information which shall comprise the verified emissions table:

    (a)   Years: in individual cells from 2005 onwards in ascending order.

    (b)   Installation identification code: in individual cells comprising the elements set out in Annex VI and in ascending order.

    (c)   Verified emissions: the verified emissions for a specified year for a specified installation shall be entered into the cell connecting that year to that installation's identification code.

2. Each Member State registry shall be capable of tabulating the following information which shall comprise the surrendered allowances table:

    (a)   Years: in individual cells from 2005 onwards in ascending order.

    (b)   Installation identification code: in individual cells comprising the elements set out in Annex VI and in ascending order.

    (c)   Surrendered allowances: the number of allowances surrendered in accordance with Articles 52, 53 and 54 for a specified year for a specified installation shall be entered into the three cells connecting that year to that installation's identification code.

3. Each Member State registry shall be capable of tabulating the following information which shall comprise the compliance status table:

    (a)   Years: in individual cells from 2005 onwards in ascending order.

    (b)   Installation identification code: in individual cells comprising the elements set out in Annex VI and in ascending order.

    (c)   Compliance status: the compliance status for a specified year for a specified installation shall be entered into the cell connecting that year to that installation's identification code. The compliance status shall be calculated in accordance with Article 55.

ANNEX III

Information concerning each operator holding account to be provided to the registry administrator

1. Points 1 to 4.1, 4.4 to 5.5 and point 7 (activity 1) of the information identifying the installation as listed in section 11.1 of Annex I to Decision 2004/156/EC.

2. The permit identification code specified by the competent authority, comprising the elements set out in Annex VI.

3. The installation identification code, comprising the elements set out in Annex VI.

4. The alphanumeric identifier specified by the operator for the account, which shall be unique within the registry.

5. Name, address, city, postcode, country, telephone number, facsimile number and e-mail address of the primary authorised representative of the operator holding account specified by the operator for that account.

6. Name, address, city, postcode, country, telephone number, facsimile number and e-mail address of the secondary authorised representative of the operator holding account specified by the operator for that account.

7. Name, address, city, postcode, country, telephone number, facsimile number and e-mail address of any additional authorised representatives of the operator holding account and their account access rights, specified by the operator for that account.

8. Evidence to support the identity of the authorised representatives of the operator holding account.

ANNEX IV

Information concerning accounts referred to in Article 11(1), (3) and (4) and person holding accounts to be provided to the registry administrator

1. Name, address, city, postcode, country, telephone number, facsimile number and e-mail address of the person requesting the opening of the person holding account.

2. Evidence to support the identity of the person requesting the opening of the person holding account.

3. The alphanumeric identifier specified by the Member State, the Commission or person for the account, which shall be unique within the registry.

4.   Name, address, city, postcode, country, telephone number, facsimile number and e-mail address of the primary authorised representative of the account specified by the Member State, the Commission or person for that account.

5.   Name, address, city, postcode, country, telephone number, facsimile number and e-mail address of the secondary authorised representative of the account specified by the Member State, the Commission or person for that account.

6.   Name, address, city, postcode, country, telephone number, facsimile number and e-mail address of any additional authorised representatives of the account and their account access rights, specified by the Member State, the Commission or person for that account.

7.   Evidence to support the identity of the authorised representatives of the account.

ANNEX V

Core terms and conditions

Structure and effect of core terms and conditions

1.   The relationship between account holders and registry administrators.

The account holder and authorised representative's obligations

2.   The account holder and authorised representative's obligations with respect to security, usernames and passwords, and access to the registry website.

3.   The account holder and authorised representative's obligation to post data on the registry website and ensure that data posted is accurate.

4.   The account holder and authorised representative's obligation to comply with the terms of use of the registry website.

The registry administrator's obligations

5.   The registry administrator's obligation to carry out account holder's instructions.

6.   The registry administrator's obligation to log the account holder's details.

7.   The registry administrator's obligation to create, update or close the account in accordance with the provisions of the Regulation.

Process procedures

8.   The process finalisation and confirmation provisions.

Payment

9.   The terms and conditions regarding any registry fees for establishing and maintaining accounts.

Operation of the registry website

10.   Provisions regarding the right of the registry administrator to make changes to the registry website.

11.   Conditions of use of the registry website.

Warranties and indemnities

12.   Accuracy of information.

13.   Authority to initiate processes.

Modification of these core terms to reflect changes to this Regulation or changes to domestic legislation

Security and response to security breaches

Dispute resolution

14.   Provisions relating to disputes between account holders.

Liability

15.   The limit of liability for the registry administrator.

16.   The limit of liability for the account holder.

Third party rights

Agency, notices and governing law

ANNEX VI

Definitions of identification codes

Introduction

1.   This Annex prescribes the elements of the following identification codes:

        (a) unit identification code;

(b) account identification code;

(c) permit identification code;

(d) account holder identification code;

(e) installation identification code;

(f) correlation identification code;

(g) transaction identification code;

(h) reconciliation identification code;

(i) project identification code.

The version of the ISO3166 codes shall be as set out in the functional and technical specifications for data exchange standards for registry systems under the Kyoto Protocol, elaborated pursuant to Decision 24/CP.8 of the Conference of the Parties to the UNFCCC.

Display and reporting of identification codes

2.   For the purpose of displaying and reporting the identification codes set out in this Annex, each element of an identification code shall be separated by a dash '-' and without spaces. Leading zeros in numeric values shall not be displayed. The separating dash '-' shall not be stored in the elements of the identification code.

Unit identification code

3.   Table VI-1 details the elements of the unit identification code. Each Kyoto unit and allowance shall be assigned a unit identification code. Unit identification codes shall be generated by registries and shall be unique throughout the registries system.

4.   A set of units shall be transmitted as a unit block defined by the starting block identifier and the ending block identifier. Every unit of a unit block shall be identical, except for their unique identifier element. Every unique identifier element of the units of a unit block shall be consecutive. When necessary to perform a transaction, keep track of, record or otherwise characterise a unit or unit block, registries or transaction logs shall create multiple unit blocks from a single unit block. When transmitting a single unit, the starting block identifier and ending block identifier shall be equal.

5.   Multiple unit blocks shall not overlap with respect to their identifier element. Multiple unit blocks in the same message shall appear in the message in ascending order of their starting block identifier.

Table VI-1: Unit Identification Code

Element	Display Order	Identifier required for the following unit types	Data Type	Length	Range or codes
Originating Registry		1 AAU, RMU, CER, ERU	A	3	ISO3166 (2 letter code), 'EU' for the Community registry
Unit Type		2 AAU, RMU, CER, ERU	N	2	0 = not a Kyoto unit 1 = AAU 2 = RMU 3 = ERU converted from AAU 4 = ERU converted from RMU 5 = CER (not ICER or tCER) 6 = tCER 7 = ICER
Supplementary Unit Type		3 AAU, RMU, CER, ERU	N	2	Blank for Kyoto units 1 = Allowance issued for the 2008-2012 period and subsequent five-year periods 2 = Allowance issued for the 2005-2007 period 3 = Force-majeure allowance
Unit Serial Block Start		4 AAU, RMU, CER, ERU	N	15	Unique numeric values assigned by registry from 1 – 999 999 999 999 999
Unit Serial Block End		5 AAU, RMU, CER, ERU	N	16	Unique numeric values assigned by registry from 1 – 999 999 999 999 999
Original Commitment Period		6 AAU, RMU, CER, ERU	N	2	0 = 2005-2007 1 = 2008-2012 ... 99
Applicable Commitment Period		7 AAU, CER, ERU	N	2	0 = 2005-2007 1 = 2008-2012 ... 99
LULUCF Activity		8 RMU, CER, ERU	N	3	1 = Afforestation and reforestation 2 = Deforestation 3 = Forest management 4 = Cropland management 5 = Grazing land management 6 = Re-vegetation
Project Identifier		9 CER, ERU	N	7	Unique numeric value assigned for project
Track		10 ERU	N	2	1 or 2
Expiry Date		11 ICER, tCER	Date		Expiration date for ICERs or tCERs

6.   Table VI-2 lists the valid initial unit type and supplementary unit type combinations. An allowance shall have a supplementary unit type regardless of the period for which it was issued and whether it has been converted from an AAU or other Kyoto unit. An AAU or other Kyoto unit that has not been converted into an allowance shall not have a supplementary unit type. On conversion of an AAU into an allowance in accordance with the provisions of this Regulation the supplementary unit type shall be set to 1. On conversion of an allowance into an AAU in accordance with the provisions of this Regulation there shall be no supplementary unit type.

Table VI-2: Valid Initial Unit Type — Supplementary Unit Type

Initial Unit Type	Supplementary Unit Type	Description
1	[not applicable]	AAU
2	[not applicable]	RMU
3	[not applicable]	ERU converted from AAU
4	[not applicable]	ERU converted from RMU
5	[not applicable]	CER (not tCER or lCER)
6	[not applicable]	tCER
7	[not applicable]	lCER
1	1	Allowance issued for the 2008-2012 period and subsequent 5-year periods and is converted from an AAU
0	2	Allowance issued for the 2005-2007 period and not converted from an AAU or other Kyoto unit
0	3	Force-majeure allowance

Account identification code

7.   Table VI-3 details the elements of the account identification code. Each account shall be assigned an account identification code. Account identification codes shall be generated by registries and shall be unique throughout the registries system. Account identification codes of accounts that were previously closed shall not be re-used.

8.   An operator holding account identification code shall be linked to one installation. An installation shall be linked to one operator holding account identification code. Holding accounts referred to in Article 11(1) and (2) shall not have an applicable commitment period, regardless of the account type.

Table VI-3: Account Identification Code

Element	Display Order	Data Type	Length	Range or codes
Originating Registry	1	A	3	ISO3166 (2 letter code), 'CDM' for the CDM registry, 'EU' for the Community registry
Account Type	2	N	3	100 = Party holding account 120 = operator holding account 121 = person holding account The remaining account types are as set out in the functional and technical specifications for data exchange standards for registry systems under the Kyoto Protocol, elaborated pursuant to Decision 24/CP.8 of the Conference of the Parties to the UNFCCC
Account Identifier	3	N	15	Unique numeric values assigned by a registry from 1 to 999 999 999 999 999
Applicable Commitment Period	4	N	2	0 for holding accounts 0-99 for retirement and cancellation accounts

9.　　Table VI-4 details the elements of the permit identification code. Each permit shall be assigned a permit identification code. Permit identification codes shall be generated by the competent authority and shall be unique throughout the registries system.

10.　　A permit identification code shall be assigned to one operator. An operator shall be assigned at least one permit identification code. A permit identification code shall be assigned to at least one installation. An installation shall have one permit identification code at any single point in time.

Table VI-4: Permit Identification Code

Element	Display Order	Data Type	Length	Range or codes
Originating Registry	1	A	3	ISO3166 (2 letter code), 'EU' for the Community registry
Permit Identifier	2	A	50	([0-9] | [A-Z] |['-']) +

Account holder identification code

11.　　Table VI-5 details the elements of the account holder identification code. Each account holder shall be assigned an account holder identification code. Account holder identification codes shall be generated by registries and shall be unique throughout the registries system. Account holder identification codes shall not be re-used for another account holder and shall not change for an account holder throughout their existence.

Table VI-5: Account Holder Identification Code

Element	Display Order	Data Type	Length	Range or codes
Originating Registry	1	A	3	ISO3166 (2 letter code), 'EU' for the Community registry
Permit Identifier	2	A	50	([0-9] | [A-Z]) +

12. Table VI–6 details the elements of the installation identification code. Each installation shall be assigned an installation identification code. Installation identification codes shall be generated by registries and shall be unique throughout the registries system. The installation identifier shall be an integer assigned as an increasing monotone sequence, starting from 1. Installation identifiers shall not contain gaps. Therefore when generating installation identifier n, a registry shall have generated every identifier in the range 1 to n-1. An installation identification code shall not be re-used for another installation and shall not change for an installation throughout its existence.

13. An installation identification code shall be assigned to one installation. An installation shall be assigned one installation identification code.

Table VI-6: Installation Identification Code

Element	Display Order	Data Type	Length	Range or codes
Originating Registry	1	A	3	ISO3166 (2 letter code), 'EU' for the Community registry
Installation Identifier	2	A	15	Unique numeric values assigned by a registry from 1 to 999 999 999 999 999

Correlation identification code

14. Table VI-7 details the elements of the correlation identification codes. Each process under Annex VIII shall be assigned a correlation identification code. Correlation identification codes shall be generated by registries and shall be unique throughout the registries system. Correlation identification codes shall not be re-used. The re-submission of a process concerning an account or verified emissions that was previously terminated or cancelled shall be assigned a new, unique correlation identification code.

Table VI-7: Correlation Identification Code

Element	Display Order	Data Type	Length	Range or codes
Originating Registry	1	A	3	ISO3166 (2 letter code), 'EU' for the Community registry
correlation Identifier	2	A	15	Unique numeric values assigned by a registry from 1 to 999 999 999 999 999

Transaction identification code

15. Each process under Annex IX shall be assigned a transaction identification code. Transaction identification codes shall be generated

by registries and shall be unique throughout the registries system. Transaction identification codes shall not be re-used. The re-submission of a process concerning a transaction that was previously terminated or cancelled shall be assigned a new, unique transaction identification code.

16.   The elements of the transaction identification codes are set out in the functional and technical specifications for data exchange standards for registry systems under the Kyoto Protocol, elaborated pursuant to Decision 24/CP.8 of the Conference of the Parties to the UNFCCC.

Reconciliation identification code

17.   Each process under Annex X shall be assigned a reconciliation identification code. Prior to the communication link between the Community independent transaction log and UNFCCC independent transaction log being established, the Community independent transaction log shall generate the reconciliation identification code when requesting reconciliation information from registries for a specified time and date. Thereafter, registries shall receive the reconciliation identification code from the UNFCCC independent transaction log. The reconciliation identification code shall be unique throughout the registries system, and all messages exchanged through all stages of a reconciliation process for a specified time and date shall use the same reconciliation identification code.

18.   The elements of the reconciliation identification codes are set out in the functional and technical specifications for data exchange standards for registry systems under the Kyoto Protocol, elaborated pursuant to Decision 24/CP.8 of the Conference of the Parties to the UNFCCC.

Project identification code

19.   Each project shall be assigned a project identification code. Project identification codes shall be generated by the executive board of the CDM for CERs and by the relevant body of the Party or the Article 6 supervisory committee in accordance with Decision 16/CP.7 of the Conference of the Parties to the UNFCCC for ERUs and shall be unique throughout the registries system.

20.   The elements of the project identification codes are set out in the functional and technical specifications for data exchange standards for registry systems under the Kyoto Protocol, elaborated pursuant to Decision 24/CP.8 of the Conference of the Parties to the UNFCCC.

ANNEX VII

List of input codes

1. This Annex defines the codes for all elements and code support tables. The version of the ISO3166 codes shall be as set out in the functional and technical specifications for data exchange standards for registry systems under the Kyoto Protocol, elaborated pursuant to Decision 24/CP.8 of the Conference of the Parties to the UNFCCC.

EU-specific codes

2. Field Name: *Activity Type*

Field Description: Numeric code indicating the activity type of an installation

Code	Description
1	Combustion installations with a rated thermal input exceeding 20 MW
2	Mineral oil refineries
3	Coke ovens
4	Metal ore (including sulphide ore) roasting or sintering installations
5	Installations for the production of pig iron or steel (primary or secondary fusion) including continuous casting
6	Installations for the production of cement clinker in rotary kilns or lime in rotary kilns or in other furnaces
7	Installations for the manufacture of glass including glass fibre
8	Installations for the manufacture of ceramic products by firing, in particular roofing tiles, bricks, refractory bricks, tiles, stoneware or porcelain
9	Industrial plants for the production of (a) pulp from timber or other fibrous materials (b) paper and board
99	Other activity opted in pursuant to Article 24 of Directive 2003/87/E

3. Field Name: *Relationship Type*

Field Description: Numeric code indicating the type of relationship between an account and a person or operator

Code	Description
1	Account holder
2	Primary authorised representative of the account holder
3	Secondary authorised representative of the account holder
4	Additional authorised representative of the account holder
5	Authorised representative of the verifier
6	Contact person for the installation

4. Field Name: *Process Type*

Field Description: Numeric code indicating the process type of a transaction

Code	Description
01-00	Issue of AAUs and RMUs
02-00	Conversion of AAUs and RMUs to ERUs
03-00	External transfer (2008-2012 onwards)
04-00	Cancellation (2008-2012 onwards)
05-00	Retirement (2008-2012 onwards)
06-00	Cancellation and replacement of tCERs and ICERs
07-00	Carry-over of Kyoto units and allowances issued for the 2008-2012 period and subsequent five-year periods
08-00	Change of expiry date of tCERs and ICERs
10-00	Internal transfer
01-51	Allowance issue (2005-2007)
10-52	Allowance issue (2008-2012 onwards)
10-53	Allowance allocation
01-54	Force-majeure allowance issue
10-55	Correction to allowances
03-21	External transfer (2005-2007)
10-01	Allowance cancellation (2005-2007)
10-02	Allowance surrender
04-03	Retirement (2005-2007)
10-41	Cancellation and replacement

5. Field Name: *Supplementary Unit Type*

Field Description: Numeric code indicating the supplementary type of a unit

Code	Description
0	No supplementary unit type
1	Allowance issued for the 2008-2012 period and subsequent five-year periods and is converted from an AAU
2	Allowance issued for the 2005-2007 period and not converted from an AAU or other Kyoto unit
3	Force-majeure allowance

6. Field Name: *Action Code*

Field Description: Numeric code indicating the action in the account update process

Code	Description
0	Add people to the account or installation
1	Update people
2	Delete people

UNFCCC codes

7.   The UNFCCC codes are set out in the functional and technical specifications for data exchange standards for registry systems under the Kyoto Protocol, elaborated pursuant to Decision 24/CP.8 of the Conference of the Parties to the UNFCCC.

ANNEX VIII

Processes concerning accounts and verified emissions with response codes

Requirements for each process

1.   The following message sequence for processes concerning an account or verified emissions shall apply:

(a)   the authorised representative of an account shall submit a request to the registry administrator of that registry;

(b)   the registry administrator shall assign a unique correlation identification code comprising the elements set out in Annex VI to the request;

(c)   prior to the communication link between the Community independent transaction log and the UNFCCC independent transaction log being established, the registry administrator shall call the appropriate operation on the Community independent transaction log account management Web service, thereafter, the registry administrator shall call the appropriate operation on the UNFCCC independent transaction log account management Web service;

(d)   the Community independent transaction log shall validate the request by calling the appropriate validation function within the Community independent transaction log;

(e)   if the request is successfully validated and thereby accepted, the Community independent transaction log shall amend the information it holds in accordance with that request;

(f)   the Community independent transaction log shall call the 'receiveAccountOperationOutcome' operation on the account management Web service of the registry which sent the request, notifying the registry as to whether the request was successfully validated and thereby accepted, or whether the request was found to contain a discrepancy and was thereby rejected;

(g)   if the request was successfully validated and thereby accepted, the registry administrator which sent the request shall amend the information held in the registry in accordance with that

the information held in the registry in accordance with that validated request; otherwise, if the request was found to contain a discrepancy and was thereby rejected, the registry administrator which sent the request shall not amend the information held in the registry in accordance with that rejected request.

Table VIII-1: Message Sequence Diagram for Processes concerning an Account or Verified Emissions

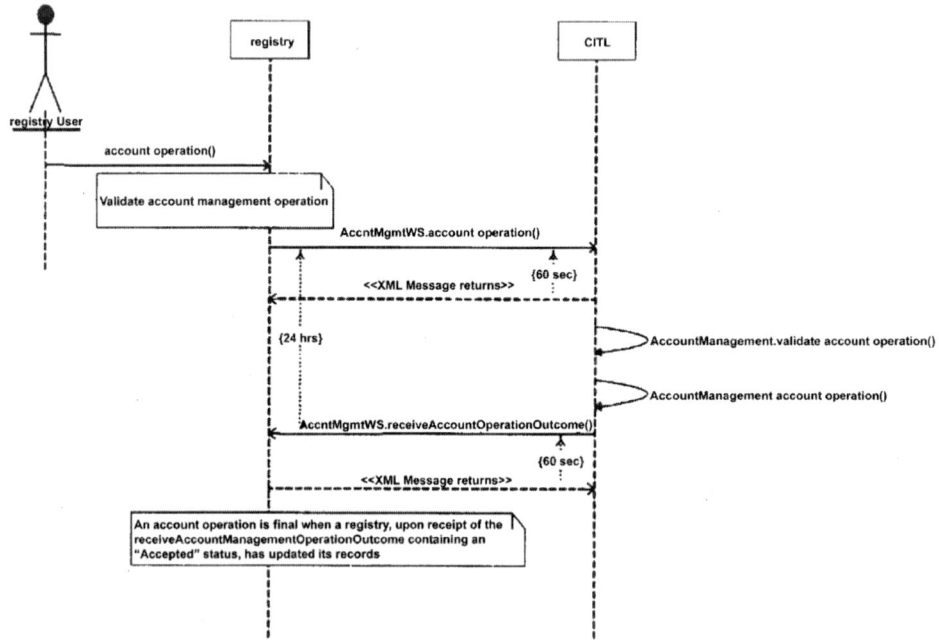

2. Prior to the communication link between the Community independent transaction log and the UNFCCC independent transaction log being established, a registry administrator sending a request should receive an acknowledgement of receipt from the Community independent transaction log within 60 seconds. Thereafter, a registry administrator sending a request should receive an acknowledgement of receipt from the UNFCCC independent transaction log within 60 seconds. A registry administrator sending a request should receive a notification of validation from the Community independent transaction log within 24 hours.

3. The status of the process during the message sequence shall be as follows:

Table VIII-2: Status Diagram for Processes concerning an Account or Verified Emissions

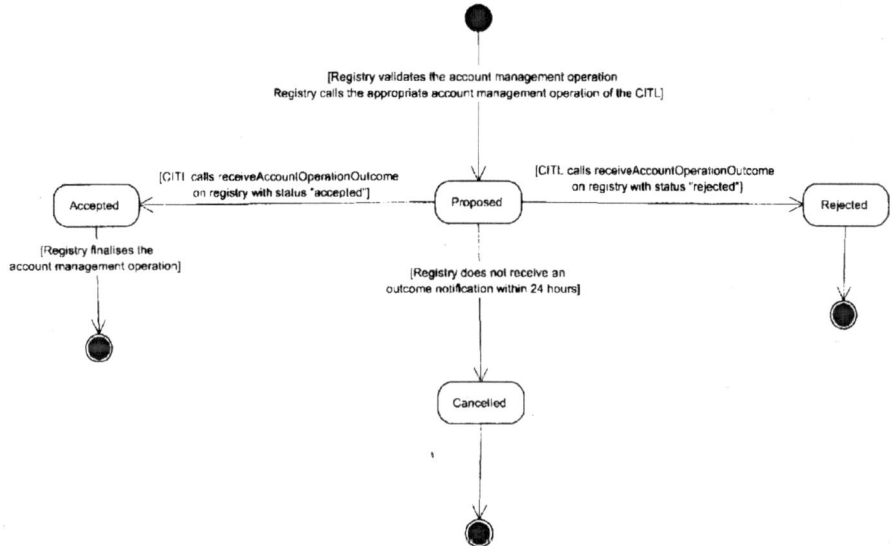

4.   The components and functions which are utilised during the message sequence are shown in table VIII-3 to VIII-18. Functions which are public shall be implemented as specified. Functions which are Privatee are for informational purposes only. The inputs of all functions have been structured to match the format and informational requirements described using web services description language, set out in the functional and technical specifications for data exchange standards for registry systems under the Kyoto Protocol, elaborated pursuant to Decision 24/CP.8 of the Conference of the Parties to the UNFCCC. An asterisk '(*)' has been used to denote the fact that an element can appear multiple times as an input.

Table VIII-3: Components and Functions for Processes concerning an Account or Verified Emissions

Component	Function	Scope
MgmtOfAccountWS	CreateAccount()	Public
	UpdateAccount()	Public
	CloseAccount()	Public
	UpdateVerifiedEmissions()	Public
	ReceiveAccountOperationOutcome()	Public
AccountManagement	ValidateAccountCreation()	Private
	CreateAccount()	Private
	ValidateAccountUpdate()	Private
	UpdateAccount()	Private
	ValidateAccountClosure()	Private
	CloseAccount()	Private
	ValidateVerifiedEmissionsUpdate()	Private
	UpdateVerifiedEmissions()	Private
DataValidation	AuthenticateMessage()	Private
	CheckVersion()	Private
	DataFormatChecks()	Private

Table VIII-4: MgmtOfAccountWS Component

Purpose	
The purpose of this component is to handle web service requests for the management of accounts and verified emissions.	
Functions exposed through Web Services	
CreateAccount()	Handles the account creation requests
UpdateAccount()	Handles the account update requests
CloseAccount()	Handles account closure requests
UpdateVerifiedEmissions()	Handles verified emissions update requests
ReceiveAccountOperationOutcome()	Receives an account operation (creation, update, ...) outcome ('accepted' or 'rejected')
Other functions	
Not applicable.	
Roles	
Community independent transaction log (for all functions) and registry (for the ReceiveAccountOperationOutcome function only)	

Table VIII-5: MgmtOfAccountWS.CreateAccount() function

Purpose	
This function receives an account creation request. The Community independent transaction log authenticates the initiating registry (Originating Registry) by calling the AuthenticateMessage() function and checks the version of the initiating registry by calling CheckVersion() function. If authentication and version checks pass, a '1' result identifier is returned without any response codes, the contents of the request are written to a file by calling the WriteToFile() function and the request is put in a queue. If authentication or version checks fail, a '0' result identifier is returned together with a single response code indicating the error cause. If the person (People) is not a natural person its name must be put in the LastName parameter. The 'PersonIdentifier' means the account holder identification code comprising the elements set out in Annex VI. The 'IdentifierInRegistry' means the alphanumeric identifier for the account as specified by the account holder pursuant to Annexes III and IV.	
Input Parameters	
From	Mandatory
To	Mandatory
CorrelationId	Mandatory
MajorVersion	Mandatory
MinorVersion	Mandatory
Account (*)	Mandatory
AccountType	Mandatory
AccountIdentifier	Mandatory
IdentifierInReg	Mandatory
CommitmentPeriod	Optional
Installation	Optional
InstallationIdentifier	Mandatory

PermitIdentifier	Mandatory
Name	Mandatory
MainActivityType	Mandatory
Country	Mandatory
PostalCode	Mandatory
City	Mandatory
Address1	Mandatory
Address2	Optional
ParentCompany	Optional
SubsidiaryCompany	Optional
EPERIdentification	Optional
Latitude	Optional
Longitude	Optional
ContactPeople (see People)	Mandatory
People (*)	Mandatory
RelationshipCode	Mandatory
PersonIdentifier	Mandatory
FirstName	Optional
LastName	Mandatory
Country	Mandatory
PostalCode	Mandatory
City	Mandatory
Address1	Mandatory
Address2	Optional
PhoneNumber1	Mandatory
PhoneNumber2	Mandatory
FaxNumber	Mandatory
Email	Mandatory
Output parameters	
Result Identifier	Mandatory
Response Code	Optional
Uses	
— AuthenticateMessage — WriteToFile — CheckVersion	
Used by	
Not applicable (called as a web service).	

Table VIII-6: MgmtOfAccountWS.UpdateAccount() function

Purpose
This function receives an account update request.
The Community independent transaction log authenticates the initiating registry (Originating Registry) by calling the AuthenticateMessage() function and checks the version of the initiating registry by calling CheckVersion() function.
If authentication and version checks pass, a '1' result identifier is returned without any response codes, the contents of the request are written to a file by calling the WriteToFile() function and the request is put in a queue.
If authentication or version checks fail, a '0' result identifier is returned together with a single response code indicating the error cause.
If the person (People) is not a natural person its name must be put in the LastName parameter.
The 'PersonIdentifier' means the account holder identification code comprising the elements set out in Annex VI.
The 'IdentifierInRegistry' means the alphanumeric identifier for the account as specified by the account holder pursuant to Annexes III and IV.

Input Parameters	
From	Mandatory
To	Mandatory
CorrelationId	Mandatory
MajorVersion	Mandatory
MinorVersion	Mandatory
Account (*)	Mandatory
AccountType	Mandatory
AccountIdentifier	Mandatory
IdentifierInReg	Optional
CommitmentPeriod	Optional
Installation	Optional
InstallationIdentifier	Optional
PermitIdentifier	Optional
Name	Optional
MainActivityType	Optional
Country	Optional
PostalCode	Optional
City	Optional
Address1	Optional
Address2	Optional
ParentCompany	Optional
SubsidiaryCompany	Optional
EPERIdentification	Optional
Latitude	Optional
Longitude	Optional
ContactPeople (see People)	Optional
People (*)	Optional
RelationshipCode	Mandatory
PersonIdentifier	Mandatory
FirstName	Mandatory
LastName	Optional
Country	Optional
PostalCode	Optional
City	Optional
Address1	Optional
Address2	Optional
PhoneNumber1	Optional
PhoneNumber2	Optional
FaxNumber	Optional
Email	Optional
Output parameters	
Result Identifier	Mandatory
Response Code	Optional
Uses	
— AuthenticateMessage	
— WriteToFile	
— CheckVersion	
Used by	
Not applicable (called as a web service).	

Table VIII-7: MgmtOfAccountWS.CloseAccount() function

Purpose
This function receives an account closure request.
The Community independent transaction log authenticates the initiating registry (Originating Registry) by calling the AuthenticateMessage() function and checks the version of the initiating registry by calling CheckVersion() function.
If authentication and version checks pass, a '1' result identifier is returned without any response codes, the contents of the request are written to a file by calling the WriteToFile() function and the request is put in a queue.
If authentication or version checks fail, a '0' result identifier is returned together with a single response code indicating the error cause.

Input Parameters	
From	Mandatory
To	Mandatory
CorrelationId	Mandatory
MajorVersion	Mandatory
MinorVersion	Mandatory
Account (*)	Mandatory
AccountIdentifier	Mandatory

Output parameters	
Result Identifier	Mandatory
Response Code	Optional

Uses
— AuthenticateMessage
— WriteToFile
— CheckVersion

Used by
Not applicable (called as a web service).

Table VIII-8: MgmtOfAccountWS.UpdateVerifiedEmissions() function

Purpose
This function receives a verified emissions update request.
The Community independent transaction log authenticates the initiating registry (Originating Registry) by calling the AuthenticateMessage() function and checks the version of the initiating registry by calling CheckVersion() function.
If authentication and version checks pass, a '1' result identifier is returned without any response codes, the contents of the request are written to a file by calling the WriteToFile() function and the request is put in a queue.
If authentication or version checks fail, a '0' result identifier is returned together with a single response code indicating the error cause.

Input Parameters	
From	Mandatory
To	Mandatory
CorrelationId	Mandatory
MajorVersion	Mandatory
MinorVersion	Mandatory
VerifiedEmissions (*)	Mandatory
Year	Mandatory
Installation (*)	Mandatory
InstallationIdentifier	Mandatory
VerifiedEmission	Mandatory

Output parameters	
Result Identifier	Mandatory
Response Code	Optional
Uses	
— AuthenticateMessage	
— WriteToFile	
— CheckVersion	
Used by	
Not applicable (called as a web service).	

Table VIII-9: MgmtOfAccountWS.ReceiveAccountOperationOutcome() function

Purpose
This function receives an account management operation outcome.
The initiating registry (Originating Registry) authenticates the UNFCCC independent transaction log (or Community independent transaction log prior to the link between the Community independent transaction log and UNFCCC independent transaction log being established) by calling the AuthenticateMessage() function and checks the version of the transaction log by calling CheckVersion() function.
If authentication and version checks pass, a '1' result identifier is returned without any response codes, the contents of the request are written to a file by calling the WriteToFile() function and the request is put in a queue.
If authentication or version checks fail, a '0' result identifier is returned together with a single response code indicating the error cause.
The response code list is populated with couples (the account or installation identifier with an adjoining response code) if the outcome is '0' for any other cause of error.

Input Parameters	
From	Mandatory
To	Mandatory
CorrelationId	Mandatory
MajorVersion	Mandatory
MinorVersion	Mandatory
Outcome	Mandatory
Response list	Optional
Output parameters	
Result Identifier	Mandatory
Response Code	Optional
Uses	
— AuthenticateMessage	
— WriteToFile	
— CheckVersion	
Used by	
Not applicable (called as a web service).	

Table VIII-10: AccountManagement Component

Purpose	
The purpose of this component is to provide the validating and update functions for the management of accounts and verified emissions.	
Functions exposed through Web Services	
Not applicable	
Other Functions	
ValidateAccountCreation()	Validates an account creation
ValidateAccountUpdate()	Validates an account update
ValidateAccountClosure()	Validates an account closure
ValidateVerifiedEmissionsUpdate()	Validates a verified emissions update
CreateAccount()	Creates accounts
UpdateAccount()	Updates accounts
CloseAccount()	Closes accounts
UpdateVerifiedEmissions()	Updates verified emissions for installations
Output parameters	
Result Identifier	Mandatory
Response Code	Optional
Uses	
— AuthenticateMessage	
— WriteToFile	
— CheckVersion	
Used by	
Not applicable (called as a web service).	

Table VIII-11: ManagementOfAccount.ValidateAccountCreation() function

Purpose	
This function validates an account creation request. If a validation test fails, the account identifier and response code are added to the response code list.	
Input Parameters	
From	Mandatory
To	Mandatory
CorrelationId	Mandatory
MajorVersion	Mandatory
MinorVersion	Mandatory
Account (*)	Mandatory
AccountType	Mandatory
AccountIdentifier	Mandatory
IdentifierInReg	Mandatory
CommitmentPeriod	Optional
Installation	Optional
InstallationIdentifier	Mandatory
PermitIdentifier	Mandatory
Name	Mandatory
MainActivityType	Mandatory
Country	Mandatory
PostalCode	Mandatory
City	Mandatory
Address1	Mandatory

Address2	Optional
ParentCompany	Optional
SubsidiaryCompany	Optional
EPERIdentification	Optional
Latitude	Optional
Longitude	Optional
ContactPeople (see People)	Mandatory
People (*)	Mandatory
RelationshipCode	Mandatory
PersonIdentifier	Mandatory
FirstName	Optional
LastName	Mandatory
Country	Mandatory
PostalCode	Mandatory
City	Mandatory
Address1	Mandatory
Address2	Optional
PhoneNumber1	Mandatory
PhoneNumber2	Optional
FaxNumber	Mandatory
Email	Optional
Output parameters	
Result Identifier	Mandatory
Response Code	Optional
Message	
Range 7101 to 7110; range 7122 to 7160.	

Table VIII-12: ManagementOfAccount.CreateAccount() function

Purpose	
This function creates accounts.	
For each account:	
Create the account and its details.	
Create all persons (People) and their details if they did not already exist and link these to the account.	
Update all information linked to persons (People) that already existed and that are linked to the account.	
Create the installation and installation details if an installation is linked to the account.	
Create all persons (People) linked to the installation (the contact person) if they did not already exist.	
Update all information linked to persons (People) that already existed and that are linked to the installation.	
Input Parameters	
From	Mandatory
To	Mandatory
CorrelationId	Mandatory
MajorVersion	Mandatory
MinorVersion	Mandatory
Account (*)	Mandatory
AccountType	Mandatory
AccountIdentifier	Mandatory
IdentifierInReg	Mandatory

CommitmentPeriod	Optional
Installation	Optional
InstallationIdentifier	Mandatory
PermitIdentifier	Mandatory
Name	Mandatory
MainActivityType	Mandatory
Country	Mandatory
PostalCode	Mandatory
City	Mandatory
Address1	Mandatory
Address2	Optional
ParentCompany	Optional
SubsidiaryCompany	Optional
EPERIdentification	Optional
Latitude	Optional
Longitude	Optional
ContactPeople (see People)	Mandatory
People (*)	Mandatory
RelationshipCode	Mandatory
PersonIdentifier	Mandatory
FirstName	Optional
LastName	Mandatory
Country	Mandatory
PostalCode	Mandatory
City	Mandatory
Address1	Mandatory
Address2	Optional
PhoneNumber1	Mandatory
PhoneNumber2	Optional
FaxNumber	Mandatory
Email	Optional
Output parameters	
Result Identifier	Mandatory
Response Code	Optional
Uses	
Not applicable	
Used by	
Not applicable (called as a web service).	

Table VIII-13: AccountManagement.ValidateAccountUpdate() function

Purpose	
This function validates an account update request. If a validation test fails, the account identifier and response code are added to the response code list.	
Input Parameters	
From	Mandatory
To	Mandatory
CorrelationId	Mandatory
MajorVersion	Mandatory
MinorVersion	Mandatory
Account (*)	Mandatory
AccountIdentifier	Mandatory
IdentifierInReg	Optional
Installation	Optional
PermitIdentifier	Optional
Name	Optional
MainActivityType	Optional
Country	Optional
PostalCode	Optional
City	Optional
Address1	Optional
Address2	Optional
ParentCompany	Optional
SubsidiaryCompany	Optional
EPERIdentification	Optional
Latitude	Optional
Longitude	Optional
ContactPeople (see People)	Optional
People (*)	Optional
Action	Mandatory
RelationshipCode	Mandatory
PersonIdentifier	Mandatory
FirstName	Optional
LastName	Optional
Country	Optional
PostalCode	Optional
City	Optional
Address1	Optional
Address2	Optional
PhoneNumber1	Optional
PhoneNumber2	Optional
FaxNumber	Optional
Email	Optional
Output parameters	
Result Identifier	Mandatory
Response Code	Optional
Uses	
Not applicable	
Used by	
Range 7102 to 7107; range 7111 to 7113; 7120; 7122; 7124; range 7126 to 7158.	

Table VIII-14: ManagementOfAccount.UpdateAccount() function

Purpose
This function updates the details of an account.
If action = 'Add':
For each link to be added:
If the person (People) existed, update its details if required.
If the person (People) did not exist, create the person (People) and link it to the account.
If action = 'Update':
For all persons (People) to update and that are linked to the account, update their details.
If action = 'Delete':
Remove the link between the person (People) and the account (for example, an additional authorised representative is removed).
If an installation is linked to the account, update the installation details if required.
Update the details of the persons (People) linked to the installation if details have been submitted (by using the same 'Add', 'Update' and 'Delete' actions).

Input Parameters	
From	Mandatory
To	Mandatory
CorrelationId	Mandatory
MajorVersion	Mandatory
MinorVersion	Mandatory
Account (*)	Mandatory
Account Type	Mandatory
AccountIdentifier	Mandatory
IdentifierInReg	Mandatory
Installation	Optional
Installation Identifier	Mandatory
PermitIdentifier	Mandatory
Name	Mandatory
MainActivityType	Mandatory
Country	Mandatory
PostalCode	Mandatory
City	Mandatory
Address1	Mandatory
Address2	Optional
ParentCompany	Optional
SubsidiaryCompany	Optional
EPERIdentification	Optional
Latitude	Optional
Longitude	Optional
ContactPeople (see People)	Mandatory
People (*)	Mandatory
RelationshipCode	Mandatory
PersonIdentifier	Mandatory
FirstName	Optional
LastName	Mandatory
Country	Mandatory
PostalCode	Mandatory
City	Mandatory
Address1	Mandatory
Address2	Optional
PhoneNumber1	Optional
PhoneNumber2	Optional

FaxNumber	Optional
Email	Optional
Output parameters	
Result Identifier	Mandatory
Uses	
Not applicable	
Used by	
Not applicable (called a web service).	

Table VIII-15: ManagementOfAccount.ValidateAccountClosure() function

Purpose	
This function validates an account closure operation. If a validation test fails, the account identifier and response code are added to the response code list.	
Input Parameters	
From	Mandatory
To	Mandatory
CorrelationId	Mandatory
MajorVersion	Mandatory
MinorVersion	Mandatory
Account (*)	Mandatory
Account Identifier	Mandatory
Output parameters	
Result Identifier	Mandatory
Response List	Optional
Messages	
7111; range 7114 to 7115; 7117; range 7153 to 7156; 7158.	

Table VIII-16: ManagementOfAccount.CloseAccount() function

Purpose	
This function closes an account or accounts by setting the end validity date of the account(s) to be closed to the current date.	
Input Parameters	
Registry	Mandatory
CorrelationId	Mandatory
MajorVersion	Mandatory
MinorVersion	Mandatory
Account (*)	Mandatory
Account Identifier	Mandatory
	Mandatory
Output parameters	
Result Identifier	Mandatory

Table VIII-17: ManagementOfAccount.ValidateVerifiedEmissionsUpdate() function

Purpose	
This function validates a verified emissions update.	
If a validation test fails, the installation identifier and response code are added to the	
Input Parameters	
From	Mandatory
To	Mandatory
CorrelationId	Mandatory
MajorVersion	Mandatory
MinorVersion	Mandatory
VerifiedEmissions (*)	Mandatory
Year	Mandatory
Installations (*)	Mandatory
InstallationIdentifier	Mandatory
VerifiedEmission	Mandatory
Output parameters	
Result Identifier	Mandatory
Response List	Optional
Messages	
Range 7118 to 7119; range 7152 to 7156; 7159.	

Table VIII-18: ManagementOfAccount.UpdateVerifiedEmissions function

Purpose	
Updates the verified emissions for the year and installation specified.	
Input Parameters	
From	Mandatory
To	Mandatory
CorrelationId	Mandatory
MajorVersion	Mandatory
VerifiedEmissions (*)	Mandatory
Year	Mandatory
Installations (*)	Mandatory
InstallationIdentifier	Mandatory
VerifiedEmission	Mandatory
Output parameters	
Result Identifier	Mandatory

Preliminary checks for each process

5.   The Community independent transaction log shall check the status of a registry for each process concerning an account or verified emissions. If the communication link between the registry and the Community independent transaction log has not been established or is temporarily suspended pursuant to Article 6(3) in respect of the requested process concerning an account or verified emissions, it shall be rejected and the response code 7005 shall be returned.

6.   Prior to the communication link between the Community independent transaction log and the UNFCCC independent transaction log being established, the Community independent transaction log shall perform

registry version and registry authentication checks, and message viability checks, on each process concerning an account or verified emissions and return the appropriate response codes if a discrepancy is detected, set out in the functional and technical specifications for data exchange standards for registry systems under the Kyoto Protocol, elaborated pursuant to Decision 24/CP.8 of the Conference of the Parties to the UNFCCC. Thereafter, each registry shall receive such response codes from the UNFCCC independent transaction log.

7.   The Community independent transaction log shall perform data integrity checks on each process concerning an account or verified emissions and return response codes in the range 7122 to 7159 if a discrepancy is detected.

Secondary checks for each process

8.   The Community independent transaction log shall perform secondary checks on each process concerning an account or verified emissions which has passed all of the preliminary checks. The secondary checks, and the adjoining response codes which are returned when a discrepancy is detected, are set out in table VIII-19.

Table VIII-19: Secondary Checks

Process description	Community independent transaction log response codes
Account creation	Range 7101 to 7110 7160
Account update	Range 7102 to 7105 Range 7107 to 7108 7111 7113 7120 7160
Account closure	7111 Range 7114 to 7115 7117
Verified emissions update	Range 7118 to 7119

ANNEX IX

Processes concerning transactions with response codes

Process types

1.   Each process concerning a transaction shall be assigned a process type consisting of an initial process type and a supplementary process type. The initial process type shall describe its category as set out in the functional and technical specifications for data exchange standards for registry systems under the Kyoto Protocol, elaborated pursuant to Decision 24/CP.8 of the Conference of the Parties to the UNFCCC. The supplementary process type shall describe its category as set out in the

provisions of this Regulation, elaborated pursuant to Directive 2003/87/ EC. The process types are set out in table IX-1.

Requirements for each process

2.   The message sequence for processes concerning a transaction, the status of the transaction and the status of the Kyoto units or allowances involved in the transaction during the message sequence, and the components and functions which are utilised during the message sequence, are set out in the functional and technical specifications for data exchange standards for registry systems under the Kyoto Protocol, elaborated pursuant to Decision 24/CP.8 of the Conference of the Parties to the UNFCCC.

Preliminary checks for each process

3.   The Community independent transaction log shall check the status of a registry for each process concerning a transaction. If the communication link between the registry and the Community independent transaction log has not been established or is temporarily suspended pursuant to Article 6(3) in respect of the requested process, it shall be rejected and the response codes 7005 or 7006 shall be returned.

4.   Prior to the communication link between the Community independent transaction log and the UNFCCC independent transaction log being established, the Community independent transaction log shall perform the following categories of preliminary checks on each process concerning a transaction:

(a)   registry version and registry authentication checks,

(b)   message viability checks,

(c)   data integrity checks,

(d)   general transaction checks, and

(e)   message sequence checks,

and return the appropriate response codes if a discrepancy is detected, as set out in the functional and technical specifications for data exchange standards for registry systems under the Kyoto Protocol, elaborated pursuant to Decision 24/CP.8 of the Conference of the Parties to the UNFCCC. Thereafter, each registry shall receive such response codes from the UNFCCC independent transaction log.

Secondary and tertiary checks for each process

5. For each process concerning a transaction which has passed all of the preliminary checks, the Community independent transaction log shall perform the following secondary checks to ascertain whether:

    (a) the Kyoto units or allowances are held in the transferring account (a discrepancy returns response code 7027);

    (b) the transferring account exists in the specified registry (a discrepancy returns response code 7021);

    (c) the acquiring account exists in the specified registry (a discrepancy returns response code 7020);

    (d) both accounts exist in the same registry for an internal transfer (a discrepancy returns response code 7022);

    (e) both accounts exist in different registries for an external transfer (a discrepancy returns response code 7023);

    (f) the transferring account is not blocked pursuant to Article 27 (a discrepancy returns response code 7025);

    (g) force majeure allowances are not being transferred (a discrepancy returns response code 7024).

6. The Community independent transaction log shall perform tertiary checks on each process concerning a transaction which has passed all of the preliminary checks. The tertiary checks, and the adjoining response codes which are returned when a discrepancy is found, are set out in table IX-1.

Table IX-1: Tertiary Checks

Process description	Process type	Community independent transaction log response Codes
Issue of AAUs and RMUs	01-00	[not applicable]
Conversion of AAUs and RMUs to ERUs	02-00	7218
External transfer (2008-2012 onwards)	03-00	Range 7301 to 7302 7304
Cancellation (2008-2012 onwards)	04-00	[not applicable]
Retirement (2008-2012 onwards)	05-00	Range 7358 to 7361
Cancellation and replacement of tCERs and ICERs	06-00	[not applicable]
Carry-over of Kyoto units and allowances issued for the 2008-2012 period and subsequent five-year periods	07-00	[not applicable]
Change of expiry date of tCERs and ICERs	08-00	[not applicable]
Internal transfer	10-00	7304 Range 7406 to 7407

Allowance issue (2005-2007)	01-51	Range 7201 to 7203 7219
Allowance issue (2008-2012 onwards)	10-52	Range 7201 to 7203 7205 7219
Allowance allocation	10-53	7202 7203 Range 7206 to 7208 7214 7216 7304 7360
Force-majeure allowance issue	01-54	7202 Range 7210 to 7211 7215 7217 7220
Correction to allowances	10-55	Range 7212 to 7213
External transfer (2005-2007)	03-21	7302 Range 7304 to 7305 Range 7406 to 7407
Allowance cancellation (2005-2007)	10-01	7212 7305
Allowance surrender	10-02	7202 7304 Range 7353 to 7356
Retirement (2005-2007)	04-03	7209 7305 7357 Range 7360 to 7362
Cancellation and replacement	10-41	(2005 to 2007) 7205 7212 7219 7360 7402 7404 Range 7406 to 7407 (2008-2012 onwards) 7202 7205 7219 7360 Range 7401 to 7402 Range 7404 to 7407

ANNEX X

Reconciliation process with response codes

Requirements for the process

1. Prior to the communication link between the Community independent transaction log and the UNFCCC independent transaction log being established, each registry shall respond to any request made by the Community independent transaction log to submit the following information for a specified time and date:

    (a)   the total number of allowances held in each account type in that registry;

(b)   the unit identification codes of any allowance held in each account type in that registry;

(c)   the transaction log and audit log history of any allowance held in each account type in that registry;

(d)   the total number of allowances held in each account in that registry;

(e)   the unit identification codes of any allowance held in each account in that registry; and

(f)   the transaction log and audit log history of each allowance held in any account in that registry.

2.   After the communication link between the Community independent transaction log and the UNFCCC independent transaction log has been established, each registry shall respond to any request made by the UNFCCC independent transaction log to submit the following information for a specified time and date:

(a)   the total number of allowances, AAUs, RMUs, ERUs, CERs (not tCERs or lCERs), lCERs and tCERs, held in each account type in that registry;

(b)   the unit identification codes of any allowance, AAU, RMU, ERU, CER (not tCER or lCER), lCER and tCER, held in each account type in that registry; and

(c)   the transaction log and audit log history of any allowance, AAU, RMU, ERU, CER (not tCER or lCER), lCER and tCER held in each account type in that registry.

3.   After the communication link between the Community independent transaction log and the UNFCCC independent transaction log has been established, each registry shall respond to any request by the UNFCCC independent transaction log made on behalf of the Community independent transaction log to submit the following information for a specified time and date:

(a)   the total number of allowances, AAUs, RMUs, ERUs, CERs (not tCERs or lCERs), lCERs and tCERs held in each account in that registry;

(b)   the unit identification codes of any allowance, AAU, RMU, ERU, CER (not tCER or lCER), lCER and tCER held in each account in that registry; and

(c)   the transaction log and audit log history of any allowance, AAU, RMU, ERU, CER (not tCER or lCER), lCER and tCER held in each account in that registry.

4.   The message sequence for the reconciliation process, the status of the reconciliation process and the status of the Kyoto units or allowances involved in the reconciliation process during the message sequence, and the components and functions which are utilised during the message sequence, are set out in the functional and technical specifications for data exchange standards for registry systems under the Kyoto Protocol, elaborated pursuant to Decision 24/CP.8 of the Conference of the Parties to the UNFCCC.

Preliminary checks for the process

5.   The Community independent transaction log shall check the status of a registry during the reconciliation process. If the communication link between the registry and the Community independent transaction log has not been established or is temporarily suspended pursuant to Article 6(3) in respect of the reconciliation process, the process shall be rejected and the response code 7005 shall be returned.

6.   Prior to the communication link between the Community independent transaction log and the UNFCCC independent transaction log being established, the Community independent transaction log shall perform registry version and registry authentication checks, message viability checks, and data integrity checks during the reconciliation process and return the appropriate response codes if a discrepancy is detected, as set out in the functional and technical specifications for data exchange standards for registry systems under the Kyoto Protocol, elaborated pursuant to Decision 24/CP.8 of the Conference of the Parties to the UNFCCC. Thereafter, each registry shall receive such response codes from the UNFCCC independent transaction log.

Secondary checks for the process

7.   The Community independent transaction log shall perform secondary checks during the reconciliation process, once the preliminary checks have been passed. The secondary checks, and the adjoining response codes which are returned when an inconsistency is detected, are set out in table X-1.

Table X-1: Secondary Checks

Process description	Community independent transaction log response codes
Reconciliation	Range: 7501 to 7524

Manual intervention

8.   If the information held in a registry has been amended in response to a process initiated but not finalised pursuant to Articles 34, 35 or 36, the Central Administrator shall instruct the registry administrator of that registry to reverse that process by amending the information held back to its original state.

If the information held in a registry has not been amended in response to a process initiated and finalised pursuant to Articles 34, 35 or 36, the Central Administrator shall instruct the registry administrator of that registry to finalise that process by amending the information held accordingly.

9.   Where the reconciliation process has identified an inconsistency, the Central Administrator shall coordinate with the registry administrator or administrators concerned in order to determine the origin of the inconsistency. The Central Administrator shall then, as necessary, either amend the information held in the Community independent transaction log or request the registry administrator or administrators concerned to make specific manual adjustments to the information held in their registry.

ANNEX XI

Administrative processes with response codes

Administrative processes

1.   The Community independent transaction log shall provide the following administrative processes:

(a)   *Transaction clean-up:* all processes under Annex IX which have been initiated but not yet terminated, completed or cancelled within 24 hours shall be cancelled. Transaction clean-up occurs on an hourly basis.

(b)   *Outstanding units:* all allowances which have not been cancelled pursuant to Articles 60 or 61 on or after 1 May 2008 and on or after 1 May in the first year of each subsequent five-year period shall be identified.

(c)   *Process status:* a registry administrator may query the status of a process under Annex IX which has been initiated by that registry administrator.

(d)   *Time synchronisation:* upon request, each registry administrator shall provide the system time of its registry in order that the consistency between the system time of a registry and the system time of the Community independent transaction log can be checked, and that the two times can be synchronised. Upon request, a registry administrator shall change the system time of its registry in order to ensure time synchronisation.

2.   After the communication link between the Community independent transaction log and the UNFCCC independent transaction log has been established, the Community independent transaction log shall only continue to provide the administrative process under paragraph 1(b).

3.   Each registry shall be capable of executing correctly the additional administrative processes provided by the UNFCCC independent transaction log, set out in the functional and technical specifications for data exchange standards for registry systems under the Kyoto Protocol, elaborated pursuant to Decision 24/CP.8 of the Conference of the Parties to the UNFCCC.

Requirements for each process

4.   The message sequence for administrative processes and the components and functions which are utilised during the message sequence are set out in the functional and technical specifications for data exchange standards for registry systems under the Kyoto Protocol, elaborated pursuant to Decision 24/CP.8 of the Conference of the Parties to the UNFCCC.

Checks for each process

5.   If, during the period referred to in paragraph 2, the Community independent transaction log detects a discrepancy under paragraph 1(a) it shall return the appropriate response codes as set out in the functional and technical specifications for data exchange standards for registry systems under the Kyoto Protocol, elaborated pursuant to Decision 24/CP.8 of the Conference of the Parties to the UNFCCC.

6.   If the Community independent transaction log detects a discrepancy under paragraph 1(b) it shall return the response code 7601.

7.   During the period referred to in paragraph 2 and where a message is received from a registry under paragraph 1(c) for a process referred to in Annex IX the Community independent transaction log shall perform the following checks:

(a)   Status of a registry: if the communication link between the registry and the Community independent transaction log has not been established or is temporarily suspended pursuant to Article 6(3) in respect of the requested process, the message shall be rejected and the response code 7005 shall be returned.

(b)   Registry version and registry authentication, message viability, and data integrity: if the Community independent transaction log detects a discrepancy, the message shall be rejected and the appropriate response codes shall be returned as set out in the functional and technical specifications for data exchange standards for registry systems under the Kyoto Protocol, elaborated pursuant to Decision 24/CP.8 of the Conference of the Parties to the

8.   During the period referred to in paragraph 2 and where a message is received from a registry under paragraph 1(d) the Community independent transaction log shall perform the following checks:

  (a)   Status of a registry: if the communication link between the registry and the Community independent transaction log has not been established or is temporarily suspended pursuant to Article 6(3) in respect of the requested process, the message shall be rejected and the response code 7005 shall be returned.

  (b)   Registry version and registry authentication, message viability, data integrity and time synchronisation: if the Community independent transaction log detects a discrepancy, the message shall be rejected and the appropriate response codes shall be returned as set out in the functional and technical specifications for data exchange standards for registry systems under the Kyoto Protocol, elaborated pursuant to Decision 24/CP.8 of the Conference of the Parties to the UNFCCC.

ANNEX XII

List of response codes for all processes

1.   The Community independent transaction log shall return response codes as part of each process, where specified in Annex VIII, Annex IX, Annex X and Annex XI. Each response code shall consist of an integer within the range 7000 to 7999. The meaning of each response code is given in table XII-1.

2.   Each registry administrator shall ensure that the meaning of each response code is maintained when displaying information in respect of a process under Annex XVI to the authorised representative who initiated that process.

Table XII-1: Community Independent Transaction Log Response Codes

Response Code	Description
7005	The current status of the initiating (or transferring) registry does not permit this process to take place.
7006	The current status of the acquiring registry does not permit this process to take place
7020	The specified account identification code does not exist in the acquiring registry.
7021	The specified account identification code does not exist in the transferring registry.
7022	The transferring account and acquiring account must be in the same registry for all transactions except external transfers.
7023	The transferring account and acquiring account must be in different registries for external transfers.
7024	Force majeure allowances cannot be transferred out of the Party holding account unless being cancelled and retired in accordance with Article 58.
7025	The transferring account is blocked for all transfers of allowances out of that account, with the exception of the surrender and cancellation and replacement processes pursuant to Articles 52, 53, 60 and 61.
7027	One or more units in the serial block are not recognised as being held by the transferring account.
7101	The account has already been created.
7102	An account must have one and only one account holder.

7103	An account must have one and only one primary authorised representative.
7104	An account must have one and only one secondary authorised representative.
7105	An installation must have one and only one contact person.
7106	The installation associated to this account is already associated to another account.
7107	The authorised representatives of the account must all be different.
7108	The alphanumeric identifier specified for the account is already specified for another account.
7109	The account type being created has not been given the correct commitment period.
7110	An operator holding account must have one and only one installation associated with that account.
7111	The specified account does not exist, and therefore it is not possible to update or close the account.
7113	It is not possible to change the account holder of a person holding account.
7114	The specified account has already been closed therefore it is not possible to close the account.
7115	The specified account still holds units and therefore it is not possible to close the account.
7117	The installation linked to the specified account is not in compliance therefore it is not possible to close the account.
7118	The specified installation does not exist and therefore it is not possible to update the verified emissions table for that installation.
7119	The specified year is a future year and therefore it is not possible to update the verified emissions table for that year.
7120	The people and their relationship with the account do not exist and therefore it is not possible to update that relationship.
7122	The correlation identifier is not in valid format or is out of range.
7124	The account alphanumeric identifier is not in valid format or is out of range
7125	The permit identifier is not in valid format or is out of range.
7126	The installation name is not in valid format or is out of range.
7127	The installation main activity is not in valid format or is out of range.
7128	The installation country is not in valid format or is out of range.
7129	The installation postal code is not in valid format or is out of range.
7130	The installation city is not in valid format or is out of range.
7131	The installation address1 is not in valid format or is out of range.
7132	The installation address2 is not in valid format or is out of range.
7133	The installation parent company is not in valid format or is out of range.
7134	The installation subsidiary company is not in valid format or is out of range.
7135	The installation EPER identification is not in valid format or is out of range.
7136	The installation latitude is not in valid format or is out of range.
7137	The installation longitude is not in valid format or is out of range.
7138	The people relationship code is not in valid format or is out of range.
7139	The person identifier is not in valid format or is out of range.
7140	The people first name is not in valid format or is out of range.
7141	The people last name is not in valid format or is out of range.
7142	The people country is not in valid format or is out of range.
7143	The people postal code is not in valid format or is out of range.
7144	The people city is not in valid format or is out of range.
7145	The people address1 is not in valid format or is out of range.
7146	The people address2 is not in valid format or is out of range.
7147	The people phonenumber1 is not in valid format or is out of range.
7148	The people phonenumber2 is not in valid format or is out of range.
7149	The people fax number is not in valid format or is out of range.
7150	The people email is not in valid format or is out of range.
7151	The people action is not in valid format or is out of range.
7152	The installation verified emission is not in valid format or is out of range.
7153	The from element is not in valid format or is out of range.
7154	The to element is not in valid format or is out of range.
7155	The major version is not in valid format or is out of range.
7156	The minor version is not in valid format or is out of range.
7157	The account type is not in valid format or is out of range.
7158	The account identifier is not in valid format or is out of range.
7159	The installation identifier is not in valid format or is out of range.
7160	It is not possible for a person holding account to have a contact person or his details, or an installation or its details (as listed in section 11.1 of Annex I to Commission Decision 2004/156/EC) associated with that account.
7201	The amount of allowances for the specified period requested to be issued exceeds the amount approved by the Commission in the national allocation plan.

7202	The acquiring account is not a Party holding account.
7203	The national allocation plan table has not been submitted to the Commission and therefore it is not possible for the issuance or allocation of allowances for the specified period to take place.
7205	The units requested to be converted into allowances must be AAUs that have been issued for a commitment period matching the commitment period for which allowances are being issued.
7206	The specified acquiring account is not the operator holding account which is associated to the specified installation.
7207	The installation does not exist in the national allocation plan table.
7208	The specified year does not exist in the national allocation plan table.
7209	The acquiring account is not the retirement account for the 2005-2007 period.
7210	Force-majeure allowances can only be issued prior to 30 June 2008.
7211	The amount of force-majeure allowances requested to be issued exceeds the amount approved by the Commission for the commitment period.
7212	The acquiring account is not the cancellation account for the 2005-2007 period.
7213	The reduction in the number of allowances exceeds the correction to the NAP as approved by the Commission.
7214	The number of allowances transferred is not strictly equal to the number foreseen in the NAP for the specified installation and specified year.
7215	The installation does not exist.
7216	The number of allowances transferred for the specified installation and specified year as foreseen in the national allocation plan has already been transferred.
7217	The specified year is not part of the period 2005-2007.
7218	The specified AAUs are allowances and therefore it is not possible to convert those AAUs into ERUs.
7219	The units requested to be issued do not have the correct allowance identification code and therefore it is not possible for the issue to take place.
7220	The units requested to be issued do not have the correct force majeure allowance identification code and therefore it is not possible for the issue to take place.
7301	Warning: approaching breach of the commitment period reserve.
7302	There is no mutual recognition agreement between the transferring registry and the acquiring registry that enables the transfer of allowances.
7304	After 30 April of the first year of the current period, allowances issued for the preceding period may only be transferred to the cancellation account or retirement account for that period.
7305	Allowances are not those issued for the 2005-2007 period.
7353	It is not possible to surrender allowances issued for the period 2005-2007 for the period 2008-2012 and subsequent five-year periods.
7354	The transferring account is not an operator holding account.
7355	It is not possible to surrender allowances issued for the current period for the previous period.
7356	Units are not eligible for surrender pursuant to Article 53.
7357	The number of allowances and force majeure allowances requested to be transferred to the retirement account is not equal to the number of allowances surrendered pursuant to Articles 52 and 54.
7358	The number of AAUs requested to be converted from allowances is not equal to the number of allowances surrendered pursuant to Article 52.
7359	The number of units requested to be transferred to the retirement account is not equal to the number of allowances surrendered pursuant to Article 52 and 53.
7360	The transferring account(s) are not Party holding account(s).
7361	Units are not eligible for retirement pursuant to Articles 58 and 59.
7262	The number of CERs requested to be transferred to the cancellation account is not equal to the number of allowances surrendered pursuant to Article 53.
7401	The number of AAUs requested to be converted into allowances is not equal to the number of allowances cancelled.
7402	Specified unit type requested to be cancelled in advance of replacement is not an allowance issued for the preceding period.
7404	The number of allowances cancelled is not equal to the number of allowances to be cancelled pursuant to Article 60(a) and 61(b).
7405	The quantity of allowances cancelled from the transferring account is not equal to the quantity of allowances transferred back to this account.
7406	The transferring account(s) must be accounts referred to in Article 11(1) and (2).
7407	The acquiring account(s) must be accounts referred to in Article 11(1) and (2).
7501	There is an inconsistency between the registry and the CITL in the operator holding account unit blocks.
7502	There is an inconsistency between the registry and the CITL in the person holding account unit blocks.
7503	Information: there are no inconsistencies between the registry and the CITL in the operator holding account unit blocks.

7504	Information: there are no inconsistencies between the registry and the CITL in the person holding account unit blocks.
7505	There is an inconsistency between the registry and the CITL in the totals of the operator holding account unit blocks.
7506	There is an inconsistency between the registry and the CITL in the totals of the person holding account unit blocks.
7507	Information: there are no inconsistencies between the registry and the CITL in the totals of the operator holding account unit blocks.
7508	Information: there are no inconsistencies between the registry and the CITL in the totals of the person holding account unit blocks.
7509	There is an inconsistency between the registry and the CITL in the Party holding account unit blocks.
7510	There is an inconsistency between the registry and the CITL in the retirement account unit blocks.
7511	There is an inconsistency between the registry and the CITL in the cancellation account unit blocks.
7512	Information: there are no inconsistencies between the registry and the CITL in the Party holding account unit blocks.
7513	Information: there are no inconsistencies between the registry and the CITL in the retirement account unit blocks.
7514	Information: there are no inconsistencies between the registry and the CITL in the cancellation account unit blocks.
7515	There is an inconsistency between the registry and the CITL in the totals of the Party holding account unit blocks.
7516	There is an inconsistency between the registry and the CITL in the totals of the retirement account unit blocks.
7517	There is an inconsistency between the registry and the CITL in the totals of the cancellation account unit blocks.
7518	Information: there are no inconsistencies between the registry and the CITL in the totals of the Party holding account unit blocks.
7519	Information: there are no inconsistencies between the registry and the CITL in the totals of the retirement account unit blocks.
7520	Information: there are no inconsistencies between the registry and the CITL in the totals of the cancellation account unit blocks.
7521	There is an inconsistency between the registry and the CITL in the replacement account unit blocks.
7522	Information: there are no inconsistencies between the registry and the CITL in the replacement account unit blocks
7523	There is an inconsistency between the registry and the CITL in the totals of the replacement account unit blocks
7524	Information: there are no inconsistencies between the registry and the CITL in the totals of the replacement account unit blocks.
7601	Reminder: the specified unit blocks of allowances issued for the previous period have not yet been cancelled pursuant to Articles 60 and 61.

ANNEX XIII

Testing procedures

1. A registry and the Community independent transaction log shall complete the following stages of testing:

    (a) Unit tests: individual components shall be tested against their specifications.

    (b) Integration tests: groups of components, comprising parts of the complete system, shall be tested against their specifications.

    (c) System tests: the system as a whole shall be tested against its specifications.

(d)   Load tests: the system shall be subjected to peaks in activity reflecting the likely demands that will be made on the system by its users.

(e)   Security testing: any security weaknesses of the system shall be identified.

2.   Individual tests for a registry carried out as part of the testing stages set out in paragraph 1 shall be conducted according to a pre-defined test plan and the results shall be documented. This documentation shall be made available to the Central Administrator on request. Any deficiencies in a registry detected during the testing stages set out in paragraph 1 shall be addressed before any testing of data exchange takes place between that registry and the Community independent transaction log.

3.   The Central Administrator shall require a registry to complete the following stages of testing:

(a)   Authentication tests: the ability of the registry to identify the Community independent transaction log, and vice versa, shall be tested.

(b)   Time synchronisation tests: the ability of the registry to establish its system time and to change its system time in order to be consistent with the system time of the Community independent transaction log and UNFCCC independent transaction log shall be tested.

(c)   Data format tests: the ability of the registry to generate messages corresponding to the appropriate process status and stage and to the appropriate format, set out in the functional and technical specifications for data exchange standards for registry systems under the Kyoto Protocol, elaborated pursuant to Decision 24/CP.8 of the Conference of the Parties to the UNFCCC, shall be tested.

(d)   Programming code and database operations tests: the ability of the registry to process messages received which correspond to the appropriate format, set out in the functional and technical specifications for data exchange standards for registry systems under the Kyoto Protocol, elaborated pursuant to Decision 24/CP.8 of the Conference of the Parties to the UNFCCC, shall be tested.

(e)   Integrated process testing: the ability of the registry to execute all processes, including all relevant statuses and stages set out in Annex VIII, Annex IX, Annex X and Annex XI, and to allow manual interventions to the database pursuant to Annex X, shall be tested.

(f)   Data logging tests: the ability of the registry to establish and maintain the records required pursuant to Article 73(2) shall be tested.

4.   The Central Administrator shall require a registry to demonstrate that the input codes referred to in Annex VII and the response codes referred to in Annex VIII, Annex IX, Annex X and Annex XI are contained within that registry's database and interpreted and used appropriately in respect of processes.

5.   The testing stages set out in paragraph 3 shall take place between the testing area of the registry and the testing area of the Community independent transaction log, established pursuant to Article 71.

6.   Individual tests carried out as part of the testing stages set out in paragraph 3 may vary to reflect the software and hardware used by a registry.

7.   Individual tests carried out as part of the testing stages set out in paragraph 3 shall be conducted according to a pre-defined test plan and the results shall be documented. This documentation shall be made available to the Central Administrator on request. Any deficiencies in a registry detected during the testing stages set out in paragraph 3 shall be addressed prior to a communication link between that registry and the Community independent transaction log being established. The registry administrator shall demonstrate that any such deficiencies have been addressed by the successful completion of the testing stages set out in paragraph 3.

ANNEX XIV

Initialisation procedures

1.   By 1 September 2004 at the latest, each Member State shall notify the Commission of the following information:

(a)   Name, address, city, postcode, country, telephone number, facsimile number and e-mail address of the registry administrator for its registry.

(b)   Address, city, postcode and country of the physical location of the registry.

(c)   The uniform resource locator (URL) and the port(s) of both the secure area and public area of the registry, and the URL and the port(s) of the testing area.

(d)   Description of the primary and backup hardware and software used by the registry, and of the hardware and software supporting the testing area pursuant to Article 68.

(e) Description of the systems and procedures for the safeguarding of all data, including the frequency with which a backup of the database is undertaken, and the systems and procedures for prompt recovery of all data and operations in the event of a disaster pursuant to Article 68.

(f) Description of the security plan of the registry established pursuant to the general security requirements under Annex XV.

(g) Description of the system and procedures of the registry in respect of change management pursuant to Article 72.

(h) Information requested by the Central Administrator to enable the distribution of digital certificates pursuant to Annex XV.

Any subsequent changes shall be promptly notified to the Commission.

2. For the period 2005-2007, each Member State shall notify the Commission of the number of force majeure allowances to be issued, subsequent to an authorisation to issue such allowances being granted by the Commission pursuant to Article 29 of Directive 2003/87/EC.

3. In advance of the 2008-2012 period and each subsequent five year period, each Member State shall notify the Commission of the following information:

(a) The total number of ERUs and CERs which operators are allowed to use for each period pursuant to Article 11a(1) of Directive 2003/87/EC.

(b) The commitment period reserve, calculated in accordance with Decision 18/CP.7 of the Conference of the Parties to the UNFCCC as 90 per cent of the Member State's assigned amount or 100 % of five times its most recently reviewed inventory, whichever is lowest. Any subsequent changes shall be promptly notified to the Commission.

National allocation plan table requirements

4. Each national allocation plan shall be submitted in accordance with the formats set out in paragraphs 5 and 7.

5. The format for submitting a national allocation plan table to the Commission is the following:

(a) Total number of allowances allocated: in a single cell the total number of allowances that are allocated for the period covered by the national allocation plan.

(b)   Total number of allowances in the new entrants reserve: in a single cell the total number of allowances that are set aside for new entrants for the period covered by the national allocation plan.

(c)   Years: in individual cells for each of the years covered in the national allocation plan from 2005 onwards in ascending order.

(d)   Installation identification code: in individual cells comprising the elements set out in Annex VI and in ascending order.

(e)   Allocated allowances: the allowances to be allocated for a specified year for a specified installation shall be entered into the cell connecting that year to that installation's identification code.

6.   The installations listed under paragraph 5(d) shall include installations unilaterally included under Article 24 of Directive 2003/87/EC and shall not include any installations temporarily excluded under Article 27 of Directive 2003/87/EC.

7.   The XML schema for submitting a national allocation plan table to the Commission is the following:

```xml
<?xml version="1.0" encoding="UTF-8"?>
<xs:schema targetNamespace="urn:KyotoProtocol:RegistrySystem:CITL:1.0:0.0" xmlns:xs="http://www.w3.org/2001/XMLSchema"
xmlns="urn:KyotoProtocol:RegistrySystem:CITL:1.0:0.0" elementFormDefault="qualified">
    <xs:simpleType name="ISO3166MemberStatesType">
        <xs:restriction base="xs:string">
            <xs:enumeration value="BE"/>
            <xs:enumeration value="GR"/>
            <xs:enumeration value="IE"/>
            <xs:enumeration value="NL"/>
            <xs:enumeration value="FI"/>
            <xs:enumeration value="DK"/>
            <xs:enumeration value="ES"/>
            <xs:enumeration value="IT"/>
            <xs:enumeration value="AT"/>
            <xs:enumeration value="SE"/>
            <xs:enumeration value="DE"/>
            <xs:enumeration value="FR"/>
            <xs:enumeration value="LU"/>
            <xs:enumeration value="PT"/>
            <xs:enumeration value="UK"/>
            <xs:enumeration value="CY"/>
            <xs:enumeration value="CZ"/>
            <xs:enumeration value="EE"/>
            <xs:enumeration value="HU"/>
            <xs:enumeration value="LV"/>
            <xs:enumeration value="LT"/>
            <xs:enumeration value="MT"/>
            <xs:enumeration value="PL"/>
            <xs:enumeration value="SK"/>
            <xs:enumeration value="SI"/>
        </xs:restriction>
    </xs:simpleType>
    <xs:simpleType name="AmountOfAllowancesType">
        <xs:restriction base="xs:integer">
            <xs:minInclusive value="1"/>
            <xs:maxInclusive value="999999999999999"/>
        </xs:restriction>
    </xs:simpleType>
    <xs:group name="YearAllocation">
        <xs:sequence>
            <xs:element name="yearInCommitmentPeriod">
                <xs:simpleType>
                    <xs:restriction base="xs:int">
```

```
                                        <xs:minInclusive value="2005"/>
                                        <xs:maxInclusive value="2058"/>
                                    </xs:restriction>
                                </xs:simpleType>
                            </xs:element>
                            <xs:element name="allocation" type="AmountOfAllowancesType"/>
                        </xs:sequence>
                    </xs:group>

                    <xs:simpleType name="ActionType">
                        <xs:annotation>
                            <xs:documentation>The action to be undertaken for the installation
A == Add the installation to the NAP
U == Update the allocations for the installation in the NAP
D == Delete the installation from the NAP
For each action, all year of a commitment period need to be given
</xs:documentation>
                        </xs:annotation>
                        <xs:restriction base="xs:string">
                            <xs:enumeration value="A"/>
                            <xs:enumeration value="U"/>
                            <xs:enumeration value="D"/>
                        </xs:restriction>
                    </xs:simpleType>
                    <xs:complexType name="InstallationType">
                        <xs:sequence>
                            <xs:element name="action" type="ActionType"/>
                            <xs:element name="installationIdentifier">
                                <xs:simpleType>
                                    <xs:restriction base="xs:integer">
                                        <xs:minInclusive value="1"/>
                                        <xs:maxInclusive value="999999999999999"/>
                                    </xs:restriction>
                                </xs:simpleType>
                            </xs:element>
                            <xs:element name="permitIdentifier">
                                <xs:simpleType>
                                    <xs:restriction base="xs:string">
                                        <xs:minLength value="1"/>
                                        <xs:maxLength value="9"/>
                                        <xs:pattern value="[A-Z0-9|'-']+"/>
                                    </xs:restriction>
                                </xs:simpleType>
                            </xs:element>
                            <xs:group ref="YearAllocation" minOccurs="3" maxOccurs="5"/>
                        </xs:sequence>
                    </xs:complexType>
                    <xs:simpleType name="CommitmentPeriodType">
                        <xs:restriction base="xs:int">
                            <xs:minInclusive value="0"/>
                            <xs:maxInclusive value="10"/>
                        </xs:restriction>
                    </xs:simpleType>
                    <xs:element name="nap">
                        <xs:complexType>
                            <xs:sequence>
                                <xs:element name="originatingRegistry" type="ISO3166MemberStatesType"/>
                                <xs:element name="commitmentPeriod" type="CommitmentPeriodType"/>

                                <xs:element name="installation" type="InstallationType" maxOccurs="unbounded">
                                    <xs:unique name="yearAllocationConstraint">
                                        <xs:selector xpath="yearInCommitmentPeriod"/>
                                        <xs:field xpath="."/>
                                    </xs:unique>
                                </xs:element>

                                <xs:element name="reserve" type="AmountOfAllowancesType"/>
                            </xs:sequence>
                        </xs:complexType>
                        <xs:unique name="installationIdentifierConstraint">
                            <xs:selector xpath="installation"/>
                            <xs:field xpath="installationIdentifier"/>
                        </xs:unique>
                    </xs:element>
                </xs:schema>
```

8.   As part of the initialisation procedures set out in the functional and technical specifications for data exchange standards for registry systems under the Kyoto Protocol, elaborated pursuant to Decision 24/CP.8 of the Conference of the Parties to the UNFCCC, the Commission shall inform the Secretariat to the UNFCCC of the account identification codes of the cancellation accounts, retirement accounts and replacement accounts of each registry.

ANNEX XV

Security standards

Communication link between the Community independent transaction log and each registry

1.   From 1 January 2005 until the communication link between the Community independent transaction log and the UNFCCC independent transaction log is established, all processes concerning allowances, verified emissions and accounts shall be completed using a communication link with the following properties:

   (a)   Secure transmission shall be achieved through the use of secure socket layer (SSL) technology with a minimum of 128 bit encryption.

   (b)   The identity of each registry shall be authenticated using digital certificates for the requests originating from the Community independent transaction log. The identity of the Community independent transaction log shall be authenticated using digital certificates for each request originating from a registry. The identity of each registry shall be authenticated using a user name and password for each request originating from a registry. The identity of the Community independent transaction log shall be authenticated using a user name and password for each request originating from the Community independent transaction log. Digital certificates shall be registered as valid by the certification authority. Secure systems shall be used to store the digital certificates and usernames and passwords, and access shall be limited. Usernames and passwords shall have a minimum length of 10 characters and shall comply with the hypertext transfer protocol (HTTP) basic authentication scheme (http://www.ietf.org/rfc/rfc2617.txt).

2.   After the communication link between the Community independent transaction log and the UNFCCC independent transaction log is established, all processes concerning allowances, verified emissions, accounts and Kyoto units shall be completed using a communication link with the properties set out in the functional and technical

specifications for data exchange standards for registry systems under the Kyoto Protocol, elaborated pursuant to Decision 24/CP.8 of the Conference of the Parties to the UNFCCC.

Communication link between the Community independent transaction log and its authorised representatives, and each registry and all authorised representatives in that registry

3. The communication link between the Community independent transaction log and its authorised representatives, and between a registry and the authorised representatives of account holders, verifiers and the registry administrator, when the authorised representatives are obtaining access from a network different from the one serving the Community independent transaction log or that registry, shall have the following properties:

    (a) Secure transmission shall be achieved through the use of secure socket layer (SSL) technology with a minimum of 128 bit encryption.

    (b) The identity of each authorised representative shall be authenticated through the use of usernames and passwords, which are registered as valid by the registry.

4. The system for issuing usernames and passwords pursuant to paragraph 3(b) to authorised representatives shall have the following properties:

    (a) At any time, each authorised representative shall have a unique username and a unique password.

    (b) The registry administrator shall maintain a list of all authorised representatives who have been granted access to the registry and their access rights within that registry.

    (c) The number of authorised representatives of the Central Administrator and registry administrator shall be kept to a minimum and access rights shall be allocated solely on the basis of enabling administrative tasks to be performed.

    (d) Any default vendor passwords with Central Administrator or registry administrator access rights shall be changed immediately after installation of the software and hardware for the Community independent transaction log or registry

    (e) Authorised representatives shall be required to change any temporary passwords they have been given upon accessing the secure area of the Community independent transaction log or registry for the first time, and thereafter shall be required to change their passwords every two months at a minimum.

(f)   The password management system shall maintain a record of previous passwords for an authorised representative and prevent re-use of the previous ten passwords for that authorised representative. Passwords shall have a minimum length of 8 characters and be a mix of numeric and alphabetical characters.

(g)   Passwords shall not be displayed on a computer screen when being entered by an authorised representative, and password files shall not be directly visible to an authorised representative of the Central Administrator or registry administrator.

Communication link between the Community independent transaction log and the general public, and each registry and the general public

5.   The public area of the website of the Community independent transaction log and the public website of a registry shall not require authentication of its users representing the general public.

6.   The public area of the Community independent transaction log website and the public area of a registry website shall not permit its users representing the general public to directly access data from the database of the Community independent transaction log or the database of that registry. Data which is publicly accessible in accordance with Annex XVI shall be accessed via a separate database.

General security requirements for the Community independent transaction log and each registry

7.   The following general security requirements shall apply to the Community independent transaction log and each registry:

(a)   A firewall shall protect the Community independent transaction log and each registry from the Internet, and shall be configured as strictly as is possible to limit traffic to and from the Internet.

(b)   The Community independent transaction log and each registry shall run regular virus scans on all nodes, workstations and servers within their networks. Anti-virus software shall be updated regularly.

(c)   The Community independent transaction log and each registry shall ensure that all node, workstation and server software is correctly configured and routinely patched as security and functional updates are released.

(d)   When necessary, the Community independent transaction log and each registry shall apply additional security requirements to ensure that the registry system is able to respond to new security threats.

ANNEX XVI

Reporting requirements of each registry administrator and the Central Administrator

Publicly available information from each registry and the Community independent transaction log

1.	The Central Administrator shall display and update the information in paragraphs 2 to 4 in respect of the registry system on the public area of the Community independent transaction log's web site, in accordance with the specified timing, and each registry administrator shall display and update this information in respect of its registry on the public area of that registry's web site, in accordance with the specified timing.

2.	The following information for each account shall be displayed in the week after the account has been created in a registry, and shall be updated on a weekly basis:

	(a)	account holder name: the holder of the account (person, operator, Commission, Member State);

	(b)	alphanumeric identifier: the identifier specified by the account holder assigned to each account;

	(c)	name, address, city, postcode, country, telephone number, facsimile number and email address of the primary and secondary authorised representatives of the account specified by the account holder for that account.

3.	The following additional information for each operator holding account shall be displayed in the week after the account has been created in the registry, and shall be updated on a weekly basis:

	(a)	points 1 to 4.1, 4.4 to 5.5 and point 7 (activity 1) of the information identifying the installation related to the operator holding account as listed in section 11.1 of Annex I to Commission Decision 2004/156/EC;

	(b)	permit identification code: the code assigned to the installation related to the operator holding account comprising the elements set out in Annex VI;

	(c)	installation identification code: the code assigned to the installation related to the operator holding account comprising the elements set out in Annex VI;

	(d)	allowances and any force majeure allowances allocated to the installation related to the operator holding account, which is part of the national allocation plan table or is a new entrant, under Article 11 of Directive 2003/87/EC.

4. The following additional information for each operator holding account for the years 2005 onwards shall be displayed in accordance with the following specified dates:

(a) verified emissions figure for the installation related to the operator holding account for year X shall be displayed from 15 May onwards of year (X+1);

(b) allowances surrendered pursuant to Articles 52, 53 and 54, by unit identification code, for year X shall be displayed from 15 May onwards of year (X+1);

(c) a symbol identifying whether the installation related to the operator holding account is or is not in breach of its obligation under Article 6(2)(e) of Directive 2003/87/EC for year X shall be displayed from 15 May onwards of year (X+1).

Publicly available information from each registry

5. Each registry administrator shall display and update the information in paragraphs 6 to 10 in respect of its registry on the public area of that registry's web site, in accordance with the specified timing.

6. The following information for each project identifier for a project activity implemented pursuant to Article 6 of the Kyoto Protocol against which the Member State has issued ERUs shall be displayed in the week after the issue has taken place:

(a) project name: a unique name for the project;

(b) project location: the Member State and town or region in which the project is located;

(c) years of ERU issuance: the years in which ERUs have been issued as a result of the project activity implemented pursuant to Article 6 of the Kyoto Protocol;

(d) reports: downloadable electronic versions of all publicly available documentation relating to the project, including proposals, monitoring, verification and issuance of ERUs, where relevant, subject to the confidentiality provisions in Decision -/CMP.1 [Article 6] of the Conference of the Parties to the UNFCCC serving as the meeting of the Parties to the Kyoto Protocol.

7. The following holding and transaction information, by unit identification code comprising the elements set out in Annex VI, relevant for that registry for the years 2005 onwards shall be displayed in accordance with the following specified dates:

(a)   the total quantity of ERUs, CERs, AAUs and RMUs held in each account (person holding, operator holding, Partyholding, cancellation, replacement or retirement) on 1 January of year X shall be displayed from 15 January onwards of year (X+5);

(b)   the total quantity of AAUs issued in year X on the basis of the assigned amount pursuant to Article 7 of Decision No 280/2004/ EC shall be displayed from 15 January onwards of year (X+1);

(c)   the total quantity of ERUs issued in year X on the basis of project activity implemented pursuant to Article 6 of the Kyoto Protocol shall be displayed from 15 January onwards of year (X+1);

(d)   the total quantity of ERUs, CERs, AAUs and RMUs acquired from other registries in year X and the identity of the transferring accounts and registries shall be displayed from 15 January onwards of year (X+5);

(e)   the total quantity of RMUs issued in year X on the basis of each activity under Article 3, paragraphs 3 and 4 of the Kyoto Protocol shall be displayed from 15 January onwards of year (X+1);

(f)   the total quantity of ERUs, CERs, AAUs and RMUs transferred to other registries in year X and the identity of the acquiring accounts and registries shall be displayed from 15 January onwards of year (X+5);

(g)   the total quantity of ERUs, CERs, AAUs and RMUs cancelled in year X on the basis of activities under Article 3, paragraphs 3 and 4 of the Kyoto Protocol shall be displayed from 15 January onwards of year (X+1);

(h)   the total quantity of ERUs, CERs, AAUs and RMUs cancelled in year X following determination by the compliance committee under the Kyoto Protocol that the Member State is not in compliance with its commitment under Article 3, paragraph 1 of the Kyoto Protocol shall be displayed from 15 January onwards of year (X+1);

(i)   the total quantity of other ERUs, CERs, AAUs and RMUs, or allowances, cancelled in year X and the reference to the Article pursuant to which these Kyoto units or allowances were cancelled under this Regulation shall be displayed from 15 January onwards of year (X+1);

(j)   the total quantity of ERUs, CERs, AAUs, RMUs and allowances retired in year X shall be displayed from 15 January onwards of year (X+1);

(k)   the total quantity of ERUs, CERs, AAUs carried over in year X from the previous commitment period shall be displayed from 15 January onwards of year (X+1);

(l)   the total quantity of allowances from the previous commitment period cancelled and replaced in year X shall be displayed from 15 May onwards of year X;

(m)   current holdings of ERUs, CERs, AAUs and RMUs in each account (person holding, operator holding, Party holding, cancellation or retirement) on 31 December of year X shall be displayed from 15 January onwards of year (X+5).

8.   The list of persons authorised by the Member State to hold ERUs, CERs, AAUs and/or RMUs under its responsibility shall be displayed in the week after such authorisations have been given, and shall be updated on a weekly basis.

9.   The total number of CERs and ERUs which operators are allowed to use for each period pursuant to Article 11a(1) of Directive 2003/87/EC shall be displayed in accordance with Article 30(3) of Directive 2003/87/EC.

10. The commitment period reserve, calculated in accordance with Decision 18/ CP.7 of the Conference of the Parties to the UNFCCC as 90 % of the Member State's assigned amount or 100 % of five times its most recently reviewed inventory, whichever is lowest, and the number of Kyoto units by which the Member State is exceeding, and therefore in compliance with, its commitment period reserve shall be displayed on request.

Publicly available information from the Community independent transaction log

11.   The Central Administrator shall display and update the information in paragraph 12 in respect of the registry system on the public area of the Community independent transaction log's web site, in accordance with the specified timing.

12.   The following information for each completed transaction relevant for the registries system for year X shall be displayed from 15 January onwards of year (X+5):

(a)   account identification code of the transferring account: the code assigned to the account comprising the elements set out in Annex VI;

(b)   account identification code of the acquiring account: the code assigned to the account comprising the elements set out in Annex VI;

(c)   account holder name of the transferring account: the holder of the account (person, operator, Commission, Member State);

(d)   account holder name of the acquiring account: the holder of the account (person, operator, Commission, Member State);

(e)   allowances or Kyoto units involved in the transaction by unit identification code comprising the elements set out in Annex VI;

(f)   transaction identification code: the code assigned to the transaction comprising the elements set out in Annex VI;

(g)   date and time at which the transaction was completed (in Greenwich Mean Time);

(h)   process type: the categorisation of a process comprising the elements set out in Annex VII.

Information from each registry to be made available to account holders

13.   Each registry administrator shall display and update the information in paragraph 14 in respect of its registry on the secure area of that registry's web site, in accordance with the specified timing.

14.   The following elements for each account, by unit identification code comprising the elements set out in Annex VI, shall be displayed on the account holder's request to that account holder only:

(a)   current holdings of allowances or Kyoto units;

(b)   list of proposed transactions initiated by that account holder, detailing for each proposed transaction the elements in paragraph 12(a) to (f), the date and time at which the transaction was proposed (in Greenwich Mean Time), the current status of that proposed transaction and any response codes returned consequent to the checks made pursuant to Annex IX;

(c)   list of allowances or Kyoto units acquired by that account as a result of completed transactions, detailing for each transaction the elements in paragraph 12(a) to (g);

(d)   list of allowances or Kyoto units transferred out of that account as a result of completed transactions, detailing for each transaction the elements in paragraph 12(a) to (g).

ANNEX 4

THE 2003 COMMISSION GUIDANCECOMMISSION OF THE EUROPEAN COMMUNITIES

Brussels, 7.1.2004
COM(2003) 830 final

COMMUNICATION FROM THE COMMISSION

on guidance to assist Member States in the implementation of the criteria listed in Annex III to Directive 2003/87/EC stablishing a scheme for greenhouse gas emission allowance trading within the Community and amending Council Directive 96/61/EC, and on the circumstances under which force majeure is demonstrated

INTRODUCTION

1.   Directive 2003/87/EC[1] provides for the establishment of a community-wide greenhouse gas emission allowance trading scheme as of 2005. Pursuant to Article 9 of the Directive, each Member State periodically has to develop a national allocation plan. These plans have to be based on objective and transparent criteria, including those listed in Annex III to the Directive. The first national allocation plans have to be published and notified to the Commission and the other Member States by 31 March 2004. For those Member States joining the Union as of 1 May 2004, the obligation to publish and notify the national allocation plan arises only with the date of accession. The Commission encourages these future Member States to publish and notify national allocation plans also by 31 March 2004.

2.   Article 9 mandates the Commission to develop guidance on the implementation of the criteria listed in Annex III by 31 December 2003. Article 29 mandates the Commission to develop guidance to describe the circumstances under which force majeure is demonstrated by the same date. The purpose of this guidance document is three-fold:

    –   First, to assist Member States in drawing up their national allocation plans, by indicating the scope of interpretation

---

[1] OJ L 275, 25.10.2003, p. 32.

of the Annex III criteria that the Commission deems acceptable;

–   Second, to support the Commission assessment of notified national allocation plans, pursuant to Article 9(3);

–   Third, to describe the circumstances under which *force majeure* is demonstrated.

3.   The Directive is a key element of the Community's climate change policy and its objective is to promote reductions of greenhouse gas emissions in a cost-effective and economically efficient manner. It is therefore important to ensure that the emissions trading scheme has a positive environmental outcome. The national allocation plans are the means to achieve this goal. This fact is reflected in the guidance developed in this document.

4.   The Commission will monitor the application of this guidance, and amend it as and when it deems necessary, in particular following any amendments to Annex III pursuant to Articles 22 and/or 30(2)(c) of the Directive.

GUIDANCE ON THE IMPLEMENTATION OF THE ANNEX III CRITERIA

5.   Annex III to Directive 2003/87/EC contains 11 criteria relating to the national allocation plans. The relationships between these criteria can be revealed by categorising them in various ways.

Table 1: Categorisation of the criteria

	Mandatory (M)/ Optional (O)	Total level	Activity/ Sector	Installation level
(1) Kyoto commitments	(M)/(O)	+		
(2) Assessments of emissions development	(M)	+		
(3) Potential to reduce emissions	(M)/(O)	+	+	
(4) Consistency with other legislation	(M)/(O)	+	+	
(5) Non-discrimination between companies or sectors	(M)	+	+	+
(6) New entrants	(O)			+
(7) Early action	(O)			+
(8) Clean technology	(O)			+
(9) Involvement of the public	(M)			
(10) List of installations	(M)			+
(11) Competition from outside the Union	(O)		+	

6. One way of categorising the criteria is on the basis of whether their implementation is mandatory or optional. A Member State has an obligation to apply all elements of criteria (2), (5), (9) and (10), and some elements of the criteria (1), (3) and (4). It can, therefore, choose whether it wants to take specific action with respect to some elements of criteria (1), (3) and (4), and the criteria (6), (7), (8) and (11). The Commission will not reject a plan if all mandatory criteria and mandatory elements of criteria are applied in a correct manner. The Commission will not reject a plan if optional criteria or optional elements of criteria are not applied. However, if these optional criteria or optional elements of criteria or additional transparent and objective criteria are applied, the Commission will assess their application. In all cases, the Commission does require information from a Member State with respect to criteria (7) and (8), even if this is only to state that a criterion has not been applied. In respect of criterion (6) a Member State must state the manner in which new entrants will be able to begin participating in the Community scheme in that Member State.

7. A second way of categorising the criteria is to distinguish between them depending on whether they are applicable to allowance allocation at the level of all covered installations, at activity or sector level, or at installation level. The Commission's interpretation is presented in Table 1.

8. The attached common format reflects the fact that criteria apply at different levels, but also that they deal with different aspects, such as technical aspects and Community legislation or policy. For the sake of clarity and in order to facilitate its use by Member States, a recommended common format for establishing and notifying the national allocation plans is attached. The common format will further assist a Member State in drawing up the plan, and, in addition, it will significantly facilitate Member States' scrutiny of each other's plans and increase the accessibility of the plans to stakeholders.

2.1 Guidance on individual criteria

9. In the following, the Commission sets out guidance on the implementation of the individual criteria. The criteria are treated individually and in the order in which they are listed in Annex III to the Directive. Cross-references are made in order to highlight relationships between different criteria. The guidance contains an introductory and an analytical section.

2.1.1. Criterion (1) – Kyoto commitments

The total quantity of allowances to be allocated for the relevant period shall be consistent with the Member State's obligation to limit its emissions pursuant to Decision 2002/358/EC and the Kyoto Protocol, taking into account, on the one hand, the proportion of overall emissions that these allowances represent in comparison with emissions from sources not covered by this Directive and, on the other hand, national energy policies, and should be consistent with the national climate change programme. The total quantity of allowances to be allocated shall not be more than is likely to be needed for the strict application of the criteria of this Annex. Prior to 2008, the quantity shall be consistent with a path towards achieving or over-achieving each Member State's target under Decision 2002/358/EC and the Kyoto Protocol.

2.1.1.1. Introduction

10. Criterion (1) makes the link between the total quantity of allowances and the Member State's individual target either under Council Decision 2002/358/EC[2] on the joint fulfilment of commitments under the Kyoto Protocol, or under the Kyoto Protocol itself. For new Member States not referred to in the Decision, their respective targets under the Kyoto Protocol is the reference point under this criterion. While the commitments established for each Member State must be met, the criterion enables a Member State to go *beyond* the "Kyoto" target. Distributing the effort to meet these targets is a "zero-sum" exercise, whereby the same result must be achieved however the effort is distributed between covered and non-covered installations and activities, as well as between covered installations.

11. Within the scope of the climate change commitment of each Member State, a Member State applying effective policies and measures to sources outside the trading scheme will necessarily be in a position to allocate more allowances to covered installations. National energy policies may also lead to adjustments of the relative contributions to the climate change commitment. If a Member State has committed itself to gradually phase-out nuclear installations on its territory, measures will have to be taken to provide the required levels of electricity. A nuclear phase-out might lead to an increase in greenhouse gas emissions, but would not justify that a Member State does not fulfil its obligations under Decision 2002/358/EC.

---

[2] OJ L 130, 15.5.2002, p. 1.

12.   The concept of the "path" reflects the fact that, before the period 2008 to 2012, Member States do not have quantitative targets but are instead required under Article 3(2) of the Kyoto Protocol to make demonstrable progress by 2005 towards meeting their quantitative commitments for 2008 to 2012. Allocations for the period 2005 to 2007 have to be mindful of the targets that will apply from 2008 to 2012. Consequently, it is understood that Member States should be making progress towards their commitments for 2008 to 2012 already in the first trading period of 2005 to 2007. The path is intended to be a trend line, not necessarily a straight one, but one that is leading towards or goes beyond the reductions and limitations called for by the Kyoto Protocol and Decision 2002/358/EC.

2.1.1.2.Analysis

13.   Criterion (1) is largely of a mandatory nature and has to be applied in determining the total quantity of allowances.

14.   While the Directive covers part of a Member State's greenhouse gas emissions, the Kyoto target applies to the total greenhouse gas emissions of a Member State. Hence, a Member State has to decide in the plan what contribution should be made by covered installations to reaching or going beyond the overall commitment in the period 2008 to 2012 and what path it will follow in the period 2005 to 2007.

15.   A Member State has to demonstrate how the chosen total quantity of allowances is consistent with reaching or over-achieving the Kyoto target, taking into account, on the one hand, the proportion of overall emissions that these allowances represent in comparison with emissions from sources not covered by this Directive and, on the other hand, national energy policies. A Member State has to present the chosen path towards reaching or over-achieving the target under Decision 2002/358/EC and the Kyoto Protocol and explain how consistency is ensured between the intended allocation and the path.

16.   In deciding the total quantity, the proportion of overall emissions of covered installations in relation to total emissions is a first element to be taken into account. A Member State should use the most recent data available to determine the proportion. In case a Member State deviates substantially from this proportion, it should give reasons for such deviations. Such reasons may include, *inter alia*, expected structural change in the economy and

national energy policy. Consistency with national energy policy may be a reason for an increase or a decrease in the proportion. A Member State phasing-out nuclear installations over the covered period may increase the proportion, if replacement is not expected to be through carbon-free alternatives. A Member State intending to increase the share of renewable energy or combined heat and power production or other forms of low-carbon or carbon-free power and heat production should decrease the proportion. The Commission recalls that under the provisions of Directive 2001/77/EC[3] on the promotion of electricity produced from renewable energy sources in the internal electricity market all Member States, including the future Member States, have committed themselves to increase the share of electricity from renewable energy sources.

17. The quantity of allowances potentially available for installations covered by the trading scheme needs to be consistent with the forecasted increases or decreases in non-covered activities. Therefore, a Member State should include clear, realistic and substantiated projections of the effectiveness of the policies aimed at non-covered activities in the national allocation plan. Furthermore, a Member State should introduce additional policies and measures to control emissions of non-covered activities in order for all relevant sectors to contribute to the achievement of the target under Decision 2002/358/EC and the Kyoto Protocol.

18. The Commission understands *"likely to be needed"* as forward-looking and linked to the projected emissions of covered installations as a whole, given that this criterion refers to the total quantity of allowances to be allocated. The Commission understands the "strict application of the criteria in this annex" to comprise the criteria with a mandatory character or containing mandatory elements – i.e. criteria (1), (2), (3), (4), and (5)[4]. In order to satisfy this requirement and fulfil all mandatory criteria and elements, a Member State should not allocate more than is needed, or warranted, by the most constraining of these criteria. It follows that any application of the optional elements of Annex III may not lead to an increase in the total quantity of allowances.

19. The chosen proportion, taking into account criteria (1), (2), (3), (4) and (5), should be multiplied by the annual average emissions allowed under Decision 2002/358/EC and, for new Member States, the Kyoto Protocol in the period 2008 to 2012. This figure could

---

[3] OJ L 283, 27.10.2001, p. 33.
[4] Criteria (9) and (10) do not relate to the determination of allocated quantities and are therefore not relevant in this context.

be scaled downwards by an appropriate factor if the Member State intends to go beyond the Kyoto target in 2008 to 2012. In order to determine the total quantity for the period 2005 to 2007 the Member State should scale this amount to the path chosen and multiply the figure by three.

20.   As a Party to the Kyoto Protocol, a Member State may use the mechanisms under Articles 6, 12, and 17 (Joint Implementation, Clean Development Mechanism, and International Emission Trading) to contribute to compliance with its commitments under the Protocol in the period 2008 to 2012. If a Member State intends to use these mechanisms it may adapt annual average emissions allowed under Decision 2002/358/EC and the Kyoto Protocol in the period 2008 to 2012. In the national allocation plan, a Member State must substantiate any such intentions to use the Kyoto mechanisms. The Commission will base its assessment notably on the state of advancement of relevant legislation or implementing provisions at the national level.

A Member State has to determine the total quantity of allowances based on the proportion of overall emissions of covered installations in relation to total emissions. A Member State should use the most recent data available to determine the proportion. In case a Member State deviates substantially from the current proportion, it should give reasons for such deviations.

A Member State must substantiate the intention to use the Kyoto mechanisms.

2.1.2. Criterion (2) – Assessments of emissions development

The total quantity of allowances to be allocated shall be consistent with assessments of actual and projected progress towards fulfilling the Member States' contributions to the Community's commitments made pursuant to Decision 93/389/EEC.

2.1.2.1.   Introduction

21.   Pursuant to Decision 93/389/EEC establishing a monitoring mechanism of Community carbon dioxide (CO_2) and other greenhouse gas emissions[5], the Commission undertakes an annual assessment of actual and projected emissions of Member States,

---

[5] OJ L 167, 9.7.1993, p.31. Decision as amended by Decision 1999/296/EC (OJ L 117, 5.5.1999, p. 35).

in total and by sector and by gas. These assessments are prepared in close cooperation with Member States. Criterion (2) is intended to ensure that the total allocation is consistent with pre-existing, publicly available and objective assessments of actual and projected emissions. The relevant reports that summarise these assessments are COM(2000)749, COM(2001)708, COM(2002)702 and COM(2003)735. The 2000 and 2001 reports cover only the existing Member States, and so are not relevant to new Member States. The 2002 and 2003 reports also cover the new Member States.

22.   Decision 93/389/EEC will be repealed and replaced in early 2004 by Decision 2004/xx/EC on the monitoring of greenhouse gas emissions and the implementation of the Kyoto Protocol[6].

2.1.2.2.   Analysis

23.   Criterion (2) is of a mandatory nature and has to be applied in determining the total quantity of allowances.

24.   The Commission conducts assessments under Decision 93/389/ EEC, in co-operation with Member States. These assessments cover recent developments of actual emissions of Member States and projected emissions during the period 2008 to 2012, in total and per sector and gas.

25.   Consistency with assessments pursuant to Decision 93/389/EEC will be deemed as ensured if the total quantity of allowances to be allocated to covered installations is not more than would be necessary taking into account actual emissions and projected emissions contained in those assessments. Consistency would not be ensured if a Member State intended to allocate a total quantity of allowances in excess of actual or projected emissions of covered installations as reported in the assessment for the relevant period.

Consistency with assessments pursuant to Decision 93/389/EEC will be deemed as ensured if the total quantity of allowances to be allocated to covered installations is not more than actual emissions and projected emissions contained in those assessments.

---

[6] This Decision, based on the Commission proposal COM(2003)51, is the subject of a first reading agreement based on the amendments adopted by the European Parliament on 21 October 2003, and is expected to enter into force early in 2004.

2.1.3. *Criterion (3) – Potential to reduce emissions*

> *Quantities of allowances to be allocated shall be consistent with the potential, including the technological potential, of activities covered by this scheme to reduce emissions. Member States may base their distribution of allowances on average emissions of greenhouse gases by product in each activity and achievable progress in each activity.*

2.1.3.1.   Introduction

26.   No definition or further determination of the term "potential" has been established, and potential should therefore not be limited to technological potential but may include, *inter alia*, economic potential. As the technical options available to reduce emissions by a tonne of carbon dioxide as well as the costs of doing so vary between activities, an allocation may be made to reflect that in some cases a reduction can be achieved at lower cost, and in other cases an equivalent reduction may be more costly. The implication is that more might be asked of activities that can make cheaper reductions, and less might be asked of activities whose reductions are expensive.

27.   The second sentence of the criterion makes explicit the possibility for Member States to use benchmarks by product in each activity and achievable progress in each activity. Under a benchmarking approach, an average of emissions per unit of output would be established, and allocations made on the basis of historic, current or expected output quantities. An installation that had lower emissions per unit of output would be given more allowances in relation to current emissions than installations whose emissions were higher per unit of output.

28.   Criterion (3) refers to the product in each activity, without defining product. It is implicitly recognised that a given activity could cover several products, so that each activity does not have to be treated as a whole. For example, achievable progress with coal-fired electricity generation is an acceptable basis for the determination of benchmarks. What is achievable by different coal-fired technologies is more limited than what may be achievable in the case of fuel-switching from coal to natural gas. However, the incentive for fuel switching to less carbon intensive fuels would not be affected.

29.   Pursuant to Article 30(2) of the Directive, the Commission shall consider in a future review the *practicality of developing Community wide benchmarks as a basis for allocation*. The Commission notes that the legislators do not consider the application of Community-wide benchmarks to be practicable for the first national allocation plan.

2.1.3.2.   Analysis

30.   Criterion (3) is mandatory in part. It has to be applied in determining the total quantity of allowances and it may be applied in determining the quantity per activity.

31.   A Member State should determine the total quantity of allowances resulting from the application of this criterion by comparing the potential of activities covered by the scheme to reduce emissions with the potential of activities not covered. The criterion will be deemed as fulfilled if the allocation reflects the relative differences in the potential between the total covered and total non-covered activities.

32.   A Member State may apply the criterion to determine separate quantities per activity. It should compare the potentials of individual activities covered by the scheme to reduce emissions against each other. If a Member State applies the criterion to determine separate quantities per activity, the criterion will be deemed as fulfilled if the allocation reflects the relative differences in the potential amongst individual covered activities.

33.   A Member State may use the relevant average emissions of covered greenhouse gases by generic product type and achievable progress in each activity to determine the quantity per activity. If a Member State chooses to do so, it should determine the actual average emissions by product using national data, and assess the average emissions by product that could be attained in the relevant period taking into account achievable progress. A Member State should indicate the applied average in the national allocation plan and justify why it considers the chosen average to be an appropriate estimate to incorporate achievable progress. The quantity of allowances per activity should be based on expected output per activity over the relevant period. A Member State should indicate the forecast used and justify why it considers the chosen forecast as the most likely development. In doing so it should also take into account recent output developments in the relevant activities.

34. In contrast to criterion (7), in which benchmarks may be applied to determine the quantity of allowances by installation, under this criterion the benchmark would be applied to determine the quantity of allowances by activity.

35. A distinction is made between technological and other potential of activities to reduce emissions. The realisation of the technological potential to reduce emissions within a trading period is limited by factors such as timing, economic viability and legal provisions.

36. Member States should consider that some measures can be implemented and will have an effect on emissions in the short term, while others may have longer lead-times and depend on investment cycles. Considering the potential of measures with a lead-time extending beyond the duration of a trading period will create an incentive for operators to act early.

37. The economic potential of activities to reduce CO_2 emissions should be based on an assessment of abatement costs per tonne of CO_2equivalent, and not on the economic viability of individual companies or installations belonging to the activity or activities concerned.

38. A Member State may make use of best available techniques reference documents (BREFs) when assessing the potential of activities. A "best" available technique is defined as a technique, which is "most effective in achieving a high general level of protection of the environment as a whole". Therefore, there is not necessarily full coherence between the use of a best available technique and the performance of an installation in terms of covered emissions.

39. In the national allocation plan, a Member State should describe the methodology it has used to assess the potential to reduce emissions. It should preferably base the assessment of the potential on a study made for the purpose of the national allocation plan. In case circumstances and timing do not allow for such a study in the process of elaborating the national allocation plan, recent existing assessments and secondary sources may be used (e.g. peer-reviewed studies). A Member State should indicate sources used and summarise the applied methodology (including major assumptions made) and results.

> A Member State has to apply the criterion to determine the total quantity of allowances. A Member State may apply the criterion to determine the quantities per activity.

2.1.4. Criterion (4) – Consistency with other legislation

> *The plan shall be consistent with other Community legislative and policy instruments. Account should be taken of unavoidable increases in emissions resulting from new legislative requirements.*

2.1.4.1.   Introduction

40.   Criterion (4) concerns the relationship between allocations under Directive 2003/87/EC and other Community legislative and policy instruments. Consistency between allowance allocations and other legislation is introduced as a requirement in order to ensure that allocation does not contravene the provisions of other legislation. In principle, no allowances should be allocated in cases where other legislation implies that covered emissions had or will have to be reduced even without the introduction of the emission trading scheme. Similarly, consistency implies that if other legislation results in increased emissions or limits the scope for decreasing emissions covered by the Directive account should be taken of this increase.

2.1.4.2.   Analysis

41.   The first sentence of the criterion is mandatory in nature, while the second one is optional.

42.   The first sentence of criterion (4) has to be applied in determining the total quantity, if the Community legislative and policy instruments affect all covered installations, or in determining the quantities for affected installations in other cases.

43.   Pursuant to the first sentence in the criterion, consistency with other Community legislative and policy instruments has to be applied in a symmetrical manner. Not only an unavoidable increase in covered greenhouse gas emissions resulting from new Community legislative and policy instruments should be taken into account, but also a decrease in covered emissions resulting from such instruments.

44.   A Member State should list all Community legislative and policy instruments it has considered and indicate which ones have been taken into account.

45. "New" legislative requirements should be understood as legislation and policy instruments that were adopted before the date of submission of the national allocation plan, and will impose relevant obligations on installations covered by the scheme after that date and before the end of the period covered by the national allocation plan. This includes the implementation of relevant parts of the *acquis communautaire* by new Member States following their accession in May 2004.

46. In order to take an unavoidable change in emissions into account, a Member State should consider, firstly, if a change in greenhouse gas emissions from covered installations is in fact due to new requirements, and, secondly, if such a change is unavoidable.

> In order to simplify administrative tasks the Commission recommends a Member State to consider a Community legislative or policy instrument only insofar as it is expected to result, per activity or in total, in a substantial increase or decrease (e.g. 10 %) of covered emissions.

2.1.5. Criterion (5) – Non-discrimination between companies or sectors

> *The plan shall not discriminate between companies or sectors in such a way as to unduly favour certain undertakings or activities in accordance with the requirements of the Treaty, in particular Articles 87 and 88 thereof.*

47. Normal state aid rules will apply.

2.1.6. Criterion (6) – New entrants

> *The plan shall contain information on the manner in which new entrants will be able to begin participating in the Community scheme in the Member State concerned.*

2.1.6.1. Introduction

48. The treatment of new entrants, i.e. installations starting operation in the course of a trading period, is one of the important design choices in any emission trading scheme. The options differ depending on the allocation method chosen for existing installations. If all allowances are sold by a government, no specific decisions are needed for new entrants. If (the majority of) allowances are, however, allocated free of charge, several options are available to integrate new entrants into the scheme.

49. The definition of new entrants in Article 3 of the Directive[7] puts new installations on an equal footing with existing installations extending capacity. The definition in relation to an updated permit applies only to the extension of an installation, and not to the entire installation, nor to the increased capacity utilisation at an existing installation.

50. The criterion foresees an informational obligation to state how new entrants will be able to begin participating in the Community scheme. The guidance outlines three options to implement the criterion against the backdrop of relevant Treaty provisions. However, the Commission will also assess any other option notified in a national allocation plan.

2.1.6.2.   Analysis

51. The obligation arising under criterion (6) will be deemed as fulfilled if a Member State explains in the national allocation plan how it intends to ensure access to allowances for new entrants. Thus the criterion will be fulfilled if a Member State indicates that it has decided to have new entrants buy all allowances on the market. There are also other options to treat new entrants. In all cases, the guiding principle is equality of treatment.

52. The EC Treaty's provisions on the right of establishment in the internal market have to be respected. It is crucial that new entrants have access to allowances, as without such access operators would be prevented from establishing a business in sectors carrying out covered activities. Guaranteeing this freedom is the essence of the second sentence of Article 11(3) of the Directive. Moreover, EU competition law would be applicable in the event that uncompetitive practices with respect to allowances were to be used to erect market entry barriers.

53. It is important to keep in mind that the new entrants issue is of a temporary nature. In principle, an installation defined as a new entrant in one trading period should no longer fall within this definition when the national allocation plan for the subsequent period is notified.

54. It follows from the definition that a new entrant is an installation for which no greenhouse gas emission permit has been issued or updated by the date the national allocation plan is notified to the

---

[7] See Article 3(h) in Directive 2003/87/EC.

Commission. A Member State may issue or update greenhouse gas emission permits with respect to installations which will start or extend operations with considerable certainty during the relevant trading period. Before issuing or updating a greenhouse gas emission permit, a Member State isrecommended to require an operator to demonstrate that it has already obtained the construction permit and any other relevant permits. Once an installation that is expected to start or extend operations during the trading period has obtained a greenhouse gas emission permit or an updated permit, it can be included in the national allocation plan and be allocated allowances in the same manner as an existing installation. The number of allowances allocated to an installation expected to operate only during part of the trading period should be proportionate to the expected duration of (extended) operations at the installation as a share of the duration of the trading period. The Member State may not withhold foreseen allowances, in case the installation does not, or not at the time intended, start or extend operations, unless it withdraws the greenhouse gas emissions permit.

55.   A Member State has at least three options to enable participation of new entrants: it may have any new entrants buy all allowances on the market, it may make use of the possibility to set aside some allowances for periodic auctioning, or it may foresee a reserve in the national allocation plan to issue allowances to new entrants free of charge.

Having new entrants buy all allowances on the market

56.   A Member State may decide to implement this criterion by having new entrants buy all allowances on the market, as any person (incl. operators) in the Community with or without covered installations may do. The Commission notes that having new entrants buy allowances on the market is in accordance with the principle of equal treatment for the following reasons. Firstly, the Commission notes that the size of the EU-wide allowance market sets the correct conditions for liquidity, which ensures that new entrants will have access to allowances. Secondly, incumbents have made their investments without having been able to take the cost of carbon into account, in contrast to new entrants, who can minimise their carbon costs through investment choices. Thirdly, new installations only fulfil the definition of a new entrant for a limited period of time, i.e. part of a trading period, and the cost of allowances for this limited period (probably less than two years in the first period) can be taken into account in the

investment and timing decisions. The Directive guarantees that as of a certain point in time the new entrant will be allocated allowances in the same manner as all other existing installations for the remainder of the lifetime of the installation.

Auctioning

57.   A Member State may enable new entrants to begin participating in the Community scheme and provide access to allowances on the basis of a periodic auction procedure. In accordance with internal market rules a Member State has to allow any person in the Community to participate in such an auction procedure. A Member State has to respect Article 10 of the Directive, pursuant to which a Member State may not auction more than 5% of the total quantity of allowances allocated during the first trading period and 10 % in the second trading period.

58.   A Member State should specify what use will be made of any allowances offered in the auction procedure but not purchased. A Member State may cancel remaining allowances and re-issue a corresponding quantity of allowances for auctioning in the subsequent period. The Commission notes that at the end of the first period this option is only available, if a Member State's national legislation provides for such re-issue (i.e. banking) of allowances in accordance with Article 13(2) second subparagraph.

59.   The Commission notes that having new entrants buy allowances in an auction is in accordance with the principle of equal treatment for the same reasons as indicated above with respect to having new entrants buy on the market.

Setting aside a reserve

60.   A Member State may provide access to allowances free of charge out of a reserve. If a reserve is set aside a Member State should indicate in the national allocation plan the size of the reserve by stating the absolute quantity of allowances out of the total quantity of allowances. The Member State should justify the size of the reserve with reference to an informed estimate of the expected number of new entrants during the trading period. Up to the quantity in the reserve, new entrants would be issued allowances free of charge according to transparent and objective rules and procedures determined in the national allocation plan. A Member State should describe the methodology by which allowances would be granted to new entrants. If such a method is used, the

Commission recommends a Member State to give applicants in possession of a recently granted or updated greenhouse gas emission permit access to allowances on a first-come first-served basis.

61. In order to respect the principle of equal treatment, the methodology that a Member State uses in order to allocate allowances to new entrants should as far as possible be the same as the one used for comparable incumbents. However, adaptations may be made for justified reasons (cf. guidance criterion (5)). Similarly, all new entrants should be treated in the same way. For instance, the Commission recommends a Member State not to create several reserves dedicated to separate activities, technologies or specific purposes, since they could result in unequal treatment between new entrants.

62. A Member State should further specify what use will be made of any allowances remaining in the reserve until the end of the period. A Member State may auction any remaining allowances, while respecting Article 10 of the Directive. As in the case of allowances offered for auctioning but not purchased, a Member State may cancel remaining allowances and re-issue a corresponding quantity of allowances into a reserve for the subsequent period. Again, the Commission notes that at the end of the first period this option is only available, if a Member State's national legislation provides for such re-issue (i.e. banking) of allowances in accordance with Article 13(2) second subparagraph.

63. A Member State should also state in the national allocation plan what transparent procedure it will follow, if new entrants apply for allowances and the reserve set aside for the period is already exhausted.

64. The Commission notes that the operation of a reserve for new entrants increases the complexity and administrative costs of the emissions trading scheme.

In case a Member State decides to set aside a reserve from which to grant allowances free of charge, the Commission recommends a Member State not to create dedicated reserves for specific activities, technologies or purposes.

2.1.7. Criterion (7) – Early action

> *The plan may accommodate early action and shall contain information on the manner in which early action is taken into account. Benchmarks derived from reference documents concerning the best available technologies may be employed by Member States in developing their National Allocation Plans, and these benchmarks can incorporate an element of accommodating early action.*

2.1.7.1. Introduction

65. The accommodation of early action is considered as desirable from a fairness point of view. Those installations that have already reduced greenhouse gas emissions in the absence of or beyond legal mandates should not be disadvantaged vis-à-vis other installations that have not undertaken such efforts. The application of this criterion necessarily implies fewer allowances available for installations that have not undertaken early action.

66. Neither the criterion nor the Directive contains a definition of early action and in which way it may be accommodated. Therefore, a Member State has a degree of freedom how to define, and whether and how to accommodate early action. This freedom is only limited by other Annex III criteria and provisions derived from the Treaty. The guidance on this criterion outlines limitations imposed by these other criteria and provisions, and contains options how to accommodate early action, if a Member State decides to do so.

67. The second sentence of the criterion builds on the reference to benchmarks in criterion (3). It repeats the possibility for Member States to use benchmarks and points out that such benchmarks are one possible option to accommodate early action. Furthermore, the reference documents developed under Directive 96/61/EC concerning integrated pollution prevention and control[8] are inferred as a potential source for developing benchmarks.

2.1.7.2. Analysis

68. Criterion (7) is optional and should, if applied, be used to determine the quantity of allowances allocated to individual installations.

69. "Early action" is to be understood as actions undertaken in covered installations to reduce covered emissions before the

---

[8] OJ L 257, 10.10.1996, p. 26.

national allocation plan is published and notified to the Commission. In line with criterion (4), only measures that operators undertook beyond requirements arising from Community legislation can qualify as early action. More stringent national legislation, applying to all covered installations in total or carrying out an activity, will be reflected in the potential to reduce emissions (cf. criterion (3)). Thus, early action is limited to reductions of covered emissions beyond reductions made pursuant to Community or national legislation, or to actions undertaken in the absence of any such legislation. A parallel can also be drawn to the Community guidelines on State aid for environmental protection, which prohibit public investment aid with respect to investments that merely bring companies into line with Community standards already adopted but not yet in force.

70.   Member States have several options to accommodate early action that operators have undertaken in existing installations. Three possible methods are elaborated below, but the Commission will also assess other methods.

Choosing an early base period

71.   The first option to accommodate early action is to base the allocation on historical emissions by applying a relatively early base period. If operators are allocated allowances corresponding to a proportion of the historical emissions from installations, those operators who have invested to reduce emissions since the base period will receive an allocation which covers a larger share of current emissions than an operator who has not made such investments. A Member State using this approach would have to verify that the difference in emission levels over time was not due to installations having implemented legal requirements.

72.   The drawback of this approach is that reliable and comparable data for emissions in an early base period may not be available, and the number of operator changes since the base period will increase with time, making it more difficult to establish reliable and complete records.

73.   An alternative is to use a recent multi-year base period and then allow an operator to choose one early year when it had higher emissions. Emissions data from one of the years in the recent period would then be replaced by data from the early year. This would increase the average annual emissions on which the allocation was based. In line with the restrictions described above,

a Member State wanting to substitute data in this way would have to verify that the difference in emission levels over time was not due to installations implementing legal requirements.

Making a two-round allocation at installation level

74.   After determining the total quantity of allowances, a share of the available allowances is set aside. The allowances set aside would be used in a second round, after an initial distribution to all installations, to give a bonus to those installations in which operators have undertaken early action. Operators would have to apply to be taken into account in the second round and would have to demonstrate that the measures they propose to be accepted as early action comply with a pre-established definition of early action. A Member State should indicate in the national allocation plan the list of measures recognised as early action, and should specify for the relevant installations which measures have been accommodated as early action and the corresponding number of allowances allocated.

Using benchmarks

75.   A Member State may accommodate early action by using benchmarks derived from reference documents concerning the best available techniques. Benchmarks accommodate early action because they imply that a more carbon efficient installation receives more allowances than a less carbon efficient one, which is not necessarily the case with a base period driven allocation formula.

76.   In contrast to criterion (3), in which benchmarks (average emissions by product incorporating achievable progress) may be applied to determine the quantity of allowances by activity, under this criterion the benchmark would be applied to determine the quantity of allowances by installation.

77.   In order to apply the benchmarking approach, a Member State should first group homogenous installations and then apply a benchmark to each of these groups. Installations in a group should be sufficiently homogenous with respect to their input or output characteristics, such that it is feasible to apply the same type of benchmark to them. If benchmarking is used to determine allocations per installation carrying out energy activities, the Commission recommends to group installations by input fuels and to apply separate input-derived benchmarks. It should be

described in the national allocation plan on what basis the grouping of installations was carried out and the respective benchmarks chosen (cf. criterion. (3)).

78.   In order to determine the allocation to an installation, the benchmark needs to be multiplied by an output value. A Member State should indicate in the national allocation plan the output values applied and justify why it considers them appropriate. A Member State may use the most recent actual output data or a forecast for the trading period, which should be clearly substantiated in the national allocation plan.

79.   Due to the ex-ante nature of the allocation decision pursuant to Article 11(1) a Member State may not base the allocation to an installation on actual output data in the trading period, i.e. data unknown at the time the national allocation plan is established but known during the course of the trading period.

80.   A benchmarking approach should not result in an allocation to installations in an activity of more allowances than determined per activity in accordance with criterion (3). A Member State would also have to verify that the installations whose emissions are lower than the benchmark have not arrived at their particular level of emissions, as a result of the implementation of legal requirements.

81.   Alternatively, a Member State may use benchmarks in a simplified way to accommodate early action. If a Member State determines allocations at installation level with a base period approach, it may use benchmarks in order to determine and apply an installation-specific correction factor to a base period driven formula. In this way the allocation to installations performing better than average is increased, while the allocation to installations performing below average is decreased. Such corrections should result in a net balance of zero across all installations concerned.

If a Member State applies this criterion, it should be used in determining the quantity of allowances allocated to individual installations. A Member State should not include measures as early action if they were taken in order to comply with legislative requirements.

If benchmarks are used to determine allocations per installation for energy activities, the Commission recommends to group installations by input fuels and to apply separate input-derived benchmarks.

2.1.8. Criterion (8) – Clean technology

> *The plan shall contain information on the manner in which clean technology, including energy efficient technologies, are taken into account.*

2.1.8.1.   Introduction

82.   This criterion enables a Member State to take account of clean technologies in setting allocations, but does not define what constitutes such clean technology.

83.   While emission trading will promote and reward the application of low-carbon technologies, this criterion is related to the criteria on potential and early action. This guidance outlines these links.

2.1.8.2.   Analysis

84.   Criterion (8) is optional and should, if applied, be used to determine the quantity of allowances allocated at installation level.

85.   Information is required from a Member State on the application of criterion (8). Thus, the criterion will be deemed as fulfilled if a Member State states that it makes no specific provisions to take clean technologies, including energy efficient technologies, into account.

86.   Criterion (8) can be seen as the extension of criterion (3) to the installation level. An installation using a clean or energy efficient technology has a lower technological reduction potential than a comparable installation not using such a technology. It follows that the use of a clean or energy efficient technology should not be rewarded under this criterion with respect to an installation belonging to an activity which has a relatively low technological reduction potential. The reduced technological reduction potential of such an installation would already have been covered in the implementation of criterion (3).

87.   Moreover, there is a link between criterion (7) on early action and criterion (8), since an early action will typically have been an investment in a clean or energy efficient technology. The Commission recommends a Member State not to apply both criteria (7) and (8) to the same installation, unless it can be shown that the early action did not consist of an investment in a clean or energy efficient technology.

88. In addition, the use of clean technologies, including energy efficient technologies, should only be taken into account under this criterion with respect to installations using such technologies before the national allocation plan is published and notified to the Commission. The Commission notes that this criterion should not be applied to clean technologies that do not result in emissions covered by the Directive.

89. By clean or energy efficient technologies, the Commission understands technologies that have resulted in lower direct emissions of covered greenhouse gases than the alternative technologies that could realistically have been deployed by the installation concerned would have led to. In determining the difference in emission levels between direct emissions from combined heat and power and an alternative technology, the latter may consist of on-site separate power and heat production.

90. As regards energy production, the Commission will accept as clean or energy efficient technologies those technologies for which it has approved State aid under the Community guidelines on State aid for environmental protection. The following list is not exhaustive:

    – High efficiency combined heat and power production. Member States may apply national definitions of "high efficiency" cogeneration production, unless such a definition has been adopted in Community law.

    – District heating, other than high efficiency combined heat and power.

91. As regards other industrial technologies than energy production, a Member State should justify why a particular technology is to be considered as clean or energy efficient. A minimum requirement is that the technology constitutes a "best available technique" as defined in Council Directive 96/61/EC of 24 September 1996 concerning integrated pollution prevention and control and that it was being used by the installation at the date of submission of the national allocation plan. However, since a "best" available technique is defined as one which is "most effective in achieving a high general level of protection of the environment as a whole", it will, in addition, have to be shown that the technique is particularly performing in limiting emissions of covered greenhouse gases.

92. Where a waste gas from a production process is used as a fuel by another operator, the distribution of allowances between the two installations is a matter for Member States to decide. For that purpose, a Member State may choose to allocate allowances to the operator of the installation transferring the waste gas, provided this is done on the basis of a pre-established criterion, compatible with the existing criteria of Annex III and the Treaty. This paragraph applies independently of whether a Member State chooses to apply criterion (7) or criterion (8) in accordance with paragraph 108.

If a Member State takes clean technology, including energy efficient technologies, into account, it should do so by applying either criterion (7) or criterion (8) but not both.

2.1.9. Criterion (9) – Involvement of the public

The plan shall include provisions for comments to be expressed by the public, and contain information on the arrangements by which due account will be taken of these comments before a decision on the allocation of allowances is taken.

2.1.9.1.   Analysis

93. This criterion is of a mandatory nature.

94. A Member State will be considered to have implemented criterion (9), if it describes in the national allocation plan how it makes the plan available to the public for comments, and how it provides for due account to be taken of any comments received. The plan should be made available in a manner which enables the public to comment on it effectively and at an early stage. This means that the public is informed, whether by public notices or other appropriate means such as electronic media, about the plan, including its text, and that also other relevant information is made available, including *inter alia* information about the competent authority to which comments or questions may be submitted.

95. A Member State should provide for a reasonable time frame for submitting comments, and co-ordinate the deadline for comments to be submitted by the public with the national decision-making procedure, so that due account can be taken of comments before the decision on the national allocation plan. "Due account" is to be understood as meaning that comments are to be taken into

account if appropriate with reference to the criteria in Annex III or to any other objective and transparent criteria applied by the Member State in the national allocation plan. A Member State should inform the Commission of any intended modifications following public participation subsequent to the publication and notification of the national allocation plan and before taking its final decision pursuant to Article 11. Feedback is to be provided, in a general form, to the public about the decision taken and the main considerations upon which it is based.

96. It should be noted that the possibility for the public to comment on the national allocation plan provided for under this criterion constitutes a second round of public consultation. Pursuant to Article 9(1) of the Directive, the comments resulting from a first round of consultation of the public on the basis of the draft plan should, where pertinent, already have been integrated into the national allocation plan prior to notification of the plan to the Commission and to the other Member States. For the overall public participation (consultation and taking account of comments) to be effective, the first round of public consultation is of particular importance. The rules described under this criterion should also be applied to the first round of consultation.

A Member State should inform the Commission of any intended modifications subsequent to the publication and notification of the national allocation plan before taking its final decision pursuant to Article 11.

2.1.10. Criterion (10) – List of installations

The plan shall contain a list of the installations covered by this Directive with the quantities of allowances intended to be allocated to each.

2.1.10.1. Introduction

97. This criterion provides for the transparency of the national allocation plan. It implies that the quantities of allowances per installation are indicated, and therefore visible to the general public, when the plan is submitted to the Commission and other Member States.

2.1.10.2. Analysis

98. This criterion will be deemed as fulfilled if a Member State has respected its obligation to list all the installations covered by the

Directive. This includes installations to be temporarily excluded in the first period pursuant to Article 27 and installations to be unilaterally included in any period pursuant to Article 24.

99.   As mentioned under criterion (5), combustion installations with a rated thermal input of more than 20 MW can be found in several sectors. A Member State should therefore indicate the main activity carried out at the site where the combustion installation is located, e.g. "paper" for a combustion installation which is part of the paper production process. A Member State should list installations by main activity, and provide subtotals of all data at activity level.

100.   A Member State has to indicate the total quantity of allowances intended to be allocated to each installation and should indicate the quantity issued in each year to each installation following Article 11(4).

101.   Article 11(4) constitutes an obligation to issue a share of the total quantity to each installation each year. Hence a Member State could issue a large proportion of the allowances in the first year(s) of a period and issue only a small share in the remaining year(s) of a period. Alternatively, a Member State could issue a small proportion of the allowances in the first year(s) of a period and issue a large share in the remaining year(s) of a period. Such approaches, in particular if taken by several Member States, may result in low market liquidity in the initial years, so that the market may fail to provide a sufficiently robust price signal. Such a signal is vital for the allowance market to provide operators of covered installations with an orientation whether to implement measures on-site or rather acquire allowances. Therefore, the Commission makes a recommendation on the proportion to be issued in each year.

102.   Furthermore, a Member State should in principle issue to all operators included in the plan equivalent, but not necessarily equal, annual shares in order to avoid undue discrimination (cf. criterion (5)).

The Commission recommends a Member State to issue each year a share that does not deviate substantially from equal proportions over the period.

2.1.11.Criterion (11) – Competition from outside the Union

The plan may contain information on the manner in which the existence of competition from countries or entities outside the Union will be taken into account.

2.1.11.1.Introduction

103. The European Union has repeatedly affirmed its commitment to respect the Kyoto target. At the same time, the European Union, at the Lisbon European Council in March 2000, set itself the strategic goal to become the most competitive and dynamic knowledge-based economy in the world, capable of sustainable economic growth with more and better jobs and greater social cohesion. Emission allowance trading is a cost-effective instrument, which allows industrial activities covered by the Directive to keep the costs of contributing to the Community's climate change commitments low. The implementation of the Kyoto Protocol will give companies in the European Union the chance for a head-start in the gradual transition to a carbon-constrained global economy, since carbon efficiency may be an important source of competitive advantage in the future, as much as labour or capital productivity today. In the short-term these commitments may imply increased costs for some companies and sectors.

2.1.11.2.Analysis

104. Criterion (11) is optional and should only be used, if applied, in determining the quantity of allowances per activity, since any effect from competition from countries or entities outside the Union would be one affecting all installations carrying out a certain activity.

105. A Member State should not use the mere existence of competition from outside the Union as a reason to apply this criterion. The Commission considers this criterion to be applicable exclusively to cases where covered installations belonging to a specific activity would be rendered significantly less competitive directly and predominantly as a result of a major difference in climate policies between the EU and countries outside the EU. In assessing any such differences in climate policy, a Member State should take into account any relevant measures that competitors outside the EU are subject to, including voluntary initiatives, technical regulation, taxes and emissions trading, and not judge solely on

the basis of whether or not the country concerned has a quantified emissions commitment and has ratified the Kyoto Protocol.

106. A Member State should not take the existence of competition from outside the Union into account in such a manner as to improve the competitive position of installations carrying out an activity *vis-à-vis* competitors outside the Union as compared to their competitive position in the absence of the EU emissions trading scheme. It should be noted that incorrect application of this criterion may constitute export aid, which is incompatible with the EC Treaty.

107. If a Member State deems it necessary to take account of competition from outside the Union, it should also consider applying other options outside the national allocation plan.

108. A Member State should keep in mind, when applying this criterion to individual activities, that where the mandatory criterion (3) is applied at activity level, installations carrying out activities with a relatively large potential to reduce emissions should still receive a smaller share of allowances in relation to emissions, compared to installations carrying out activities with a relatively small potential to reduce emissions.

109. The existence of competition should only be taken into account in the national allocation plan by a modification of the quantity of allowances per activity, without a change in the total quantity of allowances determined in accordance with the criteria (1) to (5).

If competition from outside the Union is taken into account in the national allocation plan, the criterion should only be applied in determining the quantity of allowances allocated at activity level, without a change in the total quantity of allowances.

If a Member State deems it necessary to take account of competition from outside the Union, it should also consider applying other options outside the national allocation plan.

---

3. GUIDANCE ON CIRCUMSTANCES UNDER WHICH FORCE MAJEURE IS DEMONSTRATED

Article 29

1. During the period referred to in Article 11(1), Member States may apply to the Commission for certain installations to be issued with additional allowances in cases of force majeure. The Commission shall determine whether force majeure is demonstrated, in which case it shall authorise the issue of additional and non-transferable allowances by that Member State to the operators of those installations.

2. The Commission shall, without prejudice to the Treaty, develop guidance to describe the circumstances under which force majeure is demonstrated, by 31 December 2003 at the latest.

110. In principle, allocation decisions are made by Member Sates before the beginning of the relevant trading period, thereby avoiding uncertainty in the allowance market. A limited provision allows the issuance of additional non-transferable allowances in exceptional and unforeseeable circumstances in the first period of the Community scheme.

111. Article 29 derogates from the general principle of the Community scheme, under which Member States allocate allowances before the beginning of the relevant trading period. Applications for force majeure allowances may therefore cause uncertainty in the allowance market, and, if granted, give an advantage to certain companies, which affects trade between Member States. Article 29 is therefore explicitly without prejudice to the Treaty, and the Commission will carefully consider the justification and potential effects of any application for such allowances.

112. Companies may seek insurance against various risks that could result in increased emissions, but insurance policies normally do not cover circumstances of force majeure. The Commission will not consider circumstances that could have been insured to constitute force majeure.

113. Circumstances of force majeure are, by their nature, difficult to anticipate. The Commission considers these to be exceptional and unforeseeable circumstances, which cause a substantial increase in annual direct emissions of greenhouse gases covered by Directive 2003/87/EC at an installation, which could not have been avoided even if all due care had been exercised. The circumstance

must have been beyond the control of the operator of the installation concerned and of the Member State submitting an application to the Commission under Article 29 of the Directive with respect to the installation of that operator.

114. Circumstances that the Commission may consider to be *force majeure* include notably natural disasters, war, threats of war, terrorist acts, revolution, riot, sabotage or acts of vandalism.

115. The presence of *force majeure* has to be demonstrated at installation level and on a case-by-case basis.

116. An application under Article 29 of the Directive should include, in respect of each installation, the Member State's best estimate of the increase in emissions resulting from the circumstance for which *force majeure* is pleaded and a substantiation of that estimate.

117. A Member State should submit an application under Article 29 to the Commission by 31 January the year following the year of the trading period during which the circumstance occurred for which *force majeure* is pleaded.

ANNEX

COMMON FORMAT FOR THE NATIONAL ALLOCATION PLAN 2005 TO 2007

1. DETERMINATION OF THE TOTAL QUANTITY OF ALLOWANCES

What is the Member State's emission limitation or reduction obligation under Decision 2002/358/EC or under the Kyoto Protocol (as applicable)?

What principles, assumptions and data have been applied to determine the contribution of the installations covered by the emissions trading Directive to the Member State's emission limitation or reduction obligation (total and sectoral historical emissions, total and sectoral forecast emissions, least-cost approach)? If forecast emissions were used, please describe the methodology and assumptions used to develop the forecasts.

What is the total quantity of allowances to be allocated (for free and by auctioning), and what is the proportion of overall emissions that these allowances represent in comparison with emissions from sources not covered by the emissions trading Directive? Does this proportion deviate from the current proportion of emissions from covered installations? If so, please give reasons for this deviation with reference to one or more criteria in Annex III to the Directive and/or to one or more other objective and transparent criteria.

What policies and measures will be applied to the sources not covered by the emissions trading Directive? Will use be made of the flexible mechanisms of the Kyoto Protocol? If so, to what extent and what steps have been taken so far (e.g. advancement of relevant legislation, budgetary resources foreseen)?

How has national energy policy been taken into account when establishing the total quantity of allowances to be allocated? How is it ensured that the total quantity of allowances intended to be allocated is consistent with a path towards achieving or over-achieving the Member State's target under Decision 2002/358/EC or under the Kyoto Protocol (as applicable)?

How is it ensured that the total quantity of allowances to be allocated is not more than is likely to be needed for the strict application of the criteria of Annex III? How is consistency with the assessment of actual and projected emissions pursuant to Decision 93/389/EEC ensured?

Please explain in Section 4.1 below how the potential, including the technological potential, of activities to reduce emissions was taken into account in determining the total quantity of allowances.

Please list in Section 5.3 below the Community legislative and policy instruments that were considered in determining the total quantity of allowances and state which ones have been taken into account and how.

If the Member State intends to auction allowances, please state the percentage of the total quantity of allowances that will be auctioned, and how the auction will be implemented.

2. DETERMINATION OF THE QUANTITY OF ALLOWANCES AT ACTIVITY LEVEL (IF APPLICABLE)

By what methodology has the allocation been determined at activity level? Has the same methodology been used for all activities? If not, explain why a differentiation depending on activity was considered necessary, how the differentiation was done, in detail, and why this is considered not to unduly favour certain undertakings or activities within the Member State.

If the potential, including the technological potential, of activities to reduce emissions was taken into account at this level, please state so here and give details in Section 4.1 below.

If Community legislative and policy instruments have been considered in determining separate quantities per activity, please list the instruments considered in Section 5.3 and state which ones have been taken into account and how.

If the existence of competition from countries or entities outside the Union has been taken into account, please explain how.

3. DETERMINATION OF THE QUANTITY OF ALLOWANCES AT INSTALLATION LEVEL (+ ANNEX I)

By what methodology has the allocation been determined at installation level? Has the same methodology been used for all installations? If not, please explain why a differentiation between installations belonging to the same activity was considered necessary, how the differentiation by installation was done, in detail, and why this is considered not to unduly favour certain undertakings within the Member State.

If historical emissions data were used, please state whether they have been determined in accordance with the Commission's monitoring and reporting guidelines pursuant to Article 14 of the Directive or any other set of established guidelines, and/or whether they have been subject to independent verification.

If early action or clean technology were taken into account at this level, please state so here and give details in Sections 4.2 and/or 4.3 below.

If the Member State intends to include unilaterally installations carrying out activities listed in Annex I below the capacity limits referred to in that Annex, please explain why, and address, in particular, the effects on the internal market, potential distortions of competition and the environmental integrity of the scheme.

If the Member State intends temporarily to exclude certain installations from the scheme until 31 December 2007 at the latest, please explain in detail how the requirements set out in Article 27(2)(a)-(c) of Directive 2003/87EC are fulfilled.

4.   TECHNICAL ASPECTS

4.1.   Potential, including technological potential

Has criterion (3) been used to determine only the total quantity of allowances, or also the distribution of allowances between activities covered by the scheme?

Please describe the methodology (including major assumptions made) and any sources used to assess the potential of activities to reduce emissions. What are the results obtained? How is it ensured that the total quantity of allowances allocated is consistent with the potential?

Please explain the method or formula(e) used to determine the quantity of allowances to allocate at the total level and/or activity level taking the potential of activities to reduce emissions into account.

If benchmarking was used as a basis for determining the intended allocation to individual installations, please explain the type of benchmark used, and the formula(e) used to arrive at the intended allocation in relation to the benchmark. What benchmark was chosen, and why is it considered to be the best estimate to incorporate achievable progress? Why is the output forecast used considered to be the most likely development? Please substantiate the answers.

4.2. Early action (if applicable)

If early action has been taken into account in the allocation to individual installations, please describe in which manner it is accommodated. Please list and explain in some detail the measures that were accepted as early action and what the criteria for accepting them were. Please demonstrate that the investments/actions to be accommodated led to a reduction of covered emissions beyond what followed from any Community or national legislation in force at the time the action was taken.

If benchmarks are used, please describe on what basis the grouping of installations to which the benchmarks are applied was made and why the respective benchmarks were chosen. Please also indicate the output values applied and justify why they are considered appropriate.

4.3. Clean technology (if applicable)

How has clean technology, including energy efficient technologies, been taken into account in the allocation process? If at all, which clean technology has been taken into account, and on what basis does it qualify as such? Have any energy production technologies intended to be taken into account been in receipt of approved State aid for environmental protection in any Member State? Please state whether any other industrial technologies intended to be taken into account constitute "best available techniques" as defined in Council Directive 96/61EC, and explain in what way it is particularly performing in limiting emissions of covered greenhouse gases.

5. COMMUNITY LEGISLATION AND POLICY

5.1. Competition policy (Articles 81-82 and 87-88 of the Treaty)

If the competent authority has received an application from operators wishing to form a pool and if it is intended to allow it, please attach a copy of that application to the National Allocation Plan. What percentage of the total allocation will the pool represent? What percentage of the relevant sector's allocation will the pool represent?

5.2. Internal market policy – new entrants (Article 43 of the Treaty)

How will new entrants be able to begin participating in the EU emissions trading scheme?

In the case that there will be a reserve for new entrants, how has the total quantity of allowances to set aside been determined and on what basis will the quantity of allowances be determined for each new entrant? How does the formula to be applied to new entrants compare to the formula applied to incumbents of the relevant activity? Please also explain what will happen to any allowances remaining in the reserve at the end of the trading period. What will apply in case the demand for allowances from the reserve exceeds the available quantity of allowances?

Is information already available on the number of new entrants to expect (through applications for purchase of land, construction permits, other environmental permits etc.)? Have new or updated greenhouse gas emission permits been granted to operators whose installations are still under construction, but whose intention it is to start a relevant activity during the period 2005 to 2007?

5.3. Other legislation or policy instruments

Please list other Community legislation or policy instruments that were considered in the establishment of the National Allocation Plan and explain how each one has influenced the intended allocation and for which activities.

Has any particular new Community legislation been considered to lead to an unavoidable decrease or increase in emissions? If yes, please explain why the change in emissions is considered to be *unavoidable*, and how this has been taken into account.

6. PUBLIC CONSULTATION

How is this national allocation plan made available to the public for comments?

How does the Member State provide for due account to be taken of any comments received before a decision on the allocation of allowances is taken?

If any comments from the public received during the first round of consultation have had significant influence on the national allocation plan, the Member State should summarise those comments and explain how they have been taken into account.

7. CRITERIA OTHER THAN THOSE IN ANNEX III TO THE DIRECTIVE

Have any criteria other than those listed in Annex III to the Directive been applied for the establishment of the notified National Allocation Plan? If yes, please specify which ones and how they have been implemented.

Please also justify why any such criteria are not considered to be discriminatory.

ANNEX I – LIST OF INSTALLATIONS

Please submit a matrix containing the following information:
– Identification (e.g. name, address) of each installation
– The name of the operator of each installation
– The number of the greenhouse gas emissions permit
– The unique (EPER) identifier of the installation
– The main activity, and, if applicable, other activities carried out at the installation
– Total quantity of allowances to be allocated for the period, and the annual breakdown, for each installation
– Whether the installation has been unilaterally included or temporarily excluded and whether it is part of a pool
– Annual data per installation, including emission factors if emissions data are used, which have been used in the allocation formula(e)
– A subtotal per activity of data used and number of allowances allocated

ANNEX 5

THE 2005 COMMISSION GUIDANCE

COMMISSION OF THE EUROPEAN COMMUNITIES

Brussels, 22.12.2005
COM(2005)703 final

COMMUNICATION FROM THE COMMISSION

"Further guidance on allocation plans for the 2008 to 2012 trading period
of the EU Emission Trading Scheme"

(Text with EEA relevance)

1.   INTRODUCTION

1.   This communication provides guidance to Member States for the
      design of the national allocation plans for the second trading
      period (2008 to 2012). This communication is not part of the
      ongoing review of the Emissions Trading Directive[1] "the
      Directive"), under which the Commission will produce a report
      to the European Parliament and the Council in June 2006,
      including proposals for improving the functioning of the EU
      Emissions Trading Scheme ("EU ETS"), as appropriate. In
      preparing this review, the Commission acknowledges input from
      stakeholders on a wide range of issues on the functioning and
      impact of the EU ETS.

2.   This guidance supplements the Commission's guidance of 7
      January 2004[2] for the implementation of the criteria listed in Annex
      III of the Directive. The previous guidance document contains in
      particular a technical analysis of the interpretation and interplay
      of the different criteria in Annex III and explains their role in the
      Commission's assessment of allocation plans. Key messages from
      the first guidance document are summarised in Annex 3.

---

[1] Directive 2003/87/EC of the European Parliament and of the Council of 13 October 2003
establishing a scheme for greenhouse gas emission allowance trading within the
Community and amending Council Directive 96/61/EC, OJ L 275 of 25.10.2003 p. 32–46,
as amended by Directive 2004/101/EC of the European Parliament and of the Council of
27 October 2004 amending Directive 2003/87/EC establishing a scheme for greenhouse
gas emission allowance trading within the Community, in respect of the Kyoto Protocol's
project mechanisms, OJ L 338/18 of 13.11.2004.
[2] COM (2003) 830 final

3.   The Commission considers it necessary to provide additional guidance to consistently incorporate the lessons learnt from the first allocation phase. It notes that the general nature of the criteria listed in Annex III of the Directive leaves scope for their implementation and shares the view of Member States and many stakeholders that more guidance is needed[3] to ensure more coherent allocation plans for the second trading period.

4.   In general, Member States and stakeholders also stress a preference for increasing harmonisation of allocation rules. The Commission considers it necessary to achieve more coherence in the second trading period, to the extent that the divergent progress by Member States towards their individual Kyoto targets allows for. In addition, further harmonisation is desirable beyond 2012. The Commission will consider these issues in the context of the strategic review of the EU ETS. On the basis of this review, the Commission will come forward with proposals, as appropriate, to improve the functioning of the scheme whilst safeguarding regulatory stability.

5.   The Commission urges Member States to work towards simpler plans for the second trading period. Simple allocation plans boost the understanding of the instrument among stakeholders and also increase transparency and predictability. Member States should strive to keep the second national allocation plans as simple as possible, in particular with respect to allocation methods and rules on new entrants and closures. Member States should critically assess the necessity and efficiency of rules contained in the first round national allocation plans and keep only those deemed absolutely essential.

6.   To further improve transparency of plans the Commission has developed and annexed a set of tables[4] which summarise in a standardised manner some basic information contained in a national allocation plan. The Commission regards these tables as an integral part of second round national allocation plans and expects Member States to make use of them. Furthermore, it urges Member States to continue using the Common Format[5] elaborated for the first allocation plans and, as in the first phase, will ensure a fully consistent assessment of all plans.

---

[3] On 1 December 2005, the Council invited the Commission to do its utmost to provide guidance early enough for the preparation of the second national allocation plans.
[4] See Annex 10
[5] COM(2003) 830 final, p. 25-29

2.   SUMMARY OF EXPERIENCE GAINED FROM ALLOCATION
      PLANS FOR THE FIRST PHASE (2005-2007) AND GENERAL
      LESSONS FOR THE SECOND PHASE (2008-2012)

7.   The first phase allocation process spanned about 15 months, from
      the deadline for notification on 31 March 2004 until the last
      Commission decision on 20 June 2005.This was much longer than
      envisaged in the Directive. The approval process extended well
      into the first trading period which started on 1 January 2005. The
      latenotification, approval and finalisation at national level of some
      plans introduced uncertainties not only for the respective national
      authorities and business but also for actors in the allowance market
      across Europe. This underlines the importance of timely
      notification of complete national allocation plans for the second
      allocation phase. The Commission considers that the three month
      period foreseen in Article 9(3) can only commence once a complete
      national allocation plan has been submitted. It therefore reminds
      Member States of their obligations to respect the deadline of 30
      June 2006 in order to enable the completion of the second
      allocation process, including the subsequent final national
      allocation decision, well before the start of the second trading
      period on 1 January 2008. The Commission will not accept
      amendments to national allocation plans notified after the
      deadline of 31 December 2006 specified in Article 11(2) of the
      Directive, other than those required by the respective Commission
      decision on a national allocation plan.

8.   Recognising the first phase as a learning period, the Commission
      assessed the first period allocation plans in a pragmatic manner.
      Some notable characteristics, summarised below, emerged from
      the first allocation process, resulting in a convergence of choices
      and approaches across Member States (for more details see Annex
      4):

      –   More use of emissions trading is necessary to meet the
            Kyoto targets cost-effectively.

      –   Allocations have in general been more restrictive for power
            generators than other sectors covered by the scheme.

      –   Member States experiencing considerable excess in actual
            emissions with respect to their Kyoto targets intend to
            purchase a substantial amount of Kyoto units.

–   The non-acceptance of ex-post adjustments is essential for the allowance market development.

–   Some allocation plans are more complex than necessary and not sufficiently transparent.

3.   FURTHER GUIDANCE ON SELECTED ISSUES FOR THE SECOND PERIOD NATIONAL ALLOCATION PLANS

3.1.   Progress to Kyoto targets

9.   In the 2005 progress report[6] the Commission has assessed the progress of Member States to Kyoto targets. In the comparison of 2003 actual emissions to allowed emissions in the period 2008 to 2012, a considerable number of Member States have gaps to close, some of significant magnitude. At present, it would appear that in particular Austria, Belgium, Denmark, Finland, Germany, Ireland, Italy, Luxembourg, the Netherlands, Portugal, Slovenia and Spain are not sufficiently on track towards meeting their Kyoto targets. In these Member States, more needs to be done during the second trading period to respect the Kyoto targets, which does not imply that further measures are not necessary in other Memer States as well. As it is unlikely that gaps can be closed *solely* by requiring emission reductions in the non-trading sector or relying on purchase of Kyoto units, the EU ETS needs to be used more to fully realise the potential of emissions trading.

3.2.   Setting national caps

10.   According to criterion 3 of Annex III (see Annex 2 attached to this document for the criteria of Annex III of the Directive), the quantity of allowances shall be consistent with the potential, including the technological potential, of activities covered by the scheme to reduce emissions. This means that the combination of the respective economic and technological potential to cut emissions sets an upper limit for the cap at national level.

11.   Two of the most important factors driving emissions trends are economic (GDP) growth (with higher growth leading to higher emissions) and carbon intensity (emissions per unit GDP, with reductions in carbon intensity lowering emissions). In principle, the faster an economy grows, the faster new technologies are put to use and the faster the capital stock is renewed, thereby

---

[6] Report from the Commission on the Progress towards achieving the Community's Kyoto Target of 15 December 2005, COM(2005) 655.

improving productivity and carbon intensity. The growing share of the tertiary sector and the parallel decline of the secondary sector experienced in European economies further contribute to this effect. Furthermore, the introduction of the EU ETS and the EU-wide carbon price in the trading sector will stimulate further reductions in carbon intensity.

12. Historically (in the period 1990 to 2000) carbon intensity reductions have balanced or even outweighed economic growth, which means that greenhouse gas emissions remained stable or declined. The following table indicates that this trend is likely to remain stable in the course of the ongoing decade (2000 to 2010). It needs to be emphasized that the estimates for the period 2000 to 2010 do not account for the incentives created by the first phase of the EU ETS and are therefore very likely to underestimate the actual reductions in carbon intensity during that period.

Table A: Historic and estimated GDP growth rates and carbon intensity trends[7]:

	Annual GDP Change in %	Annual Carbon Intensity* Improvement in %	Combined Net Effect on Annual Emissions Trend in %
Actual Development 1900-2000			
EU25	2.0	2.3	-0.3%
EU15	2.0	1.9	0.1%
New Member States	1.7	3.9	-2.2%
Estimated development 2000 to 2010			
EU25	2.5	2.2	0.3%
EU15	2.4	2.1	0.3%
New Member States	3.8	3.6	0.2%

Note: * Carbon intensity expresses the relation between CO_2 emissions to GDP.

13. In the analysis of the economic and technological potential to cut emissions, the Commission considers the annual GDP growth and carbon intensity reduction rates. The combined effect of these two factors gives the rate for the annual potential to reduce emissions. Starting from actual emissions in an appropriate year (e.g. 2003), assuming the trading sector to have a constant share in emissions and a similar potential to reduce emissions as the entire economy, the indicative cap consistent with criterion 3 in Annex III can be derived.

---

[7] Source: European Commission, Directorate-General for Energy and Transport, European Energy and Transport Trends to 2030, Appendix 2, January 2003, see website: http://europa.eu.int/comm/dgs/energy_transport/figures/trends_2030/index_en.htm.

14. The cap for the first phase is therefore a starting point in determining and assessing the total quantity for the second phase both at EU and Member State level. Due to criterion 1 some Member States have to lower the first period caps to respect the Kyoto target. Other Member States need to maintain their first phase caps to align the plan with the potential to reduce emissions (criterion 3). Correspondingly, the annual average EU-wide ETS cap in the second phase should be lower than the first phase cap.

15. A number of Member States have a gap to close between their 2003 actual emissions and allowed emissions according to the Kyoto target. The total gap for these Member States is 296.5 million tonnes CO_2 equivalent. This figure therefore represents the excess emissions which these Member States still need to reduce by using the instruments at their disposal to secure compliance with the Kyoto targets.

16. Member States with a gap to reach the Kyoto target should aim for a balanced mix of (i) lowering the allocation for the second phase and (ii) implementing additional measures in the non-trading sector, potentially supplemented by the (iii) government purchase of Kyoto unit credits. A balanced mix makes reductions practically more feasible and economically more efficient.

17. The table in Annex 1 lists the share of the emission trading sector expressed as the first phase allocation in comparison to 2003 actual emissions. At EU level the share amounts to some 45%. If the emissions trading sector were to contribute a proportionate share of the reduction in Member States with a gap to close, the second period total allocation in the EU-25 would be some 6% below the first period allocation, resulting in an annual average allocation of 2.063 billion allowances. To meet the Kyoto targets a reduction of less than 6 % would imply stronger efforts by the non-trading sector.

3.3. Substantiation of intended government purchase of Kyoto units

18. In view of the state of market development and constraints in the supply of Kyoto units, Member States face a considerable challenge to realise the volume of intended purchases. The decision by a Member State to purchase with public funds Kyoto units eases (as much as the purchase by companies under the terms of the Linking Directive) the need for domestic emission reductions.

19. For the above reasons, the substantiation of the intended government purchase of Kyoto units is crucial for the consistency of a national allocation plan with criterion 1 in Annex III. Therefore this was already an important element in the assessment of first period plans. Several Member States did not fully substantiate the intended purchase in the first period national allocation plans and some caps were accordingly lowered. Each Member State relying on the government purchase of Kyoto units, even if already indicated in the first round national allocation plan, thus needs to substantiate more thoroughly the intentions and demonstrate progress in realising these purchases. The Commission will base its assessment on the cumulative criteria outlined in Annex 5 and assess these aspects in a stringent manner. Where a Member State fails to satisfactorily fulfil the full set of criteria the Commission will require a proportional reduction in the cap proposed.

3.4. Substantiation of other policies and measures

20. The substantiation of the effects of implemented and additional policies and measures by Member States is crucial for a national allocation plan's consistency with criterion 1 in Annex III of the Directive. In the first period national allocation plans Member States listed a number of existing and additional policies and measures. Each Member State relying on implemented and additional policies and measures, even if already indicated in the first round national allocation plan, needs to substantiate the effects and demonstrate progress in implementing or adopting them.[8] The Commission will base its assessment on the cumulative criteria outlined in Annex 6 and assess these aspects in a stringent manner. Where a Member State fails to satisfactorily fulfil the full set of criteria the Commission will require a proportional reduction in the cap proposed.

3.5. Guidance on criterion 12 – limit on JI and CDM compliance use by operators

21. Criterion 12 of Annex III of the Directive, as amended by the Linking Directive[9], states: *"The plan shall specify the maximum*

---

[8] In this regard, the Commission stresses the importance of allocation plans to be fully consistent with Member States' obligations pursuant to Directive 2001/77/EC for the promotion of electricity produced from renewable energy sources in the internal electricity market, OJ L 283, 27.10.2001, p.33.

[9] Directive 2004/101/EC of the European Parliament and of the Council of 27 October 2004 amending Directive 2003/87/EC establishing a scheme for greenhouse gas emission allowance trading within the Community, in respect of the Kyoto Protocol's project mechanisms, OJ L 338/18 of 13.11.2004.

amount of CERs and ERUs which maybe used by operators in the
Community scheme as a percentage of the allocation of the allowances
to each installation. The percentage shall be consistent with the Member
State's supplementarity obligations under the Kyoto Protocol and
decisions adopted pursuant to the UNFCCC or the Kyoto Protocol."

22.   Criterion 12 is mandatory in the sense that the national allocation
        plan must specify the maximum amount of CERs and ERUs that
        may be used for compliance purposes by operators in the EU ETS.

23.   Criterion 12 states that the established percentage must be
        consistent with the Member State's supplementarity obligations
        under the Kyoto Protocol and decisions adopted pursuant to the
        UNFCCC or the Kyoto protocol. The Marrakesh Accords state
        that *"the use of the mechanisms shall be supplemental to domestic
        action"*[10]. No quantitative definition of the supplementarity
        obligations is available in the Kyoto Protocol or the UNFCCC or
        the decisions adopted thereunder[11]. It should also be noted that
        the meeting of the conferences of the parties to the Kyoto Protocol
        in Montreal took a range of important decisions to stimulate the
        use of CDMs, to which EU emission trading can contribute.

24.   The supplementarity requirement is applicable to aggregate
        greenhouse gas emissions of a Member State and not separately
        for individual sectors. Therefore, the intended government
        purchase of Kyoto units also needs to be taken into consideration
        when evaluating the fulfilment of this requirement.

25.   The Commission considers that Member States are free to choose
        whether to apply the limit individually in respect of each
        installation, or collectively to all installations. For greater
        flexibility, Member States are recommended to apply the limit
        for the entire trading period and collectively to all installations.

3.6.   Issues related to new entrants and closures

26.   The Commission considers it premature to draw conclusions and
        identify best practice with respect to new entrants and closures.
        Further details are specified in Annex 7.

---

[10] Decision 15/CP.7, Art 1.
[11] The Commission's proposal for the Linking Directive provided for such a quantitative
definition (COM(2003)403).

3.7. Further guidance on allocation at sector and installation level

27.   In fixing installation level allocations in the second phase, the Commission considers it necessary that Member States do not rely on first phase emissions or other first phase data. Otherwise installations which have actively reduced emissions in the first trading period are unduly disadvantaged by receiving in the second phase a smaller share of allowances than installations that have not reduced emissions during the first period.

28.   By not relying on first phase emissions or other first phase data, early action is adequately recognised, which substitutes therefore for the set-up of an early action reserve or any other means of accommodating early action.

29.   In order to reduce the complexity and administrative effort the Commission considers it inappropriate to maintain special provisions at installation level on process emissions.

30.   As already stated above, the importance of achieving a simpler design of second phase national allocation plans compared to the first phase cannot be over-emphasized. Simpler allocation rules at sector and installation level increase the transparency of the allocation process and lower the costs in particular for small- and medium-sized enterprises covered by the scheme.

3.8. Further guidance on other allocation aspects

31.   EU-wide benchmarking is not a sufficiently matured allocation method to be used for the second phase. Member states may however find appropriate use for benchmarking at national level for the installation level allocation in certain sectors and for new entrants, e.g. in the electricity sector. Experiences from such use will be examined by the Commission in the context of the review. The Commission is interested in whether the additional data requirements for benchmarking can be mastered and whether Member States consider the additional administrative effort worthwhile.

32.   The Commission stresses that Member States can make use of auctioning within the 10%-limit permitted under Article 10 of the Directive in the second trading period. Making more use of auctioning would allow Member States and the Commission to collect more experience in applying this allocation method and to inform the strategic review with practical experience. It reminds

Member States that proceeds from auctioning can, amongst others, be used to cover the administrative costs of the scheme and the government purchase of Kyoto units. Where Member States choose to auction allowances the Commission encourages them to specify the details of the auction process well in advance, preferably in the national allocation plan, especially as regards timing and quantities involved.

33.   With respect to public consultation provided for by Articles 9(1) and 11(2) and criterion 9 of Annex III of the Directive, the Commission expects Member States to provide appropriate timelines to ensure a more effective public consultation with respect to the establishment of the second phase national allocation plan. Member States should aim to conclude the public consultation pursuant to Article 11(2) and criterion 9 of Annex III in time in order to respect the deadline of 31 December 2006. As there should be less time pressure in the preparation of the second trading period than was the case for the first phase, the Commission is confident that Member States will duly comply with this requirement under their own responsibility and discretion.

4.   INTERPRETATION OF THE SCOPE OF ANNEX I OF THE DIRECTIVE

4.1.   Combustion installations

34.   With respect to the interpretation of combustion installation in Annex I of the Directive, the Commission notes that some Member States based the first phase national allocation plans on an interpretation which included all combustion processes fulfilling the specified capacity, regardless of whether the combustion process produces energy independently or as an integrated part of another production process. Other Member States applied variants of a more narrow interpretation, excluding some or all combustion processes as integrated parts of another production process.

35.   The Commission regards this situation as highly unsatisfactory. From an Internal Market perspective it needs to be avoided that the same type of installation is covered in some Member States but not in others while applying the same Directive. A consistent interpretation and coverage of combustion installations across Member States in the second trading period is vital in order to avoid significant distortions of competition throughout the Internal Market.

36. The Commission considers the interpretation of combustion installation given in Annex 8 as the appropriate one. It understands that some Member States would have to include a number of additional installations, including large installations with significant emissions as well as some of the smallest emitters. However, in the light of the following chapter, the Commission recognises that it is not useful to include additional combustion processes which are typically carried out by small installations. In order to remove inconsistencies in the second trading period, all Member States should therefore in any case include also combustion processes involving crackers, carbon black, flaring[12], furnaces[13] and integrated steelworks[14], typically carried out in larger installations causing considerable emissions. The Commission reserves the right to take all necessary measures to avoid significant distortions. Details about the Commission's interpretation of combustion installation are contained in Annex 8.

4.2 The smallest installations

37. Certain concerns have been raised by Member States and stakeholders with respect to the coverage under the Directive of the smallest installations, claiming in particular that the costs of participation for the smallest installations outweigh the benefits of being covered by the scheme. The Commission recognises that the benefits and costs of participation of certain small installations merit further consideration in the review of the EU ETS pursuant to Article 30 of the Directive.

38. The Commission underlines that some participation costs incurred by the smallest installations are "one-off" costs in the run-up to the first trading period, and will not occur in the future. In terms of recurring costs, which are largely related to the annual costs of monitoring, reporting and verifying emissions, the Commission pays particular attention to realising the potentials for cost savings for the smallest installations in the ongoing review of the monitoring and reporting guidelines. The Commission aims for entry into force of the revised guidelines by 1 January 2008, thus coinciding with the start of the second trading period.

39. Moreover, it reiterates the importance of using simpler allocation rules for the second trading period in order to benefit the smallest

---

[12] Including off-shore.
[13] Including rock wool.
[14] Including rolling mills, re-heaters, annealing furnaces and pickling.

installations and look also at other aspects, besides monitoring and allocation, to alleviate the participation costs for these installations. The Commission is confident that this will further improve the relationship between benefits and costs for such installations of participating in the EU ETS.

40. The Commission invites Member States to explore the flexibilities identified in Annex 9 in the establishment of their second phase national allocation plans. It intends to consider in the review more comprehensively the scope of the Directive with respect to the coverage of the smallest installations, including the possibility to propose an amendment to the Directive to enable the removal of some small installations from the EU ETS in the course of the the second trading period. In that context, the Commission is considering the possibility that combustion activities below a certain size threshold, such as up to 3 MW, should not be counted for the purposes of the so-called aggregation rule. The Commission is also examining the possibility of removing the part of the aggregation rule that provides for the adding together of the capacities of activities that are operated by the same operator on the same site.

Annex 1: Background data

Member State	2003 national greenhouse gas emissions	Allowed emissions annual average 2008-12 under Kyoto Protocol	ETS share[15]	First phase cap annual average 2005-07 according to Commission decisions[16]
Austria	91.6	68.3	36.00%	33
Belgium	147.7	135.8	42.60%	62.9
Cyprus	9.2	n.a.	62.00%	5.7
Czech Republic	145.4	176.8	67.10%	97.6
Denmark	74	55	45.30%	33.5
Estonia	21.4	40	88.60%	19
Finland	85.5	70.4	53.20%	45.5
France	557.2	568	28.10%	156.5
Germany	1017.5	986.1	49.00%	499
Greece	137.6	139.6	54.10%	74.4
Hungary	83.2	114.3	37.60%	31.3
Ireland	67.6	61	33.00%	22.3
Italy	569.8	477.2	40.80%	232.5
Latvia	10.5	23.3	43.40%	4.6
Lithuania	17.2	46.9	71.20%	12.3
Luxembourg	11.3	9.2	29.80%	3.4
Malta	2.9	n.a.	n.a.	2.9
Netherlands	214.8	200.3	44.40%	95.3
Poland	384	531.3	62.30%	239.1
Portugal	81.2	75.4	47.00%	38.2
Slovakia	51.7	66	59.00%	30.5
Slovenia	19.8	18.8	44.30%	8.8
Spain	402.3	329	43.40%	174.4
Sweden	70.6	75.2	32.50%	22.9
UK	651.1	657.4	37.70%	245.3
Total				**2190.8**

Note: All emission figures are in million tonnes CO_2 equivalent.

---

[15] The ETS share is calculated as the first period cap divided by 2003 national greenhouse gas emissions.
[16] These figures do not account for changes to the number of installations subsequent to the respective Commission decision (e.g. opt-ins or opt-outs of installations).

Annex 2: Criteria for national allocation plans referred to in Articles 9, 22 and 30 of Annex III of the Directive

1. The total quantity of allowances to be allocated for the relevant period shall be consistent with the Member State's obligation to limit its emissions pursuant to Decision 2002/358/EC and the Kyoto Protocol, taking into account, on the one hand, the proportion of overall emissions that these allowances represent in comparison with emissions from sources not covered by this Directive and, on the other hand, national energy policies, and should be consistent with the national climate change programme. The total quantity of allowances to be allocated shall not be more than is likely to be needed for the strict application of the criteria of this Annex. Prior to 2008, the quantity shall be consistent with a path towards achieving or over-achieving each Member State's target under Decision 2002/358/EC and the Kyoto Protocol.

2. The total quantity of allowances to be allocated shall be consistent with assessments of actual and projected progress towards fulfilling the Member States' contributions to the Community's commitments made pursuant to Decision 93/389/EEC.

3. Quantities of allowances to be allocated shall be consistent with the potential, including the technological potential, of activities covered by this scheme to reduce emissions. Member States may base their distribution of allowances on average emissions of greenhouse gases by product in each activity and achievable progress in each activity.

4. The plan shall be consistent with other Community legislative and policy instruments. Account should be taken of unavoidable increases in emissions resulting from new legislative requirements.

5. The plan shall not discriminate between companies or sectors in such a way as to unduly favour certain undertakings or activities in accordance with the requirements of the Treaty, in particular Articles 87 and 88 thereof.

6. The plan shall contain information on the manner in which new entrants will be able to begin participating in the Community scheme in the Member State concerned.

7. The plan may accommodate early action and shall contain information on the manner in which early action is taken into account. Benchmarks derived from reference documents concerning the best available technologies may be employed by Member States in developing their

National Allocation Plans, and these benchmarks can incorporate an element of accommodating early action.

8. The plan shall contain information on the manner in which clean technology, including energy efficient technologies, are taken into account.

9. The plan shall include provisions for comments to be expressed by the public, and contain information on the arrangements by which due account will be taken of these comments before a decision on the allocation of allowances is taken.

10. The plan shall contain a list of the installations covered by this Directive with the quantities of allowances intended to be allocated to each.

11. The plan may contain information on the manner in which the existence of competition from countries or entities outside the Union will be taken into account.

12. The plan shall specify the maximum amount of CERs and ERUs which may be used by operators in the Community scheme as a percentage of the allocation of the allowances to each installation. The percentage shall be consistent with the Member State's supplementarity obligations under the Kyoto Protocol and decisions adopted pursuant to the UNFCCC or the Kyoto Protocol.

Annex 3: Key messages from the first allocation guidance document

In January 2004, the Commission provided guidance to assist Member States in the preparation of the national allocation plans[17]. The guidance contained in that document on the implementation of the then eleven[18] criteria in Annex III to the Directive remainsrelevant for the second trading period 2008-2012. The Commission therefore wishes to reiterate the main elements.

Criterion (1) – Kyoto commitments

The Commission understands "likely to be needed" as forward-looking and linked to the projected emissions of covered installations as a whole, given that this criterion refers to the total quantity of allowances to be allocated. The Commission understands the reference to the "strict application of the criteria in this annex" to comprise the criteria with a mandatory character or containing mandatory elements - i.e. criteria 1, 2, 3, 4 and 5.

In order to satisfy this requirement and fulfil all mandatory criteria and elements, a Member State should not allocate more than is needed, or warranted, by the most constraining of these criteria.

It follows that any application of the optional elements of Annex III may not lead to an increase in the total quantity of allowances.

Criterion (2) – Assessments of emissions developments

Pursuant to Decision 280/2004/EC concerning a mechanism for monitoring Community greenhouse gas emissions and for implementing the Kyoto Protocol , the Commission undertakes an annual assessment of each Member State's actual emissions and projected emissions for the period 2008-2012, in total and by sector and by gas. Criterion 2 requires the total quantity of allowances to be allocated to be consistent with these assessments.

Consistency will be deemed as ensured, if the total quantity of allowances to be allocated to covered installations is not more than would be necessary taking into account actual emissions and projected emissions contained in those assessments.

---

[17] Commission Communication COM (2003) 830 final, 7.1.2004.
[18] Directive 2004/156/EC ("the Linking Directive") added a criterion 12 to Annex III to Directive 2003/87/EC.

Criterion (3) – Potential to reduce emissions

A Member State should determine the total quantity of allowances resulting from the application of criterion 3 by comparing the potential of activities covered by the scheme to reduce emissions with the potential of activities not covered.

The criterion will be deemed as fulfilled if the allocation reflects the relative differences in the potential between the total covered and non-covered activities.

Criterion (4) – Consistency with other legislation

Criterion 4 concerns the relationship between allocations under Directive 2003/87/EC and other Community legislative and policy instruments. Consistency between allowance allocations and other legislation is introduced as a requirement in order to ensure that the allocation does not contravene the provisions of other legislation.

In principle, no allowances should be allocated in cases where other legislation implies that covered emissions had or will have to be reduced even without the introduction of the emissions trading scheme. Similarly, consistency implies that if other legislation results in increased emissions or limits the scope for decreasing emissions covered by the Directive account should be taken of this increase.

Criterion (6) – New entrants

Under criterion 6, the national allocation plan should contain information on the manner in which new entrants will be able to begin participating in the emissions trading scheme in a Member State.

The guidance proposes three ways in which new entrants can begin participating in the emissions trading scheme: by buying allowances in the market, by buying them in an auction, or by receiving them for free from a reserve set aside by the Member State.

Having new entrants buy allowances in the market or in an auction is in accordance with the principle of equal treatment.

Criterion (10) – List of installations

This criterion will be deemed as fulfilled, if a Member State has respected its obligation to list all the installations covered by the Directive. A Member State has to indicate the total quantity of allowances intended to be allocated to each installation.

Annex 4: Summary of experience gained from allocation plans for the first phase (2005-2007) and general lessons for the second phase (2008-2012)

1.   **More use of emissions trading is necessary to meet the Kyoto targets cost-effectively.** Some Member States rely to a large degree on reductions in the non-trading sectors or on government purchase of Kyoto unit credits in the pursuit of their Kyoto targets. The intended government purchase of Kyoto units and the foreseen reduction efforts in the non-trading sectors have served in the first allocation phase as buffers resulting in moderate use of emission trading. In some Member States too much of the reduction effort may have been shifted to the non-trading sectors. Maintaining this imbalance would make Kyoto compliance more costly than necessary. Given that emissions trading is the most cost-effective instrument at hand, it should be used more in the second allocation round and beyond.

2.   **Allocations have in general been more restrictive for power generators than other sectors covered by the scheme.** In most Member States, the allocation to the power generating sector, in relation to projected needs, has been more restrictive, i.e. more environmentally ambitious, than the allocations to the other sectors covered by the scheme.

3.   **Member States experiencing considerable excess in actual emissions with respect to their Kyoto targets intend to purchase a substantial amount of Kyoto units.** Eight Member States announced in the first phase national allocation plans their intention to purchase with government funds in total some 500 to 600 million Kyoto units. Given the general outlook for Joint Implementation (JI) and Clean Development Mechanism (CDM), the envisaged volume will be very challenging to realise. Furthermore, the Linking Directive will add private-sector demand to government demand for such credits. The Commission considers it as a matter of priority to improve the functioning of these mechanisms.

4.   **The non-acceptance of ex-post adjustments is essential for the allowance market development.** The Commission did not approve the so-called ex-post adjustments envisaged by a number of Member States for the first trading period. This plays a vital role in the development of an efficient and liquid allowance market. The good functioning of the allowance market depends crucially on a stable and predictable allocation for the entire

trading period in order to create stable incentives for installations to reduce emissions. For compliance purposes, companies can use the full flexibility of the scheme, be it via the allowance market or via company-internal transfers across borders.

5.   **Some allocation plans are more complex than necessary and not sufficiently transparent.** In the first national allocation plans, some Member States created a complex set of special allocation rules: all Member States provided for a new entrants reserve and most also for some kind of administrative provision in the case of closure of an installation (i.e. no further allocation of allowances for the remainder of the ongoing trading period once an installation is closed). The design of new entrants and closure rules differs in detail. This contributes to a high degree of complexity and intransparency in the internal market and may result in unnecessary distortions of competition. Member States should consider simplifying all rules which they have added themselves and which are not essential for the functioning of the scheme. Simpler rules will help make national allocations plans more transparent.

Annex 5: Information requested to assess substantiation of intended government purchase of Kyoto units

Member States must substantiate the intended government purchase of Kyoto units and are requested to provide the following information in the national allocation plan:

1. indicate the amount of Kyoto units planned to be purchased for compliance with the Kyoto target and any changes in this amount compared to the first national allocation plan;

2. indicate the type of Kyoto units planned to be purchased, along with their respective projected or contracted purchase price;

3. demonstrate the existence of relevant national legislation and budget allocations;

4. provide information on the progress to date in realising the planned purchases, in particular the quantity of Kyoto units for which emission reduction purchase contracts have been signed at the time of notification of the second national allocation plan;

5. indicate the envisaged time schedule of still to be effected purchases;

6. outline the administrative arrangements put in place for realising the planned purchases, such as national programmes or purchase tenders for purchasing Kyoto units;

7. indicate details about the contributions of multilateral or private carbon purchase funds and the expected delivery of credits;

8. demonstrate the existence of contingency measures applicable in the event that planned purchases and signed purchase agreements result in the delivery of a lower than expected amount of Kyoto units.

Annex 6: Information requested to assess substantiation of other policies and measures

Member States must substantiate the effects of implemented and additional policies and measures and are requested to provide the following information in the national allocation plan:

1. indicate the **implemented** policies and measures it considers as significant in sectors not covered by the EU ETS. For sectoral framework policies implemented (e.g. rural development plan, waste management plan) the plan has to provide the individual measures included that are considered to lead to greenhouse gas emission reductions. For cross-sectoral policies and measures, the plan has to indicate in which way those measures affect emissions in the trading and non-trading sectors. The information provided has to include the year in which the implementation showed full effect;

2. indicate **additional** policies and measures not yet implemented at the time of notification which the Member State considers as significant. The plan has to present information on the status of planning or adoption of relevant legislation, agreements, incentive programmes, etc. and has to address theperiod for which full additional reduction effects are expected;

3. indicate the approximate level of current greenhouse gas emissions represented by the activity targeted by each policy or measure and include quantified annual emissions reductions for the period 2008 to 2012 for the policies and measures indicated under the two preceding bullets. If no quantitative estimation of effects is available, the plan should explain why this information could not be provided and should include additional information why the policy or measure is considered to provide significant emission reduction effects;

4. provide assumptions and methodologies used for the quantification of the effects of indicated policies and measures and provide references to sources for this information;

5. present quantitative indicators to demonstrate the effectiveness of the policy or measure under the first requirement;

6. indicate how policies and measures presented under the first two requirements are reflected in the greenhouse gas emissions projections presented in the plan;

7. indicate any developments and trends of the activities targeted by the policiesand measures provided under the first two requirements that could potentially counteract the reduction effects, e.g. increased production capacities or growing trends in consumption patterns;

8. indicate any overlapping effects among important measures (e.g. effects of cross-sectoral measures and sectoral measures on the same activity) and how such double-counting effects have been eliminated in the estimation of quantitative reduction effects.

Annex 7: Issues related to new entrants and closures

1.   The Commission notes that in the first trading period all Member States have set aside allowances for new entrants in a reserve and most adopted some form of closure provisions. The Commission did not raise objections to these administrative provisions and rules to the extent that they were not tantamount to ex-post adjustments.

2.   The Commission notes further a multitude of detailed provisions governing new entrants reserves and closures, including transfer rule arrangements, adopted by Member States in the first allocation phase. This contributes to a high degree of complexity and intransparency in the internal market and may result in distortions of competition. At this stage, there is however insufficient practical experience with regard to the practical application of these rules.

3.   For this reason, the Commission considers it premature to draw conclusions and identify best practice. In the case of new entrants' reserves and closure and transfer provisions being maintained in the second trading period, the Commission recommends Member States ensure in particular that the new entrants reserve not be replenished upon exhaustion, that allowances not allocated to closed installations be cancelled or auctioned, and that there be no allocation at projected needs to new installations.

4.   In the review report in June 2006[19], the Commission will consider alternative options (including the set-up of an EU-level new entrant reserve accompanied by EU-wide administrative rules on closure and cross-border transfer) to achieve further harmonisation with respect to new entrants and closure provisions.

---

[19] As provided for by Article 30(2) of the Directive

Annex 8: Definition of combustion installation

1. The Commission considers the interpretation including all combustion processes, i.e oxidation of fuels, fulfilling the specified capacity to be the correct interpretation of Annex I of the Directive, for the following notable reasons:

2. Firstly, the term "combustion" is used in a wide range of Community legislation including not only the Emissions Trading Directive and the IPPC-Directive, but also the LCP-Directive[20] and the Sulphur in Liquid Fuels-Directive[21]. The meaning of combustion in the context of the Emissions Trading Directive has to be interpreted within the framework of other Community legislation where definitions are included.

3. The Sulphur in Liquid Fuels-Directive in its Article 2(5) and the LCP-Directive in its Article 2(7) define 'combustion plant' as "any technical apparatus in which fuels are oxidised in order to use the heat thus generated". The LCP-Directive lists in the same Article a range of combustion plants which are specifically excluded from the scope of the LCP-Directive. The Emissions Trading Directive does not provide for such exclusion.[22]

4. Given that the Emissions Trading Directive makes no similar specific exclusions, the types of combustion installations excluded by Article 2(7) of the LCP-Directive are included within the scope of the Emissions Trading Directive where the threshold is met or exceeded.

5. Further guidance in support of this conclusion comes from Annex I of the Emissions Trading Directive itself. Annex I specifically excludes municipal and hazardous waste incineration facilities from the scope of the scheme. The combustion of e.g. hazardous waste is clearly an integrated part of the normal process undertaken by hazardous waste incinerators. If, in the absence of this specific exclusion, the Directive were to be interpreted as

---

[20] Directive 2001/80/EC on the limitation of emissions of certain pollutants into the air from large combustion plants, OJ L 309, 27.11.2001, p. 1.

[21] Directive 1999/32/EC relating to a reduction in the sulphur content of certain liquid fuels, OJ L 121, 11.05.1999, p. 13.

[22] Certain activities that are specifically excluded by the LCP-Directive are also excluded from the Emissions Trading Directive, such as "(h) any technical apparatus used in the propulsion of a vehicle, ship or aircraft" because the Emissions Trading Directive only applies to stationary technical units (Article 3(e)). The Emissions Trading Directive therefore covers neither transportation in general nor greenhouse gas emissions arising from traffic on the site of an installation.

not applying to such installations where combustion takes place as an integrated part of the installation's processes, municipal and hazardous waste installations would not need to have been specifically excluded as they would in any case have fallen outside its scope. Their specific exclusion is further confirmation that it is the presence of a combustion process with a rated thermal input exceeding 20MW that determines the Directive's coverage of stationary combustion installations.

6. It is also commonly accepted that the term "combustion installation" for the purposes of the IPPC-Directive covers not just the power generation industry but also other industries where fuels are burned. Thus the heading "Energy industries" in the context of the IPPC Directive does not imply a narrow restriction of coverage of the term "combustion installations" to combustion processes that produce energy independently, but rather also includes combustion processes taking place as an integrated part of another production process. The heading "Energy activities" used in the Emissions Trading Directive, if anything, would be broader, so at least the same conclusion would apply. This therefore provides additional support for the argument that "combustion installations" in the Emissions Trading Directive not only covers combustion installations that are part of the energy industry, but also combustion installations in other industry sectors, including sectors that are not explicitly listed in its Annex I.

7. It is well-established that industries can fall under more than one activity category of the IPPC-Directive. Integrated steel works for example carry out several Annex I activities, and refineries include combustion installations of more than 50MW. Considering the similarities between the IPPC-Directive and the Emissions Trading Directive, there is no reason to take a different approach to the interpretation of the latter in this respect. In particular, a different approach cannot be justified by the separate listing of the steel and cement industries, given that both produce substantial CO_2 emissions from (chemical) processes in addition to their emissions from combustion.

8. In the light of the above points, any installation, which includes one or more piece of stationary technical apparatus in which a combustion process takes place and that together on the same site and under the responsibility of the same operator has a rated thermal input exceeding 20MW, is therefore subject to the Emissions Trading Directive. This includes apparatus where the

heat is used in another piece of apparatus, through a medium such as electricity or steam, and apparatus where the heat resulting from combustion is used directly within that apparatus, for example, for melting, drying, flares or units providing heat input to chemical reactors. The purpose to which the product of an activity is put should not be a determining characteristic as to whether or not an installation is subject to the Directive, as this would introduce subjectivity into its scope. Energy produced by combustion may be in the form of electricity, heat, hot water or steam, and the distance between the production of energy and its eventual use is not relevant for competent authorities to decide whether or not an installation is subject to the Emissions Trading Directive.

Annex 9: Interpretation issues related to the smallest installations

1. The Commission draws Member States' attention to the fact that the so-called aggregation clause[23] contained in the second paragraph of Annex I of the Directive should be interpreted carefully so as to not cover certain small installations, without prejudice to the interpretation of such or similar wording in other Community legislation. In particular, the wording "under the same subheading" contained in this clause should be understood in the sense that a single activity falling simultaneously under several subheadings, e.g. both under "energy activities" and under a specific sectoral activity covered by Annex I of the Directive, such as "mineral industry", is considered under the more specific sectoral subheading. Multiple activities of the same type should then be aggregated on the basis of that specific sectoral subheading, and not on the basis of all of the different possible activity descriptions that could apply. There is no basis for aggregating activities that fall under a different subheading, even though they may be part of the same installation.

2. Furthermore, flexibility at the discretion of Member States comes also from the wording "and/or" in the provision governing the manufacture of ceramic products in Annex I of the Directive. If Member States want to use this flexibility the Commission notes that this provision can be interpreted in a restrictive way so as to require the simultaneous presence of all mentioned sub-elements for the second trading period, again without prejudice to the interpretation of such or similar wording in other Community legislation. In this context, the Commission draws the attention of Member States to the Declaration of the Council and the Commission of 4 September 1996[24] supporting an interpretation of the same wording contained in Annex I of the IPPC-Directive, that it is up to Member States to decide as to whether one of the two criteria or both criteria need to be fulfilled at the same time.

---

[23] "2. The threshold values given below generally refer to production capacities or outputs. Where one operator carries out several activities falling under the same subheading in the same installation or on the same site, the capacities of such activities are added together."

[24] Council Declaration of 4 September 1996 on Directive 96/61/EC of the Council on Integrated Pollution Prevention and Control, 9388/96, Interinstitutional dossier No. 00/0526 (SYN).

Annex 10: Set of NAP common format summary tables

I NAP summary table – target calculation
(Grey fields are filled out automatically)

Row	Data table	Emissions (Mt CO2eq)	Emissions (Mt CO2eq)
A		Target under Kyoto Protocol or Burden Sharing Agreement (avg. annual GHG emissions 2008-12)	
B	III	*Total GHG emissions 2003 (excluding LULUCF emissions and removals)*	
C		**Difference +/-** (row A - row B) (negative means need to reduce)	0
D	III	*Av. annual projected total GHG emissions 2008-2012 ('with measures' projection)*	
E		Difference +/- (row A - row D) (negative means need to reduce)	0
		Reduction measures *(where relevant)*	
F	V	EU emissions trading scheme	
G	VII	Additional policies and measures (other than emissions trading), including LULUCF	
H		Government purchase of Kyoto mechanisms	
I		**Total reduction measures** (row F + row G + row H)	0

IIA NAP Summary table – Basic data
(Grey fields are filled out automatically)

		1990	1991	1992	1993	1994	1995	1996	1997	1998	1999	2000	2001
A	Real GDP¹ (in billion €2000)	Absolute											
		Trend index 2003=100											
B	Emissions¹ (Mt of CO2)	Absolute											
		Trend index 2003=100											
C	Carbon intensity¹ (million tonnes CO2 / billion €)	Absolute											
		Trend index 2003=100											

		2002	2003	2004	2005	2006	2007	2008	2009	2010	2011	2012	Annual average 2008-12
A	Real GDP¹ (in billion €2000)	Absolute											
		Trend index 2003=100											
B	Emissions¹ (Mt of CO2)	Absolute											
		Trend index 2003=100											
C	Carbon intensity¹ (million tonnes CO2 / billion €)	Absolute											
		Trend index 2003=100											

[1]   Indicate source(s), separately per year where relevant.

IIB NAP Summary table – Basic data on electricity sector[1]
(Grey fields are filled out automatically)

	Year	2000	2003	2004	2005	2006	2007	2008	2009	2010	2011	2012	Average 2008-12
A	Total domestic electricity production (TWh)												
B	Imports (TWh)												
	Country 1												
	Country n												
	Other countries												
	Total Imports												
C	Exports (Twh)												
	Country 1												
	Country n												
	Other countries												
	Total Exports												
D	Electricity trade balance (TWh, total row B - total row C)												
E	Share of gas in total domestic electricity production (%)												
F	Share of oil in total domestic electricity production (%)												
G	Share of coal in total domestic electricity production (%)												
H	Share of nuclear energy in total domestic electricity production (%)												
I	Share of renewable energy, including biomass, in total domestic electricity production (%)									[2]			

[1] Indicate source(s), separately per year where relevant.
[2] This cell should also include (in parentheses) the target pursuant to Directive 2001/77/EC.

III NAP Summary table – Recent and projected greenhouse gas emissions per common reporting format sector (without taking into account additional policies and measures in Table VI)

(Grey fields are filled out automatically)

In CO2eq

Row ref	CRF subsector			2003	2004	2005	2006	2007	2008	2009	2010	2011	2012	Average annual projected emissions 2008-2012
A	1.A.1	Energy generation	GHG											
B			CO2 in ETS											
C	1.A.3	Transport	GHG											
D	1.A.4 a + b + c	Commercial and institutional, Residential, and Agricultural energy use	GHG											
E			CO2 in ETS											
F	2	Industrial processes	GHG											
G			CO2 in ETS											
I	4	Agriculture	GHG											
J	5	Land-Use Change and Forest	GHG											
K	6	Waste	GHG											
L	1.A.2 + 1.A.4	All other sectors	GHG											
M	+ 1.A.5 + 1.B + 3 + 7		CO2 in ETS											
N		Total	GHG											
O		Total in ETS	ETSCO2	Rows B + E + G + M										

IV NAP Summary table – Recent and projected CO$_2$ emissions in sectors covered by the EU emissions trading scheme

(Grey fields are filled out automatically)

	Emissions in Mt CO$_2$eq	i	ii	iii	iv	v	vi	vii	viii	ix	x	xi
	Year	2003	2004	2005	2006	2007	2008	2009	2010	2011	2012	Average annual projected emissions 2008 – 2012*
A	combustion installations total (excluding installations covered under rows B–J)											
	main activity 1											
	main activity 2											
	flaring											
	integrated steelworks											
	crackers											
	furnaces											
	main activity n											
B	mineral oil refineries											
C	coke ovens											
D	metal ore roasting, sintering, pig iron and steel producing installations											
E	cement producing installations											
F	lime producing installations											
G	glass and glass fibre producing installations											
H	ceramics producing installations											
I	pulp, paper and board producing installations											
J	Total	Rows A to J										
K	Share of EU ETS CO2 in total GHG emissions (%)	Row L / Row B in Table IIa										

* Numbers to be used in last two columns of Table V.

V NAP Summary table – Proposed allocation in relation to first period allocation (without additional policies and measures) in the sectors covered by the EU emissions trading scheme
(Grey fields are filled out automatically)

		i	ii	iii	iv	v
		2003 actual CO_2 emissions (Mt CO_2)[1]	2004 actual CO_2 emissions (Mt CO_2)	Average annual allocation 2005 - 2007	Proposed average annual allocation in 2008-2012	Proposed ETS allocation as a percentage of first period ETS
A	combustion installations total (excluding installations covered under rows B-J)					col iv / col iii
	main activity 1					
	main activity 2					
	flaring					
	integrated steelworks					
	crackers					
	furnaces					
	main activity n					
B	mineral oil refineries					
C	coke ovens					
D	metal ore roasting, sintering, pig iron and steel producing installations					
F	cement producing installations					
G	lime producing installations					
H	glass and glass fibre producing installations					
I	ceramics producing installations					
J	pulp, paper and board producing installations					
L	Total					

VI NAP Summary table – Reductions expected by policies and measures other than the EU emissions trading scheme and which have not been taken into account for the "with measures" projection presented in Table III (Mt CO$_2$eq)

	Under implementation (1)			Adopted (2)			Planned (3)		
	Expected average annual reduction (2008-12)		Full effects expected as from year	Expected average annual reduction (2008-12)		Full effects expected as from year	Expected average annual reduction (2008-12)		Full effects expected as from year
Measures	In ETS sectors	In non-ETS sectors		In ETS sectors	In non-ETS sectors		In ETS sectors	In non-ETS sectors	
i	iii		iv	v		vi	viii		ix
A									
B									
C									
D									
E									
F									
G									
H									
I									
X Subtotal									
Total	equal to row G in Table I								

(1) where the full or a substantial part of the effect can be expected, not the first year of implementation.
(2) The measure has been adopted by the final instance at the relevant local, regiona or national level, but it is not yet implemented.
(3) The measure is at least mentioned in a formal government document.

VII NAP Summary table – Government's planned use of Kyoto units (Mt CO_2eq) and status of implementation

	Planned purchase		ERUs	CERs	AAUs and others	Total
A		Total 2008-2012				Σ (equal to row H on table I)
B		Annual average				
C	Quantity of units already paid for					
D	Quantity of units contracted, but yet unpaid (delivery pending start of UN ITL) (1)					
E	Neither bought nor contracted by date of notification (A - C - D)					
F	Full budget appropriated to first commitment period (2008-12)	Currently available for 2006				
G		Committed for the future				
H	Implied future price ((F+G)/A)					

(1) Units partially paid for should be proportionally distributed between lines C and D

(Grey fields are filled out automatically)

Issues with respect to new entrants	Description of NAP provisions
Does the plan contain a new entrants' reserve?	
What is its size in absolute terms and as a percentage of the total quantity of allowances for the period?	
What use is made of allowances left over in the reserve at the end of the trading period? (cancellation, sold)	
How will new entrants be treated in case the reserve runs out of allowances before the end of the trading period? (reserve replenished, further new entrants buy in the market)	
Does the allocation to the new entrant depend on the actual choice of fuel?	
Does the allocation to the new entrant depend on the actual choice of technology?	
Does the allocation to the new entrant depend on the estimated or actual number of operating hours or does the allocation use a standard number of operating hours?	
Auctioning	
Will any allowances be auctioned?	
What share of the total quantity of allowances will be auctioned?	
Who can participate in the auction?	
What auctioning method will be used?	
When/at what intervals will the auction(s) be held?	
What quantity of allowances will be auctioned each time?	
What use will be made of the revenues?	
Will the auctions be coordinated with any auctions in other Member States?	
Closures	
Do operators have to report to the competent authority when an installation closes, and on what conditions is an installation considered to be closed?	
Does the operator continue to be issued allowances for a closed installation in the remaining years of the trading period? If the reply depends on whether the operator sets up a new entrant installation replacing the closed installation, please briefly describe the provision.	
What happens to any allowances that were intended for an installation, which will not receive them after closure? (cancellation, fed into a new entrants' reserve, auctioning)	

VIII NAP Summary table – Details on new entrants, closures and auctioning

	Power plant with a rated thermal input exceeding 20 MW	Power plant with a rated thermal input exceeding 20 MW
Maximum capacity of the actual installation	(At least 100 MW)	(At least 100 MW)
Fuel (s) used	Coal	Gas
Forecast number of operating hours/year in the period 2008 to 2012		
Annual allowance allocation in 2008 to 2012		

IX NAP Summary table – Further details on selected new entrants
X NAP Summary table - Important assumptions on annual averages

Year	EU Allowance price (in Euro)	Crude oil price (Brent) (1)	Natural gas price (1)	Coal price (1)	Exchange rate (2)	Other
2005						
2006						
2007						
2008						
2009						
2010						
2011						
2012						

(1) Use common market standard and specify, including the currency used; indicate in detail sources of data and methodologies

(2) For those Member States outside the Euro-zone

Explanatory comments on NAP Common Format summary tables

Note: Grey fields are filled in automatically when using the Excel spreadsheets.

Table I: NAP summary table – target calculation

General description:

The purpose of this table is to provide an overview of key data relevant for NAP assessment.
The gap (row C) between the Kyoto target (row A) and actual greenhouse gas emissions in 2003 (row B) is presented with necessary corresponding reduction measures (quantified in the fourth column of rows F-H, and totalled in row I). The gap is also expressed as the difference between the Kyoto target (row A) and the projected annual average total greenhouse gas emissions from 2008-2012 (row D). This figure is indicated in row E.

Specific remarks:

The second column makes a cross-reference to other data tables.

The fourth column refers to emissions or effects on emissions from measures recorded in the third column.

All rows with the exception of rows B and C contain annual averages relating to the second trading period 2008 to 2012.

Table IIa: NAP Summary table – Basic data

General description:

Table IIa gives an overview of historic and expected trends in various factors crucial to the calculation of a Member State's potential to reduce emissions: namely, real GDP (row A), greenhouse gas emissions (row B) and carbon intensity (row C).

All three factors are expressed both in absolute numbers and in a trend index, with 2003 being the base year (2003=100).

Specific remarks:

In order to have a complete picture, the Commission invites Member

States to provide annual
data from 1990 to 2012. While re-stating some date in the public domain,
Table IIa is of added value as an integral part of the NAP ensuring
transparency and easy access to this information for stakeholders and
other Member States.

Member States are required to indicate the sources of the information
used, separately per year where relevant.

For the period 2008 to 2012, the Commission prefers annual data to better
understand the development of these figures over time. In case a Member
State can justify why such annual data are not available, the Commission
would also accept the submission of only annual averages for the period
2008 to 2012, to be indicated in the respective column.

Table IIb: NAP Summary table – Basic data on electricity sector

General description:

Table IIb indicates the basic data for the electricity sector. The purpose
is to obtain a comprehensive picture of total domestic electricity
production (row A), imports (row B) and exports (row C), the electricity
trade balance (row D, constituting the difference between rows B and
C) as well as the shares of different fuels (gas, oil, coal, nuclear energy,
and renewable energy) in total domestic electricity production (rows E-
I).

Specific remarks:

Imports and exports (rows B and C) need to be disaggregated into the
most important countries to/from which the export/import takes place,
as well as a row with the remainder to other countries, and the total
figure. These figures will allow the Commission to cross-check the
plausibility of indications by individual Member States of their respective
exports and imports, which would naturally need to be compatible with
each other.

Member States are required to indicate the sources of the information
used (separately per year where relevant) and are encouraged to provide
annual data also for the period 2008 to 2012.

If a Member State can justify why such annual data are not available,
the Commission requires explanation and at least the submission of data
for a recent year and annual averages for the period 2008 to 2012.
Similarly, Member States should provide data on the fuel mix as
accurately as possible.

Naturally, the future fuel mix will depend on estimates, amongst others, of the allowance price. Member States are requested to indicate their respective estimates in the explanations in the NAP and also in Table X.

Member States should introduce also the target pursuant to Directive 2001/77/EC in Table 2b for the year 2010.

Table III: NAP Summary table – Recent and projected greenhouse gas emissions per common reporting format sector (without taking into account additional policies and measures in Table VI)

General description:

Table III relates recent and projected greenhouse gas emissions per common reporting format sector, as further specified by the numbers for the respective sub-sectors in the second column. Where indicated, the emissions should be indicated for total greenhouse gases and CO_2 in the EU ETS.

The Commission recognises the technical difficulty to complete this table but stresses the importance of bringing together the categories in the UNFCCC-based common reporting format with the categories under EU ETS reporting.

Specific remarks:

The second column indicates the sub-sectoral reference under the Common Reporting Format (CRF).

The Commission recognises that some Member States may not have all the data available to complete Table III. If a Member State can justify why such annual or sectoral data is not available, the Commission requires at least the submission of data for a recent year and annual averages for the period 2008 to 2012 for as many sectors as possible, as well as aggregate figures (total and total in ETS).

CO_2 emissions in the ETS sector depend on estimates, amongst others, on the allowance price.
Member States are requested to indicate their respective estimates in the explanations in the NAP and also in Table X.

Table IV: NAP Summary table – Recent and projected CO_2 emissions in sectors covered by the EU emissions trading scheme

General description:

Table IV looks more specifically at the recent and projected CO_2 emissions by installation or sector covered by the EU ETS, relating them to the activities mentioned in Annex I of the Directive. Certain activities have been aggregated where separate information is likely not to be available or necessary for the Commission's assessment.

Specific remarks:

Emissions from combustion installations shall be calculated without emissions from installations also covered under the specific sectors of Annex I of the Directive being indicated in rows B-J. As a matter of example, where a combustion installation is also covered by the category "installations for the production of cement clinker …" under the subheading "mineral industry" of Annex I of the Directive, emissions from that installation should fall under the entry "cement producing installations" in row E of Table IV, and should be omitted from row A "combustion installations". Moreover, emissions from these combustion installations shall be disaggregated into the most important activities to be identified by each Member State, including flaring, integrated steelworks, crackers and furnaces.

For the period 2008 to 2012, the Commission prefers annual data to better understand the development of all sectors. Where a Member State can justify the absence of such annual data for certain sectors, the Commission requires at least the submission of data for a recent year and annual averages for the period 2008 to 2012 in as many sectors as possible. If a Member State can show this to be appropriate, certain sectors may be (dis-)aggregated; in particular coke ovens (row C) with metal ore roasting, sintering, pig iron and steel producing installations (row D). Where such data are not available on an annual basis, the Commission requires a justification and at least the submission of data for a recent year as well as annual averages for the period 2008 to 2012 for as many sectors as possible, as well as aggregate figures (total and total in ETS).

The amount entered in row J, column XI correlates to Table III, row O, last column. The amount entered in row K, column XI correlates to Table III, row N, last column.

Table V: NAP Summary table – Proposed allocation in relation to first period allocation (without additional policies and measures) in the sectors covered by the EU emissions trading scheme

General description:

For installations or sectors covered by the EU ETS, Table V indicates 2003 and 2004 actual emissions (columns i and ii) as well as the proposed second period allocation in relation to first trading period allocation (columns iii and iv). Column v indicates the proposed second period allocation as a percentage of the first period allocation. The same sectoral specification is used as in Table IV.

Specific remarks:

Emissions from combustion installations shall be calculated without emissions from installations covered also under the specific sectors of Annex I of the Directive being indicated in rows B-J. As a matter of example, where a combustion installation is also covered by the category "installations for the production of cement clinker ..." under the subheading "mineral industry" of Annex I of the Directive, emissions from that installation should fall under the entry "cement producing installations" in row E of Table IV, and should be omitted from row A "combustion installations". Moreover, emissions from these combustion installations shall be disaggregated into the most important activities to be identified by each Member State, including flaring, integrated steelworks, crackers and furnaces.

For the period 2008 to 2012, the Commission prefers annual data to better understand the development of all sectors. Where a Member State can justify why such annual data is not available for all sectors, the Commission requires at least the submission of data for a recent year and annual averages for the period 2008 to 2012 in as many sectors as possible, as well as aggregate figures (total and total in ETS). If a Member State can show it to be appropriate, certain sectors may be (dis-)aggregated; in particular coke ovens (row C) with metal ore roasting, sintering, pig iron and steel producing installations (row D).

Table VI: NAP Summary table – Reductions expected by policies and measures other than the EU emissions trading scheme and which have not been taken into account for the "with measures" projection presented in Table III (Mt CO_2eq)

General description:

Table VI gives account of greenhouse gas emissions reductions expected by policies and measures other than the EU ETS, which have not been taken into account for the "with measures" projection presented in Table III.

It classifies such measures into three categories: "under implementation" (columns i-iii), "adopted" (columns iv-vi), and "planned" (columns vii-ix).

"Under implementation" means that the implementation is ongoing, and that the measure is not taken into account for the "with measures" projections presented in Table III.

"Adopted" means that the measure has been adopted by the final instance at the relevant local, regional or national level, but it is not yet implemented.

"Planned" means that the measure is at least mentioned in a formal government document, but not adopted.

Each of these three categories is again subdivided into three columns: the expected average annual reduction (2008-12), on the one hand, in ETS sectors (columns i, iv and vii), and, on the other hand, in non-ETS sectors (columns ii, v and viii). The third sub-column (iii, vi and ix, respectively) indicates the year, in which the full or a substantial part of the effects of the respective measure can be expected (not necessarily the first year of implementation).

The rows shall contain the measures to be specified in the second column.

Specific remarks:

The Commission recognises that for some measures the disaggregation of the expected reductions into those occurring outside and inside the ETS presents a technical difficulty. It is however an important element for the Commission's assessment.

Table VII: NAP Summary table – Government's planned use of Kyoto units (Mt CO_2eq) and status of implementation

General description:

Table VII gives a detailed overview on the government's planned use of Kyoto units and the status of their implementation.

It subdivides the Kyoto units into ERUs from JI projects, CERs from CDM projects, and AAUs and other units from international emissions trading. The last column indicates the total of the three types combined.

The status of implementation is presented in the rows, as follows.

Rows A and B indicate the sum across the various degrees of implementation, with row A giving the total amount in the period 2008 to 2012 and row B the annual average in that period per type of Kyoto unit and as a total. The total annual average across all three forms of Kyoto units is equal to row H of Table I.

Row C indicates the most advanced degree of implementation, i.e. the quantity of units already paid for.

Row D gives a lesser degree of implementation, which is the quantity of units contracted, but yet unpaid (delivery pending start of UN ITL). Units partially paid for should be proportionally distributed between rows C and D.

Row E relates to the quantity with the lowest degree of implementation, i.e. the units neither bought nor contracted by the date of notification (Row E = Row A – Row C – Row D).

Rows F and G give additional information on the full budget appropriated to the first commitment period (2008-12), both the one currently available for 2006 (row F) and the one committed up to 2012 (row G).

Row H indicates the implied future price of Kyoto units, which is the sum of rows F and G, divided by the total planned purchase in row A.

Specific remarks:

The Commission prefers Member States to specify the breakdown into ERUs, CERs, and AAUs and others. In case a Member State can justify why such a breakdown is not feasible, the Commission requires at least the submission of separate figures for ERUs and CERs on the one hand and AAUs and others on the other hand.

Table VIII: NAP Summary table – Details on new entrants, closures and auctioning

Table VIII contains various questions relating to important information on new entrants, auctioning and closures. The questions should be self-explanatory.

Table IX: NAP Summary table – Further details on new entrants

Table IX asks for further details on a selected new entrant, e.g. a power plant with a rated thermal input of 100 MW.

In one scenario (second column) the fuel used is coal, while in the other (third column) it is gas.

Member States are then requested to fill in row 4 (forecast number of operating hours/year in the period 2008 to 2012), where such a forecast is relevant for the allocation under the new entrants rule of the Member State, and row 5 (annual allowance allocation in 2008 to 2012).

This information will allow the Commission to better assess the standards used in the allocation to new entrants and at the same time provide for more transparency.

Table X: NAP Summary table – Important assumptions on annual averages

In Table X, Member States are requested to quantify for the years 2005-12 their key assumptions on annual average figures underlying the establishment of the NAP, in particular for:

- the EU allowance price (in Euro);

- the price for crude oil (Brent);

- the price for natural gas;

- the coal price; and

- the exchange rate (for those Member States outside the Euro-zone).

Member States should use and specify common market standards for fuel prices, including the currency used. They should indicate in detail sources of data and methodologies. This information is necessary in order to ensure comparability of data and transparency.

Member States are invited to indicate further assumptions considered important and useful for the Commission's assessment.

ANNEX 6

THE GREENHOUSE GAS EMISSIONS TRADING SCHEME REGULATIONS 2005

(unofficial consolidation)

Made   *23rd March 2005*
Laid before Parliament   *30th March 2005*
Coming into force   *21st April 2005*

ARRANGEMENT OF REGULATIONS

PART 1
GENERAL

PART 2
GREENHOUSE GAS EMISSIONS PERMITS

PART 3
ALLOWANCES

The Secretary of State, in exercise of the powers conferred upon her by subsection (2) of section 2 of the European Communities Act 1972(**1**), being a Minister designated for the purpose of that subsection in relation to greenhouse gas emission allowance trading(**2**), and subsection (1) of section 2 of the Pollution Prevention and Control Act 1999(**3**) (the "1999 Act"), having in accordance with subsection (4) of section 2 of the 1999 Act, consulted the Environment Agency, the Scottish Environment Protection Agency, such bodies or persons appearing to her to be representative of the interests of local government, industry, agriculture and small businesses respectively as she considers appropriate and such other bodies and persons as she considers appropriate, hereby makes the following Regulations:

---

(1) 1972 c. 68. As regards Scotland, see also section 57(1) of the Scotland Act 1998 (c. 46), which provides that, despite the transfer to the Scottish Ministers by virtue of that Act of functions in relation to observing and implementing obligations under Community law, any function of a Minister of the Crown in relation to any matter shall continue to be exercisable by him as regards Scotland for the purposes specified in section 2(2) of the European Communities Act 1972.

(2)   S.I. 2004/1984.

(3)   1999 c.24.

PART 1

GENERAL

Citation and commencement

1. These Regulations may be cited as the Greenhouse Gas Emissions Trading Scheme Regulations 2005 and shall come into force on 21st April 2005.

Interpretation

2. — (1) In these Regulations —

"1995 Act" means the Environment Act 1995(**4**);

"address" means, in relation to electronic communications, any number or address used for the purposes of such communications;

"allowance" has the meaning given to it in Article 3 of the Directive;

"appeal body" means the body to which an appeal may be made under regulations 32 or 33;

"appropriate authority" means (except where regulation 27(15) or 32(14) apply) —

(i)   in relation to an installation which is (or will be) situated in England and an offshore installation, the Secretary of State;

(ii)   in relation to an installation (other than an offshore installation) which is (or will be) situated in Scotland, the Scottish Ministers;

(iii)   in relation to an installation (other than an offshore installation) which is (or will be) situated in Wales, the National Assembly for Wales; and

(iv)   in relation to an installation (other than an offshore installation) which is (or will be) situated in Northern Ireland, the Department of the Environment;

"approved national allocation plan" means, in respect of a scheme phase, a national allocation plan specified in approved NAP regulations as the approved national allocation plan for that scheme phase;

"approved NAP regulations" means, in relation to a scheme phase, regulations made by the Secretary of State under section 2(2) of the European Communities Act 1972 specifying as the approved

----

(**4**) 1995 c. 25.

national allocation plan a national allocation plan developed for that scheme phase which has not been rejected by the European Commission or in relation to which the European Commission has accepted amendments in accordance with Article 9(3) of the Directive;

"CER" means certified emission reduction and has the meaning given to it in Article 3 of the Directive;

"change in operation" means, in relation to an installation, a change in the nature, functioning or scope of the installation which—

(i)   affects any information included in the greenhouse gas emissions permit pursuant to regulation 9(8)(d); or

(ii)   might, in the opinion of the regulator, require any monitoring and reporting condition to be amended;

"chief inspector" means the chief inspector constituted under regulation 8(3) of the Northern Ireland Regulations;

"the Directive" means Directive 2003/87/EC of the European Parliament and of the Council establishing a scheme for greenhouse gas emission allowance trading within the Community and amending Council Directive 96/61/EC(5), as amended by Directive 2004/101/EC;

"electronic communication" has the same meaning as in the Electronic Communications Act 2000(6);

"ERU" means an emission reduction unit and has the meaning given to it in Article 3 of the Directive;

"enforcement notice" has the meaning given by regulation 29(1);

"excluded installation" means an installation in respect of which the operator holds a valid certificate served under regulation 11(6);

"greenhouse gas emissions" means the release of greenhouse gases into the atmosphere from sources in an installation;

"greenhouse gas emissions permit" means a permit granted under regulation 9;

---

(5) OJ No. L 275, 25.10.03, p.32. The Directive is amended by Directive 2004/101/EC, OJ No. L 338, 13.11.2004, p.18.
(6) 2000 c.7; the definition of electronic communication in section 15(1) was amended by the Communications Act 2003 (c. 21), section 406(1) and Schedule 17, paragraph 158.

"greenhouse gases" has the meaning given to it in Article 3 of the Directive;

"installation" means (except where it appears in Schedule 1)—

(i)   a stationary technical unit where one or more Schedule 1 activities are carried out; and

(ii)   any other location on the same site where any other directly associated activities are carried out which have a technical connection with the activities carried out in the stationary technical unit and which could have an effect on greenhouse gas emissions and pollution,

and references to an installation include references to part of an installation;

"issue" means, in relation to allowances, the transfer in the registry of allowances allocated in respect of an installation from the party holding account to the operator holding account relating to that installation;

"monitoring and reporting condition" means a condition of a greenhouse gas emissions permit imposed pursuant to regulation 10(2) (but excluding conditions imposed pursuant to regulation 10(2)(c));

"the Monitoring and Reporting Decision" means Commission Decision 2004/156/EC establishing guidelines for the monitoring and reporting of greenhouse gas emissions pursuant to Directive 2003/87/EC of the European Parliament and of the Council(7);

"national allocation plan" means a plan developed in accordance with Articles 9, 10 and 30(3) of and Annex III to the Directive;

"new entrant" has the meaning given in Article 3 of the Directive;

"new entrant reserve" means any reserve of allowances provided for in the approved national allocation plan for distribution to new entrants and shall not include any allowances removed from the reserve by virtue of a correction to the national allocation plan table made under Article 38(2) or 44(2) of the Registries Regulation;

"Northern Ireland Regulations" means the Pollution Prevention and Control Regulations (Northern Ireland) 2003(8);

"notice of surrender" has the meaning given by regulation 16(5);

---

(7) OJ No. L 59 , 26.02.04, p.1.
(8) S.R. (NI) 2003 No 46, amended by S.I. 2003/496, S.I.2003/3311; there is another amending instrument which is not relevant.

"offshore installation" means an installation which is (or will be) situated in the area (together with places above and below it) comprising—

(i)   those parts of the sea adjacent to England and Wales from the low water mark to the landward baseline of the United Kingdom territorial sea;

(ii)   the United Kingdom territorial sea apart from those areas comprised in any controlled waters within the meaning of section 30A(1) of the Control of Pollution Act 1974(**9**); and

(iii)   those areas of sea in any area for the time being designated under section 1(7) of the Continental Shelf Act 1964(**10**);

"Offshore Regulations" means the Offshore Combustion Installations (Prevention and Control of Pollution) Regulations 2001(**11**);

"operator" means, subject to paragraph (2), in relation to an installation, the person who has control over its operation;

"Planning Appeals Commission" means the Planning Appeals Commission established under Article 110 of the Planning (Northern Ireland) Order 1991(**12**);

"project activity" has the meaning given to it in Article 3 of the Directive;

"Registries Regulation" means Commission Regulation 2216/2004 of 21 December 2004 for a standardised and secured system of registries pursuant to Directive 2003/87/EC of the European Parliament and of the Council and Decision 280/2004/EC of the European Parliament and the Council(**13**) and, unless the context otherwise requires, expressions used in these Regulations which are also used in the Registries Regulation have the same meaning as they have in the Registries Regulation;

"regulator" means—

(**9**) 1974 c. 40; section 30A(1) was inserted by section 169 of, and Schedule 23 to, the Water Act 1989 (c. 15).
(**10**) 1964 c. 29; section 1(7) was amended by the Oil and Gas (Enterprise Act) 1982 (c. 23), section 37 and Schedule 3, paragraph 1.
(**11**) S.I. 2001/1091, amended by S.I. 2003/3311.
(**12**) S.I. 1991/1220 (N.I.11); relevant amending instruments are S.I. 1999/663, 2003/430 (N.I.8).
(**13**) OJ No. L 386, 29.12.2004, p.1

(i)   in relation to an installation (other than an offshore installation) which is (or will be) situated in England and Wales, the Environment Agency;

(ii)   in relation to an installation (other than an offshore installation) which is (or will be) situated in Scotland, the Scottish Environment Protection Agency;

(iii)   in relation to an installation (other than an offshore installation) which is (or will be) situated in Northern Ireland, the chief inspector; and

(iv)   in relation to an offshore installation, the Secretary of State;

"reportable emissions" means, in relation to an installation, the total specified emissions (expressed in tonnes of carbon dioxide equivalent) which arise from the Schedule 1 activities carried out in that installation; and "annual reportable emissions" means, subject to regulation 10(4), the reportable emissions arising during any scheme year;

"responsible authority" has the meaning given by regulation 11(12);

"retention notice" has the meaning given by regulation 24(8);

"revocation notice" has the meaning given by regulation 17(1);

"Schedule 1 activity" means an activity falling within a description in Schedule 1;

"scheme phase" means—

(i)   the three year period beginning on 1st January 2005;

(ii)   the five year period beginning on 1st January 2008; or

(iii)   each subsequent five year period;

"scheme year" means a year beginning with 1st January;

"specified emissions" means, in relation to any Schedule 1 activity, the greenhouse gas emissions specified in Schedule 1 in relation to that activity;

"tonne of carbon dioxide equivalent" has the meaning given to it in Article 3 of the Directive;

"variation notice" has the meaning given by regulation 14(9);

"working day" means any day other than a Saturday, Sunday, Good Friday, Christmas

Day or day which is a bank holiday within the meaning of the Banking and Financial

Dealings Act 1971(**a**).

(2)   For the purposes of these Regulations—

(a)   where an installation has not been put into operation, the person who will have control over the operation of the installation when it is put into operation shall be treated as the operator of the installation;

(b)   where an installation has ceased to be in operation, the person who holds the greenhouse gas emissions permit which relates to the installation shall be treated as the operator of the installation; and

(c)   where a permit holder has ceased to be the operator of an installation to which a greenhouse gas emissions permit relates references to the operator shall be read as references to the permit holder.

Application of these Regulations to the Crown

3. —   (1) Subject to the provisions of this regulation, these Regulations bind the Crown.

(2) No contravention by the Crown of any provision of these Regulations shall make the Crown criminally liable under regulation 38 but the High Court, or in relation to an installation in Scotland the Court of Session, may, on the application of a regulator, declare unlawful any act or omission of the Crown which constitutes such a contravention.

(3) Notwithstanding paragraph (2), the provisions of these Regulations apply to persons in the service of the Crown as they apply to other persons.

(4) If the Secretary of State certifies that it appears to her, as respects any Crown premises and any specified powers of entry exercisable under section 108 of the Environment Act 1995(**14**) or regulation 27 of the Northern Ireland Regulations in relation to functions conferred or imposed by these Regulations, that it is requisite or expedient that, in the interests of national security, the powers of entry should not be exercisable in relation to those premises, those powers shall not be exercisable in relation to those premises; and

---

(**14**) 1995 c. 25; section 108 was amended by the Pollution Prevention and Control Act 1999 (c. 24), section 6(2) and Schedule3, in relation to England and Wales by the Pollution Prevention and Control Regulations 2000(S.I.2000/1973), regulation 39 and Schedule 10, Part 1, paragraphs 14 and 16 and the Anti-Social Behaviour Act 2003, section 55(6), (7), (8) and (9)and in relation to Scotland by the Pollution Prevention and Control (Scotland) Regulations 2000 (S.S.I. 2000/323), regulation 36, Schedule 10, Part 1, paragraph 5(1) and (3).

in this paragraph "specified" means specified in the certificate and "Crown premises" means premises held or used by or on behalf of the Crown.

(5) The following persons shall be treated as if they were the operator of the installation concerned for the purpose of any notice served or given or any proceedings instituted in relation to an installation operated or controlled by any person acting on behalf of the Royal Household, the Duchy of Lancaster or the Duke of Cornwall or other possessor of the Duchy of Cornwall—

(a) in relation to an installation operated or controlled by a person acting on behalf of the Royal Household, the Keeper of the Privy Purse;

(b) in relation to an installation operated or controlled by a person acting on behalf of the Duchy of Lancaster, such person as the Chancellor of the Duchy appoints in relation to that installation; and

(c) in relation to an installation operated or controlled by a person acting on behalf of the Duchy of Cornwall, such person as the Duke of Cornwall, or the possessor for the time being of the Duchy of Cornwall, appoints in relation to that installation.

Notices

4. (1) Any notice or other document served or given under these Regulations by an appropriate authority, a responsible authority, the Secretary of State or a regulator shall be in writing or if the person to be served with or given any such notice or document has provided an address for service using electronic communications, by electronic communications.

(2) Any such notice or other document may be served on or given to a person by—

(a) leaving it at his proper address;

(b) sending it by post to him at that address; or

(c) where an address for service using electronic communications has been given by that person, sending it using electronic communications to that person at that address.

(3) Any such notice or other document may—

(a) in the case of a body corporate (other than a limited liability partnership), be served on the secretary or clerk of that body;

    (b)   in the case of a limited liability partnership, be served on a member; or

    (c)   in the case of a partnership (other than a limited liability partnership), be served on or given to a partner or person having the control or management of the partnership business.

(4) For the purpose of this regulation and of section 7 of the Interpretation Act 1978(**15**) (service of documents by post) in its application to this regulation, the proper address of any person on or to whom any such notice or other document is to be served or given shall be his last known address, except that—

    (a)   in the case of a body corporate (other than a limited liability partnership) or its secretary or clerk, it shall be the address of the registered or principal office of that body;

    (b)   in the case of a limited liability partnership or a member of a limited liability partnership, it shall be the registered or principal office of that partnership;

    (c)   in the case of a partnership (other than a limited liability partnership) or person having the control or management of the partnership business, it shall be the principal office of the partnership,

and for the purposes of this paragraph the principal office of a company registered outside the United Kingdom or of a partnership carrying on business outside the United Kingdom shall be its principal office within the United Kingdom.

(5) If the person to be served with or given any such notice or document has specified an address in the United Kingdom other than his proper address within the meaning of paragraph (4) as the one at which he or someone on his behalf will accept notices or documents of the same description as that notice or document, that address shall also be treated for the purposes of this regulation and section 7 of the Interpretation Act 1978 as his proper address.

(6) Where a notice or document is served or given using electronic communications, the service is deemed to be effected by properly addressing and transmitting the electronic communication.

Applications

5. —   (1) A regulator may require any application or type of application made to it under any provision of these Regulations to be made on a form made available by the regulator.

(**15**) 1978 c. 30.

(2) A form made available by a regulator under paragraph (1) shall specify the information required by the regulator to determine the application, which shall include any information required to be contained in the application by the provision of these Regulations under which the application is made.

(3) Where a regulator makes available a form under paragraph (1) in relation to the making of applications to it under a provision of these Regulations any application made to it under that provision shall be made on that form.

(4) Any application made under these Regulations may, with the agreement of the regulator, be sent to the regulator electronically.

(5) Where an application which is required to be accompanied by a fee is sent electronically, the fee may be sent to the regulator separately from the application, but the application shall not be treated as having been received by the regulator until the fee has also been received.

(6) An application made under these Regulations may be withdrawn at any time before it is determined.

(7) In its application to regulation 11 (excluded installations), paragraphs (1) to (6) shall apply as if any reference to "the regulator" were a reference to "the responsible authority".

(8) In its application to regulation 27 (pooling), paragraphs (1) to (6) shall apply as if any reference to "the regulator" were a reference to "the appropriate authority".

Functions of the regulator: Northern Ireland

6. Any functions conferred or imposed by these Regulations on the chief inspector may be delegated by him to any inspector appointed under regulation 8(1) of the Northern Ireland Regulations.

PART 2

GREENHOUSE GAS EMISSIONS PERMITS

Requirement for greenhouse gas emissions permit to carry out Schedule 1 activities

7. No person shall carry out a Schedule 1 activity resulting in specified emissions, except under and to the extent authorised by a greenhouse gas emissions permit.

Applications for greenhouse gas emissions permits

8. — (1) An application for a greenhouse gas emissions permit shall be made to the regulator in accordance with this regulation and shall, except where the application relates to an offshore installation, be accompanied by the fee prescribed in respect of the application in Schedule 5.

(2) An application under paragraph (1) shall contain the following information—

(a)   the name of the applicant, his telephone number and postal address (including post code) and, if different, any address to which correspondence relating to the application should be sent and, if the applicant is a body corporate, its registered number, the postal address of its registered or principal office and, if that body corporate is a subsidiary of a holding company (within the meaning of section 736 of the Companies Act 1985(**16**) or, in relation to Northern Ireland, article 4 of the Companies (Northern Ireland) Order 1986(**17**)), the name of the ultimate holding company and the postal address of its principal office;

(b)   the postal address of the site of the installation and its national grid reference (or for offshore installations equivalent information identifying the installation and its location), a description of that site and the location of the installation on that site, and, for installations other than offshore installations, the name of any local authority in whose area the site is situated;

(c)   a description of the installation and the Schedule 1 activities to be carried out in the installation including a description of the technology used;

(d)   the raw and auxiliary materials used in carrying out Schedule 1 activities in the installation, the use of which is likely to lead to specified emissions;

(e)   the sources of specified emissions from the Schedule 1 activities carried out in the installation;

(f)   a description of the measures which are planned to monitor and report specified emissions in accordance with the Monitoring and Reporting Decision;

(g)   a description, including the reference number, of any environmental licence issued in relation to the installation;

(**16**) 1985 c. 6; section 736 was substituted by section 144(1) of the Companies Act 1989 (1989 c. 40).
(**17**) S.I. 1986/1032 (N.I.6), amended by S.I. 1990/1504 (N.I.10); there is other amending legislation but none of it is relevant.

(h)   any additional information which the applicant wishes the regulator to take into account in considering the application; and

(i)   a non-technical summary of the information referred to in sub-paragraphs (c) to (h).

(3) For the purpose of paragraph (2)(g), "environmental licence" means —

(a)   an authorisation under Part I of the Environmental Protection Act 1990(**18**) or the Industrial Pollution Control (Northern Ireland) Order 1997(**19**); or

(b)   a permit granted under —

(i)   the Pollution Prevention and Control (England and Wales) Regulations 2000(**20**);

(ii)   the Pollution Prevention and Control (Scotland) Regulations 2000(**21**);

(iii)   the Offshore Regulations; or

(iv)   the Northern Ireland Regulations.

(4) Where an application is for a greenhouse gas emissions permit to operate more than one installation the application shall contain the information required by paragraph (2) in relation to each installation.

(5) The regulator may, by notice to the applicant, require him to furnish such further information specified in the notice, within the period so specified, as the regulator may require for the purpose of determining the application and if the applicant fails to furnish the specified information within the period specified, the application shall, if the regulator gives notice to the applicant that it treats the application as having been withdrawn, be deemed to have been withdrawn at the end of that period.

Determination of applications and grant of greenhouse gas emissions permits

9.—   (1) The regulator shall give notice of its determination of an application for a greenhouse gas emissions permit within a period

---

(**18**) 1990 c. 43.
(**19**) S.I. 1997/2777 (N.I.18), to which there are amendments not relevant to these Regulations.
(**20**) S.I. 2000/1973, relevant amending instruments are S.I. 2002/1559, S.I. 2003/3311; there are other amending instruments but none is relevant.
(**21**) S.S.I. 2000/323, relevant amending instruments are S.S.I. 2003/146, 2003/170, 2003/235, 2004/26 and 2005/101 and S.I. 2003/3311.

of two months beginning on the date on which it received the application or within such longer period as may be agreed in writing with the applicant.

(2) For the purpose of calculating the period of two months mentioned in paragraph (1) no account shall be taken of any period beginning with the date on which notice is served on the applicant under regulation 8(5) and ending on the date on which the applicant furnishes the information specified in the notice.

(3) If a regulator fails to give notice of its determination of an application for a greenhouse gas emissions permit within the period allowed by or under paragraph (1), the application shall, if the applicant notifies the regulator that he treats the application as having been refused, be deemed to have been refused at the end of that period.

(4) Subject to paragraph (6), where an application is duly made to the regulator, the regulator shall either grant the greenhouse gas emissions permit subject to the conditions required or authorised to be imposed by regulation 10 or refuse the application.

(5) Where a regulator grants a greenhouse gas emissions permit—

(a) in relation to an offshore installation; or

(b) in response to an application for a greenhouse gas emissions permit which was made before the date on which these Regulations enter into force,

the fee prescribed in Schedule 5 in respect of the grant of the permit shall be payable within a period of 28 days beginning on the date on which the regulator serves a notice on the operator requesting payment of the fee.

(6) A greenhouse gas emissions permit shall not be granted if the regulator—

(a) considers that the applicant will not be the operator of the installation concerned after the grant of the permit; or

(b) is not satisfied that the applicant will ensure that the installation is operated so as to comply with the monitoring and reporting conditions which would be included in the permit.

(7) A greenhouse gas emissions permit may authorise the operation of more than one installation on the same site operated

by the same operator but may not otherwise authorise the operation of more than one installation.

(8)   A greenhouse gas emissions permit authorising the operation of an installation shall include—

(a)   the name and postal address of the operator and, if different, any address to which correspondence should be sent;

(b)   the postal address of the site of the installation and its national grid reference (or for offshore installations equivalent information identifying the installation and its location);

(c)   a description of the site and the location of the installation on that site; and

(d)   a description of the installation, the Schedule 1 activities to be carried out in the installation and the specified emissions from those activities.

(9)   Where—

(a)   the provisions of a greenhouse gas emissions permit have been varied under regulation 14 or by a retention notice or affected by a partial transfer under regulation 15; or

(b)   there is more than one greenhouse gas emissions permit applying to installations on the same site operated by the same operator,

the regulator may replace the permit with a consolidated permit applying to the same Schedule 1 activities and subject to the same conditions as the permit being replaced.

Conditions of greenhouse gas emissions permits

10.—   (1) There shall be included in a greenhouse gas emissions permit such conditions as the regulator considers appropriate and in particular such conditions as the regulator considers appropriate to comply with paragraphs (2) to (6).

(2)   A greenhouse gas emissions permit shall include conditions concerning the monitoring and reporting of specified emissions from the installation to which it relates and, in particular—

(a)   conditions to ensure that any specified emissions from the Schedule 1 activity to which it relates are monitored and reported in accordance with the Monitoring and Reporting Decision, including conditions—

(i)   specifying the monitoring methodology and frequency; and

(ii)   requiring the operator to submit reports of the annual reportable emissions to the regulator and concerning the timing of such reports;

(b)   a requirement that all reports submitted pursuant to conditions imposed under sub-paragraph (ii) are verified in accordance with the criteria set out in Annex V of the Directive and that the regulator is informed of the results of any such verification; and

(c)   a requirement that an operator notifies the regulator as soon as he becomes aware of any factor which might prevent him from complying with any of the conditions included in a greenhouse gas emissions permit pursuant to this paragraph.

(3)   A greenhouse gas emissions permit shall contain conditions to ensure that the operator surrenders allowances equal to the annual reportable emissions from the installation within four months of the end of the scheme year during which those emissions arose.

(4)   A greenhouse gas emissions permit shall provide that for the purpose of assessing compliance with a condition imposed pursuant to paragraph (3) in relation to a recovery year, the annual reportable emissions from the installation in that year shall be deemed to be increased by an amount equal to the amount of annual reportable emissions in respect of which the operator failed to comply with that condition in the non-compliance year.

(5)   For the purposes of paragraph (4)—

(a)   "a non-compliance year" shall be a scheme year in respect of which an operator fails to comply with a condition of the permit imposed pursuant to paragraph (3); and

(b)   "the recovery year" shall be the scheme year following the non-compliance year, or where the non-compliance results from an error in the report submitted by an operator under a monitoring and reporting condition, the scheme year in which the error is discovered.

(6)   A greenhouse gas emissions permit shall contain a condition stating that in relation to any period for which the installation is an excluded installation (the "exclusion period")—

(a)   the operator shall be deemed to be in compliance with any conditions imposed pursuant to paragraphs (2) and (3); and

(b)   the operator shall be required to notify the regulator of any change in operation during the exclusion period, at least 2 months before the end of that exclusion period or within 10 days of the revocation of a certificate served under regulation 11(6) (except in relation to a part of the installation in respect of which a new certificate is issued in accordance with regulation 11(11)(b)).

(7)   Subject to paragraph (8), where an operator makes a change in operation to an excluded installation, the greenhouse gas emissions permit which relates to that installation shall, for the duration of the period for which the installation is an excluded installation, be deemed to authorise the change in operation.

(8)   Paragraph (7) shall not prevent an operator from making an application under regulation 14(2) for a variation of the provisions of a greenhouse gas emissions permit which relates to an excluded installation.

(9)   Regulators shall periodically review the conditions of greenhouse gas emissions permits and may do so at any time.

Excluded installations

11.—   (1) Where the European Commission has provided for the temporary exclusion of an installation under Article 27(2) of the Directive, the operator of the installation may apply to the responsible authority for a certificate stating that the installation is an excluded installation.

(2)   The Secretary of State shall, within 7 days beginning on the date of a notification by the European Commission of a decision under Article 27(2) of the Directive providing for the temporary exclusion of an installation or of the publication of the decision in the Official Journal of the European Union, whichever is the earlier, publish or, as the case may be, further publish the decision in such manner as she considers appropriate.

(3)   Subject to paragraph (4), an application under paragraph (1) shall be made before the expiry of the period of two months beginning with the date on which the Secretary of State publishes or, as the case may be, further publishes a European Commission decision in accordance with paragraph (2), whichever is the later.

(4)   A responsible authority may accept an application after the date by which an application is required to be made under paragraph (3).

(5)   An application under paragraph (1) shall—

(a)   identify the installation in question;

(b)   contain the name and postal address of the operator and, if different, any address to which correspondence should be sent;

(c)   contain a copy of the greenhouse gas emissions permit relating to the installation identified in sub-paragraph (a);

(d)   identify the regulator which granted that permit; and

(e)   identify the national policy by virtue of which the European Commission has provided for the temporary exclusion of the installation.

(6)   Where an application is duly made, the responsible authority shall serve on the regulator and the operator a certificate which shall—

(a)   identify the installation;

(b)   identify the operator and the regulator of that installation;

(c)   state the date from which the installation is excluded and the duration of the exclusion and identify any period for which it is deemed to be excluded in accordance with paragraph (7); and

(d)   specify any conditions applying to the exclusion.

(7) Where a decision of the European Commission under Article 27(2) of the Directive provides for an installation to be temporarily excluded from a date prior to the date of that decision, the certificate served under paragraph (6) in respect of that installation may provide that for the purposes of compliance with any conditions of the greenhouse gas emissions permit imposed pursuant to regulation 10(2) or (3) the installation shall be treated as if it were an excluded installation from the date provided for in the decision of the European Commission.

(8) Where an application for a certificate under paragraph (1) relates to an installation in respect of which an allocation for the first scheme phase has been made under regulation 21(1)(b), or if applicable (1)(c), or under regulation 22(13)(b) or (18) the responsible authority, shall before serving a certificate under paragraph (6) take such steps as are necessary to ensure that—

(a)   no allowances will be issued in respect of the installation to which the application relates in respect of any whole scheme year for which the European Commission has provided for the installation to be excluded;

(b) the amount of allowances to be issued in respect of the installation to which the application relates in respect of any scheme year in which the European Commission provides for the installation to be excluded for only part of the year shall be reduced in proportion to the part of the year to which the exclusion relates; and

(c) the total quantity of allowances to be issued in the scheme phase is reduced by the number of allowances which will not be issued in accordance with sub-paragraph (a) and (b).

(9) Where an operator fails to comply with the conditions referred to in paragraph (6)(d), the responsible authority may serve a notice on the operator and the regulator revoking the certificate served under paragraph (6).

(10) Where the regulator effects a transfer or partial transfer under regulation 15 of a greenhouse gas emissions permit which relates to an excluded installation, the regulator shall notify the responsible authority of the transfer or partial transfer and provide a copy of the updated permit and any new permit granted which relates to that installation.

(11) Where the regulator notifies the responsible authority in accordance with paragraph (10)—

(a) in the case of a transfer of the whole greenhouse gas emissions permit, if the responsible authority is satisfied that the installation will continue to be covered by the national policy identified in the application under paragraph (1), the responsible authority shall serve a notice on the operator and regulator including a copy of the certificate served under paragraph (6) and specifying the change of operator;

(b) in the case of a partial transfer, the responsible authority shall revoke the certificate served under paragraph (6) and if the responsible authority is satisfied that any part of the installation will continue to be covered by the national policy identified in the application under paragraph (1), the responsible authority shall serve on the operator and the regulator of that part a certificate under paragraph (6) in relation to that part;

(c)   in any other case, the responsible authority shall serve a notice on the operator and the regulator revoking the certificate served under paragraph (6).

(12) For the purposes of this regulation, the "responsible authority" means, subject to paragraph (13), the person who is responsible for the national policy by virtue of which the European Commission has provided for the temporary exclusion of the installation.

(13) If there is doubt as to who is responsible for a particular national policy, the Secretary of State shall decide who is to be considered to be responsible for the policy for the purposes of this regulation.

Proposed change in operation

12.— (1) Subject to paragraph (4), where an operator of an installation who holds a greenhouse gas emissions permit in respect of the Schedule 1 activities carried out in the installation proposes to make a change in operation the operator shall, at least 14 days before making the change, notify the regulator.

(2)   A notification under paragraph (1) shall be in writing and shall contain a description of the proposed change in operation including a brief explanation of whether and, if so, why it—

(a)   affects any information included in the greenhouse gas emissions permit pursuant to regulation 9(8)(d); or

(b)   might require any monitoring and reporting condition to be amended.

(3)   A regulator shall, by notice served on the operator, acknowledge receipt of any notification received under paragraph (1).

(4)   Paragraph (1) shall not apply where—

(a)   a change in operation is to be made more than 2 months before the end of the period for which the installation to which the change relates is an excluded installation; or

(b)   the operator applies under regulation 14(2) for the variation of the conditions of his greenhouse gas emissions permit before making the proposed change in operation and the application contains a description of that change.

Commencement of a Schedule 1 activity

13.— (1) Before the latest of—

(a)   1ˢᵗ April in each year;

(b)   the expiry of a period of 14 days beginning on the date of the grant of a greenhouse gas emissions permit under regulation 9(1); or

(c)   the expiry of a period of 14 days beginning with the date on which these Regulations enter into force,

the operator of an installation in respect of which a greenhouse gas emissions permit has been granted but which has not been put into operation may notify the regulator that it does not intend to put the installation into operation on or before 31ˢᵗ March in the following year ("the notified non-operation year").

(2)   Where an operator which has notified the regulator under paragraph (1) puts the installation into operation in a notified non-operation year, it shall notify the regulator that the installation has been put into operation before the expiry of a period of 14 days beginning on the day on which the installation is put into operation.

Variation of provisions of greenhouse gas emissions permits

14.— (1) The regulator may at any time vary any provision of a greenhouse gas emissions permit (including the extent to which the permit authorises a Schedule 1 activity) and shall do so if it appears to the regulator at that time, whether as a result of a review under regulation 10(9) or otherwise, that regulation 9(8) or 10 requires provisions to be included in the permit which are different from the subsisting provisions.

(2) An operator of an installation who holds a greenhouse gas emissions permit in respect of the Schedule 1 activity carried out in that installation may apply to the regulator for the variation of the provisions of his permit.

(3) An application under paragraph (2) shall be made in accordance with paragraph (5) and shall, subject to paragraph (4), be accompanied by the fee prescribed in respect of the application in Schedule 5.

(4) Where an application under paragraph (2) relates to an offshore installation, the fee prescribed in Schedule 5 in respect of the application shall be payable within the period of 28 days beginning on the date on which the regulator serves a notice on the operator requesting payment of the fee.

(5) An application under paragraph (2) shall contain the following information—

(a)   the name of the operator, his telephone number and postal address (including post code) and, if different, the address to which correspondence should be sent;

(b)   the postal address of the site of the installation to which the greenhouse gas emissions permit relates and its national grid reference (or, for offshore installations, equivalent information identifying the installation and its location);

(c)   if relevant, a description of the proposed change in operation requiring the variation and a statement of any change as respects the matters dealt with in regulation 8(2)(c) to (f) which would result if the proposed change in operation were made;

(d)   an indication of the variations of the provisions of the greenhouse gas emissions permit which the operator wishes the regulator to make; and

(e)   any additional information which the operator wishes the regulator to take into account in considering his application.

(6) The regulator may, by notice to the operator, require him to furnish such further information specified in the notice, within the period so specified, as the regulator may require for the purpose of determining the application; and if the operator fails to furnish the specified information within the period specified in the notice, the application shall, if the regulator gives notice to the operator that it treats the application as having been withdrawn, be deemed to have been withdrawn at the end of that period.

(7) Where an application is duly made to the regulator under paragraph (2), the regulator shall determine whether to vary the provisions of the greenhouse gas emissions permit and shall give notice of its determination within two months beginning with the day on which the regulator received the application or within such longer period as may be agreed in writing with the operator.

(8) For the purpose of calculating the periods mentioned in paragraph (7) no account shall be taken of any period beginning with the date on which notice is served on an operator under paragraph (6) and ending on the date on which the operator furnishes the information specified in the notice.

(9) Where the regulator decides to vary the provisions of the greenhouse gas emissions permit, whether on an application under paragraph (2) or otherwise, it shall serve a notice on the operator (a "variation notice") specifying the variations of the provisions of the permit and the date or dates on which the variations are to take effect and, unless the notice is withdrawn, the variations specified in the notice shall take effect on the date or dates so specified.

(10) A variation notice served under paragraph (9) shall, unless served for the purpose of determining an application under paragraph (2), require the operator to pay, within such period as may be specified in the notice, the fee prescribed in respect of the variation notice in Schedule 5.

(11) Where the regulator decides on an application under paragraph (2) not to vary the provisions of the greenhouse gas emissions permit, it shall give notice of its decision to the operator.

(12) If the regulator fails to give notice of its determination of an application under paragraph (2) within the period allowed by or under paragraphs (7) and (8), the application shall, if the operator notifies the regulator that he treats the application as having been refused, be deemed to have been refused at the end of that period.

Transfer of greenhouse gas emissions permits

15.— (1) Subject to paragraph (2), where the operator of an installation wishes to transfer, in whole or in part, his greenhouse gas emissions permit to another person ("the proposed transferee") the operator and the proposed transferee shall jointly make an application to the regulator to effect the transfer.

(2) A greenhouse gas emission permit which relates to an installation in which a Schedule 1 activity is no longer carried out may not be transferred.

(3) An application under paragraph (1) shall, subject to paragraph (5), be accompanied by the fee prescribed in respect of the transfer in Schedule 5 and shall contain the following information—

(a) the operator's and the proposed transferee's telephone number and postal address and, if different, any address to which correspondence relating to the application should be sent; and

(b) the postal address of the site of the installation to which the greenhouse gas emissions permit relates and its national grid reference (or, for offshore installations, equivalent information identifying the installation and its location).

(4) Where the operator wishes to transfer only part of his greenhouse gas emissions permit (a "partial transfer"), an application under paragraph (1) shall —

(a) identify the Schedule 1 activity or part of a Schedule 1 activity to which the transfer applies (the "transferred activity");

(b) identify the installation in which that transferred activity is carried out (the "transferred unit");

(c) where an application for an allocation of allowances from the new entrant reserve has been made under regulation 22(1) in respect of the installation to which the permit relates and either —

(i) the regulator has not determined the application in accordance with regulation 22(13); or

(ii) the number of allowances allocated under regulation 22(13)(b) or (18) is less than the number of allowances determined under regulation 22(13)(a),

specify whether the application under regulation 22(1) relates to the transferred unit;

(d) where the installation to which the permit relates is included in a pool in accordance with regulation 27, specify whether the installation, or in the case of a partial transfer, the transferred unit, should continue to be included in the pool; and

(e) where an application for an allowances from the late installation element of the new entrant reserve has been made under regulation 22A in respect of the installation to which the permit relates and the Secretary of State has not determined the application in accordance with regulation 22A(5), specify whether the application under regulation 22A relates to the transferred unit.

(5) Where an application under paragraph (1) relates to an offshore installation, the fee prescribed in Schedule 5 in respect of the application shall be payable within the period of 28 days beginning on the date on which the regulator serves a notice on the operator requesting payment of the fee.

(6) The regulator shall effect the transfer unless the regulator considers that—

(a)   the proposed transferee will not be the operator of the transferred unit after the transfer is effected; or

(b)   the proposed transferee will not ensure that the installation is operated so as to comply with any monitoring and reporting condition.

(7) The regulator shall effect a transfer under this regulation by—

(a)   in the case of a partial transfer—

(i)   issuing a new greenhouse gas emissions permit to the proposed transferee which–

(aa)   applies to the transferred activity;

(bb)   identifies the transferred unit; and

(cc)   includes the conditions required by paragraph (8); and

(ii)   reissuing the original greenhouse gas emissions permit to the operator, updated to record the transfer and varied to–

(aa)   identify the Schedule 1 activity to be carried out in the installation after the transfer and the specified emissions from that activity;

(bb)   describe the installation after the transfer; and

(cc)   specify the conditions applying after the transfer as required by paragraph (8);

(b)   in the case of a transfer of the whole greenhouse gas emissions permit, reissuing to the operator and the proposed transferee the permit updated to include the name and other particulars of the proposed transferee as the operator of the transferred unit,

and the transfer shall take effect from such date as may be agreed with the applicants and specified in the updated greenhouse gas emissions permit and, in the case of a partial transfer, the new greenhouse gas emissions permit.

(8) In the case of a partial transfer effected under this regulation, the conditions included in the new greenhouse gas emissions permit and the original greenhouse gas emissions permit after the transfer shall be the same as the conditions included in the original permit immediately before the transfer in so far as they are relevant, respectively, to any installation to which the new

permit relates or the original permit continues to relate but subject to such variations as, in the opinion of the regulator, are necessary to take account of the transfer.

(9) If within the period of two months beginning with the date on which the regulator receives an application under paragraph (1), or within such longer period as the regulator and the applicants may agree in writing, the regulator has neither effected the transfer nor given notice to the applicants that it has rejected the application, the application shall, if the applicants notify the regulator in writing that they are treating the application as having been refused, be deemed to have been refused at the end of that period or that longer period, as the case may be.

(10) The regulator may, by notice, require the operator or the proposed transferee to furnish such further information specified in the notice, within the period so specified, as the regulator may require for the purpose of determining an application under this regulation.

(11) Where a notice is served on an operator or proposed transferee under paragraph (10)–

(a)   for the purpose of calculating the period of two months mentioned in paragraph (9), no account shall be taken of the period beginning with the date on which the notice is served and ending on the date on which the information specified in the notice is furnished; and

(b)   if the specified information is not furnished within the period specified, the application shall, if the regulator gives notice to the operator and proposed transferee that it treats the application as having been withdrawn, be deemed to have been withdrawn at the end of that period.

(12)   Where a regulator effects the transfer of a greenhouse gas emissions permit in accordance with paragraph (7)(b), the regulator shall notify the registry administrator of the transfer.

Applications to surrender a greenhouse gas emissions permit

16.— (1) Where an operator has ceased carrying out in an installation all of the Schedule 1 activities authorised by a greenhouse gas emissions permit in relation to that installation, the operator shall apply to the regulator to surrender the permit.

(2) An application under paragraph (1) shall be made before the expiry of a period of one month beginning on the date on which the operator ceased to carry out the activity or activities in the

installation to which the greenhouse gas emissions permit relates or by the date of entry into force of approved NAP regulations in relation to the first scheme phase, whichever is the later.

(3) Paragraph (1) shall not apply where—

(a)   an approved national allocation plan provides for all allowances allocated under these Regulations in respect of any installation in which a Schedule 1 activity is no longer carried out to continue to be issued to the operator of such installation during the scheme phase to which the approved national allocation plan relates;

(b)   an approved national allocation plan provides that, if conditions specified in that plan are met, an operator which ceases to carry out a Schedule 1 activity in an installation may retain the allowances, or a proportion of those allowances, allocated in respect of the installation under these Regulations and the operator has, before the expiry of a period of one month beginning on the date on which the operator ceased to carry out the Schedule 1 activities or of the date on which the approved NAP Regulations in relation to the scheme phase for which the allowances are allocated enter into force, whichever is the later, made an application to retain its allocation or a proportion of its allocation under regulation 24(1); or

(c)   the greenhouse gas emissions permit applies to more than one installation, the operator continues to carry out a Schedule 1 activity in one of those installations, and the operator has applied to vary the permit so that it no longer applies to the installation in which he has ceased to carry out a Schedule 1 activity.

(4) An application under paragraph (1) shall, subject to paragraph (5), be accompanied by the fee prescribed in respect of the application in Schedule 5, and shall contain the operator's telephone number and postal address and, if different, any address to which correspondence relating to the application should be sent.

(5) Where an application under paragraph (1) relates to an offshore installation, the fee prescribed in Schedule 5 in respect of the application shall be payable within the period of 28 days beginning on the date on which the regulator serves a notice on the operator requesting payment of the fee.

(6) Where an application is duly made under paragraph (1), the regulator shall within two months give the operator and, where the surrender relates to an installation included in a pool in accordance with regulation 27, the appropriate authority, notice of its determination of the application (which, if approving the application, shall be known as a "notice of surrender") and the notice shall take effect, subject to regulation 32(10), on the date specified in the notice.

(7) A notice of surrender of the permit shall require the operator, in relation to the scheme year in which the notice of surrender takes effect, to—

(a)   except where a notice of surrender relates to an excluded installation which was an excluded installation for the whole of the scheme year up to the date on which the notice takes effect, submit to the regulator by the date specified in the notice a report specifying the reportable emissions from the beginning of the scheme year in which the notice of surrender takes effect until the date on which the notice of surrender takes effect (excluding any period for which the installation was an excluded installation) and to ensure that such report is verified in accordance with the relevant monitoring and reporting conditions; and

(b)   by the date specified in the notice surrender allowances equal to—

(i)   the reportable emissions specified in a report referred to in sub-paragraph (7);

(ii)   where an operator has failed to comply with a condition of a greenhouse gas emissions permit imposed pursuant to regulation 10(3) in respect of the last scheme year for which the date for surrendering allowances in accordance with that condition has passed, the annual reportable emissions in respect of which the operator failed to comply with that condition in that year;

(iii)   where a notice of surrender is served in a scheme year in which an error in the report submitted by an operator under a monitoring and reporting condition in relation to any earlier scheme year has been discovered, the annual reportable emissions in respect of which, as a result of that error, the operator failed to comply with the condition of a greenhouse gas emissions permit imposed pursuant to regulation 10(3) in respect of the scheme year to which the error relates;

(iv)   where a supplementary decision has been made under regulation 25(2) or (7), the total number of allowances which on the date on which the notice of surrender is served have been issued in respect of the installation which would not have been issued if the statement referred to in regulation 25(1)(a) or 25(7)(a) had not been false or misleading; and

(v)   where an operator has failed to comply with paragraph (1), the total number of allowances which on the date on which the notice of surrender is served have been issued in respect of the installation which would not have been issued if the operator had complied with paragraph 16(1).

(8) The report referred to in paragraph (7) shall be prepared and verified in accordance with the monitoring and reporting conditions in the greenhouse gas emissions permit to which the application to surrender relates.

(9) From the date on which the notice of surrender takes effect, the greenhouse gas emissions permit shall cease to have effect to authorise the carrying out of a Schedule 1 activity and to require the monitoring of emissions but any conditions of the permit shall continue to have effect so far as they are not superseded by the requirements of the notice pursuant to paragraph (7) until the regulator certifies either that the requirements of paragraph (7) and any conditions of the greenhouse gas emissions permit imposed pursuant to regulation 10(3) have been complied with or that there is no reasonable prospect of further allowances being surrendered by the operator in respect of the installation to which the notice relates.

(10) From the scheme year following the scheme year in which the notice of surrender takes effect, for the purposes of assessing compliance with a condition of the permit imposed pursuant to regulation 10(3), the reportable emissions of the installation, before any increase in accordance with regulation 10(4), shall be deemed to be zero.

(11) Except where paragraph (12) applies, where the regulator certifies in accordance with paragraph (8) that there is no reasonable prospect of further allowances being surrendered by the operator it shall notify the registry administrator.

(12) Where the regulator certifies in accordance with paragraph (8) that there is no reasonable prospect of further allowances being

surrendered by the operator because the operator holding account has been closed in accordance with Article 17(1) of the Registries Regulation, regulation 39 shall apply as if the failure to surrender sufficient allowances to comply with any conditions of the greenhouse gas emissions permit imposed pursuant to regulation 10(3) in a previous scheme year prior to the date on which the operator holding account is closed, were a further failure to comply with a condition imposed pursuant to regulation 10(3).

(13) The requirements specified in a notice of surrender pursuant to paragraph (7) shall be treated as if they were monitoring and reporting conditions.

(14) The requirements specified in a notice of surrender pursuant to paragraph (7)(b) shall be treated as if they were conditions of the greenhouse gas emissions permit imposed pursuant to regulation 10(3) and the number of allowances required to be surrendered by the notice of surrender were the annual reportable emissions of the installation in respect of the scheme year to which the notice relates.

(15) Where an installation fails to comply with the requirements of a notice of surrender included pursuant to paragraph (7), the regulator shall notify the registry administrator.

(16) The regulator may, by notice to the operator, require him to furnish such further information specified in the notice, within the period so specified, as the regulator may require for the purpose of determining an application under this regulation.

(17) Where a notice is served on an operator under paragraph (16) for the purpose of calculating the period of two months mentioned in paragraph (5), no account shall be taken of the period beginning with the date on which the notice is served and ending on the date on which the information specified in the notice is furnished.

Revocation of greenhouse gas emissions permits

17.— (1) Subject to paragraph (3), the regulator may at any time revoke a greenhouse gas emissions permit by serving a notice ("a revocation notice") on the operator and where the revocation relates to an installation included in a pool in accordance with regulation 27, on the appropriate authority.

(2) Without prejudice to the generality of paragraph (1) the regulator shall serve a notice under paragraph (1) where an operator fails to comply with an obligation under regulation 16(1).

(3) Where an approved national allocation plan provides for allowances allocated in respect of an installation in which a Schedule 1 activity is no longer carried out to continue to be issued during the scheme phase to which the approved national allocation plan relates, the regulator shall not revoke the greenhouse gas emissions permit which relates to that installation until after 28th February in the last scheme year in that scheme phase.

(4) A revocation notice shall specify the date on which the notice shall, subject to regulation 32(10), take effect, which shall be at least 28 days after the date on which the notice is served.

(5) A revocation notice shall specify that the operator is, in relation to the scheme year in which the revocation takes effect, required to —

(a)   except where a revocation notice relates to an excluded installation which was an excluded installation for the whole of the scheme year up to the date on which the notice takes effect, submit to the regulator by the date specified in the notice a report specifying the reportable emissions from the beginning of the scheme year in which the revocation notice takes effect until the date on which the revocation notice takes effect on the operator and to ensure that such report is verified in accordance with the relevant monitoring and reporting conditions; and

(b)   by the date specified in the notice surrender allowances equal to —

(i)   the reportable emissions specified in a report referred to in sub-paragraph (a);

(ii)   where an operator has failed to comply with a condition of a greenhouse gas emissions permit imposed pursuant to regulation 10(3) in respect of the last scheme year for which the date for surrendering allowances in accordance with that condition has passed, the annual reportable emissions in respect of which the operator failed to comply with that condition in that year;

(iii)   where a revocation notice is served in a scheme year in which an error in the report submitted by an operator under a monitoring and reporting condition in relation to any earlier scheme year has been discovered, the annual reportable emissions in respect of which, as a result of that error, the operator failed to comply with the condition of a greenhouse gas emissions permit imposed pursuant to regulation 10(3) in respect of the scheme year to which the error relates;

(iv) where a supplementary decision has been made under regulation 25(2) or (7), the total number of allowances which on the date on which the revocation notice is served have been issued in respect of the installation which would not have been issued if the statement referred to in regulation 25(1)(a) or 25(7)(a) had not been false or misleading; and

(v) where the notice is served in accordance with paragraph (2), the total number of allowances which on the date on which the notice is served have been issued in respect of the installation which would not have been issued if the operator had complied with regulation 16(1).

(6) Where a revocation notice is served in accordance with paragraph (2), regulation 24(11) or regulation 25(3), the revocation notice shall require the operator to pay, within such period as may be specified in the notice, the fee prescribed in respect of a revocation notice in Schedule 5.

(7) From the date on which the revocation notice takes effect, the greenhouse gas emissions permit shall cease to have effect to authorise the carrying out of a Schedule 1 activity and to require the monitoring of emissions but the conditions of the permit shall continue to have effect in so far as they are not superseded by the requirements of the notice specified pursuant to paragraph (5) until the regulator certifies that either the requirements of the notice specified pursuant to paragraph (5) and any conditions of the greenhouse gas emissions permit imposed pursuant to regulation 10(3) have been complied with, or that there is no reasonable prospect of further allowances being surrendered by the operator in respect of the installation to which the notice relates.

(8) From the scheme year following the scheme year in which the revocation notice takes effect, for the purposes of assessing compliance with a condition of the permit imposed pursuant to regulation 10(3), the reportable emissions of the installation, before any increase in accordance with regulation 10(4), shall be deemed to be zero.

(9) Except where paragraph (10) applies, where the regulator certifies in accordance with paragraph (7) that there is no reasonable prospect of further allowances being surrendered by the operator it shall notify the registry administrator.

(10) Where the regulator certifies in accordance with paragraph (7) that there is no reasonable prospect of further allowances being surrendered by the operator because the operator holding account has been closed in accordance with Article 17(1) of the Registries Regulation, regulation 39 shall apply as if the failure to surrender sufficient allowances to comply with any conditions of the greenhouse gas emissions permit imposed pursuant to regulation 10(3) in a previous scheme year prior to the date on which the operator holding account is closed, were a further failure to comply with a condition imposed pursuant to regulation 10(3).

(11) The requirements specified in a revocation notice pursuant to paragraph (5)(a) shall be treated as if they were monitoring and reporting conditions.

(12) The requirements specified in a revocation notice pursuant to paragraph (5)(b) shall be treated as if they were conditions of the greenhouse gas emissions permit imposed pursuant to regulation 10(3) and the number of allowances required to be surrendered were the annual reportable emissions of the installation in respect of the scheme year to which the notice relates.

(13) Where an installation fails to comply with the requirements of a revocation notice included pursuant to paragraph (5), the regulator shall notify the registry administrator.

(14) A regulator which has served a revocation notice may, before the date on which the revocation takes effect, withdraw the notice.

Fees and charges

18.— (1) An operator which holds a greenhouse gas emissions permit shall pay a charge for the subsistence of such permit in accordance with Schedule 5.

(2) If an operator has failed to pay a charge referred to in paragraph (1), the regulator may revoke the greenhouse gas emissions permit under regulation 17(1).

(3) The provisions of Schedule 5 shall apply until such time as they are superseded by the provisions of a charging scheme made—

(a) in respect of installations (other than offshore installations) situated in England, Wales or Scotland, under section 41 of the 1995 Act; or

(b) in respect of offshore installations, under regulation 19.

(4) A charging scheme made under section 41 of the 1995 Act which supersedes the provisions of Schedule 5, or any such scheme made under regulation 19, shall specify which of those provisions it supersedes.

(5) Where a fee or charge prescribed in Schedule 5 is superseded in accordance with paragraph (3), a reference to a fee or charge prescribed in Schedule 5 shall be read as a reference to a fee or charge prescribed in a charging scheme which supersedes that fee or charge.

Charging scheme for offshore installations

19.— (1) The Secretary of State may make, and from time to time revise, a scheme prescribing —

(a)   fees payable in respect of, or of applications for, a greenhouse gas emissions permit for an offshore installation;

(b)   fees payable in respect of, or of applications for, the variation, transfer and surrender of such permits;

(c)   charges payable in respect of the subsistence of such permits;

(d)   charges payable in respect of, or in respect of applications for, the allocation of allowances to an operator of an offshore installation;

(e)   charges payable in respect of, or in respect of applications for, the retention of allowances by an operator of an offshore installation ceasing to carry on an activity to which they relate;

(f)   charges payable in respect of the revocation of a greenhouse gas emissions permit for an offshore installation; and

(g)   charges payable in respect of the subsistence of an account required to be held in the registry established under regulation 26 by an operator of an offshore installation.

(2) The fees and charges prescribed in a scheme under paragraph (1) shall be paid to the Secretary of State.

(3) The Secretary of State shall, on making or revising a scheme under paragraph (1), lay a copy of the scheme or of the revisions made to the scheme or, if she considers it more appropriate, the scheme as revised, before each House of Parliament.

(4) A scheme under paragraph (1) may, in particular —

(a) make different provision for different cases, including different provision in relation to different persons in different circumstances or localities;

(b) allow for reduced fees or charges payable in respect of greenhouse gas emissions permits granted to the same operator;

(c) provide for the times at which and the manner in which the payments required by the scheme are to be made (subject to the requirements in these Regulations as to times at which payment is required); and

(d) make such incidental, supplementary and transitional provisions as appears to the Secretary of State to be appropriate.

(5) The Secretary of State shall take such steps as she considers appropriate for bringing the provisions of any charging scheme made by her which is for the time being in force to the attention of persons likely to be affected by it.

(6) In this regulation "prescribed" means specified in, or determined under, a scheme made under this regulation.

PART 3

ALLOWANCES

National Allocation Plans

20. — (1) Subject to regulation 46, the Secretary of State shall develop a national allocation plan in respect of the second scheme phase and in respect of each subsequent scheme phase.

(2) The Secretary of State shall send to the Scottish Ministers, the National Assembly for Wales and the Department of the Environment —

(a) a copy of the national allocation plan developed under paragraph (1), at least 18 months before the beginning of the scheme phase in respect of which it is developed;

(b) information on whether the European Commission has accepted or rejected a national allocation plan or any aspect of a plan as soon as practicable after being advised of such acceptance or rejection; and

(c) any amendment to the national allocation plan proposed by the Secretary of State as soon as practicable after its communication to the European Commission.

(3) The Secretary of State shall publish in England the national allocation plan developed for each scheme phase, at least 18 months before the beginning of the relevant phase.

(4) The Secretary of State shall publish in England information on whether the European Commission has accepted or rejected a national allocation plan or any aspect of a plan as soon as practicable after being advised of such acceptance or rejection.

(5) Where the European Commission rejects a national allocation plan or any aspect of such plan under Article 9(3) of the Directive and the Secretary of State proposes an amendment to the plan, the Secretary of State shall publish in England the amendment as soon as practicable after its communication to the European Commission.

(6) The Scottish Ministers shall publish in Scotland any plan, information or amendment sent to them by the Secretary of State under paragraph (2) as soon as practicable after it is received.

(7) The National Assembly for Wales shall publish in Wales any plan, information or amendment sent to it by the Secretary of State under paragraph (2) as soon as practicable after it is received.

(8) The Department of the Environment shall publish in Northern Ireland any plan, information or amendment sent to it by the Secretary of State under paragraph (2) as soon as practicable after it is received.

Allocation and issue of allowances

21.— (1) Subject to regulation 46, for the second scheme phase and each subsequent scheme phase, the Secretary of State shall decide upon—

(a) the total quantity of allowances to be allocated for that phase;

(b) the allocation of allowances in respect of each installation including the number of those allowances to be issued in each scheme year in that phase; and

(c) where there is more than one greenhouse gas emissions permit relating to an installation, the division of the allowances allocated in respect of that installation under sub-paragraph (b) between each part of the installation to which a separate greenhouse gas emissions permit relates.

(2) Decisions under paragraph (1) shall—

(a)   be based upon the national allocation plan for the relevant phase which has not been rejected by the European Commission or in relation to which the European Commission has accepted amendments in accordance with Article 9(3) of the Directive; and

(b)   take due account of comments from the public in accordance with the provisions of the national allocation plan.

(3) The Secretary of State shall publish in England a decision under paragraph (1), at least twelve months before the beginning of the scheme phase to which the decision relates.

(4) The Secretary of State shall notify the Scottish Ministers, the National Assembly for Wales and the Department of the Environment of her decision under paragraph (1).

(5) Where the European Commission has provided for additional allowances to be allocated in respect of an installation, or installations of any description, under Article 29(1) of the Directive, the Secretary of State shall instruct the registry administrator to issue allowances in accordance with Article 43(1) of the Registries Regulation.

(6) Where—

(a)   an approved national allocation plan provides that where conditions specified in the approved national allocation plan are met allowances allocated under these Regulations in respect of an installation for a particular scheme year in the scheme phase to which the plan relates should be issued to the operator of the installation after 28th February; and

(b)   such conditions are met in respect of an installation,

the relevant decision maker shall notify the registry administrator and the operator that the second sentence of Article 40 or 46 of the Registries Regulation applies to that installation in respect of the scheme year in which the notice is served.

(7) A notice under paragraph (6) shall, in accordance with the approved national allocation plan, either—

(a)   specify the date after 28th February on which, in accordance with the approved national allocation plan, allowances should be issued in respect of the installation; or

(b)   where the approved national allocation plan provides for such date to be determined by reference to when specified conditions have been met, specify the conditions and indicate that the relevant decision maker will further notify the registry administrator when those conditions have been met.

(8) Where paragraph (7)(b) applies, the relevant decision maker shall periodically, or where requested by a notice served on the relevant decision maker by the operator, assess whether the conditions specified in the notice under paragraph (6) have been met and shall notify the operator and the registry administrator when the conditions have been met.

(9) Where a notice is served under paragraph (6), the reference to "later date" in Article 40 or 46 of the Registries Regulation shall be to the date specified in accordance with paragraph (7)(a) or to the date of a notice under paragraph (8).

(10) For the purposes of this regulation, "relevant decision maker" means—

(a)   where the conditions referred to in paragraph (6)(a) relate to an installation in respect of which an application for temporary exclusion under Article 27(2) of the Directive has been made, the responsible authority;

(b)   where the conditions referred to in paragraph (6)(a) relate to a supplementary decision of the Secretary of State under regulation 25(7), the Secretary of State;

(c)   in all other cases, the regulator.

Allocation by auction or sale

21A.– (1) Before 1st May 2008, the Secretary of State may decide to allocate allowances which meet the conditions in paragraph (2) by way of auction or sale.

(2) The allowances must—

(a)   be for the first scheme phase;

(b)   be contained within the new entrant reserve; and

(c)   qualify for auction or sale in accordance with the approved national allocation plan for that phase.

(3) Subject to regulation 46, if the Secretary of State decides under paragraph (1) to allocate allowances by way of auction or sale, he may enter into an agreement with an account holder ("the purchaser") to allocate allowances to that person in exchange for payment.

(4) Where the Secretary of State—

(a)   agrees to allocate allowances to the purchaser in accordance with this regulation; and

(b)   has received the agreed payment from the purchaser by the agreed date,

he shall, within 7 days of receiving the payment, serve a notice on the purchaser and the registry administrator allocating those allowances to the purchaser.

(5) The Secretary of State may take one of the steps specified in paragraph (6) if—

(a)   he has agreed to allocate allowances to the purchaser in accordance with this regulation; and

(b)   he has not received the agreed payment from the purchaser by the agreed date.

(6) The steps are—

(a)   not to allocate those allowances to the purchaser, by serving a notice on him to that effect; or

(b)   to recover the outstanding payment from the purchaser summarily as a civil debt.

(7) If in accordance with paragraph (6)(b) the Secretary of State recovers the outstanding payment from the purchaser before 10th April 2008, he shall, within 7 days of receiving such payment, serve a notice on the purchaser and the registry administrator allocating the allowances that were the subject of their agreement to the purchaser.

(8) A notice under paragraph (4) or (7) shall specify—

(a)   the purchaser;

(b)   the purchaser's holding account;

(c)   the number of allowances to be allocated to the purchaser,

(d)   where all of the allowances are not to be issued in the same year, the number of allowances to be issued in each remaining year or part year of the first scheme phase; and

(e)   the date, or where all of the allowances are not to be issued in the same year the dates in each relevant year, by which the allowances are to be issued (the first of which shall be no later than 14 days after the date of the notice).

(9) Where a notice is served on the registry administrator in accordance with paragraph (4) or (7), the registry administrator shall transfer the quantity of allowances specified in the notice from the party holding account to the account specified in the notice.

(10) The registry administrator shall transfer allowances in accordance with paragraph (9) by the date specified in the notice in accordance with paragraph (8)(e).

(11) The registry administrator shall transfer allowances in accordance with the internal transfer process set out in Annex IX to the Registries Regulation.

(12) Where appointed to do so by the Secretary of State, the Environment Agency or the Scottish Environment Protection Agency may exercise any of the Secretary of State's functions under this regulation (other than the power of appointment under this paragraph) subject to any limitations imposed by the Secretary of State when making the appointment.

Application for an allocation from the new entrant reserve

22. — (1) Where an approved national allocation plan provides for a new entrant reserve in the scheme phase to which it relates, an operator of an installation may apply to the regulator for an allocation of allowances in respect of that installation from the new entrant reserve.

(2) Subject to paragraph (3), an application under paragraph (1) shall be combined with an application for a greenhouse gas emissions permit under regulation 8(1) or an application for a variation of a greenhouse gas emissions permit under regulation 14(2).

(3) Paragraph (2) shall not apply where an application under paragraph (1) relates to—

(a) an installation in respect of which an operator made an application for a greenhouse gas emissions permit under regulation 8(1) before the date on which the approved NAP Regulations enter into force; or

(b) a change in operation in respect of which an operator made an application for a variation of a greenhouse gas emissions permit under regulation 14(2) before the date on which the approved NAP Regulations enter into force.

(4) An application under paragraph (1) shall contain such information as the regulator may reasonably require for the purpose of determining the application in accordance with the provisions of the approved national allocation plan and, except where paragraph (6) or (7) applies, shall be accompanied by the fee prescribed in Schedule 5 in respect of such application.

(5) Subject to paragraph (6), where—

(a) before the date on which approved NAP Regulations for the first scheme phase enter into force; and

(b) in the case of an application relating to an installation which was put into operation on or after 27th February 2005 or a change in operation which was made on or after 27th February 2005, on or after 10th January 2005,

an operator has made to the regulator an application which if made after the date on which the approved NAP Regulations enter into force would have complied with paragraph (1), it shall be deemed to be an application under paragraph (1) made on the date of entry into force of the approved NAP Regulations and where more than one application is deemed to have been received on that day, the applications shall be deemed to have been received on that day in the order in which they were originally received by the regulator.

(6) Except where paragraph (7) applies, paragraph (5) shall only apply to an application in respect of which the operator has before the expiry of a period of 15 working days beginning on the date on which these Regulations enter into force paid to the regulator the fee prescribed in Schedule 5 in respect of an application under paragraph (1).

(7) Where an application under paragraph (1), or deemed to have been made under paragraph (1) in accordance with paragraph (5), relates to an offshore installation, the fee prescribed in Schedule 5 in respect of the application shall be payable within the period of 28 days beginning on the date on which the regulator serves a notice on the operator requesting payment of the fee.

(8) If an operator fails to comply with paragraph (7) the regulator may refuse the application.

(9) The regulator may, by notice to the applicant, require him to furnish such further information specified in the notice, within the period so specified, as the regulator may require for the purpose of determining the application and if the applicant fails to furnish the specified information within the period specified,

the application shall, if the regulator gives notice to the applicant that it treats the application as having been withdrawn, be deemed to have been withdrawn at the end of that period.

(10) Subject to paragraph (11), the regulator shall give notice under paragraph (13) of its determination of an application under paragraph (1) within a period of two months beginning on the later of—

(a)   the date on which the application under paragraph (1) was received by the regulator; or

(b)   the date of entry into force of the approved NAP Regulations for the first scheme phase,

or within such longer period as may be agreed in writing with the applicant.

(11) For the purposes of calculating the period of two months mentioned in paragraph (10) no account shall be taken of—

(a)   any period beginning with the date on which notice is served on the applicant under paragraph (9) and ending on the date on which the applicant furnishes the information specified in the notice;

(b)   any period beginning with the date on which notice is served on the applicant under regulation 8(5) or 14(6) in respect of an application with which an application under paragraph (1) is combined in accordance with paragraph (2).

(12) If the regulator fails to give notice of its determination of an application under paragraph (1) within the period allowed by paragraph (10), the application shall, if the operator notifies the regulator that it treats the application as having been refused, be deemed to have been refused.

(13) Where an application under paragraph (1) is duly made to the regulator, the regulator shall, in accordance with the provisions of the approved national allocation plan—

(a)   determine the eligible allocation subject to such conditions as the regulator considers appropriate to comply with the approved national allocation plan;

(b)   determine the eligible allocation and, subject to paragraph (14), allocate allowances to the operator in respect of the installation to which the application relates; or

(c) reject the application,

by serving a notice on the operator and, in the case of a determination under sub-paragraph (b), on the registry administrator.

(14) Where the eligible allocation is greater than the number of available allowances, a notice under paragraph (13)(b) shall allocate the available allowances and the regulator may make additional allocations of allowances under paragraph (13)(b) if additional allowances subsequently become available in the new entrant reserve in accordance with the approved national allocation plan until the number of allowances allocated under paragraph (13)(b) equals the eligible allocation.

(15) A notice under paragraph (13)(a) shall–

(a) specify the conditions applying to the determination;

(b) require the operator to notify the regulator if any of the information provided in the application or provided to the regulator in response to a request for further information under paragraph (9) changes; and

(c) where the approved national allocation plan provides for allowances from the new entrant reserve to be reserved for an installation in respect of which an application under paragraph (1) has been made, indicate whether, and, if so, how many allowances have been reserved in respect of the installation.

(16) Where it appears to the regulator, whether as a result of a notification in accordance with a condition imposed pursuant to paragraph (13)(a) or otherwise, that the approved national allocation plan requires a variation of any of the provisions of the notice served on the operator under paragraph (13)(a), the regulator shall serve a further notice under paragraph (13)(a) on the operator varying the provisions of the previous notice.

(17) Where it appears to the regulator, whether as a result of a notification in accordance with a condition imposed pursuant to paragraph (13)(a) or otherwise, that it would no longer be consistent with the approved national allocation plan for an allocation of allowances to be made in respect of an installation to which a notice under paragraph (13)(a) relates, the regulator shall serve a further notice on the operator rejecting the application.

(18) Subject to paragraph (19), where the regulator is satisfied that all the conditions specified in a notice under paragraph (13)(a) are met, the regulator shall serve a notice on the operator and the registry administrator allocating allowances in respect of the installation.

(19) Where the eligible allocation is greater than the number of available allowances, a notice under paragraph (18) shall allocate the available allowances and the regulator may make additional allocations of allowances under paragraph (18) if additional allowances subsequently become available in the new entrant reserve in accordance with the approved national allocation plan until the number of allowances allocated under paragraph (18) equals the eligible allocation.

(20) A notice under paragraph (13)(b) or (18) shall specify —

(a)   the operator and installation identification code of the installation in respect of which the allocation is made and the permit identification code of the greenhouse gas emissions permit which relates to that installation;

(b)   the allocation of allowances to the operator in respect of the installation including the number of allowances to be issued in each remaining year or part year of the phase in relation to which the allocation is made and the date on which the allowances will be issued in the year in which the notice is served.

(21) A notice under paragraph (13)(b) or (18) shall be an instruction to the registry administrator for the purposes of Article 42 or 48 of the Registries Regulation.

(22) Where an application under regulation 15(1) specifies that an application under paragraph (1) in respect of the installation relates to the transferred unit, any allocation of allowances under paragraph (13)(b) or (18) made after the transfer takes effect shall be allocated to the proposed transferee.

(23) For the purposes of this regulation —

"eligible allocation" means the amount of allowances which may be allocated in respect of an installation under the provisions of the approved national allocation plan subject to the number of available allowances;

"available allowances" means allowances in the new entrant reserve which are available for allocation in accordance with the approved national allocation plan;

"new entrant reserve" does not include any late installation element (as defined in an approved national allocation plan) of the new entrant reserve;".

"proposed transferee" and "transferred unit" shall have the same meaning as in regulation 15(1) and (4).

Application for an allocation for late installations

22A. — (1) Before 1st March 2007, an operator of an installation may apply to the Secretary of State for an allocation of allowances from the element of the new entrant reserve which, in accordance with an approved national allocation plan, has been set aside for late installations.

(2) An application shall contain such information as the Secretary of State may reasonably require for the purpose of determining the application in accordance with the approved national allocation plan.

(3) Where an operator has made to the Secretary of State an application before 6th April 2006 which, if made on or after that date would have complied with paragraph (2), it shall be deemed to be an application made and received by the Secretary of State on 6th April 2006.

(4) The Secretary of State may, by notice to the applicant, require him to furnish to her, within 5 working days or such longer period as may be specified in the notice, such further information as she may require for the purpose of determining the application; and if the applicant fails to furnish the further information within the relevant period, the application shall, if the Secretary of State gives notice to the applicant that she treats the application as having been withdrawn, be deemed to have been withdrawn at the end of that period.

(5) The Secretary of State shall approve an application unless —

(a)   the installation in question was put into operation on or after 1st January 2004;

(b)   the installation in question is not the subject of a greenhouse gas emissions permit;

(c)   the installation in question was the subject of a greenhouse gas emissions permit before 1st January 2005;

(d)   the installation in question was allocated allowances in a decision made under regulation 19(1)(b) of the Greenhouse Gas Emissions Trading Scheme Regulations2003(**22**); or

---

(**22**) S.I. 2003/3311. Regulation 47 revokes these regulations subject to certain savings.

(e)   the application is made more than 30 working days after whichever is the later of the installation being granted a greenhouse gas emissions permit under regulation 9, and 6th April 2006,

in which case she shall refuse the application.

(6) The Secretary of State shall notify the operator of her determination of the application by serving a notice upon him, and if she has approved the application, the notice shall state the eligible allocation for the installation to which the application relates.

(7) For the purposes of calculating the period of 30 working days mentioned in paragraph (5)(e) no account shall be taken of any period beginning with the date on which the applicant requests information from the Secretary of State relating to the manner in which he should compile and submit information as part of an application under paragraph (1), and ending on the date on which the Secretary of State furnishes that information.

(8) The Secretary of State shall give notice of her determination of an application as required by paragraph (6) —

(a)   within a period of 20 working days beginning with the date on which the application is received by the Secretary of State; or

(b)   within such longer period as may be agreed in writing with the applicant.

(9) For the purposes of calculating the period of 20 working days mentioned in paragraph (8) no account shall be taken of any period beginning with the date on which notice is served on the applicant under paragraph (4) and ending on the date on which the applicant furnishes the information specified in the notice.

(10) If the Secretary of State fails to give notice of her determination of an application within the period allowed by paragraph (8), the application shall, if the operator notifies the Secretary of State that he treats the application as having been refused, be deemed to have been refused.

(11) Subject to paragraphs (12) to (15), the Secretary of State shall allocate allowances equal to the eligible allocation in respect of each application that she has approved within a period of 15 working days beginning with the day on which she served a notice on the operator under paragraph (6) by serving a notice on the operator and the registry administrator.

(12) Where the Secretary of State becomes aware that if she were to approve all of the applications that she has received under paragraph (1) of this regulation but which she has not determined, the total of—

(a)   the eligible allocations relating to those applications; and

(b)   the eligible allocations relating to all of the applications which she has approved but in respect of which she has not yet allocated allowances,

would exceed the number of available allowances, she shall not allocate any allowances until she has determined all of those applications.

(13) Where paragraph (12) applies, once the Secretary of State has determined all of the applications that she has received, for the purpose of allocating allowances in respect of those applications she shall prioritise those applications according to the order in which an application for a greenhouse gas emissions permit in relation to the installation which is the subject of the application was duly made, giving the highest priority to the installation in relation to which the earliest application for a greenhouse gas emissions permit was duly made.

(14) Where paragraph (12) applies, the Secretary of State shall allocate allowances equal to the eligible allocation in respect of each application that she has approved in the order of priority established under paragraph (13), until she has available allowances to cover the next eligible allocation in respect of an application in part only, in which case she shall allocate the remainder of the available allowances in respect of that application, and shall not allocate any allowances in respect of any subsequent applications.

(15) If—

(a)   by virtue of paragraph (14) the Secretary of State has not allocated the eligible allocation in respect of an application that she has approved; and

(b)   in accordance with an approved national allocation plan, she has subsequently added further allowances to the late installation element of the new entrant reserve,

she shall repeat the process set out in paragraphs (13) and (14) as if paragraph (12) applied, except that when allocating in relation to an application in respect of which she has already allocated part of the eligible allocation, she shall only allocate the remainder of the eligible allocation.

(16) The Secretary of State shall allocate allowances under paragraph (14) by serving a notice on the operator and the registry administrator.

(17) A notice under paragraph (11) or (16) shall specify—

(a)   the operator and installation identification code of the installation in respect of which the allocation is made and the permit identification code of the greenhouse gas emissions permit which relates to that installation; and

(b)   the allocation of allowances to the operator in respect of the installation including, subject to regulations 22C and 23(2), the number of allowances to be issued in each remaining year or part year of the phase in relation to which the allocation is made and the date on which the allowances will be issued in the year in which the notice is served.

(18) A notice under paragraph (11) or (16) shall set out the date or dates upon which allowances are to be issued to the operator and such notice shall be an instruction from the competent authority to the registry administrator for the purposes of Article 42 or 48 of the Registries Regulation.

(19) Where an application under regulation 15(1) specifies that an application under paragraph (1) in respect of the installation relates to the transferred unit, any allocation of allowances under this regulation made after the transfer takes effect shall be allocated to the transferee.

(20) For the purposes of this regulation the eligible allocation shall be calculated in accordance with regulation 22B.

(21) For the purposes of this regulation—

"available allowances" means allowances—

(a)   in the late installation element of the new entrant reserve as set out in an approved national allocation plan, or

(b)   which have been added to that element of the reserve by the Secretary of State, which have been neither allocated in accordance with this regulation nor removed from the reserve by virtue of a correction to the national allocation plan table made under Article 38(2) or 44(2) of the Registries Regulation; and "transferred unit" has the same meaning as in regulation 15(4).

Eligible allocations

22B.—(1) For the purposes of regulation 22A the eligible allocation shall be the allocation for the year in which the permit is issued plus—

(a) where the installation first became the subject of a greenhouse gas emissions permit in 2005, twice the annual allocation;

(b) where the installation first became the subject of a greenhouse gas emissions permit in 2006, the annual allocation;

(c) where the installation first became the subject of a greenhouse gas emissions permit in any other year, zero.

(2) For the purposes of paragraph (1), the allocation for the year in which the permit is issued shall be calculated using the following formula:

$$Z = \frac{D}{365} \times X$$

Where:

Z is the allocation for the year in which the permit is issued;

X is the annual allocation calculated in accordance with paragraphs (3) or (4);

D (permitted days) is the number of days comprised in the period beginning on the date upon which the installation became the subject of a greenhouse gas emissions permit and ending at the end of the same calendar year.

(3) Except where paragraph (4) applies, the annual allocation shall, for the purposes of paragraph (1), be calculated using the following formula:

$$X = \left(\frac{T - A}{Y - 1}\right) \times S \times L$$

Where:

X is the annual allocation;

T (total baseline emissions) is the total verified emissions of the installation in question

during the period 1998 to 2003 inclusive;

A (lowest year emissions) is the verified emissions of the installation in question for the year in which its emissions were lowest during the period 1998 to 2003 inclusive, excluding years in which its emissions were zero;

Y (number of years) is the number of years in which the verified emissions for the

installation in question were greater than zero;

S (sector reduction factor) is the number listed as the sector reduction factor in the table in

Schedule 7 adjacent to the sector in which, in accordance with the approved national allocation plan the Secretary of State considers the installation should be classified;

L (late installation reduction factor) is—

(a) where an application for a greenhouse gas emissions permit in respect of the installation was made on or before 31st August 2005, and a statement of verified emissions for the installation was submitted to the Secretary of State on or before 28th February 2005, 100%;

(b) where an application for a greenhouse gas emissions permit in respect of the installation was made on or before 31st August 2005, but a statement of verified emissions for the installation was not submitted to the Secretary of State on or before 28th February 2005, 90%;

(c) where an application for a greenhouse gas emissions permit in respect of the installation is made after 31st August 2005, 75%.

(4) For the purposes of paragraph (1), where the installation in question was put into operation on or after 1st January 2003, the annual allocation shall be calculated using the following formula:

$$B \times N \times L$$

Where:

X is the annual allocation;

B (benchmarked allocation) is the annual allocation for which the installation would have been eligible had it been a new entrant, calculated in accordance with the approved national allocation plan;

N (new entrant reserve reduction) is the number listed as the new entrant reserve reduction factor in the table in Schedule 7 adjacent to the sector in which, in accordance with the approved national allocation plan the Secretary of State considers the installation should be classified;

L (late installation reduction factor) is—

(a)   where an application for a greenhouse gas emissions permit in respect of the installation was made on or before 31st August 2005, and a statement of verified emissions for the installation was submitted to the Secretary of State on or before 28th February 2005, 100%;

(b)   where an application for a greenhouse gas emissions permit in respect of the installation was made on or before 31st August 2005, but a statement of verified emissions for the installation was not submitted to the Secretary of State on or before 28th February 2005, 90%;

(c)   where an application for a greenhouse gas emissions permit in respect of the installation is made after 31st August 2005, 75%.

(5) In this regulation, "verified emissions" means the reportable emissions from the installation in question for the period 1998 to 2003 inclusive, as set out in the application made under regulation 22A and verified in accordance with the guidance issued by the Secretary of State on 5th August 2005 entitled "EU Emissions Trading Scheme Guidance on Annual Verification, version 1(**23**)

Power to withhold allowances

22C.—(1) Where—

(a)   an approved national allocation plan provides that where conditions specified in the plan are met, allowances allocated under these Regulations in respect of an installation for a particular scheme year in the scheme phase to which the plan relates should not be issued to the operator of the installation on or before 28th February of that scheme year; and

(b)   such conditions are met in respect of an installation in relation to which allowances have been allocated under regulation 22 or 22A,

---

(**23**) Available on the Defra website at http://defraweb/environment/climatechange/ trading/eu/permits/pdf/annverifguide.pdf

the regulator shall serve a notice on the registry administrator and the operator stating that for the scheme year in question, the allocation of allowances shall be withheld.

(2) A notice under paragraph (1) shall in accordance with the approved national allocation

plan either—

(a)   specify the date after 28th February on which, in accordance with the approved national allocation plan, allowances should be issued in respect of the installation; or

(b)   where the approved national allocation plan provides for such date to be determined by reference to when specified conditions have been met, specify the conditions and indicate that the regulator will further notify the registry administrator when those conditions have been met.

(3) Where a notice under paragraph (1) contains the information required by subparagraph (2)(b), the regulator shall periodically, or when requested by a notice served on the regulator by the operator, assess whether the conditions specified in the notice under paragraph (1) have been met and shall notify the operator and the registry administrator when the conditions have been met.

(4) A notice under paragraph (1) containing the information required by sub-paragraph (2)(a) shall be an instruction to the registry administrator for the purposes of Article 42 or 48 of the Registries Regulation and shall supersede a previous instruction issued for the installation and scheme year in question.

(5) A notice under paragraph (1) containing the information required by sub-paragraph (2)(b) shall revoke a previous instruction to the registry administrator for the purposes of Article 42 or 48 of the Registries Regulation for the installation and scheme year in question.

(6) A notice served by the regulator in accordance with paragraph (3) shall be an instruction to the registry administrator for the purposes of Article 42 or 48 of the Registries Regulation for the installation and scheme year in question.

Allowance allocation where permit varied, surrendered or revoked

23.— (1) Where a greenhouse gas emissions permit is varied under regulation 14 so that it no longer applies to an installation in which the operator has ceased to carry out a Schedule 1 activity, surrendered under regulation 16 or revoked under regulation 17(1)—

(a)   the regulator shall notify the Secretary of State; and

(b)   where an allocation of allowances in respect of the installation which has ceased to be the subject of a permit has been made under regulation 22 or 22A for the scheme phase in which the permit is varied, surrendered or revoked or the subsequent scheme phase, the regulator shall take such steps as it considers necessary to ensure that no further allowances are issued in respect of that installation from the date on which the variation notice, the notice of surrender or the revocation notice takes effect.

(2) Where the regulator notifies the Secretary of State that a greenhouse gas emissions permit has been varied so that it no longer applies to an installation in which the operator has ceased to carry out a Schedule 1 activity, surrendered or revoked, the Secretary of State shall take such steps as she considers necessary to ensure that no further allowances are issued in respect of the installation which has ceased to be the subject of a permit from the date on which the variation notice, the notice of surrender or the revocation notice takes effect.

Applications to retain allocation

24.— (1) Where the approved national allocation plan provides that, if conditions specified in the approved national allocation plan are met, an operator which ceases to carry out a Schedule 1 activity in an installation may retain the allowances, or a proportion of the allowances, allocated in respect of that installation under these Regulations, the operator may apply to the regulator to retain its allocation.

(2)   An application under paragraph (1) shall contain such information as the regulator may reasonably require for the purpose of determining the application in accordance with the provisions of the approved national allocation plan.

(3)   Subject to paragraph (4), the regulator shall give notice under paragraph (7) of its determination of an application which is duly made under paragraph (1) within a period of two months beginning on the day on which the regulator received the application or within such longer period as may be agreed in writing with the applicant.

(4)   For the purposes of calculating the period of two months mentioned in paragraph (3) no account shall be taken of any period beginning with the date on which notice is served on the applicant under paragraph (6) and ending on the date on which the applicant furnishes the information specified in the notice.

(5) If the regulator fails to give notice of its determination of an application under paragraph (1) within the period allowed by paragraph (3), the application shall, if the operator notifies the regulator that it treats the application as having been refused, be deemed to have been refused.

(6) The regulator may, by notice to the applicant, require him to furnish such further information specified in the notice, within the period so specified, as the regulator may require for the purpose of determining the application and if the applicant fails to furnish the specified information within the period specified, the application shall, if the regulator gives notice to the applicant that it treats the application as having been withdrawn, be deemed to have been withdrawn at the end of that period.

(7) Where an application is duly made under paragraph (1), the regulator shall, in accordance with the provisions of the approved national allocation plan—

(a)   accept the application and provide, subject to such conditions as the regulator considers appropriate, for either–

(i)   all the allowances allocated in respect of the installation for the scheme phase to which the application relates to be retained, or

(ii)   a proportion of the allowances allocated in respect of the installation for the scheme phase to which the application relates to be retained; or

(b)   refuse the application,

by serving a notice on the operator.

(8) A notice under paragraph (7)(a) (a "retention notice") shall—

(a)   specify such variations to the provisions of the greenhouse gas emissions permit as the regulator considers appropriate which shall take effect from the date on which the notice takes effect;

(b)   specify the number of hours which the regulator required to determine the application under paragraph (1) and shall require the operator to pay, within such period as may be specified in the notice, the fee prescribed in Schedule 5 in respect of the determination of the application.

(9) Where—

(a)   a retention notice provides for only a proportion of the allowances allocated under regulation 22(13)(b) or (18) in respect of the installation to be retained, the regulator shall take such steps as are necessary to ensure that from the date on which the retention notice takes effect only such proportion of the allowances as is specified in the notice are issued in respect of the installation;

(b)   a retention notice provides for only a proportion of the allowances allocated under regulation 21(1)(b) or, if applicable, (1)(c) in respect of the installation to be retained, the regulator shall notify the Secretary of State who shall take such steps as she considers necessary to ensure that from the date on which the retention notice takes effect only such proportion of the allowances as is specified in the notice are issued in respect of the installation.

(10)   Subject to regulation 32(10), where a retention notice provides for a proportion of the allowances allocated in respect of the installation to which the application relates to be retained, the provision of the notice specifying the proportion of the allowances to be retained shall take effect from the expiry of a period of 15 working days beginning on the date of the notice.

(11)   Where—

(a)   an application under paragraph (1) is withdrawn by the operator or refused by the regulator; or

(b)   any conditions specified in a notice under paragraph (7)(a) are no longer met, the regulator shall revoke the greenhouse gas emissions permit in accordance with regulation 17(1).

Supplementary decisions by the regulator or the Secretary of State

25.—   (1) Subject to paragraph (3), paragraph (2) shall apply where—

(a)   a person has made a false statement–

(i)   in connection with an application under regulation 22(1); or

(ii)   in connection with an application under regulation 24(1); and

(b)   the statement resulted in an over-allocation in respect of an installation.

(2) Where this paragraph applies, the regulator may make a supplementary decision by serving a notice on the operator.

(3) Where a person has made a false statement in connection with an application under regulation 24(1) and the application would, in accordance with the approved national allocation plan, have been refused if the statement had not been false or misleading, the regulator shall revoke the greenhouse gas emissions permit in accordance with regulation 17(1).

(4) A supplementary decision under paragraph (2) shall—

(a)   identify the false statement and specify the amount of the over-allocation;

(b)   set out the steps which the regulator will take in accordance with paragraph (5) and, if applicable, paragraph (6), and, subject to regulation 32(10), shall take effect from the expiry of a period of two months beginning on the date of the notice.

(5) Subject to paragraph (6), where the regulator makes a supplementary decision under paragraph (2), the regulator shall take such steps as it considers necessary to ensure that the amount of allowances issued pursuant to an allocation under regulation 22(13)(b) or (18) in respect of the installation to which the false statement relates are reduced by the amount of the over-allocation.

(6) Where paragraph (2) applies and–

(a)   all allowances allocated under regulation 22(13)(b) or (18) in respect of the installation for the scheme phase to which the false statement relates have been issued; or

(b)   the amount of such allowances which have not been issued is less than the amount of the over-allocation, the regulator shall notify the Secretary of State specifying the amount of over-allocation or, where the regulator has made a supplementary decision under paragraph (2), the remaining over-allocation.

(7) Subject to regulation 46, where—

(a)   a person has made a false statement in response to a request for information from the Secretary of State for the purposes of developing a national allocation plan under regulation 20(1) or making a decision under regulation 21(1) and the statement has resulted in an over-allocation in respect of the installation; or

(b)   the regulator notifies the Secretary of State in accordance with paragraph (6), or

(c)   a person has made a false statement in connection with an application under regulation 22A and the statement resulted in an over-allocation in respect of an installation, the Secretary of State may make a supplementary decision by serving notice on the operator.

(8) A supplementary decision under paragraph (7) shall—

(a)   identify the false statement and specify the amount of the over-allocation;

(b)   set out the steps which the Secretary of State will take in accordance with paragraph (9), and, subject to regulation 33(6), shall take effect from the expiry of a period of two months beginning on the date of the notice.

(9) Where the Secretary of State makes a supplementary decision under paragraph (7), the Secretary of State shall take such steps as she considers necessary to ensure that the number of allowances issued in respect of the installation to which the false statement relates for—

(a)   the phase to which the false statement relates; and

(b)   where the number of allowances allocated in respect of the installation in that scheme phase which have not been issued is less than the amount of the over-allocation or, if applicable, the remaining over-allocation, a subsequent scheme phase, are reduced by the amount of the over-allocation or, if applicable, the remaining over-allocation.

(10) The steps which may be taken by the Secretary of State under paragraph (9) may include—

(a)   where the supplementary decision relates to an installation situated in England (other than an offshore installation), directing the regulator; or

(b)   where the supplementary decision relates to an installation, other than an offshore installation, situated in Scotland, Wales or Northern Ireland, arranging for the appropriate authority in relation to that installation to direct the regulator, to take such steps as are necessary to reduce the amount of allowances issued pursuant to an allocation under regulation 22(13)(b) or (18).

(11) As soon as practicable after the Secretary of State makes a supplementary decision under paragraph (7), she shall publish her supplementary decision under paragraph (7) in England and notify the Scottish Ministers, the National Assembly for Wales and Department of the Environment of the decision.

(12) The Scottish Ministers shall publish in Scotland any decision notified to them under paragraph (11) as soon as practicable on notification.

(13) The National Assembly for Wales shall publish in Wales any decision notified to them under paragraph (11) as soon as practicable on notification.

(14) The Department of the Environment shall publish in Northern Ireland any decision notified to them under paragraph (11) as soon as practicable on notification.

(15) A regulator which has served a notice under paragraph (2) or the Secretary of State who has served a notice under paragraph (7), may before the date on which the notice takes effect withdraw the notice.

(16) For the purposes of this regulation–

"false statement" means a statement which is which is false or misleading in a material particular;

"over-allocation" means—

(i)   the number of allowances by which an allocation under regulation 21(1)(b), or if applicable, (1)(c) or regulation 22(13)(b) or (18) in respect of an installation in the scheme phase to which a false statement relates exceeds the number of allowances which would have been allocated in accordance with the approved national allocation plan for that scheme phase if the statement had not been false or misleading;

(ii)   in relation to an application under regulation 24(1), the number of allowances which are retained by the operator which would not have been retained in accordance with the approved national allocation plan if the statement had not been false or misleading;

(iii)   in relation to an allocation of allowances made under regulation 22A, the difference between the number of allowances allocated in respect of the application made under that regulation, and the number of allowances that would have been allocated if the statement had not been false or misleading;

"remaining over-allocation" means the difference between the over-allocation and the amount by which the number of allowances to be issued in respect of an installation is reduced by the supplementary decision under paragraph (2).

Registry

26. — (1) Subject to regulation 46, the Secretary of State shall establish a registry in accordance with the requirements of Article 19 of the Directive and the Registries Regulation.

(2) The Environment Agency shall —

(a)   maintain the registry in accordance with the requirements of Article 19 of the Directive;

(b)   act as registry administrator for the purposes of the Registries Regulation and these Regulations.

(3) Subject to paragraph (4), the regulator shall be the competent authority for the purposes of the Registries Regulation.

(4) The Secretary of State shall be the competent authority for the purposes of Articles 15, 38, 41, 43, 44, 47, 59 and 60 of the Registries Regulation.

(5) The Secretary of State shall be the relevant body for the purposes of Articles 12(1), 13, 50(1) and 63(1) of and paragraph 19 of Annex VI to the Registries Regulation.

(6) It shall be the duty of the operator to comply with the requirements of Article 15(1) and (3) of the Registries Regulation.

(7) A holder of an account in the Registry may nominate an additional authorised representative for that account in accordance with Article 23(2) of the Registries Regulation.

(8) Subject to paragraph (15), where an operator fails to comply with a condition imposed pursuant to regulation 10(3) in respect of an installation, the registry administrator shall ensure that an operator may not transfer any allowances out of the operator holding account for that installation until the compliance status figure for that installation calculated in accordance with Article 55 of the Registries Regulation is greater than or equal to zero.

(9) If a person has failed to comply with any terms and conditions agreed in accordance with Article 19(4) of the Registries Regulation, the registry administrator may–

(a) prevent, subject to paragraph (15), the transfer of any allowances out of any person holding accounts in the registry held in the name of that person until the person complies with the terms and conditions, or

(b) if the person continues to fail to comply with the terms and conditions, serve a notice on the person indicating that, subject to regulation 32(10), any person holding accounts in the registry held in his name will be closed on the expiry of such period as may be specified in a notice.

(10) The registry administrator may, before the date on which the notice under paragraph (9)(b) takes effect, withdraw the notice.

(11) Subject to paragraph (15), if an operator has failed to comply with any terms and conditions agreed in accordance with Article 15(4) of the Registries Regulation, the registry administrator may prevent the transfer of any allowances out of any operator holding account in the registry held in the name of that operator until the operator complies with the terms and conditions.

(12) An application for the creation of a person holding account under Article 19(1) of the Registries Regulation shall be accompanied by the fee prescribed in respect of the application in Schedule 5.

(13) An application by an organisation which wishes to be able to be appointed by an account holder as his additional authorised representative, the holder of a person holding account or a verifier —

(a) to be included in the registry;

(b) to appoint an individual as a new user in the registry; or

(c) to authorise an individual to use the registry on his behalf,

shall be accompanied by such fee as may be prescribed in respect of the application in Schedule 5.

(14) Where—

(a) an operator fails to submit to the regulator the report required to be submitted to the regulator by the terms of a notice of surrender included pursuant to regulation 16(7) or by the terms of a revocation notice included pursuant to regulation 17(5)(a) or the report submitted is incomplete; or

(b) the report submitted to the regulator in accordance with the terms of a notice of surrender included pursuant to regulation 16(7) or by the terms of a revocation notice included pursuant to regulation 17(5)(a) or part of such report cannot be verified in accordance with the relevant monitoring and reporting conditions, the registry administrator shall ensure that, subject to paragraph (15), the operator or, where the installation is covered by a notice under regulation 27(10)(b) authorising a pool, the pool administrator, may not transfer allowances out of the operator holding account for the installation to which the notice of surrender or the revocation notice relates until the report has been submitted to the regulator and has been verified in accordance with the terms of a notice of surrender included pursuant to regulation 16(7) or by the terms of a revocation notice included pursuant to regulation 17(5)(a) or the regulator has notified a determination in accordance with regulation 30(3).

(15) Paragraphs (8), (9)(a), (11) and (14) shall not prevent—

(a)   the surrender of allowances in accordance with Article 52 or 54 of the Registries Regulation; or

(b)   the cancellation and replacement of allowances in accordance with Articles 60 and 61 of the Registries Regulation.

(16) Where a registry administrator prevents the transfer of allowances out of an account under paragraphs (8), (9)(a), (11) or (14) or Article 27 of the Registries Regulation, the registry administrator shall notify the account holder specifying the reason why and the period during which transfers will be prevented.

(17) An operator holding account and a person holding account shall be capable of holding all ERUs and CERs.

Pooling

27.—   (1) One or more operators of installations to which this regulation applies may make a joint application to the appropriate authority to form a pool for the second scheme phase.

(2) This regulation applies to installations which carry out activities which—

(a)   fall within the same description in Schedule 1; and

(b)   do not fall within any description in Annex I of Council Directive 96/61/EC concerning integrated pollution prevention and control(**24**).

(24) OJ No. L 257, 10.10.1996, p.26; to which there are amendments not relevant to this regulation.

(3) An application under paragraph (1) shall be made at least 6 months before the start of the scheme phase in which the operators wish to form a pool and shall—

(a)   identify the installations to be included in the pool;

(b)   contain the names and postal addresses of the operators of those installations and, if different, any addresses to which correspondence should be sent;

(c)   contain a copy of the greenhouse gas emissions permit in respect of each of those installations and identify the regulator which granted that permit;

(d)   nominate a person to act as pool administrator and contain a declaration from that person that he is willing to act as pool administrator; and

(e)   contain evidence that the pool administrator will be able to fulfil the obligations in paragraph (12).

(4) Where an application is duly made under paragraph (1) and the appropriate authority considers it appropriate to allow the pool—

(a)   where the Secretary of State is not the appropriate authority, the appropriate authority shall send a copy of the application to the Secretary of State; and

(b)   the Secretary of State shall submit the application to the European Commission.

(5) The appropriate authority shall notify—

(a)   the operator of each installation to be included in the pool;

(b)   the regulator or regulators for the installations to be included in the pool; and

(c)   the person nominated to act as pool administrator under paragraph (3)(d), whether it considers it appropriate to allow the pool.

(6) If the European Commission rejects the application within three months of the date it receives the application—

(a)   where the Secretary of State is not the appropriate authority, the Secretary of State shall notify the appropriate authority; and

(b)   the appropriate authority shall notify–

(i)   the operator of each installation to be included in the pool;

(ii)   the regulator or regulators for the installations to be included in the pool; and

(iii)   the person nominated to act as pool administrator under paragraph (3)(d), that the application has been rejected and of the reasons given by the European Commission for the rejection.

(7) Where operators are notified under paragraph (6) that the European Commission has rejected their application, they may within a period of four weeks beginning on the date of the notice under paragraph (6)(b) submit an amended application to the appropriate authority.

(8) If the appropriate authority considers that the amended application addresses the reasons given by the European Commission for rejection of the application—

(a)   where the Secretary of State is not the appropriate authority, the appropriate authority shall send a copy of the amended application to the Secretary of State; and

(b)   the Secretary of State shall submit the amended application to the European Commission.

(9) The appropriate authority shall notify—

(a)   the operator of each installation to be included in the pool;

(b)   the regulator or regulators for the installations to be included in the pool; and

(c)   the person nominated to act as pool administrator under paragraph (3)(d), of whether it considers that the amended application addresses the reasons given by the European Commission for rejection of the application.

(10) If the European Commission does not reject the application within three months of the date it receives the application or accepts an amended application submitted under paragraph (8)(b)—

(a)   where the Secretary of State is not the appropriate authority, the Secretary of State shall notify the appropriate authority; and

(b)   the appropriate authority shall serve a notice authorising the pool on–

(i)   the operator of each installation to be included in the pool;

(ii)   the regulator for each installation to be included in the pool;

(iii)   the person nominated to act as pool administrator under paragraph (3)(d); and

(iv)   the registry administrator.

(11)   A notice under paragraph (10)(b) shall—

(a)   identify the installations included in the pool;

(b)   identify the person who will act as pool administrator;

(c)   specify any conditions applying to the approval of the pool; and

(d)   specify the phase for which the pool is approved.

(12) For the duration of the period for which a group of installations are covered by a notice under paragraph (10)(b) authorising the pool—

(a)   the operator of each installation referred to in paragraph (11)(a) shall ensure that the pool administrator is nominated as primary authorised representative for the operator holding account for the installation;

(b)   the pool administrator shall surrender allowances equal to the annual reportable emissions from all the installations within the pool for which he is acting as pool administrator (as increased by any condition of a greenhouse gas emissions permit relating to an installation included in the pool imposed pursuant to regulation 10(4)) within four months of the end of the scheme year during which those emissions arose; and

(c)   regulation 39 shall apply to a pool administrator who fails to comply with the obligation in sub-paragraph (b) as it applies to an operator who fails to comply with a condition of a greenhouse gas emissions permit imposed pursuant to regulation 10(3).

(13) An operator of an installation which is included in a notice authorising a pool in accordance with paragraph (11)(a) shall be deemed, unless the notice is revoked, to be in compliance with any condition of a greenhouse gas emissions permit imposed pursuant to regulation 10(3).

(14) Where the pool administrator fails to pay a civil penalty under regulation 39 by the due date determined in accordance with regulation 41(3), the appropriate authority shall serve a notice on

the persons specified in paragraph (10)(b) providing for the notice under paragraph (10)(b) authorising the pool to be revoked, subject to paragraph 33(6), from the date specified in the notice.

(15) For the purposes of this regulation, where an application to form a pool relates to installations in more than one country of the United Kingdom, the appropriate authority in relation to the application and any subsequent pool shall, subject to regulation 46, be the Secretary of State.

(16) Where a notice of surrender or a revocation notice is served in respect of an installation which is included in a pool, the appropriate authority shall serve on the persons on which the notice under paragraph (10)(b) in respect of the pool was served, a further notice amending the notice under paragraph (10)(b) to remove the installation from the list of installations included in the pool from the date on which the notice of surrender or revocation notice takes effect.

(17) Where the regulator effects a transfer under regulation 15 of a greenhouse gas emissions permit which relates to an installation included in a pool, the appropriate authority shall serve on the persons on which the notice under paragraph (10)(b) in respect of the pool was served, a further notice amending the notice under paragraph (10)(b) from the date on which the transfer takes effect to take account of the transfer in accordance with the provisions of the application for transfer under regulation 15(4)(d).

(18) Where —

(a) the Secretary of State is the appropriate authority by virtue of paragraph (15);

(b) no agreement has been reached under regulation 46 in relation to a decision under paragraph (4); and

(c) the deadline in paragraph (3) for making an application to form a pool has passed, the Secretary of State shall forthwith serve a notice on those persons referred to in paragraph (5) indicating that, as no agreement has been reached, it is not considered appropriate to allow the pool and providing that the operators of installations included in the application which are situated in the same country of the United Kingdom may make a new application under paragraph (1) to the appropriate authority within two weeks of the date of the notice under this paragraph.

(19) Where —

(a)   the Secretary of State is the appropriate authority by virtue of paragraph (15);

(b)   no agreement has been reached under regulation 46, in relation to a decision under paragraph (8); and

(c)   a period of four weeks from the date of submission of an amended application to the Secretary of State under paragraph (7) has expired, the Secretary of State shall forthwith serve a notice on those persons referred to in paragraph (9) indicating that it has not been agreed that the amended application addresses the reasons given by the European Commission for rejection of the application to form a pool and providing that the operators of installations included in the application which are situated in the same country of the United Kingdom may submit a further amended application under paragraph (8) to the appropriate authority within two weeks of the date of the notice under this paragraph.

Use of CERs and ERUs

27A.—(1) Subject to paragraphs to (5), an operator of an installation may, in accordance with Article 53 of the Registries Regulation, use any combination of—

(a)   CERs from project activities;

(b)   ERUs from project activities; and

(c)   allowances,

to comply with a requirement to surrender allowances imposed pursuant to regulation 10(3), 16(7)(b), 17(5)(b) or 40(2)(a).

(2) An operator may not use ERUs to comply with an obligation relating to emissions in the first scheme phase.

(3) An operator may not use CERs or ERUs generated from—

(a)   nuclear facilities; or

(b)   land use, land use change and forestry activities.

(4) In relation to the second and subsequent scheme phases, an operator may only use CERs and ERUs up to the limit provided for in the approved national allocation plan for that scheme phase.

(5) In relation to a scheme phase in which a partial transfer of a greenhouse gas emissions permit is effected in accordance with regulation 15(7)(a), an operator may not use CERs or ERUs where the obligation relates to the transferred unit.

PART 4

ENFORCEMENT

Duty of regulator to enforce compliance with monitoring and reporting conditions

28. While a greenhouse gas emissions permit is in force it shall be the duty of the regulator to take such action under these Regulations as may be necessary for the purpose of ensuring that the monitoring and reporting conditions are complied with.

Enforcement notices

29.— (1) If the regulator is of the opinion that an operator has contravened, is contravening or is likely to contravene any monitoring and reporting condition, the regulator may serve on him a notice (an "enforcement notice").

(2) An enforcement notice shall—

(a) state that the regulator is of the opinion that an operator has contravened, is contravening or is likely to contravene any monitoring and reporting condition;

(b) specify the matters constituting the contravention or the matters making it likely that the contravention will arise, as the case may be;

(c) specify the steps that must be taken to comply with the monitoring and reporting condition or, to the extent possible, to remedy any failure to comply with the monitoring and reporting condition, as the case may be; and

(d) specify the period within which those steps must be taken.

(3) The regulator may withdraw an enforcement notice at any time.

Power of the regulator to determine reportable emissions

30.— (1) Where—

(a) an operator serves notice on the regulator in accordance with a condition of the greenhouse gas emissions permit imposed pursuant to regulation 10(2)(c) notifying it of factors that might prevent him from complying with the monitoring and reporting conditions of the permit and requesting the regulator to determine all or part of the annual reportable emissions from the installation or, in respect of the surrender or revocation of a greenhouse gas emissions permit, the reportable emissions from the installation for the period specified in regulation 16(7) or 17(5)(a);

(b)   an operator fails to comply with the conditions included in a greenhouse gas emissions permit pursuant to regulation 10(2)(a)(ii) or 10(2)(b); or

(c)   an operator fails to comply with the requirements included in a notice of surrender pursuant to regulation 16(7) or in a revocation notice pursuant to regulation 17(5)(a),

the regulator shall determine the reportable emissions from the installation in the relevant period and the regulator's determination of the reportable emissions shall be treated as the reportable emissions from that installation for the period to which the determination relates.

(2) When determining annual reportable emissions under paragraph (1) the regulator shall take account of the Monitoring and Reporting Decision and the requirements set out in Annex V of the Directive.

(3) The regulator shall notify any determination under paragraph (1) to the operator of the installation.

(4) A notice under paragraph (3) shall be served on the registry administrator and shall be an instruction to the registry administrator for the purposes of Article 51(2) of the Registries Regulation.

(5) Where a regulator makes a determination under paragraph (1) it may recover the cost of making that determination from the operator concerned.

Powers of entry: offshore installations

31.— (1) The Secretary of State may authorise in writing any person who appears suitable to her to exercise, in accordance with the terms of that authorisation, any of the powers specified in paragraph (2) in respect of offshore installations for the purposes of—

(a)   determining whether the requirements, restrictions or prohibitions imposed by or under these Regulations are being, or have been, complied with;

(b)   discharging one or more of the functions conferred or imposed upon the Secretary of State by or under these Regulations; or

(c)   determining whether and, if so, how such a function should be discharged.

(2) The powers exercisable under paragraph (1) are the powers in paragraphs (a) to (k) of regulation 13(2) of the Offshore Regulations and subject to paragraphs (3) and (4) of that regulation.

(3) Regulation 18(1)(f) of the Offshore Regulations shall apply to a failure to comply with an obligation imposed pursuant to a power exercisable under paragraph (1) as it applies to a failure to comply with an obligation imposed pursuant to regulation 13(2) of the Offshore Regulations.

PART 5

APPEALS

Appeals against a decision of, or a notice served by, the regulator or registry administrator

32.— (1) Subject to paragraph (5), the following persons, namely—

(a)   a person who has been refused the grant of a greenhouse gas emissions permit under regulation 9;

(b)   a person who has been refused the variation of the provisions of a greenhouse gas emissions permit on an application under regulation 14(2);

(c)   a person who is aggrieved by the provisions of his greenhouse gas emissions permit following an application under regulation 8 or by a variation notice following an application under regulation 14(2);

(d)   a person whose application under regulation 15(1) for a regulator to effect the transfer of a greenhouse gas emissions permit has been refused or who is aggrieved by the provisions of his greenhouse gas emissions permit to take account of such a transfer;

(e)   a person whose application under regulation 16(1) to surrender a greenhouse gas emissions permit has been refused or who is aggrieved by the terms of the notice of surrender; or

(f)   a person who is aggrieved by the regulator's determination of reportable emissions under regulation 30,

may appeal to the appropriate authority or, where an appeal relates to an offshore installation, to the regulator.

(2) Subject to paragraph (5), a person on whom a variation notice is served, other than following an application under regulation 14(2), or on whom a revocation notice or an enforcement notice is served may appeal to the appropriate authority or, where an appeal relates to an offshore installation, to the regulator.

(3) Subject to paragraph (5),

(a)   an operator who is aggrieved by a notice served by the regulator under regulation 21(6);

(b)   an operator whose application for an allocation from a new entrant reserve under regulation 22(1) is refused or who is aggrieved by the provisions of a notice under regulation 22(13)(a) or (b), (17) or (18), or regulation 22C;

(c)   an operator whose application under regulation 24(1) to retain an allocation of allowances is refused or who is aggrieved by the terms of a retention notice;

(d)   an operator who is aggrieved by the terms of a notice under regulation 25(2); or

(e)   an operator on which a notice under regulation 41(2)(b) has been served,

may appeal to the appropriate authority or, where an appeal relates to an offshore installation, to the regulator.

(4) Subject to paragraph (5)—

(a)   a person who is aggrieved by a decision to prevent transfers out of his account in accordance with regulation 26(8), (9)(a), (11) or (14) or under Article 27 of the Registries Regulation; or

(b)   a person who is aggrieved by a notice under regulation 26(9)(b),

may appeal to the appropriate authority.

(5) Paragraphs (1) to (4) shall not apply where the decision or notice, as the case may be, implements a direction of the appropriate authority given under paragraph (7) or regulation 42 or of the Secretary of State given under regulation 44.

(6) An appeal under paragraph (3)(a) may include an appeal against a decision of the regulator referred to in paragraph (1)(a), (b) or (c).

(7) Except where an appeal under paragraph (1), (2) or (3) relates to an offshore installation, in determining an appeal under this regulation against a decision of a regulator or a notice served by a regulator the appropriate authority may—

(a)   affirm the decision or notice;

(b)   quash all or part of the decision or notice;

(c)   vary the decision or notice; or

(d)   give directions to the regulator or, in the case of an appeal under paragraph (4), to the registry administrator, in relation to the exercise of its functions under these Regulations in relation to the subject matter of the appeal.

(8) Where an appeal made under paragraph (1), (2) or (3) relates to an offshore installation, the regulator shall reconsider its decision and may affirm, reverse or vary its decision.

(9) Where an appeal is brought under—

(a)   paragraph (1)(c) or (d) in relation to the provisions attached to a greenhouse gas emissions permit;

(b)   paragraph (1)(f);

(c)   paragraph (2) against a variation notice or an enforcement notice;

(d)   paragraph (3)(a);

(e)   paragraph (3)(a) against the provisions of a notice under regulation 22(13)(a) or (b) or (18), or regulation 22C; or

(f)   paragraph (4)(a),

the bringing of the appeal shall, subject to paragraph (12), not have the effect of suspending the operation of the decision or notice.

(10) Where an appeal is brought under—

(a)   paragraph (2) against a revocation notice or under paragraph (1)(e) against the provisions of a notice of surrender;

(b)   paragraph (3)(c) against the terms of a retention notice, paragraph (3)(d) or (e); or

(c)   paragraph (4)(b),

the bringing of the appeal shall have the effect of suspending the operation of the notice pending the final determination or the withdrawal of the appeal.

(11) Where an appeal is brought under paragraph (3)(c) against a refusal to accept an application under regulation 24(1), the application shall not be deemed to have been refused for the purposes of regulation 24(11)(a) unless the refusal is affirmed on appeal or the appeal is withdrawn.

(12) Where an appeal is brought under paragraph (1)(f) against a determination of reportable emissions the determination shall not be used for the purpose of checking compliance with a condition included in a greenhouse gas emissions permit pursuant to regulation 10(3) pending the final determination or the withdrawal of the appeal.

(13) Regulation 10 shall apply where the appropriate authority, in exercising any of the powers in paragraph (7), gives directions as to the conditions to be attached to a greenhouse gas emissions permit as they would apply to the regulator when determining the conditions of the permit.

(14) For the purposes of appeals under this regulation, the appropriate authority in relation to installations (other than offshore installations) situated in Northern Ireland shall be the Planning Appeals Commission.

Appeals for reconsideration of decisions

33.— (1) A person who is aggrieved by a decision or notice under regulation 27(4), (8) or (14) or the terms of a notice pursuant to regulation 27(16) or (17) may appeal to the appropriate authority.

(2) A person who is aggrieved by —

(a) a certificate or notice served under regulation 11(6), (9) or (11); or

(b) a notice served by the responsible authority under regulation 21(6),

may appeal to the responsible authority.

(3) A person who is aggrieved by —

(a) a supplementary decision under regulation 25(7); or

(b) a notice served by the Secretary of State under regulation 21(6), or 22A(5)

may appeal to the Secretary of State.

(4) Where an appeal is made under this regulation, the appeal body shall reconsider its decision and may affirm, reverse, or vary its decision.

(5) Where an appeal is made under—

(a)   paragraph (1) against a notice under regulation 27(16) or (17); or

(b)   paragraph (2) or (3)(b),

the bringing of the appeal shall not have the effect of suspending the certificate or notice.

(6) Where an appeal is made under—

(a)   paragraph (3)(a); or

(b)   paragraph (1) against a notice under regulation 27(14),

the decision or notice to which the appeal relates shall not take effect pending the final determination or withdrawal of the appeal.

Procedure for appeals under regulations 32 and 33

34.— (1) Except where paragraph (4) applies and subject to paragraph (5), Schedule 2 shall have effect in relation to the making and determination of appeals under regulations 32 or 33.

(2) Except where paragraph (4) applies, the appeal body may—

(a)   appoint any person to exercise on its behalf, with or without payment, the function of determining an appeal under regulation 32 or 33 or any matter or question involved in such an appeal; or

(b)   refer any matter or question involved in an appeal under regulation 32 or 33 to such person as it may appoint for the purpose, with or without payment.

(3) Schedule 3 shall have effect with respect to appointments under paragraph (2)(a).

(4) Where an appeal under regulation 32 relates to an installation (other than an offshore installation) situated in Northern Ireland, Schedule 4 shall have effect in relation to the making and determination of the appeal.

(5) Where an appeal is made under regulation 32(4), references in Schedule 2 to the regulator shall be taken to be references to the registry administrator.

PART 6

INFORMATION

Information

35.— (1) For the purposes in paragraph (4), an appropriate authority, a responsible authority or the Secretary of State may, by notice served on a regulator, require the regulator to furnish such information about the discharge of its functions as a regulator as the appropriate authority, responsible authority or the Secretary of State may require.

(2) For the purposes in paragraph (4), an appropriate authority, a responsible authority, the Secretary of State or a regulator may, by notice served on any person, require that person to furnish such information as is specified in the notice, in such form and within such period following service of the notice or at such time as is so specified.

(3) The information which a person may be required to furnish by a notice served under paragraph (2) includes information, which, although it is not in the possession of that person or would not otherwise come into the possession of that person, is information which it is reasonable to require that person to compile for the purpose of complying with that notice.

(4) The purposes referred to in paragraphs (1) and (2) are—

(a)   the purpose of the discharge of the relevant body's functions under these regulations;

and

(b)   the purpose of applying, seeking to apply, or assessing whether to seek to apply emission allowance trading to activities, installations and greenhouse gases which are not listed in Schedule 1 in accordance with Article 24 of the Directive.

(5) Where the Secretary of State is entitled to serve a notice on a person under paragraph (2)—

(a)   in relation to England and Wales, the regulator; and,

(b)   in relation to Scotland and Northern Ireland, the regulator or the Environment Agency,

may serve that notice for the purpose of assisting the Secretary of State.

(6) The Secretary of State may use any information which she holds or has obtained for the purposes of these Regulations, and may share such information with other government bodies, for the purpose of preparing and publishing national energy and emissions statistics, including the preparation and publication of a national inventory.

(7) In this regulation, "national inventory" means the estimation of anthropogenic emissions of greenhouse gases by sources and removals of all greenhouse gases by sinks not controlled by the Montreal Protocol(25) under Article 4(1)(a) of the UNFCCC(26).

Publication of a list of operators subject to penalties

36. As soon as possible after the expiry of the period of 4 months after the end of each scheme year of each scheme phase, the regulator shall publish a list of the names of operators who are liable to a civil penalty under regulation 39 or 40.

National Security

37.— (1) No information included in a national allocation plan developed under regulation 20(1), in a decision under regulation 21(1) or in a supplementary decision under regulation 25(7) shall be published, if, in the opinion of the Secretary of State, the inclusion of that information, or information of that description, would be contrary to the interests of national security.

(2) No information shall be included in the list published under regulation 36 if in the opinion of the Secretary of State, the inclusion of that information , or information of that description, would be contrary to the interests of national security.

(3) The Secretary of State may, for the purpose of—

(a) ensuring that information to which paragraph (1) applies is not published; or

(b) securing the exclusion from the list published under regulation 36 of the information to which paragraph (2) applies,

give to the appropriate authorities for installations (other than offshore installations) situated in Scotland, Wales and Northern Ireland directions specifying information, or descriptions of

---

(25)Montreal Protocol on Substances that Deplete the Ozone Layer adopted at Montreal on 16th September 1987. This is available at http://hq.unep.org/ozone/Montreal-Protocol/Montreal-Protocol2000.shtml.

(26)United Nations Framework Convention on Climate Change signed in New York on 9th May 1992. This is available at http://unfccc.int/resource/docs/convkp/conveng.pdf

information, which shall not be published or shall be excluded from the list published under regulation 36.

(4) The other appropriate authorities referred to in paragraph (3) shall notify the Secretary of State of any information which is not published or which is excluded from a list published in accordance with regulation 36 in pursuance of directions under paragraph (3).

PART 7

OFFENCES AND CIVIL PENALTIES

Offences

38.— (1) It is an offence for a person—

(a) to contravene regulation 7;

(b) to fail to comply with or to contravene a condition of a greenhouse gas emissions permit (except where regulation 39 or 40 apply to such failure to comply or contravention);

(c) to fail to comply with regulation 12(1), 13(2) or 16(1) or to fail to comply with a condition of a notice under regulation 22(13)(a) imposed pursuant to regulation 22(15)(b);

(d) to fail to comply with the requirements of an enforcement notice;

(e) to fail, without reasonable excuse, to comply with any requirement imposed by a notice under regulation 16(16) or 35(2);

(f) to make a statement which he knows to be false or misleading in a material particular, or recklessly to make a statement which is false or misleading in a material particular, where the statement is made—

(i) in purported compliance with a requirement imposed by a notice under regulation 8(5), 14(6), 15(10), 16(16), 22(9), 24(6) or 35(2);

(ii) for the purpose of obtaining the grant of a greenhouse gas emissions permit to himself or any other person, or the variation, transfer or surrender of a greenhouse gas emissions permit;

(iii) for the purpose of obtaining a certificate under regulation 11;

(iv) for the purpose of obtaining a notice authorising a pool under regulation 27;

(v) as part of the verification of a report required under a monitoring and reporting condition;

(vi) for the purpose of obtaining an allocation from a new entrant reserve provided for in the approved national allocation plan;

(vii) for the purpose of retaining the allocation or a proportion of the allocation of allowances in respect of an installation which ceases to carry out a Schedule 1 activity;

(g) intentionally to make a false entry in any record required to be kept under the condition of a greenhouse gas emissions permit;

(h) with intent to deceive, to forge or use a document issued or authorised to be issued under a condition of a greenhouse gas emissions permit or required for any purpose under a condition of such a permit or to make or to have in his possession a document so closely resembling any such document as to be likely to deceive;

(i) to fail to comply with Article 15(1), 15(3) or 19(3) of the Registries Regulation.

(2) A person guilty of an offence under paragraph (1) shall be liable—

(a) on summary conviction, to a fine not exceeding the statutory maximum or to imprisonment for a term not exceeding three months;

(b) on conviction on indictment, to a fine or to imprisonment for a term not exceeding two years or to both.

(3) Where an offence under this regulation is committed by—

(a) a body corporate (other than a limited liability partnership) and is proved to have been committed with the consent or connivance of, or to have been attributable to any neglect on the part of, any director, manager, secretary, or other similar officer of the body corporate or a person who was purporting to act in any such capacity;

(b) a limited liability partnership and is proved to have been committed with the consent or connivance of, or to have been attributable to any neglect on the part of, any member of the limited liability partnership or a person who was purporting to act as such; or

(c)   a partnership in Scotland (other than a limited liability partnership) (a "Scottish partnership") and is proved to have been committed with the consent or connivance of, or have been attributable to any neglect on the part of, any partner or a person who was purporting to act as such,

that person as well as the body corporate, the limited liability partnership or the Scottish partnership, as the case may be, shall be guilty of that offence and shall be liable to be proceeded against and punished accordingly.

(4) Where the affairs of a body corporate (other than a limited liability partnership) are managed by its members, paragraph (3) shall apply in relation to the acts or defaults of a member in connection with his functions of management as if he were a director of the body corporate.

(5) Where the commission by any person of an offence under this regulation is due to the act or default of some other person, that other person may be charged with and convicted of the offence by virtue of this paragraph whether or not proceedings for the offence are taken against the first-mentioned person.

Civil penalties: excess emissions

39.— (1) Any operator who fails to comply with a condition imposed pursuant to regulation 10(3) in respect of an installation shall be liable to a penalty.

(2) The amount of the penalty to which the operator of an installation is liable under paragraph (1) shall be the excess emissions of the installation multiplied by the excess emissions penalty.

(3) For the purpose of paragraph (2)—

(a)   "excess emissions" means, in respect of an installation, the amount in tonnes of carbon dioxide equivalent by which the annual reportable emissions from the installation exceeded the number of allowances surrendered for that installation;

(b)   "excess emissions penalty" means—

(i)   in respect of excess emissions which relate to reportable emissions which were released between 1st January 2005 and 31st December 2007, 40 Euro; and

(ii) in respect of excess emissions which relate to reportable emissions which were released on or after 1st January 2008, 100 Euro.

(4) In relation to paragraph (3)(b), the reference to an amount in Euro shall be taken to be a reference to the sterling equivalent of that number of Euro, converted by reference to the rate of conversion published in the C series of the Official Journal of the European Communities in September of the scheme year preceding that in which the liability for the penalty arose.

Civil penalties: understatement of reportable emissions

40. — (1) Subject to paragraph (2), where in relation to the application for surrender of a greenhouse gas emissions permit under regulation 16(1) or the revocation of a greenhouse gas emissions permit under regulation 17(1), the report submitted in accordance with the requirements included in a notice of surrender or a revocation notice pursuant to regulation 16(7) or 17(5)(a) understates the reportable emissions from the installation to which the report relates, the operator shall be liable to a penalty equal to the amount of the understatement of reportable emissions multiplied by the excess emissions penalty under regulation 39(3)(b) which applied to excess emissions in the year in which the understatement was made.

(2) Conduct falling with paragraph (1) shall not give rise to liability to a penalty under this regulation if the person who made the understatement—

(a) surrenders allowances equal to the amount of the understatement; and

(b) satisfies the regulator, that he did not knowingly or recklessly understate the reportable emissions from the installation.

Civil penalties: general

41. — (1) In this regulation "civil penalty" means any penalty which—

(a) is imposed by or under these Regulations; and

(b) arises otherwise than in consequence of a person's conviction for a criminal offence.

(2) Where a person is liable to a civil penalty, the regulator shall—

(a) assess the amount due by way of penalty; and

(b) notify the person liable to the penalty of that amount.

(3) Subject to regulation 32(10), a penalty shall be due on the day (the "due date") following the expiry of a period of two months beginning on the date on which the person is notified by the regulator under paragraph (2)(b) and shall be paid to the regulator.

(4) Where a regulator makes an assessment under paragraph (2) of any penalty to which a person is liable the amount of that penalty shall carry interest for the period which—

(a)   begins on the due date; and

(b)   ends with the day before the day on which the assessed penalty is paid.

(5) Interest under this regulation shall be payable at a rate of one percentage point above LIBOR on a day to day basis.

(6) For the purposes of paragraph (5), "LIBOR" means the sterling three months London interbank offered rate in force during the period between the due date and the date on which the penalty is paid.

(7) Where an amount has been assessed and notified to any person under paragraph (2), the amount and any interest incurred under paragraph (4) shall be recoverable on demand.

(8) The regulator shall notify the appropriate authority of any penalty assessed under paragraph (2) and shall pass any civil penalties and any interest paid to it to the appropriate authority.

PART 8

APPROPRIATE AUTHORITY'S POWERS

Directions to regulators

42.—   (1) Subject to paragraph (5), an appropriate authority may give directions to the regulator of a general or specific character with respect to the carrying out of any of their functions under these Regulations in relation to installations for which it is the appropriate authority.

(2) Without prejudice to the generality of the power conferred by paragraph (1), a direction under that paragraph may direct a regulator–

(a)   to exercise any of their powers under these Regulations or to do so in such circumstances as may be specified in the directions or in such manner as may be so specified; or

(b) not to exercise those powers, or not to do so in such circumstances or such manner as may be specified in the directions.

(3) Any direction given under these Regulations shall be in writing and may be varied or revoked by a further direction.

(4) It shall be a duty of a regulator to comply with any direction which is given to it under these Regulations.

(5) This regulation shall not apply in respect of offshore installations.

Guidance to regulators

43.— (1) An appropriate authority may issue guidance to a regulator with respect to the carrying out of any of their functions under these Regulations in relation to installations for which it is the appropriate authority.

(2) A regulator, in carrying out any of its functions under these Regulations, shall have regard to any guidance issued by the appropriate authority under this regulation.

PART 9

SECRETARY OF STATE'S POWERS

Directions to registry administrator

44.— (1) Subject to paragraph 46, the Secretary of State may give directions to the registry administrator of a general or specific character with respect to the carrying out of any of its functions under these Regulations or the Registries Regulation.

(2) Without prejudice to the generality of the power conferred by paragraph (1), a direction under that paragraph may direct the registry administrator–

(a) to exercise any of its powers under these Regulations or the Registries Regulation or to do so in such circumstances as may be specified in the directions or in such manner as may be so specified; or

(b) not to exercise those powers, or not to do so in such circumstances or such manner as may be specified in the directions.

(3) Any direction given under this regulation shall be in writing and may be varied or revoked by a further direction.

(4) It shall be the duty of the registry administrator to comply with any direction which is given to it under this regulation.

Guidance to the registry administrator

45.— (1) Subject to regulation 46, the Secretary of State may issue guidance to the registry administrator with respect to the carrying out of any of its functions under these Regulations or the Registries Regulation.

(2) The registry administrator, in carrying out any of its functions under these Regulations or the Registries Regulation, shall have regard to any guidance issued by the Secretary of State under this regulation.

PART 10

SUPPLEMENTARY

Agreement of Scottish Ministers, the National Assembly for Wales and the Department for Environment

46.— (1) Subject to paragraphs (2), (3) and (4), any power of the Secretary of State under regulation 20(1), 21(1), 22A(5), 25(7), 26(1), 27(15), 44 and 45 is exercisable—

(a)   in so far as it relates to installations situated in Scotland (other than offshore installations), only with the agreement of the Scottish Ministers;

(b)   in so far as it relates to installations situated in Wales (other than offshore installations), only with the agreement of the National Assembly for Wales; and

(c)   in so far as it relates to installations situated in Northern Ireland (other than offshore installations), only with the agreement of the Department of the Environment.

(2) The Secretary of State may exercise a power referred to in paragraph (1) in relation to installations situated in Scotland where—

(a)   no agreement has been reached with the Scottish Ministers;

(b)   she considers that it is necessary to do so to ensure that the United Kingdom complies with its obligations under the Directive; and

(c)   she serves a notice on the Scottish Ministers stating that she has decided to exercise a power referred to in paragraph (1) in relation to Scotland.

(3) The Secretary of State may exercise a power referred to in paragraph (1) in relation to installations situated in Wales where—

(a)   no agreement has been reached with the National Assembly for Wales;

(b)   she considers that it is necessary to do so to ensure that the United Kingdom complies with its obligations under the Directive; and

(c)   she serves a notice on the National Assembly for Wales stating that she has decided to exercise a power referred to in paragraph (1) in relation to Wales.

(4) The Secretary of State may exercise the power under paragraph (1) in relation to installations situated in Northern Ireland where—

(a)   no agreement has been reached with the Department of the Environment;

(b)   she considers that it is necessary to do so to ensure that the United Kingdom complies with its obligations under the Directive; and

(c)   she serves a notice on the Department of the Environment stating that she has decided to exercise a power referred to in paragraph (1) in relation to Northern Ireland.

(5) Subject to paragraph (8), the power of the Secretary of State under regulation 21A(1) (but not any power of the Environment Agency or the Scottish Environment Protection Agency which arises by virtue of regulation 21A(12)) is exercisable only with the consent of the Scottish Ministers, the National Assembly for Wales and the Department of the Environment ("the devolved administrations").

(6) Subject to paragraph (8), the Secretary of State's power of appointment under regulation 21A(12), including the power to impose limitations on such an appointment, is exercisable only with the consent of the devolved administrations.

(7) Consent under paragraph (5) means consent to the manner in which the Secretary of State intends to conduct an auction or sale of allowance.

(8) The Secretary of State may exercise the power in regulation 21A(1) or (12) where—

(a)   consent of any one or more of the devolved administrations has not been given;

(b) he considers that it is necessary to exercise the power to ensure that the United Kingdom complies with its obligations under the Directive; and

(c) he serves a notice on each of the devolved administrations which has not given its consent stating that he has decided to exercise the power without such consent.

PART 11

REVOCATION AND CONSEQUENTIAL AMENDMENTS

Revocation and savings provisions

47.— (1) Subject to paragraphs (2) and (3), the following Regulations are revoked—

(a) The Greenhouse Gas Emissions Trading Scheme Regulations 2003(**27**) (the "2003 Regulations"); and

(b) The Greenhouse Gas Emissions Trading Scheme (Amendment) Regulations 2004(**28**).

(2) Regulation 18 of the 2003 Regulations shall continue to apply until the obligations under that regulation is fulfilled in respect of the first phase.

(3) Paragraphs (1) to (4) and (12) to (18) of regulation 19 of the 2003 Regulations shall continue to apply until the obligations under those paragraphs are fulfilled in respect of the first phase.

(4) In this regulation, "first phase" means the period referred to in regulation 18(2)(a) of the 2003 Regulations.

Consequential amendments

48. The enactments mentioned in Schedule 6 shall have effect with the amendments there specified (being amendments consequential on provisions of these Regulations).

Elliot Morley

Minister of State

Department for Environment, Food and Rural Affairs

(**27**) S.I. 2003/3311
(**28**) S.I. 2004/3390

Regulations 2(1) and 7

SCHEDULE 1

Activities

PART 1: ACTIVITIES AND SPECIFIED EMISSIONS

Activities	*Specified emissions*
1. Energy Activities	
1.1 Activities of combustion installations with a rated thermal input exceeding 20 megawatts (excluding hazardous or municipal waste installations).	Carbon dioxide
1.2 Activities of mineral oil refineries.	Carbon dioxide
1.3 Activities of coke ovens.	Carbon dioxide
2. Production and processing of ferrous metals	
2.1 Activities of metal ore (including sulphide ore) roasting and sintering installations.	Carbon dioxide
2.2 Activities of installations for the production of pig iron or steel (primary or secondary fusion), including continuous casting, with a capacity of more than 2.5 tonnes per hour.	Carbon dioxide
3. Mineral Industries	
3.1 Activities of installations for the production of cement clinker in rotary kilns with a production capacity of more than 500 tonnes per day.	Carbon dioxide
3.2 Activities of installations for the production of lime in rotary kilns or other furnaces with a production capacity of more than 50 tonnes per day.	Carbon dioxide
3.3 Activities of installations for the manufacture of glass including glass fibre where the melting capacity of the plant is more than 20 tonnes per day.	Carbon dioxide
3.4 Activities of installations for the manufacture of ceramic products (including roofing tiles, bricks, refractory bricks, tiles, stoneware or porcelain) by firing in kilns where–	Carbon dioxide
(i) the kiln production capacity is more than 75 tonnes per day; or	
(ii) the kiln capacity is more than 4m³ and the setting density is more than 300 kg/m³.	
4. Other activities	
4.1 Activities of industrial plants for the production of pulp from timber or other fibrous materials.	Carbon dioxide
4.2 Activities of industrial plants for the production of paper and board with a production capacity of more than 20 tonnes per day.	Carbon dioxide

PART 2: INTERPRETATION OF SCHEDULE 1

1. The following rules apply for the interpretation of Part 1 of this Schedule.

2. An activity shall not be taken to be an activity falling within Part 1 if it is carried out for research, development or testing of new products or processes.

3.— (1) This paragraph applies for the purpose of determining whether an activity carried out in a stationary technical unit falls within the description of an activity in Part 1 which refers to capacity.

(2) Where a person carries out several activities falling within the same description in Part 1 in different parts of the same stationary technical unit or in different stationery technical units on the same site, the capacities of each part or unit, as the case may be, shall be added together and the total capacity shall be attributed to each part or unit for the purpose of determining whether the activity carried out in each part or unit falls within a description in Part 1.

(3) For the purposes of sub-paragraph (2), no account shall be taken of capacity when determining whether activities fall within the same description.

Regulation 34(1)

SCHEDULE 2

Appeals (other than appeals to which Schedule 4 applies)

1.— (1) A person who wishes to appeal to the appeal body under regulation 32 or 33 shall give to the appeal body written notice of the appeal together with the documents specified in sub-paragraph (2) and shall at the same time send to the regulator a copy of that notice together with copies of the documents specified in sub-paragraph (2)(a) and (f).

(2) The documents mentioned in sub-paragraph (1) are—

(a)   a statement of the grounds of appeal;

(b)   a copy of any relevant application;

(c)   a copy of any relevant greenhouse gas emissions permit;

(d)   a copy of any relevant correspondence between the appellant and–

(i)   in the case of an appeal under regulation 32, the regulator;

(ii)   in the case of an appeal under regulation 33(1), the appropriate authority;

(iii)   in the case of an appeal under regulation 33(2), the responsible authority; or

(iv)   in the case of an appeal under regulation 33(3), the Secretary of State;

(e)   a copy of any decision or notice which is the subject matter of the appeal; and

(f)   a statement indicating whether the appellant wishes the appeal to be in the form of a hearing or to be disposed of on the basis of written representations.

(3) An appellant may withdraw an appeal by notifying the appeal body in writing and shall send a copy of that notification to the regulator.

2.—   (1) Subject to sub-paragraph (2), notice of appeal in accordance with paragraph 1 is to be given—

(a)   in the case of an appeal under regulation 32(1)(a) to (d) or under regulation 32(1)(e) against the refusal of an application to surrender a greenhouse gas emissions permit, before the expiry of the period of six months beginning with the date of the decision or deemed decision which is the subject matter of the appeal;

(b)   in the case of an appeal under regulation 32(1)(e) against a notice of surrender or 32(2) against a revocation notice, before the date on which the notice of surrender or the revocation notice takes effect;

(c)   in the case of an appeal under regulation 32(2) against a variation notice or an enforcement notice, of an appeal under regulation 32(3)(a) or 33(1) or (2)(b), or of an appeal under regulation 33(3)(b) other than an appeal which relates to a notice issued under regulation 22A(5), before the expiry of the period of two months beginning with the date of the notice which is the subject matter of the appeal;

(d)   in the case of an appeal under regulation 32(1)(f) against a determination of reportable emissions, before the expiry of the period of two months beginning with the date of the notice which is the subject matter of the appeal;

(e)   in the case of an appeal under regulation 32(3)(a) or (c), before the expiry of the period of 15 working days beginning with the date of the decision which is the subject matter of the appeal;

(f)   in the case of an appeal under regulation 32(3)(d) or (e), 32(4), or 33(3)(a), before the expiry of the period of two months beginning with the date of the decision which is the subject matter of the appeal;

(g)   in the case of an appeal under regulation 33(2), before the expiry of the period of 2 months beginning with the date of service of the certificate or notice which is the subject matter of the appeal

(h)   in the case of an appeal under regulation 33(3)(b) which relates to a notice issued under regulation 22A(5), before the expiry of a period of 10 working days beginning with the date of the decision which is the subject matter of the appeal.

(2) The appeal body may in a particular case allow notice of appeal to be given after the expiry of the periods mentioned in sub-paragraph (1)(a) or (c).

3.—   (1) In the case of an appeal under regulation 32, the regulator shall, within 14 days of receipt of the copy of the notice of appeal sent in accordance with paragraph 1, give notice of it to any person who appears to the regulator to have a particular interest in the subject matter of the appeal.

(2) In the case of an appeal under regulation 33, the appeal body shall, within 14 days of receipt of the copy of the notice of appeal sent in accordance with paragraph 1, give notice of it to any person who appears to the appeal body to have a particular interest in the subject matter of the appeal.

(3) A notice under sub-paragraph (1) or (2) shall—

(a)   state that notice of appeal has been given;

(b)   state the name of the appellant and the location of the installation concerned;

(c)   describe the decision or notice to which the appeal relates;

(d)   state that representations with respect to the appeal may be made to the appeal body in writing by any recipient of the notice within a period of 21 days beginning with the date of the notice and that copies of any representations so made will be furnished to the appellant and to the regulator; and

(e)   state that if a hearing is to be held wholly or partly in public, a person mentioned in sub-paragraph (1) or (2) will be notified of the date of the hearing.

(4) The regulator shall, within 14 days of sending a notice under sub-paragraph (1), notify the appeal body of the persons to whom and the date on which the notice was sent.

(5) In the event of an appeal under regulation 32 being withdrawn, the regulator shall give notice of the withdrawal to every person to whom notice was given under sub-paragraph (1).

(6) In the event of an appeal under regulation 33 being withdrawn, the appeal body shall give notice of the withdrawal to every person to whom notice was given under sub-paragraph (2).

4.—   (1) Before determining an appeal, the appeal body may afford the appellant and, where applicable, the regulator an opportunity of appearing before and being heard by a person appointed by it (the "person holding the hearing") and it shall do so in any case where a request is duly made by the appellant or, where applicable, the regulator to be so heard.

(2) A hearing held under sub-paragraph (1) may, if the person holding the hearing so decides, be held wholly or partly, in private.

(3) Where the appeal body causes a hearing to be held under sub-paragraph (1) it shall give the appellant and, if applicable, the regulator at least 28 days notice (or such shorter period of notice as they may agree) of the date, time and place fixed for the holding of the hearing.

(4) In the case of a hearing which is to be held wholly or partly in public, the appeal body shall, at least 21 days before the date fixed for the holding of the hearing—

(a)   publish a copy of the notice mentioned in sub-paragraph (3) in a newspaper circulating in the locality in which the installation is operated; and

(b)   serve a copy of that notice on every person mentioned in paragraph 3(1) who has made representations in writing to the appeal body.

(5) The appeal body may vary the date fixed for the holding of any hearing and sub-paragraphs (3) and (4) shall apply to the variation of a date as they applied to the date originally fixed.

(6) The appeal body may also vary the time or place for the holding of a hearing and shall give such notice of any such variation as appears to him to be reasonable.

(7) The persons entitled to be heard at a hearing are the appellant and, if applicable, the regulator.

(8) Nothing in sub-paragraph (7) shall prevent the person holding the hearing from permitting any other persons to be heard at the hearing and such permission shall not be unreasonably withheld.

(9) After the conclusion of a hearing, the person holding the hearing shall make a report in writing to the appeal body which shall include his conclusions and his recommendations or his reasons for not making any recommendation.

(10) Paragraph 4(5) and (6) of Schedule 3 shall apply to hearings held under this paragraph as if references to the appointed person in those paragraphs were references to the person holding the hearing under this paragraph.

5. — (1) Where an appeal under regulation 32 (other than an appeal which relates to an offshore installation) is to be disposed of on the basis of written representations, the regulator shall submit any written representations to the appeal body not later than 28 days after receiving a copy of the documents mentioned in paragraph 1(2)(a) and (f).

(2) The appellant shall make any further representations by way of reply to any representations from the regulator not later than 17 days after the date of submission of those representations by the regulator.

(3) Any representations made by the appellant or the regulator shall bear the date on which they are submitted to the appeal body.

(4) When the regulator or the appellant submits any representations to the appeal body they shall at the same time send a copy of them to the other party.

(5) The appeal body shall send to the appellant and the regulator a copy of any representations made to it by the persons mentioned in paragraph 3(1) and shall allow the appellant and the regulator a period of not less than 14 days in which to make representations on them.

(6) The appeal body may in a particular case —

(a)   set later time limits than those mentioned in this paragraph;

(b)   require exchanges of representations between the parties in addition to those mentioned in paragraphs (1) and (2).

6.—   (1) The appeal body shall give notice to the appellant of its determination of the appeal and shall provide him with a copy of any report mentioned in paragraph 4(9).

(2) The appeal body shall at the same time send—

(a)   a copy of the documents mentioned in sub-paragraph (1) to the regulator; and

(b)   a copy of its determination of the appeal to any person mentioned in paragraph 3(1) who made representations to the appeal body and, if a hearing was held, to any other person who made representations in relation to the appeal at the hearing.

7. Where a determination of the appeal body on an appeal is quashed in proceedings before any court, the appeal body—

(a)   shall send to the persons notified of its determination under paragraph 6 a statement of the matters with respect to which further representations are invited for the purposes of its further consideration of the appeal;

(b)   shall afford to those persons the opportunity of making, within 28 days of the date of the statement, written representations in respect of those matters; and

(c)   may, as it thinks fit, cause a hearing to be held or reopened and, if it does so, paragraphs 4(2) to (10) shall apply to the hearing or the reopened hearing as they apply to a hearing held under paragraph 4(1),

and paragraph 6 shall apply to the re-determination of the appeal as it applies to the determination of an appeal.

Regulation 34(3)

SCHEDULE 3

Delegation of Appellate Functions

1. In this Schedule —

"appointed person" means a person appointed under regulation 34(2)(a);

"appointment", in the case of any appointed person, means appointment under regulation 34(2)(a).

2. An appointment must be in writing and —

(a)   may relate to any particular appeal, matters or questions specified in the appointment or to appeals, matters or questions of a description so specified;

(b)   may provide for any function to which it relates to be exercisable by the appointed person either unconditionally or subject to the fulfilment of such conditions as may be specified in the appointment; and

(c)   may, by notice in writing to the appointed person, be revoked at any time by the appeal body in respect of any appeal, matter or question which has not been determined by the appointed person before that time.

3. Subject to the provisions of this Schedule, an appointed person shall, in relation to any appeal, matter or question to which his appointment relates, have the same powers and duties as the appeal body, other than any function of appointing a person for the purpose —

(a)   of enabling persons to appear before and be heard by the person so appointed; or

(b)   of referring any question or matter to that person.

4.—   (1) If either of the parties to the appeal, matter or question expresses a wish to appear before and be heard by the appointed person, the appointed person shall give both of them an opportunity of appearing and being heard.

(2) Whether or not a party to an appeal, matter or question has asked for an opportunity to appear and be heard, the appointed person —

(a)   may hold a local inquiry or other hearing in connection with the appeal, matter or question; and

(b)   shall if the appeal body so directs, hold a local inquiry in connection with an appeal, matter or question.

(3) Where an appointed person holds a local inquiry or other hearing by virtue of this Schedule, an assessor may be appointed by the appeal body to sit with the appointed person at the inquiry or hearing and advise him on any matters arising, notwithstanding that the appointed person is to determine the appeal, matter or question.

(4) Subject to paragraphs (5) and (6), the costs of a local inquiry held under this Schedule shall be defrayed by the appeal body.

(5) Subject to sub-paragraph (6), subsections (2) to (5) of section 250 of the Local Government Act 1972(**29**) (local inquiries: evidence and costs) shall apply to hearings held under this Schedule by an appointed person as they apply to inquiries caused to be held under that section by a Minister, but with the following modifications, that is to say—

(a)   with the substitution in subsection (2) (evidence) for the reference to the person appointed to hold the inquiry of a reference to the appointed person;

(b)   with the substitution in subsection (4) (recovery of costs of holding the inquiry) for the references to the Minister causing the inquiry to be held of references to the appeal body;

(c)   with the substitution for the reference in that subsection to a local authority of a reference to the regulator;

(d)   with the substitution in subsection (5) (orders as to the costs of the parties) for the reference to the Minister causing the inquiry to be held of a reference to the appeal body.

(6) In the case of an appeal to the Scottish Ministers, subsections (3) to (8) of section 210 of the Local Government (Scotland) Act 1973(**30**) (which relates to the costs of and holding of local inquiries) shall apply to hearings held under this Schedule by an appointed person as they apply to inquiries held under that section, but with the following modifications, that is to say—

(**29**) 1972 c. 70; section 250 has been amended by the Statute Law (Repeals) Act 1989 (c. 43), Schedule 1, Part IV, the Criminal Justice Act 1982 (c. 48), sections 37, 38 and 46 and the Housing and Planning Act 1986 (c. 63), Schedule 12, Part III.

(**30**) 1973 c. 65, section 210 was amended by the Criminal Procedure (Scotland) Act 1975 (c. 21), sections 289F and 289G (which were inserted into that Act by the Criminal Justice Act 1982 (c. 48), section 54) and the Housing and Planning Act 1986, Schedule 11, paragraph 39.

(a)   with the substitution in subsection (3) (notice of inquiry) for the reference to the person appointed to hold the inquiry of a reference to the appointed person;

(b)   with the substitution in subsection (4) (evidence) for the reference to the person appointed to hold the inquiry and, in paragraph (b), the reference to the person holding the inquiry of references to the appointed person;

(c)   with the substitution in subsection (6) (expenses of witnesses etc) for the references to the Minister causing the inquiry to be held of a reference to the appointed person or the Scottish Ministers;

(d)   with the substitution in subsection (7) (expenses)—

(i)   for the first reference to the Minister of a reference to the Scottish Ministers; and

(ii)   for the second reference to the Minister of a reference to the appointed person or the Scottish Ministers;

(e)   with the substitution in subsection (7A) (recovery of entire administrative expense)–

(i)   for the first reference to the Minister of a reference to the appointed person or the Scottish Ministers;

(ii)   in paragraph (a), for the reference to the Minister of a reference to the Scottish Ministers; and

(iii)   in paragraph (b), for the reference to the Minister holding the inquiry of a reference to the Scottish Ministers;

(f)   with the substitution in subsection (7B) (power to prescribe daily amount)–

(i)   for the first reference to the Minister of a reference to the Scottish Ministers;

(ii)   in paragraphs (a) and (c), for the references to the person appointed to hold the inquiry of references to the appointed person; and

(iii)   in paragraph (d), for the reference to the Minister of a reference to the appointed person or the Scottish Ministers; and

(g)   with the substitution in subsection (8) (certification of expenses)—

    (i)   for the words "the Minister has", of the words "the Scottish Ministers have";

    (ii)   for the reference to him and the reference to the Crown of references to the appointed person or the Scottish Ministers.

5.—   (1) Where under paragraph 2(c) the appointment of the appointed person is revoked in respect of any appeal, matter or question, the appeal body shall, unless it proposes to determine the appeal, matter or question itself, appoint another person under regulation 34(2)(a) to determine the appeal, matter or question instead.

(2) Where such a new appointment is made, the consideration of the appeal, matter or question, or any hearing in connection with it, shall be begun afresh.

(3) Nothing in sub-paragraph (2) shall require any person to be given an opportunity of making fresh representations or modifying or withdrawing representations already made.

6.—   (1) Anything done or omitted to be done by an appointed person in, or in connection with, the exercise or purported exercise of any function to which the appointment relates shall be treated for all purposes as done or omitted to be done by the appeal body in its capacity as such.

(2) Sub-paragraph (1) shall not apply —

(a)   for the purposes of so much of any contract made between the appeal body and the appointed person as relates to the exercise of the function; or

(b)   for the purposes of any criminal proceedings brought in respect of anything done or omitted to be done by an appointed person in, or in connection with, the exercise or purported exercise of any function to which the appointment relates.

Regulation 34(4)

SCHEDULE 4

Appeals under regulation 32: Northern Ireland

1.— (1) A person who wishes to appeal to the Planning Appeals Commission ("the appeals commission") under regulation 32 shall give to the appeals commission written notice of the appeal together with the documents specified in sub-paragraph (2) and shall at the same time send to the regulator a copy of that notice together with copies of the documents specified in sub-paragraphs (2)(a) and (f).

(2) The documents mentioned in sub-paragraph (1) are—

(a)   a statement of the grounds of appeal;

(b)   a copy of any relevant application;

(c)   a copy of any relevant greenhouse gas emissions permit;

(d)   a copy of any relevant correspondence between the appellant and the regulator;

(e)   a copy of any decision or notice which is the subject matter of the appeal; and

(f)   a statement indicating whether the appellant wishes the appeal to be in the form of a hearing or to be disposed of on the basis of written representations.

(3) An appellant may withdraw an appeal by notifying the appeal body in writing and shall send a copy of that notification to the regulator.

2.— (1) Subject to sub-paragraph (2), notice of appeal in accordance with paragraph 1 is to be given—

(a)   in the case of an appeal under regulation 32(1)(a) to (e), before the expiry of the period of six months beginning with the date of the decision or deemed decision which is the subject matter of the appeal;

(b)   in the case of an appeal under regulation 32(2) against a revocation notice, before the date on which the revocation notice takes effect;

(c)   in the case of an appeal under regulation 32(2) against a variation notice or an enforcement notice or an appeal under regulation 32(3)(a), before the expiry of the period of two months beginning with the date of the notice which is the subject matter of the appeal;

(d)   in the case of an appeal under regulation 32(1)(f) against a determination of reportable emissions, before the expiry of the period of two months beginning with the date of the notice which is the subject matter of the appeal;

(e)   in the case of an appeal under regulation 32(3)(a) or (c), before the expiry of the period of 15 working days beginning with the date of the decision which is the subject matter of the appeal;

(f)   in the case of an appeal under regulation 32(3)(d) or (e), 32(4), before the expiry of the period of 2 months beginning with the date of the decision which is the subject matter of the appeal.

(2) The appeals commission may in a particular case allow notice of appeal to be given after the expiry of the periods mentioned in sub-paragraph (1)(a) or (c).

3.—   (1) The regulator shall, within 14 days of receipt of the copy of the notice of appeal sent in accordance with paragraph 1, give notice of it to any person who appears to the regulator to have a particular interest in the subject matter of the appeal.

(2) A notice under sub-paragraph (1) shall—

(a)   state that notice of appeal has been given;

(b)   state the name of the appellant and the location of the installation concerned;

(c)   describe the application or greenhouse gas emissions permit to which the appeal relates;

(d)   state that representations with respect to the appeal may be made to the appeals commission in writing by any recipient of the notice within a period of 21 days beginning with the date of the notice and that copies of any representations so made will be furnished to the appellant and to the regulator; and

(e)   state that if a hearing is to be held wholly or partly in public, a person mentioned in sub-paragraph (1) or (2) who makes representations with respect to the appeal and any person mentioned in sub-paragraph (1) will be notified of the date of the hearing.

(3) The regulator shall, within 14 days of sending a notice under sub-paragraph (1), notify the appeals commission of the persons to whom and the date on which the notice was sent.

(4) In the event of an appeal under regulation 32 being withdrawn, the regulator shall give notice of the withdrawal to every person to whom notice was given under sub-paragraph (1).

4.— (1) The appeals commission shall determine the appeal and paragraphs (1), (3), (4) and (5) of Article 111 of the Planning (Northern Ireland) Order 1991 shall apply in relation to the determination of the appeal as they apply in relation to the determination of an appeal under that Order.

(2) If either party to the appeal so requests, the appeals commission shall afford to each of them an opportunity of appearing before and being heard by the appeals commission.

(3) A hearing held under sub-paragraph (2) may, if the appeals commission so decides, be held wholly or partly, in private.

Regulation 8(1), 9(5), 14(3), (4) and (10), 15(3) and (5), 16(5) and (6),
17(6), 18(1), (3), (4) and (5), 22(4), (6) and (7), 24(8) and 26(12)

SCHEDULE 5

Fees And Charges

PART 1

Fees in relation to the grant, variation, transfer, surrender and revocation
of a greenhouse gas emissions permit

1.—   (1) The following fees are prescribed and shall be payable to the
regulator—

(a)   in respect of an application for a greenhouse gas emissions
permit under regulation 8(1) (other than in respect of an
offshore installation)–

(i)   in respect of an installation emitting less than 50kt per year,
£1230;

(ii)   in respect of an installation emitting at least 50 and no more
than 500kt per year, £2300;

(iii) in respect of an installation emitting more than 500kt per
year, £5490;

(b)   in respect of the grant of a greenhouse gas emissions permit
authorising the operation of an offshore installation–

(i)   in respect of an installation emitting less than 50kt per year,
£1230;

(ii)   in respect of an installation emitting at least 50 and no more
than 500kt per year, £2300;

(iii) in respect of an installation emitting more than 500kt per
year, £5490;

(c)   in respect of the grant of the greenhouse gas emissions
permit where regulation 9(5)(b) applies—

(i)   in respect of an installation emitting less than 50kt per year,
£700;

(ii)   in respect of an installation emitting at least 50 and no more
than 500kt per year, £1770;

(iii) in respect of an installation emitting more than 500kt per
year, £4960;

(d)   in respect of an application for the variation of the provisions of a greenhouse gas emissions permit under regulation 14(2) (except where the regulator considers that a variation relates to minor changes or changes of a purely administrative nature), £240;

(e)   in respect of a variation notice varying the provisions of a greenhouse gas emissions permit otherwise than on an application under regulation 14(2) (except where the regulator considers that a variation relates to minor changes or changes of a purely administrative nature), £240.

(f)   in respect of an application under regulation 15(1) to transfer a greenhouse gas emissions permit, in whole or in part, £240;

(g)   in respect of an application under regulation 16(1) to surrender a greenhouse gas emissions permit, £620;

(h)   in respect of the revocation of a greenhouse gas emissions permit pursuant to regulation 17(2), 24(11) or 25(3), £620.

(2) A fee prescribed under paragraph 1 in respect of a variation notice or in respect of a revocation notice shall be payable by the date specified in the notice.

PART 2

Fees in respect of the allocation of allowances

2.—   (1) The following fees are prescribed and shall be payable to the regulator—

(a)   in respect of an application for an allocation from the new entrant reserve under regulation 22(1), £1030.

(b)   in respect of a retention notice or a notice under regulation 24(7)(b), £115 multiplied by the number of hours specified in the notice in accordance with regulation 24(8).

(2) A fee prescribed under sub-paragraph (1)(b) shall be payable by the date specified in the notice to which it relates.

PART 3

Registry fees

3.—   (1) The following fees are prescribed and shall be payable to the regulator—

(a)   in respect of an application for the creation of a person holding account under Article 19(1) of the Registries Regulation, £175;

(b)   in respect of an application by—

(i)   a verifier; or

(ii)   an organisation which wishes to be able to be appointed by an account holder as his additional authorised representative,

to be included in the registry, £175

(c)   subject to sub-paragraph (2), in respect of an application by the holder of a person holding account to appoint an individual as a new user in the registry, whom he may subsequently appoint as one of his authorised representatives in accordance with Article 19(3) of the Registries Regulation, £50;

(d)   subject to sub-paragraph (2), in respect of an application by—

(i)   a verifier; or

(ii)   an organisation which has been included in the registry so that it is able to be appointed by an account holder as his additional authorised representative,

to appoint an individual as a new user in the registry, whom he may subsequently authorise to use the registry on his behalf, £50.

(e)   subject to sub-paragraph (2), in respect of an application by—

(i)   a verifier; or

(ii)   an organisation which has been included in the registry so that it is able to be appointed by an account holder as his additional authorised representative,

to authorise an individual, who has not previously been appointed as a new user in the registry, to use the registry on his behalf, £50.

(2) Where, in respect of a verifier or an organisation, the application referred to in (1)(b) included an application to authorise only one individual to use the registry in its behalf, paragraphs (1)(d) and (1)(e) shall not apply to the verifier or organisation's first subsequent application—

(a)   to appoint an individual as a new user in the registry; or

(b)   to authorise an individual to use the registry on its behalf.

PART 4

Subsistence Charges

4. Subject to paragraphs 6, 9 and 13 of this Schedule, the charge payable by an operator to the regulator prescribed for the subsistence of a greenhouse gas emissions permit for the financial year 2005/2006 shall be as shown in Table 1 and shall be payable in accordance with paragraph 10 of this Schedule.

Table 1

Charge for the financial year 2005/2006

		Estimated 2005 emissions or, where applicable, the estimated annual specified emissions from the installation to which the greenhouse gas emissions permit relates-		
		less than 50 kilotonnes per year	*at least 50 and no more than 500 kilotonnes per year*	*greater than 500 kilotonnes per year*
Charge if on date on which these Regulations enter into force, the total number of installations published by the Secretary of State in accordance with paragraph **Error! Reference source not found.** of this Schedule is-	less than 500	£2,540	£3,390	£4,230
	500 to 599	£2,280	£3,050	£3,810
	600 to 699	£2,110	£2,820	£3,520
	700 to 799	£1,990	£2,650	£3,320
	800 to 899	£1,900	£2,530	£3,170
	900 to 999	£1,830	£2,440	£3,050
	1000 to 1099	£1,750	£2,350	£2,900
	1100 to 1199	£1,720	£2,300	£2,870
	1200 or more	£1,690	£2,250	£2,810

5. Subject to paragraphs 6, 9 and 13 of this Schedule, the charges payable by an operator prescribed for the subsistence of a greenhouse gas emissions permit for the financial year 2006/2007 and for each subsequent financial year shall be as shown in Table 2 and shall be payable in accordance with paragraph 10 of this Schedule.

Table 2

Charge for the financial year 2006/2007 and subsequent financial years

		Amount of annual specified emissions or, where applicable, estimated annual specified emissions from the installation to which the greenhouse gas emissions permit relates-		
		less than 50 kilotonnes per year	*at least 50 and no more than 500 kilotonnes per year*	*greater than 500 kilotonnes per year*
Charge if on 1st April of the financial year to which the charge relates, the total number of installations published by the Secretary of State in accordance with paragraph **Error! Reference source not found.** of this Schedule is-	less than 500	£2,915	£3,765	£4,605
	500 to 599	£2,553	£3,323	£4,083
	600 to 699	£2,341	£3,051	£3,751
	700 to 799	£2,190	£2,850	£3,520
	800 to 899	£2,076	£2,706	£3,346
	900 to 999	£1,988	£2,598	£3,208
	1000 to 1099	£1,893	£2,493	£3,043
	1100 to 1199	£1,850	£2,430	£3,000
	1200 or more	£1,815	£2,375	£2,935

6. The charge prescribed for the subsistence of a greenhouse gas emissions permit under paragraph 4 or 5 of this Schedule shall not be payable in respect of a greenhouse gas emissions permit relating to—

(a)   an installation which is for the duration of the financial year to which the charge relates an excluded installation;

(b)   a planned installation which is not put into operation during the financial year to which the charge relates; or

(c)   an installation in respect of which a retention notice has been served prior to 1st April in the financial year to which the charge relates and is not revoked during that financial year.

7. Subject to paragraph 8 of this Schedule, the Secretary of State shall before the expiry of a period of 7 days beginning, in relation to the financial year 2005/2006, on the date on which these Regulations enter into force or, in relation to the financial year 2006/2007 and each subsequent financial year, on 1st April 2006 and 1st April in each subsequent financial year—

(a)   calculate the total number of installations on 1st April in that financial year; and

(b)   publish in such manner as she considers appropriate the total number of installations calculated under sub-paragraph (a) and the appropriate charges for the financial year as set out in relation to the financial year 2005/2006 in Table 1 in paragraph 4 of this Schedule or in relation to subsequent financial years, in Table 2 in paragraph 5 of this Schedule.

8. Where on the date of entry into force of these Regulations—

(a)   the Secretary of State has made an application for an installation to be temporarily excluded under Article 27(2) of the Directive and the European Commission has not made a decision determining the application; or

(b)   the date determined in accordance with regulation 11(3) by which an application under regulation 11(1) must be made has not passed ("the application deadline"),

paragraphs 4 and 7 of this Schedule shall apply as if references to the date on which these Regulations enter into force were to the date of the expiry of a period of 7 days beginning on the application deadline or, where the European Commission refuses the application, the date on which the European Commission notifies its decision.

9. Where during a financial year—

(a)   a greenhouse gas emissions permit is granted in relation to an installation under regulation 9(4);

(b)   an installation ceases to be an excluded installation;

(c)   a planned installation is put into operation; or

(d)   a permit is partially transferred in accordance with regulation 15,

the charge payable under paragraph 4 or 5 of this Schedule in respect of the subsistence of the greenhouse gas emissions permit relating to the installation (or in the case of a partially transferred permit the transferred unit) for the remainder of that financial year shall be a proportion of the charge shown in Table 1 or Table 2 calculated on a daily basis for the remainder of the financial year commencing on the date of the grant of the greenhouse gas emissions permit, the date on which the installation ceased to be an excluded installation, the date on which the planned installation is put into operation, or the date upon which the transfer took effect, as appropriate.

10. Subject to paragraph 11 of this Schedule–

(a)   the charge prescribed under paragraph 4 of this Schedule shall be payable on the expiry of a period of 28 days beginning on the date on which notice of the estimated 2005 emissions or, where paragraph 9 applies in relation to the installation, the estimated annual specified emissions and the charge is sent by the regulator to the operator;

(b)   the charge prescribed under paragraph 5 of this Schedule shall be payable on the expiry of a period of 28 days beginning on the date on which notice of the charge and, in relation to a charge for a financial year in which paragraph 9 applies in relation to the installation and the following financial year, the estimated annual specified emissions, is sent by the regulator to the operator.

11. The operator of an installation may notify the regulator that it wishes to pay the charges prescribed under paragraphs 4 and 5 of this Schedule in instalments.

12. Where an operator notifies the regulator under paragraph 11 —

(a)   in the financial year in which the notice is given, the charge shall be payable in equal instalments payable on the first day of each quarter remaining in the financial year or, if later, on the expiry of a period of 28 days beginning on the date on which notice of—

(i)   in relation to a charge for the financial year 2005/2006, the estimated 2005 emissions or, where paragraph 9 applies in relation to the installation, the estimated annual specified emissions; and

(ii)   the charge,

is sent by the regulator to the operator;

(b)   in subsequent financial years, the charge shall be payable in four equal instalments payable on the first day of each quarter in the financial year or, if later, on the expiry of a period of 28 days beginning on the date on which notice of—

(i)   in relation to a charge for a financial year in which paragraph 9 applies in relation to the installation and the following financial year, the estimated annual specified emissions; and

(ii)   the charge,

is sent by the regulator to the operator.

13. Where during a financial year a greenhouse gas emissions permit is surrendered under regulation 16 or revoked under regulation 17(1) or the installation to which the permit relates becomes an excluded installation, the regulator shall make a refund to the operator of a proportion of the charge payable under paragraph 4 or 5 of this Schedule in respect of the remainder of that financial year calculated as follows —

    (a)   if the charge has been paid for the whole financial year, a refund calculated on a daily basis for the remainder of the financial year commencing on the date on which the notice of surrender or revocation notice takes effect or the date of service of the certificate under regulation 11(6), as appropriate; or

    (b)   if the charge has been paid only for the quarter in which the surrender or revocation occurs, a refund calculated on a daily basis for the remainder of that quarter commencing on the date on which the notice of surrender or revocation notice takes effect or the date of service of the certificate under regulation 11(6), as appropriate.

14.— (1) The registry administrator shall within 14 days of the date on which the Secretary of State publishes the total number of installations in accordance with paragraph 7(b), notify the regulator of the element of the charge prescribed under paragraph 4 or 5 which relates to the subsistence of the operator holding account in the registry ("the operator registry charge").

(2) The regulator shall pass on to the registry administrator any operator registry charge which it receives.

15. For the purposes of this Schedule —

"annual specified emissions" means the annual reportable emissions from the installation in the scheme year which ended in the financial year prior to the financial year to which the charge relates;

"estimated 2005 emissions" means a reasonable estimate, in the opinion of the regulator, of the reportable emissions likely to be emitted from the installation in the calendar year 2005;

"estimated annual specified emissions" means, in relation to a financial year in which paragraph 9 applies in relation to the installation and the following financial year, a reasonable estimate, in the opinion of the regulator of the reportable emissions likely to be emitted from the installation in the year beginning on the date on which the permit is granted, the installation ceases to be an excluded installation or the planned installation is put into operation;

"financial year" means a year beginning on 1ˢᵗ April and ending on 31ˢᵗ March;

"planned installation" means an installation in respect of which an operator has notified the regulator under regulation 13(1);

"quarter" means a three month period beginning with 1ˢᵗ April, 1ˢᵗ July, 1ˢᵗ October and 1ˢᵗ January;

"total number of installations" means the number of installations covered by greenhouse gas emissions permits in the United Kingdom excluding–

(i)   any excluded installations or planned installations;

(ii)   any installation in respect of which a retention notice has been served; and

(iii) any installations included in the European Commission's provision for temporary exclusion under Article 27(2) of the Directive which have applied for a certificate of temporary exclusion in accordance with regulation 11(1).

Regulation 48

SCHEDULE 6

Consequential Amendments

The Environment Act 1995

1.— (1) The 1995 Act shall be amended in accordance with this paragraph.

(2) After section 41, insert a new section 41A as follows —

"41A Charges in respect of greenhouse gas emissions permits

(1) Without prejudice to subsections (1)(b) and (2) of section 41 above, the following charges may be prescribed under that section as respects permits ("greenhouse gas emissions permits") granted under the Greenhouse Gas Emissions Trading Scheme Regulations 2005 ("the regulations") —

(a) charges in respect of, or in respect of an application for, the allocation of allowances to an operator;

(b) charges in respect of, or in respect of an application for, the retention of allowances by an operator ceasing to carry on an activity to which they relate;

(c) charges in respect of the revocation of a greenhouse gas emissions permit;

(d) charges in respect of the subsistence of an account required to be held in the trading scheme registry by an operator ("operator registry charges").

(2) If the Agency —

(a) proposes to prescribe operator registry charges, or to amend any provision for such charges included in a charging scheme, and

(b) notifies SEPA of its proposals,

the Agency and SEPA shall each include in a charging scheme (subject to approval by the Secretary of State under section 42(2) below) provision giving effect to the proposals.

(3) If the Agency revises any proposals of which it has given notification under subsection (2) above, and notifies SEPA accordingly, the obligations imposed by that subsection apply in relation to the proposals as revised.

(4) A notification under subsection (2) or (3) above shall include details of the amount of the proposed charges.

(5) SEPA shall pass on to the Agency any operator registry charges that it receives.

(6) A charging scheme made by the Agency may require the payment to the Agency of such charges as may from time to time be prescribed in respect of—

(a)   the creation of an account in the trading scheme registry, other than one that is required to be held by an operator;

(b)   the subsistence of such an account;

(c)   the updating of information provided to the Agency in relation to such an account.

(7) In this section—

"allowance" and "operator" have the same meaning as in the regulations;

"charging scheme" and "prescribed" have the same meaning as in section 41;

"trading scheme registry" means the registry established under the regulations."

(3)   In sub-section (1) of section 56 of the 1995 Act (interpretation of Part 1), in each of the two definitions of "environmental licence" insert after paragraph (j)—

"(k)   a greenhouse gas emissions permit granted under the Greenhouse Gas Emissions Trading Scheme Regulations 2005".

The Pollution Prevention and Control Act 1999

2.—   (1) The Pollution Prevention and Control Act 1999 shall be amended in accordance with sub-paragraph (2).

(2) In Schedule 1—

(a)   after paragraph 9, insert a new paragraph 9A as follows—

"9A.

(1)   Authorising the Secretary of State to make schemes for the charging by regulators of charges, as respects greenhouse gas emissions permits in relation to offshore installations, corresponding to those that may be prescribed under section 41 (read with section 41A) of the Environment Act 1995.

(2)   Subsections (2) to (5) of section 41A of that Act apply in relation to the Secretary of State and a charging scheme made by virtue of this paragraph as they apply in relation to the Scottish Environment Protection Agency and a charging scheme made by that Agency under the 1995 Act.

(3)   In this paragraph "greenhouse gas emissions permit" and "offshore installation" have the same meaning as in the Greenhouse Gas Emissions Trading Scheme Regulations 2005."

(b)   in after paragraph 24, after "paragraph 9" insert ", 9A".

Regulation 22B

SCHEDULE 7

Table of Sector and new entrant reserve reduction factors

Sector	Sector reduction factor	New Entrant Reserve reduction factor
Aluminium -- AFED	109.51%	97.9%
Textiles -- BATC	69.63%	99.3%
Brewing -- BBPA	91.55%	98.7%
Cement -- BCA	96.52%	90.1%
Ceramics -- BCC-F	103.21%	97.1%
Ceramics -- BCC-M	89.58%	99.3%
Ceramics -- BCC-N	95.12%	95.1%
Ceramics -- BCC-R	82.94%	99.3%
Ceramics -- BCC-W	92.79%	97.1%
Glass -- BGMC	95.56%	92.7%
Lime -- BLA	110.12%	98.4%
Poultry -- BPMF2	87.77%	98.5%
Rubber -- BRMA-T	89.65%	97.0%
Cement -- non-CCA	0%	0%
Ceramics -- non-CCA	98.73%	99.3%
Chemicals -- non-CCA	90.24%	98.2%
Chemicals -- CIA	88.33%	88.3%
Coal Mining -- non-CCA	49.99%	95.7%
Cathode Ray Tubes -- CRT	107.49%	93.8%
Dairies -- DIAL	93.78%	98.5%
Engineering & Vehicles -- non-CCA	48.87%	99.3%
Mineral Wool -- EUR	116.02%	83.7%
Food & Drink -- FDF	91.97%	95.4%
FDT -- CIA	93.28%	98.5%
FDT -- non-CCA	83.43%	96.2%
Glass -- non-CCA	107.07%	98.7%
Iron & Steel -- non-CCA	0%	0%

Lime -- UKSA	100.73%	99.3%
Malting -- MAGB	105.45%	95.4%
Non-Ferrous -- NFA	87.09%	94.9%
Nuclear Fuel -- CIA	90.59%	99.3%
Nuclear Fuels -- non-CCA	107.06%	99.3%
Offshore	93.71%	91.9%
Other Non-metallic -- non-CCA	113.23%	99.3%
Other Oil & Gas	92.44%	81.7%
Power Stations	72.08%	95.4%
Pulp & Paper -- non-CCA	0%	96.6%
Refineries	106.70%	98.0%
Refineries -- CIA	104.96%	99.3%
Aerospace -- SBAC	88.10%	87.2%
Semiconductors -- SC	52.67%	93.8%
Spirits -- SEEC	98.61%	99.3%
Services	83.58%	97.1%
Vehicle Manufacture -- SMMT	96.23%	99.0%
Foundries -- T2010	0%	100%
Pulp & Paper -- TPF	100.24%	97.9%
Rendering -- UKRA	24.12	59.7%
Iron & Steel	104.64%	85.8%
Wood & Wood Products -- non-CCA	0%	0%
Wood Board -- WPIF	108.55%	97.3%
Aluminium -- AFED	109.51%	98.9%
Textiles -- BATC	69.63%	99.3%"

EXPLANATORY NOTE

(This note is not part of the Regulations)

These Regulations consolidate the Greenhouse Gas Emissions Trading Scheme Regulations 2003 (S.I. 2003/3311) (the "2003 Regulations") and the Greenhouse Gas Emissions Trading Scheme (Amendment) Regulations 2004 (S.I. 2004/3390) (the "2004 Regulations") with amendments. In addition to minor and drafting changes, the Regulations include substantive amendments to make provision for the treatment of new installations and installations which cease to carry out an activity listed in Schedule 1 of the Regulations, to provide for subsistence charges in relation to permits and to take account of the entry into force of the Commission Regulation 2216/2004 for a standardised and secured system of registries pursuant to Directive 2003/87/EC of the European Parliament and the Council and of the Council Decision 280/204/EC of the European Parliament and of the Council (the "Registries Regulation") and of the Commission Decision 2004/156/EC establishing guidelines for the monitoring and reporting of greenhouse gas emissions pursuant to Directive 2003/87/EC of the European Parliament and of the Council.

The Regulations provide the framework for a greenhouse gas emissions trading scheme for the purpose of implementing Directive 2003/87/EC of the European Parliament and the Council establishing a scheme for greenhouse gas emission allowance trading within the Community and amending Council Directive 96/61/EC (the "Emissions Trading Directive"). The Regulations apply to the United Kingdom. *Regulation 2* provides for the Scottish Ministers, Department of the Environment in Northern Ireland and National Assembly for Wales to act as appropriate authority for installations situated in their area (other than installations falling within the definition of offshore installation).

The Regulations control emissions of carbon dioxide from any of the activities listed in *Schedule 1 to the Regulations. Part 2 of Schedule 1* sets out rules for the interpretation of *Part 1 of Schedule 1*.

Part 1 of the Regulations (regulations 1 to 6) sets out general provisions. *Regulation 2* contains definitions including designating the regulators for installations under the scheme. The other regulations in Part 1 deal with general matters such as the service of notices under the Regulations.

Part 2 deals with the need for a permit to operate an installation covered by the Regulations *(regulation 7)*, the procedure for granting permits and the contents of permits *(regulations 8 to 10)* and the treatment of permits once granted *(regulations 14 to 18)*. The conditions of permits *(regulation 10)* must ensure that the emissions of the installation are properly monitored and reported and that the operator surrenders within 4 months of the end of each scheme year allowances equal to the annual reportable emissions from the installation during that year.

Regulation 11 enables an installation in respect of which the European Commission has provided for temporary exclusion to apply for a certificate excluding it from the scheme. *Regulations 14 to 17* deal with the variation, transfer, surrender and revocation of permits.

Regulation 18 provides for a charge to be payable in respect of the subsistence of a permit. The subsistence charge and other charges in relation to the scheme are set out in *Schedule 5*. Regulation 18 provides for the provisions of the Schedule to be superseded by charging schemes adopted by the regulators. *Schedule 6* provides for amendments to the Environment Act 1995 to ensure that the Environment Agency and the Scottish Environment Protection Agency are able to adopt charging schemes to supersede the charges in Schedule 5. *Regulation 19* is made under section 2(1) of the Pollution Prevention and Control Act 1999 and provides for the Secretary of State to make a charging scheme for specified charges in relation to offshore installations. *Schedule 6* provides for an amendment to the Pollution Prevention and Control Act 1999 to enable future regulations to extend regulation 19 to cover all the types of charge provided for in Schedule 5.

Part 3 deals with the allocation of allowances. *Regulation 20* requires the Secretary of State to develop a national allocation plan for each phase of the scheme and *Regulation 21* provides for the Secretary of State to make a final allocation decision. *Regulations 22 to 24* set the framework for the allocation of allowances to installations which obtain a permit after the national allocation plan has been submitted to the European Commission and for the treatment of installations which cease to be covered by the scheme. *Regulation 26* makes provisions consequential to the Registries Regulation. *Regulation 27* enables operators of certain installations to apply to form a pool.

Part 4 (regulations 28 to 31) contains the enforcement powers under the Regulations. *Part 5 (regulations 32 to 34) and Schedules 2 to 4* provide for appeals against decisions of the regulator and for appeals for the appropriate authority, responsible authority or the Secretary of State to reconsider decisions under the Regulations. *Part 6 (regulations 35 to 37)* sets out information gathering powers and publicity requirements. *Part 7 (regulations 38 to 41)* sets out offences for contraventions of the Regulations and civil penalties where an operator fails to surrender sufficient allowances to cover its specified emissions. *Part 8 (regulations 36 and 37)* enables the appropriate authority to give directions and guidance to regulators and *Part 9 (regulations 44 and 45)* enables the

Secretary of State to give directions and guidance to the registry administrator.

Part 10 (regulation 46) identifies powers under the regulations which can be exercised only with the agreement of the devolved administrations in relation to installations situated in their area (other than installations falling within the definition of offshore installations). This includes the power to develop a national allocation plan under regulation 20 and to decide upon the allocation of allowances under regulation 21. There is a default power for the Secretary of State to act where no agreement is reached if it is necessary to ensure that the United Kingdom complies with its obligations under the Emissions Trading Directive.

Part 11 (regulations 47 and 48) and Schedule 6 revoke the 2003 Regulations and the 2004 Regulations subject to savings and introduce the consequential amendments.

A regulatory impact assessment has been prepared and placed in the library of each House of Parliament. Copies can be obtained from National Climate Change Policy Division, Department for the Environment, Food and Rural Affairs, Ashdown House, Victoria Street, London SW1.

ANNEX 7

THE UK GREENHOUSE GAS EMISSIONS TRADING SCHEME 2002
(unofficial consolidation)

(as amended by the UK Greenhouse Gas Emissions Trading (Amendment) Scheme 2004, the UK Greenhouse Gas Emissions Trading (Amendment) Scheme 2005, the UK Greenhouse Gas Emissions Trading (Amendment) (No. 2) Scheme 2005, the UK Greenhouse Gas Emissions Trading (Amendment) Scheme 2006 and the UK Greenhouse Gas Emissions Trading (Amendment) Scheme 2007).

ARRANGEMENT OF SCHEME

PART A
INTRODUCTION

Citation and purposes of scheme
A1.— (1) This scheme may be cited as the UK Greenhouse Gas Emissions Trading Scheme 2002.

(2) The scheme is made by the Secretary of State for the purposes of—

(a)   achieving reductions in emissions of greenhouse gases in a cost-effective manner;

(b)   facilitating compliance with the UK's obligations under the UN Framework Convention on Climate Change and the Kyoto Protocol; and

(c)   implementing the UK's climate change programme.

Interpretation

A2.   Schedule 1 (Interpretation) shall have effect.

Participation in the scheme
A3.   A person may participate in the scheme as—

(a)   a direct participant by entering into a direct participant agreement with the Secretary of State and taking on an emissions reduction target in accordance with Part C of this scheme;

(b)   a group participant by entering into a group participant agreement with the Secretary of State and taking on an emissions reduction target in accordance with Part F of this scheme;

(c)   a trading participant by entering into a trading participant agreement and opening a trading account in the Registry in accordance with Part B of this scheme; and

(d)   a CCA participant by submitting a registration form to the Secretary of State in accordance with Part D of this scheme.

PART B
EMISSIONS TRADING REGISTRY

Emissions Trading Registry
B1.— (1) The Secretary of State will establish and maintain an Emissions Trading Registry ("the Registry").

(2) The Registry will contain a number of accounts which will be used to record the allocation, holding, transfer, cancellation and retirement of allowances.

(3)   An allowance for the purpose of the scheme is a unit of account, representing one tonne of carbon dioxide equivalent, which is used for determining compliance with emissions limitation commitments.

The Registry website
B2.— (1) The Registry will be in electronic form and will be accessible through the Registry website.

(2) The Secretary of State will take all reasonable steps to ensure that the Registry website will be available for access 24 hours a day 7 days a week.

(3) The Secretary of State will publish a user manual which may be updated from time to time setting out the detailed procedures for access to and use of the Registry website.

(4) The Secretary of State may suspend access to the Registry website if—

(a)   the Secretary of State considers that it is necessary to close or suspend provision of any of the services on the Registry website for the purposes of repair, maintenance, development or any other reason; or

(b)   access to the Registry website or operation of the any of the services is interrupted for reasons beyond the control of the Secretary of State.

(5) If the Secretary of State becomes aware that it is necessary to close or suspend access to the Registry for a prolonged period, she will put in place alternative arrangements for dealing with accounts and transactions.

(6) The Secretary of State will not be liable for any loss suffered as a result of the interruption of the Registry website.

Registry accounts
B3.— (1) An allowance will only be allocated or transferred to a person who holds an account in the Registry.

(2) There are the following types of account—

(a)   compliance accounts;

(b)   trading accounts;

(c)   the national cancellation account (see rule B15); and

(d)   the national retirement account (see rule B16).

(3) All compliance and trading accounts will be located in either the absolute sector or the relative sector.

Trading accounts
B4.— (1) A direct participant, a CCA participant or a trading participant may open a number of trading accounts in accordance with the procedure set out in the user manual.

(2) A participant opening a trading account may choose whether it is to be located in either the absolute sector or the relative sector but, once an account is opened, the sector in which it is located cannot be changed.

Trading participants

B5. —(1) Any person wishing to open a trading account who is not a direct participant or a CCA participant must apply to enter the scheme as a trading participant.

(2) An application to become a trading participant must be submitted to the Secretary of State in such manner as she may specify together with such information as she may require.

(3) Upon receipt of a complete application under paragraph (2), the Secretary of State will decide whether to enter into a trading participant agreement with that person and the Secretary of State shall inform the applicant of that decision.

Account users

B6.— (1) Once an account has been opened, all transactions and communications in relation to the account will be undertaken by or through either principal account users or secondary account users, appointed in accordance with the user manual.

(2) A participant may change the principal account user or secondary account user for any of his accounts in accordance with the user manual.

Secure area of the Registry website

B7.— (1) Access by account users to the Registry will be provided via the secure area of the Registry website which can only be accessed by users satisfying security checks in accordance with the user manual.

(2) Account users may only access the Registry website using the technology specified in the user manual and are responsible for making sure that their computers and other equipment can be used with the software in operation on the Registry website.

(3) Account users will be provided with security information (user names, passwords etc.) in accordance with the user manual in order to enable them to access the secure area of the Registry website and to undertake transactions in relation to the accounts to which they have access.

(4) Account users must keep all security information provided in relation to their appointed accounts secret and prevent anyone who is not authorised to access the accounts from doing so.

(5) An account user must notify the Secretary of State as soon as possible if he suspects that—

(a)   an unauthorised person knows any of his security information;

(b)   an unauthorised person has tried or intends to access the secure area of the Registry website;

(c)   he has forgotten or lost his security information; or

(d)   any security information has been kept in a form that may be accessible to an unauthorised person.

(6) Until she receives a notification under paragraph (5), the Secretary of State will be entitled to assume that all actions taken in relation to an account via the secure area of the Registry website have been taken by an authorised account user.

(7) The Secretary of State will take all reasonable steps to ensure that unauthorised access to the secure area of the Registry website does not occur.

(8) The Secretary of State may suspend an account user's access to any of the accounts to which he would otherwise have access if the Secretary of State has reason to believe that the account user has—

(a)   failed to comply with paragraph (4) or (5);

(b)   attempted to access accounts which he is not authorised to access;

(c)   attempted to introduce a virus or other harmful programme into the Registry software system or any other files, programmes or records held in the Registry;

(d)   repeatedly attempted to access an account using the wrong security information; or

(e)   otherwise failed to comply with the user manual.

(9) The procedure for lifting a suspension under paragraph (8) will be set out in the user manual.

Account identification and information

B8.   (1) The Secretary of State will assign a unique account name and identifying number to each account opened in the Registry.

(2) The account number will include the following information—

(a)   the account type (compliance or trading);

(b)   whether the account is located in the absolute sector or the relative sector; and

(c)   such other information as the Secretary of State determines.

Serial numbers for allocated allowances

B9. Upon the allocation of allowances to a compliance account, the Secretary of State will assign to each allowance a unique serial number that will comprise the following information—

(a)   country of origin (GB for all allowances issued under this scheme);

(b) Kyoto commitment period (0 for all pre-Kyoto commitment period allowances issued under this scheme);

(c) unit (a number unique to the allowance for each year in which allowances are allocated under this scheme);

(d) signifier of the unit type (allowance or credit);

(e) vintage.

Account balance

B10. The holdings from time to time of allowances in a compliance account or trading account is known as the account balance and will reflect —

(a) all allowances allocated under rules B12, C5 and D4;

(b) all changes in allowance holdings as a result of allowance transfers under rule B13;

(c) all allowances cancelled under rule B15; and

(d) all allowances retired under rule C7 and D5.

Transaction log

B11. The Secretary of State will publish a transaction log containing the following information in relation to all allowance allocations, transfers, cancellations and retirements —

(a) date of the transaction;

(b) the accounts involved in the transaction; and

(c) serial numbers of the allocated, transferred, cancelled or retired allowances.

Conversion of renewables obligation certificates

B12. — (1) If the registered holder of renewable obligations certificates wishes to convert certificates held by him into allowances, he may do so by following the procedure set out below.

(2) The registered holder must notify the Secretary of State of the request by providing her with the following details —

(a) the identification number of the certificates in question;

(b) the number of allowances to which the certificates are equivalent; and

(c) the details of the account at the Registry into which he would like the equivalent allowances allocated.

(3) The registered holder must submit a request to the Gas and Electricity Markets Authority asking it to substitute the Secretary of State as the registered holder for the certificates in accordance with the relevant order under sections 32 to 32C of the Electricity Act 1989[1] and, within 5 business days of receiving a notice under

---

[1] 1989 c.29. Section 32 of the Electricity Act 1989 was substituted for the section 32 originally enacted by section 62 of the Utilities Act 2000 (c.27). Sections 32A to 32C of the Electricity Act 1989 were inserted by sections 63 to 65 respectively of the Utilities Act 2000.

paragraph (2), the Secretary of State will submit the relevant request in accordance with the order.

(4) Within 5 business days of receiving notification from the Gas and Electricity Markets Authority that the substitution has taken place, the Secretary of State will request that the certificates be deleted from the relevant register established by the Gas and Electricity Markets Authority.

(5) Within 5 business days of receiving notification from the Gas and Electricity Markets Authority that the certificates have been deleted, the Secretary of State will allocate allowances into the account specified at the rate of 0.43 allowances for each certificate deleted, rounding down to the nearest whole number.

(6) The Secretary of State will notify the principal account user of the account in accordance with the user manual that the conversion has been completed and of the amount of allowances allocated to the account.

Transfers

B13.—(1) All requests for transfers of allowances must be submitted in accordance with the user manual.

(2) Subject to paragraph (3), a transfer request will be accepted if—

(a) the information required by the user manual is accurate and complete;

(b) the transferor account includes sufficient allowances to proceed with the transfer request; and

(c) neither the transferor account nor the transferee account are for the time being suspended under rule B20, C4 or D7.

(3) In the case of a transfer of allowances from a relative sector account to an absolute sector account, the transfer request will only be accepted if under rule B14(4) the relative sector gateway is open but, if the relative sector gateway is shut for the time being, the transfer request will be rejected unless the person submitting the transfer request has made an election in accordance with the user manual for the transfer request to enter the queuing mechanism under rule B14(8).

(4) Where a transfer request is not accepted because it does not satisfy any of the requirements of paragraphs (2) and (3), the person submitting the request will be notified in accordance with the user manual.

(5) Once a transfer request has been accepted by the Registry, the transfer will be completed automatically in accordance with the user manual by the deletion of each transferred allowance from the transferor account and the allocation of that allowance to the transferee account in accordance with the details in the transfer request.

(6) When completing a transfer under paragraph (5), allowances will be deleted and allocated in the following order (unless otherwise specified in the transfer request) —

(a)   allowances from the earliest vintage first; and

(b)   within vintages, in the order in which the allowances were allocated or transferred into the transferor account beginning with the latest allocated or transferred.

(7) Completion of the transfer takes place when the relevant entries appear in the accounts of both the transferor and the transferee.

(8) Each party to a transfer will be notified of the completion of the transfer in accordance with the user manual.

Relative sector gateway

B14.—(1) The relative sector gateway ("the gateway") will operate on the principle that there will be no net transfer of allowances out of the relative sector into the absolute sector.

(2) Transfers of allowances from relative sector accounts into absolute sector accounts will only be completed when the gateway is open.

(3) The Registry software will monitor all transfers into and out of the relative sector and cumulative figures will be maintained for (A) the total quantity of allowances transferred from absolute sector accounts to relative sector accounts and (B) the total quantity of allowances transferred from relative sector accounts to absolute sector accounts.

(4) The gateway will be open when the total quantity of allowances transferred from absolute sector accounts to relative sector accounts is greater than the total quantity of allowances transferred from relative sector accounts to absolute sector accounts (i.e. when A exceeds B).

(5) The gateway will close when the total quantity of allowances transferred from absolute sector accounts to relative sector accounts is equal to the total quantity of allowances transferred from relative sector accounts to absolute sector accounts (i.e. when A equals B).

(6) The following information about the status of the gateway will appear on the Registry website —

(a)   whether the gateway is open or closed;

(b)   if open, the maximum number of allowances which can be transferred through it before it closes;

(c)   if closed, the total number of allowances in the transfer requests in the queuing mechanism.

(7) While the gateway is closed —

(a)   no further transfers from relative sector accounts to absolute sector accounts will be completed;

(b)   requests to transfer will enter the queuing mechanism under paragraph (8) if the transferor made the appropriate election when submitting the transfer request; and

(c)   if no such election was made, the transfer request will be rejected and the transferor notified.

(8) The queuing mechanism will operate in the following way—

(a)   subject to the queuing process set out in the user manual, all pending transfer requests will be processed in the order in which they were submitted;

(b)   the principal or secondary account user of the transferor account will be notified of—

(i)   the transfer request's place in the queue;

(ii)   the total number of allowances involved in the transfer requests which are to be processed ahead of it in the queue; and

(iii)   the time and date at which the gateway was closed.

(9) Subject to paragraph (11), transfer requests which are in the queue but have not been processed may be withdrawn at any time in accordance with the user manual.

(10) If the gateway is open and a transfer request from a relative sector account to an absolute sector account (including a transfer request in the queuing mechanism) would if completed trigger the closure of the gateway in accordance with paragraph (5), a partial transfer of exactly the number of allowances which triggers closure under that paragraph will be accepted, provided the transferor made the appropriate election when submitting the transfer request but not otherwise.

(11) Once a transfer has been split under paragraph (10), the remainder of the transfer request may not be withdrawn (other than in exceptional circumstances in accordance with the user manual with the agreement of the Secretary of State).

National cancellation account

B15.—(1) The Secretary of State will establish a national cancellation account which will hold all allowances cancelled by participants under this rule.

(2) All cancellation requests must be submitted in accordance with the user manual.

(3) Subject to rule B18, an allowance which has been transferred into the national cancellation account cannot be subject to any further transaction.

National retirement account

B16.—(1) The Secretary of State will establish a national retirement account which will hold all allowances retired under rule C7 and D5.

(2) Only the Secretary of State may transfer allowances into the national retirement account.

(3) Subject to rule B18, an allowance which has been transferred into the national retirement account cannot be subject to any further transaction.

Banking

B17.—(1) All allowances (other than those which have been retired or cancelled) will be carried over from one commitment year to the next and, at the end of the commitment period, will be carried over until the process set out in paragraph (2) is carried out.

(2) After 31st March 2007, but no later than 31st December 2007—

(a)   any allowances held by a direct participant in—

(i)   his compliance account after the final reconciliation process is completed; and

(ii)   any trading account of his at the final reconciliation deadline will be dealt with in accordance with paragraph (3);

(b)   any allowances held by a trading participant at the final reconciliation deadline will be dealt with in accordance with paragraph (5); and

(c)   any allowances held by a CCA participant will be dealt with in accordance with rule D7.

(3) Allowances falling within paragraph (2)(a) will—

(a)   up to the unrestricted banking limit (calculated in accordance with paragraph (4)), be carried over into the first Kyoto Protocol compliance period[2] in an account in the national registry [3]; and

(b)   in excess of the limit, be carried over as described in sub-paragraph (a) subject to a banking restriction imposed by the Secretary of State.[4]

(4) The unrestricted banking limit under paragraph (3) equals the net amount by which the participant has reduced his emissions beyond his annual targets over the commitment period calculated as follows:

---

[2] A five year period starting on 1st January 2008.

[3] The Secretary of State will issue guidance on the future form of the national registry no later than 31st December 2005.

[4] The restriction will operate by cancelling a percentage of the allowances eligible to be carried over. The exact percentage will be announced by the Secretary of State no later than 31st December 2005.

tA – tC – tVE

where—

tA is the total number of allowances allocated to the participant over the commitment period; tC is the net adjustment under Schedule 3 in allowances resulting from any changes; and tVE is the total of the direct participant's verified emissions over the commitment period.

(5) Allowances falling within paragraph (2)(b) will be carried over into the first Kyoto Protocol compliance period in an account in the national registry subject to a banking restriction imposed by the Secretary of State.[5]

(6) Any allowances carried over under paragraphs (1), (3) and (5) will either maintain their original serial numbers or will be exchanged for first Kyoto Protocol Commitment Period serial numbers on a like for like basis.

Correction of account errors

B18.—(1) The Secretary of State may correct any error or omission in any account in relation to the allocation, transfer, retirement or cancellation of allowances.

(2) Participants must notify the Secretary of State as soon as possible if they find any error, omission, failure or delay in relation to the allocation, transfer, retirement or cancellation of allowances.

(3) If the Secretary of State makes a correction under paragraph (1), she will notify the concerned participants and account users of the correction, giving her reasons.

(4) If a participant or account user who is aggrieved by a correction makes representations within 5 days of receipt of a notice under paragraph (3) to the Secretary of State, she will reconsider the correction and will notify the concerned participants and account users of the result of her reconsideration, giving her reasons.

Closure of trading accounts

B19.—(1) A principal account user or secondary account user may instruct the Secretary of State to close a trading account in accordance with the user manual if there are no allowances held in the account at that time.

(2) If a trading account—

(a)   has not held any allowances since it was opened and shows no activity for a period of 12 months or more; or

(b)   does not hold any allowances and shows no activity for a period of 30 months or more,

---

[5] See footnote (4).

the Secretary of State will notify the relevant participant and the principal account user of her intention to close the account and will close the account unless a valid transfer request is made to transfer allowances into the account or the participant or principal account user demonstrates to the Secretary of State's satisfaction a good reason why the inactive account should not be closed.

(3) Where under the terms of a trading participant agreement or direct participant agreement, the Secretary of State may terminate such agreement for reasons of insolvency or for any other reason, the Secretary of State may and without further notice to the participant—

(a)   access any trading account held in his name in order to transfer allowances to another account (including the national cancellation account); and

(b)   close or suspend or suspend in part any trading account held in his name, in accordance with the terms of such agreement.

Suspension of trading accounts

B20.— (1) The Secretary of State may suspend any trading account if—

(a)   the participant (or any of his directors, partners or controllers) is subject to an ongoing investigation relating to finance and markets or the formation and activities of a body corporate or other business or professional entity by any authority which may lead to conviction, or a legal or regulatory sanction, for an offence or other act involving fraud, dishonesty or professional misconduct;

(b)   the Secretary of State has reason to believe that the participant or one of the account users with access to the account in question has failed to comply with any of the rules in this Part (including in particular rule B7(4) or (5)); or

(c)   the Secretary of State has reason to believe that an unauthorised person may attempt to access the account (for example where notified under rule B7(5)).

(2) A suspension imposed under paragraph (1) will continue until the Secretary of State is satisfied that—

(a)   in a case falling within paragraph (1)(a), there is no continuing risk to the integrity of the scheme (for example if the person under investigation were to be acquitted);

(b)   in a case falling within paragraph (1)(b), the failure to comply has been remedied and is not likely to recur; or

(c)   in a case falling within paragraph (1)(c), there is no further risk of unauthorised access to the account (for example when the security information relating to the account user and, if necessary, all account users with access to the account has been changed and all relevant parties notified).

(3) Subject to paragraph (4), the effect of suspending a trading account under paragraph (1) is that no transfers may be made into or out of the account until the suspension is lifted.

(4) The Secretary of State may upon receipt of an application made in accordance with the user manual allow transfers from any suspended trading account of a direct participant or a CCA participant to be made to his compliance account.

PART C
DIRECT PARTICIPANTS

Benefits

C1. — (1) Subject to rules C3 and G1, the Secretary of State will in respect of each commitment year —

(a)   make an incentive payment to each direct participant calculated in accordance with rule C2; and

(b)   allocate allowances to his compliance account in accordance with rule C5; if the conditions specified in paragraph (2) are fulfilled in relation to that year.

(2) The conditions are that on the reconciliation deadline for the relevant year the direct participant —

(a)   complies with his annual emissions limitation commitment under rule C6(1);

(b)   complies with his obligations under rule C8; and

(c)   is a party to a valid and continuing direct participant agreement.

Incentive payments

C2. — (1) The amount of a direct participant's incentive payment for any commitment year will be equal to the clearing price multiplied by his annual target for that year.

(2) The payment will be made within 28 days of the direct participant submitting a verification opinion to the Secretary of State but not before the first banking day after the reconciliation deadline.

Repayment of incentive payments

C3. — (1) Any direct participant whose total verified emissions during the commitment period exceed the total of —

(a)   the number of allowances retired from his compliance account prior to the final reconciliation deadline; plus

(b)   the allowances held in his compliance account on that deadline,

shall forthwith repay with interest any incentive payments which he has received from the Secretary of State and any such payment will be recoverable as a civil debt.

(2) Compliance with paragraph (1) will be assessed by the Secretary of State on the final reconciliation deadline.

(3) Subject to paragraph (4), any direct participant withdrawing from the scheme or ceasing for whatever reason to be a direct participant before the termination date shall be liable to repay with interest any incentive payments which he has received from the Secretary of State and any such payment will be recoverable as a civil debt.

(4) Where a direct participant's targets are reduced in accordance with Schedule 3 as a result of a joint-venture partner withdrawing consent for the inclusion of a source in the scheme, the direct participant shall be liable to repay with interest the corresponding proportion of any incentive payments which he has received from the Secretary of State, calculated in accordance with paragraph (5) and any such payment shall be recoverable as a civil debt.

(5) The amount to be repaid (A) under paragraph (4) is the result of the following calculation—

$$A = \frac{B_S \times I_P}{B_O}$$

Where—

B_S is the figure for the baseline emissions of the source for which consent has been withdrawn;

I_P is the total amount of incentive payments received by the participant prior to the withdrawal;

B_O is the participant's baseline prior to the withdrawal;

(5A) Where a direct participant's source list is amended in accordance with Schedule 3 as a result of the removal of an incorrectly included source, the direct participant shall be liable to repay with interest the corresponding proportion of any incentive payments which he has received from the Secretary of State, calculated in accordance with paragraph (5B) and any such payment shall be recoverable as a civil debt.

(5B) The amount to be repaid (A) under paragraph (5A) is the result of the following calculation—

$$A = \frac{B_S \times I_P}{B_O}$$

Where—

B_S is the figure for the baseline emissions of the source for which was incorrectly included;

I_P is the total amount of incentive payments received by the participant prior to the withdrawal;

B_O is the participant's baseline prior to the incorrect inclusion;

(5C) Paragraph 7(5)(a) of Schedule 2 shall apply to determine whether a source has been incorrectly included for the purposes of paragraphs (5A) and (5B).

(6) Interest is payable on any incentive payment to be repaid under this rule—

(a)   from the date on which the incentive payment was made by the Secretary of State until payment;

(b)   at an annual interest rate equal to the daily rate of 1 per cent above the BBA Sterling one month LIBOR rate, as published in the Financial Times on the first business day of the month; and

(c)   computed on the basis of the actual number of days elapsed.

Compliance accounts

C4.— (1) The Secretary of State will open a compliance account for each direct participant on the basis of information provided by the direct participant and such other information as the Secretary of State may reasonably require.

(2) Only one compliance account will be opened for each direct participant.

(3) Only allowances held in a direct participant's compliance account will be used to assess compliance with his annual emissions limitation commitments under rule C6(1).

(4) The compliance accounts of direct participants will be located in the absolute sector.

(5) The Secretary of State may suspend the compliance account of a direct participant if—

(a)   the participant (or any of his directors, partners or controllers) is subject to an ongoing investigation by any authority which may lead to conviction, or a legal or regulatory sanction, for an offence or other act involving fraud, dishonesty or professional misconduct relating to finance and markets or the formation and activities of a body corporate or other business or professional entity;

(b)   the Secretary of State has reason to believe that the participant or one of the account users with access to the account in question has failed to comply with any of the rules in this Part, including in particular rule B7(4) or (5); or

(c)   the Secretary of State has reason to believe that an unauthorised person may attempt to access the account (for example where notified under rule B7(5).

(6) A suspension imposed under paragraph (5) will continue until the Secretary of State is satisfied that—

(a)   in a case falling within paragraph (1)(a), there is no continuing risk to the integrity of the scheme (for example if the person under investigation were to be acquitted);

(b)   in a case falling within paragraph (1)(b), the failure to comply has been remedied and is not likely to recur; or

(c)   in a case falling within paragraph (1)(c), there is no further risk of unauthorised access to the account (for example when the security information relating to the account user and, if necessary, all account users with access to the account has been changed and all relevant parties notified).

(7) The effect of suspending an account under paragraph (5) will be that—

(a)   subject to paragraph (8), no allowances can be transferred into or out of the account in question until the suspension is lifted; and

(b)   allowances may be retired from the account as part of the reconciliation process described in rule C6.

(8) The Secretary of State may upon receipt of an application made in accordance with the user manual allow transfers to be made into a suspended compliance account.

(9) Where under the terms of a direct participant agreement, the Secretary of State may terminate the agreement for reasons of an insolvency event or for any other reason, the Secretary of State may without notice to the participant—

(a)   access any compliance account held in his name in order to transfer allowances to another account (including the national cancellation account); and

(b)   close or suspend or suspend in part any compliance account held in his name, in accordance with the terms of such agreement.

Allowances

C5.—(1) Allowances will be allocated to a direct participant on or before the following date—

(a)   in the first commitment year, the later of—

(i)   1st April 2002 or

(ii)   15 days after the direct participant submits a verification opinion in respect of his original baseline in accordance with Part 3 of Schedule 2; and

(b)   in subsequent commitment years, 15 days after he submits a verification opinion in respect of his annual emissions in accordance with Part 3 of Schedule 2.

(2) Subject to paragraph 2A and rule C6(4) and (5), the number of allowances allocated to the participant will be equal to his verified baseline minus (his annual target multiplied by the appropriate multiplier for the relevant commitment year shown in the following Table), rounded to the nearest whole number.

TABLE

Commitment Year	Multiplier
2002	1
2003	2
2004	3
2005	4
2006	5

(2A) Where, in 2005 or 2006 —

(a)   a direct participant's source list contains a source in respect of which a certificate of temporary exclusion has been served under regulation 11 of the Greenhouse Gas Emissions Trading Scheme Regulations 2005[6] or an equivalent provision in any superseding legislation, and the certificate is in force; and

(b)   the Secretary of State has made her decision under sub-paragraph (1)(b) of regulation 19 of the Greenhouse Gas Emissions Trading Scheme Regulations 2003[7],

the number of allowances to be allocated to the participant in respect of that source shall be calculated in accordance with the methodology set out in the document, "UK application to provide for temporary exclusion from the EU Emissions Trading Scheme for Direct Participants in the UK Emissions Trading Scheme" as revised in May 2004, as read with paragraph (2) of this rule.

(2B) Where, in 2005 or 2006 —

(a)   a direct participant's source list contains a source in respect of which a certificate of temporary exclusion has been served under regulation 11 of the Greenhouse Gas Emissions Trading Scheme Regulations 2005 or an equivalent provision in any superseding legislation; and

---

[6] S.I. 2005/925.

[7] S.I. 2003/3311. These Regulations have been revoked, save that regulation 19(1) shall continue to apply until the obligations under that regulations are fulfilled in respect of the period 2005-2007 – see regulation 47 of the Greenhouse Gas Emissions Trading Scheme Regulations 2005 (S.I. 2003/3311).

    (b)   the Secretary of State has made her decision under sub-paragraph (1)(b) of regulation 19 of the Greenhouse Gas Emissions Trading Scheme Regulations 2003 after allowances have been allocated in accordance with paragraph (2),

the Secretary of State shall cancel allowances from the compliance account of the participant in question equivalent to the difference between the number of allowances allocated and the number of allowances that would have been allocated in accordance with paragraph (2A) had paragraph (2A) applied at the time of allocation.

(2C) If, in relation to any direct participant, the Secretary of State is unable to comply with paragraph (2B) due to there not being sufficient allowances in the participant's compliance account, she shall notify the participant of —

    (a)   the number of allowances that she was required to cancel under paragraph (2B);

    (b)   the number of allowances that she has cancelled under paragraph (2B); and

    (c)   the number of outstanding allowances.

(2D) A direct participant shall cancel the number of outstanding allowances within 21 days of the Secretary of State sending a notice to him under paragraph (2C).

(2E) If a direct participant fails to comply with paragraph (2D), the direct participant shall be deemed to have failed to comply with his emissions limitation commitment as set out in rule C6(1) for that commitment year, and his excess emissions for the purpose of rule C6 shall be increased by the number of outstanding allowances.

(2F) For the purposes of this rule, "outstanding allowances" means the number of allowances that the Secretary of State is obliged to cancel under paragraph (2B) in relation to a direct participant minus the number of allowances cancelled under both paragraphs (2B) and (2D) in relation to that participant.

(3) Allowances may also be allocated to a direct participant following acquisition of a source

(including a substitute source in accordance with Part 1 of Schedule 3).

(4) Any allowance allocated to the participant under this rule will be allocated to his compliance account.

Annual emissions limitation commitment

C6.— (1) A direct participant shall ensure that for any commitment year his reconciliation balance for that year is equal to or greater than his verified emissions during that year.

(2) The reconciliation process set out in rule C7 will be carried out in order to assess whether a direct participant has complied with his emissions limitation commitment under paragraph (1).

(3) If a direct participant fails to comply with paragraph (1)—

(a)   for a commitment year beginning prior to the coming into force of legislation which provides a statutory basis for financial penalties for non-compliance with the scheme, paragraph (4) shall apply; and

(b)   for a commitment year beginning after the coming into force of legislation which provides a statutory basis for financial penalties for non-compliance with the scheme, paragraph (5) shall apply.

(4) In any case falling within paragraph (3)(a)—

(a)   the Secretary of State will not make any incentive payment to the participant for the commitment year in question; and

(b)   the number of allowances which will be allocated to the participant's compliance account under rule C5 for the following commitment year will be reduced by 1.3 allowances for each tonne of excess emissions as defined in paragraph (6), rounding to the nearest whole number any calculation which does not result in a whole number.

(5) In any case falling within paragraph (3)(b)—

(a)   the Secretary of State will not make any incentive payment to the participant for the commitment year in question;

(b)   the participant will pay the Secretary of State a financial penalty of £30 for each tonne of excess emissions as defined in paragraph (6); and

(c)   the number of allowances which will be allocated to the participant's compliance account under rule C5 for the following commitment year will be reduced by one allowance for each tonne of excess emissions as defined in paragraph (6).

(6) For the purpose of paragraphs (5) and (6), a direct participant's excess emissions will be the amount in tonnes of carbon dioxide equivalent of verified emissions, during the year for which compliance with paragraph (1) is being assessed, for which allowances were not retired under rule C7.

(7) Any financial penalty incurred under paragraph (5)(b) will be payable to the Secretary of State from the date on which the reconciliation process is completed for the commitment year in question, together with interest running from that date.

(8) Interest is payable on any financial penalty under paragraph (7)—

(a)   from the date on which the reconciliation process was completed;

(b)   at an annual interest rate equal to the daily rate of 1 per cent above the BBA Sterling one month LIBOR rate, as published in the Financial Times on the first business day of the month; and

(c)   computed on the basis of the actual number of days elapsed.

Reconciliation process

C7.— (1) An allowance will only be valid for the purpose of assessing compliance with rule C6(1) under this rule if it has a vintage which is either earlier than or the same year as the commitment year for which compliance is being assessed.

(2) The total quantity of valid allowances held in the compliance account of each direct participant at the reconciliation deadline ("the reconciliation balance") will be calculated as soon as practicable after the earlier of the date of delivery of the participant's verification opinion under Part 3 of Schedule 2 or the reconciliation deadline.

(3) If in any year access to the Registry is suspended under rule B2(4) for more than 8 consecutive hours during the 30 day period prior to the reconciliation deadline, the reconciliation deadline will be delayed by twice the duration of the suspension.

(4) The Secretary of State will notify each direct participant of his reconciliation balance once it has been calculated.

(5) As soon as practicable after he has both calculated a direct participant's reconciliation balance and received a verification opinion from that direct participant, the Secretary of State will reconcile the holdings of valid allowances in the participant's compliance account against his verified emissions by retiring allowances in accordance with paragraphs (6) to (8).

(6) The Secretary of State will retire allowances in the following order—

(a)   allowances from the earliest vintage first; and

(b)   within vintages, in the order in which they were allocated or transferred into the compliance account beginning with the earliest allocated or transferred.

(7) The Secretary of State will transfer all allowances which she retires to the national retirement account.

(8) The Secretary of State will retire allowances until either the number of allowances retired is equal to the total verified emissions for the relevant compliance period or until no more valid allowances remain in the compliance account.

(9) If the reconciliation balance of a direct participant is equal to or exceeds his total verified emissions, the Secretary of State will notify the direct participant that he is in compliance with his emissions limitation commitment under rule C6(1).

(10) If the total verified emissions of a direct participant exceed the final compliance account balance, the Secretary of State will notify the direct participant that he is not in compliance with his emissions limitation commitment under rule C6(1) and rule C6(3) will apply.

Monitoring, reporting and verification

C8. For each commitment year, a direct participant must—

(a) submit a duly completed verification opinion for his annual emissions in accordance with Part 3 of Schedule 2 on or before the reconciliation deadline;

(b) comply with the other requirements of Schedule 2 (monitoring, reporting and verification); and

(c) comply with the requirements of Schedule 3 (changes of operation and source list errors).

Failure to submit a verification opinion

C9.— (1) Where a direct participant has not complied with the requirements of rule C8 but—

(a) the only failure consists of a delay in submitting a verification opinion for that commitment year;

(b) the direct participant has by the reconciliation deadline notified the Secretary of State that the reason for the delay is the occurrence of a force majeure event or there is some other good explanation for the delay and provided her with satisfactory evidence in support; and

(c) the direct participant has since submitted the verification opinion, then he will be treated as if he had complied with those requirements.

(2) In any case falling within paragraph (1)—

(a) allowances will be allocated under rule C5(1) within 15 days; and

(b) any incentive payment will be paid under rule C2(2) within 28 days, after receipt by the Secretary of State of the participant's verification opinion.

(3) If a participant fails to comply with rule C8(a) and it is not a case falling within paragraph (1) of this rule,

(a) no allowances will be allocated to the participant for the next commitment year under rule C5(2);

(b) the reconciliation process in rule C7 will be carried out, and rule C6(3) to C6(7) shall apply, as if the participant's verified emissions for the commitment year in question were equal to his baseline, except that any reduction under rule C6(4)(b) or C6(5)(c) will be applied to the current balance of the participant's compliance account.

(4) Where a direct participant is only able to submit a qualified verification opinion in relation to its baseline or annual emissions, due to inadequate or incomplete emissions data for any of his sources in the case of a force majeure event or for any other reason, the Secretary of State may—

(a) treat the participant as having complied with rule C8(a);

(b) reduce the allocation of allowances to the participant by an amount not exceeding the baseline emissions of the relevant source; and

(c) reduce the incentive payments to the participant by a fraction not exceeding (the baseline emissions of the relevant source divided by the participant's baseline).

Other failures to comply with monitoring, reporting and verification requirements

C10.—(1) If in any commitment year a direct participant fails to comply with rule C8(b) or (c), the Secretary of State may take any of the following steps—

(a) declare that the participant's statement of emissions or verification opinion for that commitment year is invalid in whole or in part;

(b) withhold or delay in whole or in part any incentive payment to the participant;

(c) withhold or delay in whole or in part the allocation of any allowance to the participant;

(d) require the participant to amend his source list, baseline and targets, and make any associated transfer or cancellation of allowances, in accordance with Schedule 3.

(2) In deciding how to exercise her powers under paragraph (1), the Secretary of State will have regard to the nature and seriousness of the failure to comply, taking into account the purposes of the scheme.

(3) Where the Secretary of State considers that a participant has failed to comply with rule C8(b) or (c), she will serve a notice on the participant—

(a) identifying the obligation which she considers has not been complied with;

(b) setting out her reasons; and

(c) setting out the action she is minded to take under paragraph (1).

(4) Where a participant is aggrieved with a notice from the Secretary of State under paragraph (8), the participant may invoke the dispute resolution procedure in Schedule 4.

PART D
CCA PARTICIPANTS

Introduction

D1. — (1) This part of the scheme sets out the way in which operators and sector associations may participate in emissions trading for the purposes of their umbrella and underlying agreements.

(2) In particular —

(a)   rule D2 sets out the registration procedure to be followed by an operator or a sector association wishing to become CCA participants;

(b)   rule D3 explains how compliance accounts will be opened for CCA participants;

(ba)   rule D3A sets out the requirements in cases of an EU ETS overlap;

(c)   rule D4 sets out the circumstances in which allowances will be allocated to a CCA participant and the procedures for making the appropriate adjustment to the relevant CCA target;

(d)   rule D5 sets out the circumstances in which allowances will be retired from the compliance account of a CCA participant and the procedures for making the appropriate adjustment to the relevant CCA target;

(e)   rule D6 sets out the circumstances in which allowances held in the accounts of a CCA participant will be carried over into the first Kyoto Protocol commitment period;

(f)   rule D7 sets out the rules for suspending compliance accounts of a CCA participant;

(g)   rule D8 sets out the circumstances in which a CCA participant may cease to be a participant in the scheme; and

(h)   rule D9 explains how the scheme applies to agreements falling within paragraph 47 of Schedule 6 to the Finance Act 2000.[8]

(3) An operator may participate in this scheme in relation to any target unit listed in an underlying agreement to which it is a party, other than target units which form part of a trading group.

(4) A sector association may participate in this scheme in relation to any target unit which forms part of a trading group whose identity is provided for in an umbrella agreement to which it is a party.[9]

---

[8] The remainder of this part has been drafted on the basis that the majority of CCA participants will be participating in respect of combinations of umbrella and underlying agreements falling within paragraph 48 of Schedule 6 to the Finance Act 2000.

[9] An umbrella agreement may not provide for the identification of more than one trading group.

(5) For the purposes of this scheme, a "CCA participant" means—

(a)   an operator which has submitted a duly completed registration form under rule D2(1); or

(b)   a sector association which has submitted a duly completed registration form under rule D2(3).

Registration

Operators

D2.— (1) An operator who wishes to become a CCA participant must submit a registration form to the Secretary of State in such manner as she may specify together with—

(a)   the target unit identifiers for all target units listed in the underlying agreement in relation to which the operator wishes to participate in the scheme; and

(b)   the operator's consent to the publication of the information set out in rule G2(2) and (3A);

(c)   such other information as the Secretary of State may reasonably require.

(2) If an operator is a party to underlying agreements which apply to facilities identified in more than one umbrella agreement, paragraph (1) must be complied with in respect of each umbrella agreement.

Sector associations

(3) A sector association who wishes to become a CCA participant must submit a registration form to the Secretary of State in such manner as she may specify together with—

(a)   the target unit identifers for the facilities in the trading group;

(b)   the consents necessary to vary the relevant umbrella agreement to provide for trading by the trading group;

(c)   the sector association's consent to the publication of the information set out in rule G2(2) and (3A); and

(d)   such other information as the Secretary of State may reasonably require.

(4) A sector association may only participate in relation to more than one trading group if it is party to more than one umbrella agreement and each agreement provides for the identification of a trading group. In such a case, paragraph (3) must be complied with in respect of each trading group.

Compliance accounts
Operators

D3.—(1) Compliance accounts for operators will be opened in accordance with this rule on the basis of information provided in the registration form under rule D2(1) and any other information which the Secretary of State may reasonably require.

(2) A separate compliance account will be opened for each target unit identified under paragraph D2(1) and the account will be linked to the target unit for the purposes of this part of the scheme.

(3) A compliance account linked to a target unit with an absolute CCA target will be located in the absolute sector.

(4) The compliance account linked to a target unit with an relative CCA target will be located in the relative sector.

(5) Where an operator applies to vary the CCA target of a target unit linked to a compliance account under the terms of the relevant underlying agreement from an absolute to a relative CCA target or vice versa, the operator must provide the Secretary of State with details of the compliance account at the time of its application and must transfer any allowances from that account into another account in the Registry.

(6) The Secretary of State will not approve any application referred to in paragraph (5) until all allowances have been transferred out of the compliance account linked to the target unit whose target is the subject of the application.

(7) Where an application referred to in paragraph (5) is approved, the existing compliance account will be closed and a new compliance account linked to the target unit will be opened in the same sector as the new target.

Sector associations

(8) Compliance accounts for sector associations will be opened in accordance with this rule on the basis of the information provided in the registration form under rule D2(3) and any other information which the Secretary of State may reasonably require.

(9) A single compliance account will be opened for each trading group on whose behalf the sector association is participating.

(10) The compliance account of a sector association will be located in the relative sector.

EU-ETS overlap

D3A.—(1) This rule applies where—

    (a)   (i) in a case falling within rule D1(3), an operator; or
           (ii) in a case falling within rule D1(4), an operator of a facility within a trading group, is (or has been) an EU ETS participant in respect of an EU ETS installation comprised

within a facility forming part of a target unit or trading group; and

(b)   in relation to that facility, an EU ETS overlap occurs (or has occurred).

(2) An EU ETS overlap occurs to the extent that any emissions from the facility are also reportable emissions from the relevant EU ETS installation.

Notification of EU ETS overlap

(3) For the purpose of assessment of any EU ETS overlap during each target period —

(a)   the operator; or

(b)   where at the time of notification the relevant target unit for the facility in question is part of a trading group, the sector association, shall, in respect of each scheme year specified in the table below, provide the Secretary of State with the notification required by paragraph (4) on or by each notification deadline —

Scheme year	Notification deadline
2005	7 February 2007
2007	7 February 2009
2009	7 February 2011

(4) In respect of the relevant scheme year, the operator or sector association (as the case may be) shall, in accordance with Schedule 5A, specify —

(a)   the EU ETS overlap; and

(b)   in relation to that overlap —

(i)   the EU ETS overachievement;

(ii)   the EU ETS underachievement; or

(iii)   that there has been neither.

(5) For the purpose of paragraph (4)(b) —

(a)   an EU ETS overachievement occurs where the overlap emissions emitted during the scheme year are less than the number of EU ETS allowances allocated to the relevant EU ETS installation for that year and which are attributable to those emissions, the difference between the two constituting the overachievement; and

(b)   an EU ETS underachievement occurs where the overlap emissions emitted during the scheme year are greater than the number of EU ETS allowances allocated to the relevant EU ETS installation for that year and which are attributable to those emissions, the difference between the two constituting the underachievement.

Reserve powers of Secretary of State

(6) The Secretary of State may require the operator or sector association to—

(a)   obtain a verification opinion in respect of its notification; or

(b)   review or amend its notification.

Effect of notification

(7) Paragraphs (8) to (10) apply for the purpose of adjusting the CCA target or trading group target applying to the facility in question so that the target takes account of any EU ETS overachievement or EU ETS underachievement notified.

(8) In the case of an EU ETS overachievement, except where the operator or the sector association has already retired a number of EU ETS allowances at least equal to its overachievement, the notification shall, for the purpose of the emissions trading provisions of—

(a)   the underlying agreement which applies to the facilities making up the target unit applicable to the overlap emissions at the time of the adjustment; or

(b)   the umbrella agreement which provides for the identification of the relevant trading group applicable to the overlap emissions at the time of the adjustment,

be deemed to be the transfer from that target unit or trading group target of an equivalent number of carbon emission allowances, in accordance with a trading scheme established by the Secretary of State which applies to the relevant agreement, and the number of allowances thus deemed to have been transferred shall constitute a negative balance in relation to that target unit or trading group.

(9) In all cases of an EU ETS underachievement, the notification shall, for the purposes set out in paragraphs (8)(a) and (b), be deemed to be the transfer to that target unit or trading group target of an equivalent number of carbon emission allowances, in accordance with a trading scheme established by the Secretary of State which applies to the relevant agreement, and the number of allowances thus deemed to have been transferred shall constitute a positive balance in relation to that target unit or trading group.

(10) Where a notification is amended, that amendment shall be deemed to be the transfer of the commensurate number of carbon emission allowances to reflect the amendment, in accordance with the principles of paragraphs (8) and (9).

Guidance

(11) The operator or sector association shall have regard to any guidance issued by the Secretary of State in respect of the application of this rule.

Interpretation

(12) For the purposes of this rule and Schedule 5A —

(a)   "EU ETS installation" is an "installation" within the meaning of regulation 2(1) of the Greenhouse Gas Emissions Trading Scheme Regulations 2005(a);

(b)   an "EU ETS participant" means an "operator" within the meaning of that regulation who also falls within the ambit of paragraph (1)(a) above;

(c)   "reportable emissions" and "scheme year", have the meaning set out in that regulation; and

(d)   "overlap emissions" are emissions falling within paragraph (2)."

Allocation of allowances for over-achievement of CCA targets

D4. — (1) A CCA over-achievement which occurs in relation to an operator or a trading group may be converted into allowances in accordance with this rule.

(2) For the purpose of this rule, "a CCA over-achievement" occurs when —

(a)   in relation to an operator, the carbon emitted or energy used by a target unit during a target period is less than the CCA target (as adjusted in accordance with rule D3A(7) to (10)) for that target unit; and

(b)   in relation to a trading group, the carbon emitted or energy used by the facilities which comprise that trading group during a target period is less than the trading group target (as adjusted in accordance with rule D3A(7) to (10)).

Operators

(3) If —

(a)   a CCA over-achievement occurs in relation to a target unit;

(b)   the operator of the target unit was a CCA participant in respect of the target period in which the over-achievement occurred;

(c)   a compliance account has been opened and linked to that target unit under rule D3; and

(d)   the operator has complied with the notification requirements in paragraph (5); the notification of the over-achievement will, for the purpose of the emissions trading provisions of the underlying agreement which applies to the facilities making up that target unit, be deemed to be the transfer of carbon emission allowances from that target unit in accordance with a trading scheme established by the Secretary of State which applies to that agreement and the number of allowances so reported shall be deemed to be a negative balance in relation to that target unit.

(4) Where—

(a)   the conditions specified in paragraph (3)(a), (b), (c) and (d) are satisfied; and

(b)   the operator has complied with the application procedure under paragraph (7);

the Secretary of State will within 15 days of receiving the application either allocate the appropriate number of allowances to the compliance account linked to the target unit to which the over-achievement related or, if she reasonably considers that the CCA over-achievement has not been adequately demonstrated, require the operator to provide such further information in support of his claim as she may specify and, where it is relevant, the notification procedure under rule D3A(3) and (4).

(5) An operator wishing to convert a CCA over-achievement into allowances or to preserve the over-achievement for possible future conversion under this rule ("ring-fencing") must notify the relevant sector association and the Secretary of State of the over-achievement by the reporting deadline set by his sector association, by providing the information specified in paragraph (6).

(6) The information required by paragraph (5) is as follows—

(a)   facility numbers of the facility or facilities making up the target unit in respect of which the over-achievement has occurred;

(b)   the amount in the currency of the CCA target by which it has been over-achieved and the number of allowances to which the over-achievement is equivalent; and

(c)   any other information needed to calculate the adjustment to be made to the CCA target under the emissions trading provisions of the underlying agreement which applies to the target unit.

(7) An application to convert a CCA over-achievement into allowances must be submitted by the operator to the Secretary of State and copied to the relevant sector association prior to 31 December 2012 and must be accompanied by—

(a)   a statement of over-achievement prepared in accordance with paragraph 3 of Schedule 6; and

(b)   an unqualified verification opinion from a verifier prepared in accordance with paragraph 4 of Schedule 6.

(8) For the purpose of paragraphs (5) and (7), the relevant sector association is the sector association specified in the umbrella agreement which applies to the facilities comprising the target unit to which the over-achievement relates.

Sector associations

(9) If —

(a)   a CCA over-achievement occurs in relation to a trading group;

(b)   the sector association whose umbrella agreement provides for the identification of the trading group was a CCA participant in respect of the target period in which the overachievement occurred;

(c)   a compliance account has been opened for that trading group under rule D3; and

(d)   the sector association has complied with the notification requirements in paragraph (11);

the notification of the over-achievement will, for the purpose of the emissions trading provisions of the umbrella agreement which provides for the identification of the trading group, be deemed to be the transfer of carbon emission allowances from that trading group in accordance with a trading scheme established by the Secretary of State which applies to that agreement and the number of allowances so notified shall be deemed to be a negative balance in relation to that trading group.

(10) Where —

(a)   the conditions specified in paragraph (9)(a), (b)(c) and (d) are satisfied; and

(b)   the sector association has complied with the application procedure under paragraph (13)

the Secretary of State will within 15 days of receiving the application either allocate the appropriate number of allowances to the compliance account linked to the trading group or, if she reasonably considers that the CCA over-achievement has not been adequately demonstrated, require the sector association to provide such further information in support of his claim as she may specify and, where it is relevant, the notification procedure under rule D3A(3) and (4).

(11) A sector association wishing to convert a CCA over-achievement into allowances or to preserve that over-achievement for possible future conversion under this rule ("ring-fencing") must notify the Secretary of State of the over-achievement by the

CCA reconciliation deadline by providing the information specified in paragraph (12).

(12) The information required by paragraph (11) is as follows-

(a) the trading group target in respect of which the over-achievement has occurred;

(b) the amount in the currency of the trading group target by which it has been over-achieved and the number of allowances to which that is equivalent;

(c) any other information needed to calculate the adjustment to be made to the trading group target under the emissions trading provisions of the umbrella agreement identifying the trading group.

(13) An application to convert CCA over-achievement into allowances must be submitted by the sector association to the Secretary of State prior to 31 December 2012 and must be accompanied by—

(a) a statement of over-achievement prepared in accordance with paragraph 3 of Schedule 6; and

(b) an unqualified verification opinion prepared in accordance with paragraph 4 of Schedule 6.

General

(14) The vintage of any allowance allocated under this rule will be the year in which the allowance is allocated, except that—

(a) the vintage of any allowance allocated between 1 January and 31 March will be the previous year; and

(b) the vintage of any allowance allocated after 31 December 2012 will be "2012".

(15) Any over-achievement relating to a target unit with a relative target which has been ringfenced in accordance with paragraph (5) or (11) but for which an application for conversion into allowances is not received prior to 1st January 2008 will be discounted prior to conversion by a percentage factor to be announced by the Secretary of State no later than 31 December 2005.

(16) Where—

(a) a CCA over-achievement has been ring-fenced in accordance with paragraph (5) or (11); and

(b) prior to submitting an application under paragraph (7) or (13), either—

(i) the CCA target to which the over-achievement related has been varied from an absolute to a relative CCA target or vice versa; or

(ii) the operator or sector association has ceased to be a CCA participant under rule D8(4);

the Secretary of State will allocate allowances which relate to the over-achievement of the original CCA target not into a compliance account as provided for in paragraph (4) or (10) but instead into a trading account in the same sector as the original CCA target, the details of which are to be provided in the application submitted under paragraph (7) or (13).

Increasing CCA targets by retiring allowances from compliance accounts

D5.—(1) An operator may increase the CCA target (as adjusted in accordance with rule D3A(7) to (10)) for a target unit and a sector association may increase a trading group target (as adjusted in accordance with rule D3A(7) to (10)) by retiring allowances in accordance with this rule.

Operators

(2) If—

(a) the operator of a target unit is a CCA participant in relation to a target period for which the CCA target of a target unit is to be adjusted;

(b) a compliance account has been opened and linked to that target unit under rule D3;

(c) the operator has supplied the Secretary of State with the information specified in paragraph (3); and

(d) allowances have been retired from the compliance account under paragraph (7);

the retirement of those allowances will, for the purpose of the emissions trading provisions of the underlying agreement which applies to the facilities making up that target unit, be deemed to be the transfer of carbon emission allowances to that target unit in accordance with a trading scheme established by the Secretary of State which applies to that agreement and the amount of allowances retired shall be deemed to be a positive balance in relation to that target unit.

(3) An operator seeking to increase the CCA target of a target unit must provide the Secretary of State with the following information on or before the CCA reconciliation deadline—

(a) the target unit identifier;

(b) the amount by which the target is to be increased and the number of allowances to which that amount is equivalent (i.e. the number to be retired);

(c) the details of the compliance accounts to which the target unit is linked; and

(d)   the earliest date on which any retirement may take place (which may not be later than the CCA reconciliation deadline).

(4) To the extent to which any of the information required under paragraph (3) has been supplied to the Secretary of State by the relevant sector association, there is no requirement for the operator to supply that information separately.

(5) Subject to paragraph (17) and (17A), the "CCA reconciliation balance" for each target unit specified in a CCA retirement request will be the total quantity of allowances held in the compliance account linked to that target unit on the CCA reconciliation deadline or an earlier date agreed between the relevant sector association and the Secretary of State (which shall not be earlier than the date specified by the operator under paragraph (3)(d)).

(6) If the CCA reconciliation balance for a target unit is less than the number of allowances specified in the information supplied to the Secretary of State under paragraph (3) or (4), no allowances will be retired and there will be no increase to the CCA target of that target unit.

(7) If the CCA reconciliation balance for a target unit is equal to or greater than the number of allowances specified in the information supplied to the Secretary of State under paragraph (3) or (4), allowances will be retired from the compliance account linked to the target unit in the following order—

(a)   allowances with the earliest vintage first; and

(a)   within vintages, in the order in which they were allocated or transferred into the compliance account beginning with the earliest allocated or transferred.

Sector associations

(8) If—

(a)   a sector association is a CCA participant in relation to a target period for which a trading group target is to be adjusted;

(b)   a compliance account has been opened for a trading group under rule D3;

(c)   the sector association has supplied the Secretary of State with the information specified in paragraph (9); and

(d)   allowances have been retired from the compliance account under paragraph (13);

the retirement of those allowances will, for the purpose of the emissions trading provisions of the umbrella agreement which provides for the identification of that trading group, be deemed to be the transfer of carbon emission allowances to that trading group in accordance with a trading scheme established by the

Secretary of State which applies to that agreement and the amount of allowances retired shall be deemed to be a positive balance in relation to that trading group.

(9) A sector association seeking to increase the CCA target of a trading group must provide the Secretary of State with the following information on or before the CCA reconciliation deadline—

(a)   the amount by which the target is to be increased and the number of allowances to which that amount is equivalent (i.e. the number to be retired);

(b)   the details of the compliance accounts for the trading group; and

(c)   the earliest date on which any retirement may take place (which may not be later than the CCA reconciliation deadline).

(10) To the extent to which any of the information required under paragraph (9) has already been supplied to the Secretary of State by the sector association, there is no requirement for the sector association to supply that information again.

(11) Subject to paragraph (17) and (17A), the "CCA reconciliation balance" for a trading group for which a CCA retirement request has been submitted will be the total quantity of allowances held in the trading group's compliance account on the CCA reconciliation deadline or an earlier date agreed between the sector association and the Secretary of State.

(12) If the CCA reconciliation balance for a trading group is less than the number of allowances specified in the information supplied to the Secretary of State under paragraph (9) or (10), no allowances will be retired and there will be no increase to the trading group target.

(13) If the CCA reconciliation balance for a trading group is equal to or greater than the number of allowances specified in the information supplied to the Secretary of State under paragraph (9) or (10), allowances will be retired from the compliance account of the trading group in the following order—

(a)   allowances with the earliest vintage first; and

(b)   within vintages, in the order in which they were allocated or transferred into the compliance account beginning with the earliest allocated or transferred.

General

(14) All allowances retired under this rule will be transferred to the national retirement account.

(15) Subject to paragraph (16), the "CCA reconciliation deadline" means, in respect of any target period, 12.00 noon on the next business day after 6 February following the end of the target period.

(16) If in any year access to the Registry is suspended under rule B2(4) for more than 8 consecutive hours during the 30 day period ending on the CCA reconciliation deadline, the CCA reconciliation deadline will be delayed by twice the duration of the suspension and, if the extended deadline is later than 5.00pm, to 12.00 noon on the next business day.

(17) The Secretary of State may allow allowances, which were transferred into a relevant compliance account after the CCA reconciliation deadline but prior to 31st March following the deadline, to be included in a CCA reconciliation balance for the purposes of this rule, where a CCA participant submits an application in writing setting out the reasons why the Secretary of State should allow the allowances to be retired, supported by such further information as the Secretary of State may require.

(17A) Notwithstanding paragraph (17), where on or after 31st March following the CCA reconciliation deadline —

(a)   an error has been identified in information supplied by a CCA participant or a sector association to the Secretary of State, and

(b)   the CCA participant or the sector association, as the case may be, has satisfied the Secretary of State that the error was innocently made,

the Secretary of State may allow allowances which have been transferred into a relevant compliance account, to be included in a CCA reconciliation balance for the purposes of this rule.

(17B) For the purposes of paragraph (17A), a CCA participant shall be deemed to have complied with paragraph (3) or (9) even if —

(a)   the information required under paragraph (3) or (9) is provided after the reconciliation deadline; or

(b)   the date specified as the earliest date under paragraph (3) or (9) on which any retirement may take place is later than the CCA reconciliation deadline.

(18) For the purpose of paragraphs (4) and (5), the relevant sector association is, subject to rule D9(2), the sector association specified in the umbrella agreement which applies to the facilities comprising the target unit whose target is to be adjusted.

Banking

D6.—(1) After 31st March 2007 but no later than 31st December 2007, the following rules shall apply to the carrying over of allowances into the first Kyoto Protocol compliance period in accounts in the national registry [10]—

(a) any allowances held by an operator participating in relation to a target unit with an absolute CCA target will be dealt with in accordance with paragraphs (2) to (5); and

(b) any allowances held by an operator participating in relation to a target unit with a relative CCA target or by a sector association will be dealt with in accordance with paragraphs (6) and (7).

(2) Subject to paragraph (3), allowances held by the operator in the compliance account linked to that target unit or in any trading account will be carried over into the first Kyoto Protocol compliance period.

(3) Subject to paragraph (4), the maximum number of allowances which may be carried over under paragraph (2) is equal to the total number of allowances allocated under rule D4 since 1st January 2002 in respect of the performance of that target unit.

(4) In the case of a target unit which previously had a relative CCA target, only those allowances which relate to the over-achievement of an absolute CCA target will be taken into account in calculating the figures referred to in paragraph (3).

(5) Any allowances in excess of the limit in paragraph (3) will be subject to a banking restriction imposed by the Secretary of State[11] before being carried over into the first Kyoto Protocol compliance period.

(6) Subject to paragraph (7), allowances in any account will be subject to a banking restriction imposed by the Secretary of State[12] before being carried over into the first Kyoto Protocol compliance period .

(7) In the case of the operator of a target unit which previously had an absolute CCA target, allowances equal in number to those which were allocated under rule D4 in respect of the overachievement of that absolute target may be carried over into the first Kyoto Protocol compliance period without restriction.

---

[10] The Secretary of State will issue guidance on the future form of the national registry no later than 31st December 2005.

[11] The restriction will operate by cancelling a percentage of the allowances eligible to be carried over. The exact percentage
will be announced by the Secretary of State no later than 31st December 2005.

[12] See previous footnote.

Suspension of compliance accounts

D7.—(1) The Secretary of State may suspend any or all of the compliance accounts of a CCA participant if—

(a) the participant (or any of his directors, partners or controllers) is subject to an ongoing investigation by any authority which may lead to conviction, or a legal or regulatory sanction, for an offence or other act involving fraud, dishonesty or professional misconduct relating to finance and markets or the formation and activities of a body corporate or other business or professional entity;

(b) the Secretary of State has reason to believe that the participant or one of the account users with access to the account in question has failed to comply with any of the rules in Part B, including in particular rule B7(4) or (5);

(c) the Secretary of State has reason to believe that an unauthorised person may attempt to access the account (for example where notified under rule B7(5)); or

(d) an operator or sector association fails to notify the Secretary of State as required by rule D3A(3) and (4).

(2) A suspension imposed under paragraph (1) will continue until the Secretary of State is satisfied that—

(a) in a case falling within paragraph (1)(a), there is no continuing risk to the integrity of the scheme (for example if the person under investigation were to be acquitted);

(b) in a case falling within paragraph (1)(b), the failure to comply has been remedied and is not likely to recur; or

(c) in a case falling within paragraph (1)(c), there is no further risk of unauthorised access to the account (for example when the security information relating to the account user and, if necessary, all account users with access to the account has been changed and all relevant parties notified).

(3) The effect of suspending an account under paragraph (1) is that—

(a) subject to paragraph (4), no allowances can be transferred into or out of the account in question until the suspension is lifted; and

(b) allowances may be retired from the account under rule D5.

(4) The Secretary of State may upon receipt of an application made in accordance with the user manual allow transfers to be made into a suspended compliance account.

Ceasing to be a CCA participant

D8.— (1) A CCA participant may cease to be a participant in the scheme in the circumstances set out in this rule.

(2) The Secretary of State may (but is not obliged to) serve a termination notice on a CCA participant if—

(a)   the CCA participant is in the reasonable opinion of the Secretary of State guilty of any unlawful conduct or wilful conduct amounting to an abuse of the scheme or the Registry or in relation to any Registry account; or

(b)   the CCA participant or any of its directors, partners or controllers is, in relation to finance and markets or the formation and activities of a body corporate or other business or professional entity, convicted of any criminal offence, or subject to legal or regulatory sanction, for fraud, dishonesty or professional misconduct.

(3) Where a termination notice is served on a CCA participant under paragraph (2)—

(a)   all compliance accounts and trading accounts held by the participant will be suspended;

(b)   if there are allowances remaining in any compliance or trading account held in the participant's name, the Secretary of State will transfer the remaining allowances into an account held in the name of a third party, provided that—

(i)   the Secretary of State reasonably considers that the allowances were not acquired as a result of the event giving rise to the termination; and

(ii)   the CCA participant notifies the Secretary of State of the account to which any remainder is to be transferred within two months of the service of the termination notice;

(c)   if the conditions set out in sub-paragraph (b) are not satisfied, the Secretary of State may cancel any allowances remaining in the account; and

(d)   following the transfer or cancellation of the remaining allowances, the Secretary of State will close all compliance accounts or trading accounts held in the participant's name.

(4) Where a climate change agreement to which a CCA participant is a party is terminated and a termination notice under paragraph (2) has not been served, paragraphs (3)(a) to (d) will apply to compliance and trading accounts held in the participant's name save that—

(a)   if the participant has successfully applied to become a trading participant under rule B5, paragraphs (3)(a) to (d) shall not apply to its trading accounts; and

(b)   if the participant is party to another climate change agreement and that agreement has not been terminated, paragraphs (3)(a) to (d) shall only apply to those compliance accounts which relate to the climate change agreement which has been terminated.

(5) A CCA participant may serve a withdrawal notice on the Secretary of State if there are no allowances held in any of its compliance accounts and —

(a)   all of its trading accounts have been closed in accordance with rule B19(1); or

(b)   it has successfully applied to become a trading participant under rule B5.

(6) Where a withdrawal notice under paragraph (5) is served, the participant will cease to be a CCA participant and the Secretary of State will close its compliance accounts.

(7) Any notice under this rule should be served in accordance with the terms of the climate change agreement to which the CCA participant concerned is a party.

Agreements falling within Finance Act 2000, Schedule 6, paragraph 47

D9. — (1) A person who is a party to an agreement falling within paragraph 47 of Schedule 6 to the Finance Act 2000 (a "paragraph 47 agreement") may participate in this scheme as a CCA participant as if it were an operator.

(2) Where a party to a paragraph 47 agreement wishes to participate as a CCA, the scheme shall be interpreted, unless the context otherwise requires, as follows —

(a)   references to an operator mean the person who is the party to the paragraph 47 agreement other than the Secretary of State;

(b)   references to an underlying agreement mean the paragraph 47 agreement;

(c)   the requirements under rule D4(5) and (7) to notify the relevant sector association do not apply;

(d)   the references to the relevant sector association in rule D5(4) and (5) mean the person who is the party to the paragraph 47 agreement;

(e)   the references to an umbrella agreement in the definitions of certification period and facility in Schedule 1 are deemed to be to the paragraph 47 agreement; and

(f)   the references to an umbrella agreement in Schedule 6 shall not apply.

PART E
PROJECT PARTICIPANTS

E1. [This Part has been left intentionally blank]

PART F
GROUP PARTICIPANTS

F1. — (1) The purpose of this part is to provide a mechanism whereby two or more persons may participate in the scheme as members of a group represented in the scheme by a single group participant.
(2) A group participant means a person who has entered into a group participant agreement under this rule.
(3) A group participant agreement means an agreement in the form specified by the Secretary of State entered into between a group participant and the Secretary of State by virtue of which the group participant participates in the scheme.
(4) A group participant agreement will contain appropriate provisions to provide that the scheme will apply to a group participant substantially as if he were a direct participant with a direct participant agreement, subject to any additional provisions necessary to ensure that the purposes of the scheme are fulfilled and to give effect to the following principles:

Restriction on use of compliance account

(a)   the compliance account of the group participant will be suspended immediately after opening and no allowances may be transferred into or out of the account except by the Secretary of State on application from the group participant or an account user nominated by him;

(b)   transfers of allowances into the group participant's compliance account will be allowed without restriction on application to the Secretary of State;

(c)   transfers of allowances out of the group participant's compliance account will be allowed on application to the Secretary of State only after the reconciliation process described in rule C6 has been completed and then only if and to the extent that the group participant's reconciliation balance exceeded the total verified emissions of the group for the relevant commitment year;

Accountability for group members

(d)   the group participant will be liable for any act or omission of any group member which amounts to non-compliance with any requirements of the scheme and in particular of Schedule 2 as if that act or omission were his own;

(e)   a failure by a group member to comply with the requirements of Schedule 2 will not in itself be regarded as a good explanation for a delay in submitting a verification opinion under rule C9(1)(b);

(f)   the group participant must demonstrate to the satisfaction of the Secretary of State that the Secretary of State will be able to pursue the group members directly if the group participant is unable to discharge his liabilities under the scheme;

Management control

(g)   for the purposes of the scheme in general, in particular Part 1 of Schedule 2, the group participant will be deemed to have management control of the sources over which the group members have management control, save that the source list of a group participant will be required to indicate which group members control which sources;

(h)   any change in management control over a source in a group participant's source list which falls under Part 1 of Schedule 3 will be dealt with as if that change were a change for which the group participant was responsible;

Withdrawal

(i)   the withdrawal of a group member from a group will be treated as the closure of the sources controlled by that group member in accordance with Part 1 of Schedule 3; and

(j)   if the group participant withdraws from the scheme, the group participant will be required to repay to the Secretary of State any incentive payments made to it except if the group members become members of another group under this part or become direct participants by entering into direct participant agreements in their own right.

PART G
MISCELLANEOUS

Power to amend or revoke the scheme

G1.— (1) This scheme may be amended or revoked by the Secretary of State at any time by another scheme made by the Secretary of State.

(2) Without prejudice to the generality of paragraph (1), the powers under that paragraph may be exercised —

(a)   for the purposes of correcting errors in the scheme;

(b)   for improving the functioning of the scheme in the light of experience;

(c)   for extending the scheme—

(i)   to additional direct participants as a result of any auction held after the first auction;

(ii)   to parties to climate change agreements,

(iii)   to projects;

(iv)   to provide for international trading;

(d)   for transferring responsibilities of the Secretary of State to an emissions trading authority or other body;

(e)   for the purpose of complying with any decision taken by the Commission in relation to State Aids approval in relation to the scheme;

(f)   in connection with any future legislation dealing with greenhouse gases, including legislation authorising the imposition of penalties; and

(g)   for making any incidental, supplemental or transitional provisions.

(3) Subject to paragraph (4), the Secretary of State will consult such participants as she considers may be affected by the exercise of her powers under paragraph (1) before exercising those powers.

(4) Paragraph (3) does not apply in relation to any exercise of powers under paragraph (1)—

(a)   in relation to the provisions related to auctions;

(b)   for the purpose of complying with any decision taken by the Commission in relation to State Aids approval in relation to the scheme; or

(c)   in any case where the Secretary of State considers that the matter is so urgent that it is inappropriate to consult.

Access to information

G2.— (1) By participating in the scheme, participants consent to the publication by the Secretary of State of the following information.

(2) All participants consent to the publication of the following details in relation to all accounts at the Registry—

(a)   the name of the participant;

(b)   the account name;

(c)   the account number; and

(c)   the name and contact details of the principal account users; and

(d)   the transaction log described in rule B11.

(3) A direct participant consents to the publication of—

(a)   his overall and annual targets;

(b)   the number of allowances allocated to him in each commitment year under rule C5;

(c)   his verified baseline in aggregate form;

(d)   his aggregate emissions in tCO2e for each commitment year; and

(e)   whether he has complied with his annual emissions limitation commitment under rule C6(1) or to what extent he has failed to do so.

(3A) A CCA participant consents to the publication of—

(a)   the number of allowances allocated to compliance accounts held in his name under rule D4; and

(b)   the number of allowances retired from compliance accounts held in his name under rule D5.

(4) The information in paragraphs (2)(e), (3)(d) and (e) and (3A)(a) and (b) will be published annually.

(5) The Secretary of State will be entitled to disclose, without the participant's consent, any other information relating to the participant's participation in the scheme in the following circumstances—

(a)   where the disclosure is made under and in accordance with the terms of any legislation;

(b)   where the disclosure is made to a relevant authority for the purposes of the authority's functions; or

(c)   where the disclosure is required to be made in the course of legal proceedings.

(6) The Secretary of State will consult the participant before making any disclosure under paragraph (5) where she considers it appropriate in the circumstances to do so.

(7) Once the Secretary of State has disclosed information in accordance with paragraph (5) if the information is already in the public domain, she may publish the information without the consent of the participant.

(8) Any participant who objects to the release of any information (other than that falling within the scope of paragraph (2), (3) or (3A)) supplied to the Secretary of State for the purposes of the scheme must notify the Secretary of State at the time the information is supplied, setting out reasons for objecting to its release.

(9) Save as provided for by paragraphs (2), (3), (5) and (7), the Secretary of State will only disclose information relating to the participant's participation in the scheme with the consent of the participant.

(10) A relevant authority for the purpose of paragraph (5) is—

(a)   either House of Parliament (including any committee);

(b)   the European Commission;

(c)   the relevant environmental regulator under Part I of the Environmental Protection Act 1990 or regulations made under section 2 of the Pollution Prevention and Control Act 1999 or corresponding legislation for Northern Ireland;

(d)   an auditor appointed under Part 4 of Schedule 2;

(e)   an adjudicator appointed under Schedule 4.

(11) The Secretary of State will take steps to prevent any person whom she appoints under Part 4 of Schedule 2 from disclosing information relating to the participation of any person in this scheme and obtained in carrying out the audit to any one other than the Secretary of State, except to the extent needed to carry out the audit.

Dispute resolution

G3. Schedule 4 (Dispute Resolution) shall have effect.

Notices

G4.— (1) Save as otherwise provided in the scheme, all notices or other communications shall be in writing and in English and shall be given by letter delivered by hand against receipt, sent by prepaid registered or recorded delivery post, or sent by facsimile or email, and will be deemed to have been received—

(a)   in the case of delivery by hand, when delivered against receipt;

(b)   in the case of recorded delivery prepaid post, on the day following the recorded date of delivery; or

(c)   in the case of facsimile or email, on the day following acknowledgement of receipt by the addressee's facsimile receiving equipment or email system.

(2) In proving the giving of a notice it shall be sufficient to prove that the notice was left, or that the envelope containing the notice was correctly addressed and was posted, or that the facsimile was correctly addressed and was despatched and despatch of the transmission was confirmed and confirmed as having been sent to the number referred to in paragraph (3), or that the email was correctly addressed and delivery to the addressee's email address referred to in paragraph (3) was confirmed.

(3) Save as otherwise provided in the scheme, all notices or other communications shall be sent to the address, facsimile number or email address and marked for the attention of the addressee's representative as set out in the direct participant agreement or trading participant agreement.

Guidance

G5.— (1) The Secretary of State may from time to time give guidance in connection with the practical operation of the scheme.

(2) Guidance under paragraph (1) may be given from time to time in order to assist in the interpretation of the scheme.

(3) Participants, verifiers and all other persons acting under or by virtue of this scheme shall act in accordance with any guidance issued under this rule which is for the time being in force.

Transitional provisions

G6. Any action taken by the Secretary of State or any other person in anticipation of this scheme in any period before it came into force shall have the same effect as if the scheme had been in force throughout that period.

MADE ON BEHALF OF THE SECRETARY OF STATE FOR ENVIRONMENT, FOOD AND RURAL AFFAIRS

Sarah Hendry
Head of Global Atmosphere Division
(Authorised by the Secretary of State to act in this regard on her behalf)

SCHEDULE 1
INTERPRETATION

1. The Interpretation Act 1978 shall apply for the interpretation of this scheme as it applies to the interpretation of subordinate legislation.

2. Where guidance about the interpretation of this scheme has been issued under rule G5(2), the scheme shall be interpreted in accordance with that guidance.

3. In this scheme, unless the context otherwise requires, expressions used have the meaning given to them below —

'absolute sector' means one of the two parts into which the Registry is divided (the other being the relative sector) and in which trading and compliance accounts are located in accordance with rule B4(2), C4(4) or D3;
'absolute CCA target' means a CCA target which is expressed in either carbon emitted or energy used without reference to throughput;
'account' means an account in the Registry;
'account balance' has the meaning given to it in rule B10;
'allowance' has the meaning given to it in rule B1(3);
'annual emissions' means, in relation to a direct participant, the total emissions from all the sources in the participant's source list in any commitment year;
'annual emissions limitation commitment' means the obligation set out in rule C6(1);

'annual target' means, in relation to a direct participant, the participant's original annual target (being the amount by which the participant must reduce his annual emissions in each commitment year), determined in accordance with Part 1 of Schedule 2, as amended from time to time in accordance with Schedule 3;

'auction' means an auction conducted in accordance with Schedule 5;

'auction guidance' means the UK Emissions Trading Scheme Auction Guidance dated 25 January 2002 (ETS (02)02);

'baseline' means, in relation to a direct participant, the participant's original baseline calculated in accordance with Part 1 of Schedule 2, as amended from time to time in accordance with Schedule 3;

'baseline emissions' means, in relation to a source, the average annual emissions over the baseline period of that source;

'baseline period' means the three year period 1998-2000 (inclusive), or 1999-2000 (inclusive), or 2000 where a direct participant satisfies his verifier that verifiable emissions data is not available for a source for the earlier years;

'bidder's manual' means the manual distributed by the Secretary of State under paragraph 3 of Schedule 5 setting out detailed instructions on how to use the auction software system and the procedures for the electronic entry of bids in the auction;

'business day' means a day other than —

(a) a Saturday or Sunday;

(b) a Bank Holiday within the meaning of the Banking and Financial Dealings Act 1971 (other than a day which is a bank holiday only in Scotland or Northern Ireland);

'cancel' means, in relation to an allowance, the transfer of an allowance into the national cancellation account in accordance with rule B15;

'carbon dioxide equivalent' or 'CO2e' means

(a) in relation to carbon dioxide the actual quantity of those emissions; and;

(b) in relation to any other greenhouse gas the quantity of carbon dioxide which has the same global warming potential as those emissions (as specified by the Intergovernmental Panel on Climate Change in their Second Assessment Report "1995 IPCC GWP values");

'CCA overachievement' has the meaning given to it in rule D4(2);

'CCA participant' has the meaning given to it in rule D1(5);

'CCA reconciliation balance' has the meaning given to it, in relation to a target unit, in rule D5(5) and, in relation to a trading group, in rule D5(11);

'CCA reconciliation deadline' has the meaning given to it in rule D5(15);

'CCA target' means, in relation to an operator, the target set for a target unit in the relevant underlying agreement and, in relation to a sector association, the trading group target provided for in the relevant umbrella agreement;

'certification period' means, in relation to a target unit or a trading group, the certification period set out or provided for in the relevant umbrella agreement;

'change of operation' has the meaning given to it in Part 1 of Schedule 2;

'clearing price' means the price per tonne of carbon dioxide equivalent to be paid to direct participants in respect of emissions reductions set in an auction held in accordance with Schedule 5;

'climate change agreement' means an umbrella agreement or an underlying agreement;

'commitment period' means 1st January 2002 to 31st December 2006;

'commitment year' means any calendar year in the commitment period;

'compliance account' means an account at the Registry held by a direct participant or CCA participant into which allowances are allocated and from which allowances are retired by the Secretary of State through the reconciliation process and which may also be used for holding and transferring
allowances;

'direct emissions' means emissions released from a source;

'direct participant' means a party to a direct participant agreement who has taken on a target in accordance with Part C of the scheme;

'direct participant agreement' means the agreement entered into between a direct participant and the
Secretary of State by virtue of which the direct participant participates in the scheme;

'emissions' means emissions, measured in tonnes of carbon dioxide equivalent, of one or more greenhouse gases, including both direct and indirect emissions;

"Emissions Trading Directive" Directive 2003/87/EC of the European Parliament and of the Council establishing a scheme for greenhouse gas emission allowance trading within the Community and amending Council Directive 96/61/EC;

'emissions trading provisions' means those paragraphs of the relevant climate change agreement which set out how CCA targets are to be adjusted to take account of emissions trading[13];

"excluded emissions" emissions from sources in the source list which, in relation to any Schedule 1 activity, are emissions specified in that Schedule in relation to that activity, and in respect of which a certificate of temporary exclusion has been served under regulation 11 of the Greenhouse Gas Emissions Trading Scheme Regulations 2003 (S.I. 2003/3311) or an equivalent provision in any superseding legislation;

'facility' means, subject to rule D9(2), a facility to which an underlying agreement applies;

---

[13] In an underlying agreement, these are typically but not in every case paragraphs 1.3 to 1.5 of Schedule 2.

'force majeure event' means, in relation to a direct participant, trading participant or the Secretary of State, any event or circumstance beyond the reasonable control of that person and which results in or causes the failure of that person to perform any of its obligations under the scheme, including strike, lockout or other industrial disturbance, war, threat of war, terrorist act , blockade, revolution, riot, insurrection, sabotage, act of vandalism, lightning, fire, storm, flood, earthquake or drought;

'Framework Document' means the "Framework Document for the UK Emissions Trading Scheme" published by DEFRA in August 2001;

'greenhouse gases' means carbon dioxide(CO2), methane (CH4), nitrous oxide (N2O), hydrofluorocarbons (HFCs), perfluorocarbons (PFCs), and sulphur hexafluoride (SF6) and any other gas added to the list in Annex A to the Kyoto Protocol from time to time;

'incentive payment' means a payment made or to be made to a direct participant in accordance with rule C2;

'indirect emissions' means, in relation to a source, emissions attributable to the consumption of electricity which is not generated on the same site as that source;

'insolvency event' has the meaning given to it in the relevant direct participant agreement or trading participant agreement;

'Kyoto Protocol' means the Protocol to the United Nations Framework Convention on Climate Change, adopted at a Conference of the parties in Kyoto, Japan, 10 December 1997, as amended or implemented by subsequent Conferences of the Parties to the Protocol or otherwise and 'first Kyoto Protocol commitment period' means 1st January 2008 to 31st December 2012;

'legislation' means primary or secondary national legislation and includes directives and other legislation of the European Community;

"Monitoring and Reporting Decision" Commission Decision 2004/156/ EC establishing guidelines for the monitoring and reporting of greenhouse gas emissions pursuant to Directive 2003/87/EC of the European Parliament and of the Council;

'national cancellation account' means the account established under rule B15 into which allowances are transferred for the purpose of cancellation;

'national registry' means the registry established by the UK in accordance with its obligations under the Kyoto Protocol;

'national retirement account' means the account established under rule B16 into which allowances are transferred for the purpose of retirement;

'operator' means, subject to rule D9(2), a person who is a party to an underlying agreement other than the Secretary of State;

'original baseline' means, in respect of a direct participant, the total baseline emissions of all sources in a direct participant's approved source list;

'overall target' means in relation to a direct participant, the participant's original overall target, determined in accordance with paragraph 8 of

Part 1 of Schedule 2, as amended from time to time in accordance with Schedule 3;

'principal account user' means the person appointed in accordance with the user manual to act as aprincipal account user on behalf of a participant;

'queuing mechanism' means the mechanism operated under rule B14(8);

'reconciliation balance' has the meaning given to it in rule C7(2);

'reconciliation deadline' means, in respect of each commitment year, subject to rule C7(3), 31st March following the end of that commitment year and 'final reconciliation deadline' means 31st March 2007;

'reconciliation process' means the process of retiring allowances for the purpose of compliance described in rule C7 and 'final reconciliation process' means the reconciliation process which takes place after the end of the commitment period;

'Registry' means the Emissions Trading Registry established by the Secretary of State under rule B1(1);

'relative CCA target' means a CCA target which is expressed in either carbon emitted per unit of throughput or energy used per unit of throughput;

'relative sector' means one of the two parts into which the Registry is divided (the other being the absolute sector) and in which trading and compliance accounts are located in accordance with rules B4(2) and D3;

'relative sector gateway' means the restriction on transfer of allowances from the relative sector to the absolute sector which operates in accordance with rule B14;

'renewable obligation certificate' means a renewable obligation certificate issued by the Gas and Electricity Markets Authority in accordance with the relevant order under the sections 32 to 32C of the Electricity Act 1989;

'Reporting Guidelines'means the "Guidelines for the Measurement and Reporting of Emissions in the UK Emissions Trading Scheme" published by DEFRA in August 2001, as revised from time to time;

'reporting protocol' means the method approved by the Secretary of State by which a direct participant must measure and calculate his emissions;

'retire' means, in relation to an allowance, the transfer of an allowance by the Secretary of State into the national retirement account for the purpose of compliance under rules C7 and D5;

'ring-fencing' has the meaning given to it, in relation to a target unit, in rule D4(5) and, in relation to a trading group, in rule D4(11);

"Schedule 1 activity" An activity falling within a description in Schedule 1 to the Emissions Trading Directive;

'scheme' means this scheme as amended from time to time;

'Secretary of State' means the Secretary of State for Environment, Food and Rural Affairs;

'secondary account user' means the person appointed in accordance with the user manual to act as a secondary account user on behalf of a participant;

'sector association' means a person who is a party to an umbrella agreement other than the Secretary of State;

'site' has the meaning given to it in Part 1 of Schedule 2;

'source' has the meaning given to it in Part 1 of Schedule 2;

'source list' means, in relation to a direct participant, the list of sources in his approved source list, prepared in accordance with Part 1 of Schedule 2, and included in his direct participant agreement, as amended from time to time in accordance with Schedule 3;

'source list error' has the meaning given to it in Part 1 of Schedule 2;

'targets' means, in relation to a direct participant, the participant's overall target and his annual target;

'target period' means, in relation to a target unit, the target period specified in the relevant underlying agreement and, in relation to a trading group, the target period specified in the relevant umbrella agreement;

'target unit' means a facility or group of facilities with a CCA target in an underlying agreement which only applies to that facility or that group and 'target unit identifier' means the unique identifying number allocated to a target unit by the Secretary of State;

'throughput' means input or output depending on how it is calculated under the terms of the relevant climate change agreement;

'trading account' means an account in the Registry held by a trading participant or any other participant for the purposes of allocating, holding and transferring allowances;

'trading group' means a trading group whose identity is provided for in an umbrella agreement;

'trading group target' means the trading group target defined in the umbrella agreement which provides for the identity of the trading group;

'trading participant' means a person who has opened a trading account in accordance with rule B5 but is not a direct participant;

'trading participant agreement' means an agreement entered into between a trading participant and the Secretary of State by virtue of which the trading participant may hold an account at the Registry;

'transaction log' means the record of allocations, transfers, cancellations and retirements of allowances described in rule B11;

'umbrella agreement' means an umbrella agreement within the meaning of paragraph 48 of Schedule 6 of the Finance Act 2000;

'underlying agreement' means, subject to rule D9(2), an underlying agreement within the meaning of paragraph 48 of Schedule 6 of the Finance Act 2000;

'user manual' means the manual published by the Secretary of State setting out the detailed procedures for access to and use of the Registry website, as updated from time to time;

'verification opinion' means a signed opinion provided by a verifier under, in relation to a direct participant, Part 3 of Schedule 2 or, in relation to a CCA participant, paragraph 4 of Schedule 6; and a verification opinion is 'qualified' if it is submitted with qualifications about the emissions or energy use to which it refers;

'verified baseline' means a baseline verified by a verifier in accordance with Part 3 of Schedule 2 as amended in accordance with Schedule 3;

'verified emissions' means, in relation to a direct participant, emissions from the sources on his source list which are measured, verified and reported in accordance with Part 1 to 3 of Schedule 3;

'verifier' means a person accredited by the UK Accreditation Service (UKAS) to assess the accuracy of emissions and energy use in accordance with Part 3 of Schedule 2 and paragraph 4 of Schedule 6; and

'vintage' means, subject to rule D4(14) and the exception below, the year in which the allowance is issued, and the vintage of an allowance represents the first commitment year for which an allowance may be used for the purposes of complying with a direct participant's emissions limitation commitment. The exception is that all allowances allocated to direct participants under rule C5(1)(a) will have a vintage of "2002".

SCHEDULE 2

MONITORING, REPORTING AND VERIFICATION (DIRECT PARTICIPANTS)

PART 1
SOURCE LISTS, BASELINES, TARGETS

Preparation and approval of source lists
1.   A direct participant's source list must be prepared in accordance with —
    (a)   the methodology set out in the Framework Document and the Reporting Guidelines;
    (b)   Schedule 3, in relation to any change of operation during the baseline period; and
    (c)   any guidance provided by the Secretary of State from time to time.

2.—   (1) A direct participant must submit his source list to the Secretary of State for approval.

(2) The participant must keep a written record of—

(a)   the methodology used in preparing his source list; and

(b)   the decisions taken when preparing his source list, including the reasons for excluding particular sources from the list.

(3) The record must be submitted to the Secretary of State at the same time as the participant's source list.

(4) A direct participant must provide the Secretary of State with such additional information as she requires in relation to the approval of the source list.

3.—   (1) For the purposes of this Schedule—

(a)   'point source' means any separately identifiable point from which greenhouse gases are emitted;

(b)   'site' means (subject to sub-paragraphs (2) and (3) below)—

(i)   a building or other substantial structure; or

(ii)   a stationary technical unit; and

(c)   'source' means a point source or a collection of point sources of the same type on the same site.

(2) A participant may determine to include within a site any additional building, structure or plant in which a directly associated activity is carried out which has a technical or functional connection with the activities carried out in the building, structure or unit forming the site.

(3) A record of any determination under sub-paragraph (2) must be kept and submitted to the Secretary of State under paragraph 2(2) and (3).

(4) The Secretary of State may in approving any source list under paragraph 4 of this Schedule modify or depart from the definition of "site" in sub-paragraphs (1)(b) and (2) if she considers that—

(a)   there would be practical difficulty in applying the definition in the circumstances of that case; or

(b)   it is appropriate to do so to protect the environmental integrity or viability of the scheme or to prevent potential abuse of the scheme.

4.—   (1) Where the Secretary of State approves a direct participant's source list, she will confirm her approval in writing to the participant.

(2) The Secretary of State's approval of the participant's source list may be subject to instructions to the direct participant about how the baseline must be calculated and may include additional comments, information or explanation needed to identify and describe the relevant sources (including any determination made

by the participant under paragraph 3(2) or any modification or departure from the definition of site under paragraph 3(4)).

(3) Once approved by the Secretary of State, a source list may not be amended otherwise than in accordance with Schedule 3.

Calculation of original baseline

5.— (1) A direct participant must calculate his original baseline on the basis of the aggregate baseline emissions from all sources in his approved source list and in accordance with—

(a) the methodology set out in the Framework Document and in the Reporting Guidelines and associated protocols;

(b) any instructions or comments made in the source list approval; and

(c) Schedule 3 in relation to—

(i) any change of operation during the baseline period or since the date of the participant's source list approval; or

(ii) the discovery of any source list error since the date of the participant's source list approval; and

(d) any guidance given by the Secretary of State from time to time.

(2) In complying with sub-paragraph (1), a direct participant must—

(a) identify the appropriate protocol for calculating all emissions relevant to his source list;

(b) use in his calculations the appropriate emissions factors (default, activity-specific, etc.) where more than one factor is set out in the relevant protocol; and

(c) use the activity data (fuel used, production input or output etc.) appropriate to enable him to undertake the calculations envisaged in the relevant protocol.

(3) A direct participant's baseline may not be amended otherwise than in accordance with Schedule 3.

Targets

6.— (1) A direct participant's overall target will be determined through an auction held by the Secretary of State in accordance with Schedule 5.

(2) The participant's annual target will be calculated by dividing his overall target by the number of commitment years to which it refers (i.e. five years for all direct participants entering the scheme in the first commitment year).

(3) A direct participant's targets may not be amended otherwise than in accordance with Schedule 3.

Amendment of source lists, baselines and targets

7.— (1) Where a change of operation occurs, or a direct participant becomes aware of a source list error, during any commitment year, he must in the circumstances specified in Schedule 3—

(a) amend his source list, baseline and targets; and

(b) make any associated requests to transfer or cancel allowances, in accordance with that Schedule, before the end of that commitment year, except for where the change of operation occurred in 2004, in which case, on or before 20th January 2005.

(3) A 'change of operation' means any of the circumstances described in Part 1 of Schedule 3.

(4) A 'source list error' occurs where a source which has been included in or omitted from a source list incorrectly.

(5) A source has been—

(a) incorrectly included where it would have been omitted had the source list been properly prepared in accordance with paragraph 1; and

(b) incorrectly omitted where it would have been included had the source list been properly prepared in accordance with paragraph 1.

(6) Except where paragraph (7) applies, for the purpose of calculating a direct participant's annual emissions under paragraph 8 of this Schedule, any amendments made to a direct participant's source list, baseline and targets pursuant to Schedule 3 shall take effect from 1st January of the commitment year in which the change of operation took place or the source list error was discovered.

(7) Where a change of operation takes place after 30th June in any commitment year, for the purpose of calculating a direct participant's annual emissions under paragraph 8 of this Schedule, any amendments made to a direct participant's source list, baseline and targets pursuant to Schedule 3 shall take effect from 1st July of that commitment year.

PART 2
ANNUAL EMISSIONS

Calculation of annual emissions

8.— (1) A direct participant must calculate his annual emissions for each commitment year , in accordance with—

(a) the methodology set out in the Framework Document and the Reporting Guidelines and associated protocols;

(b)   Schedule 3, in relation to any change of operation or discovery of any source list error during that commitment year; and

(c)   any guidance given by the Secretary of State from time to time.

(2) In complying with sub-paragraph (1), a direct participant must—

(a)   identify the appropriate protocol for calculating all emissions relevant to his participation in the scheme;

(b)   use in his calculations the appropriate emissions factors (default, activity-specific, etc.) where more than one factor is set out in the relevant protocol; and

(c)   record and check the activity data (fuel used, production input or output etc.) necessary to enable him to undertake the calculations envisaged in the relevant protocol.

PART 3
VERIFICATION AND REPORTING

Verification of original baseline

9.—   (1) A direct participant must have his original baseline verified by a verifier in accordance with the following provisions of this paragraph.

(2) A direct participant must provide the verifier with the following information—

(a)   his approved source list, a copy of his source list approval and any supporting information provided to the Secretary of State;

(b)   his baseline emissions data, together with details of the procedures, decisions, measurements and calculations relevant to the calculation of the baseline under paragraph 5;

(c)   details of—

(i)   any change of operation which has taken place during the baseline period or since the date of his source list approval;

(ii)   any source list error which has been discovered since the date of his source list approval;

(iii)   any consequential amendments made to his source list, baseline and targets in accordance with Schedule 3;

(d)   any other information the participant is required to provide to his verifier under Schedule 3; and

(e)   such other information as the verifier requires.

(3) The direct participant must make a written declaration to the verifier confirming that the baseline emissions data and other

information submitted is, to the best of the participant's knowledge, accurate and complete.

(4) The declaration must be signed by the direct participant or a director or manager or other designated representative of the direct participant.

10.— (1) The verifier must assess the direct participant's original baseline on the basis of the source list approval (including any aggregation of point sources approved by the Secretary of State in the source list approval) and in accordance with—

(a)   the Framework Document and the Reporting Guidelines and associated protocols;

(b)   guidance to verifiers issued by UKAS;

(c)   Schedule 3; and

(d)   any other guidance given by the Secretary of State from time to time.

(2) The verifier must take account of all relevant information (whether or not provided by the participant).

(3) The direct participant must make any corrections to his baseline required by the verifier to correct any material error found by the verifier.

(4) The direct participant must require his verifier to complete a verification opinion in relation to his original baseline in accordance with the UKAS guidance to verifiers.

Reporting of original baseline

11.— (1) A direct participant must submit to the Secretary of State a statement of baseline emissions together with a signed verification opinion in relation to his original baseline, before he will be allocated allowances under rule C5 for the first commitment year.

(2) The statement of baseline emissions must set out the following information—

(a) the direct participant's verified original baseline;

(b) details of—

(i) any change of operation during the baseline period or since the date of the participant's source list approval letter;

(ii) any source list error which has been discovered since the date of the participant's source list approval letter; and

(iii) any consequential amendments to the participant's source list, baseline and targets made in accordance with Schedule 3;

(c) any other information which the participant is required to provide to the Secretary of State under Schedule 3; and

(d) any other information reasonably requested by the Secretary of State.

Verification of annual emissions

12. A direct participant must have the annual emissions , and any amendments to his baseline and targets, for each commitment year, verified by a verifier in accordance with the paragraphs 13 to 16.

13.— (1) The participant must provide the verifier with the following information—

    (a)   his approved source list, a copy of his source list approval and any supporting information from the Secretary of State, and his verified baseline;

    (b)   the total annual emissions from the sources on his source list, together with details of the data, procedures, decisions, measurements and calculations on which this figure is based;

    (c)   details of—

    (i)   any change of operation which took place during the relevant commitment year;

    (ii)   any source list error which has been discovered during the relevant commitment year; and

    (iii)   any consequential amendments made to his source list, baseline and targets in accordance with Schedule 3;

    (d)   any other information which the participant is required to provide to the verifier under Schedule 3; and

    (e)   such other information as the verifier requires.

(2) The direct participant must make a written declaration to the verifier confirming that the annual emissions data and other information submitted is, to the best of the participant's knowledge, accurate and complete.

(3) The declaration must be signed by the direct participant or a director or manager or other designated representative of the direct participant.

14.— (1) The verifier must assess the direct participant's emissions data for the relevant commitment year on the basis of his current source list and in accordance with—

    (a)   the Framework Document and the relevant Reporting Guidelines and associated protocols;

    (b)   guidance to verifiers issued by UKAS;

    (c)   Schedule 3; and

    (d)   any other guidance given by the Secretary of State from time to time.

(2) The verifier must take account of all relevant information (whether or not provided by the participant) when he is verifying the participant's annual emissions and any amendments to the participant's baseline or targets.

Reporting of annual emissions

15.— (1) A direct participant must submit to the Secretary of State a statement of his annual emissions together with a signed verification opinion in relation to his annual emissions, by the reconciliation deadline for each commitment year.

(2) The statement of annual emissions must set out the following information—

(a)   total emissions for the relevant commitment year;

(b)   details of—

(i)   any change of operation which took place during the relevant commitment year;

(ii)   any source list error which was discovered during the relevant commitment year;

(iii)   any consequential amendments to the participant's source list, baseline and targets made in accordance with Schedule 3;

(iv)   any associated transfer, cancellation or allocation of allowances;

(c)   any other information which the participant is required to provide to the Secretary of State under Schedule 3; and

(d)   any other information reasonably requested by the Secretary of State.

(3) In addition to the foregoing obligations, a direct participant shall separately monitor his excluded emissions in accordance with the Monitoring and Reporting Decision.

Refusal to issue a verification opinion

16. A direct participant must notify the Secretary of State if any verifier employed by him has at any time refused to issue a verification opinion in any commitment year and must provide the Secretary of State with any reasons given by the verifier for the refusal.

PART 4
RETENTION OF AND ACCESS TO INFORMATION

Retention of information

17.— (1) A direct participant must retain proper records relating to his participation in the scheme sufficient to demonstrate his compliance with the requirements of this Schedule.

(2) These records must be sufficient to allow a third party to audit the participant's baseline and annual emissions at any time during the operation of the scheme.

(3) In particular, a direct participant must retain—

(a)   evidence of the decisions taken in relation to the compilation of his source list and the calculation of his baseline;

(b)   the activity data and other information on which any calculation of the emissions from the sources in his source list was based, including documents justifying the selection of protocols and emissions factors used in such calculations;

(c)   any information relevant to adjustments to his baseline, source list and target which take place in accordance with Schedule 3, including information used to decide whether or not such adjustments need to be made; and

(d)   information and data not required under sub-paragraphs (a) to (c) but which is required by the verifier to verify the data submitted by the participant to the Secretary of State.

(4) All such records must be retained for the purposes of compliance with this Schedule for at least two years after the end of the final commitment year.

(5) Where the volume of activity data makes such retention impractical, a participant may retain the information in an aggregated form following satisfactory verification of the data, provided that this satisfies the condition in sub-paragraph (2).

(6) A direct participant must record and manage the information retained under this rule in a systematic manner and must employ an effective data management system as provided for in section 6 of the Reporting Guidelines.

Investigation and audit

18. A direct participant must comply with any request from the Secretary of State for information in connection with his compliance with the requirements of this Schedule (including information provided to verifiers in connection with the verification of his baseline or annual emissions) within such period as may be specified by the Secretary of State (not being less than 10 working days).

19. The Secretary of State may appoint an auditor to undertake an independent audit of the information provided by the direct participant, and the participant must cooperate with the auditor (including keeping proper records and making them available for inspection when required by the auditor).

20.— (1) Without prejudice to the foregoing provisions, a direct participant shall provide to the Secretary of State any data which—

(a)   relates to emissions from any source which is or has at any time been included in the direct participant's source list; and

(b)   has been requested in writing by the Secretary of State, within 20 working days of the Secretary of State making the request.

(2) If a direct participant fails to comply with paragraph (1) but within 20 days of the request notifies the Secretary of State that there is good reason for his failure to comply, and provides her with satisfactory evidence in support, the Secretary of State shall take this into account in deciding how to exercise her powers under rule C10.

SCHEDULE 3

CHANGE OF OPERATION AND SOURCE LIST ERRORS

PART 1
CHANGE OF OPERATION

Transfer of management control over sources between direct participants

1.—   (1)Where a direct participant divests management control over a source in his source list to another direct participant, paragraph 2 shall apply.

(2) "Management control", in relation to a source, has the meaning given in Annex A to the Framework Document.

(3) A direct participant divests management control over a source in his source list where, due to a sale, lease or restructuring of his operations, or the out-sourcing of an activity, or any similar transaction, he no longer exercises management control over that source.

(4) A direct participant acquires management control over a source where, due to a purchase, lease or restructuring of his operations, or the in-sourcing of an activity, or any similar transaction, he starts to exercise management control over that source.

(5) The divestment or acquisition of management control over a source shall be deemed to have occurred on—

(a)   the contractual date of completion;

(aa)   where management control of the source is divested by virtue of paragraph 3(1)(d), 1st January 2005 or, if later, the date upon which the Schedule 1 activity commenced.

(b) where paragraphs (a) and (aa) does not apply, the date on which the divestment or acquisition took legal effect; or

(c) where paragraphs (a), (aa) and (b) do not apply, the date on which the divestment or acquisition took practical effect.

2.— (1) If the change threshold is triggered in relation to the participant who divested the source, the participant's source list, baseline and targets shall be amended in accordance with paragraph 17.

(2) If the change threshold is triggered in relation to the participant who acquired the source, that participant's source list, baseline and targets shall be amended in accordance with paragraph 18.

(3) Where the change threshold is triggered in relation to either or both participants, and allowances have been allocated for the commitment year in which the source was divested, then—

(a) if the change threshold is triggered in relation to both participants, the divesting participant must transfer to the acquiring participant the number of allowances calculated in accordance with paragraph 19;

(b) if the change threshold is triggered only in relation to the divesting participant, then that participant must notify the Secretary of State of the fact, and on receipt of such notification, the Secretary of State shall cancel the number of allowances calculated in accordance with paragraph 19;

(c) if the change threshold is triggered only in relation to the acquiring participant—

(i) that participant must notify both his verifier and the Secretary of State, as part of his verification and reporting obligations under Part 3 of Schedule 2; and

(ii) the Secretary of State will allocate to the participant allowances equivalent to the number calculated under paragraph (a) above.

(4) If the change threshold is not triggered in relation to a participant, then—

(a) the participant's source list, baseline and targets shall not be amended on that occasion; but

(b) if the change threshold is triggered by a subsequent divestment or acquisition, the amendments that would have been made if the change threshold had been triggered on the first occasion shall then be made (along with any other amendments required by this Schedule).

(5) The "change threshold", in this paragraph and in paragraphs 4, 11 and 12, has the meaning given in paragraph 16, and the change threshold is triggered in the circumstances set out in that paragraph.

Divestment of management control over sources other than to a direct participant

3.— (1) Where—

    (a)   a direct participant divests management control over a source in his source list to a person who is not a direct participant;

    (b)   a direct participant closes a source in his source list;

    (c)   a direct participant's joint-venture partner withdraws consent to the inclusion in the scheme of a source in his source list, or

    (d)   a source, whose emissions are not excluded emissions, is an installation, or part of an installation, carrying out a Schedule 1 activity, paragraphs 4 to 6 shall apply.

(2) For the purposes of this Schedule, each of the circumstances described in sub-paragraph (1)(b), (c) and (d) shall be treated as a "divestment" of the source.

(3) Paragraph 1(2), (3) and (5) shall apply in relation to sub-paragraph (1).

(4) A direct participant "closes" a source where the operation of that source ceases, and the source shall be deemed to be closed on that date.

4.— (1) If the change threshold is triggered in relation to the direct participant—

    (a)   the participant's source list, baseline and targets shall be amended in accordance with paragraph 17; and

    (b)   if allowances have been allocated for the commitment year in which the source was divested, the participant shall notify the Secretary of State of the fact, and on receipt of such notification, the Secretary of State shall cancel the number of allowances calculated in accordance with paragraph 19.

(2) If the change threshold is not triggered in relation to the direct participant, then—

    (a)   the participant's source list, baseline and targets shall not be amended on that occasion; but

    (b)   if the change threshold is triggered by a subsequent divestment or acquisition, the amendments that would have been made if the change threshold had been triggered on the first occasion shall then be made (along with any other amendments required by this Schedule).

5.— (1) Where paragraph 3(1)(a) applies, the person who acquired the source may apply to enter the scheme as a direct participant in relation to that source.

(2) Before he may participate as a direct participant in the scheme, the person must—

(a)   submit a direct participant application to the Secretary of State stating that he wishes to take over the direct participant's source list, baseline and targets in relation to that source; and

(b)   if the application is successful, enter into a direct participant agreement with the Secretary of State.

(3) The source list, baseline and targets of a successful applicant under this paragraph shall be determined as follows—

(a)   his source list shall comprise the acquired source;

(b)   his baseline shall be the baseline emissions of the acquired source; and

(c)   his targets shall be the reduction in the targets of the divesting participant.

6.—   (1) Where paragraph 3(1)(c) applies, the direct participant must notify both his verifier and the Secretary of State as part of his verification and reporting obligations under Part 3 of Schedule 2 for the commitment year in which the divestment occurred.

(2) The notice must—

(a)   identify the relevant source;

(b)   state that the joint-venture partner has withdrawn consent to the inclusion of the source in the Scheme; and

(c)   be signed by the joint-venture partner.

Substantial closure or divestment of a source

7.—   (1) Where a direct participant—

(a) divests a substantial part of a source in his source list to a person who is not a direct participant; or

(b) closes a substantial part of a source in his source list (including as a result of a force majeure event),

the Secretary of State may require the participant to treat the divestment or closure in the same way as the divestment or closure of a source under paragraph 3(1)(a) or (b), where she considers it appropriate having regard to the purposes of the scheme set out in rule A1.

(2) In such cases, the part of the source that has been closed or divested shall be treated as a "source" for the purposes of this Schedule.

(3) The Secretary of State may issue guidance from time to time on the application of this paragraph.

(4) A direct participant may seek an opinion from the Secretary of State regarding the application of this paragraph in any particular case.

(5) Where this paragraph applies, the direct participant must notify and provide details to both his verifier and the Secretary of State as part of his verification and reporting obligations under Part 3 of Schedule 2 for the commitment year in which the closure or divestment occurred.

(6) The notice must—

(a)   identify the relevant source;

(b)   describe the activity performed by the source;

(c)   where applicable, describe the effect of the force majeure event on the source; and

(d)   attach a copy of any opinion received from the Secretary of State pursuant to subparagraph (4) above.

Substantial increase in emissions from a source

8.—   (1) Where the quantity of emissions from a source in a direct participant's source list substantially increases as a result of a force majeure event, the Secretary of State may allow the participant to treat the affected source in the same way as the closure of a source under paragraph 3(1)(b), where it would be unreasonable to require the participant to retain that source in his source list.

(2) The Secretary of State may issue guidance from time to time on the application of this paragraph.

(3) A direct participant may seek an opinion from the Secretary of State regarding the application of this paragraph in any particular case.

(4) Where this paragraph applies, the direct participant must notify and provide details to both his verifier and the Secretary of State as part of his verification and reporting obligations under Part 3 of Schedule 2 for the commitment year in which the force majeure event occurred.

(5) The notice must—

(a)   identify the relevant source;

(b)   describe the activity performed by the source;

(c)   describe the effect of the force majeure event on the source; and

(d)   attach a copy of any opinion received from the Secretary of State pursuant to subparagraph (3) above.

Acquisition of management control over sources other than from a direct participant

9.—   (1) Where a direct participant—

(a)   acquires management control over a source from a person who is not a direct participant; or

(b)   opens a new source, the participant's source list, baseline and targets shall not be amended.

(2) For the purposes of this Schedule, the opening of a new source shall be treated as an "acquisition" of the source.

(3) A "new source" means a source that did not exist until after 1st January 2000, and a direct participant shall be deemed to have opened a new source on the date the source becomes operational.

Acquisition by a direct participant of substitute sources

10.— (1) Where a direct participant divests management control over a source in his source list in the circumstances described in paragraphs 1 or 3, but continues the activity performed by the divested source by acquiring a substitute source, then paragraphs 11 to 13 shall apply.

(2) Subject to sub-paragraph (3), for the purposes of sub-paragraph (1) above, a direct participant acquires a substitute source by—

(a)   acquiring management control over a source in the circumstances described in paragraph 1 or 9;

(b)   re-opening a closed source;

(c)   resuming the normal operation of a source affected by a force majeure event and falling under paragraph 8(1); or

(d)   any similar transaction or restructuring of operations.

(3) For the purposes of this paragraph, if a direct participant opens or re-opens a source, then insofar as the source emits, in relation to any Schedule 1 activity, any of the greenhouse gas emissions specified in that Schedule in relation to that activity, then the direct participant shall not be taken as acquiring a substitute source.

11.— (1) If the divestment under paragraph 10(1) triggers the change threshold in relation to the direct participant, then (whether the substitute source was acquired in the same commitment year or a subsequent commitment year) sub-paragraphs (2) and (3) shall apply.

(2) In respect of the divested source—

(a)   the participant's source list, baseline and targets shall be amended in accordance with paragraph 17;

(b)   if allowances have been allocated for the commitment year in which the source was divested—

(i)   if the source was divested to another direct participant, then paragraph 2(3)(a) and (b) applies.

(ii)   if the source was divested other than to a direct participant, then paragraph 4(1)(b) applies.

(3) In respect of the substitute source, if the acquisition of the source triggers the change threshold in relation to the direct participant—

(a)   the participant's source list, baseline and targets, shall be amended as follows—

(i)   add the substitute source to the participant's source list;

(ii)   add back into the participant's baseline the amount subtracted under sub-paragraph (2)(a);

(iii)   add back into the participant's targets the amounts subtracted under sub-paragraph (2)(a).

(b)   if allowances have been allocated for the commitment year in which the substitute source was acquired—

(i)   if the substitute source was acquired from another direct participant and the change threshold is triggered in relation to that other (divesting) participant—

(aa)   the divesting participant shall transfer to the participant the number of allowances calculated in accordance with paragraph 19; and

(bb)   where necessary, the participant shall notify the Secretary of State of the number of excess allowances, and on receipt of such notification, the Secretary of State shall cancel those allowances, or additional allowances will be allocated to the participant by the Secretary of State, so that his total allowance allocation for that commitment year under rule C5(2) reflects the acquisition of the substitute source;

(ii)   if the substitute source is acquired from another direct participant but the change threshold is not triggered in relation to that other (divesting) participant, the Secretary of State will allocate allowances to the participant so that his total allowance allocation for that commitment year under rule C5(2) reflects the acquisition of the substitute source;

(iii)   if the substitute source is acquired other than from a direct participant, then the Secretary of State will allocate allowances to the participant so that his total allowance allocation for that commitment year under rule C5(2) reflects the acquisition of the substitute source.

12.—   (1) If the divestment under paragraph 10(1) does not trigger the change threshold in relation to the direct participant, then (whether the substitute source was acquired during the same commitment year or a subsequent commitment year)—

(a)   the participant's source list, baseline and targets shall not be amended on that occasion; but

(b)   if the change threshold is triggered by a subsequent divestment or acquisition, the amendments that would have been made if the change threshold had been triggered on the first occasion shall then be made (along with any other amendments required by this Schedule).

(2) If the acquisition under paragraph 10(2) does not trigger the change threshold in relation to the direct participant, then—

(a)   the participant's source list, baseline and targets shall not be amended on that occasion; but

(b)   if the change threshold is triggered by a subsequent divestment or acquisition, the amendments that would have been made if the change threshold had been triggered on the first occasion shall then be made (along with any other amendments required by this Schedule).

13.—   (1) Where paragraph 10 applies, the direct participant must notify both his verifier and the Secretary of State, as part of his verification and reporting obligations under Part 3 of Schedule 2 for the commitment year in which the substitute source was acquired.

(2) The notice must—

(a)   identify the divested source;

(b)   describe the activity performed by that source;

(c)   state that the activity has been continued by a substitute source; and

(d)   identify the substitute source acquired.

13A.—(1) If, in relation to any direct participant, the Secretary of State is unable to comply with her obligations to cancel allowances under this Schedule due to there not being sufficient allowances in the participant's compliance account she shall notify the participant of—

(a)   the number of allowances that she was required to cancel under this Schedule;

(b)   the number of allowances that she has cancelled under this Schedule; and

(c)   the number of outstanding allowances.

(2) A direct participant shall cancel the number of outstanding allowances within 21 days of the Secretary of State sending a notice to him under sub-paragraph (1).

(3) If a direct participant fails to comply with paragraph (2), the direct participant shall be deemed to have failed to comply with his emissions limitation commitment as set out in rule C6(1) for that commitment year, and his excess emissions for the purpose of rule C6 shall be increased by the number of outstanding allowances.

(4) For the purposes of this paragraph, "outstanding allowances" means the number of allowances that the Secretary of State is obliged to cancel under this Schedule in relation to a direct participant minus the number of allowances cancelled by the Secretary of State pursuant to that obligation, and the number of allowances cancelled in accordance with sub-paragraph (2).

PART 2
SOURCE LIST ERRORS

Incorrectly included sources
14.— (1) If, at any time after the approval of his source list, a direct participant becomes aware that a source has been incorrectly included in his approved source list —
- (a) the participant's source list, baseline and targets shall be amended in accordance with paragraph 17; and
- (b) if allowances have been allocated for the commitment year in which the amendments are made, the direct participant must cancel the number of allowances calculated in accordance with paragraph 19, as if the incorrectly included source were a divestment of a source.

(2) The definition of an "incorrectly included" source in paragraph 7(5)(a) of Schedule 2 applies in respect of sub-paragraph (1).

Incorrectly omitted sources
15.— (1) If, at any time after the approval of his source list, a direct participant becomes aware that a source has been incorrectly omitted from his approved source list —
- (a) subject to sub-paragraph (2), the participant's source list and baseline shall be amended in accordance with paragraph 18(2) and (3), as if the incorrectly omitted source were an acquisition of a source; but
- (b) the participant's targets shall not be amended.

(2) The participant's baseline shall not be amended if the baseline emissions from the omitted source are less than the size threshold (being the lesser of 10,000 tCO2e or 1% of the participant's approved source list).

(3) The definition of an "incorrectly omitted" source in paragraph 7(5)(b) of Schedule 2 applies in respect of sub-paragraph (1).

PART 3
CALCULATING AMENDMENTS TO SOURCE LIST, BASELINE AND TARGETS

Change Threshold

16.— (1) The change threshold referred to in Part 1 of this Schedule is the lesser of 25,000 tCO2e or 2.5% of the direct participant's verified original baseline.

(2) The change threshold is triggered in relation to a direct participant where the baseline emissions of a source divested or acquired by the participant, combined with the aggregate baseline emissions of any other sources previously divested or acquired by the participant, equals or exceeds the change threshold, expressed as follows:

$$\left| \sum_{i=1}^{n} \left(A_i + D_i \right) \right| \geq CT$$

where—

$| \; |$ means the absolute value of the quantity;

$\sum$ means the sum of;

$\sum_{i=1}^{n} A_i$ is the sum of the baseline emissions from all acquired sources (a positive figure);

$\sum_{i=1}^{n} D_i$ is the sum of the baseline emissions from all divested sources (a negative figure).

Amendments to source list, baseline and targets upon divestment of a source

17.— (1) Where a direct participant divests a source, his source list, baseline and targets shall be amended where required by this Schedule as follows.

(2) Remove the divested source from the participant's source list.
(3) Calculate the participant's amended baseline (B_1) using the
following formula—

$$B_1 = B_0 - B_S$$

where—
B_0 is the participant's baseline;
B_S is the baseline emissions of the divested source.

(4) Calculate the participant's amended overall target (OT_1) using
the following formula—

$$OT_1 = OT_0 - OT_S$$

where—
OT_0 is the participant's overall target;
OT_S is the portion of OT_0 associated with the divested source,
calculated as follows—

$$OT_S = OT_0 \times \frac{B_S}{B_0}$$

(5) Calculate the participant's amended annual target (AT_1) using
the following formula—

$$AT_1 = \frac{OT_1}{Y}$$

where—
Y is the number of commitment years to which the overall target
refers (i.e. five for direct participants who entered the scheme in
the first commitment year).
(6) Where a direct participant divests management control of a
source after 30th June in a commitment year the calculations in
sub-paragraphs (3) to (5) shall be performed in two stages—
(a)   the calculations to determine the amended baseline and
        targets for the commitment year in which the source was
        divested shall be performed as set out in sub-paragraphs
        (3) to (5) but using "½BS" instead of "BS", and
(b)   the calculations to determine the baseline and targets for
        subsequent years shall be performed as set out in sub-
        paragraphs (3) to (5).".

Amendments to source list, baseline and targets upon acquisition of a source

18.— (1) Where a direct participant acquires a source from another direct participant, his source list, baseline and targets shall be amended where required by this Schedule as follows.

(2) Add the acquired source to the participant's source list.

(3) Calculate the participant's amended baseline (B_1) using the following formula—

$$B_1 = B_0 + B_S$$

where—

B_0 is the participant's baseline;

B_S is the baseline emissions of the acquired source.

(4) Calculate the participant's amended overall target (OT_1) using the following formula—

$$OT_1 = OT_0 = OT_S$$

where—

OT_0 is the acquiring participant's overall target;

OT_S is the portion of the divesting participant's OT_0 associated with the acquired source (calculated under paragraph 17(4)).

(5) Calculate the participant's amended annual target (AT_1) as follows—

$$AT_1 = \frac{OT_1}{Y}$$

where—

Y is the number of commitment years to which the overall target refers (i.e. five for direct participants who entered the scheme in the first commitment year).

(6) Where a direct participant acquires a source after 30th June in a commitment year the calculations in sub-paragraphs (3) to (5) shall be performed in two stages—

(a) the calculations to determine the amended baseline and targets for the commitment year in which the source was acquired shall be performed as set out in sub-paragraphs (3) to (5) but using "$\frac{1}{2}B_S$" instead of "B_S", and

(b) the calculations to determine the baseline and targets for subsequent years shall be performed as set out in sub-paragraphs (3) to (5).".

Transfer or cancellation of allowances

19.– (1) The number of allowances to be transferred or cancelled by a direct participant under paragraphs 2(3), 4, 11 and 14 shall be calculated as follows—

$$N = A \times \frac{B_S}{B_0}$$

where—

N is the number of allowances (which shall be rounded to the nearest whole number);

A is the participant's total allowance allocation, prior to the divestment, for the commitment year in which the source was divested;

B_S is the baseline emissions of the divested source;

B_0 is the participant's baseline.

(2) Where a direct participant divests a source after 30th June in a commitment year, the calculation in sub-paragraph (1) shall be performed using "$\frac{1}{2}B_S$" instead of "B_S"

SCHEDULE 4
DISPUTE RESOLUTION PROCEDURE

Scope

1.— (1) This Schedule sets out the procedure to be adopted in the event of any dispute that may arise out of or relate to the scheme, a direct participant agreement or a trading participant agreement (other than a dispute relating to Schedule 5 (Auction)).

(2) For the purposes of this schedule, 'parties' means the Secretary of State and the participant or participants who are aggrieved by or seeking to challenge an act, omission or decision of the Secretary of State.

Mediation

2. The parties will use their best efforts to discuss in good faith and to settle any dispute subject to this Schedule prior to referring the matter to an adjudicator under paragraph 3.

Adjudication

3.— (1) If any dispute cannot be resolved through discussion and agreement under paragraph 2, either party may refer it to an appointed adjudicator who is appropriately qualified to consider the dispute in question.

(2) In default of agreement between the parties as to the appointment of an adjudicator, the Secretary of State will nominate an adjudicator who in her opinion is appropriately qualified.

(3) The liability to pay the costs of the appointment of an adjudicator and of the adjudication proceedings shall be determined by the adjudicator whose decision in that regard shall be binding on the parties, but not the legal and professional costs of either party, which shall be paid by the party incurring them.

4. Each party will have 20 working days from the date that it receives such notice to make representations to the adjudicator, and must copy any representations made to the other party.

5. The nominated adjudicator will determine the procedure which is to be followed in relation to the adjudication, but in doing so—

(a) he must take account of the wishes of the parties and the scheme timetable;

(b) he may take expert advice, after consulting the parties, to assist him with the adjudication;

(c) he must ensure that each party has an adequate opportunity to respond to any representations made by the other party, and any evidence that he proposes to take into account;

(d) he may request further information from either party to the dispute, in which case the information provided must be copied to the other party;

(e) he may hold an oral hearing into the matter and decide on the procedure for such a hearing; and

(f) he may impose or extend any time limit for any action to be taken by either party and may proceed with the adjudication in such manner as he considers appropriate in the circumstances if a time limit is not complied with.

6.— (1) The adjudicator will as a condition of his appointment undertake, on the basis of the representations provided to him and any additional information he considers to be relevant, to make a report, giving his findings of fact, his reasons and his recommendation.

(2) The adjudicator will send a copy of his report to each party.

Confidentiality

7.— (1) Unless otherwise agreed any adjudication will be conducted and any adjudication report issued on an open basis.

(2) If all parties consent—

    (a)   the discussions and (with the consent of the adjudicator) the adjudication or any part of them will be conducted; and

    (b)   the adjudication report will be issued in private and upon such terms as may be agreed as to the non-disclosure of any matter.

Consequences

8.— (1) To the extent that any dispute involves a challenge to any decision, act or omission of the Secretary of State, she will reconsider such decision, act or omission in the light of the adjudicator's report.

(2) The Secretary of State will notify any participant affected of the action she proposes to take upon reconsideration and of her reasons.

9. Subject to paragraphs 3(3) and 8 above, the adjudicator's report and recommendation will not be binding on either party, and each party will be free to accept or reject the adjudicator's report or recommendation and to commence any legal proceeding in respect of the matters in dispute.

10. If the parties accept the adjudicator's recommendation or otherwise settle the dispute between them, such agreement shall be recorded in writing and once signed on their behalf shall be final and binding on the parties.

11. Subject to paragraph 7(2), either party may refer to and produce the adjudicator's report in evidence in any subsequent legal proceedings relating to the dispute.

12. Nothing in this Schedule shall restrict either party's freedom to commence legal proceedings if necessary to preserve confidentiality or any proprietary right or remedy.

SCHEDULE 5

AUCTION

Purpose

1. The purpose of the auction is to provide a procedure to determine—
(a) which of the direct participants are to continue in the scheme;
(b) the overall target of each of the direct participants; and

(c) the clearing price (used to calculate the amount of incentive payments under rule C2(1)).

Timing

2.—(1) The auction will start on the auction date and continue until the Secretary of State announces that the clearing price has been reached in accordance with paragraph 12(2).

(2) Direct participants will be notified of the auction date and the schedule of auction rounds in accordance with paragraph 3.

(3) The Secretary of State may change the auction schedule without the need for any consultation and will notify direct participants of any changes by announcing it on the auction software system or by any other means she considers appropriate.

Preparation for the auction

3. By the later of 4 days before the auction date or 2 business days after a direct participant enters into a direct participant agreement, the Secretary of State will notify the direct participant of—

(a) the auction date;

(b) the auction schedule setting out the dates and times at which the first round and any subsequent rounds will start and finish;

(c) the internet address of the auction software system;

(d) the unique auction logins and auction passwords which the direct participant will use to access the auction software system and participate in the auction;

(e) a bidder's manual setting out detailed instructions on how to use the auction software system and the procedures for the electronic entry of bids;

(f) the Secretary of State's telephone and fax numbers for the purpose of alternative bidding arrangements under paragraph 10; and

(g) the telephone number of a dedicated auction help line.

Structure of the auction

4.—(1) Subject to paragraph 10, the auction will be conducted electronically on the auction software system in accordance with this Schedule and the auction guidance[14].

(2) The auction will consist of a series of rounds and shall start with the first round on the date and at the time stated in the auction schedule notified to the direct participant under paragraph 3.

(3) Each round will consist of the period of time announced in the auction schedule during which all direct participants must enter a valid bid in accordance with the auction guidance and the bidder's manual or retire from the auction.

---

[14] UK Emissions Trading Scheme Auction Guidance dated 25 January 2002 (ETS (02)02).

(4) The budget for the auction (as defined in the auction guidance) will be determined by the Secretary of State after the first round of the auction and will be announced on the auction software system.

Access to the auction software system

5.—(1) Direct participants may only access the auction software system using the technology specified in the bidder's manual and are responsible for making sure that their computers and other equipment can be used with the auction software system.

(2) Direct participants will be provided with user names and passwords in accordance with paragraph 3(d) in order to enable them to access and submit bids using the auction software system and to submit bids using the alternative arrangements set out in paragraph 10.

(3) Direct participants and their authorised representatives must keep all user names and passwords secret to prevent anyone who is not authorised from submitting bids.

(4) A direct participant must notify the Secretary of State as soon as possible if he suspects that—

(a) an unauthorised person knows any of his user names or passwords;

(b) an unauthorised person has tried or intends to access the auction software system;

(c) he has forgotten or lost any of his user names or passwords; or

(d) any user names or passwords have been kept in a form that may be accessible to an unauthorised person.

(5) Until she receives a notification under sub-paragraph (4), the Secretary of State will be entitled to consider that all bids submitted in relation to a direct participant have been made by an authorised person.

(6) The Secretary of State will take all reasonable steps to ensure that unauthorised access to the auction software system does not occur.

Bids

6.— (1) Bids shall be submitted in accordance with the auction guidance and the bidder's manual.

(2) The validity of any bid shall be assessed automatically by the auction software system and, in the event of an invalid bid, an error message will be displayed.

Withdrawing bids

7.— (1) A bid may be withdrawn at any time before the end of a round in accordance with the auction guidance and the procedure set out in the bidder's manual.

(2) Withdrawal of a bid shall not prevent the bidder from submitting another bid in that round.

Retirement

8.— (1) Any bidder may, during any round, notify his retirement from the auction by submitting a zero bid in accordance with the auction guidance and the bidder's manual.

(2) A bidder who has retired from the auction will not be entitled to bid in any later bidding round.

Failure to enter a bid

9.— (1) If a direct participant fails to enter a valid bid before the end of a round, the Secretary of State may attempt to contact the direct participant by telephone, using the number specified in the direct participant agreement.

(2) If contacted under sub-paragraph (1), a direct participant may—

(a) indicate that he wishes to participate in the auction in which case the Secretary of State may extend the time for submission of the bid and may accept bids using the alternative arrangements set out in paragraph 10; or

(b) confirm that he wishes to retire from the auction, in which case the Secretary of State will retire the direct participant from the auction and he will not be allowed to enter any further bids in subsequent rounds.

(3) If no contact is made under sub-paragraph (1), the Secretary of State will automatically retire the direct participant from the auction, in which case he will not be allowed to enter any further bids in subsequent rounds.

Alternative arrangements for bidding

10.— (1) Where at any time during the auction a direct participant is unable to participate in the auction by using the auction software system, he must notify the Secretary of State by telephone using the number notified under paragraph 3(f).

(2) If contacted under sub-paragraph (1), the Secretary of State may allow the participant to participate in the auction by telephone, in which case any communication must be accompanied by one of the auction usernames and passwords notified to the participant under paragraph 3(d).

(3) If any direct participant participates in the auction by telephone, the Secretary of State may make a tape recording of any conversation with that direct participant.

(4) If a participant who has contacted the Secretary of State under sub-paragraph (1) cannot provide one of the auction usernames and passwords, he may participate in the auction by fax using the number notified under paragraph 3(f), in which case any communication must be accompanied by the specified signature

of one of the persons authorised under the direct participant agreement.

(5) The Secretary of State may communicate additional information about the auction (including notification of invalid bids) to a direct participant participating in the auction under sub-paragraph (2) or (4) by whatever means she considers appropriate.

(6) In the event of the auction software system failing to operate, the Secretary of State will notify direct participants of an alternate internet address by whatever means she considers appropriate.

Cancellation

11.— (1) The Secretary of State may at any time decide to cancel the auction or postpone it.

(2) The Secretary of State will notify direct participants of any decision under sub-paragraph (1) by announcing it on the auction software system or by any other means she considers appropriate.

Final round

12.— (1) After each round the Secretary of State will consider the bids received in accordance with the auction guidance in order to determine whether the clearing price has been reached.

(2) If the Secretary of State decides that the clearing price has been reached, she will announce the clearing price on the auction software system.

(3) The overall targets for all direct participants submitting valid bids in the round of the auction in which the clearing price is reached will be calculated in accordance with the auction guidance.

Exclusion

13.— (1) If a direct participant breaches one of more of the rules set out in sub-paragraphs (2) to (4), the Secretary of State may, at her discretion, notify that direct participant of his exclusion from the auction.

(2) A direct participant must not submit to the Secretary of State any information affecting her decision to enter into a direct participant agreement which the direct participant knows to be false or misleading.

(3) A direct participant must not compromise or attempt to compromise the security of the auction software system and in particular must comply with paragraphs 5(3) and (4).

(4) A direct participant must not convey or incite another person to convey any confidential information directly or indirectly relating to any proposed bid which might have an effect on the

way in which any other direct participant proposes to participate in the auction.

Dispute resolution

14. Any question arising in the course of the auction about the application of this Schedule or the auction guidelines shall be decided by the Secretary of State and her decision shall be final and conclusive on that question.

SCHEDULE 5A
Notification of an EU ETS overlap

Introduction

1.— (1) This Schedule prescribes the information an operator or sector association is required to submit to the Secretary of State under rule D3A(4).

(2) For the purposes of this Schedule, references to an "operator" are to be taken as a reference to a sector association in cases falling within rule D3A(3)(b).

The EU ETS overlap

2. In respect of the CCA facility and target, the operator shall—

- (a) identify the facility, including the facility number; and
- (b) specify the target unit applying to the facility, including in particular—
- (i) where the target is a relative target, details of throughput;
- (ii) where the target unit has an energy target, details of the fuel split from the facility and the relevant carbon emission factors used.

3. In respect of each EU ETS installation the emissions of which constitute the EU ETS overlap with the facility in question, the operator shall—

- (a) identify the relevant EU ETS installation, and its National Allocation Plan ID;
- (b) confirm—
- (i) the number of EU ETS allowances issued to that installation; and
- (ii) the level of emissions emitted from that installation, in respect of the scheme year notified.

4. The operator shall, in accordance with the methodology set out in any guidance issued by the Secretary of State under rule D3A(11), specify—

(a)   the overlap emissions emitted from the facility in question; and

(b)   the extent to which the reportable emissions from the relevant EU ETS installation do not constitute overlap emissions in the particular case, providing the Secretary of State (and, where relevant, any verifier) with the necessary data and calculations to justify its notification in respect of sub-paragraphs (a) and (b).

EU ETS overachievement or underachievement

5. The operator shall, in accordance with the methodology set out in any guidance issued by the Secretary of State under rule D3A(11), specify —

(a)   (i) the amount of the EU ETS overachievement; or
(ii) the amount of the EU ETS underachievement, expressed, in either case, in terms of the number of EU ETS allowances to which that overachievement or underachievement is equal; or

(b)   where there has been neither, that fact, providing the Secretary of State (and, where relevant, any verifier) with the necessary data and calculations to justify its notification in respect of this paragraph.

Retirement of EU ETS allowances

6. The operator shall confirm that it has already retired a number of EU ETS allowances at least equal to its EU ETS overachievement, if this is the case.

Other information

7. In addition to supplying the information required by this Schedule, the operator shall provide any other information —

(a)   which is requested by the Secretary of State; or

(b)   which it considers would be relevant to establish an EU ETS overlap, an EU ETS overachievement or an EU ETS underachievement (or to establish that neither has occurred) in the particular case.

SCHEDULE 6

MONITORING, REPORTING AND VERIFICATION OF CCA OVER-ACHIEVEMENTS

Introduction
1. A CCA participant which wishes to convert a CCA over-achievement into allowances under rule D4 must comply with this Schedule.

Monitoring
2.— (1) For the target period in which the CCA over-achievement occurred, the CCA participant must calculate in accordance with the documents and other materials listed in sub-paragraph (2)—

(a)   for an operator converting the CCA over-achievement of a target unit,

(i)   the throughput of the target unit; and

(ii)   either the energy used by or the emissions from the target unit in tonnes of carbon dioxide equivalent;

(b)   for a sector association converting the CCA over-achievement of a trading group—

(i)   the throughput of all facilities in the trading group; and

(ii)   either the energy used by or the emissions from all facilities in the trading group in tonnes of carbon dioxide equivalent.

(2) The documents and other materials referred to in sub-paragraph (1) are—

(a)   the umbrella agreement and the underlying agreement which apply to the target unit or the facilities in the trading group to which the CCA over-achievement relates;

(b)   CCA guidance documents which relate to emissions trading, including in particular CCA3 and CCA10; and

(c)   any further guidance given by the Secretary of State from time to time.

(3) In making the calculation referred to in sub-paragraph (1), no account may be taken of—

(a)   any tolerance bands specified in an underlying agreement;

(b)   any adjustments made to a CCA target in accordance with the terms of an underlying agreement to account for product mix or throughput; or

(c)   any increase in energy use or emissions which was disregarded in accordance with the terms of an underlying agreement because it was the result of unexpected energy supply disruptions.

Reporting

3.— (1) An application to convert a CCA over-achievement into allowances under rule D4(7) or (13) must be accompanied by a statement of over-achievement prepared in accordance with this paragraph.

(2) For an operator, a statement of over-achievement must—

(a)   set out the total amount by which the CCA target for the target unit to which the application relates has been exceeded during the target period, denominated in tonnes of carbon dioxide equivalent;

(b)   certify that the operator has not submitted a previous claim in respect of the same target unit over the same target period;

(c)   certify that the target unit is not part of a trading group and that the over-achievement has and will not be used as part of the over-achievement of a trading group; and

(d)   certify that the operator has complied with the notification requirements of rule D4(5) in respect of the over-achievement.

(3) For a sector association, a statement of over-achievement must—

(a)   set out the total amount by which the trading group target to which the application relates has been exceeded during the target period, denominated in tonnes of carbon dioxide equivalent;

(b)   certify that the sector association has not submitted a previous claim in respect of the same trading group over the same target period;

(c)   certify that no operator has submitted a claim in respect of any part of the CCA overachievement which is the subject of the application; and

(d)   certify that the sector association has complied with the notification requirements of rule D4(11) in respect of the over-achievement.

(4) If an application under rule D4(7) or (13) relates to CCA over-achievements occurring in more than one target period, the statement of over-achievement must include a breakdown of the total over-achievement showing the individual over-achievement in each target period and, in a situation falling within rule D4(16)(b)(i), how the over-achievement related to the absolute and relative CCA targets concerned.

Verification

4.— (1) A CCA participant wishing to convert a CCA over-achievement into allowances must have its compliance with the requirements of rule D4 and paragraphs 2 and 3 of this Schedule verified by a verifier in accordance with this paragraph.

(2) The participant must provide the verifier with the following information—

(a) for an operator—

(i) the total energy use or emissions from the target unit during the target period to which the over-achievement relates and, for target units with relative targets, the relevant output data;

(ii) details of the data, procedures, decisions, measurements and calculations on which the figures referred to in sub-paragraph (a) are based;

(iii) the facility number of each facility in the target unit;

(iv) the CCA target for the target unit for the relevant target period; and

(v) such other information as the verifier requires;

(b) for a sector association—

(i) the total energy use or emissions from the facilities in the trading group during the target period to which the over-achievement relates and, for target units with relative targets, the relevant output data;

(ii) details of the data, procedures, decisions, measurements and calculations on which the figures referred to in sub-paragraph (a) are based;

(iii) the facility numbers of the facilities in the trading group;

(iv) the trading group target for the relevant target period; and

(v) such other information as the verifier requires.

(3) The participant must provide the verifier with a written declaration signed by a director manager or other designated representative of the participant confirming that the information which it has provided is, to the best of the participant's knowledge, accurate and complete.

(4) The verifier must assess the information provided by the participant in accordance with—

(a) the provisions of the relevant umbrella and underlying agreements;

(b) CCA guidance documents which relate to emissions trading, including in particular CCA3 and CCA10;

(c) guidance to verifiers issued by UKAS; and

(d)   any other guidance given by the Secretary of State from time to time; and must take account of all relevant information (whether or not provided by the participant).

(5) A verifier shall only provide the participant with an unqualified verification opinion, where a verifier is satisfied that the statement of over-achievement prepared by the participant under paragraph 3 accurately reflects the extent of the over-achievement and that the requirements of rule D5 and this Schedule have been complied with.

(6) Where a verifier cannot provide the participant with an unqualified verification opinion, it may nonetheless provide a qualified verification opinion but such an opinion will not satisfy the requirements of rule D4(7)(b) or (13)(b).

(7) A CCA participant must inform both the Secretary of State and any verifier which it subsequently instructs of any previous refusal by a verifier to provide an unqualified verification opinion or of any qualifications in a verification opinion in respect of any CCA over-achievement, together with any reasons given by the verifier for the refusal or qualifications.

Retention of information

5.—   (1) Compliance with this paragraph will be treated by the Secretary of State as satisfying the obligation of a CCA participant under the underlying or umbrella agreement to which it is a party to keep proper records in respect of an application to convert a CCA over-achievement into allowances.

(2) The participant must keep sufficient records to allow any person appointed by the Secretary of State under an umbrella or underlying agreement an to undertake an independent audit of any application under rule D4.

(3) In particular, the participant must retain—

(a)   the activity data and other information on which any calculation of his emissions in a target period was based; and

(b)   information and data not required under sub-paragraph (a) but which is required by the verifier to verify the data submitted by the participant to the Secretary of State.

(4) All such records must be retained for the purposes of compliance with this Schedule for at least six years from the date on which they are created.

(5) A CCA participant must record and manage the information retained under this rule in a systematic manner and must employ an effective data management system.

ANNEX 8

UK GREENHOUSE GAS EMISSIONS TRADING SCHEME 2002: DIRECT PARTICIPANT AGREEMENT

[Limited Company version]

THIS AGREEMENT is made this   day of   2002

BETWEEN

(1) The Secretary of State for Environment, Food and Rural Affairs; and
(2) [NAME] a company incorporated in _____ whose registered office address is [ADDRESS] ("The Direct Participant")

RECITALS

WHEREAS:

a) The Secretary of State is establishing a Greenhouse Gas Emissions Trading Scheme for the purposes of achieving reductions in emissions of greenhouse gases in a cost-effective manner and for the other purposes set out in Rule A1 of the Scheme.

b) The Secretary of State is empowered by virtue of section 153 of the EnvironmentalProtection Act 1990 to provide financial assistance in the form of the incentive payment provided for in the Scheme.

c) The European Commission has decided that the scheme is compatible with the state aid rules and has given approval for the making of incentive payments and the allocation of Allowances by the Secretary of State to Direct Participants in the Scheme. Both parties acknowledge that the modification or withdrawal of such approval wholly or in part will or will be likely to necessitate the revocation of or substantial amendment to the Scheme.

d) The Secretary of State is proposing at the earliest opportunity to introduce a Billproviding a statutory basis for financial penalties for failure to comply with the Scheme. The Direct Participant acknowledges that, unless he has previously withdrawn from the Scheme in accordance with the Rules and the terms of this

Agreement, he may be exposed to fines or other penalties imposed by statute for failure to comply with his obligations under the Scheme.

e)   It is acknowledged by the parties that they wish to benefit from early experience of greenhouse gas trading by participating in the Scheme and that the Scheme is innovative. It is also acknowledged that before the expiry of this Agreement the Scheme will be subject to review and that the Rules and the terms of this Agreement may be amended or revised if it appears necessary to the Secretary of State for the achievement of the purposes of the Scheme, including for example to permit new and late entrants to join the Scheme, or to accommodate international trading . The Secretary of State's policy in relation to future domestic controls on emissions is that the Direct Participant will not be penalised for actions taken to reduce its emissions as a result of its participation in the scheme and that the Direct Participant's baselines and targets under the scheme will be taken into account in developing any future controls.

f)   The Direct Participant has submitted to the Secretary of State an application form seeking to participate in the Scheme and a Source List.

g)   The Secretary of State has agreed to permit the Direct Participant to participate in the Scheme.

NOW IT IS AGREED AS FOLLOWS

1.   DEFINITIONS AND INTERPRETATION

1.1.   In this Agreement, unless the context otherwise requires the following words have the meaning given to them below:

"Agreement" means (as amended from time to time in accordance with the provisions hereof) this document and all annexes to it;

"Application Form" means the document enclosed at Annex 1 to this Agreement and the documents and information submitted by the Direct Participant to the Secretary of State in connection with it;

"Commencement Date" means the date set out at the top of this document;

"Insolvency Event" means, in relation to the Direct Participant, any of the following events:

(a)   it goes into or is the subject of a petition for compulsory winding up;

(b)   it passes a resolution for voluntary winding up;

(c)   an administrator, administrative receiver or receiver is appointed or takes possession over the whole or any part of its assets;

(d)   it has entered into or made a proposal for or appointed any person as a nominee in respect of a voluntary arrangement with its creditors under Part I of the Insolvency Act 1986, or has proposed or entered into any scheme of arrangement or composition with its creditors under section 425 of the Companies Act 1985; or

(e)   anything analogous to, or having a substantially similar effect to, any of the circumstances specified in paragraph (a) to (d) above occurs in relation to the Direct Participant in any jurisdiction;

"Parties" means the parties to this Agreement whose names appear above;

"Scheme" means the UK Greenhouse Gas Emissions Trading Scheme 2002, as amended from time to time;

"Source List" means the document set out at Annex 2 to this Agreement as amended from time to time; and

"Termination Date" means 31st March 2007.

1.2.   Words and expressions which are defined in Schedule 1 to the Scheme shall bear the same meaning where they appear in this Agreement.

1.3.   The rules of interpretation referred to in Schedule 1 to the Scheme shall apply to this Agreement.

1.4.   The Scheme is incorporated into and forms part of this Agreement as if set out here and, unless the context otherwise requires, references in the Scheme to a direct participant shall mean the Direct Participant herein and references to a source list shall mean the Source List set out in Annex 2 of this Agreement.

COMMENCEMENT

2.1   This Agreement shall come into full force and effect on the Commencement Date.

2.2   Upon the Commencement Date the Scheme and the terms of this Agreement shall be given effect and be made binding upon each of the parties.

3. DURATION

3.1   Subject to the provisions of clauses 8 and 9 below and subject always to the exercise by the Secretary of State of her rights and powers under Rule G1 of the Scheme, this Agreement shall continue in full force and effect and be binding on the parties until the Termination Date.

UNDERTAKINGS

4.1   Subject to the limitations and restrictions set out in this Agreement and in the Scheme and to the exercise by the Secretary of State of any of her rights or powers under this Agreement or under the Scheme, and in reliance upon the representations and warranties of the Direct Participant set out in Clause 5, the Secretary of State shall at the times and in the manner provided for in the Scheme:-

(a)   open and maintain accounts at the Registry in the name of the Direct Participant (in accordance in particular with Part B and Rule C4 of the Scheme);

(b)   allocate Allowances to a compliance account in the name of the Direct Participant (in accordance in particular with Rule C5);

(c)   pay to the Direct Participant incentive payments (in accordance in particular with Rules C1 and C2);

(d)   apply the Scheme to other participants as necessary to promote the purposes of the Scheme; and

(e)   comply with the other obligations imposed upon her by the Scheme so far as the same directly concern the Direct Participant.

4.2   The Direct Participant shall at the times and in the manner provided in the Scheme:-

(a)   undertake through the Auction an annual emissions reduction target for each commitment year (in accordance in particular with Schedule 5);

(b)   monitor, calculate, report and submit emissions statements and verification opinions in respect of his baseline and his annual emissions (in accordance in particular with Schedule 2);

(c)   hold in his compliance account on the reconciliation deadline for each commitment year a reconciliation balance which is equal to or greater than his verified emissions during that year (in accordance in particular with Rule C6(1));

(d)   ensure that the sum of (i) his reconciliation balance on the final reconciliation deadline and (ii) the total number of allowances retired from his compliance account prior to the final reconciliation deadline and are equal to or exceed his total verified emissions during the commitment period (in accordance in particular with Rule C3(1));

(e)   repay to the Secretary of State forthwith any incentive payments received by him together with interest in the circumstances provided for in the Scheme (in accordance in particular with Rule C3); and

(f)   comply with the other obligations imposed upon him by the Scheme.

4.3   Subject to any express provision in the Scheme as to the time within which the Direct Participant shall notify the Secretary of State of any matter, he shall do so as soon as reasonably possible after becoming aware of the occurrence of the notifiable event.

4.4   Further the Direct Participant shall promptly notify the Secretary of State as soon as he becomes aware:-

(a)   of any material change in the information contained in sections 2 and 4 of the Application Form; and

(b)   that one of the Insolvency Events has occurred in relation to the Direct Participant.

5. REPRESENTATIONS AND WARRANTIES

5.1 The Direct Participant represents and warrants to the Secretary of State that :-

(a) the information contained in, and the documents and information submitted to the Secretary of State in connection with the Application Form are true and accurate and not misleading;

(b) in so far as any information contained in or provided pursuant to the Application Form has materially changed, the Direct Participant has informed the Secretary of State of such change;

(c) he has power to enter into this Agreement and has obtained all necessary approvals to do so;

(d) to the best of his knowledge there is no legal, regulatory, contractual or other restriction upon his entering into and performing his obligations under the Agreement;

(e) the signatory or signatories below is/are duly authorised with full power and authority to execute this Agreement on behalf of the Direct Participant;

(f) he has had adequate access to and has considered the Scheme so far as it applies to direct participants before entering into this Agreement; and

(g) he has prepared his Source list in accordance with the requirements of Schedule 2 of the Scheme.

6. AMENDMENT AND VARIATION

6.1 The Scheme may be amended by the Secretary of State from time to time or may be revoked by any subsequent scheme made by the Secretary of State in accordance with the provisions of Rule G1 of the Scheme.

6.2 Any such amendment or revocation shall take effect without the requirement for further signature or acknowledgement on the part of the Direct Participant.

6.3 The Secretary of State shall keep a record of all such amendments and revocations.

6.4   No other amendments or variations to this Agreement shall take effect unless made in
writing, signed by authorised representatives of each of the parties.

7.   ASSIGNMENT AND TRANSFER

7.1   The Secretary of State may at any time assign or transfer all or any part of her rights and obligations under this Agreement and the Scheme to an agent appointed by her for the purposes of administering the Scheme and the Direct Participant consents to all such dealings.

7.2   The Direct Participant shall not assign, charge, transfer or otherwise deal with this Agreement without prior the written consent of the Secretary of State, who may refuse such consent in her discretion.

8.   TERMINATION

8.1   This Agreement will automatically terminate:

(a)   forthwith if the Direct Participant retires from the auction;

(b)   upon the Termination Date.

8.2   If any of the events in clause 8.3 (each an "Event of Default") occurs in relation to the Direct Participant, the Secretary of State may (but shall not be obliged to) terminate this Agreement by a notice in writing (such notice to take effect as specified in the notice).

8.3   The Events of Default referred to in clause 8.2 are the following:

(a)   an Insolvency Event occurs in respect of the Direct Participant;

(b)   the Direct Participant is in the reasonable opinion of the Secretary of State guilty of any unlawful conduct or wilful conduct amounting to abuse of the Scheme or the Registry, or in relation to any Registry account or the Auction process; and

(c)   any the Direct Participant or any of its directors, partners or controllers is, in relation to finance and markets or the formation and activities of a body corporate or other business or professional entity, convicted of any criminal offence, or subject to legal or regulatory sanction, for fraud, dishonesty or professional misconduct.

8.4   Upon termination of this Agreement by the Secretary of State for any Event of Default,

(a)   the Direct Participant shall be liable to repay forthwith any incentive monies it has received together with interest thereon calculated on the basis set out in rule C3(6) of the Scheme;

(b)   all allowances held in trading accounts in the name of the Direct Participant will be automatically transferred to the Direct Participant's compliance account, the trading accounts will be closed and the compliance account will be suspended under rule C4(9);

(c)   the Secretary of State will cancel allowances in the Direct Participant's compliance account up to the total number of allowances allocated to him under Rule C5 for the commitment year in which the termination is to take effect, and, if no allowances remain in the account after the cancellation, will close the account;

(d)   if any allowances remain in the compliance account following the cancellation under sub-paragraph (c), the Secretary of State will transfer the remaining allowances to an account held in the name of a third party, provided that the Direct Participant (i) notifies the Secretary of State of the account to which any remainder is to be transferred within two months of the service of the termination notice and (ii) has discharged any liability under sub-paragraph (a); and

(e)   following the transfer of all remaining allowances, or if the Direct Participant fails to notify a transfer, under sub-paragraph (d), the Secretary of State will cancel any remaining allowances in the compliance account and close the account

8.5   Termination of this Agreement shall not affect any rights of the parties accrued up to and including the date of termination. In particular but without limitation, the Direct Participant shall remain liable to the Secretary of State under Rule C3 of the Scheme to repay incentive monies and interest notwithstanding termination and to pay any financial penalties for which he is liable Rule C6 under the Scheme.

8.6   The provisions of Rule G2 (access to information), Part 4 of Schedule 2 (retention of information) and Rule G3 of the Scheme (dispute resolution) shall survive termination of the Agreement.

9.   WITHDRAWAL

9.1   The Direct Participant may withdraw from this Agreement and cease to be a direct participant in the Scheme if (but only if) all of the following conditions are satisfied:-

(a)   if the Direct Participant intends to withdraw from the Scheme at the end of a commitment year:

   (i)   the Direct Participant has served on the Secretary of State a notice of intention to withdraw not less than 28 days before the effective date of withdrawal;

   (ii)   the Direct Participant has repaid to the Secretary of State in cleared funds before the effective date of withdrawal all incentive monies received together with interest thereon calculated in accordance with the provisions of rule 34(4) of the Scheme;

   (iii)   the Direct Participant has submitted an verification opinion in relation to his emissions for the complete commitment year immediately preceding withdrawal;

   (iv)   the Direct Participant has complied with any outstanding requests by the Secretary of state or any person appointed by her under Part 4 of Schedule 2 to the Scheme and any audit has been completed to the satisfaction of the Secretary of State;

   (v)   as at the date of withdrawal the Direct Participant holds in his compliance account a reconciliation balance for the relevant commitment year which is equal to or greater than his verified emissions during that year and the Secretary of State has retired allowances equal to his verified emissions; and

   (vi)   the Secretary of State has indicated her consent to withdrawal – which consent will not be unreasonably withheld;

(b)   if the Direct Participant intends to withdraw from the Scheme during a commitment year:

   (i)   conditions (i), (ii), (iv) and (vi) under sub-paragraph (a) apply; and

(ii)   the Direct Participant holds in his compliance account at the date on which he served his notice of intention to withdraw a balance which is equal to or greater than the total number of allowances allocated to him under Rule C5 for that year and the Secretary of State has cancelled allowances equal to his allocation.

9.2   The effective date of withdrawal shall be the first date upon which the above conditions are satisfied.

9.3   As soon as the Secretary of State receives a notice of intention to withdraw under clause 9.1 above, all allowances in trading accounts held in the name of the Direct Participant will be automatically transferred to the Direct Participant's compliance account, the trading accounts will be closed and the compliance account will be suspended under rule C4(9), subject to the transfer of allowances into the account under rule C4(8).

9.4   Upon withdrawal, if any allowances remain in the compliance account following the cancellation process under clause 9.1(a)(v) or (b)(ii), the Secretary of State will allow the Direct Participant to transfer the remaining allowances to an account in the name of a third party, or to a trading account held in the name of the Direct Participant if the Direct Participant has before the effective date of withdrawal entered into a trading participant agreement with the Secretary of State.

9.5   When the account balance of the Direct Participant's compliance account is zero, the Secretary of State will close the compliance account.

9.6   Withdrawal from this Agreement shall not affect any rights of the parties accrued up to and including the date of withdrawal.

9.7   The provisions of Rule G2 (access to information) and G3 (dispute resolution) of the Scheme shall survive notwithstanding the withdrawal of the Direct Participant.

10.   FORCE MAJEURE

10.1   Subject to any express provision under the Scheme or elsewhere in this Agreement, neither party shall be liable for any delay in performing, or failure to perform, any of its obligations under this Agreement or the Scheme, where such performance is rendered impossible by reason of a force majeure event.

10.2   The party claiming a force majeure event shall promptly notify the other party in writing of the reason for the delay or failure to perform the necessary obligation, and its likely duration and shall use all reasonable efforts to remedy its inability to perform.

10.3   Subject to that party's compliance with clause 10.2, its performance under this Agreement and the Scheme will be suspended for the period that the force majeure event subsists.

11.   CONSEQUENCES OF BREACH

11.1   To the extent permitted by law, and subject to any express provisions elsewhere in this Agreement or in the Scheme :-

(a)   the Secretary of State's liability to the Direct Participant whether for failure to perform its obligations under this Agreement or the Scheme, in tort, for negligence, breach of statutory duty or otherwise shall not in any event affect the aggregate amount of incentive payment to which the Direct Participant would have been entitled but for such failure;

(b)   neither party shall be liable to the other in contract, tort, negligence, breach of statutory duty or otherwise for any loss, damage, costs or expenses of any nature whatsoever incurred or suffered by that other party of an indirect or consequential nature including without limitation any economic loss or other loss of turnover, profits, business, business opportunity or goodwill.

11.2   The Direct Participant acknowledges that damages may not be an adequate remedy for failure to comply with his obligations set out in Part 3 and Schedule 2 of the Scheme and accepts that the Secretary of State is entitled to injunctive relief and/or an order for specific performance of those obligations.

12.   SET OFF

12.1   The Direct Participant shall pay all sums due to the Secretary of State under the terms of this Agreement or the Scheme without any discount, deduction, set-off or counterclaim whatsoever.

13.   DISPUTE RESOLUTION

13.1   Schedule 4 of the Scheme (Dispute Resolution) shall have effect in relation to any dispute that may arise out of or relate to this Agreement, the Scheme or any breach of it.

14.   NOTICES

14.1   All notices and other communications between Parties under or in connection with the Agreement and the Scheme must be given in accordance with Rule G4.

14.2   The address and contact details for the Secretary of State are set out Part 1 of Annex 3 to this Agreement. The address and contact details for the Direct Participant are set out in Part 2 of Annex 3 to this Agreement.

14.3   The address, contact details and specimen signatures of those persons authorised by the Direct Participant to participate in the auction in accordance with Schedule 5 of the Scheme are set out in Part 3 of Annex 3 of this Agreement.

15.   GENERAL

15.1   This Agreement shall be governed by and construed in accordance with English law and each party agrees to submit to the jurisdiction of the Courts of England and Wales. The submission to such jurisdiction shall not (and shall not be construed so as to) limit the right of the Secretary of State to take proceedings against the Direct Participant in any other court of competent jurisdiction, nor shall the taking of proceedings in any other court of competent jurisdiction preclude the taking of proceedings in any other jurisdiction whether concurrently or not.

15.2   If the whole or any part of any provision of this Agreement or the Rules is declared invalid or illegal or unenforceable by the courts of any jurisdiction to which it is subject, such invalidity, unenforceability or illegality shall not prejudice or affect the remainder of that provision or any other provisions in the Scheme or in this Agreement (in so far as the offending parts can be severed) which shall continue in full force and effect notwithstanding such invalidity, unenforceability or illegality.

15.3   For the purposes of the Contracts (Rights of Third Parties) Act 1999 this Agreement is not intended to, and does not, give any person who is not a party to it any right to enforce any of its provisions.

15.4   No omission or delay on the part of either party in enforcing or exercising any rights powers or remedies under the Agreement shall impair such right power or remedy or be construed as a waiver thereof or of any other right power or remedy.

15.5   This Agreement together with the Scheme contains the whole of the agreement between the Parties and the Parties confirm that they have not entered into this agreement on the basis of any representations that are not expressly incorporated in this Agreement and the Scheme.

Signed by the Parties by their Authorised Signatories:

For and on behalf of the Secretary of State

Director/Authorised person

For and on behalf of

ANNEX 1

Application Form

ANNEX 2

Source list

ANNEX 3

Contact details

Part 1: Notices to the Secretary of State

Address for notices:	Global Atmosphere Division
	Department for Environment, Food and Rural Affairs
	3/F3 Ashdown House
	123 Victoria Street
	London SW1E 6DE
Facsimile:	020 7944 5219
Telephone:	020 7944 5933
email:	ets@defra.gsi.gov.uk
Attention:	Emissions Trading Scheme

Part 2: Notices to the Direct Participant

Address for notices:	*[Insert]*
Facsimile:	*[Insert]*
Telephone:	*[Insert]*
email:	*[Insert]*
Attention:	*[Insert]*

Part 3: Auction notices and authorised persons

Address for auction notices:	*[Insert]*
Facsimile:	*[Insert]*
Telephone:	*[Insert]*
email:	*[Insert]*
Authorised Person 1:	*[Insert]*
Signature	*[Insert]*

Authorised Person 2:   *[Insert]*
Signature   *[Insert]*
Authorised Person 3:   *[Insert]*
Signature   *[Insert]*

ANNEX 9

EXTRACT FROM SCHEDULE 6 TO THE FINANCE ACT 2000 (PARAGRAPHS 44 TO 52)

44

(1)   Where the Secretary of State gives a certificate to the Commissioners stating that, for a period specified in the certificate, a facility is to be taken as being covered by a climate change agreement, the Commissioners shall publish a notice in respect of the facility.

(2)   Such a notice shall—

  (a)   state the day on which it is published,
  (b)   identify the facility or facilities in respect of which it is published,
  (c)   for each facility—
      (i)   set out the first and last days of the period specified for the facility in the Secretary of State's certificate, and
      (ii)   indicate the effect of sub-paragraph (3), and
  (d)   indicate that the notice may be varied by later notices.

(3)   For the purposes of this Schedule, a reduced-rate supply is a taxable supply in respect of which the following conditions are satisfied—

  (a)   the first condition is that the taxable commodity supplied by the supply is supplied to a facility identified in a notice published under sub-paragraph (1);
  (b)   the second condition is that the supply is made at a time falling in the period that begins with the later of—
      (i)   the first day set out for the facility under sub-paragraph (2)(c), and
      (ii)   the day on which the notice is published, and ends with the last day set out for the facility under sub-paragraph (2)(c).

(4)   Sub-paragraph (3) has effect subject to paragraph 45.

(5)   The Commissioners may, for the purposes of sub-paragraph (3), by regulations make provision for determining whether any taxable commodity is supplied to a facility.

(6)   The provision that may be made by regulations under sub-paragraph (5) includes, in particular, provision for a taxable commodity of any description specified in the regulations to be taken as supplied to a facility only if the commodity is delivered to the facility.

Reduced-rate supplies: variation of notices under paragraph 44

45

(1)   This paragraph applies where the Secretary of State, after having given in respect of a facility such a certificate as is mentioned in paragraph 44(1) ("the original certificate"), gives a certificate (a "variation certificate") to the Commissioners stating—

> (a)   that, throughout the period ("the original period") specified for the facility in the original certificate, the facility is to be taken as not being covered by a climate change agreement; or
>
> (b)   that, for so much of the original period as falls on or after a day specified in the variation certificate (being a day falling within the original period), the facility is to be taken as no longer being covered by a climate change agreement.

(2)   Where the Commissioners receive a variation certificate in respect of a facility before they have published a notice under paragraph 44(1) in response to the original certificate so far as relating to the facility, their obligation to publish a notice under paragraph 44(1) in respect of the facility shall have effect as an obligation to publish such a notice in response to the original certificate as varied by the variation certificate.

(3)   Where the Commissioners receive a variation certificate but sub-paragraph (2) does not apply, they shall publish a notice (a "variation notice") that—

> (a)   states the day on which it is published,
>
> (b)   identifies the facility or facilities in respect of which it is published,
>
> (c)   sets out, for each facility in respect of which the statement in the variation certificate is of the type described in sub-paragraph (1)(b), the date specified for the facility in the variation certificate, and
>
> (d)   for each facility, indicates the effect of sub-paragraphs (4) to (7) as they apply in the case of the facility.

(4)   Sub-paragraphs (5) to (7) set out the effect of a variation notice being published in respect of a facility.

(5)   If—

    (a)   the statement in the variation certificate in respect of the facility is of the type described in sub-paragraph (1)(a), and

    (b)   the day on which the variation notice is published falls before the beginning of the original period, the notice ("the original notice") published under paragraph 44(1) in response to the original certificate has effect as if the facility had never been identified in it.

(6)   If—

    (a)   the statement in the variation certificate in respect of the facility is of the type described in sub-paragraph (1)(a), and

    (b)   the day on which the variation notice is published falls during the original period, the original notice has effect as if the last day set out for the facility under paragraph 44(2)(c) were the day on which the variation notice is published.

(7)   If the statement in the variation certificate in respect of the facility is of the type described in sub-paragraph (1)(b), the original notice has effect as if the last day set out for the facility under paragraph 44(2)(c) were the later of—

    (a)   the day on which the variation notice is published, and

    (b)   the day set out in the variation notice for the facility under sub-paragraph (3)(c).

Climate change agreements

46

In this Schedule "climate change agreement" means—

    (a)   an agreement that falls within paragraph 47, or

    (b)   a combination of agreements that falls within paragraph 48.

Climate change agreements: direct agreement with Secretary of State

47

(1)   An agreement (including one entered into before the passing of this Act) falls within this paragraph if it is an agreement—

    (a)   entered into with the Secretary of State,

    (b)   expressed to be entered into for the purposes of the reduced rate of climate change levy,

    (c)   identifying the facilities to which it applies,

    (d)   to which a representative of each facility to which it applies is a party,

    (e)   setting, or providing for the setting of, targets for the facilities to which it applies,

    (f)   specifying certification periods (as to which see paragraph 49(1)) for the facilities to which it applies, and

    (g)   providing for five-yearly (or more frequent) reviews by the Secretary of State of targets set by or under the agreement for those facilities and for giving effect to outcomes of such reviews.

(2)   In this paragraph and paragraph 48 "representative", in relation to a facility to which an agreement applies, means—

    (a)   the person who is the operator of the facility at—

        (i)   the time the agreement is entered into, or

        (ii)   if later, the time the facility last became a facility to which the agreement applies, or

    (b)   a person authorised by that operator to agree to the facility being a facility to which the agreement applies.

Climate change agreement: combination of umbrella and underlying agreements

48

(1)   A combination of agreements falls within this paragraph if the following conditions are satisfied.

(2)   The first condition is that the combination is a combination of—

    (a)   an umbrella agreement (including one entered into before the passing of this Act), and

  (b)   an agreement (including one entered into before the passing
         of this Act) that, in relation to the umbrella agreement, is
         an underlying agreement.

(3)   The second condition is that between them the two agreements—

  (a)   set, or provide for the setting of, targets for the facilities to
         which the underlying agreement applies,
  (b)   specify certification periods (as to which see paragraph
         49(1)) for the facilities to which the underlying agreement
         applies, and
  (c)   provide for five-yearly (or more frequent) reviews by the
         Secretary of State of targets set by or under the agreements
         for those facilities and for giving effect to outcomes of such
         reviews.

(4)   For the purposes of this paragraph an "umbrella agreement" is
an agreement—

  (a)   entered into with the Secretary of State,
  (b)   expressed to be entered into for the purposes of the reduced
         rate of climate change levy,
  (c)   identifying the facilities to which it applies, and
  (d)   to which a representative of each facility to which it applies
         is a party.

(5) For the purposes of this paragraph an agreement is an "underlying
agreement" in relation to an umbrella agreement if it is an agreement—

  (a)   expressed to be entered into for the purposes of the
         umbrella agreement,
  (b)   entered into—
         (i)   with the Secretary of State, or
         (ii)   with a party to the umbrella agreement other than
                the Secretary of State,
  (c)   approved by the Secretary of State if he is not a party to it,
  (d)   identifying which of the facilities to which the umbrella
         agreement applies are the facilities to which it applies, and
  (e)   to which a representative of each facility to which it applies
         is a party.

(6) In the case of a climate change agreement that is a combination of
agreements that falls within this paragraph, references to the facilities

to which the climate change agreement applies are references to the facilities to which the underlying agreement applies.

Climate change agreement: supplemental provisions

49

(1)   The first certification period specified by a climate change agreement for a facility to which it applies shall begin with the later of—

> (a)   the date on which the agreement, so far as relating to the facility, is expressed to take effect, and
> (b)   1st April 2001;

> and each subsequent certification period so specified shall begin immediately after the end of a previous certification period.

(2)   Where a climate change agreement (the "new agreement") applies to a facility to which another climate change agreement previously applied, the first certification period specified by the new agreement for the facility shall be—

> (a)   a period beginning as provided by sub-paragraph (1), or
> (b)   a period that—
> > (i)   begins earlier than that, and
> > (ii)   is a period that was a certification period specified for the facility by any climate change agreement that previously applied to the facility.

A period such as is mentioned in paragraph (b) includes a period beginning, or beginning and ending, before the date on which the new agreement, so far as relating to the facility, is expressed to take effect.

(3)   For the purposes of giving certificates such as are mentioned in paragraphs 44(1) and 45(1), the Secretary of State may take a facility as being covered by a climate change agreement for a period if the facility is one to which the agreement applies and either—

> (a)   that period is the first certification period specified by the agreement for the facility, or
> (b)   that period is a subsequent certification period for the facility and it appears to the Secretary of State that progress made in the immediately preceding certification period

towards meeting targets set for the facility by the agreement or by a climate change agreement that previously applied to the facility is, or is likely to be, such as under the provisions of the agreement in question is to be taken as being satisfactory.

(4)   For the purposes of sub-paragraph (3)(b) a climate change agreement may (in particular) provide that progress towards meeting any targets for a facility is to be taken as being satisfactory if, in the absence (or partial absence) of any such progress required under the agreement, alternative requirements provided for by the agreement are satisfied.

(5)   For the purposes of sub-paragraphs (2) and (3), the circumstances in which a facility to which a climate change agreement applies is one to which another such agreement previously applied include those where the facility is—

(a)   a part, or a combination of parts, of a facility to which another such agreement previously applied,

(b)   a combination of two or more such facilities,

(c)   any combination of parts of such facilities, or

(d)   any combination of such facilities and parts of such facilities.

(6)   Paragraphs 47 and 48 and sub-paragraph (4) above are not to be taken as meaning that an agreement, or combination of agreements, containing provision in addition to any mentioned in those paragraphs and that sub-paragraph is not a climate change agreement.

(7) For the purposes of paragraphs 47 and 48 and this paragraph "target", in relation to a facility to which a climate change agreement applies, means a target relating to—

(a)   energy, or energy derived from a source of any description, used in the facility or an identifiable group of facilities within which the facility falls, or

(b)   emissions, or emissions of any description, from the facility or such a group of facilities;

and for this purpose "identifiable group" means a group that is identified in the agreement or that at any relevant time can be identified under the agreement.

(8)   Nothing in this Schedule is to be taken as requiring the Secretary of State to—

    (a)   enter into any climate change agreement,

    (b)   enter into a climate change agreement with any particular person or persons, in respect of any particular facility or facilities or on any particular terms, or

    (c)   approve any, or any particular, proposed climate change agreement.

Facilities to which climate change agreements can apply

50

(1)   This paragraph applies where, in connection with concluding or varying a climate change agreement, it falls to be determined whether a facility is to be, or is to continue to be, identified in the agreement as a facility to which the agreement applies.

(2)   For the purposes of such a determination "facility" is (subject to any regulations under sub-paragraph (3) or (4)) to be taken as meaning—

    (a)   an installation covered by paragraph 51; or

    (b)   a site on which there is or are—

          (i)   such an installation or two or more such installations,

          (ii)   a part, or parts, of such an installation,

          (iii) a part, or parts, of each of two or more such installations, or

          (iv)   any combination of such installations and parts of such installations.

(3)   The Secretary of State may by regulations make provision for an installation covered by paragraph 51 to be taken to be a facility for those purposes only if—

    (a)   the taxable commodities supplied to the installation by taxable supplies are intended to be burned (or, in the case of electricity, consumed)—

          (i)   in the installation, or

          (ii)   on the site where the installation is situated but not in the installation, and

    (b)   the amounts of taxable commodities, and of any other commodities specified in the regulations, subject to each of those intentions are such that any conditions specified in the regulations are satisfied.

(4)   The Secretary of State may by regulations make provision for a site to be taken to be a facility for those purposes only if—

(a)   the taxable commodities supplied to the site by taxable supplies are intended to be burned (or, in the case of electricity, consumed)—
   (i)   in installations on the site that are covered by paragraph 51 (or in parts of such installations), or
   (ii)   on the site but not in any such installation (or part of such an installation), and
(b)   the amounts of taxable commodities, and of any other commodities specified in the regulations, subject to each of those intentions are such that any conditions specified in the regulations are satisfied.

(5)   Regulations under sub-paragraph (3) or (4) may make provision for deeming, for the purposes of the regulations, commodities to be intended to be burned (or, in the case of electricity, consumed) in circumstances specified in the regulations.

(6)   In this paragraph and paragraph 51 "installation" means a stationary technical unit.

Energy-intensive installations

51

(1) An installation is covered by this paragraph if it falls within any one or more of the descriptions of installation set out in the Table.

(2) Sub-paragraph (2A) applies where—

(a)   an installation falls within any one or more of those descriptions, and
(b)   there is, on the same site as the installation, a location at which ancillary activities are carried out.

(2A) The installation (taken alone) is not covered by this paragraph, but the combination—
(a)   of the installation and that location, or

(b)   where there is more than one such location, of the installation and all of those locations, is to be taken as being an installation covered by this paragraph.

(2B)   In sub-paragraph (2) "ancillary activities" means activities that—

(a)   are directly associated with any of the primary activities carried out in the installation,

(b)   have a technical connection with those primary activities, and

(c)   could have an effect on environmental pollution.

(3)   Sub-paragraphs (1) to (2B) are subject to any regulations under paragraph 52.

(4)   . . .

(5)   . . .

(6)   In sub-paragraph (2B)—

"environmental pollution" has the same meaning as in the Pollution Prevention and Control Act 1999;

"primary activity", in relation to an installation falling within any one or more of the descriptions of installation set out in the Table, means an activity the carrying out of which at the installation results in the installation falling within one or more of those descriptions.

Table: Descriptions of Energy-Intensive Installations

Installations regulated under the Pollution Prevention and Control (England and Wales) Regulations 2000 (SI 2000/1973)

1   Part A installations.

Installations that would be so regulated but for a threshold or exception

2   Installations that would be Part A installations but for—
(a)   a relevant numeric threshold, or
(b)   a relevant exception.

Installations that would be so regulated if certain modifications were made to the Regulations

3   Installations that would be Part A installations if the relevant modifications were made.

Corresponding installations in Scotland and Northern Ireland

4   Installations that are situated in Scotland or Northern Ireland, but if situated in England and Wales—
      (a)   would be Part A installations,
      (b)   would be Part A installations but for—
           (i)   a relevant numeric threshold, or
           (ii)   a relevant exception, or
      (c)   would be Part A installations if the relevant modifications were made.

Interpretation of entries 1 to 4

5

(1)   In this entry "the Schedule" means Schedule 1 to the Pollution Prevention and Control (England and Wales) Regulations 2000.

(2)   In entries 1 to 4—
      (a)   "Part A installation" has the meaning given in Part 3 of the Schedule;
      (b)   "relevant exception" means—
           (i)   the exception in paragraph (b)(i) of Part A(1) of Section 2.1 of Part 1 of the Schedule,
           (ii)   the exceptions in paragraph (c) of Part A(1) of Section 5.1 of Part 1 of the Schedule for activities falling within Part B of that Section and for the incineration of specified hazardous waste in an exempt incineration plant, or
           (iii)   the exception in paragraph (e) of Part A(1) of Section 5.1 of Part 1 of the Schedule for incineration as part of a Part B activity in so far as this exception relates to the activities referred to in paragraphs (a) and (b) of Part B of that Section;
      (c)   "the relevant modifications" means the omission of the following provisions of Part 1 of the Schedule:
           (i)   the final twelve words of paragraph (b) of Part A(1) of Section 4.4;
           (ii)   the final twelve words of paragraph (b) of Part A(1) of Section 4.5;

(iii) paragraph 1 of the Interpretation of Part A(1) of Section 5.4;

(iv) the final fourteen words of paragraph (c) of Part A(1) of Section 6.1;

(v) the final fourteen words of paragraph (c) of Part A(1) of Section 6.4; and

(vi) the final fourteen words of paragraph (f)(ii) of Part A(1) of Section 6.8; and

(d) "relevant numeric threshold" means a numeric threshold specified in any of the following provisions of Part 1 of the Schedule:

(i) paragraphs (c) and (d) of Part A(1) of Section 2.1;

(ii) Part A(2) of Section 2.1;

(iii) paragraph (b) of Part A(1) of Section 2.2;

(iv) Part A(1) of Section 2.3;

(v) paragraph (b) of Part A(1) of Section 3.1;

(vi) paragraph (b) of Part A(2) of Section 3.1;

(vii) paragraph (b) of Part A(1) of Section 3.3;

(viii) Part A(2) of Section 3.3;

(ix) paragraph (a) of Part A(1) of Section 3.4;

(x) Part A(2) of Section 3.6;

(xi) paragraphs (c) and (d) of Part A(1) of Section 4.1;

(xii) paragraphs (d) and (e) of Part A(1) of Section 5.1;

(xiii) Part A(1) of Section 5.2;

(xiv) Part A(1) of Section 5.3;

(xv) paragraph (c) of Part A(1) of Section 5.4;

(xvi) paragraph (b) of Part A(1) of Section 6.1;

(xvii) Part A(1) of Section 6.3;

(xviii) paragraphs (a) and (b) of Part A(1) of Section 6.4;

(xix) Part A(2) of Section 6.4;

(xx) Part A(2) of Section 6.7;

(xxi) paragraphs (a) to (e) of Part A(1) of Section 6.8;

(xxii) Part A(2) of Section 6.8; and

(xxiii) Part A(1) of Section 6.9; and

(e) any reference to a part of the United Kingdom includes the territorial waters adjacent to that part.

Power to vary the installations covered by paragraph 51

52

(1) The Treasury may make provision by regulations for varying the installations covered by paragraph 51.

ANNEX 10

EXAMPLE UMBRELLA CLIMATE CHANGE
AGREEMENT

UMBRELLA CLIMATE CHANGE AGREEMENT FOR THE [   ] SECTOR
THIS AGREEMENT is made the

BETWEEN :

(1)   the Secretary of State for Environment, Food and Rural Affairs
       ("the Secretary of State"); and

(2)   [   ] ("the sector association").

1.   RECITALS

1.1   Section 30 of, and Schedule 6 to, the Finance Act 2000 makes
provision for a new tax known as the climate change levy. The levy will
be charged on the supply of taxable commodities. Paragraph 42(1)(c) of
Schedule 6 provides that the amount payable by way of levy on the
supply of taxable commodities shall be 20% of the full rate if the supply
is a reduced-rate supply.

1.2   Paragraphs 44 to 52 of Schedule 6 set out the circumstances in
which a supply is a reduced-rate supply. To be a reduced-rate supply a
supply has to be supplied to a facility which is certified by the Secretary
of State as being covered by a climate change agreement.

1.3   Paragraph 46(b) of Schedule 6 provides that a climate change
agreement may consist of a combination of agreements that falls within
paragraph 48 of that Schedule. Paragraph 48 provides that the
combination is a combination of an umbrella agreement and an
underlying agreement.

1.4   This agreement is an umbrella agreement entered into for the
purposes of the reduced rate of climate change levy. It is not intended
to give rise to contractual obligations between the parties.

1.5   The sector association is a representative (as defined in paragraph
47(2) of Schedule 6) of each facility to which this agreement applies.

1.6   The provision for a reduction in the climate change levy in respect of taxable commodities supplied to facilities covered by a climate change agreement is subject to approval by the European Commission under the state aid rules. Approval is currently being considered by the Commission. This agreement and the underlying agreements are entered into on the assumption that state aids approval will apply to the provision for a reduction in the climate change levy throughout the duration of the agreements. Clause 9.3 makes provision for the termination of this agreement if state aids approval is not granted by the Commission or ceases to apply. Paragraph 7 of Schedule 6 makes provision for the variation of this agreement if the terms of the state aids approval are significantly different at any time during this agreement from the assumed state aid approval terms as defined in that paragraph of that Schedule.

2.   INTERPRETATION AND NOTICES

2.1   In this agreement, unless the context otherwise requires-

"acceptable currency" means a currency described in paragraph 5 of Schedule 2 and references to an absolute target, a relative target, a carbon target or an energy target shall be construed in accordance with that paragraph;

"certification period" means, in relation to a facility to which this agreement applies, a period set out in Part 2 of Schedule 1 for that facility; and "first certification period" and "subsequent certification periods" means, in relation to such a facility, the first certification period and the subsequent certification periods set out in that Part of that Schedule for that facility respectively;

"facility" shall be construed in accordance with clause 3;

"fuel" means coal, coke, gas oil, heavy fuel oil, petrol, liquid petroleum gas, jet kerosene, ethane, naphtha, refinery gas, petroleum coke, natural gas and electricity;

"notice" includes any document whether in paper or electronic form;

"operator" means the operator of a facility to which this agreement applies who enters into an underlying agreement applying to the facility or a person who enters into such an agreement on the operator's behalf;

"sector" means the sector consisting of facilities which in accordance with clause 3.3 belong to the [   ] sector;°›:

"sector target" means the targets for the sector set by clause 5.2 as varied from time to time;

"served" includes copied;

"target period", in relation to a sector, has the meaning given by paragraph 1.1 of Schedule 2 and, in relation to a facility, has the meaning given by paragraph 1.1 of Schedule 2 to the relevant underlying agreement;

"target unit" has the meaning given by clause 2.1 of the relevant underlying agreements;

"termination notice" means a notice served by the Secretary of State on the sector association under clause 6.13 or 9.3 or paragraph 2.13, 4.13 or 6.12 of Schedule 6 to terminate this agreement;

"throughput" means input or output depending on how it is calculated under paragraph 4 of Schedule 2 to the relevant underlying agreement;

"underlying agreement" means an agreement applying to one or more of the facilities identified in Part 1 of Schedule 1 between the Secretary of State and the operator of the facilities to which it applies which is expressed to be entered into for the purposes of this agreement;

"variation certificate" means a variation certificate under paragraph 45 of Schedule 6 to the Finance Act 2000; and

"working day" means any day other than a Saturday, Sunday, Christmas Day, Good Friday or a day falling on a bank holiday in any part of the United Kingdom.

2.2   Any notice served under this agreement shall be in writing.

2.3   A notice served on the sector association may be served by sending it by post to:

```
                                                              
                                                              
                                                              
```

or electronically to:

```
                                                              
```

2.4   A notice served on the Secretary of State may be served by sending it by post to:

> The Climate Change Agreements Secretariat,
> SEP,
> Defra,
> Ashdown House,
> 123 Victoria Street,
> London SW1E 6DE

or electronically to:

> levy.agreements@defra.gsi.gov.uk

2.5   A notice served on an operator may be served in accordance with clause 2.3 of the relevant underlying agreement.

3.   FACILITIES TO WHICH THIS AGREEMENT APPLIES

3.1   This agreement applies to the facilities identified in Part 1 of Schedule 1 but subject to the following provisions of this agreement.

3.2   A facility is only eligible at any time for inclusion in this agreement if, and to the extent that, at that time-

> (a)   it is a facility within the meaning of paragraph 50(2) to (6) of Schedule 6 to the Finance Act 2000;
>
> (b)   it belongs to the [   ] sector; and
>
> (c)   it is not included in another agreement falling within paragraph 47 of Schedule 6 to the Finance Act 2000 or another combination of agreements falling within paragraph 48 of that Schedule.

3.3   A facility belongs to the [   ] sector if it is a facility which manufactures [   ].

3.4   A facility which is eligible for inclusion in this agreement may be added to the list of facilities identified in Part 1 of Schedule 1 by varying this agreement in accordance with paragraph 1 of Schedule 6.

3.5   A facility which is identified in the list of facilities in Part 1 of Schedule 1-

(a)   may be removed from that list by varying this agreement in accordance with paragraph 1 of Schedule 6; and

(b)   shall be removed from that list in the circumstances specified in paragraph 3.3 of that Schedule.

4.   CERTIFICATION PERIODS

4.1   The certification periods for a facility to which this agreement and an underlying agreement apply are the periods set out in Part 2 of Schedule 1 in relation to that facility.

5.   TARGETS AND CURRENCIES

5.1   The targets set for a facility to which this agreement and an underlying agreement apply consist of -

(a)   the sector targets set by clause 5.2; and

(b)   the targets set for the facility in the relevant underlying agreement.

5.2   The sector targets for the facilities to which this agreement applies are the targets set out in paragraph 1.1 of Schedule 2.

5.3   The sector targets for facilities to which this agreement applies shall be varied at the end of 2002 in accordance with the procedure set out in paragraph 5 of Schedule 6 to take account of any variations under clause 4.2 of the relevant underlying agreements of any targets set by those agreements.

5.4   The Secretary of State shall carry out a review at the end of the year 2004 of the sector targets for the final three target periods and shall carry out a further review at the end of the year 2008 of the sector target for the final target period.

5.5   Any such review shall be to ensure that the sector targets being reviewed continue to represent the potential for cost effective energy savings taking account of any changes in technical or market circumstances.

5.6   In carrying out any such review the Secretary of State shall consult with the sector association and take account of its representations on the review.

5.7   The sector targets shall be varied, where appropriate, to take account of the results of any such review in accordance with the procedure set out in paragraph 6 of Schedule 6.

5.8   A similar approach to that adopted when the original sector targets were set shall be adopted when the targets are revised following a review under clause 5.4.

5.9   The acceptable currencies for sector targets, the conversion conventions for those currencies and the requirement for those currencies to be co-ordinated with the currency of targets in the underlying agreements are set out in Part 2 of Schedule 2.

6.   OBLIGATIONS OF THE SECTOR ASSOCIATION

6.1   The sector association shall encourage their members who operate facilities within the sector to enter into underlying agreements with the Secretary of State.

6.2   The sector association shall not impose unreasonable requirements on non-members who operate facilities within the sector and wish to enter into underlying agreements with the Secretary of State.

6.3   The sector association shall not, in particular, impose unreasonable charges on operators or potential operators (whether members or non-members) in respect of the negotiation of this agreement or underlying agreements or the carrying out of its obligations under this agreement.

6.4   The sector association shall meet with the Secretary of State by the end of February in alternate years commencing with the year 2002 to review the operation of this agreement.

6.5   Subject to clause 6.6, by the end of January in alternate years commencing with the year 2003 the sector association shall, in relation to the most recently completed target period, supply the Secretary of State with -

> (a)   the information specified in Part 1 of Schedule 3; and
>
> (b)   if the sector target is not met, the information specified in Part 2 of that Schedule.

6.6   If the sector association is unable to comply with clause 6.5 because of the failure of any operator to comply with clause 6.5 of the relevant underlying agreement, the association shall -

> (a)   serve notice on the Secretary of State informing him or her that this is the case and giving the facility number of the facility concerned; and
>
> (b)   supply him or her with the information which would have been required if the facility had not been a facility to which this agreement applies.

6.7   Where the Secretary of State receives notice under clause 6.6(a) and is supplied with the information required by clause 6.6(b), he or she may terminate the relevant underlying agreement by notice under clause 9.3 of that agreement and vary the sector targets in accordance with paragraph 9 of Schedule 6.

6.8   The Secretary of State may at any time serve a notice on the sector association requesting it to supply him or her with such information as he or she may require in connection with his or her functions under Schedule 6 to the Finance Act 2000 within such period as may be specified in the notice (not being less than 10 working days or, if information is required from an operator, 15 working days).

6.9   Where the Secretary of State serves a notice under clause 6.8, the sector association shall supply the information requested within the period specified in the notice.

6.10   The sector association may serve a notice on an operator requesting it to supply the association with such information as it requires to comply with a request from the Secretary of State for information under clause 6.8.

6.11   The Secretary of State may appoint a person to undertake an independent audit of information provided by the sector association (and information provided by operators to the association) and the sector association shall co-operate with any person so appointed; and, for that purpose, the sector association shall keep proper records and make them available for inspection when required by the auditor.

6.12.   Where the Secretary of State considers that the sector association has failed to comply with an obligation under this clause, he or she may serve a notice on the sector association identifying -

> (a)   the obligation which he or she considers has not been complied with; and

> (b)   the steps that he or she considers the sector association should take to comply with that obligation.

6.13   Where the Secretary of State serves a notice under clause 6.12 and the sector association fails to take the steps set out in the notice, he may terminate this agreement by serving a termination notice on the sector association.

6.14   A termination notice under clause 6.13 shall specify the date on which this agreement ceases to have effect (which shall be at least 10 working days after the date on which the notice is served) and this agreement shall cease to have effect on the date specified in the termination notice unless the notice is withdrawn before that date. Where this agreement ceases to have effect in this way no new certificate will be given under clause 7 for any further certification periods and any existing certificates given under that clause will be terminated by a variation certificate.

7.   CERTIFICATION OF FACILITIES BY THE SECRETARY OF STATE

7.1   Subject to clause 7.8 and 7.9, the Secretary of State shall certify that a facility to which this agreement and an underlying agreement apply is covered by a climate change agreement for the first certification period for that facility.

7.2   Subject to clause 7.8 and 7.9, the Secretary of State shall certify that a facility to which this agreement and an underlying agreement apply is covered by a climate change agreement for a subsequent certification period for that facility if it appears to him or her that progress made in the immediately preceding certification period towards meeting the targets set for the facility is satisfactory.

7.3   For the purpose of clause 7.2, progress made in an immediately preceding certification period towards meeting the targets set for a facility is to be taken as being satisfactory if -

    (a)   the sector target for the target period falling within the immediately preceding certification period has been met; or

    (b)   the target set for the facility in the relevant underlying agreement for that target period has been met and -

        (i) there is not a tolerance band in the underlying agreement in relation to that target; or

        (ii) there is such a tolerance band but the target has been met without needing to take account of it; or

        (iii)   there is such a tolerance band, the target has only been met by taking account of it and the circumstances are as set out in clause 7.4(a); or

    (c)   the circumstances are as set out in clause 7.4(a) and (b).

7.4   The circumstances mentioned in clause 7.3 are that -

    (a)   the qualitative requirements set for the facility in the relevant underlying agreement for the purpose of this clause have been met; and

    (b)   the target set for the facility in the relevant underlying agreement for the target period falling within the immediately preceding certification period has not been met because of a relevant constraint or requirement which had a major impact on the performance of the operator of the facility and prevented the target from being achieved.

7.5   In clause 7.4(b) "relevant constraint or requirement" means, in relation to a target -

(a)   a constraint or requirement imposed by or under town and country planning, environmental, health and safety or food hygiene legislation;

(b)   a constraint or requirement imposed on the construction or operation of a combined heat and power plant under section 14 of the Energy Act 1976 or section 36 of the Electricity Act 1989; or

(c)   a constraint imposed by the gas or electricity network,

where the constraint or requirement is inconsistent with an assumption used for setting or reviewing the target.

7.6   In clause 7.5 "constraint" includes a delay.

7.7   Where the Secretary of State certifies that a facility to which this agreement applies is covered by a climate change agreement for a certification period, he or she shall give a certificate to the Commissioners of Customs and Excise stating that for that period the facility is to be taken as being covered by a climate change agreement and shall copy that certificate to the sector association and to the operator of the facility. A copy of any relevant variation certificate given to the Commissioners of Customs and Excise shall also be copied to the sector association and the operator of the facility.

7.8   The Secretary of State will not certify that a facility to which this agreement applies is covered by a climate change agreement (or where a certificate has already been issued, will issue a variation certificate) if -

(a)   the facility is by virtue of clause 3.2 not eligible for inclusion in this agreement; or

(b)   the conditions for issuing a certificate under clause 7.1 or 7.2 are not met; or

(c)   the facility is excluded from this agreement under paragraph 1.3 or 3.3 of Schedule 6.

7.9   The Secretary of State may decide not to certify that a facility to which this agreement applies is covered by a climate change agreement (or where a certificate has already been issued, decide to issue a variation certificate) if -

(a)   the sector association has failed to supply him or her with the information required in relation to that facility or the information supplied is incomplete or inaccurate;

(b)   the sector association or the operator of the facility has failed -

(i) to co-operate with a person appointed under clause 6.11 in relation to the carrying out of an audit of information provided in relation to that facility; or

(ii) to keep proper records for that purpose or make them available for inspection by the auditor;

(c)   the operator of the facility has failed to comply with any obligation in clause 6 of the underlying agreement applying to the facility; or

(d)   the sector association or the operator fails to meet its share of the costs of an adjudication in relation to which it is a party.

7.10   Subject to clause 7.12, where the Secretary of State is minded not to certify that a facility to which this agreement applies is covered by a climate change agreement (or where a certificate has already been issued, he or she is minded to issue a variation certificate), he or she shall serve a notice on the sector association and the operator of the facility stating that he or she is so minded and setting out his or her reasons.

7.11   Where the Secretary of State serves a notice under clause 7.10, the sector association or the operator may, not more than 10 working days following receipt of that notice, serve a notice on the Secretary of State (copied to the sector association or the operator, as the case may be) setting out why the server of the notice considers that the Secretary of State should certify that the facility is covered by a climate change agreement (or not issue a variation certificate) and, where such a notice is served, the dispute procedure set out in Schedule 4 shall apply.

7.12   Clause 7.10 shall not apply where -

(a) a termination notice has been served either under this agreement or under the relevant underlying agreement in relation to the facility; or

(b) the facility concerned is no longer a facility to which an underlying agreement applies.

8.   CONFIDENTIALITY

8.1   The Secretary of State shall be entitled to publish, without the sector association's consent, this agreement and a list of the facilities in Part 1 of Schedule 1 which are certified under clause 7 as being covered by a climate change agreement.

8.2   The Secretary of State shall be entitled to disclose, without the sector association's consent, any other information relating to this agreement in the following circumstances -

(a) where the disclosure is made under and in accordance with the terms of any legislation;

(b) where the disclosure is made to a relevant authority for the purposes of-

(i) the Secretary of State's functions under Schedule 6 to the Finance Act 2000; or

(ii)the authority's functions; or

(c) where the disclosure is made in the course of legal proceedings.

8.3   In clause 8.2(a) "legislation" means primary or secondary legislation and includes legislation of the European Community.

8.4   The Secretary of State shall consult the sector association before making any disclosure under clause 8.2 where he or she considers it appropriate in the circumstances to do so.

8.5   Save as provided for in clause 8.1 and 8.2, the Secretary of State shall only disclose information relating to this agreement with the consent of the sector association.

8.6   The relevant authorities referred to in clause 8.2(b) are -

(a) either House of Parliament (including any committee);

(b) the European Commission;

(c) the Commissioners of Customs and Excise;

(d) the relevant environmental regulator for a facility under Part I of the Environmental Protection Act 1990 or regulations made under section 2 of the Pollution Prevention and Control Act 1999 or corresponding legislation for Northern Ireland;

(e) any person appointed by the Secretary of State under clause 6.11 to undertake an audit of information provided by the sector association; and

(f) the authorities charged with regulating under the Competition Act 1998.

8.7   The Secretary of State shall take steps to prevent any person, whom he or she appoints under clause 6.11 to undertake an audit, from disclosing information obtained in carrying out the audit to any one other than the Secretary of State, except to the extent needed to carry out the audit.

9.   DURATION OF THIS AGREEMENT

9.1   Subject to clause 9.2, this agreement shall continue in force from the date on which it is made until 31st March 2013.

9.2   This agreement may be terminated before 31st March 2013 –

(a) by notice served by the sector association on the Secretary of State; or

(b) by a termination notice served by the Secretary of State on the sector association in accordance with clause 6.13 or 9.3 or paragraph 2.13, 4.13 or 6.12 of Schedule 6.

9.3   If state aids approval for the reduction in climate change levy in respect of taxable commodities supplied to facilities covered by this agreement is not granted by the European Commission or ceases to apply, the Secretary of State may terminate this agreement by serving a termination notice on the sector association.

9.4   A termination notice under clause 9.3 shall specify the date on which this agreement ceases to have effect (which shall be at least 10 working days after the date on which the notice is served) and this agreement shall cease to have effect on the date specified in the termination notice unless the notice is withdrawn before that date. Where this agreement ceases to have effect in this way, no new certificate will be given under clause 7 for any further certification periods and any existing certificates given under that clause will be terminated by a variation certificate.

10.   VARIATION OF AGREEMENT

10.1   Subject to clause 10.2, any provision of this agreement may be varied by agreement of the Secretary of State and the sector association in accordance with Part 1 of Schedule 6.

10.2   The sector targets and the currency of the targets shall be varied in accordance with Part 2 of Schedule 6 but, save as otherwise expressly provided in this agreement, no variation of the sector targets or the currency of the targets shall be made.

Signed by authority of
the Secretary of State

Signed on behalf of the
sector association

SCHEDULE 1

FACILITIES TO WHICH THIS AGREEMENT APPLIES AND CERTIFICATION PERIODS

PART 1
LIST OF FACILITIES

Facility number	Operator of facility	Address of facility	Description of facility

PART 2
CERTIFICATION PERIODS

1   The first certification period for the facilities listed above is 1st April 2001 to 31st March 2003.

2   The subsequent certification periods for those facilities are –

(a)   1st April 2003 to 31st March 2005;

(b)   1st April 2005 to 31st March 2007;

(c)   1st April 2007 to 31st March 2009;

(d)   1st April 2009 to 31st March 2011; and

(e)   1st April 2011 to 31st March 2013.

SCHEDULE 2

SECTOR TARGETS AND CURRENCIES

PART 1

SECTOR TARGETS

1.   Sector targets

1.1   The sector targets are set out in the following Table.

Target period	Sector target
1st October 2001 to 30th September 2002	
1st October 2003 to 30th September 2004	
1st October 2005 to 30th September 2006	
1st October 2007 to 30th September 2008	
1st October 2009 to 30th September 2010	

1.2   The sector target for a target period shall be adjusted to reflect any adjustments made under paragraph 1.3 (adjustment of targets to take account of emissions trading) of Schedule 2 to the relevant underlying agreements in relation to the facilities in the sector.

1.3   For the purpose of determining whether a sector target has been met, the sector shall be treated as consisting of the facilities to which this agreement and the underlying agreements apply.

1.4   For the purpose of determining whether a sector target has been met, the units of energy used by the facilities in the sector, the units of carbon emitted from those facilities and the throughput of those facilities shall be calculated in accordance with the following paragraphs of this Part of this Schedule.

1.5   Where the target period for a target unit under a relevant underlying agreement is different from the target period for the sector targets but overlaps with that period, the figures for the target unit which are relevant to that overlapping period shall be used for the purposes of determining under those paragraphs whether the relevant sector target has been met and this agreement shall apply in relation to that target unit with appropriate modifications.

2.   Calculation of units of energy used by sector

2.1   The total number of units of energy used by the facilities in the sector during a target period shall be the sum of the units of energy used by those facilities during that target period calculated in accordance with paragraph 2 of Schedule 2 to the relevant underlying agreements.

2.2   The units of energy used shall be measured in kilowatt hours.

3.   Calculation of carbon emissions from sector

3.1   The total number of units of carbon emitted from the facilities in the sector during a target period shall be the sum of the units of carbon emitted from those facilities during that target period calculated in accordance with paragraph 3 of Schedule 2 to the relevant underlying agreements.

3.2   The units of carbon emitted shall be expressed as kilogrammes of mass carbon (rather than carbon dioxide).

4.   Calculation of throughput of sector

4.1   The total throughput of the facilities in the sector to which this agreement and the underlying agreements apply during a target period shall be the sum of the throughput of those facilities during that target period calculated in accordance with paragraph 4 of Schedule 2 to the relevant underlying agreements.

PART 2

CURRENCIES

5.   Acceptable currencies

5.1   The acceptable currencies for the sector targets in Part 1 of this Schedule are –

(a)   for an absolute carbon target, carbon emitted during the target period;

(b) for an absolute energy target, energy used during the target period;

(c) for a relative carbon target, carbon emitted during the target period per unit of throughput during that period; and

(d) for a relative energy target, energy used during the target period per unit of throughput during that period.

6.   Conversion conventions for currencies

6.1   The conversion conventions for the currencies set out in paragraph 5 are as follows.

Converting an energy target to a carbon target

6.2   To convert an energy target to a carbon target multiply the energy target by the relevant carbon emission factors as follows:

> Carbon target = Energy target x ((assumed percentage of electricity x electricity carbon emission factor) + (assumed percentage of gas x gas carbon emission factor) + (assumed percentage of coal x coal carbon emission factor) + (assumed percentage of oil x oil carbon emission factor) + (assumed percentage of other fuels x relevant carbon emission factors for those fuels)).

Converting a carbon target to an energy target

6.3   To convert a carbon target to an energy target divide the carbon target by the relevant carbon emission factors as follows:

> Energy target = Carbon target / ((assumed percentage of electricity x electricity carbon emission factor) + (assumed percentage of gas x gas carbon emission factor) + (assumed percentage of coal x coal carbon emission factor) + (assumed percentage of oil x oil carbon emission factor) + (assumed percentage of other fuels x relevant carbon emission factors for those fuels)).

Converting a relative target to an absolute target

6.4   To convert a relative target to an absolute target multiply the relative target by the assumed throughput during the target period to get the absolute target.

Converting an absolute target to a relative target

6.5   To convert an absolute target to a relative target divide the absolute target by the assumed throughput during the target period to get the relative target.

6.6   For the purpose of this paragraph -

(a)   the carbon emission factors for each fuel shall be the carbon emission factors for those fuels set out in the Table in paragraph 3.1 of Schedule 2 to the relevant underlying agreements;

(b)   the assumed percentage of a fuel is the percentage of the total number of units of energy used by the facilities to which the target applies during the relevant target period which are assumed to be units of energy from that fuel and shall be agreed by the Secretary of State and the sector association before each application of the convention or, in default of agreement, specified by the Secretary of State in a notice served on the sector association; and

(c)   the assumed throughput for the relevant target period shall be agreed by the Secretary of State and the sector association before each application of the convention or, in default of agreement, specified by the Secretary of State in a notice served on the sector association.

7.   Co-ordination of currencies in this agreement with underlying agreements

7.1   The currency of the sector targets shall be co-ordinated with the currency of the targets in the underlying agreements as follows –

(a)   if more than 50% of underlying agreement targets in a target period are absolute targets, the sector targets for the following target periods shall also be absolute targets;

(b) if more than 50% of the underlying agreement targets in a target period are relative targets, the sector targets for the following target periods shall also be relative targets;

(c)   if more than 50% of the underlying agreement targets in a target period are carbon targets, the sector targets for the following target periods shall also be carbon targets; and

(d)   if more than 50% of the underlying agreement targets in a target period are energy targets, the sector targets for the following target periods shall also be energy targets.

7.2   For the purposes of paragraph 7.1, the percentage of targets for a target period which are absolute or relative targets or carbon or energy targets shall be determined by reference to the total units of energy used by the relevant facilities in that period calculated in accordance with paragraph 2 of Schedule 2 to the relevant underlying agreements.

SCHEDULE 3

INFORMATION TO BE SUPPLIED TO THE SECRETARY OF STATE

PART 1

INFORMATION REQUIRED WHETHER OR NOT SECTOR TARGET IS MET

The information required by clause 6.5(a) is as follows -

1.   The total number of units of energy used by relevant facilities during the relevant target period calculated in accordance with paragraph 2 of Schedule 2 with a sufficient breakdown of that information to determine whether the currencies of the sector targets need to be changed under paragraph 7 of that Schedule.

2.   If the sector target for the relevant target period is a carbon target, the total number of units of carbon emitted from the relevant facilities during that period calculated in accordance with paragraph 3 of Schedule 2.

3.   The total throughput for the relevant facilities in the sector for the relevant target period calculated in accordance with paragraph 4 of Schedule 2.

4.   The adjustment to be made to the sector target for the relevant target period in accordance with paragraph 1.2 of Schedule 2.

5.   For each target unit with an absolute target for the relevant target period, the throughput of that target unit during that period calculated in accordance with paragraph 4 of Schedule 2 to the relevant underlying agreement.

PART 2

INFORMATION REQUIRED ONLY IF SECTOR TARGET IS NOT MET

The information required by clause 6.5(b) is as follows -

6.   For each target unit, the total number of units of energy used during the relevant target period by the target unit in relation to each type of fuel calculated in accordance with paragraph 2 of Schedule 2 to the relevant underlying agreement.

7.   For each target unit with a carbon target for the relevant target period, the total number of units of carbon emitted from the target unit during that period calculated in accordance with paragraph 3 of Schedule 2 to the relevant underlying agreement.

8.   For each target unit with a relative target for the relevant target period, the throughput of that target unit during that period calculated in accordance with paragraph 4 of Schedule 2 to the relevant underlying agreement.

9.   For each target unit where the target is to be adjusted under paragraph 1.2 or 1.3 of Schedule 2 to the relevant underlying agreement, the information needed to calculate the adjustment.

10.   For each operator which relies on clause 7.4(a), a copy of its energy plan and a description of the steps taken to implement the plan.

11.   For each operator which relies on clause 7.4(b), details of the relevant constraint or requirement and of its impact on the performance of the operator.

SCHEDULE 4

CERTIFICATION DISPUTE RESOLUTION

1.   Where the Secretary of State receives a notice under clause 7.11 he shall, not more than 10 working days after receiving the notice, serve notice on the sector association and the operator stating whether or not he intends to change his proposed decision and, if not, setting out his reasons.

2.   Where the Secretary of State serves notice on the sector association and the operator under paragraph 1 that he or she does not intend to change his or her proposed decision and there is a dispute on the facts, the sector association or the operator may, not more than 10 working days after service of the notice, refer the dispute for adjudication to an institution named on the list prepared under paragraph 1 of Schedule 5 and that Schedule shall apply in relation to the adjudication. A notice stating that there has been any such referral shall be served by the party making the referral on the other parties to the dispute.

3.   Each party to the dispute shall have 20 working days from the date on which they receive notice from the nominated adjudicator of the address to which representations should be sent to make representations to the nominated adjudicator. Any such representations shall be copied by the party making the representations to the other parties to the dispute.

4.   The adjudicator may request further information from any of the parties to the dispute. Any such request shall be in writing, shall specify the date by which the information is required and shall be copied to the other parties to the dispute. Any information provided in response to such a request shall be copied by the party providing the information to the other parties to the dispute.

5.   The adjudicator will, on the basis of the representations provided to him or her and any additional information he or she considers to be necessary, make a finding on the disputed questions of fact and notify the parties of that finding. The adjudicator's finding on the disputed questions of fact shall be binding on the parties.

6.   Not more than 5 working days after the receipt of the adjudicator's finding, the Secretary of State shall serve notice on the sector association and the operator stating whether or not he or she intends to change his or her proposed decision and, if not, setting out his reasons.

SCHEDULE 5

ADJUDICATORS

1.   A person who wishes to refer a dispute on the facts to an institution for adjudication may refer that dispute to one of the institutions named in a list agreed by the sector association and the Secretary of State or, in default of agreement, prepared by the Secretary of State and served on the sector association.

2.   Where a dispute is referred to one of the institutions named in the list mentioned above, that institution shall nominate a person to carry out the adjudication.

3.   The nominated adjudicator shall notify the parties of his appointment and shall inform them of the address to which representations are to be sent.

4.   Subject to express provisions of this agreement on procedural matters in relation to the adjudication, it shall be for the nominated adjudicator to determine the procedure which is to be followed in relation to the adjudication but in doing so-

(a)   he or she shall take account of the wishes of the parties;

(b)   he or she may take expert advice, after consulting the parties, to assist him with the adjudication;

(c)   he or she shall ensure that each party has an adequate opportunity to respond to any representations made by the other party, and any evidence, that he proposes to take into account; and

(d)   he or she may impose or extend any time limit for any action to be taken by any party to the dispute and may proceed with the adjudication in such manner as he or she considers appropriate in the circumstances if a time limit is not complied with.

5.   The parties to the adjudication shall bear their own costs.

6.   The cost of the adjudication shall be shared equally between the parties.

SCHEDULE 6

VARIATIONS

PART 1

GENERAL VARIATIONS

1.   General variations

1.1   Where a party to this agreement wishes to vary any of its provisions other than the sector targets or the currency of the targets, it may serve a notice on the other party setting out the variation and the reasons for the variation.

1.2   A party which receives a notice under paragraph 1.1 shall have 20 working days after its receipt to respond in writing.

1.3   Where the variation is agreed this agreement shall be varied accordingly.

1.4   Any notice served under paragraph 1.1 or 1.2 and any variation agreed shall be copied to the operator of every facility to which this agreement applies.

PART 2

VARIATION OF SECTOR TARGET PROVISIONS

2. Variation of sector targets due to inclusion or exclusion by agreement of a facility

2.1   Subject to paragraph 2.3(b), sector targets shall be varied in accordance with the following provisions of this paragraph where -

(a)   in the case of relative and absolute targets, a facility is included in the list of facilities in Part 1 of Schedule 1 by means of a variation of this agreement in accordance with paragraph 1 and that facility is identified in an underlying agreement;

(b)   in the case of relative targets, a facility has been excluded from the list of facilities in Part 1 of Schedule 1 by means of a variation of this agreement in accordance with paragraph 1 and that facility was identified in an underlying agreement.

2.2   Any variation of sector targets required by paragraph 2.1(a) or (b) shall be effected at the end of the first target period which ends after the inclusion or exclusion (the "relevant target period").

2.3   The sector targets for the relevant target period and for any subsequent target periods shall be varied to reflect -

(a)   in the case of absolute targets, all inclusions in the sector of facilities not previously taken into account;

(b)   in the case of relative targets, all inclusions in and exclusions from the sector of facilities not previously taken into account where these affect the overall product mix but not otherwise.

2.4   Where a variation to sector targets is required by this paragraph, the Secretary of State shall, not more than 5 working days after receiving the information mentioned in paragraph 3 of Schedule 3, serve a notice on the sector association and on all operators stating the variation that he proposes should be made to the sector targets.

2.5   The sector association and operators shall each have 5 working days after receipt of a notice under paragraph 2.4 to serve notice of its response on the Secretary of State.

2.6   If the Secretary of State and the sector association agree on a variation of the sector targets, the targets shall be varied accordingly.

2.7   If the sector association and the Secretary of State fail to agree on a variation of the sector targets within the period of 5 working days mentioned in paragraph 2.5 and there is a dispute on the facts, either party may refer the dispute for adjudication to an institution named on the list prepared under paragraph 1 of Schedule 5 and that Schedule shall apply in relation to the adjudication.

2.8   Notice of any such referral shall be served by the party making the referral on the other party.

2.9   Each party to the dispute shall have 5 working days, from the date on which they receive notice from the nominated adjudicator of the address to which representations should be sent, to make representations to the nominated adjudicator. Any such representations shall be copied by the party making the representations to the other party to the dispute.

2.10   The adjudicator may request further information from either party to the dispute. Any such request shall be in writing, shall specify the date by which the information is required and shall be copied to the other party to the dispute. Any information provided in response to such a request shall be copied by the party providing the information to the other party to the dispute.

2.11   The adjudicator will, on the basis of the representations provided to him or her and any additional information he or she considers to be necessary, make a finding on the disputed questions of fact and notify the parties of that finding.

2.12   The adjudicator's finding on the disputed questions of fact shall be binding on the parties but it shall be for the parties to agree, in the light of such finding, on the variation required and if agreement is reached, the targets shall be varied accordingly.

2.13   Where, despite any finding made by the adjudicator, the Secretary of State and the sector association fail to agree the variation of the sector targets, the Secretary of State may terminate this agreement by serving a termination notice on the sector association.

2.14   A termination notice under paragraph 2.13 shall specify the date on which this agreement ceases to have effect (which shall be at least 10 working days after the date on which the notice is served) and shall be accompanied by a statement setting out the Secretary of State's reasons for not agreeing with the sector association on the variation required.

2.15   Where the Secretary of State serves a termination notice under paragraph 2.13, this agreement shall cease to have effect on the date specified in the notice unless the notice is withdrawn before that date. Where this agreement ceases to have effect in this way, no new certificate will be given under clause 7 for any further certification periods and any existing certificates given under that clause will be terminated by a variation certificate. Where a termination notice is withdrawn, the targets shall be varied to reflect any agreement reached between the Secretary of State and the sector association.

3.   Variation of agreement and sector targets to reflect ineligibility exclusions

3.1   This agreement shall be varied in accordance with the following provisions of this paragraph-

> (a)   to exclude a facility (or part of it) which has been excluded from a relevant underlying agreement on the basis that it was not eligible for inclusion in that agreement; and

(b)   to vary the sector targets to take account of the exclusion.

3.2   The Secretary of State shall serve notice on the sector association and the operators specifying the facility or part to be excluded and setting out the variations which he or she considers should be made to this agreement to take account of the exclusion, including any variation of the sector targets, and inviting representations from the sector association and the operators within 20 working days.

3.3   After considering any representations made the Secretary of State shall serve a further notice on the sector association and the operators specifying the facility or part to be excluded and setting out the variations which he or she considers should be made to this agreement to take account of the exclusion, including any variation of the sector targets, and this agreement shall be varied in accordance with that notice without any further action being required. Where this agreement is varied in this way, a variation certificate will be issued if necessary to reflect the exclusion.

4.   Variation of absolute sector targets due to fall of throughput

4.1   Where the annual level of throughput of the sector during a target period (the "relevant target period") calculated in accordance with paragraph 4 of Schedule 2 is less than 90 per cent of the annual throughput of the sector when the sector targets were set, the sector target for the relevant target period shall be varied to take account of that fall in throughput if they are absolute targets.

4.2   For the purpose of paragraph 4.1, where the relevant target period is the second target period, the annual level of throughput of the sector when the sector targets were set is the level of throughput of facilities in the sector-

(a)   if there has not been a relevant variation, in the year 1999; or

(b)   if there has been a relevant variation, in the target period which immediately preceded the variation or, if there has been more than one such variation, the last one.

"Relevant variation" means any variation other than one under paragraph 5 or 7 and does not include an adjustment under paragraph 1.2 of Schedule 2.

4.2A For the purpose of paragraph 4.1, where the relevant target period is one of the final three target periods, the annual level of throughput of the sector when the sector targets were set is the level of throughput of facilities in the sector either: -

(a) as agreed between the Secretary of State and the sector association during the target review under clause 5.3 for the relevant target period, or if there has been more than one such review, during the last target review; or

(b) the assumed throughput for the relevant target period as agreed or specified in accordance with paragraph 6.6(c) of Schedule 2 to the umbrella agreement,

whichever is the latest.

4.3   Any variation of an absolute sector target required by paragraph 4.1 shall be effected at the end of the relevant target period. The absolute sector target for the relevant target period shall be varied.

4.4   Where a variation to the sector target is required by paragraph 4.1, the Secretary of State shall, not more than 5 working days after receiving the information mentioned in paragraph 3 of Schedule 3, serve a notice on the sector association and all operators stating the variation which he proposes should be made to the sector target.

4.5   The sector association or an operator shall respond, not more than 5 working days after receipt of a notice under paragraph 4.4, by serving a notice of its response on the Secretary of State.

4.6   If the Secretary of State and the sector association agree on a variation of the sector target, the target shall be varied accordingly.

4.7   If the sector association and the Secretary of State fail to agree on a variation of the sector target within the period of 5 working days mentioned in paragraph 4.5 and there is a dispute on the facts, either party may refer the dispute for adjudication to an institution named on the list prepared under paragraph 1 of Schedule 5 and that Schedule shall apply in relation to the adjudication.

4.8   Notice of any such referral shall be served by the party making the referral on the other party.

4.9   Each party to the dispute shall have 5 working days, from the date on which they receive notice from the nominated adjudicator of the address to which representations should be sent, to make representations to the nominated adjudicator. Any such representations shall be copied by the party making the representations to the other party to the dispute.

4.10   The adjudicator may request further information from either party to the dispute. Any such request shall be in writing, shall specify the date by which the information is required and shall be copied to the other party to the dispute. Any information provided in response to such a request shall be copied by the party providing the information to the other party to the dispute.

4.11   The adjudicator will, on the basis of the representations provided to him or her and any additional information he or she considers to be necessary, make a finding on the disputed questions of fact and notify the parties of that finding.

4.12   The adjudicator's finding on the disputed questions of fact shall be binding on the parties but it shall be for the parties to agree, in the light of such finding, on what variation is required and if agreement is reached, the target shall be varied accordingly.

4.13   Where, despite any finding made by the adjudicator, the Secretary of State and the sector association fail to agree the variation of the sector target, the Secretary of State may terminate this agreement by serving a termination notice on the sector association.

4.14   A termination notice under paragraph 4.13 shall specify the date on which this agreement ceases to have effect (which shall be at least 10 working days after the date on which the notice is served) and shall be accompanied by a statement setting out the Secretary of State's reasons for not agreeing with the sector association on the variation required.

4.15   Where the Secretary of State serves a termination notice under paragraph 4.13, this agreement shall cease to have effect on the date specified in the notice unless the notice is withdrawn before that date. Where this agreement ceases to have effect in this way no new certificate will be given under clause 7 for any further certification periods and any existing certificates given under that clause will be terminated by a variation certificate. Where a termination notice is withdrawn, the target shall be varied to reflect any agreement reached between the Secretary of State and the sector association.

5.   Variations of sector targets following combined heat and power appraisals

5.1   The Secretary of State shall serve notice on the sector association and relevant operators setting out the variations which he or she considers should be made to sector targets to take account of any variations under clause 4.2 of the relevant underlying agreements of any targets set by those agreements and inviting representations from the sector association and the operators within 20 working days.

5.2   After considering any representations made the Secretary of State shall serve a further notice on the sector association and the operators setting out the variations he or she intends to make and the sector targets shall be varied in accordance with that notice without any further action being required.

6.   Variations of sector targets following a review at end of 2004 and 2008

6.1   Where the Secretary of State carries out a review under clause 5.4 and he or she considers that it is appropriate to vary the sector targets reviewed to take account of the results of the review, he or she shall serve a notice of the appropriate variations on the sector association setting out -

(a)   the revised sector targets in an acceptable currency; and

(b)   his or her reasons for considering those variations to be appropriate.

6.2   The sector association shall, not more than 20 working days after receipt of a notice under paragraph 6.1, serve notice of its response on the Secretary of State.

6.3   If the Secretary of State and the sector association fail to agree on the variations and there is a dispute on the facts, either party may refer the dispute for adjudication to an institution named on the list prepared under paragraph 1 of Schedule 5 and that Schedule shall apply in relation to the adjudication.

6.4   Notice of any such referral shall be served by the party making the referral on the other party.

6.5   Each party to the dispute shall have 20 working days from the date on which they receive notice from the nominated adjudicator of the address to which representations should be sent to make representations to the nominated adjudicator. Any such representations shall be copied by the party making the representations to the other party to the dispute.

6.6   The adjudicator may request further information from either party to the dispute. Any such request shall be in writing, shall specify the date by which the information is required and shall be copied to the other party to the dispute. Any information provided in response to such a request shall be copied by the party providing the information to the other party to the dispute.

6.7   The adjudicator will, on the basis of the representations provided to him or her and any additional information he considers to be necessary, make a finding on the disputed questions of fact and notify the parties of that finding.

6.8   The adjudicator's finding on a disputed questions of fact shall be binding on the parties but it shall be for the parties to agree, in the light of that finding, what variations to the sector targets are required to ensure that the targets continue to represent the potential for cost effective energy savings taking account of any changes in technical or market circumstances.

6.9   Where the Secretary of State and the sector association agree the variations to the sector targets, the sector association shall serve a notice on the Secretary of State setting out the variations that it proposes should be made to the targets in the underlying agreements to take account of those variations.

6.10   If the Secretary of State and the sector association agree to the variations that should be made to the targets in the underlying agreements to take account of the sector target variations, the Secretary of State shall serve a notice on each operator setting out the agreed variations to the sector targets and to the targets in that operator's underlying agreement and inviting representations on those variations from the operator within 20 working days.

6.11   If, having considered any representations made by the operators under paragraph 6.10, the Secretary of State and the sector association are in agreement on the variations that should be made to the sector targets and the targets in the underlying agreements, the sector targets shall be varied accordingly and the Secretary of State shall serve a notice on each operator setting out the variations to the targets in that operator's underlying agreement as agreed with the sector association.

6.12   The Secretary of State may serve a termination notice on the sector association terminating this agreement if -

(a)   the Secretary of State and the sector association fail to agree on the variation of the sector targets or the targets in the underlying agreements; or

(b)   the sector association fails to serve a notice under paragraph 6.9.

6.13   A termination notice under paragraph 6.12 shall specify the date on which this agreement ceases to have effect, which shall be at least 10 working days after the date on which the notice is served. A termination notice under paragraph 6.12(a) shall be accompanied by a statement setting out the Secretary of State's reasons for not agreeing with the sector association on the variations required.

6.14   Where the Secretary of State serves a termination notice under paragraph 6.12, this agreement shall cease to have effect on the date specified in the notice unless the notice is withdrawn before that date. Where this agreement ceases to have effect in this way no new certificate will be given under clause 7 for any further certification periods and any existing certificates given under that clause will be terminated by a variation certificate. Where a termination notice is withdrawn, the targets shall be varied to reflect any agreement reached between the Secretary of State and the sector association.

7. Variations of sector targets on the basis of a significant difference in the terms of the state aids approval

7.1   If the terms of the state aids approval given in relation to the reduction in climate change levy are significantly different at any time during this agreement from the assumed state aid approval terms, then either party to this agreement who wishes to vary the sector targets to take account of that significant difference may serve a notice on the other party setting out the variation and the reasons for the variation.

7.2   A party which receives a notice under paragraph 7.1 shall have 20 working days from its receipt to respond in writing.

7.3   Where the Secretary of State and the sector association agree the variations to the sector targets, the sector association shall serve a notice on the Secretary of State setting out the variations that it proposes should be made to the targets in the underlying agreements to take account of those sector target variations.

7.4   If the Secretary of State and the sector association agree to the variations that should be made to the targets in the underlying agreements to take account of the sector target variations, the Secretary of State shall serve a notice on each operator setting out the agreed variations to the sector targets and to the targets in that operator's underlying agreement and inviting representations from the operator within 20 working days on those variations.

7.5   If, having considered any representations made by the operators under paragraph 7.4, the Secretary of State and the sector association are in agreement on the variations that should be made to the sector targets and the targets in the underlying agreements, the sector targets shall be varied accordingly and the Secretary of State shall serve a notice on each operator setting out the variations to the targets in that operator's underlying agreement as agreed with the sector association.

7.6   In paragraph 7.1 "assumed state aid approval terms" means state aids approval for -

(a)   the amount payable by way of climate change levy on a reduced-rate supply being only 20 per cent of the amount that would be payable if the supply were neither a half-rate supply nor a reduced-rate supply;

(b)   the exemption from the climate change levy provided for by paragraph 15 of Schedule 6 to the Finance Act 2000 (supplies to combined heat and power stations);

(c)   the exemption from the climate change levy provided for by paragraph 18 of that Schedule (supply not used as fuel);

(d)   the exemption from the climate change levy provided for by paragraph 19 of that Schedule (electricity from renewable sources); and

(e)   the provision of enhanced capital allowances for investment in energy efficiency measures.

8.   Variation of the currency of a sector target following a change in the currency of underlying agreements

8.1   Where a change of the currency in underlying agreements results in the currency of the sector targets in this agreement no longer being co-ordinated with the currency of the underlying agreements in the way referred to in paragraph 7 of Schedule 2, the currency in the sector targets shall be varied to restore the co-ordination.

8.2   The Secretary of State shall serve a notice on the sector association setting out the variation of the currency of the sector targets required to bring the sector targets back into co-ordination with the targets in the underlying agreements. The variation shall be in accordance with the conversion conventions set out in paragraph 6 of Schedule 2.

8.3   The sector targets will then be varied in accordance with that notice on the date specified in the notice without further action being required.

9.   Variation of sector targets where operator has failed to supply information needed to decide whether or not targets have been met

9.1   Where clause 6.7 applies and the Secretary of State proposes to vary the sector targets for the relevant period and subsequent periods, he or she shall serve notice specifying the proposed variations on the sector association.

9.2   Where the sector association receives notice under paragraph 9.1, it shall have 20 working days from receipt of the notice to make representations to the Secretary of State.

9.3   After considering any representations made in accordance with paragraph 9.2, the Secretary of State may serve a further notice on the sector association specifying the variations; and the targets shall be varied in accordance with that notice without any further action being required.

ANNEX 11

EXAMPLE UNDERLYING CLIMATE CHANGE AGREEMENT

Underlying Climate Change Agreement For The [ ] Sector

THIS AGREEMENT is made the

BETWEEN :

(1)   the Secretary of State for the Environment, Food and Rural Affairs("the Secretary of State"); and

(2)   ("the operator").

(2) | ("the operator").

1.   RECITALS

1.1   Section 30 of, and Schedule 6 to, the Finance Act 2000 makes provision for a new tax known as the climate change levy. The levy will be charged on the supply of taxable commodities. Paragraph 42(1)(c) of Schedule 6 provides that the amount payable by way of levy on the supply of taxable commodities shall be 20% of the full rate, if the supply is a reduced-rate supply.

1.2   Paragraphs 44 to 52 of Schedule 6 set out the circumstances in which a supply is a reduced-rate supply. To be a reduced-rate supply a supply has to be supplied to a facility which is certified by the Secretary of State as being covered by a climate change agreement.

1.3   Paragraph 46(b) of Schedule 6 provides that a climate change agreement may consist of a combination of agreements that falls within paragraph 48 of that Schedule. Paragraph 48 provides that the combination is a combination of an umbrella agreement and an underlying agreement.

1.4   This agreement is an underlying agreement entered into for the purposes of the umbrella agreement. It is not intended to give rise to contractual obligations between the parties.

1.5   The operator is a representative (as defined in paragraph 47(2) of Schedule 6) of each facility to which this agreement applies.

2.   INTERPRETATION AND NOTICES

2.1   In this agreement, unless the context otherwise requires-

"acceptable currency" means a currency described in paragraph 5 of Schedule 2 to the umbrella agreement and references to an absolute target, a relative target, a carbon target or an energy target shall be construed in accordance with that paragraph;

"energy plan" has the meaning given by paragraph 1.1 of Schedule 3;

"facility" shall be construed in accordance with clause 3 of the umbrella agreement;

"facility number", in relation to a facility, means the number of that facility specified in Part 1 of Schedule 1 to the umbrella agreement;

"fuel" means coal, coke, gas oil, heavy fuel oil, petrol, liquid petroleum gas, jet kerosene, ethane, naphtha, refinery gas, petroleum coke, natural gas and electricity;

"notice" includes any document whether in paper or electronic form;

"sector association" means the [   ] sector

"sector target" has the same meaning as in the umbrella agreement;

"served" includes copied;

"target period" has the meaning given by paragraph 1.1 of Schedule 2;

"target unit" means a facility or a group of facilities with a target set under this agreement which only applies to that facility or that group;

"termination notice" means a notice served by the Secretary of State on the operator under clause 6.12 or 9.3 or paragraph 3.13, 4.11 or 5.13 of Schedule 5 to terminate this agreement;

"throughput" means input or output depending on how it is calculated under paragraph 4 of Schedule 2;

"umbrella agreement" means the agreement entered into between the Secretary of State and the sector association for the purposes of the reduced rate of climate change levy;

"underlying agreement" means an agreement applying to one or more facilities identified in Part 1 of Schedule 1 to the umbrella agreement which is expressed to be entered into for the purpose of the umbrella agreement;

"variation certificate" means a variation certificate under paragraph 45 of Schedule 6 to the Finance Act 2000; and

"working day" means any day other than a Saturday, Sunday, Christmas Day, Good Friday or a day falling on a bank holiday in any part of the United Kingdom.

2.2   Any notice served under this agreement shall be in writing.

2.3   A notice served on the operator may be served by sending it by post to

or electronically to:

2.4   A notice served on the Secretary of State may be served by sending it by post to:

> The Climate Change Agreements Team,
> Defra,
> Zone 6/E4,
> Ashdown House,
> 123 Victoria Street,
> London SW1E 6DE,

or electronically to:

> levy.agreements@defra.gsi.gov.uk

2.5   A notice served on the sector association may be served by sending it by post to:

[.]

or electronically to:

[ ]

3.   FACILITIES TO WHICH THIS AGREEMENT APPLIES

3.1   This agreement applies to the facilities identified in Schedule 1 but subject to the following provisions of this agreement.

A facility is only eligible at any time for inclusion in this agreement if, and to the extent that, at that time-

it is a facility within the meaning of paragraph 50(2) to (6) of Schedule 6 to the Finance Act 2000;

it belongs to the [   ] sector; and

it is identified in Part 1 of Schedule 1 to the umbrella agreement and it is not included in another underlying agreement or an agreement falling within paragraph 47 of Schedule 6 to the Finance Act 2000 or another combination of agreements falling within paragraph 48 of that Schedule.

3.3   A facility belongs to the [   ] sector if it is a facility which manufactures [   ].

4.   TARGETS

4.1   The targets and tolerance bands (if any) for the facilities to which this agreement applies are set out in paragraph 1.1 of Schedule 2. Paragraphs 1.2 to 1.5 of that Schedule provide for adjustments to take account of product mix and/or throughput and emissions trading. The rules for determining whether or not a target is met are set out in the remaining provisions of that Schedule.

4.2   The targets for the facilities to which this agreement applies shall be reviewed, and if necessary varied, at the end of the year 2002 in accordance with the procedure set out in paragraph 4 of Schedule 5 to ensure that they fully reflect any cost effective energy savings which could be achieved through the use, or additional use, of combined heat and power.

4.3   Where the sector targets in the umbrella agreement are varied following a review under clause 5.4 of that agreement, the targets set under this agreement shall, where appropriate, be varied to take account of the variation of the sector targets in accordance with the procedure set out in paragraph 6 of Schedule 5.

5.   QUALITATIVE REQUIREMENTS

5.1   For the purpose of clause 7 of the umbrella agreement, the qualitative requirements for the facilities to which this agreement applies are set out in Schedule 3.

6.   OBLIGATIONS OF OPERATOR

The operator shall serve a notice on the Secretary of State (copied to the sector association) immediately it has reason to believe that a facility listed in Schedule 1 (or any part of it) might not be eligible by virtue of clause 3.2 for inclusion in this agreement.

6.2   The operator shall, for each facility to which this agreement applies, supply the Secretary of State with the name of a person who can be contacted in respect of the facility together with that person's postal address, telephone number, fax number and e-mail address.

6.3   If there is any change in the information mentioned in clause 6.2, the operator shall serve a notice on the Secretary of State specifying the change.

6.4   The Secretary of State may by notice request the operator to supply him with the following information in relation to any facility to which this agreement applies-

(a)   the reference number of every meter recording units of energy supplied to the facility; and

(b)   for each such meter, whether it is intended to use all of the energy recorded by that meter in that facility,

and the operator shall comply with such a request.

6.5   By the end of January in alternate years commencing with the year 2003, the operator shall supply the sector association with the information specified in Schedule 4 in relation to the most recently completed target period.

6.6   The operator shall comply with a request from the sector association for information under clause 6.10 of the umbrella agreement not more than 5 working days after receipt of the notice served under that clause.

6.7   The operator shall co-operate with any person appointed by the Secretary of State under clause 6.11 of the umbrella agreement to undertake an independent audit of the information provided by the operator to the sector association and, for that purpose, the operator shall keep proper records and make them available for inspection when required by the auditor.

6.8   The operator shall ensure that an appraisal is carried out of the potential for use, or additional use, of combined heat and power in the facilities to which this agreement applies.

6.9   The appraisal shall be carried out in accordance with a procedure specified by the Secretary of State in a notice served on the operator.

6.10   The operator shall submit a report to the Secretary of State setting out the results of the appraisal by no later than 30th September 2002.

6.11   Where the Secretary of State considers that the operator has failed to comply with any of its obligations under clause 6.8 to 6.10, he may serve a notice on the operator identifying the failure in question and the steps he considers the operator should take to comply with that obligation.

6.12   Where the Secretary of State serves a notice under clause 6.11 and the operator fails to take the steps set out in the notice, he may terminate the agreement by serving a termination notice on the operator.

6.13   A termination notice under clause 6.12 shall specify the date on which the agreement ceases to have effect (which shall be at least 10 working days after the notice is served) and the agreement shall cease to have effect on the date specified in the termination notice unless the notice is withdrawn before that date. Where this agreement ceases to have effect in this way no new certificate will be given under clause 7.1 for any further certification periods and any existing certificates given in respect of such facilities under that clause will be terminated by a variation certificate.

7.   CERTIFICATION OF FACILITIES BY SECRETARY OF STATE

7.1   The Secretary of State will certify that a facility to which this agreement   applies is covered by a climate change agreement in the circumstances set out in clause 7 of the umbrella agreement.

8.   CONFIDENTIALITY

8.1   The Secretary of State shall be entitled to publish, without the operator's consent, the list of the facilities in Schedule 1 and a list of the facilities in that Schedule which are certified under clause 7.1 as being covered by a climate change agreement.

8.2   The Secretary of State shall be entitled to disclose, without the operator's consent, any other information relating to this agreement in the following circumstances -

(a)   where the disclosure is made under and in accordance with the terms of any legislation;

(b)   where the disclosure is made to a relevant authority for the purposes of-

(i)   the Secretary of State's functions under Schedule 6 to the Finance Act 2000; or

(ii)   the authority's functions; or

(c)   where the disclosure is made in the course of legal proceedings.

8.3   In clause 8.2(a) "legislation" means primary or secondary legislation and includes legislation of the European Community.

8.4   The Secretary of State shall consult the operator before making any disclosure under clause 8.2 where he considers it appropriate in the circumstances to do so.

8.5   Save as provided for in clause 8.1 and 8.2, the Secretary of State will only disclose information relating to this agreement with the consent of the operator.

8.6   The relevant authorities referred to in clause 8.2(b) are -

(a)   either House of Parliament (including any committee);

(b)   the European Commission;

(c)   the Commissioners of Customs and Excise;

(d)   the relevant environmental regulator for a facility under Part I of the Environmental Protection Act 1990 or regulations made under section 2 of the Pollution Prevention and Control Act 1999 or corresponding legislation for Northern Ireland;

(e)   a person appointed by the Secretary of State under clause 6.11 of the umbrella agreement to undertake an audit of information; and

(f)   the authorities charged with regulating under the Competition Act 1998.

8.7   The Secretary of State shall take steps to prevent any person, whom he appoints under clause 6.11 of the umbrella agreement to undertake an audit, from disclosing information relating to this agreement and obtained in carrying out the audit to any one other than the Secretary of State, except to the extent needed to carry out the audit.

9.   DURATION OF THIS AGREEMENT

9.1   Subject to clause 9.2 and 9.6, this agreement shall continue in force from the date on which it is made until 31st March 2013.

9.2   This agreement may be terminated before 31st March 2013 -

(a)   by notice served by the operator on the Secretary of State; or

(b)   by a termination notice served by the Secretary of State on the operator in accordance with clause 6.12 or 9.3 or paragraph 3.13, 4.11 or 5.13 of Schedule 5.

9.3   Where the operator is in breach of clause 6.5 and clause 6.7 of the umbrella agreement applies, the Secretary of State may terminate this agreement by serving a termination notice on the operator.

9.4   A termination notice under clause 9.3 shall specify the date on which this agreement ceases to have effect (which shall be at least 10 working days after the date on which the notice is served).

9.5   Where the Secretary of State serves a termination notice under clause 9.3, this agreement shall cease to have effect on the date specified in the notice unless the notice is withdrawn before that date. Where this agreement ceases to have effect in this way, no new certificate will be given under clause 7.1 for any further certification periods and any existing certificates given in respect of such facilities under that clause will be terminated by a variation certificate.

9.6   This agreement shall terminate if the umbrella agreement is terminated.

10.   VARIATION OF AGREEMENT

10.1   Subject to clause 10.2, any provision of this agreement may be varied by agreement of the Secretary of State and the operator in accordance with Part 1 of Schedule 5.

10.2   The list of facilities in Schedule 1 and the targets in Schedule 2 and the currency of those targets shall be varied in accordance with Part 2 of Schedule 5 but, save as otherwise expressly provided in this agreement, no variation of the list of facilities, the targets or the currency of the targets shall be made.

Signed by authority of   Signed on behalf of the
the Secretary of State   operator

SCHEDULE 1

FACILITIES TO WHICH THIS AGREEMENT APPLIES

1.   This agreement applies to the facilities which have the following facility numbers -

SCHEDULE 2

TARGETS

FACILITY TARGETS

1.   Targets

1.1   The targets and tolerance bands for facilities to which this agreement applies are set out in the following Table.

Facility number	Target period	Target	Tolerance band
	1st January 2003 to 31st December 2004		
	1st January 2005 to 31st December 2006		
	1st January 2007 to 31st December 2008		
	1st January 2009 to 31st December 2010		
	1st January 2003 to 31st December 2004		
	1st January 2005 to 31st December 2006		
	1st January 2007 to 31st December 2008		
	1st January 2009 to 31st December 2010		

Adjustment of relative targets to take account of product mix and/or throughput

1.2   Where -

(a)   a relative target for a target unit does not have a tolerance band;

(b)   a procedure for adjusting the target to take account of product mix and/or throughput has been agreed between the operator and the Secretary of State; and

(c)   the operator serves notice on the Secretary of State not more than 10 working days after a target period ending before 1st January 2007 specifying that it wishes the target for that period to be adjusted under this paragraph, the target shall be adjusted in accordance with that procedure.

Adjustment of targets to take account of emissions trading

1.3   The target for a target unit shall be adjusted in relation to a target period where carbon emission allowances have been transferred to or from the target unit in accordance with a trading scheme established or approved by the Secretary of State which applies to this agreement and there is a positive or negative balance in relation to that target unit for that period.

1.4   For the purpose of paragraph 1.3 –

(a)   there is a positive balance in relation to a target unit if for that period there have only been transfers to the target unit or the number of carbon emission allowances transferred to the target unit exceeds the number of carbon emission allowances transferred from it;

(b)   there is a negative balance in relation to a target unit if for that period there have only been transfers from the target unit or the number of carbon emission allowances transferred from the target unit exceeds the number of carbon emission allowances transferred to it.

1.5   The target in relation to the target unit shall be increased for that period if there is a positive balance, and shall be reduced if there is a negative balance -

> (a)   if the target is an absolute energy target, by an amount equal to B/C;

> (b)   if the target is a relative energy target, by an amount equal to B/(C x T);

> (c)   if the target is an absolute carbon target, by an amount equal to B;

> (d)   if the target is a relative carbon target, by an amount equal to B/T,

where -

B is the balance mentioned in paragraph 1.4(a) or 1.4(b) expressed in kilogrammes;

C is the carbon emission factor for the fuel used in the target unit in the relevant target period set out in the Table in paragraph 3.1 or, if more than one fuel is used in that period, the weighted average of the carbon emission factors set out in that Table for the fuels so used (the weighted average shall be determined by reference to the total number of units of energy calculated in accordance with paragraph 2 for each fuel used in the target unit in that period); and

T is the throughput of the target unit for the relevant target period calculated in accordance with paragraph 4.1.

1.6   For the purpose of determining whether a target has been met, the units of energy used by the facilities to which this agreement applies, the units of carbon emitted from those facilities and the throughput of those facilities shall be calculated in accordance with the following paragraphs of this Schedule.

2.   Calculation of units of energy used by a facility

2.1   The number of units of energy used by a facility to which this agreement applies shall be measured in kilowatt hours and calculated as follows.

Fossil fuels

2.2   The units of fossil fuels used shall be calculated on a gross calorific value basis. No correction shall be applied to account for the energy used in the extraction, processing and supply of the fossil fuels to a facility.

General electricity imports

2.3   Where a facility imports electricity other than from a CHP plant (see paragraphs 2.4 and 2.5), a dedicated electricity generator (see paragraph 2.6) or new renewable energy supplies (see paragraph 2.10), the units of electricity used shall be multiplied by a factor of 2.6.

Energy from a plant which converts an energy input into both heat and power (a "combined heat and power plant" or "CHP plant")

2.4   Where energy from a CHP plant is used, the units of energy used shall be calculated on the basis of the units of energy input to the CHP plant not the units of energy produced by the CHP plant.

2.5   Where a facility has a CHP plant and all of the energy from the CHP plant is used within the facility, the facility shall be treated as using all of the units of energy input to the plant and no allocation of those units is required. Where a facility has a CHP plant and some of the heat or electricity from the plant is exported from the facility, or where a facility imports heat or electricity which is generated by a CHP plant which is not part of the facility, the energy input to the CHP plant shall be allocated to each user of the heat or electricity as follows –

(1) First, allocate energy input to the CHP plant to the heat and electricity produced using the following formulae -

Heat Energy =
$$\frac{\text{Fuel Input} \times \text{Heat Output}}{(2 \times \text{Electricity Output}) + \text{Heat Output}}$$

Electricity Energy =
$$\frac{2 \times \text{Fuel Input} \times \text{Electricity Output}}{(2 \times \text{Electricity Output}) + \text{Heat Output}}$$

where -

"Heat Energy" is the energy allocated to the heat produced;

"Electricity Energy" is the energy allocated to the electricity produced;

"Fuel Input" is the fuel supplied to the CHP plant;

"Heat Output" is the quantity of heat produced; and

"Electricity Output" is the quantity of electricity produced.

(Fuel and heat units should be expressed on a Gross Calorific Value basis. Energy units should be consistent throughout (ideally kWh). Where absorption cooling is used to produce a cooling supply, the cooling output should be metered and divided by the average coefficient of performance (COP) of the cooling system to estimate the heat used.)

(2) Then apportion the energy input to each user of the heat and electricity as follows–

–   allocate the heat energy to each user of the heat in proportion to the quantity of heat from the CHP plant which each consumes;

–   allocate the electricity energy to each user of the electricity in proportion to the quantity of electricity from the CHP plant which each consumes.

Thus, if the heat is distributed to a number of users such that –

Heat Output = Heat_1 + Heat_2 + Heat_n

or the electricity is distributed to a number of users such that -

Electricity Output = Electricity_1 + Electricity_2 + Electricity_m

Then –

$$\text{Heat Energy}_1 = \frac{\text{Heat}_1}{\text{Heat Output}} \times \text{Heat Energy, and}$$

$$\text{Electricity Energy}_1 = \frac{\text{Electricity}_1}{\text{Electricity Output}} \times \text{Electricity Energy}$$

where Heat Energy_1 and $\text{Electricity Energy}_1$ are the quantities of energy allocated to user 1, and likewise for the other users.

(3) If some of the electricity is exported to the public supply (and not directly to a known user), allocate credit for this electricity to each of the heat users as follows -

– multiply the quantity of exported electricity by 2.6;

– subtract from this the energy apportioned to the exported electricity from the CHP plant, which gives the saving that has been made;

– divide this saving amongst each of the users of the heat from the CHP plant on a pro-rata basis to the quantity of heat which each uses; and

– subtract the pro-rata saving from each of the heat figures calculated in accordance with step (2) above, to get a revised heat figure.

Thus, if electricity were exported to the public supply instead of being supplied to user m in step (2) above, the revised heat figures would be –

Revised Heat Energy_1 =

$$\text{Heat Energy}_1 - \frac{((\text{Exported Electricity} \times 2.6) - \text{Electricity Energy}_m) \times \text{Heat}_1}{\text{Heat Output}}$$

---

[1] Where there are imports or exports of steam, calculations shall be based on the fuel or fuels used to generate the steam.

Revised Heat Energy$_2$ =

Heat Energy$_2$ – ((Exported Electricity x 2.6) - Electricity Energy$_m$) x Heat$_2$
$$\text{Heat Output}$$

Revised Heat Energy$_n$ =

Heat Energy$_n$ – ((Exported Electricity x 2.6) - Electricity Energy$_m$) x Heat$_n$
$$\text{Heat Output}$$

where "Exported Electricity" is the quantity of exported electricity expressed as the number of units of electricity exported.

(If there were no electricity exports to the public supply, then user 1 would report total energy from the plant or generator as Heat Energy$_1$ plus Electricity Energy$_1$. If there were electricity exports, user 1 would report total energy from the plant or generator as Revised Heat Energy$_1$ plus Electricity Energy$_1$.)

Electricity from a dedicated electricity generation plant

2.6   Where all of the electricity from a dedicated electricity generation plant located within a facility is used within that facility, then all of the energy input into the plant shall be allocated to the facility. Where there is more than one user of the electricity from the plant, the total energy input to the plant shall be allocated to the consumers of the electricity on a pro-rata basis.

Steam

2.7   Imported or exported steam shall be accounted for by taking the enthalpy of the steam and dividing by the efficiency of the system which generates the steam and distributes it to the user's facility boundary (in order to account for the total energy used to produce the steam).

(Account should be taken of steam pressure – for example, where sites import high pressure steam and return it at a lower pressure.)

Renewable energy

2.8   Renewable energy used to produce heat (for example, energy from coppicing) shall not be counted as part of a facility's energy use.

2.9   Electricity produced on site from a renewable source (for example, a wind turbine) shall not be counted as part of a facility's energy use.

2.10   Electricity which is certified by the supplier as being from new renewable energy supplies and as being electricity which the supplier is not relying upon for the purpose of fulfilling any obligation imposed upon it by any enactment in relation to the generation of such electricity shall not be counted as part of a facility's energy use.

2.11   For the purpose of paragraph 2.10 "new renewable energy supplies" means –

wind energy;
hydro power, excluding hydro power from plants exceeding 10MW output;
tidal power;
wave energy;
photovoltaics;
photoconversion;
geothermal hot dry rock;
geothermal aquifers;
municipal and industrial wastes;
landfill gas;
agriculture and forestry wastes; and
energy crops.

2.12   The Secretary of State may by notice served on the sector association vary the list in paragraph 2.11.

2.13   Electricity supplied from other renewable energy supplies shall be accounted for along with other conventional electricity supplies with the factor of 2.6 applied to convert it to primary energy.

Energy from waste

2.14   The use of energy from waste-derived fuels shall not be counted as part of a facility's energy use.

Fuel used as a chemical feedstock

2.15   Fuels used as a chemical feedstock and embodied in a chemical product shall not be counted as part of a facility's energy use. However, fuels which are used as reductants shall be counted.

Electrolysis

2.16 All energy used for electrolysis shall be counted as part of a facility's energy use

Energy from exothermic reactions
2.17 Energy from exothermic reactions not involving fossil fuels shall not be counted as part of a facility's energy use.

Unexpected energy supply disruptions
2.18 An increase in the number of units of energy used in a facility during the relevant target period (compared with the number of units which would otherwise have been used in that period) shall be disregarded if it was the result of -

(a) an unexpected disruption in the supply of energy to the facility; or

(b) an unexpected total failure of a dedicated electricity generation plant located within the facility and -

(i) the total failure continued for at least 240 hours; and

(ii) the operator is able to demonstrate from his records (or in some equally satisfactory manner) that the plant was being properly operated at the relevant time and has been properly maintained.

2.19 For the purposes of paragraph 2.18 an event is only unexpected if it is the kind of event which was not taken into account in setting the target.

3. Calculation of carbon emissions from a facility

3.1 The total number of units of carbon emitted from a facility to which this agreement applies during a target period shall be calculated by multiplying the units of energy used from each fuel used in the facility during the relevant target period calculated in accordance with paragraph 2 by the relevant carbon emission factor set out below for that fuel.

Fuel[1] Carbon Emission Factor (kg/kWh)
Electricity   0.0453
Coal   0.0817

---

[1] Where there are imports or exports of steam, calculations shall be based on the fuel or fuels used to generate the steam.

Coke 0.1170
Gas Oil   0.0680
Heavy Fuel Oil   0.0709
Petrol 0.0655
LPG   0.0627
Jet Kerosene   0.0655
Ethane   0.0545
Naphtha   0.0709
Refinery Gas   0.0545
Petroleum Coke   0.0927
Natural Gas 0.0518

3.2   The units of carbon emitted shall be expressed as kilogrammes of mass carbon (rather than carbon dioxide).

Process carbon emissions

3.3   Carbon emissions from industrial processes shall not be counted as part of a facility's carbon emissions unless they result from combustion or oxidation of fossil fuels.

3.4   Carbon emissions from electrodes shall not be counted as part of a facility's carbon emissions.

Unexpected energy supply disruptions

3.5   An increase in the number of units of carbon emitted from a facility during the relevant target period (compared with the number of units which would otherwise have been emitted in that period) shall be disregarded if it was the result of -

(a)   an unexpected disruption in the supply of energy to the facility; or

(b)   an unexpected total failure of a dedicated electricity generation plant located within the facility and -

(i)   the total failure continued for at least 240 hours; and

(ii)   the operator is able to demonstrate from his records (or in some equally satisfactory manner) that the plant was being properly operated at the relevant time and has been properly maintained.

3.6   For the purposes of paragraph 3.5 an event is only unexpected if it is the kind of event which was not taken into account in setting the target.

4.   Calculation of throughput of a facility

4.1   The throughput of a facility to which this agreement applies during a target period shall be calculated by applying the following accounting conventions -

SCHEDULE 3

QUALITATIVE REQUIREMENTS

The qualitative requirements for the facilities to which this agreement applies are as follows.

1.   Preparation and implementation of energy plan

1.1   The operator shall prepare and implement a document (an "energy plan") setting out the energy policy for each facility to which this agreement applies. The degree of detail required in the energy plan and the approach to be taken shall depend on the size of the facility, its energy use and the pre-existing systems in place.

1.2   The energy plan shall be designed to demonstrate the operator's commitment to continuous improvement in energy efficiency at the facility concerned in order to meet the targets in this agreement set for the facility. It shall, in particular -

   (a)   set out a programme for regular surveys of the main energy consuming parts of the facility to identify means of reducing energy use by -

      (i)   improved 'house-keeping' measures;

      (ii)   improved management and control of processes;

      (iii) installation of better processes, either by retrofits or new build; and

      (iv)   increased use of combined heat and power where appropriate;

(b)   commit the operator, when capital investments are planned in relation to the facility, to giving energy efficiency due regard in the selection and configuration of plant, and adopting the most energy efficient equipment available when the marginal cost is justifiable;

(c)   provide for adequate resourcing of energy management in relation to the facility and for measurable objectives; and

(d)   identify the Director or Senior Manager with overall responsibility for the energy plan and its implementation.

1.3   The energy plan shall also contain management policies designed to facilitate the implementation of the policies in the plan. The management policies shall, in particular, set out–

(a)   who is responsible for energy use in the facility and for the implementation of the policies in the energy plan;

(b)   appropriate methods for communication to ensure that the policies and procedures are understood and that management commitment to them is visible;

(c)   training plans, both for energy managers and the workforce as appropriate;

(d)   procedures for appropriate planned and emergency maintenance of equipment, and for its replacement; and

(e)   procedures for assessing the cost-effectiveness of an energy saving measure, which should take a view of savings over the lifetime of the measure.

2.   Monitoring and control

2.1   The operator shall set up a system for monitoring and controlling progress in the implementation of the energy plan and identifying improvement actions.

2.2   The monitoring and control system shall–

(a)   measure the principal energy flows within the facility;

(b)   report energy use in appropriate units to operating managers at a frequency appropriate to the quantity of energy consumed;

(c)   provide standards of performance that managers are charged with achieving;

(d)   allow the review of achievements against standards in order to identify where action needs to be taken; and

(e)   allow the review of standards periodically so that they may be tightened as performance improves.

3.   Reporting

3.1   The operator shall produce management reports on energy use and management (progress against objectives, conclusions from regular reviews, etc) in a way appropriate to the size and complexity of the facility.

3.2   The reports shall include–

(a)   progress reports as necessary or as required by the appropriate senior management body (e.g. Board) in order to ensure adequate control and review of objectives; and

(b)   frequent reports for operational management control.

4.   Review

4.1   The operator shall carry out regular reviews of the energy plan and its implementation.

4.2   The reviews shall include -

(a)   consideration of the plan's policy (its aims and objectives, scope, adequacy);

(b)   comparisons of quantitative performance against targets;

(c)   comparisons with benchmark data (where available); and

(d)   reviews of the barriers to the implementation of energy efficiency improvements, and proposals for addressing these as far as is possible.

4.3   The operator shall vary the energy plan, where appropriate, to take account of the results of the reviews.

SCHEDULE 4

INFORMATION TO BE SUPPLIED TO SECTOR ASSOCIATION

The information required by clause 6.5 is as follows -

1.   For each target unit, the total number of units of energy used during the relevant target period for each type of fuel calculated in accordance with paragraph 2 of Schedule 2.

2.   For each target unit with a carbon target for the relevant target period, the total number of units of carbon emitted from the target unit during that period calculated in accordance with paragraph 3 of Schedule 2.

3.   For each target unit, the throughput of that target unit during the relevant target period calculated in accordance with paragraph 4 of Schedule 2.

4.   For each target unit where the target is to be adjusted under paragraph 1.2 or 1.3 of Schedule 2, the information needed to calculate the adjustment.

5.   If the operator relies on clause 7.4(a) of the umbrella agreement, a copy of its energy plan and a description of the steps taken to implement the plan.

6.   If the operator relies on clause 7.4(b) of the umbrella agreement, details of the relevant constraint or requirement and details of its impact on the performance of the operator.

SCHEDULE 5

VARIATIONS

PART 1

GENERAL VARIATIONS

1. Variations of provisions other than Schedules 1 and 2

1.1 Where a party to this agreement wishes to vary any of its provisions (other than the list of facilities in Schedule 1 or the targets in Schedule 2 or the currency of those targets) it may serve a notice on the other party setting out the variation and the reasons for the variation.

1.2 A party which receives a notice under paragraph 1.1 shall have 20 working days after its receipt to respond in writing.

1.3 Where the variation is agreed this agreement shall be varied accordingly.

PART 2

VARIATION OF LIST OF FACILITIES AND TARGETS

2. Including facilities

A facility which is listed in Part 1 of Schedule 1 to the umbrella agreement shall be included in the list of facilities in Schedule 1 to this agreement if it is eligible for inclusion in this agreement by virtue of clause 3.2 and-

(a) the operator has served a notice on the Secretary of State stating that it wishes the facility to be included;

(b) the Secretary of State and the operator have agreed to its inclusion and the variations to this agreement which are to be made to reflect the inclusion of the facility, including any variations altering existing targets or specifying new targets which will apply to the facility; and

(c) the Secretary of State has served notice on the operator specifying the agreed variations,

and on the date fixed by the Secretary of State's notice the facility shall be included in the list of facilities in Schedule 1 to this agreement and the variations specified in the Secretary of State's notice shall have effect.

3. Excluding facilities

3.1 A facility may be excluded from the list of facilities in Schedule 1 in accordance with the following provisions.

3.2   Where the operator wishes to exclude a facility from the list of facilities in Schedule 1, the operator shall serve a notice on the Secretary of State stating that this is the case, identifying the facility and setting out the reasons for the exclusion and shall copy the notice to the sector association. The notice shall also state whether it is intended to apply for the facility's inclusion in another underlying agreement within 12 months.

3.3   Where the Secretary of State wishes to exclude a facility in whole or in part from the list of facilities in Schedule 1 on the basis that it is not eligible by virtue of clause 3.2 for inclusion in this agreement, he shall serve a notice on the operator stating that this is the case, identifying the facility or part and setting out the reasons for the exclusion and shall copy the notice to the sector association.

3.4   The exclusion of the whole or part of a facility from a target unit consisting of a group of facilities shall require the variation of the targets applying to that unit if -

(a)   it is not eligible by virtue of clause 3.2 for inclusion in this agreement; or

(b)   a notice in relation to the facility has been served under paragraph 3.2 and it is intended to apply for its inclusion in another underlying agreement within 12 months.

3.5   The sector association may, not more than 10 working days after receiving a copy of a notice under paragraph 3.2 or 3.3, make written representations to the Secretary of State (copied to the operator).

3.6   If the Secretary of State and the operator agree on the exclusion of the facility or part of it from the list of facilities in Schedule 1 and, where paragraph 3.4 requires the variation of a target, on any variation of the target, this agreement shall be varied accordingly.

3.7   If -

(a)   the Secretary of State and the operator fail to agree as mentioned in paragraph 3.6 not more than 10 working days after service of the notice under paragraph 3.2 or 3.3, as the case may be; and

(b)   there is a dispute on the facts,

either party may refer the dispute for adjudication to an institution named on the list prepared under Schedule 5 to the umbrella agreement and that Schedule shall apply in relation to the adjudication.

3.8   Notice of any such referral shall be served by the party making the referral on the other party.

3.9   Each party to the dispute shall have 20 working days from the date on which they receive notice from the nominated adjudicator of the address to which representations should be sent to make representations to the nominated adjudicator. Any such representations shall be copied by the party making the representations to the other party to the dispute.

3.10   The adjudicator may request further information from the parties. Any such request shall be in writing, shall specify the date by which the information is required and shall be copied to the other party to the dispute. Any information provided in response to such a request shall be copied by the party providing the information to the other party to the dispute.

3.11   The adjudicator will, on the basis of the representations provided to him and any additional information he considers to be necessary, make a finding on the disputed questions of fact and notify the parties of that finding.

3.12   The adjudicator's finding on the disputed questions of fact shall be binding on the parties, but it shall be for the parties to agree, in the light of any such finding, on whether or not a facility is eligible in whole or in part for inclusion in this agreement and on any variation to a target required by this paragraph and if agreement is reached, this agreement shall be varied accordingly.

3.13   Where, despite any finding made by the adjudicator, the Secretary of State and the operator fail to agree as mentioned in paragraph 3.6, the Secretary of State may terminate this agreement by serving a termination notice on the operator.

3.14   A termination notice under paragraph 3.13 shall specify the date on which this agreement ceases to have effect (which shall be at least 10 working days after the date on which the notice is served) and shall be accompanied by a statement setting out the Secretary of State's reasons for not agreeing with the operator on the variation required.

3.15 Where the Secretary of State serves a termination notice under paragraph 3.13, this agreement shall cease to have effect on the date specified in the notice unless the notice is withdrawn before that date. Where this agreement ceases to have effect in this way, no new certificate will be given under clause 7.1 for any further certification periods and any existing certificates given in respect of such facilities under that clause will be terminated by a variation certificate. Where a termination notice is withdrawn, this agreement shall be varied to reflect any agreement reached between the Secretary of State and the operator.

4.   Review of targets following combined heat and power appraisal

4.1   Following receipt of the operator's report under clause 6.10, the Secretary State shall review the targets for the facilities to which this agreement applies to determine whether they need to be varied to ensure that they fully reflect any cost effective energy savings which could be achieved through the use, or additional use, of combined heat and power.

4.2   The targets shall be varied, if it is appropriate to do so in the light of the review, in accordance with the following procedure.

4.3   The Secretary of State shall serve a notice on the operator setting out the proposed variations of the targets, his reasons for the proposals and inviting the operator to agree the variations within 10 working days from the date of the notice or, if he disagrees, to submit any representations to him within that period.

4.4   If the operator and the Secretary of State agree on the variations which should be made, this agreement shall be varied accordingly.
it:
4.5   If no agreement is reached within 20 working days from the date of the notice and there is a dispute on the facts, either party may refer the dispute for adjudication to an institution named on the list prepared under Schedule 5 to the umbrella agreement and that Schedule shall apply in relation to the adjudication.

4.6   Notice of any such referral shall be served on the other party by the party making the referral.

4.7   Each party to the dispute shall have 20 working days from the date on which they receive notice from the nominated adjudicator of the address to which representations should be sent to make representations to the nominated adjudicator. Any such representations shall be copied by the party making the representations to the other party to the dispute.

4.8   The adjudicator may request further information from the parties. Any such request shall be in writing, shall specify the date by which the information is required and shall be copied to the other party to the dispute. Any information provided in response to such a request shall be copied by the party providing the information to the other party to the dispute.

4.9   The adjudicator will, on the basis of the representations provided to him and any additional information he considers to be necessary, make a finding on the disputed question of fact and notify the parties of that finding.

4.10   The adjudicator's finding on a disputed question of fact shall be binding on the parties, but it shall be for the parties to agree, in the light of any such finding, on any variation of the targets and if agreement is reached, this agreement shall be varied accordingly.

4.11   Where, despite any finding made by the adjudicator, the Secretary of State and the operator fail to agree as mentioned in paragraph 4.10, the Secretary of State may terminate this agreement by serving a termination notice on the operator.

4.12   A termination notice under paragraph 4.11 shall specify the date on which this agreement ceases to have effect (which shall be at least 10 working days after the date on which the notice is served) and shall be accompanied by a statement setting out the Secretary of State's reasons for not agreeing with the operator on the variation required.

4.13   Where the Secretary of State serves a termination notice under paragraph 4.11 this agreement shall cease to have effect on the date specified in the notice unless the notice is withdrawn before that date. Where this agreement ceases to have effect in this way, no new certificate will be given under clause 7.1 for any further certification periods and any existing certificates given in respect of such facilities under that clause will be terminated by a variation certificate. Where a termination notice is withdrawn, this agreement shall be varied to reflect any agreement reached between the Secretary of State and the operator.

5.   Variation of absolute targets due to fall of throughput

5.1   Where the annual level of throughput of a target unit during a target period (the "relevant target period") calculated in accordance with paragraph 4 of Schedule 2 of the umbrella agreement is less than 90 per cent of the annual throughput of that target unit when the targets applying to the unit were set, the target for the relevant target period shall be varied to take account of that fall of throughput if it is an absolute target.

5.2   For the purpose of paragraph 5.1, where the relevant target period is the second target period, the annual level of throughput of the target unit when the targets applying to that unit were set is the level of throughput -

(a)   if there has not been a relevant variation, in the year 1999 or such other year as the Secretary of State and the operator may agree for this purpose; or

(b)   if there has been a relevant variation, in the target period which immediately preceded the variation or, if there has been more than one such variation, the last one.

"Relevant variation" means any variation other than one under paragraph 4 or 7 and does not include an adjustment under paragraph 1.2 or 1.3 of Schedule 2.

5.2A For the purpose of paragraph 5.1, where the relevant target period is one of the final three target periods, the annual level of throughput of the target unit when the targets applying to that unit were set is either:

(a)   the level of throughput as agreed between the Secretary of State and the sector association in relation to that target unit during the target review under clause 5.3 of the umbrella agreement for the relevant target period, or if there has been more than one such review, during the last target review; or

(b)   the assumed throughput of the target unit for the relevant target period as agreed or specified in accordance with paragraph 6.6(c) of Schedule 2 to the umbrella agreement,

whichever is the latest.

5.2B Where relevant target period is the second target period, if:

(a) there has been a relevant variation due to a fall in the annual throughput of the target unit during the first target period; and

(b) the operator of the target unit is a Small or Medium-Sized Enterprise and annual throughput during the second target period is greater than annual throughput during the first target period; or

(c) the operator of the target unit is not a Small or Medium-Sized Enterprise and the annual throughput of the target unit during the second target period is more than 110% of the annual throughput of the target unit during the first target period, the target shall be varied to take account of that rise in throughput if it is an absolute target.

"Small or Medium-Sized Enterprise" means an enterprise:

(i) with fewer than 250 employees; and

(ii) which has an annual turnover not exceeding EUR 50 million; or

(iii) an annual balance sheet total not exceeding EUR 43 million.

5.2C For the purpose of 5.2B, the target shall not be varied to the extent that it is less demanding that the target as originally agreed on the date on which this agreement was entered into.

5.3 Any variation of an absolute target required by paragraph 5.1 or 5.2B shall be effected at the end of the relevant target period. The absolute target for the relevant target period shall be varied.

5.4 Where a variation is required by paragraph 5.1 or 5.2B, the Secretary of State shall, not more than 5 working days after receiving from the sector association details of the throughput of the relevant target unit which it is required to provide under paragraph 5 of Schedule 3 to the umbrella agreement, serve a notice on the operator stating the variation which he proposes should be made to the target and shall copy the notice to the sector association.

5.5 On receiving a notice under paragraph 5.4, the sector association may, not more than 10 working days after receiving the notice, make written representations to the Secretary of State (copied to the operator).

5.6   If the Secretary of State and the operator agree on the variation of the target, this agreement shall be varied accordingly.

5.7   If the Secretary of State and the operator fail to agree as mentioned in paragraph 5.6 not more than 10 working days after service of the notice under paragraph 5.4 and there is a dispute on the facts, either party may refer the dispute for adjudication to an institution named on the list prepared under paragraph 1 of Schedule 5 to the umbrella agreement and that Schedule shall apply in relation to the adjudication.

5.8   Notice of any such referral shall be served by the party making the referral on the other party.

5.9   Each party to the dispute shall have 20 working days from the date on which they receive notice from the nominated adjudicator of the address to which representations should be sent to make representations to the nominated adjudicator. Any such representations shall be copied by the party making the representations to the other party to the dispute.

d :5.10   The adjudicator may request further information from the parties. Any such request shall be in writing, shall specify the date by which the information is required and shall be copied to the other party to the dispute. Any information provided in response to such a request shall be copied by the party providing the information to the other party to the dispute.

5.11   The adjudicator will, on the basis of the representations provided to him and any additional information he considers to be necessary, make a finding on the disputed question of fact and notify the parties of that finding.

5.12   The adjudicator's finding on a disputed questions of fact shall be binding on the parties, but it shall be for the parties to agree, in the light of any such finding, on any variation to the target required by this paragraph and if agreement is reached, this agreement shall be varied accordingly.

5.13   Where, despite any finding made by the adjudicator, the Secretary of State and the operator fail to agree as mentioned in paragraph 5.6, the Secretary of State may terminate this agreement by serving a termination notice on the operator.

5.14   A termination notice under paragraph 5.13 shall specify the date on which this agreement ceases to have effect (which shall be at least 10 working days after the date on which the notice is served) and shall be accompanied by a statement setting out the Secretary of State's reasons for not agreeing with the operator on the variation required.

5.15   Where the Secretary of State serves a termination notice under paragraph 5.13, this agreement shall cease to have effect on the date specified in the notice unless the notice is withdrawn before that date. Where this agreement ceases to have effect in this way, no new certificate will be given under clause 7.1 of the umbrella agreement for any further certification periods and any existing certificates given in respect of such facilities under that clause will be terminated by a variation certificate. Where a termination notice is withdrawn, this agreement shall be varied to reflect any agreement reached between the Secretary of State and the operator.

6.   Variations of targets following a review at end of 2004 and 2008

6.1   Where, following a review under clause 5.4 of the umbrella agreement, the Secretary of State and the sector association agree on the variations that should be made to the targets set under this agreement to take account of agreed variations to the sector targets, the Secretary of State shall serve a notice on the operator setting out the agreed variations and inviting representations from the operator on those variations within 20 working days, as provided for in paragraph 6.10 of Schedule 6 to the umbrella agreement.

6.2   Where, having considered any such representations, the Secretary of State serves a notice on the operator under paragraph 6.11 of Schedule 6 to the umbrella agreement setting out variations to the targets set under this agreement to take account of the variations of the sector targets, the targets set under this agreement shall be varied in accordance with that notice without any further action being required.

7.   Variation of targets following a significant difference in the terms of the state aids approval

7.1   Where, under paragraph 7 of Schedule 6 to the umbrella agreement, the Secretary of State and the sector association agree on the variations that should be made to the targets set under this agreement to take account of agreed variations to the sector targets to take account of a significant difference in state aids approval, the Secretary of State shall serve a notice on the operator setting out the agreed variations and inviting representations from the operator on those variations within 20 working days, as provided for in paragraph 7.4 of that Schedule to the umbrella agreement.

7.2   Where, having considered any such representations, the Secretary of State serves a notice on the operator under paragraph 7.5 of Schedule 6 to the umbrella agreement setting out variations to the targets set under this agreement to take account of the variations of the sector targets, the targets set under this agreement shall be varied in accordance with that notice without any further action being required.

8.   Variation of the currency of a target on application of operator

8.1   Subject to paragraph 8.2, the currency of a target may be varied by agreement of the Secretary of State and the operator from one acceptable currency to another.

8.2   Variations to the currency of a target may be made at any time before the beginning of the target period. However, over the course of the agreement the operator may apply to make only -

(a)   two changes between absolute and relative targets, in the case of a target unit whose targets were initially set as absolute targets, or one change between absolute and relative targets, in the case of a target unit whose targets were initially set as relative targets; and

(b)   one change between absolute energy targets and absolute carbon targets or between relative energy targets and relative carbon targets.

8.3.   If the operator wishes to vary the currency of a target, it shall serve a notice on the Secretary of State setting out its reasons.

8.4   Where the operator and the Secretary of State agree, the currency of the target will be changed in accordance with the conversion conventions set out in paragraph 6 of Schedule 2 to the umbrella agreement (but for this purpose the assumed percentage of each fuel shall be as agreed by the Secretary of State and the operator).